57

Mobile Multimedia Broadcasting Standards

Fa-Long Luo
Editor

Mobile Multimedia Broadcasting Standards

Technology and Practice

Springer

Editor
Fa-Long Luo
Anyka, Inc. & Element CXI
1815 McCandless Drive
San Jose, CA 95035-8046
USA
f.luo@ieee.org

ISBN: 978-0-387-78262-1 e-ISBN: 978-0-387-78263-8
DOI: 10.1007/978-0-387-78263-8

Library of Congress Control Number: 2008924074

Printed on acid-free paper

springer.com

Preface

Mobile multimedia broadcasting compasses a broad range of topics including radio propagation, modulation and demodulation, error control, signal compression and coding, transport and time slicing, system on chip real-time implementation in hardware, software and system levels. The major goal of this technology is to bring multimedia enriched contents to handheld devices such as mobile phones, portable digital assistants, and media players through radio transmission or internet protocol (IP) based broadband networks. Research and development of mobile multimedia broadcasting technologies are now explosively growing and regarded as new killer applications. A number of mobile multimedia broadcasting standards related to transmission, compression and multiplexing now coexist and are being extensively further developed. The development and implementation of mobile multimedia broadcasting systems are very challenging tasks and require the huge efforts of the related industry, research and regulatory authorities so as to bring the success.

From an implementation design and engineering practice point of view, this book aims to be the first single volume to provide a comprehensive and highly coherent treatment for multiple standards of mobile multimedia broadcasting by covering basic principles, algorithms, design trade-off, and well-compared implementation system examples. This book is organized into 4 parts with 22 chapters.

The first part of the book consists of seven well-organized chapters to mainly deal with system, implementation, compatibility and comparison of all the coexisting standards related to mobile TV and multimedia broadcasting including T-DMB, DAB, DVB-H/T, CMMB, Media-FLO, ISDB-T and WiMAX, ATSC digital TV and NTSC analog TV.

Part 2 is devoted to fundamental principles, algorithms, implementation, design and testing for baseband processing in mobile multimedia broadcasting. Organized into six chapters, this part presents the link layer, transport mechanisms, basic modulation schemes, error control methods, and channel coding techniques employed in multiple standards.

The third part consists of five chapters to cover all the aspects related to the compression, transmission, error concealment, quality assessment, and real-time implementation of video coding in broadcasting systems with emphasis on H.264 and AVS-M.

Four chapters on the standards for audio coding, classification, and surround effects are organized in the last part of this book. The general principles and algorithms in audio coding are first presented. Then, the overview of China's DRA audio coding standard and MPEG-4 AAC standard family (AAC, High Efficiency AAC and High Efficiency AAC Version 2) is given. The last chapter explains the general concepts behind spatial audio coding and then discusses the specific aspects of MP3 Surround and MPEG Surround which are playing a very important role in digital audio/multimedia broadcasting systems for multichannel contents.

It is hoped that this book could serve as a complete and invaluable reference for engineers, researchers, broadcasters, manufacturers, network operators, software developers, content providers, service providers, and regulatory bodies for the delivery of television, data service and multimedia enriched contents to mobile systems. This book is also very suitable as a textbook for graduate students in electronics engineering, communications, networking and computer sciences.

Silicon Valley, California, USA *Fa-Long Luo, Ph.D.*
August 2008

List of the Standards Covered in this Book

Transmission and Multiplexing Standards:

1. DVB-H
2. DVB-T
3. DAB
4. Media-FLO
5. T-DMB
6. CMMB
7. DMB-T/H
8. ISDB-T
9. WiMax
10. A-VSB (ATSC)
11. NTSC (Analog TV)
12. MPEG-2 TS
13. DAB-IP

Video Coding Standards:

1. H.264 (MPEG-4 AVC)
2. MPEG-2
3. AVS-M

Audio Coding Standards:

1. MPEG-4 AAC
2. MPEG-4 HE-AAC
3. MPEG-4 HE-AAC V2
4. DRA
5. MPEG-Surround

Contents

Part I
System, Implementation, Compatibility and Comparison of Multiple Standards

Chapter 1
Fundamentals of DVB-H Broadcasting Transmission and Reception

Wout Joseph and Luc Martens

1.1 Introduction

The digital broadcasting standard DVB-H (Digital Video Broadcasting – Handheld) enables a high data rate broadcast access for hand-held terminals (e.g., portable, pocket-size battery-operated phones) [5–8]. The broadband downstream channel features a useful data rate of up to several Mbps and may be used for audio and video streaming applications, file downloads, and many other kinds of services.

The DVB-H technology is an extension of DVB-T (Digital Video Broadcasting – Terrestrial) [6] and takes the specific properties of typical hand-held terminals into account. The three main new physical-layer techniques that have been introduced for DVB-H are time slicing, MPE–FEC (Multi-Protocol Encapsulation–Forward Error Correction), and the 4K mode [5,6]. First, DVB-H uses time slicing, a power-saving algorithm based on the time-multiplexed transmission of different services. This technique results in a battery power-saving effect and allows soft handover if one moves from a network cell to another one. Secondly, for reliable transmission in poor signal reception conditions, an enhanced error-protection scheme, called MPE–FEC, is introduced. Thirdly, the 4K mode (next to the 2K and 8K mode of DVB-T) for OFDM (orthogonal frequency division multiplexing) is defined, addressing the specific needs of hand-held terminals. The 4K mode aims to offer an additional trade-off between transmission cell size and mobile reception capabilities, providing an additional degree of flexibility for DVB-H network planning for single-frequency networks (SFNs). These techniques make DVB-H a very promising standard for broadcast services requiring high data rates for hand-held devices and offer extended possibilities for content providers and network operators.

W. Joseph (✉) and L. Martens
Ghent University/IBBT, G. Crommenlaan 8, box 201, 9050 Gent, Belgium
e-mail: wout.joseph@intec.ugent.be, luc.martens@intec.ugent.be

F.-L. Luo (ed.), *Mobile Multimedia Broadcasting Standards: Technology and Practice*,
DOI: 10.1007/978-0-387-78263-8_1,

The outline of this chapter is as follows. Several characteristics of a DVB-H system, which are needed for coverage planning, are discussed in Sect. 1.2. Section 1.3 presents coverage planning and throughput versus range for a DVB-H network using several path loss (PL) models. Link budget calculations are elaborated for a realistic example. Next, Sect. 1.4 describes propagation measurements of a DVB-H signal and the methodology to develop a PL model. The methodology is illustrated using results of the Flemish DVB-H trial. In Sect. 1.5, the parameter evaluation and performance analysis of a DVB-H system based on measurements are discussed. DVB-H network design for indoor reception is described in Sect. 1.6. This section includes the calculation of the required number of base stations (BS) for good indoor coverage for a region and the relation with the required carrier-to-noise ratio. Finally, conclusions are formulated in Sect. 1.7.

1.2 Selection of Characteristics of DVB-H System

Characteristics and parameters of a DVB-H system are selected. This enables coverage planning and link budget calculations in Sect. 1.3.

1.2.1 Coverage Classes

DVB-H services will be provided in different circumstances. Coverage for DVB-H will be demanding since reception is expected in different conditions (the hand-held terminal is moving, body loss, no line of sight, etc.).

Two types of reception are defined in [8]: *portable* and *mobile* reception. Portable antenna reception is defined as the reception at no speed or very low speed (walking speed), while mobile antenna reception is defined as the reception at medium to high speed. Four different receiving conditions for portable and mobile reception are then defined:

- **Class A**: hand-held portable outdoor reception
  This class contains outdoor reception no less than 1.5 m above ground level at very low or no speed. The receiver is assumed to be portable with an attached or built-in antenna.
- **Class B**: hand-held portable indoor reception at ground floor
  This class contains indoor reception no less than 1.5 m above floor level in rooms at very low or no speed. Furthermore, reception on the ground floor and windows in the external walls of the building is assumed. The receiver is assumed to be portable with an attached or built-in antenna.
- **Class C**: integrated car antenna mobile reception
  This class contains outdoor reception (no less than 1.5 m above ground level) with a moving DVB-H terminal. An example of this class is an antenna integrated in a car.

- **Class D**: hand-held mobile reception (i.e., terminals are used within a moving vehicle)
  This class contains reception inside cars or vehicles (e.g., bus, train, etc.). Again reception in a car or vehicle no less than 1.5 m above ground level is assumed.

Portable antenna reception will in practice take place under a great variety of conditions (outdoor, indoor, ground floor, first floor, upper floors). The hand-held receiver will be probably moved (walking speed). However, for planning purposes, a simplification could be considered for both classes A and B: the portable receiver is not moved during reception and large objects near the receiver are also not moved [8].

Coverage in a small area (typically 100 m^2) is classified as follows:

- "**Good**", if at least 95% of the receiving locations at the edge of the area are covered for portable reception and 99% of the receiving locations within it are covered for mobile reception.
- "**Acceptable**", if at least 70% of the locations at the edge of the area are covered for portable reception and 90% of the receiving locations within it are covered for mobile reception.

Those percentages apply to the edge of the coverage area, the average value of the area is then a larger value. Furthermore, in [8], a receiving location (with area 0.5 m^2) is regarded as covered if the required carrier-to-noise and carrier-to-interference values are achieved for 99% of the time.

1.2.2 Frequency Band Selection

When planning a network, the operator or broadcaster has to make a choice between the available frequency bands. The selection of the frequency band to be used has a major effect on the dimensioning and planning of the network. DVB-H aims to use the so-called broadcast bands: VHF band III (174–230 MHz) or UHF band IV (470–598 MHz) or UHF band V (598–862 MHz), using the standardized 5, 6, 7 or 8 MHz channel bandwidths. Some remarks have to be made when these bands are used.

VHF Band III: the propagation, Doppler characteristics, and building penetration characteristics in this band are exceptionally good. The wavelengths λ in the VHF band (>1 m) imply a large size receiving antenna, which would be difficult to integrate in a small hand-held terminal. Accordingly, except for receiving systems embedded in vehicles, this band is not attractive for handset manufacturers.

UHF Bands IV–V: the propagation and building penetration characteristics remain acceptable to offer large coverage. The Doppler shifts accepted by receivers in these bands correspond to a speed of 250–500 km/h and the size of antenna is suitable for integration. With the exception of the upper part of band V, where GSM900 transmissions could interfere with the DVB-H reception, these bands are the preferred bands for delivering services to mobile handsets from a technical standpoint. However, bands IV and V are heavily congested as broadcasters simulcast Analogue TV and Digital Terrestrial TV services there. In some parts of Europe, access to these bands could be delayed until 2010–2020 when the analogue TV switch off will occur and suitable UHF channels are released.

This chapter will focus on UHF bands IV and V, as those bands are used for the Flemish DVB-H trial in Ghent, Belgium. Furthermore, a channel width of 8 MHz will be considered.

1.2.3 Building Penetration and Vehicle Entry Loss

1.2.3.1 Building Penetration Loss

Portable DVB-H reception will take place at outdoor and indoor locations. The field strength at indoor locations will be attenuated significantly depending on the materials and the construction of the house. Several measurements have been carried out to verify real values of attenuation. A large spread of building penetration losses have been measured [8]. The range of the obtained penetration loss is between 7 and 15 dB. As building penetration loss, [8] assumes a median value of 11 dB with a standard deviation of 6 dB in the UHF band.

The shadowing margin (Sect. 1.3.2) at *indoor* locations is the combined result of the outdoor variation and the variation factor due to building attenuation.

1.2.3.2 Vehicle Entry Loss

For mobile reception inside cars or any other vehicle, entry loss (also called vehicle penetration loss) must be taken into account. For planning purposes, the European Telecommunications Standards Institute (ETSI) proposes an entry loss of 7 dB in case of Class D mobile reception [8]. The standard deviation for vehicle entry loss is neglected in [8].

Only limited data about vehicle entry loss is available. In [20], the mean value of the penetration loss for a car is about 8 dB with an associated standard deviation of 2–3 dB. Measurements and simulations of the vehicle penetration loss at 600, 900, 1800, and 2400 MHz are presented in [30]. The measured average penetration loss varies from 3.2 to 23.8 dB depending on frequency, illuminated vehicle side, and in-vehicle antenna orientation. Vehicle penetration loss tends to follow a lognormal distribution. At, e.g., 600 MHz vehicle penetration losses of about 14.7 dB are obtained with a standard deviation of 7.9 dB [30]. Again, the shadowing margin for reception *inside* cars is the combined result of the outdoor variation and the variation factor due to vehicle entryloss.

Penetration loss into trains is described in [1,2,10]. Propagation into trains is difficult due to the high penetration losses, severe shadowing within the deep cuttings for trains, and due to high velocities causing large Doppler frequencies. The propagation loss into a cutting can be significant. This may result in lost connections. For old Belgian trains with nonmetallized windows, losses up to 22 dB were registered at 2 GHz. For Belgian trains with metallized windows (most modern trains) the penetration loss increased to values between 20 and 32 dB at 2 GHz. For Swiss trains with metallized windows, a mean attenuation of 20 dB (900 MHz) and 22 dB (1900 MHz) is obtained [1]. Measurement campaigns in France and Germany revealed that metallized windows introduced a loss of 25–30 dB [2]. Standard deviations between 2 and 4 dB are mentioned in [1, 10].

1.2.3.3 Influence of Body of Person, People Walking Around

The influence of people walking around the receiving antenna has also been estimated. The signal level variations (10% and 90% value) ranged from +2.6 dB to −2.6 dB. These variations are relatively small and it does not seem necessary to take them into account for planning purposes according to [8].

1.2.4 Receiver Characteristics

1.2.4.1 Receiver Antenna Gain

The antenna solution in a small hand-held terminal has to be an integral part of the terminal construction and will therefore be small when compared to the wavelength. For planning purposes the values of Table 1.1 for the receiver gain G_r (dBi or dBd) can be used for bands IV and V in class A and B reception according to ETSI [8]. It will be necessary to distinguish between class C and class D reception. In class D reception a hand-held terminal is used with the same antenna gain as in classes A and B. In class C reception, a vehicular built-in antenna is used with a greater gain than for hand-held terminals. The practical standard antenna for vehicle reception is a $\lambda/4$ monopole, which uses the metallic roof as ground plane. For passive antenna systems the values in Table 1.1 can be used for planning purposes [8].

Table 1.1 Antenna gain in dBd for hand-held reception for planning purposes [8]

Band	G_r (dBd)	
	Class A, B, D	Class C
Band IV	−12	−2
Band V	−7	−1

1.2.4.2 Receiver Sensitivity

The minimum signal input level to the receiver (Rx) is influenced by the required *C*/*N* ratio (carrier-to-noise ratio required by the system). In [8] *C*/*N* values for different modulation schemes are listed and the receiver noise figure (*F*) has been chosen as 6 dB for the frequency bands IV to V. The minimum receiver sensitivities ($P_{\min}$(dBW)) can be calculated from the receiver noise input power (P_n(dBW)) by using the following formulas:

$$P_n = F + 10\log(k \times T_0 \times B) \quad (1.1)$$

$$P_{\min} = P_n + C/N \quad (1.2)$$

where k is Boltzmann's constant (1.38×10^{-23} Ws/K), T_0 the absolute temperature (290 K), and B the receiver noise bandwidth (Hz). The equivalent minimum receiver field strength $E_{\min}$ (dBμV/m) can be calculated as follows:

$$E_{\min} = P_{\min} - A_a + L_f + 145.8 \quad (1.3)$$

where $A_a = G_r$ (dBd) $+ 10\log(1.64 \cdot \lambda^2/4\pi)$ is the effective antenna aperture in (dBm^2) [4, 8, 24] and L_f is the feeder loss (dB). Tables 1.2 and 1.3 show the Rx sensitivity (Rx sens) and equivalent minimum receiver field strength for the *C*/*N* values required by the specified modulation types according to ETSI TR 102 377 [8] for portable and mobile reception, respectively (MFER = Multi-Protocol Encapsulation Frame Error Rate, PER = Packet Error Rate). For the required *C*/*N* in mobile conditions (C and D) an additional 3 dB could be taken into account [8, 29]. This margin permits higher speeds.

1.2.5 Example of Selected Parameters and Scenarios

The objective is to provide coverage in a certain area. A scenario with a transmitting base station (Tx) in a suburban environment is investigated. The height of this Tx is $h_{Tx} = 64$ m. An ERP = 5 kW (ERP = Equivalent Radiated Power) is considered.

Table 1.2 Rx sensitivity and equivalent minimum receiver field strength $E_{\min}$ for the subscriber terminal (*static Rayleigh channel*, BW = 8 MHz, *F* = 6 dB, *portable*, band V, PER = 10^{-4} [8])

Modulation	Required *C*/*N* (dB)	Rx sens (dBm)	$E_{\min}$ (dBμV/m)
QPSK 1/2	5.4	−93.8	43.9
QPSK 2/3	8.4	−90.8	46.9
16-QAM 1/2	11.2	−88.0	49.7
16-QAM 2/3	14.2	−85.0	52.7
64-QAM 1/2	16.0	−83.2	54.5
64-QAM 2/3	19.3	−79.9	57.8

Table 1.3 Rx sensitivity and equivalent minimum receiver field strength E_{min} for the subscriber terminal (*typical urban channel* for typical reference receiver, BW = 8 MHz, $F = 6$ dB, *mobile*, band V, MFER <5% [8])

Modulation	Min *C/N* (dB)	Rx sens ($C/N + 3$ dB) (dBm)	E_{min} (dBμV/m): C ($C/N + 3$ dB)	E_{min} (dBμV/m): D ($C/N + 3$ dB)
QPSK 1/2	9.5	−86.7	45.0	51.0
QPSK 2/3	12.5	−83.7	48.0	54.0
16-QAM 1/2	15.5	−80.7	51.0	57.0
16-QAM 2/3	18.5	−77.7	54.0	60.0
64-QAM 1/2	20.5	−75.7	56.0	62.0
64-QAM 2/3	24.0	−72.2	59.5	65.5

Table 1.4 Characteristics of the selected DVB-H system

Parameter	Value
Frequency	Channel 37: 602 MHz
Mode	4K
Modulation	Adaptive QPSK, 16-QAM, 64-QAM
Channel bandwidth BW	8 MHz
ERP	5 kW
Tx feeder loss L_f	0 dB
Tx height h_{Tx}	64 m
Receiver antenna gain G_r	Table 1.1 (in dBd or dBi)
Rx feeder loss	0 dB
Rx height h_{Rx}	1.5 m
CPE	Classes A, B, C, D
C/N	Tables 1.2 and 1.3
Coverage requirement	Good / acceptable (portable: 95% / 70%) Good / acceptable (mobile: 99% / 90%)

The height of the receiver (Rx) is $h_{Rx} = 1.5$ m. For broadcast planning purposes this is the common value, taking into account that the differences between planning at 1 and 2 m are negligible.

CPE (customer premises equipment) is considered for coverage classes A, B, C, and D. Because the 4K mode is exclusively defined for use in DVB-H systems we investigate the performance of this 4K mode. Furthermore, a channel BW = 8 MHz is assumed. Table 1.4 summarizes the selected parameters of the considered DVB-H system.

1.3 Coverage Planning for DVB-H

Coverage planning and link budget calculations for a DVB-H network using several PL models will be discussed in this section. First, different PL models are proposed. Next, shadowing and fading margins, which have to be taken into account, are discussed. Link budget calculations for an example based on Sect. 1.2.5 are elaborated. Finally, delay spread and interference are discussed and throughput for a DVB-H system versus the range is analyzed.

1.3.1 Path Loss Models

To calculate the range of the system (and select the optimal parameter setting), PL models are needed. The PL between a pair of antennas is the ratio of the transmitted power to the received power. It includes all of the possible elements of loss associated with interaction between the propagating wave and any objects between the transmitting and receiving antennas [28]. The parameter PL is used for the estimation of the coverage of a system. Different models to obtain the PL have been developed. Models that are applicable for the estimation of the coverage of DVB-H systems are here discussed: the ITU-R P.1546 model [15], the Cost 231 Hata-model [12, 23], and the Ghent model [25, 26].

1.3.1.1 ITU-R P.1546 Model

The ITU-R P.1546 model [15] uses tabulated field-strength values for 1 kW effective radiated power (ERP) at nominal frequencies of 100, 600, and 2000 MHz, respectively, as a function of various parameters. Propagation curves have been plotted using these values. From the field values the PL can be calculated. Some curves refer to land paths, others refer to sea paths. Interpolation or extrapolation of the values obtained for the nominal frequency values should be used to obtain field-strength values for any given frequency. The propagation curves represent the field-strength values exceeded for 50%, 10%, and 1% of the time. The ITU-R P.1546 model [15] is not valid for field strengths exceeded for percentage times outside the range from 1% to 50%.

The curves are based on measurement data mainly relating to mean climatic conditions in temperate regions containing cold and warm seas, e.g., the North Sea and the Mediterranean Sea. The land path curves were prepared from data obtained mainly from temperate climates as encountered in Europe and North America. The sea-path curves were prepared from data obtained mainly from the Mediterranean and the North Sea regions.

Although this model is often used, it has some disadvantages. First, this prediction method uses a receiving height of 10 m [15]. For portable reception, an antenna height of 10 m above ground level is not realistic. For this reason a receiving antenna height of 1.5 m above ground level (outdoor) or above floor level (indoor) has been assumed [8]. Therefore correction factors, noted as height losses, must be introduced. Height loss L_{h} values are provided for some type of environments for 500 and 800 MHz in [8] and [4]. The height loss can also be calculated using [15]. Table 1.5 provides these height losses for different types of environments.

Table 1.5 Receiving antenna height loss for different environments [15]

Frequency (MHz)	Height loss L_h (dB)		
	Rural	Suburban	Urban
Band IV	11	16	22
Band V	13	18	24

Further, the model uses tabulated values at three frequencies, specific distances (not smaller than 1 km and up to 1000 km), and Tx heights (10–1200 m). Interpolation (provided by [15]) is thus necessary.

1.3.1.2 Cost 231 Hata-Model

The most widely used PL model is the Hata–Okumura model [12, 23]. This model is valid for the 500–1500 MHz frequency range. There exists an elaboration on the Hata-Okumura model that extends the frequency range (up to 2100 MHz). This model is not suitable for lower base station antenna heights (valid for $h_{\mathrm{Tx}} = 30-200$ m), and hilly or moderate-to-heavy wooded terrain. More information about this model can be found in [12, 23, 28].

1.3.1.3 Ghent Model

In [25] and [26], the PL is determined using measurements of an actual DVB-H signal at 602 MHz in a suburban environment. This model will be noted as the Ghent model. In Sect. 1.4 the experimental development of a PL model will be discussed and the Ghent model will be used as an example to explain the methodology. The PL is modeled according to a lognormal shadowing model. It was shown in [25] that the variation around the mean PL is well described by a lognormal distribution. This model is valid from 70 m up to 14 km, for BS heights around 60 m, and frequencies of 600 MHz in a suburban environment.

Figure 1.1 shows the PL at 602 MHz for the ITU and Ghent models for a suburban environment. The height of the BS (h_{BS}) is 60 m and the height of the receiving antenna (h_{Rx}) is 2.85 m. The height correction of [15] can then be used to obtain values at $h_{\mathrm{Rx}} = 1.5$ m. Up to 1.96 km, the Ghent model delivers the highest path losses, resulting in the most restrictive model. From that distance on, the ITU model is more restrictive (the environment around Ghent is less dense suburban).

The relation between the equivalent field strength E (dBμV/m) and the PL is as follows [15]:

$$\mathrm{PL} = 139 - E + 20 \log f \qquad (1.4)$$

where E is the field strength for 1 kW ERP and f the frequency in MHz.

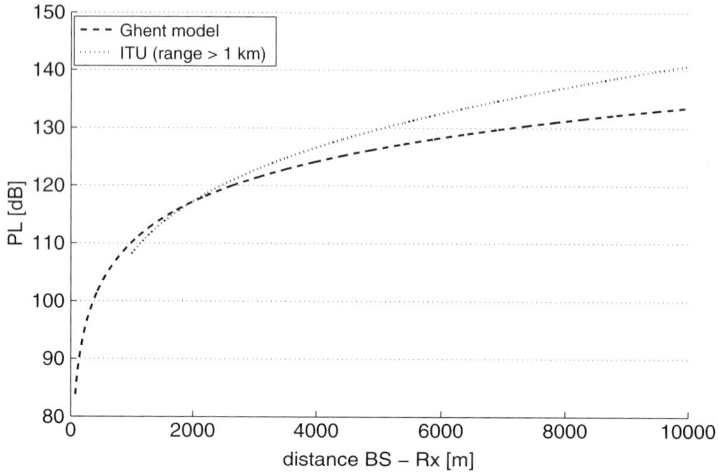

Fig. 1.1 PL at 602 MHz as a function of the distance from the BS for ITU and Ghent models (suburban environment, $h_{\mathrm{BS}} = 60\,\mathrm{m}$ and $h_{\mathrm{Rx}} = 2.85\,\mathrm{m}$)

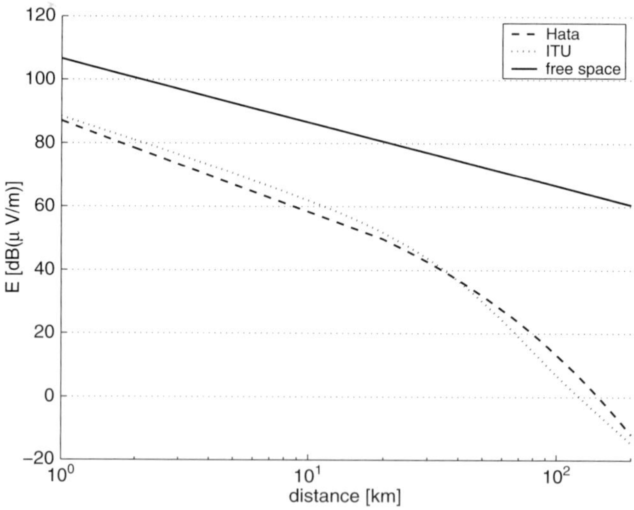

Fig. 1.2 Field strength at 500 MHz as a function of the distance from the Tx for different models (ERP = 1 kW, $h_{\mathrm{Tx}} = 300\,\mathrm{m}$ and $h_{\mathrm{Rx}} = 1.5\,\mathrm{m}$)

Example

Figure 1.2 shows the field strengths E (dB(μV/m)) at 500 MHz from 1 to 200 km at $h_{\mathrm{Rx}} = 1.5\,\mathrm{m}$ calculated with the ITU-R P.1546 model ($L_{\mathrm{h}} = 16\,dB$, $t = 50\%$) and the COST 231 Hata-model for an outdoor receiver in a suburban environment. No penetration loss is taken into account. The height of the Tx (h_{Tx}) is 300 m and an input power of 1 kW ERP is assumed. Also the free-space field is shown in this figure. Due to the multipath environment the field strengths are much lower than in free space.

The equivalent PL – calculated using (1.4) – is shown in Fig. 1.3. Due to the multipath environment the PL is much higher than in free space. The Hata and ITU model correspond reasonably well (differences of about 5 dB). The PL is of course higher for larger distances between Tx and Rx.

Figure 1.4 shows the PL as a function of the height of the base station from 10 to 300 m. The distance between Tx and Rx is 10 km and $h_{Rx} = 1.5$ m. The ITU

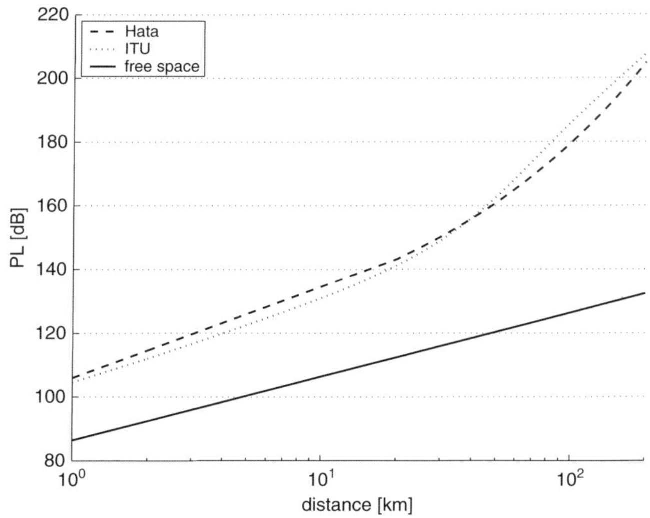

Fig. 1.3 PL at 500 MHz as a function of the distance from the Tx for different models ($h_{Tx} = 300$ m and $h_{Rx} = 1.5$ m)

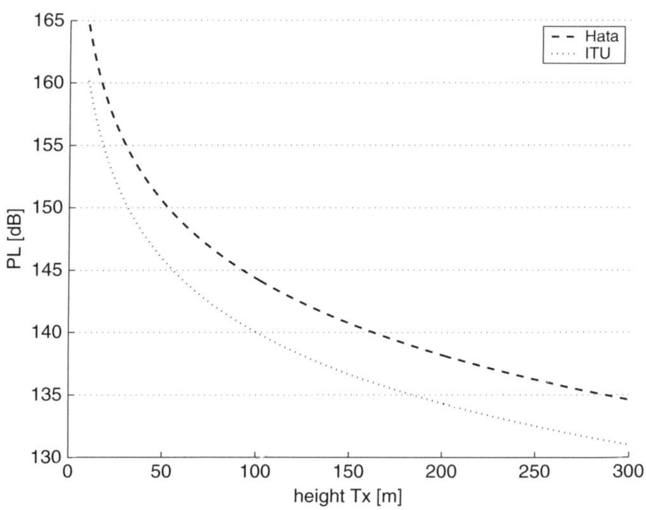

Fig. 1.4 PL at 500 MHz as a function of the height of the Tx (distance BS – Rx = 10 km, $h_{Rx} = 1.5$ m)

model delivers a PL that is about 5 dB lower than the Hata-model for the different heights. It is obvious that placing the BS higher results in a lower PL and thus a larger coverage. Changing the height of the Tx has a large influence on the PL: e.g., at $h_{\mathrm{Tx}} = 10\,\mathrm{m}$ the PL is 160 dB, while at $h_{\mathrm{Tx}} = 300\,\mathrm{m}$ the PL is only 131 dB (ITU model).

1.3.2 Shadowing and Fade Margins

The shadowing margin and fade margin have an important role in the link budget. They depend on the coverage and reliability requirements of the operator for the system. A service could have to be provided at 90% of all locations within a cell with 99% reliability.

In this chapter all ranges will be calculated for good reception. This "good" reception is defined as follows: at least 95% of receiving locations at the edge of the area are covered for portable reception and 99% of receiving locations within it are covered for mobile reception [8].

The standard deviation of the ITU model is used for determining the shadowing margin (noted as location correction factor in [8]) for *outdoor* locations (portable and mobile). The shadowing margin at *indoor* locations is the combined result of the outdoor variation and the variation factor due to building attenuation. These distributions are expected to be uncorrelated. The standard deviation of the indoor field strength distribution can therefore be calculated by taking the root of the sum of the squares of the individual standard deviations.

At UHF, the outdoor and indoor macroscale standard deviations are 5.5 and 6 dB according to [8, 15], respectively. The combined value is then 8.3 dB. Table 1.6 shows the shadowing margins considered here and based on ETSI [8]. We assume here a building penetration loss of 11 dB as proposed in [8] (Sect. 1.2.3) and a vehicle entry loss of 8 dB with standard deviation of 2.5 dB [20] (Sect. 1.2.3). The combined value of the outdoor and vehicle entry deviation is then 6 dB. The shadowing margin for mobile reception in a car (class D) is then 14 dB for 99% coverage.

The shadowing effect follows a lognormal distribution. The coverage requirement described here is for locations at the *edge* of the cell. Sometimes it may be more appropriate to design the system for a coverage probability over the whole cell [28].

Table 1.6 Shadowing margins based on ETSI guideline TR 102 377 [8]

Reception mode	Coverage requirement (%)	Shadowing margin (dB)	
		Outdoor	Indoor
Portable	>95	9	14
	>70	3	4
Mobile	>99	13	14
	>90	7	7

For the calculation of shadowing margins, the following formula for the location variability σ_L can be used [28]:

$$\sigma_L = 0.65\,(\log(f_c))^2 - 1.3\log(f_c) + A \tag{1.5}$$

where $A = 5.2$ in the urban case and 6.6 in the suburban case. f_c is the frequency in MHz. These values apply to macrocells. At, e.g., 500 MHz for suburban and urban regions, standard deviations of 7.8 and 6.4 dB are obtained, respectively. These values are higher than the 5.5 dB of the ITU-R P.1546 model [15].

The fade margin takes the yearly availability of the system into account. The link availability will be affected by clear-air multipath and rain multipath fading. The ITU-R P.530 model [16] described in ECC report 33 [3] can be used to determine the fade margin. A fade margin of 10 dB results in a yearly availability of 99.995% for a cell radius of 10 km.

1.3.3 Prediction of Range of a DVB-H System: Example

In this section an example of range prediction for a DVB-H system will be elaborated. The parameters of Sect. 1.2.5 will be used. The calculation and tabulation of signal powers, gains, losses, and *C/N* for a complete communication link is called a link budget, which is a useful approach for the basic design of a communication system. To determine the coverage range of the DVB-H system we use the parameters of Tables 1.2 – 1.6 and formulas (1.1), (1.2), and (1.3).

We analyze a receiver in band V (channel 37: 602 MHz) and assume good coverage (e.g., a coverage requirement of 95% for coverage class A). An ERP = 5 kW is assumed. We investigate the scenario described in Sect. 1.2.5 (Table 1.4, h_{Tx} = 64 m). The environment is suburban.

Table 1.7 summarizes the ranges of the different modulation schemes for the scenario under consideration (four coverage classes, good coverage) using the ITU (t = 50% [8]) and Hata-model. The range that can be obtained depends on the type of modulation. QPSK 1/2 results in ranges up to 3.7 km for suburban regions (class B,

Table 1.7 Ranges (km) for the considered scenario

Modulation	Ranges (km)							
	ITU				Hata			
	A	B	C	D	A	B	C	D
QPSK 1/2	10.5	3.7	7.8	2.8	8.9	3.0	6.2	2.2
QPSK 2/3	8.8	3.0	6.4	2.2	7.2	2.4	5.1	1.8
16-QAM 1/2	7.4	2.5	5.3	1.8	6.0	2.0	4.1	1.5
16-QAM 2/3	6.1	2.0	4.3	1.4	4.9	1.6	3.4	1.2
64-QAM 1/2	5.4	1.7	3.7	1.2	4.3	1.4	2.9	1.1
64-QAM 2/3	4.3	1.4	2.9	1.0	3.4	1.2	2.3	<1.0

ITU) while 64-QAM 2/3 is only possible up to 1.4 km (class B, ITU). Outdoor portable reception results in the largest ranges, while indoor (class B) and mobile (class D) reception result in the lowest ranges due to additional building and vehicle penetration loss, respectively. Table 1.7 shows that for, e.g., 16-QAM 1/2 ranges of 7.4 km (class A, outdoor), 2.5 km (class B, indoor), 5.3 km (class C, outdoor vehicle), and 1.8 km (class D, inside vehicle) can be obtained using the ITU model. The Hata-model delivers higher PL values for the considered scenario than the ITU model. Therefore the ranges obtained with the Hata-model are lower (e.g., for class B from 1.2 to 3.0 km). Thus one has to be careful when using only the ITU model, which may be too optimistic.

Calculations for reception into trains can also be done. The range for the considered DVB-H system for reception in trains (class D) is calculated for a coverage requirement 90%. The ratio *C/N* is calculated without a 3 dB margin. A train entry loss of 26 dB and standard deviation of 3 dB is considered [1, 2, 10]. The shadowing margin for a coverage requirement of 90% and for the combined standard deviation is then 8.1 dB. This reception mode results in ranges (ITU model) of 1.4 km for QPSK 1/2, 1.2 km for QPSK 2/3, and lower than 1 km for the higher modulation schemes. Reception in trains will thus be very difficult.

1.3.4 Estimation of Throughput

In this section first the delay spread is estimated and interference is analyzed. Next, suitable OFDM parameters are selected taking into account the delay spread and interference. Finally, the physical-layer throughput is determined.

1.3.4.1 Estimation of Delay Spread

The phenomenon of delay spread is due to multipath scattering. In order to avoid inter-symbol interference (ISI) and inter-carrier interference (ICI), a cyclic prefix (CP) or guard time T_g (guard interval GI) is introduced in front of every data part of an OFDM symbol. It is therefore necessary to choose a CP larger than the maximum delay spread.

The delay profile is characterized by τ_{rms} (root-mean-square (rms) delay spread of the entire delay profile). It was found that the rms delay spread for omnidirectional antennas [11] follows a lognormal distribution and that the median of this distribution grows as some power of distance. The model of [11] is the following:

$$\tau_{rms} = T_1\, d^{\varepsilon} y \tag{1.6}$$

where τ_{rms} is the rms delay spread, d is the distance in km, T_1 is the median value of τ_{rms} at d = 1 km, ε is an exponent that lies between 0.5 and 1.0 and y is a lognormal variate with standard deviation between 2 and 6 dB. The parameters T_1 and ε of the

Table 1.8 Rms delay spread τ_{rms} for a range up to 10 km using the model of [11]

Environment	T_1 (μs)	ε	Median τ_{rms} (μs) at 10 km	Maximum τ_{rms} (μs) in 90% at 10 km
Rural	0.1	0.5	0.3	1.3
Suburban	0.3	0.5	1.0	2.4
Urban micro	0.4	0.5	1.3	2.6
Urban macro	0.7	0.5	2.2	4.6
Mountain	0.5	1.0	5.0	11.6

model at 1 km are shown in Table 1.8 [11]. The delay spread remains unchanged for frequencies above 30 MHz, since the wavelengths become much smaller than human-made architectural structures [14, 19].

Table 1.8 also lists the median delay spread values obtained by simulations with the model of (1.6) for different types of environment for a cell range of 10 km and the 90th percentile of the delay spread values at 10 km. Table 1.8 shows thus that 90% of the delay spread values are smaller than 3 μs for suburban regions. The delay spread is the smallest for rural areas and has the largest values for mountainous areas. As the terrain is mostly flat in the considered suburban environment (example Sect. 1.3.3), a maximum delay spread of 3 μs is an acceptable assumption for this environment. In [19] a majority (95%) of the measured delay spread values is less than 1.75 μs.

1.3.4.2 Interference

The two main interference sources that constrain the cell size of a DVB-H network are inner and outer interference. The inner interference is the interference generated by the transmitters in the SFN. The outer interference is the interference coming from other SFNs or MFNs (multifrequency networks) that operate at the same frequency.

The maximum acceptable echo delay depends on the used guard interval. When the echo delay of the signal is higher than the guard interval then interference occurs. Thus as long as the distance between the transmitters in the SFN is less than a value R_g then there is no inner interference in a network with only two transmitters. Otherwise, inner interference will occur. The maximum distance R_g is defined by the following formula [8]:

$$R_g = c \cdot T_g \tag{1.7}$$

where c is the velocity of light and T_g is the guard time.

Due to multipath environments, different signals will be received with different delays and interference is inevitable. Table 1.9 shows for different guard intervals or ratios T_g/T_u (T_u = useful time) the maximum distance (see (Eq. 1.7)) between the transmitters in an SFN for which minimal inner interference may appear [6,8]. Minimal interference means that if the distances between the transmitters were larger

Table 1.9 R_g and T_g for the different DVB-H modes [8]

T_g/T_u	2K mode		4K mode		8K mode	
	$T_g(\mu s)$	$R_g(km)$	$T_g(\mu s)$	$R_g(km)$	$T_g(\mu s)$	$R_g(km)$
1/4	56	16.8	112	33.6	224	67.2
1/8	28	8.4	56	16.8	112	33.6
1/16	14	4.2	28	8.4	56	16.8
1/32	7	2.1	14	4.2	28	8.4

Table 1.10 Choice of OFDM parameters for 4K mode and channel bandwidth (BW) of 8 MHz (c = velocity of light)

OFDM parameters	Value	Choice BW = 8 MHz
Mode	2K, 4K, 8K	4K
Number of carriers N_{FFT} (FFT-size)	2048, 4096, 8192	4096
Modulated carriers K	1705, 3409, 6817	3409
Elementary period T	7/64	7/64
Useful time T_u	224, 448, 896 μs	448 μs
Carrier spacing Δf	$1/T_u$	2.232 kHz
T_g/T_u	1/32, 1/16, 1/8, 1/4	1/4
Guard time T_g	$T_u/4$	112 μs
Symbol time T_s	$T_u + T_g$	560 μs
Max distance of Tx R_g	$c \cdot T_g$	33.6 km

than the ones in Table 1.9 – the other corresponding network parameters remain the same – the network itself will incur more interference. For the 4K mode, this SFN radius is two times larger than for the 2K mode and half of the radius for the 8K mode [8]. A trade-off between the coverage range and interference has to be made. The larger the range, the higher the percentage of the cell area that will receive interference from other transmitters of the same SFN.

1.3.4.3 OFDM Parameters

We analyze the configuration of the DVB-H system of Table 1.4. We consider thus the 4K mode for an 8 MHz channel. We choose the guard time T_g equal to 1/4th of the useful time. The guard time is then equal to 112 μs and is thus larger than $\tau_{rms} = 3\,\mu s$. The maximum distance R_g is 33.6 km for this guard time. The chosen OFDM parameters are shown in Table 1.10 [6].

1.3.4.4 Physical Layer (PHY) Bit Rates

Using the OFDM parameters of Table 1.10 (BW = 8 MHz), the physical layer (PHY) bit rates can be calculated for the different modulations and coding rates [6].

Table 1.11 shows these raw bit rates on the PHY level. The throughput varies from 5 to 20 Mbps. The spectral efficiency varies from 0.6 to 2.5 bit/s/Hz for the considered parameters.

1.3.5 Throughput and Coverage Results

Figure 1.5 shows the PHY throughput for the suburban scenario (Table 1.4, 4K mode, 8 MHz channel) as a function of the distance at 602 MHz. For this figure we combine the coverage distances obtained from Sect. 1.3.3 with the bit rates from Table 1.11. The theoretical PHY throughput for the four coverage classes is shown. Each marker on the curve of a class corresponds with a modulation and coding rate. QPSK 1/2 corresponds with the highest ranges while 64-QAM 2/3 will reach the shortest distances. For example, at a range of 7.4 km a throughput of

Table 1.11 PHY modes and bit rates for 4K mode, $T_g/T_u = 1/4$ for BW = 8 MHz [6]

Modulation	Coding rate	PHY bit rate (Mbps)	Spectral efficiency (bit/s/Hz)
QPSK	1/2	4.98	0.62
QPSK	2/3	6.64	0.83
16-QAM	1/2	9.95	1.24
16-QAM	2/3	13.27	1.66
64-QAM	1/2	14.93	1.87
64-QAM	2/3	19.91	2.49

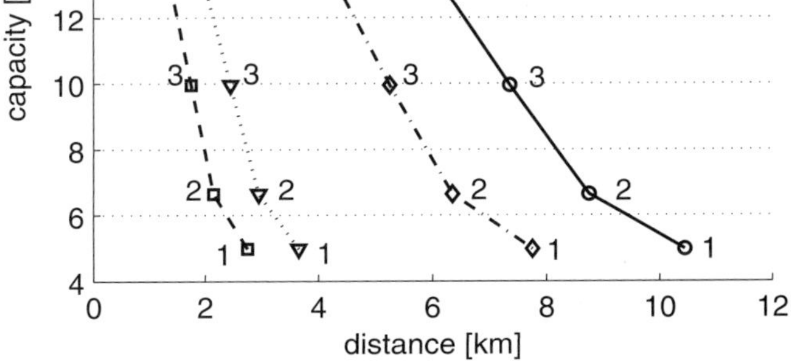

Fig. 1.5 Throughput for 8 MHz channel as a function of the distance for the four coverage classes (ERP = 5 kW, h_{BS} = 64 m, and h_{Rx} = 1.5 m, 1 = QPSK 1/2, 2 = QPSK 2/3, 3 = 16-QAM 1/2, 4 = 16-QAM 2/3, 5 = 64-QAM 1/2, 6 = 64-QAM 2/3)

10 Mbps can be obtained for class A using the ITU model. The indoor performance is limited to much smaller ranges: e.g., 10 Mbps is only possible up to 1.8 km in a vehicle (class D) and up to 2.5 km inside houses (class B). No MPE–FEC protection is assumed in Fig. 1.5.

1.4 Coverage Measurements for a Path Loss Model

In this section, the measurement methodology for the development of a PL model is discussed. As an example, propagation measurements, performed at 602 MHz in a suburban environment in Ghent, Belgium, are presented. From the experimental data a statistical PL model is derived. This model can be used for coverage estimation of DVB-H networks.

1.4.1 Measurement Site and Measurement Equipment

The base station (BS) antenna is located on the roof of a building at a height of $h_{\mathrm{BS}} = 64$ m. The measurement area is mostly suburban. The considered DVB-H system is based on the ETSI standards [5]. The azimuth pattern of the BS antenna is omnidirectional. The vertical 3 dB-beamwidth equals 6°. The gain of the antenna is 7.5 dBd. A DVB-H signal with modulation scheme 16-QAM 1/2 and a bandwidth (BW) of 7.61 MHz is injected in the transmitting antenna (Tx). The used frequency is 602 MHz. The ERP of the BS antenna equals 37.76 dBW.

As Rx, an omnidirectional Jaybeam antenna type 7511 is used. The measurements are performed at 2.85 m above ground level. The receiver antenna is vertically polarized and its gain is 0 dBd. Figure 1.6 shows the receiver part of measurement

Fig. 1.6 Vehicle with Rx used for the data acquisition

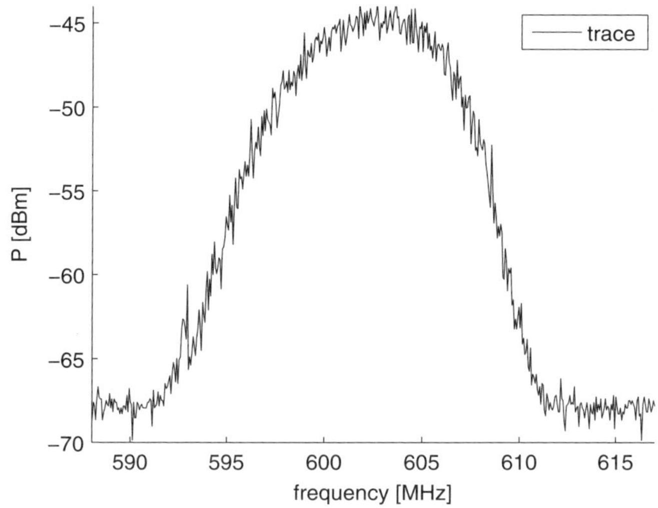

Fig. 1.7 Trace of the DVB-H signal measured by the spectrum analyzer at non-line-of-sight location

system. The measurements are performed with a Rohde & Schwarz FSEM30 spectrum analyzer (SA) with a frequency range from 20 Hz up to 26.5 GHz. The output of the SA is sampled and stored on a laptop. Optimal settings for measurements with the SA have been determined: a center frequency of 602 MHz, frequency span of 30 MHz, and a resolution bandwidth of 5 MHz. The measurement value also depends on the detector mode: the rms mode is used, as proposed in [17]. The measurement positions are acquired with a GPS (Global Positioning System) device. Using a car, the Rx is moved with constant speed in the environment. Figure 1.7 shows a trace of the DVB-H signal (bandwidth of 7.61 MHz), measured with the SA with the proposed settings at a non-line-of-sight location in Ghent, Belgium.

1.4.2 Data Processing

First, the noise floor is determined for each measurement track. Then an additional margin of 5 dB is added. Samples which are below the noise floor plus this margin are discarded. This margin is a compromise between sensitivity of the measurement system and noise elimination. In this way not all samples far from the BS are discarded. On the other hand, 5 dB is high enough to separate signal from noise. Figure 1.8 shows a data track, the noise level, and the additional margin. To remove the fluctuations of the fast fading, the received signal strength is averaged and sampled according to the Lee criterium [21], i.e., about 50 samples for each 40 wavelengths. The sampling of the measurement points depends upon the velocity of the car. Using [17] and [21] and the sampling of the measurement system,

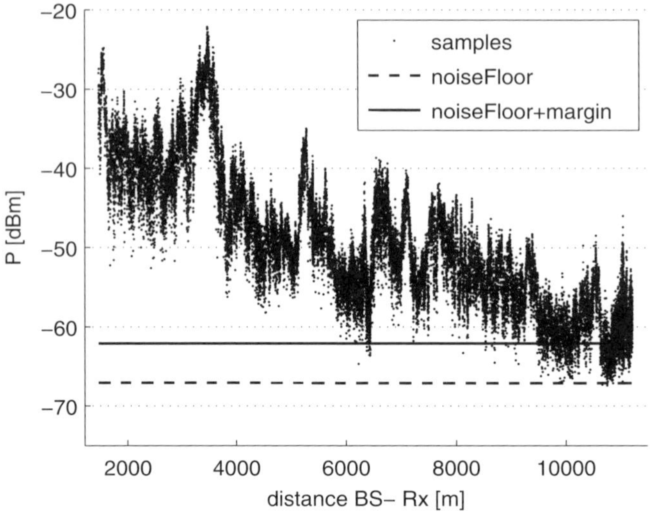

Fig. 1.8 Example of measurement track and noise floor at 602 MHz as a function of the distance BS − Rx ($h_{BS} = 64$ m and $h_{Rx} = 2.85$ m)

a maximum speed of 25 km/h has to be used for the acquisition of the measurement data. A constant velocity of 20 km/h is used. For each track, the environmental conditions (wide/narrow street, residential/suburban, high/low buildings) were determined. We consider distances from about 100 to 14000 m from the BS.

1.4.3 Path Loss

PL is defined in Sect. 1.3.1 [28]. The PL is modeled according to a lognormal shadowing model:

$$\text{PL} = P_0 + 10\, n \log(d/d_0) + \chi \tag{1.8}$$

where d is the distance between BS and Rx in m, d_0 is a reference distance in m, and n is the PL exponent. d_0 was chosen 100 m here. Furthermore, χ is the shadowing fading variation and has a standard deviation σ. A fit with two parameters, P_0 and n, was performed. The rms deviation of the measurement points was minimized with a linear regression fit. Figure 1.9 shows the measurements and fit of PL at $h_{Rx} = 2.85$ m as a function of the distance BS − Rx. The parameter P_0 equals 86.8 dB. We obtain a PL exponent $n = 2.34$ and a standard deviation of 6.18 dB.

To investigate the correctness of the model, the cdf (cumulative distribution function, i.e., Prob [deviation < abscissa]) of the difference between the experimental data and the PL model is analyzed. This cdf is then being fit using a cdf of a normal distribution (in dB). Again the rms deviation is minimized with a linear regression fit, where the mean value and the standard deviation are adjusted.

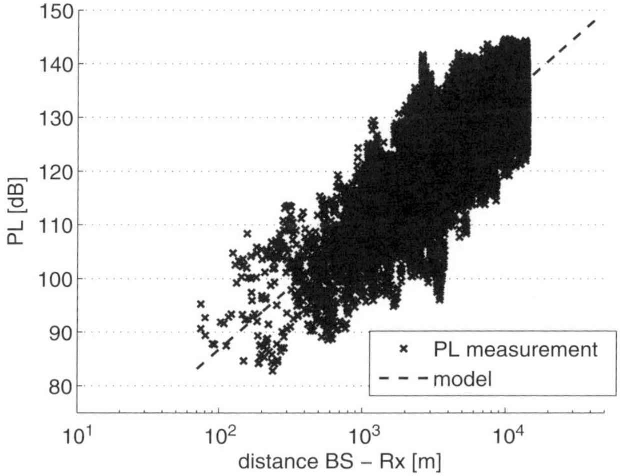

Fig. 1.9 Scatter plot and linear regression fit of PL at 602 MHz as a function of the distance BS – Rx ($h_{BS} = 64$ m and $h_{Rx} = 2.85$ m)

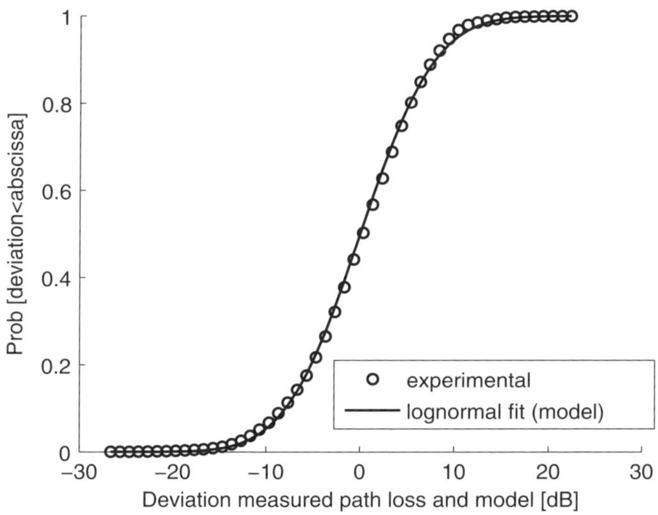

Fig. 1.10 Cumulative distribution of experimental data and linear regression fit

The standard deviation σ_{fit} of the normal fit equals 6.17 dB, which agrees excellently with the 6.18 dB of the experimental data, indicating the correctness of the model. Figure 1.10 shows the cdf of the experimental data and the linear regression fit. To further assess lognormality of the samples, a statistical goodness-of-fit test was performed. A Kolmogorov–Smirnov (K–S) test was conducted on the samples. The measurements passed the K–S test at a significance level $\alpha = 0.05$.

1.4.4 Equivalent Field Strength

Based on [15] and [17], the power measurements in Ghent are converted to corresponding values of the equivalent electric field strength E (dBμV/m):

$$E = V_o + k + A, \tag{1.9}$$

with V_o the output voltage of the antenna in dBμV, k the antenna factor in dB(1/m), and A the attenuation of the antenna signal path in dB. A, in this case, is the cable loss between Rx and SA.

V_o can be calculated as follows:

$$V_o = 20 \cdot \log \left(10^6 \cdot \sqrt{50 \cdot \frac{10^{\frac{P_r}{10}}}{1000}} \right). \tag{1.10}$$

k is defined as

$$k = 20 \cdot \log(f) - 29.7707 - G_r, \tag{1.11}$$

with f the frequency under consideration (in MHz) and G_r the gain of the receiver antenna.

In Fig. 1.11 a map of the measurement environment is shown together with the field-strength values. The field strengths are divided into five categories. The locations marked in blue, green, yellow, or red indicate to which coverage class that location belongs [7]. The corresponding minimal electric field strength necessary to obtain good reception at 2.85 m (height correction [18]) is 64.5, 69.5, 80.5, and 84.5 dBμV/m for classes A, C, B, and D, respectively [7]. The locations marked in black (except for the BS, marked with a black dot) do not have a good reception in either of the four coverage classes.

From the map it is clear that near the BS, reception quality is best and the further from the BS, the lower the field strengths are. In more open areas such as north of the BS though, the electric field decreases slower than in denser areas, such as south west from the BS.

Finally, we calculate the range for 16-QAM 1/2 for indoor reception of the system considered in Sects. 1.2.5 and 1.3.3 but with the PL model of (1.8). A range of 3.2 km is obtained for class B reception using the Ghent model compared to a range of 2.5 and 2.0 km using the ITU and Hata-model, respectively.

1.5 Parameter Evaluation and Performance Analysis of DVB-H System

In this section, physical-layer measurements for a DVB-H system in realistic channel conditions are investigated. This system is based on the specifications and guidelines of ETSI [5–8]. The goal is to evaluate the performance of the system for

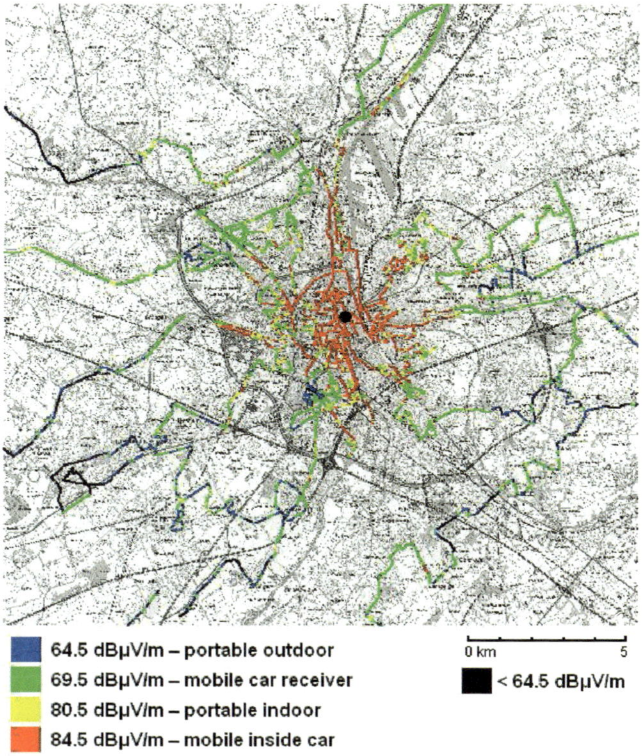

Fig. 1.11 Measured equivalent electric field strengths in Ghent Belgium (black dot indicates the location of the base station, $h_{\mathrm{Tx}} = 64\,\mathrm{m}$, $h_{\mathrm{Rx}} = 2.85\,\mathrm{m}$)

different modulation schemes, guard intervals, and MPE–FEC rates. This analysis will enable better estimations of the performance of the DVB-H system and an optimized DVB-H performance model (range, influence of different parameter settings). Different reception conditions (indoor walking, indoor standing, outdoor walking, car 20 km/h, car 70 km/h, car 120 km/h, train, tram, and bus) will be discussed and performance for each of these scenarios is analyzed.

1.5.1 Transmitting Network

Many results presented in this section are obtained from the Flemish DVB-H trial during 2007 in Ghent, Belgium. Therefore the transmitting network in Ghent is here presented. Figure 1.12 shows a map of Ghent where the Flemish DVB-H trial network is situated and the location of the three BS marked with large black dots (other labels will be explained further). The environment is rural and suburban. The operation frequency is 602 MHz with a bandwidth of 8 MHz. In Belgium, there are

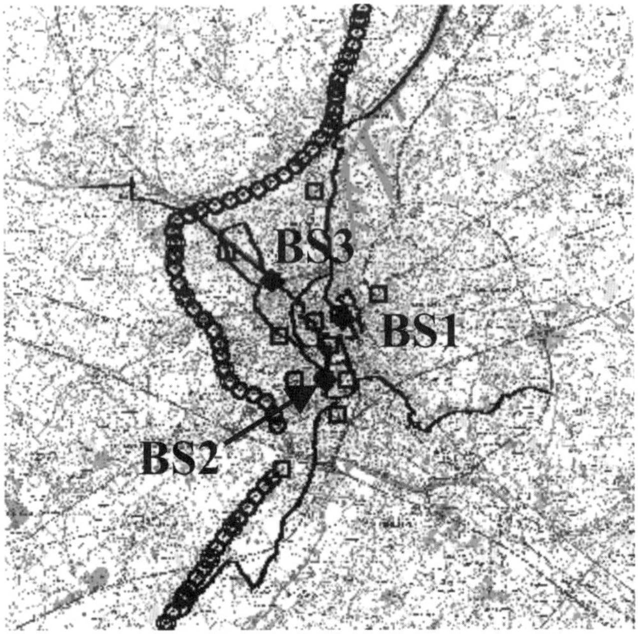

Fig. 1.12 Map of Ghent with the three transmitting antennas (large black dots), the selected buildings for the indoor portable measurements (squares), and the routes for the mobile measurements (the routes at 20 km/h are shown with a full black line, the routes at 70 km/h are shown with circles)

almost no urban regions (urban is defined here as an area as the center of New York, only parts of the town Brussels in Belgium are urban). Cities as Ghent can be classified as suburban, the region around the city center can be classified as rural. All transmitting antennas are omnidirectional. The heights of these BS are $h_{BS} = 64$ m (BS1), $h_{BS} = 63$ m (BS2), and $h_{BS} = 57$ m (BS3), respectively. The EIRP (EIRP = Effective Isotropic Radiated Power) used for these BS is 39.0, 40.9, and 36.6 dBW, respectively.

1.5.2 Reception Scenarios

DVB-H services will be provided in different circumstances. Coverage for DVB-H will be demanding since reception is expected in different conditions (the terminal is moving, body loss, no line-of-sight conditions, etc.). The measurements discussed here have been performed for portable outdoor (class A), portable indoor (class B), and mobile reception (class D).

We define nine different scenarios. Each scenario has been assigned an identification number (ID). Table 1.12 summarizes the different scenarios, each with a brief description. In the following sections, the different scenarios will be explained.

Table 1.12 Investigated scenarios during trial

Class	Scenario	ID	Description
A	Outdoor walking	I	Twenty routes
B	Indoor walking	II	Thirteen buildings, measurements on each floor (rooms, corridors, and staircases)
B	Indoor standing	III	Thirteen buildings, 1 minute each floor
D	Car 20 km/h	IV	Six routes, 70.5 km
D	Car 70 km/h	V	Three routes, 37.5 km
D	Car 120 km/h	VI	Two routes, 50 km
D	Train	VII	Station Gent St-Pieters → Wondelgem
D	Tram	VIII	Station Gent Sint-Pieters → Evergem
D	Bus	IX	Wondelgem → station Gent Sint-Pieters

1.5.2.1 Portable Reception

For class A (scenario I), 20 routes in 11 different parts of the city have been investigated (Table 1.12). The routes selected for class A measurements are situated near the buildings investigated for class B. For class B (II and III), 13 buildings have been selected. Indoor walking (II) as well as indoor standing (III) have been investigated. In Fig. 1.12 the different buildings are marked with squares. For scenarios II and III, measurements have been performed during 1 min on each floor in every building.

1.5.2.2 Mobile Reception

For class D (scenarios IV–IX), measurements have been performed for different mobile scenarios (Table 1.12): reception inside a train (VII), a tram (VIII), a bus (IX), and a car (IV–VI). For reception inside a car, different reception velocities have been investigated: 20 km/h (IV), 70–90 km/h (V) and 100–120 km/h (VI). Figure 1.12 shows a map of Ghent with routes of different mobile scenarios of class D.

1.5.3 Parameters Used to Analyze Performance

This section defines the parameters used to analyze the performance of the DVB-H system. First, MpegLock and MpegDataLock are explained. Next, parameters corresponding with MPE tables, transmission quality, and signal quality parameters are described. Finally, MPE–FEC gain is explained.

Basic definitions

- MpegLock: If MpegLock is "on," the TS (transport stream) synchronization is achieved.

- MpegDataLock: If MpegDataLock is "on," the TS (transport stream) synchronization is achieved and the TS packet is valid.

Parameters corresponding with MPE tables

- *%Lock*: the percentage of the time that the logged parameters MpegLock and MpegDataLock are both "on." When both are "on," it is possible to receive MPE tables.
- *%Not locked*: the percentage of the time that at least one of the logged parameters MpegLock and MpegDataLock is "off."

$$\%Not\ Locked = 100 - \%Lock \tag{1.12}$$

- *%Correct tables*

$$= \frac{number\ of\ correct\ MPE\ tables\ received \times 100}{number\ of\ received\ tables} \tag{1.13}$$

- *%Incorrect tables* = MFER
- *%Corrected tables*

$$= \frac{number\ of\ corrected\ MPE\ tables\ received \times 100}{number\ of\ received\ tables} \tag{1.14}$$

- *%Valid reception* is the percentage of the time that the receiver is locked and receives either correct, or corrected tables:

$$\%Valid\ reception = 100 - \left[\%Not\ locked + \left(\frac{\%Lock \times \%Incorrect\ tables}{100}\right)\right] \tag{1.15}$$

Transmission quality criteria

- *%Locations*: the ratio of the number of valid tables received in a 3 dB range for E (dBμV/m) / 1 dB range for $\frac{C}{N+I}$ (dB) and the total number of tables received in that 3 dB / 1 dB range. This definition can also be found in [7].
- *%Locations QEF ok*: the ratio of the number of samples that meet the Quasi-Error-Free criterion and the total number of samples recorded (in a 3 dB range for E/1 dB range for $\frac{C}{N+I}$). Samples meet the Quasi-Error-Free criterion when the recorded BER (bit error rate) is smaller than $2 \cdot 10^{-4}$ after the Viterbi decoder [29].
- *%Locations ESR5 ok*: the ratio of the number of samples that meet the Erroneous Second Ratio 5% criterion and the total number of samples recorded (in a 3 dB range for E/1 dB range for $\frac{C}{N+I}$). Samples meet the Erroneous Second Ratio 5% criterion when there is a maximum of 1 erroneous second in 20 s. This corresponds to a recorded BER smaller than $2 \cdot 10^{-3}$ after the Viterbi decoder [29].

Signal quality requirements

- $\frac{C}{N+I}|_{MFER5\%}$: the minimal value of $\frac{C}{N+I}$ (dB) for which the MFER is at most 5%, or *%Locations* is at least 95% *after* MPE–FEC correction.
- $\frac{C}{N+I}|_{FER5\%}$: the minimal value of $\frac{C}{N+I}$ (dB) for which the FER is at most 5%, or *%Locations* is at least 95% *before* MPE–FEC correction.
- $E|_{MFER5\%}$: the minimal value of E (dBμV/m) for which the MFER is at most 5%, or *%Locations* is at least 95% *after* MPE–FEC correction.
- $E|_{FER5\%}$: the minimal value of E (dBμV/m) for which the FER is at most 5%, or *%Locations* is at least 95% *before* MPE–FEC correction.

MPE–FEC gain

- $\triangle C_{X\%}$: the reduction in dB for $\frac{C}{N+I}$ obtained by using MPE–FEC, while maintaining the same reception quality

$$\triangle C_{X\%} = \frac{C}{N+I}|_{FERX\%} - \frac{C}{N+I}|_{MFERX\%} \tag{1.16}$$

- $\triangle E_{X\%}$: the reduction in dB for E obtained by using MPE–FEC, while maintaining the same reception quality

$$\triangle E_{X\%} = E|_{FERX\%} - E|_{MFERX\%} \tag{1.17}$$

1.5.4 Measurement Methodology

1.5.4.1 Investigated Transmission Schemes for Optimization

The parameters that have been tuned are modulation, coding rate, guard interval GI, and MPE–FEC coding rate level. For each set or combination of parameters we define a number. A transmission scheme will be noted as follows: "FFT mode, guard interval, modulation and inner code rate, MPE–FEC rate." Table 1.13 gives an overview of the 14 investigated transmission schemes and their corresponding physical bit rate. The parameters (in Table 1.13) that are varied are shown in bold. Six parameters sets (no. 1–6) have been selected to investigate MPE–FEC rates ranging from 67/68 to 1/2. For six transmission schemes, different modulation schemes and inner coding rates from QPSK 1/2 to 64-QAM 2/3 are selected (no. 7–11 and no. 2). Finally, to study the influence of the guard interval (1/32 up to 1/4) four schemes are selected as shown in Table 1.13.

1.5.4.2 Measurement Method

The measurements are performed with a DVB-H tool implemented on a PCMCIA card with a small receiver antenna [26, 27]. The gain of the antenna is −5 dBi. The PCMCIA card is plugged into a laptop, which is used to collect and process the

Table 1.13 Transmission schemes and corresponding physical bit rate, investigated to determine the influence of MPE–FEC, modulation scheme (and inner code rate), and the guard interval

No	Transmission scheme	PHY bit rate (Mbps)
1	4K, 1/8, 16-QAM 1/2 **MPE–FEC 67/68**	10.90
2	4K, 1/8, 16-QAM 1/2 **MPE–FEC 7/8**	9.68
3	4K, 1/8, 16-QAM 1/2 **MPE–FEC 5/6**	9.22
4	4K, 1/8, 16-QAM 1/2 **MPE–FEC 3/4**	8.30
5	4K, 1/8, 16-QAM 1/2 **MPE–FEC 2/3**	7.37
6	4K, 1/8, 16-QAM 1/2 **MPE–FEC 1/2**	5.53
7	4K, 1/8, **QPSK 1/2** MPE–FEC 7/8	4.84
8	4K, 1/8, **QPSK 2/3** MPE–FEC 7/8	6.45
2	4K, 1/8, **16-QAM 1/2** MPE–FEC 7/8	9.68
9	4K, 1/8, **16-QAM 2/3** MPE–FEC 7/8	12.91
10	4K, 1/8, **64-QAM 1/2** MPE–FEC 7/8	14.52
11	4K, 1/8, **64-QAM 2/3** MPE–FEC 7/8	19.36
12	4K, **1/4**, 16-QAM 1/2 MPE–FEC 7/8	8.71
2	4K, **1/8**, 16-QAM 1/2 MPE–FEC 7/8	9.68
13	4K, **1/16**, 16-QAM 1/2 MPE–FEC 7/8	10.25
14	4K, **1/32**, 16-QAM 1/2 MPE–FEC 7/8	10.55

measurements later. Every 0.5 s, a sample is recorded, while the receiver is either locked or unlocked [26,27]. A locked receiver can receive DVB-H frames, which are either correct or incorrect. Incorrect tables can be corrected by the MPE–FEC code. The tool logs parameters as $\frac{C}{N+I}$ (carrier-to-interference-noise ratio), FER (Frame Error Rate), MFER (Multi-Protocol Encapsulation FER), and electric-field strength. Location and speed are recorded with a GPS device.

1.5.5 Results for One Reception Scenario and One Transmission Scheme: Car 20 km/h

In this section, the performance of DVB-H for scenario IV (driving in a car at 20 km/h) is investigated for transmission scheme 2: 4K, 1/8, 16-QAM 1/2, MPE–FEC 7/8 (Table 1.13).

Six routes at 20 km/h have been chosen: one in each wind direction, one through the city center and one around the city (Sect. 1.5.2). The total length of the routes for scenario IV is 70.5 km (Table 1.12).

1.5.5.1 %Locations MFER, FER

Figure 1.13 shows the percentage of locations with valid tables as a function of the recorded electric-field strength E (dBμV/m) for reception in a car driving at 20 km/h.

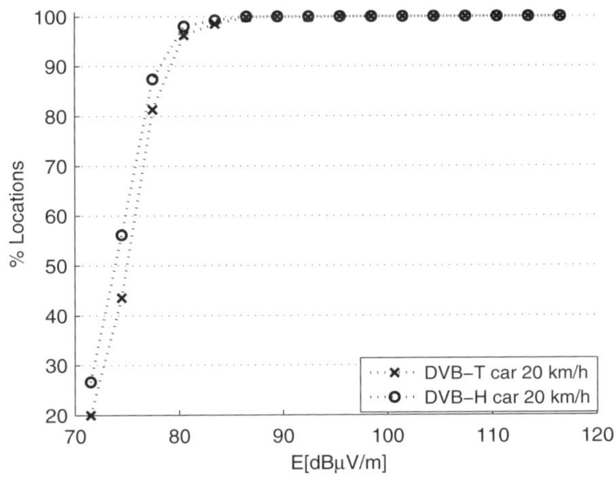

Fig. 1.13 Percentage of locations with valid tables as a function of the recorded electric-field strength before (DVB-T) and after (DVB-H) MPE–FEC correction for reception inside car (20 km/h) for 4K, 1/8, 16-QAM 1/2, MPE–FEC rate 7/8

Table 1.14 Minimal electric-field and $\frac{C}{N+I}$ values for FER and MFER equal to 0%, 1%, and 5% (car at 20 km/h)

E (dbμV/m)	0%	1%	5%
FER	95.5	84.63	80.25
MFER	86.5	82.88	79.65
$\frac{C}{N+I}$ (dB)	0%	1%	5%
FER	21.5	16.03	14.76
MFER	16.5	15.30	12.94

For each 3 dB range for E, a minimum of 20 tables is required so that *%Locations* can be accurately computed for this bin [7].

Table 1.14 shows $\frac{C}{N+I}|_{MFERX\%}$, $\frac{C}{N+I}|_{FERX\%}$, $E|_{MFERX\%}$ and $E|_{FERX\%}$ for $X = 0$, 1, and 5. For MFER 5%, a $\frac{C}{N+I}$ of 12.94 dB and E of 79.65 dBμV/m are required. In [7, 8] (Table 1.3 in Sect. 1.2.4), 18.5 dB for $\frac{C}{N+I}$ is assumed. The 18.5 dB of [7,8] includes a 3 dB margin and is calculated for higher driving velocities. For that reason, the $\frac{C}{N+I}|_{MFER5\%}$ value is higher than the one presented here. The required electric-field strength E and required $\frac{C}{N+I}$ is higher if a higher percentage of correct tables is needed. Also, for an equal reception quality, the required E and $\frac{C}{N+I}$ values are higher without MPE–FEC than with MPE–FEC.

1.5.5.2 %Locations ESR5, QEF

Figure 1.14 shows the percentage of recorded measurements for which the BER was lower than $2 \cdot 10^{-3}$ (ESR5) and $2 \cdot 10^{-4}$ (QEF) (a) as a function of the recorded $\frac{C}{N+I}$

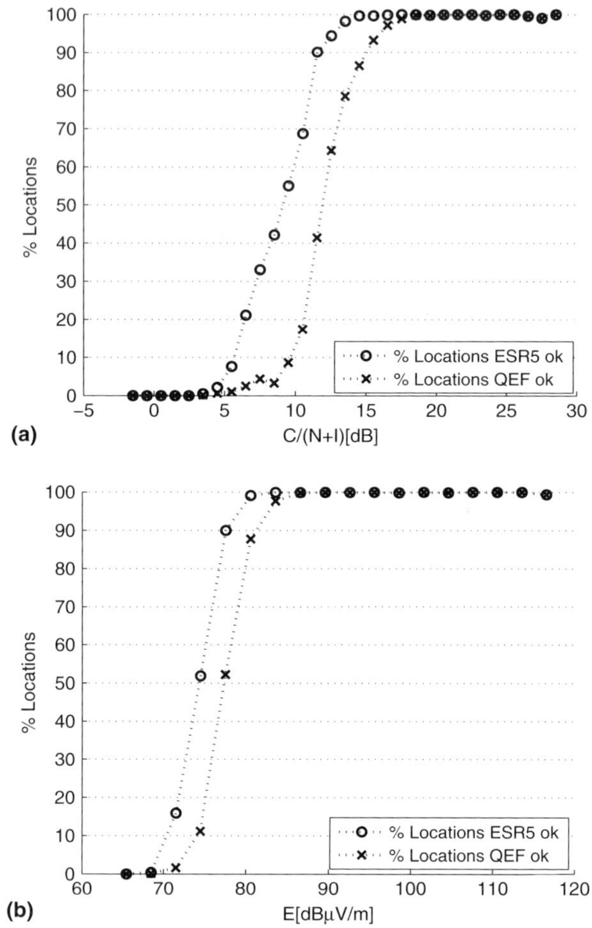

Fig. 1.14 Percentage of recorded measurements for which the BER is lower than $2 \cdot 10^{-3}$ (ESR5) and $2 \cdot 10^{-4}$ (QEF) **(a)** as a function of the received $\frac{C}{N+I}$ and **(b)** as a function of the received electric-field strength for reception inside a car driving at 20 km/h

and (b) as a function of the recorded electric-field strength for reception inside a car driving at 20 km/h. The minimal values where 95% of the locations satisfy the ESR5 and QEF criterion are 12.65 and 15.94 dB, respectively, for $\frac{C}{N+I}$, and 79.15 and 82.69 dBμV/m, respectively, for E. The ESR5 criterion corresponds very well with the $\frac{C}{N+I}$ and E values for MFER 5% (Table 1.14).

1.5.6 Comparison of Different Reception Scenarios

In this section, the performance of a DVB-H system for different scenarios for transmission scheme 2 (4K, 1/8, 16-QAM 1/2, MPE–FEC 7/8) is compared.

Figure 1.15a shows the percentage of the time that the receiver was not locked (*%Not locked*) and the percentage of the received tables that were incorrect (*%Incorrect tables*) for the different scenarios (transmission scheme 2). Figure 1.15b shows the percentage of correct tables, corrected tables, and incorrect tables for the different scenarios. Figure 1.15a and b shows that car 70 km/h, car 120 km/h, and train are the most difficult reception conditions: the receiver is less locked, and more incorrect tables are received. Nonetheless, one should be careful with this comparison of scenarios, because different scenarios consist of a different number of measurement points and are measured on different locations.

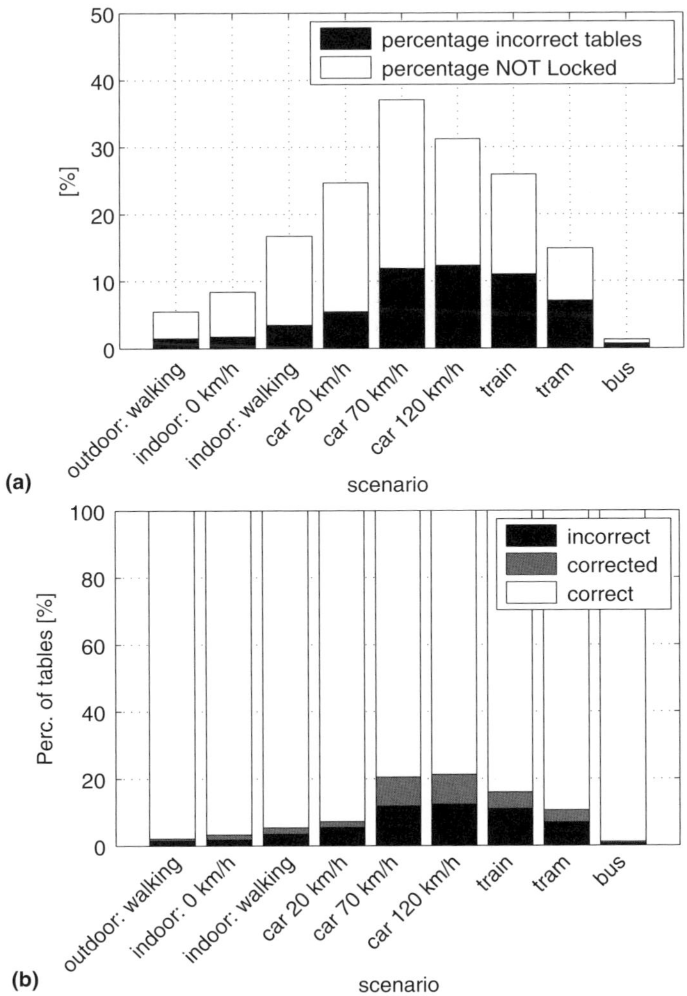

Fig. 1.15 **(a)** % Not Locked, %Incorrect tables and **(b)** %Correct tables, %Corrected tables, %Incorrect tables for the different scenarios (transmission scheme 2)

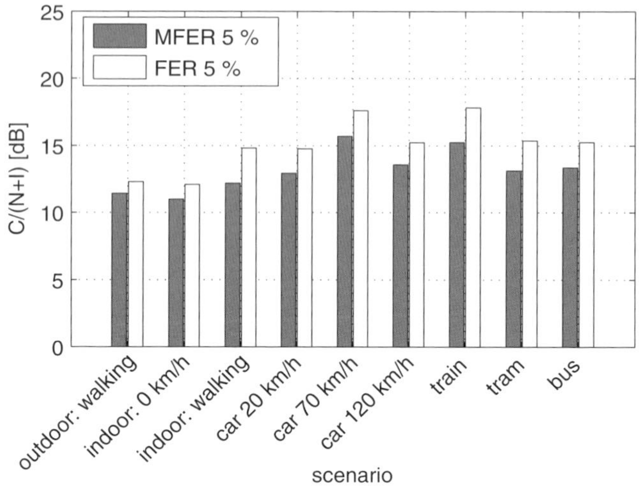

Fig. 1.16 $\frac{C}{N+I}|_{\mathrm{MFER5\%}}$ and $\frac{C}{N+I}|_{\mathrm{FER5\%}}$ for the different scenarios (transmission scheme 2)

When the analysis (*%Locations*) of Sect. 1.5.5 is performed for all scenarios and investigated schemes, one can compare the MFER 5% and FER 5% values. Figure 1.16 shows $\frac{C}{N+I}|_{MFER5\%}$ and $\frac{C}{N+I}|_{FER5\%}$ for the different scenarios. It shows that the more difficult the reception conditions are (higher speed, reception in trains, ...), the higher the required $\frac{C}{N+I}$ is. For 120 km/h, however, the required $\frac{C}{N+I}$ is lower than for 70 km/h, in contrast to what may be expected. The reason is the choice of the routes for the different scenarios: not the worst-case conditions have been investigated for the Doppler shift at 120 km/h. The very northern and southern part of the route at 70 km/h leads straight towards/away from the nearest transmitter, resulting in worst-case Doppler shifts, in contrary to the 120 km/h route (Fig. 1.12). It has to be mentioned that the choice of the routes for 120 km/h is limited due to speed limits. Hence, the actual experienced Doppler shift might be higher for the route at 70 km/h than for the route at 120 km/h. The values will be discussed in Sect. 1.5.7 and compared with other schemes.

1.5.7 Influence of MPE–FEC Coding Rate

The MPE–FEC feature interleaves the data and adds a Reed-Solomon Code in order to make the DVB-H signal more robust. The influence of the MPE–FEC coding rate is also investigated in [7,9]. The schemes of Table 1.13 are used (Sect. 1.5.4) to analyze experimentally the influence of the MPE–FEC. Investigated coding rate levels are 1/2, 2/3, 3/4, 5/6, 7/8, and 67/68. A decreasing level of protection corresponds to a higher coding rate level. The coding rate level 67/68 corresponds to almost no MPE–FEC protection. In this section the parameters 4K, 1/8, 16-QAM 1/2 are fixed and the MPE–FEC rate is varied.

Table 1.15 $\triangle C_{5\%}$ values for different MPE–FEC rates for different scenarios (4K, 1/8, modulation scheme fixed at 16-QAM 1/2)

Scenario	Gain $\triangle C_{5\%}$ (dB) for MPE–FEC					
	67/68	7/8	5/6	3/4	2/3	1/2
Indoor walking	0.57	2.62	1.26	0.82	2.11	2.73
Car 20 km/h	0.12	1.82	1.13	1.94	1.96	2.70
Car 70 km/h	0.38	1.91	1.36	2.96	1.62	2.68
Average (I-VIII)	0.13	1.85	1.31	1.71	1.89	2.49

Table 1.15 shows the value of the MPE–FEC gain $\triangle C_{5\%}$ for the different MPE–FEC rates (reception inside car at 20 and 70 km/h and indoor walking reception), determined using the method of Sect. 1.5.4.2 [26,27]. As expected, Table 1.15 shows that the more MPE–FEC coding is used (lower MPE–FEC rate), the higher the gain in $\frac{C}{N+I}$ (dB). On average (last line of Table 1.15) over scenarios I–VIII, $\triangle C_{5\%}$ varies from 0.13 dB (67/68) to 2.49 dB (1/2). The MPE–FEC rate 7/8 provides a somewhat larger gain than expected but corresponds reasonably with data of [31] where gains from 1.6 dB (at Doppler frequency of 10 Hz, i.e., portable) up to 4.3 dB (at about 250 z) are obtained for a typical urban (TU6) channel [8]. The values of Table 1.15 are based on actual measurements (realistic channel). For 67/68, the $\frac{C}{N+I}$ gain is very limited and maximally about 0.6 dB. Finally, for 1/2 maximal $\frac{C}{N+I}$ gains of about 2.7 dB are obtained.

In [13], an MPE–FEC gain (constellation 16-QAM 1/2, MPE–FEC rate 3/4) of 1.5 dB for a car driving at 30 km/h has been obtained, when measuring an IP PER (Packet Error Rate) of 5%. This corresponds reasonably well to the MPE–FEC gains obtained in Table 1.15 for an MPE–FEC rate of 3/4 (a gain of 1.94 dB for a car driving at 20 km/h). In [9] it is stated that already in portable situations (Doppler frequency below 10 Hz), the effect of the MPE–FEC rate allows to reach improved $\frac{C}{N+I}$. The effect of the MPE–FEC gradually improves the $\frac{C}{N+I}$ for higher Doppler frequencies.

1.5.7.1 % Valid Reception

Table 1.16 shows the mean value of *%Correct tables*, *%Corrected tables*, *%Incorrect tables*, *%Lock*, and *%Valid reception* for the different MPE–FEC rates. Each value is an average over the different scenarios (I–IX) (Sect. 1.5.2). Table 1.16 represents average reception in Ghent under a random reception condition. Table 1.16 shows that *%Correct tables* is more or less constant, while *%Corrected tables* increases as the MPE–FEC rate increases. This corresponds with a decrease of *%Incorrect tables*. Table 1.16 also shows an increase in the *%Valid reception* (defined in Sect. 1.5.3) value when the MPE–FEC coding increases. For MPE–FEC 67/68, the value is 83.56%, while for MPE–FEC 1/2 this value increases up to 87.91%.

Table 1.16 Average value over all scenarios of the percentage of correct, corrected, and incorrect tables, %Lock and %Valid reception for the different MPE–FEC rates (modulation scheme 16-QAM 1/2)

MPE–FEC rate	%Correct tables	%Corrected tables	%Incorrect tables	%Lock	%Valid reception
67/68	93.98	0.37	5.64	88.56	83.56
7/8	91.19	3.25	5.56	88.72	83.79
5/6	91.90	2.56	5.53	88.83	83.92
3/4	91.52	3.71	4.77	88.50	84.28
2/3	92.02	4.55	3.43	91.61	88.47
1/2	92.21	5.05	2.73	90.38	87.91

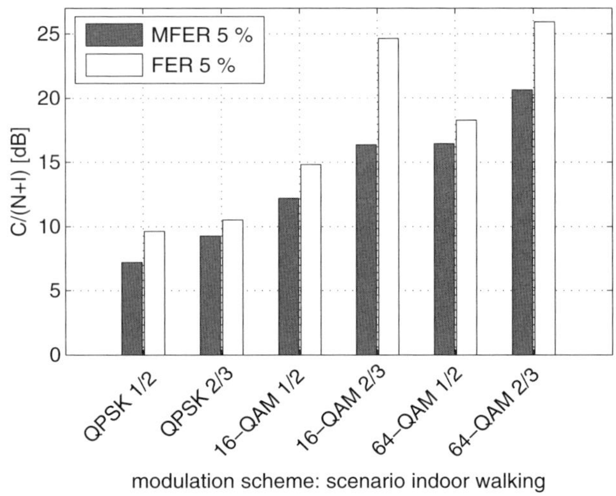

Fig. 1.17 $\frac{C}{N+I}|_{\mathrm{MFER}5\%}$ and $\frac{C}{N+I}|_{\mathrm{FER}5\%}$ values for different modulation schemes for indoor walking reception (4K, 1/8, MPE–FEC fixed at 7/8)

1.5.8 Influence of Modulation Scheme

In Sect. 1.5.4 (Table 1.13) the investigated transmission schemes are shown. Figures 1.17 and 1.18 show the $\frac{C}{N+I}|_{MFER5\%}$ and $\frac{C}{N+I}|_{FER5\%}$ values for the investigated modulation schemes for indoor walking reception and for reception inside a car at 20 km/h [26, 27]. These figures show that the higher the modulation scheme, the higher the required $\frac{C}{N+I}$ values, e.g., in Fig. 1.18 (scenario IV car 20 km/h) for QPSK 1/2, a $\frac{C}{N+I}|_{MFER5\%}$ value of 7.28 dB is required, while for 64-QAM 2/3, 20.28 dB is required. The required MFER 5% values (gray bars in Fig. 1.17) are of course lower than the FER 5% values (white bars in Figs. 1.17 and 1.18), due to the MPE–FEC feature. In Fig. 1.18, the FER 5% value is missing for 64-QAM 2/3: reception without MPE–FEC coding is not possible with the receiver (Sect. 1.5.4). The required $\frac{C}{N+I}$ value is higher than the value that can be measured with the used receiver.

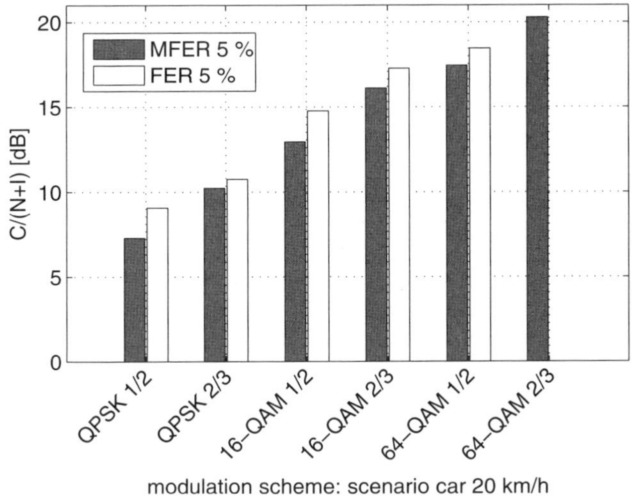

Fig. 1.18 $\frac{C}{N+I}|_{\text{MFER5\%}}$ and $\frac{C}{N+I}|_{\text{FER5\%}}$ values for different modulation schemes for reception inside a car at 20 km/h (4K, 1/8, MPE–FEC fixed at 7/8)

Table 1.17 The required $\frac{C}{N+I}|_{\text{MFER5\%}}$ for the considered modulation schemes for different scenarios (4K, 1/8, MPE–FEC fixed at 7/8)

Scenario	Modulation scheme					
	QPSK	QPSK	16-QAM	16-QAM	64-QAM	64-QAM
	1/2	2/3	1/2	2/3	1/2	2/3
Indoor Walking	7.19	9.27	12.20	16.38	16.45	20.65
Car 20 km/h	7.28	10.23	12.94	16.11	17.45	20.28
Car 70 km/h	8.20[a]	12.00[a]	15.70	16.60	19.55	23.40*
Average (I-VIII)	7.07	9.82	13.15	15.92	16.91	19.78

[a] data of [31].

Table 1.17 shows the $\frac{C}{N+I}|_{FER5\%}$ and $\frac{C}{N+I}|_{MFER5\%}$ values for different scenarios and for all modulation schemes. It shows again that the higher the modulation scheme, the higher the required $\frac{C}{N+I}$ value. On average (last line of Table 1.17) over scenarios I–VIII, $\frac{C}{N+I}|_{MFER5\%}$ varies from 7.07 dB (QPSK 1/2) to 19.78 dB (64-QAM 2/3). When the inner coding changes from 1/2 to 2/3, the $\frac{C}{N+I}|_{MFER5\%}$ value is 2 to 3 dB higher. When the constellation is 16-QAM instead of QPSK, the $\frac{C}{N+I}|_{MFER5\%}$ value is about 6 dB higher. When changing from 16-QAM to 64-QAM, an increase in $\frac{C}{N+I}$ of about 4 dB is needed for indoor walking to maintain the same reception quality.

In [8, 9] required $\frac{C}{N+I}$ values of 15.5, 18.5, 20.5, and 23.4 dB for 16-QAM 1/2, 16-QAM 2/3, 64-QAM 1/2, and 64-QAM 2/3, respectively, are presented for mobile reception. These values correspond reasonably well with our measurements for the scenario of car 70 km/h (see Table 1.17), where the obtained values are 15.70, 16.60, 19.55 dB, and an unmeasurable value (the required value for 64-QAM 2/3 is higher than the value that can be measured with the used receiver). For 64-QAM 2/3 a value of 23.4 dB [9, 31] for mobile reception is listed. For QPSK 1/2 and QPSK 2/3 for car 70 km/h also values of [9,31] are taken. The differences between the values from our measurements and those of [9, 31] may be caused by the lower Doppler shift in our measurements, compared to the mobile (higher velocity) situation in [8, 9, 31]. It should also be noted that in [8, 9], the MPE–FEC rate is probably 3/4 and the channel is typical urban (TU6 channel), while the values presented here are for the MPE-FEC rate of 7/8 for measurements in an actual (realistic) channel.

1.5.8.1 %Valid Reception

Table 1.18 shows the mean value of *%Correct tables*, *%Corrected tables* and *%Incorrect tables*, *%Lock* and *%Valid reception* for the different modulation schemes. Each value is an average over the different scenarios (I–IX). This table shows that the higher the modulation scheme, the lower *%Valid reception* (and *%Lock*). A modulation scheme with more inner coding performs slightly better than the nearest lower modulation scheme with less inner coding, e.g., 83.79% for 16-QAM 1/2 versus 79.21% for QPSK 2/3. The average *%Valid reception* values vary from 55.12% (64-QAM 2/3) to 95.43% (QPSK 1/2). QPSK requires a lower $\frac{C}{N+I}$ than 64-QAM to obtain equal reception quality.

1.5.9 Influence of Guard Interval

Scenario IV "car 20 km/h" has been investigated to study the influence of the guard interval. Table 1.19 shows $\frac{C}{N+I}|_{MFER5\%}$, $\frac{C}{N+I}|_{FER5\%}$, and the maximal deviation for the different guard intervals. Table 1.19 shows that these values remain relatively constant for the different guard intervals, indicating that the influence of the guard interval on these values is small (maximally about 1 dB).

Table 1.18 Average value over all scenarios of the percentage of correct, corrected, and incorrect tables, %Lock and %Valid reception for the different modulation schemes

Modulation scheme	%Correct tables	%Corrected tables	%Incorrect tables	%Lock	%Valid reception
QPSK 1/2	96.71	1.06	2.23	97.61	95.43
QPSK 2/3	89.20	3.46	7.34	85.49	79.21
16-QAM 1/2	91.19	3.25	5.56	88.72	83.79
16-QAM 2/3	82.39	5.94	11.67	80.64	71.22
64-QAM 1/2	85.33	5.36	9.31	80.51	73.01
64-QAM 2/3	67.42	13.07	19.52	68.49	55.12

Table 1.19 $\frac{C}{N+I}|_{\mathrm{MFER5\%}}$ and $\frac{C}{N+I}|_{\mathrm{FER5\%}}$ values for different guard intervals for car 20 km/h (4K, 16-QAM 1/2, 7/8)

Parameter	Guard interval				Max deviation
	1/4	1/8	1/16	1/32	
$\frac{C}{N+I}|_{\mathrm{FER5\%}}$	14.77	14.76	15.16	15.09	0.40
$\frac{C}{N+I}|_{\mathrm{MFER5\%}}$	13.36	12.94	14.06	13.22	1.12

1.6 DVB-H Broadcast Network Design for Indoor Reception

This section describes a methodology to determine the required number of BS for good indoor reception in a certain area. As an example, indoor reception for Flanders (13,522 km^2) will be investigated. The methodology can of course be applied to other regions.

We will make use of the existing models [15] and coverage models developed using propagation measurements in the DVB-H network in Ghent [25,26]. Different categories of BS are defined and different scenarios are proposed. The number of BS for "good indoor reception" will then be determined for Flanders. Finally, the number of BS as a function of the *C*/*N* ratio (carrier-to-noise ratio) is determined and a formula for this relationship is proposed.

1.6.1 Selection of Parameters of Base Stations

Channel 37, corresponding to 602 MHz (band V), is considered. This channel is used during the Flemish DVB-H trial in Ghent, Belgium. Furthermore, a channel width BW of 8 MHz is considered. The required number of BS will be calculated for good indoor coverage for a throughput of about 10 Mbps [26]. This throughput is also used for the trial in Ghent, as described in [26]. The following settings are considered: 4K mode, a guard interval of 1/8, modulation scheme 16-QAM 1/2, and MPE–FEC coding rate of 7/8. The resulting bit rate is then 9.68 Mbps. For the estimation of the MPE–FEC gain, the data of [9, 26, 31] is used for portable reception. An MPE–FEC gain of 3.0 dB is considered for portable and mobile reception [26, 31]. The network dimensioning is executed for these settings.

1.6.2 Receiver

For the network dimensioning we consider the reference receiver Rx defined in [8] (Sect. 1.2.4). Different reception classes have been adopted: only class B (portable indoor reception, Sect. 1.2.1) will be considered here. For planning purposes, a receiver gain G_r of −4.86 dBi can be used for band V according to ETSI [8] (antenna in a small hand-held terminal). The minimum signal input level to the receiver is determined by the required C/N ratio listed in [8, 29]. The receiver noise figure (F) has been chosen as 6 dB for the frequency bands IV to V (reference receiver). For 16-QAM 1/2 the following parameters are used according to ETSI TR 102 377 [8] for portable reception (BW = 8 MHz, band V, static Rayleigh channel, PER = Packet Error Rate = 10^{-4} [29]): C/N = 11.2 dB, Rx sensitivity of −88.0 dBm, and equivalent minimum receiver field strength E_{min} of 49.7 dBμV/m (Table 1.2 in Sect. 1.2.4).

1.6.3 Area to Be Covered: Example Flanders

The surface of Flanders to be covered is 13,522 km^2 [22]. There are 13 cities in Flanders, namely Antwerpen and Ghent and 11 regional cities: Aalst, Brugge, Genk, Hasselt, Kortrijk, Leuven, Mechelen, Oostende, Roeselare, St-Niklaas, and Turnhout. The surface of these cities is 668.49 km^2. Also coverage for Brussels is necessary. The surface of this area (Brussels – Capital Region) is 162 km^2 [22].

1.6.4 Categories of Base Stations and Scenarios

1.6.4.1 Categories

We define three categories of BS:

- **Category 1**: low towers/low power, height of base station $h_{\mathrm{BS}} = 35$ m, ERP of 2 kW
- **Category 2**: medium towers/medium power, $h_{\mathrm{BS}} = 60$ m, ERP of 5 kW
- **Category 3**: high tower/high power, $h_{\mathrm{BS}} = 150$ m, ERP of 20 kW

1.6.4.2 Scenarios

Using these categories, five different scenarios are proposed: three scenarios with 100% of the BS from one category, a realistic scenario based on the available antenna sites of the public broadcaster VRT (Flemish Radio and Television Network), supplemented by low power stations, and a scenario assuming building additional medium power infrastructure. For all investigated scenarios, the network is first developed for rural coverage. Additionally, suburban/urban coverage for the 13 cities and Brussels is provided.

Scenario 1

For scenario 1, Flanders and Brussels are covered for 100% using BS of category 1 (height h_{BS} = 35 m and ERP = 2 kW). Figure 1.19 shows this scenario.

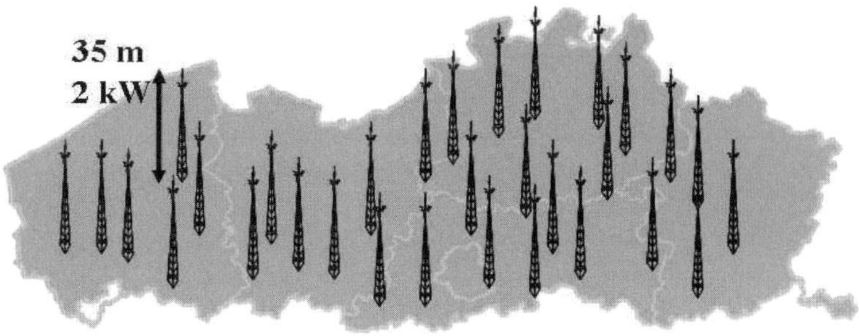

Fig. 1.19 Scenario 1

Fig. 1.20 Scenario 2

Fig. 1.21 Scenario 3

Scenario 2

For scenario 2, Flanders and Brussels are covered for 100% using BS of category 2 (height $h_{\mathrm{BS}} = 60$ m and ERP = 5 kW). Figure 1.20 shows this scenario.

Scenario 3

For scenario 3, Flanders and Brussels are covered for 100% using BS of category 3 (height $h_{\mathrm{BS}} = 150$ m and ERP = 20 kW). Figure 1.21 shows this scenario.

Table 1.20 Assumed number of BS of categories 2 and 3 for scenario 4

Category	Flanders rural	Cities suburban	Brussels suburban
1	?	?	?
2	15	5	0
3	5	1	1

Table 1.21 Assumed number of BS of categories 2 and 3 for scenario 5

Category	Flanders rural	Cities suburban	Brussels suburban
1	?	?	?
2	50	13	0
3	5	1	1

Scenario 4

Scenario 4 is based upon available antenna sites of VRT in Flanders. Scenario 4 uses existing medium and high infrastructure ($h_{BS} \geq 60$ m). In Table 1.20, we assume 15 BS of categories 2 and 5 BS of category 3 for rural coverage. Furthermore, we assume 5 BS of category 2 and 1 BS of category 3 in the regional cities. Finally, for Brussels there is 1 BS of category 3. The remaining number of BS of category 1 required for the coverage is then calculated.

Scenario 5

The realistic scenario 5 assumes building additional high infrastructure ($h_{BS} \geq$ 60 m). In Table 1.21 we assume 50 BS of category 2 and 5 BS of category 3 for rural coverage. Furthermore, we assume 13 BS of category 2 (1 for every regional city) and 1 BS of category 3 in the regional cities. Finally, for Brussels there is 1 BS of category 3. The remaining number of BS of category 1 required for the coverage is then calculated.

1.6.5 Required Number of Base Stations

1.6.5.1 Range of a Single Base Station

To calculate the range of the system, PL models are needed (Sect. 1.3.1). To calculate the range of BS of category 2, the Ghent model of [25, 26] is used for suburban coverage (P_0 = 86.8 dB, n = 2.34, and σ = 6.18). The rural coverage is calculated using the rural ITU R P.1546 model [15]. The coverage of BS of categories 1 and 3 is also calculated using the ITU-R P.1546 model [15]. The range of a single BS is then calculated for the reference receiver of [7,8]. The ranges R for $h_{Rx} = 1.5$ m are summarized (together with the used models) in Table 1.22 for reception class B.

The covered area is assumed to be a circle or a hexagon (21% less coverage). Using hexagons, empty spaces are avoided. The covered area is then calculated for a circle as follows:

$$covered\ area_{1BS} = \pi \cdot R^2 \tag{1.18}$$

The covered area for a hexagon is calculated as follows:

$$covered\ area_{1BS} = \frac{3}{2} \cdot \sqrt{3} \cdot R^2 \tag{1.19}$$

The number of BS #BS for a certain area with surface S is then:

$$\#BS = S/covered\ area_{1BS} \tag{1.20}$$

1.6.5.2 Calculations for Different Scenarios

The number of BS obtained for indoor reception for scenarios 1–5 is shown in Table 1.23. For suburban coverage in 13 cities, a minimum of 13 base station antennas is assumed when the #BS obtained from the calculations is smaller than 13 (e.g., for scenario 3). For scenario 1 (100% category 1), the largest number of BS is of course obtained. The number of BS is much higher for hexagons compared to circles

Table 1.22 Range of base station for different categories (class B reception, 4K mode, a guard interval of 1/8, modulation scheme 16-QAM 1/2, and MPE–FEC coding rate of 7/8)

Category	Terrain	Model	Range R (km)
1	Rural	ITU	2.45
	Suburban	ITU	1.65
2	Rural	ITU	4.15
	Suburban	Ghent model	3.35
3	Rural	ITU	10.20
	Suburban	ITU	7.25

Table 1.23 #BS for different scenarios

Scenario	#BS	
	Circle	Hexagon
1	816	986
2	274	332
3	47	65
4	653	823
5	563	733

due to the less covered area of a single BS. From Table 1.23, it can be concluded that a large number of BS are required for the realistic scenario 4 (653 for circles) and scenario 5 (563 for circles). The #BS for scenario 5 is lower than for scenario 4 because in scenario 5 additional medium infrastructure would be built.

1.6.5.3 Number of Base Stations Versus *C/N*

The number of BS required for indoor DVB-H coverage in Flanders as a function of the *C/N* is here determined. A high number of calculations were performed to obtain #BS for *C/N* values from 5 to 20 dB. Up to now the reference receiver with *C/N* = 11.2 dB for class B was considered. The following settings and parameters were considered:

- Class B
- 16 QAM 1/2, MPE–FEC 7/8
- Guard interval 1/8
- MPE-gain = 3.0 dB, [9, 26, 31]
- Coverage for Flanders and Brussels
- *C/N*: varied from 5 to 20 dB

As an example, Fig. 1.22 shows the number of BS as a function of the *C/N* for scenario 2 using circles and hexagons. The figure shows that the number of BS and thus the costs are very sensitive to the *C/N* value. The dependence is exponentially: for a *C/N* of 13.8 dB for class B reception (value of a realistic handheld), 395 BS for scenario 2 for circles are needed (compared to 274 for the reference receiver with *C/N* = 11.2 dB).

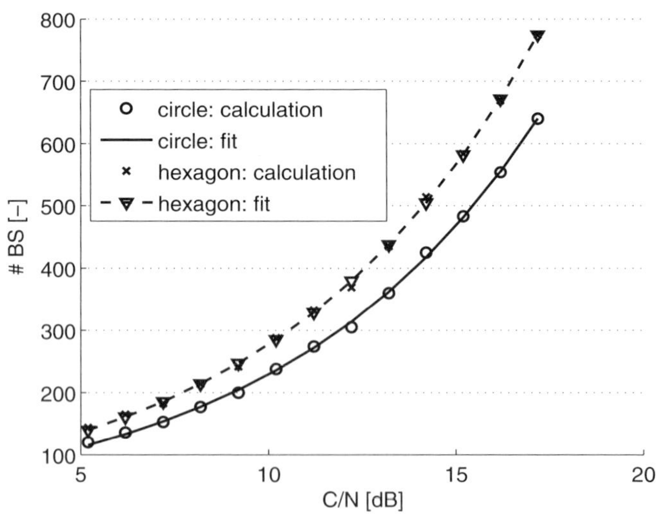

Fig. 1.22 Calculated and fitted number of BS as a function of the *C/N* for scenario 2

Table 1.24 Values of parameters *a*, *b*, and *c* to calculate #BS in Flanders for different scenarios

Scenario	Area	a	b	c
1	Circle	150.62	0.1508	0
	Hexagon	181.71	0.1510	0
2	Circle	55.09	0.1426	0
	Hexagon	66.52	0.1427	0
3	Circle	10.73	0.1335	0
	Hexagon	12.85	0.1339	0
4	Circle	78.13	0.1841	27
	Hexagon	106.25	0.1781	27
5	Circle	44.08	0.2136	70
	Hexagon	67.60	0.2014	70

Therefore we modeled the number of BS for the different scenarios as follows:

$$\#BS = a \cdot e^{b \cdot C/N} + c \tag{1.21}$$

where a, b, and c are the parameters of the model. Parameter c is of course zero for scenarios 1 to 3, as no constant number of existing or additional medium or high infrastructure is assumed. Figure 1.22 shows the results of the fits for scenario 2. For the different scenarios, the average deviation between the model and the calculations is smaller than 2.5%. Using this formula and the *C*/*N* of a terminal, one can make an estimate for the required number of BS for coverage in Flanders. Table 1.24 shows the values for the parameters a, b, and c. Using these parameters, the required number of BS for the different scenarios can easily be calculated.

1.7 Conclusions

DVB-H is a very promising standard for broadcast services requiring high data rates and offers extended possibilities for content providers and network operators. Several characteristics of a DVB-H system, which are needed for coverage planning, have been discussed in this chapter. Coverage planning and throughput versus range for a DVB-H network are presented. Next, propagation measurements of a DVB-H signal and the methodology to develop a PL model are discussed.

A methodology to perform and analyze physical-layer measurements for a DVB-H system is presented. Portable scenarios (indoor/outdoor walking, indoor standing) as well as mobile scenarios (car at different velocities, train, tram, bus) have been described. Measurement results have been presented for different modulation schemes, MPE–FEC rates, and guard intervals. The gain in $\frac{C}{N+I}$ obtained by MPE–FEC coding varies from 0.13 dB (MPE–FEC 67/68) to 2.49 dB (MPE–FEC 1/2). Lower modulation schemes require lower $\frac{C}{N+I}$ values: the $\frac{C}{N+I}$ requirement for MFER 5% varies from 7.07 dB (QPSK 1/2) to 19.78 dB (64-QAM 2/3). When changing the inner code rate from 1/2 to 2/3, the $\frac{C}{N+I}$ requirement increases 2 to 3 dB. 16-QAM modulation requires 6 dB more signal strength than QPSK to obtain equal reception quality and 64-QAM about 4 dB more than 16-QAM.

The number of BS required for good indoor DVB-H coverage is calculated for a physical-layer bit rate of about 10 Mbps. The region of Flanders has been chosen as an example. Three different categories of BS have been defined and five different scenarios have been proposed. Technical trial results are used to obtain realistic coverage models. The number of BS (and thus the cost) is very sensitive to the *C*/*N* value. This dependence has been modeled.

Acknowledgements The authors wish to thank David Plets and Leen Verloock for their effort during the DVB-H tests. The authors also thank the IBBT-MADUF project, co-funded by the IBBT (Interdisciplinary institute for BroadBand Technology), a research institute founded by the Flemish Government in 2004, and the involved companies and institutions. W. Joseph is a Post-Doctoral Fellow of the FWO-V (Research Foundation – Flanders).

References

1. Abrishamkar F, Irvine J (2000) Comparison of current solutions for the provision of voice services to passengers on high speed trains. 52nd Vehicular Technology IEEE VTS-Fall Conference Proceedings: 2068–2075
2. COST 231 Final Report (1991) Digital Mobile Radio Towards Future Generation Systems. COST Telecom Secretariat, Brussels, Belgium
3. ECC Report 33 (2003) The analysis of the coexistence of FWA cells in the 3.4–3.8 GHz band. Electronic Communication Committee (ECC) [Online]. Available: www.ero.dk
4. ECC Report 49 (2004) Technical criteria of digital video broadcasting terrestrial (DVB-T) and terrestrial digital audio broadcasting (TDAB) allotment planning. Electronic Communication Committee (ECC), Copenhagen, Denmark [Online]. Available: www.ero.dk
5. ETSI, EN 302 304 v1.1.1 (2004) Digital Video Broadcasting (DVB); Transmission System for Handheld Terminals (DVB-H). European Telecommunications Standards Institute (ETSI), Sophia Antipolis Cedex, France
6. ETSI, EN 300 744 v1.5.1 (2004) Digital Video Broadcasting (DVB); Framing Structure, Channel Coding and Modulation for Digital Terrestrial Television. European Telecommunications Standards Institute (ETSI), Sophia Antipolis Cedex, France
7. ETSI, TR 102 401 v1.1.1 (2005) Digital Video Broadcasting (DVB); Transmission to Handheld Terminals (DVB-H); Validation Task Force Report. European Telecommunications Standards Institute (ETSI), Sophia Antipolis Cedex, France
8. ETSI, TR 102 377 v1.1.1 (2005) Digital Video Broadcasting (DVB); DVB-H Implementation Guidelines. European Telecommunications Standards Institute (ETSI), Sophia Antipolis Cedex, France
9. Faria G, Hendriksson JA, Stare E, Talmola P (2006) DVB-H: digital broadcast services to handheld devices. Proc. IEEE 94(1): 194–209
10. Goller M (1195) Application of GSM in high speed trains: measurements and simulations. IEE Colloquium Radiocommun. Transportation 5: 1–7
11. Greenstein L, Erceg V, Yeh Y, Clark M (1997) A new path-gain/delayspread propagation model for digital cellular channels. IEEE Trans. Veh. Technol. 46(2): 477–485
12. Hata M (1980) Empirical formula for propagation loss in land mobile radio services. IEEE Trans. Veh. Technol. 29: 317–325

13. Himmanen H and Jekola T (2007) DVB-H field trials: Studying radio channel characteristics. IEEE International Symposium on Broadband Multimedia Systems and Broadcasting, Orlando, Florida, USA
14. Hoymann C, Puttner M, Forkel I (2003) The HIPERMAN standard a performance analysis. Proceedings of the IST Mobile & Communication Summit, Aveiro, Portugal IST:827–831
15. ITU-R Recommendation P.1546 (2003–2005) Method for point-to-area predictions for terrestrial services in the frequency range 30 MHz to 3000 MHz. International Telecommunication Union ITU, Geneva, Switzerland
16. ITU-R Recommendation P.530-10 (2001) Propagation data and prediction methods required for the design of terrestrial line-of-sight systems. International Telecommunication Union ITU, Geneva, Switzerland
17. ITU-R Recommendation SM.1708 (2005) Field-strength measurements along a route with geographical coordinate registrations, Geneva, Switzerland
18. ITU-R Recommendation P.370-7 (1994–1995) VHF and UHF propagation curves for the frequency range from 30 MHz to 1000 MHz, Geneva, Switzerland
19. Kepler JF, Krauss T, Mukthavaram S (2002) Delay spread measurements on a wideband MIMO channel at 3.7 GHz. Proc. IEEE Int. VT Symp. Vancouver, Canada, VTC 2002-Fall 2:2498-2502
20. Kostanic I, Hall C, McCarthy J (1998) Measurements of the vehicle penetration loss characteristics at 800 MHz. Proc. 48th IEEE Int. VT Symp: 1-4
21. Lee WCY (1993) Mobile Communications Design Fundamentals. John Wiley & Sons Inc, New York, NY
22. Nationaal Instituut voor Statistiek NIS, Algemene Directie Statistiek en Economische Informatie (2007) Fysische geografie – algemeen. [Online]. Available: http://statbel.fgov.be/figures/d110 nl.asp
23. Okumura Y, Ohmori E, Kawano T, Fukua K (1968) Field strength and its variability in UHF and VHF land-mobile radio service. Rev. Elec. Commun. Lab. 16(9): 825–873
24. Owens T, Zhang C, Itagaki T, Outters J, Lauterjung J, Martucci M, Bouquet D, Mazieres B, Prudent J, Christ P, Gaspard I, Ritscher S, Zimmermann G, Christ P (2005) Deliverable 6.1: Radio spectrum, traffic engineering and resource management Tech. Rep. [Online]. Available: www.ist-instinct.org
25. Plets D, Joseph W, Tanghe E, Verloock L, Martens L (2007) Analysis of propagation of actual DVB-H signal in a suburban environment. 2007 IEEE International Symposium on Antennas and Propagation, Honolulu, Hawaii, USA, APS: Paper No. 1386
26. Plets D, Joseph W, Martens L, Deventer E, Gauderis H (2007) Evaluation and validation of the performance of a DVB-H network. 2007 IEEE International Symposium on Broadband Multimedia Systems and Broadcasting, Orlando, Florida, USA. Available on CDROM
27. Plets D, Joseph W, Verloock L, Tanghe E, Martens L, Deventer E, Gauderis H (2007) Evaluation of performance characteristics of a DVB-H network for different reception conditions. 57th Annual IEEE Broadcast Technology Society Symposium, Washington DC, USA, ABS: Paper No. 12-07
28. Saunders S. R. (1999) Antennas and Propagation for Wireless Communication Systems. John Wiley & Sons Inc, New York, NY
29. Schramm R (2004) DVB-T C/N values for portable single and diversity reception. EBU Technical review
30. Tanghe E, Joseph W, Verloock L, Martens L (2008) Evaluation of Vehicle Penetration Loss at Wireless Communication Frequencies IEEE Trans. Veh. Techn. (57): 2036–2041
31. TeamCast (2007) DVB-H calculator [Online]. Available: http://www.teamcast.com/en/maj-e/c2a2i9177/support/dvb-h-calculator

Chapter 2
Digital Video Broadcast: Systems and Implementations

Yue Zhang, Kok-Keong Loo, John Cosmas, and Yong-Hua Song

2.1 Development of DVB Infrastructure

Terrestrial broadcasting has brought entertainment and media information to mass audiences around the world for nearly a century. In the last two decades, the demand for consumption of multimedia services anywhere anytime has increased greatly. Because the spectrum resource is limited, a new spectrum efficient broadcasting technology is required to meet this demand. Digital broadcast technologies, such as the European Digital Video Broadcast – Terrestrial (DVB-T), moved terrestrial broadcast into the digital age. While analog broadcast can only deliver one programme per channel, digital broadcast namely DVB-T allows multiprogramme broadcasting where 2 to 4 standard definition TV (SDTV) programmes can be transmitted with time division multiplexing in a single 8 MHz channel. The bandwidth of DVB-T (from 12 to 24 Mbit/s) can be allocated to offer different TV qualities, such as enhanced definition television (EDTV, requiring about 10 to 12 Mbit/s per programme) or high definition TV (HDTV, requiring about 24 Mbit/s per programme) [1, 2]. In a nutshell, DVB-T has brought a higher quality service to TV program and it helps by reducing the use of spectrum, thus allowing other promising wireless technologies to diversify its applications. This leads to the development of emergence standard, Digital Video Broadcast – Handheld (DVB-H), that takes DVB-T a step further by making mobile reception of digital broadcasting possible with small, handheld devices [3].

The governments of different European countries have made their plans for analog to digital TV switchover and fully deploying DVB-T/H services. United Kingdom, in early 2007, switched off analog TV in one Welsh community as an

Y. Zhang (✉), K.-K. Loo, and J. Cosmas
School of Engineering and Design, Brunel University, UB8 3PH, UK
e-mail: Yue_Zhang@ieee.org

Y.-H. Song
Department of Electrical Engineering and Electronics, University of Liverpool, L69 3GJ, UK

F.-L. Luo (ed.), *Mobile Multimedia Broadcasting Standards: Technology and Practice*,
DOI: 10.1007/978-0-387-78263-8_2,

experiment. It is expected that the analog TV is completely phased out by the end of 2012 [4] nationwide. Meanwhile, analog and digital TV continue to co-exist in such a way to enable a soft transition from analog to digital. This period is known as the simulcast phase that will entail increased spectrum needs.

2.2 DVB Standard: Physical Layer

The DVB physical layer includes outer coder/decoder, inner coder/decoder, QAM mapping, frame adaptation and TPS insertion, OFDM and up/down converter as shown in Fig. 2.1 [5,6]. GIR block in Fig. 2.1 means guard interval removal.

The functional blocks are discussed in the following sections.

2.2.1 Scrambler/Descrambler

The MPEG2 Transport Stream (MPEG2-TS) is a source input to DVB physical layer that is organized in fixed length of 188-byte packet (see Fig. 2.3a) [7]. Every first byte of the MPEG2-TS 188-byte packet is a sync-word with a known value 47_{Hex}. In order to ensure the energy dispersal of binary data, a scrambler is used to randomize the MPEG2-TS input stream as shown in Fig. 2.2. The scrambler is a 16-bit Pseudo Random Binary Sequence (PRBS) with generator polynomial, $1+X^{14}+X^{15}$. The PRBS register is loaded with initial sequence "100 1010 1000 0000,"" and is re-initialized at the start of every group of 8 MPEG2-TS packets (see Fig. 2.3b). The first sync byte in a group of eight packets is inverted from 47_{HEX} to B8_{HEX} for identification of every randomized group of packets. In other words,

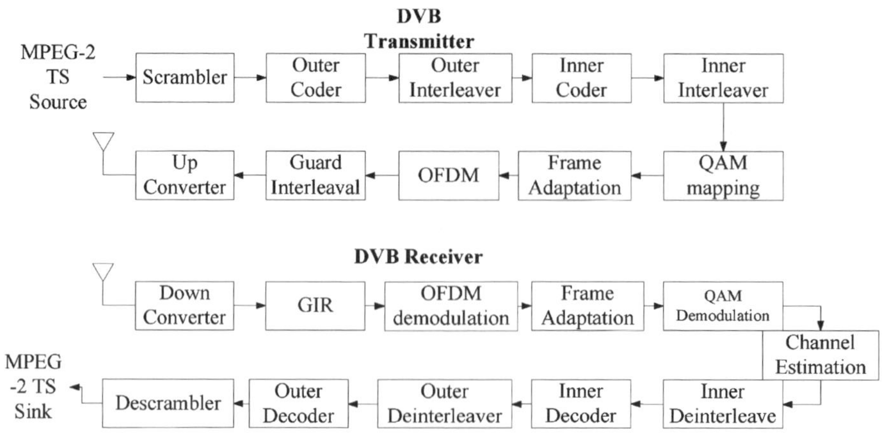

Fig. 2.1 Block diagram of DVB system

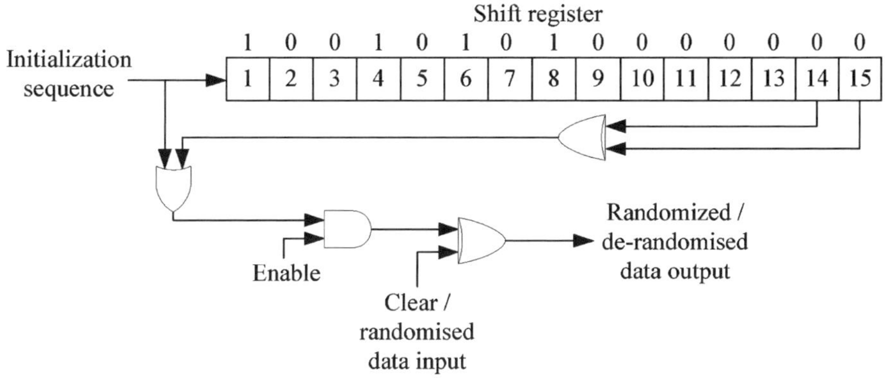

Fig. 2.2 Scrambler schematic diagram

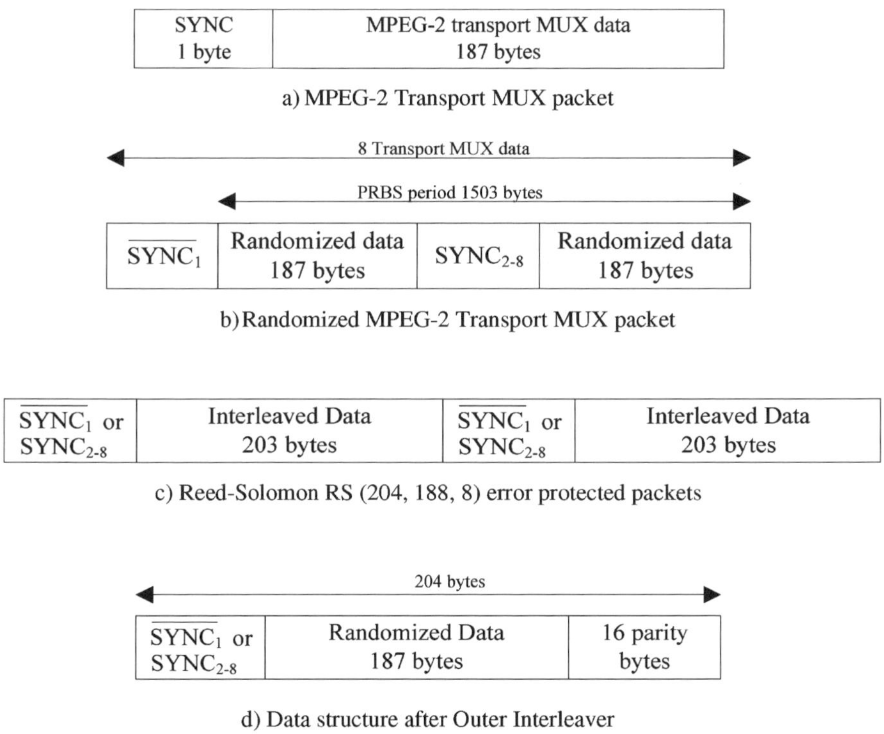

Fig. 2.3 Outer coding and outer interleaving process

the PRBS should have a period of 1503 bytes. The randomization process should stay active even if the modulator input bit-stream is nonexistent or noncompliant with the MPEG2-TS format (1 sync byte + 187 packet bytes). Reverse operation is carried out in descrambler at the receiver in order to obtain the sink MPEG-2 TS.

2.2.2 Outer Coder and Outer Interleaver

Reed-Solomon (RS) encoder is used as outer coder to generate 16-byte parity block for each randomized transport packet (188-byte) fed by the scramble. In essence, 51 null bytes are intentionally added to the input randomized transport packet such that a NASA-standard RS(255, 239, $t = 8$) mother code can be used. The null bytes are discarded after the coding procedure leading to an RS codeword of $N = 204$ including the 16-byte parity block appended at the end of each input as shown in Fig. 2.3c. The RS code uses a generator polynomial, $g(x) = (x+\lambda^0)(x+\lambda^1)(x+\lambda^2)\cdots(x+\lambda^{15})$, where $\lambda = 02_{\text{HEX}}$, and a primitive field generator polynomial is $p(x) = x^8+x^4+x^3+x^2+1$, that has the capability of correcting up to 8-byte random error. Following the RS coder, the error protected packets are fed into a 12-byte depth convolutional byte-wise interleaver. The interleaved data structure is composed of interleaved error protected packets and is delimited by the MPEG2-TS sync bytes (inverted or non-inverted) which preserves the periodicity of 204 bytes (see Fig. 2.3d).

2.2.3 Inner Coding

In DVB system, a mother convolutional code of rate 1/2 with 64 states is used to create a range of punctured convolutional codes (see Fig. 2.4). The generator polynomials of the mother convolutional codes are $G_1 = X+X^1+X^2+X^3$ for X output stream, and $G_2 = X+X^2+X^3+X^5$ for Y output stream. With mother code of rate 1/2, the input bitstream is doubled by the convolutional coder and the output is reduced by using puncturing method to provide an appropriate punctured rate such as 2/3, 3/4, 5/6, and 7/8. For example, to obtain rate 2/3, one of each 4 bits data will be punctured. Similarly, to obtain rate 3/4, two of each 6 bits data will be punctured. At receiver, the punctured bits are padded with zeroes such that the decoding is carried as if rate 1/2 code. In essence, the punctured bits are lost bits which increase system BER.

2.2.4 Inner Interleaver

The inner interleaver is a group of joined blocks of demultiplexer, bit-interleaver, and symbol-interleaver as shown in Fig. 2.5. Firstly, the output of convolutional encoder is split into six substreams. The splitting scheme is defined as a mapping of the input bits onto the output bits in one of six substreams. Each substream is then processed by a separate bit-interleaver. Bit-interleaver maps 128-bit input block onto 128-bit output block. The outputs of the six bit-interleavers are sequentiallygrouped

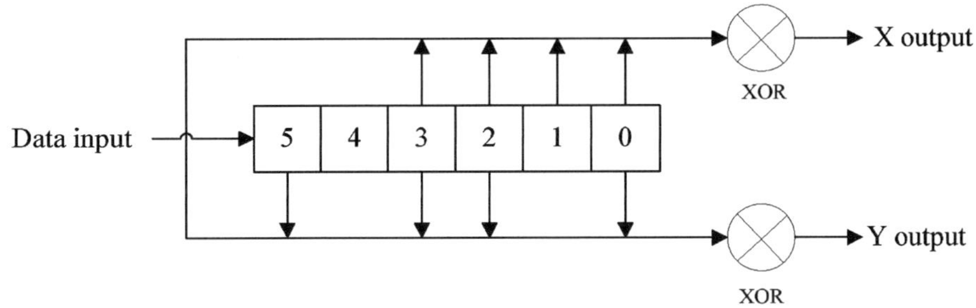

Fig. 2.4 Convolutional code generator

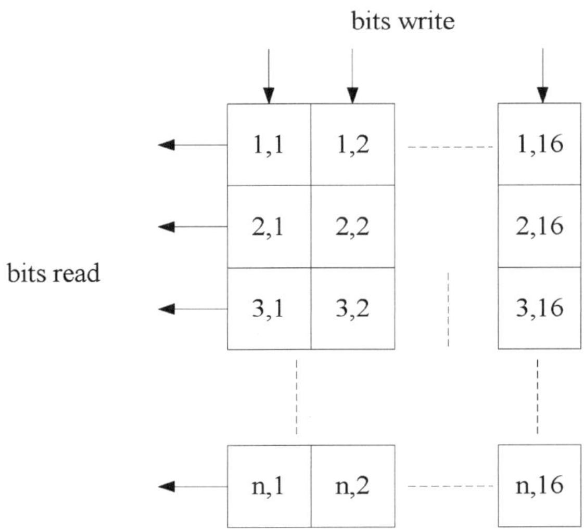

Fig. 2.5 Inner interleaver

to form six-bit symbols. These six-bit symbols are again interleaved by symbol-interleaver. In 2K mode, the symbol-interleaver interleaves 12 sets of 126, six-bit symbols, and in 8K mode, it interleaves 48 sets of 126 symbols. Furthermore, the in-depth interleaver used in 4K mode, which interleaves 24 sets of 126 symbols, is used for DVB-H systems. In addition to the 2K and 8K transmission modes provided originally by the DVB-T standard, the 4K mode with its in-depth interleaver brings additional flexibility in network design by trading off mobile reception performance and size of SFN (single frequency network) networks.

DVB-H is principally a transmission system allowing reception of broadcast information on single antenna hand-held mobile devices. In the DVB-T system, the 2K mode is known to provide better mobile reception performance than the 8K mode, due to the larger inter-carrier spacing it implements. However, the associated guard interval duration of the 2K mode OFDM symbols is very short. This makes the 2K mode only suitable for small size SFN. However, 4K OFDM symbol has a longer duration and longer guard interval than 2K mode. This makes 4K mode suitable for medium size SFN networks. It can increase the spectral efficiency for SFN network planning. As for 8K mode, the symbol duration of 4K mode is shorter than in the 8K mode and so the channel estimation can be done more frequently in the 4K mode demodulator. Therefore, it provides a better mobile performance than 8K mode, although not as high as with the 2K transmission mode, it is enough for the use of DVB-H scenarios. In this scenario, 4K mode provides a good trade-off between the two important aspects of the system: spectral efficiency for the DVB-H network designers and high mobility for the DVB-H consumers.

2.2.5 Frame Adaptation, TPS, and QAM Map

In an OFDM-based system, a group of complex symbols are modulating with frequency carriers of the equivalent length to create an OFDM symbol. In DVB, transmitted signal is organized in frames; a frame is composed of 68 OFDM symbols, and 4 frames constitute a super-frame. Besides carrying data, OFDM symbol also carries scattered pilots, continual pilots, and TPS (transmission parameter signaling) carriers. For example, in 2K mode, an OFDM frame comprises 1705 carriers accommodating 1512 data carriers, 131 scattered pilots, 45 continual pilots, and 17 TPS carriers. In 8K mode, an OFDM frame has 6817 carriers (6048 data carriers, 177 continual pilots, 524 scattered pilots, 68 TPS carriers). The power of pilot carrier is 16/9 times stronger than data carrier, whereas TPS carrier is transmitted at normal power as a data carrier. While TPS carriers are used for signaling various transmission parameters, the pilot signals are used in frame adaptation (see Fig. 2.1) in order to provide synchronization to the DVB frame structure through various methods such as frame synchronization, frequency synchronization, time synchronization, and channel estimation. Note that the DVB receiver must be synchronized and equalized such that the received signal can be decoded successfully and to gain access to the information held by the TPS pilots. All data carriers in an OFDM frame are modulated using QPSK, 16-QAM, 64-QAM, nonuniform 16-QAM, or nonuniform 64-QAM constellations as shown in Fig. 2.6.

2.2.6 OFDM Structure

OFDM is a proven technique for achieving high data rate whilst overcoming multipath fading in wireless communication [8–10]. OFDM can be thought of as a hybrid of multicarrier modulation (MCM) and frequency shift keying (FSK) modulation. Orthogonality amongst the carriers is achieved by separating them by a frequency which is an integer multiple of the inverse of symbol duration of the parallel bit streams thus minimizing intersymbol interference (ISI). Carriers are spread across the complete channel, fully occupying it and hence using the bandwidth very

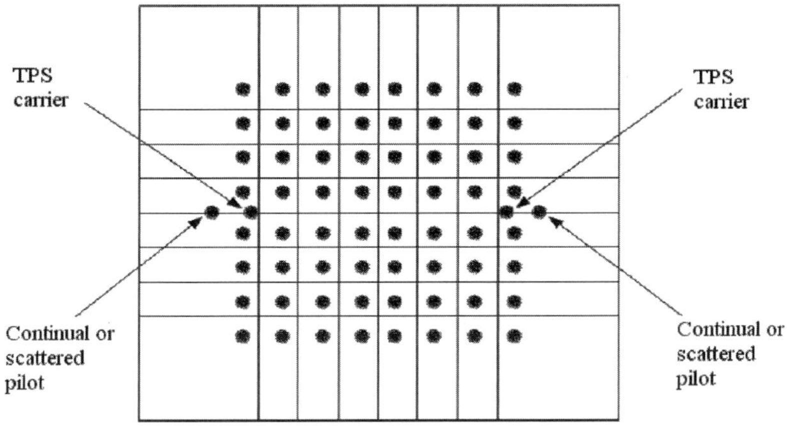

Fig. 2.6 Continual pilots, scattered pilots and TPS carriers in the DVB constellation diagram

efficiently. The OFDM receiver uses adaptive bit loading techniques based on a dynamic estimate of the channel response, to adapt its processing and compensate for channel propagation characteristics and achieves near ideal capacity. The most important advantage of OFDM systems over single carrier systems is obtained when there is frequency-selective fading. The signal processing in the receiver is rather simple; multiplying each subcarrier by a complex transfer factor thereby equalizing the channel response compensates distortion of the signals [11]. It is not feasible for conventional single carrier transmission systems to use this method. However some disadvantages of OFDM systems are that the peak-to-average power ratio (PAPR) of OFDM is higher than for a single carrier system and OFDM is sensitive to a flat fading channel.

The OFDM signal consists of N orthogonal subcarriers modulated by N parallel data streams. Denoting the frequency and complex source symbol of the kth subcarrier as f_k and d_k, respectively, the baseband representation of an OFDM is

$$x(t) = \frac{1}{\sqrt{N}} \sum_{k=0}^{N-1} d_k e^{j2\pi f_k t}, \quad 0 \leq t \leq NT \tag{2.1}$$

where d_k is typical taken from a PSK or QAM symbol constellation, and NT is the duration of the OFDM symbol. The subcarrier frequencies are equally spaced at $f_k = k/NT$. The OFDM signal in Eq. 2.1 can be derived by using a single IFFT operation rather than using a bank of oscillators. Therefore the OFDM symbol can be represented:

$$x[n] = \frac{1}{\sqrt{N}} \sum_{k=0}^{N-1} d_k W_N^{kn}, \quad 0 \leq n \leq N-1 \tag{2.2}$$

where $W_N = e^{j2\pi/N}$.

2.2.7 Differences Between DVB-T and DVB-H

There are major three differences between DVB-T and DVB-H which are the 4K mode, in-depth interleaver, and TPS-bit signaling used in the DVB-H [8].

The 4K mode is architecturally compatible with existing DVB-T infrastructure, requiring only minor changes in the modulator. It constitutes a new FFT size added to the native DVB-T 2K and 8K FFT sizes, all other parameters being the same. As the C/N performance with the three channel models (AWGN, Rice and Rayleigh) is FFT size independent, so the 4K size provides the same performance as the other two modes in these channels. The real feature of the 4K mode is the performance enhancement in mobile reception. The current DVB-T standard provides excellent mobile performance with 2K mode, but with 8K mode the performance is unsatisfactory, especially with reasonable receiver complexity. The 4K mode provides roughly two times better Doppler performance than 8K mode. Compared to an existing 2K/8K DVB-T receiver, the addition of the 4K mode and of the in-depth symbol interleaver does not require extra memory, significant amounts of logic, or extra power.

Secondly, there is an in-depth interleaver in 4K mode DVB-H [6] systems. The longer symbol duration of the 4K mode makes it more resilient to impulsive interference. For a given amount of noise power in a single impulse noise, power is averaged over 4096 subcarriers by the FFT in the demodulator.

Furthermore, TPS-bit signaling provides robust multiplex level signaling capability to the DVB-H transmission system. TPS is known to be a robust signaling channel as a TPS lock in a demodulator can be achieved with a low C/N value. Two formerly unused TPS bits have been allocated for DVB-H signaling. One is a time-slicing indicator to signal that at least one time-sliced DVB-H service is available in the transmission channel. The other is an MPE-FEC indicator to signal that at least one DVB-H service in the transmission channel is protected by MPE-FEC. The DVB OFDM configuration for 2K, 8K and 4K modes is provided in Tables 2.1–2.3, respectively.

Table 2.1 OFDM parameters for the 2K mode

Parameter	2K mode			
Elementary period T	$7/64\,\mu s$			
Number of carriers	1705			
Value of carrier number K_{min}	0			
Value of carrier number K_{max}	1704			
Duration Tu	$224\,\mu s$			
Carrier spacing 1/Tu	4464 Hz			
Spacing between carriers K_{min} and K_{max}	7.61 MHz			
Allowed guard interval Δ/Tu	1/4	1/8	1/16	1/32
Duration of symbol part Tu	$2048 \times T$; $224\,\mu s$			
Duration of symbol interval Δ	$512 \times T$; $56\,\mu s$	$256 \times T$; $28\,\mu s$	$128 \times T$; $14\,\mu s$	$64 \times T$; $7\,\mu s$
Symbol duration $T_s = \Delta + \text{Tu}$	$280\,\mu s$	$252\,\mu s$	$238\,\mu s$	$231\,\mu s$

Table 2.2 OFDM parameters for the 4K mode

Parameter	4K mode			
Elementary period T	$7/64\,\mu s$			
Number of carriers	3409			
Value of carrier number K_{min}	0			
Value of carrier number K_{max}	3408			
Duration Tu	$448\,\mu s$			
Carrier spacing 1/Tu	2232 Hz			
Spacing between carriers K_{min} and K_{max}	7.61 MHz			
Allowed guard interval Δ /Tu	1/4	1/8	1/16	1/32
Duration of symbol part Tu	$4096 \times T$; $448\,\mu s$			
Duration of symbol interval Δ	$1024 \times T$; $112\,\mu s$	$512 \times T$; $56\,\mu s$	$256 \times T$; $28\,\mu s$	$128 \times T$; $14\,\mu s$
Symbol duration $T_s = \Delta + \text{Tu}$	$560\,\mu s$	$504\,\mu s$	$476\,\mu s$	$462\,\mu s$

Table 2.3 OFDM parameters for the 8K mode

Parameter	8K mode			
Elementary period T	$7/64\,\mu s$			
Number of carriers	6817			
Value of carrier number K_{min}	0			
Value of carrier number K_{max}	6816			
Duration Tu	$896\,\mu s$			
Carrier spacing 1/Tu	1116 Hz			
Spacing between carriers K_{min} and K_{max}	7.61 MHz			
Allowed guard interval Δ /Tu	1/4	1/8	1/16	1/32
Duration of symbol part Tu	$8192 \times T$; $896\,\mu s$			
Duration of symbol interval Δ	$2048 \times T$; $224\,\mu s$	$1024 \times T$; $112\,\mu s$	$512 \times T$; $56\,\mu s$	$256 \times T$; $28\,\mu s$
Symbol duration $T_s = \Delta + \text{Tu}$	$1120\,\mu s$	$1008\,\mu s$	$952\,\mu s$	$924\,\mu s$

2.2.8 Compatibility Between DVB-H and DVB-T

The new DVB-H components are compatible to the existing DVB-T systems [6]. First of all, as for time-slicing and MPE-FEC, they do not raise any incompatibility issues and are fully compatible with the existing DVB physical layer such as DVB-T, S, and C. Additionally time-slicing and MPE-FEC in DVB-H modify the MPE protocol in a fully backward compatible way. Allocating bytes of the MAC address fields, located in the MPE section header, is fully supported by the DVB-SI standard. Furthermore, DVB-H signaling is fully backward compatible as there are bits reserved for future use and these are ignored by the DVB-T receivers.

However, the proposed new 4K mode and in-depth symbol interleaver of DVB-H affects the compatibility with the current DVB-T physical layer specification. In system level, the new 4K mode could be considered just as an interpolation of the existing 2K and 8K mode, requiring only an additional parameter in the DVB-T system and a little control logic in the equipment. For receiver level, it is obvious that current 2K or 8K receivers will be unable to receive 4K signals, but this is not a severe restriction as any new DVB-H network using the 4K mode would be targeted towards new services and new types of hand portable terminals. The only restriction in this case arises when sharing the multiplex between traditional DVB-T and DVB-H services. The standard allows new 4K capable receivers to receive both 2K and 8K transmissions.

2.3 Hardware Implementation of DVB Transmitter

The DVB transmitter (or receiver) system is, generally, composed of digital system and analog front-end as shown in Fig. 2.7. The digital system for transmitter mainly consists of baseband signal processing and digital up-conversation. As far as the baseband signal processing is concerned, the DVB bandwidth and the transmission capacity can be tailored by changing only the system clock namely the *elementary period*, *T* of 7/64us (see Tables 2.1 to 2.3). The system clock is equivalent to 9.14MSPS which also means 9.14M symbols sampled per second. The analog front-end circuit for the transmitter mainly consists of signal amplification, bandwidth filtering, and conversion from immediate frequency (IF) to the radio frequency (RF) of interest.

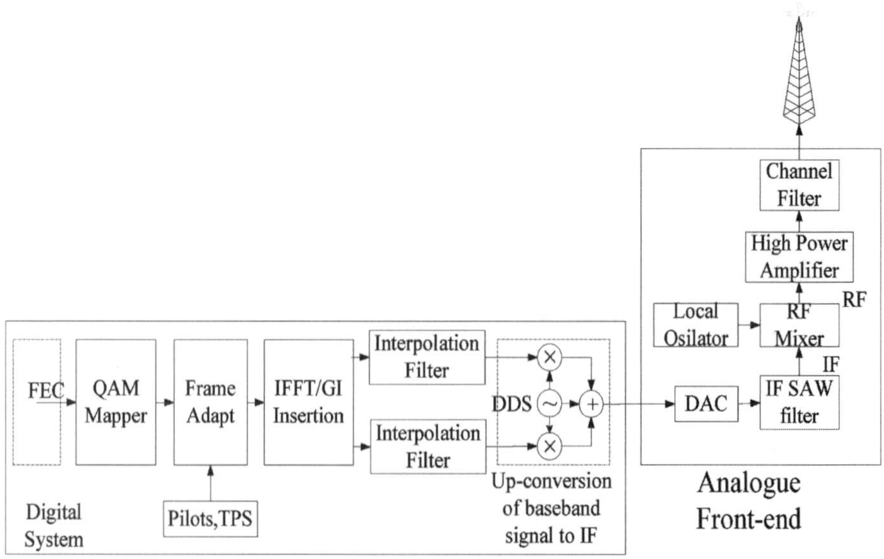

Fig. 2.7 Hardware design for DVB transmitter

2.3.1 Digital System for DVB Transmitter

It is suggested that Xilinx Virtex-4 XtremeDSP FPGA development board (see Fig. 2.8) is ideal for implementation of the digital part of the DVB transmitter (or receiver).

Virtex-4 incorporates up to 200K logic cells, 500 MHz performance and system architecture that is able to deliver twice the density, twice the performance and half the power consumption of previous generation of Xilinx FPGAs. Virtex-4 has 192 DSP slices and new DSP algorithms allow higher levels of DSP integration. Most baseband signal processing is performed in Virtex-4 (see Fig. 2.9).

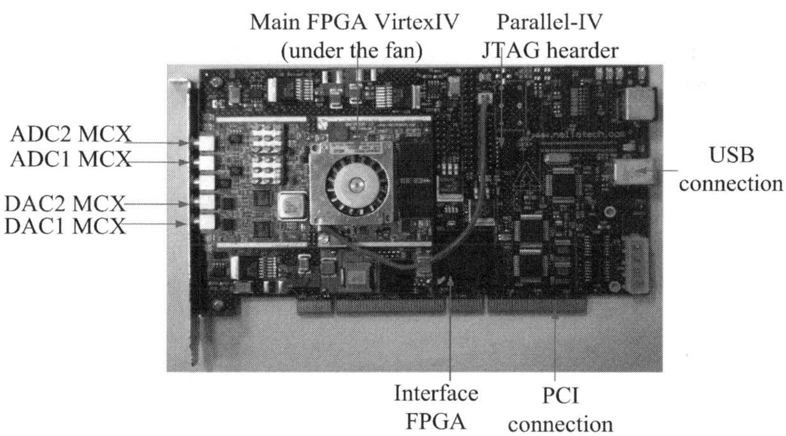

Fig. 2.8 Xilinx Xtreme DSP FPGA development platform

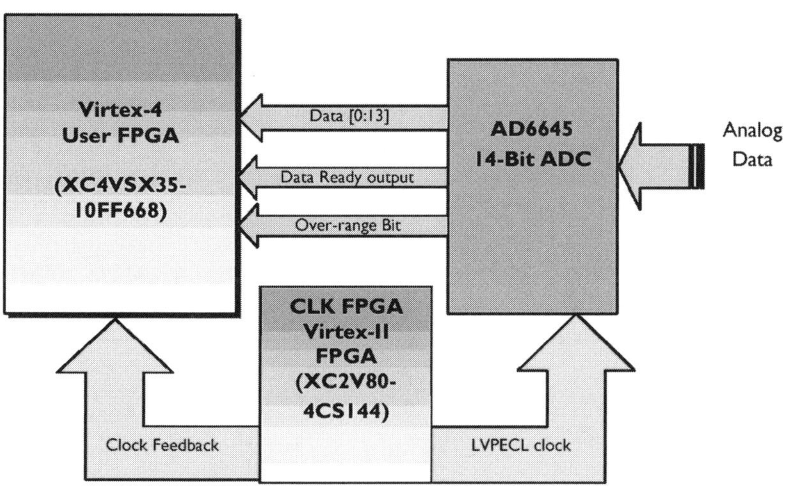

Fig. 2.9 The DACs and FPGAs interfacing for DVB transmitter

The Virtex-2 performs as a time/clock control for the Virtex-4 FPGA and AD9772A DAC. The DAC offers 14-bit resolution and a maximum conversion rate of 160 MSPS. Additional control signals exist between the DAC and the FPGA, i.e. "clock feedback" and "LVPECL" enables full control of the system clock and the DAC output sampling. The clock feedback controls the clocking of the main FPGA, in other words, system clock. The LVPECL clock from Virtex-2 controls the clocking of DAC output sampling. Having these features, the functional blocks in Fig. 2.7 can be easily implemented on this FPGA platform.

After the forward error correction (FEC) block, the error-controlled data stream is distributed uniformly over the entire channel and sent to the QAM mapping and frame adaptation. Through the QAM mapping, all the payload carriers are then mapped depending on whether hierarchical or nonhierarchical modulation is used. This results in two tables, namely that for the real part Re(f) and that for the imaginary part Im(f). However, they also contain gaps into which the pilots and the TPS earners are then inserted by the frame adaptation block. Therefore, the complete tables, comprising 2K, 4K or 8K values, respectively, are then fed into the heart of the DVB modulator, the IFFT block. Here, the 2K UK mode is considered for the implementation in FPGA.

IFFT with 2048 points are used to get the time domain of the transmitter signals. The pipelined streaming I/O solution is adopted in the FFT and IFFT implementations, as shown in Fig. 2.10.

The pipelined streaming I/O pipelines several radix-2 butterfly-processing engines to offer continuous data processing [12]. Each processing engine has its own memory banks to store the input and intermediate data. The core has the ability to simultaneously perform transform calculations on the current frame of data, load input data for the next frame of data, and unload the results of the previous frame of data. This architecture supports unscaled and scaled fixed point arithmetic methods. In the scaled fixed-point mode, the data is scaled after every pair of radix-2 stages.

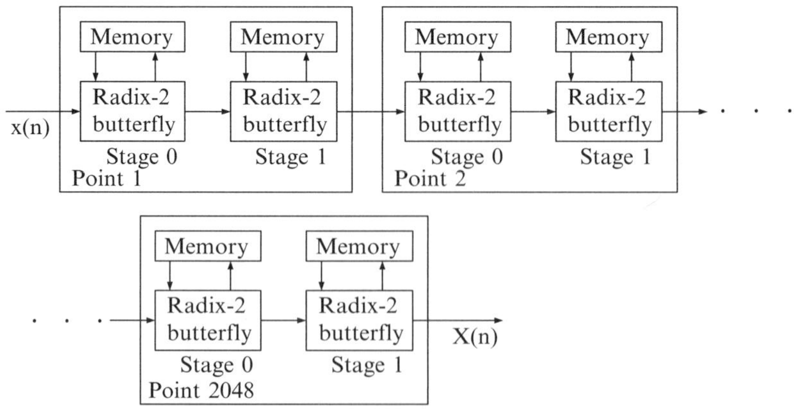

Fig. 2.10 IFFT implementation structure

After IFFT block, the OFDM signal is separated into real and imaginary part in the time domain. The 2048 values respectively are stored in buffers organized along the lines of the pipeline principle.

After the IFFT block, the digital fixed-point signals in time domain are alternately written into one buffer whilst the other one is being read out. During read-out, the end of the buffer is read out first as a result of which the guard interval is formed as shown in Fig. 2.11.

The signals then are sent to the interpolation filter to get high sampling rate data. The interpolation filter can be divided into three stages as shown in Fig. 2.12 [13].

Figure 2.12 illustrates a block diagram of the DUC (digital up-converter) core. The DUC is a digital circuit which implements the conversion of a complex digital baseband signal to a real passband signal. The input complex baseband signal is sampled at a relatively low sampling rate, typically the digital modulations symbol rate 9.14 MSPS in DVB systems. The baseband signal is filtered and converted to a

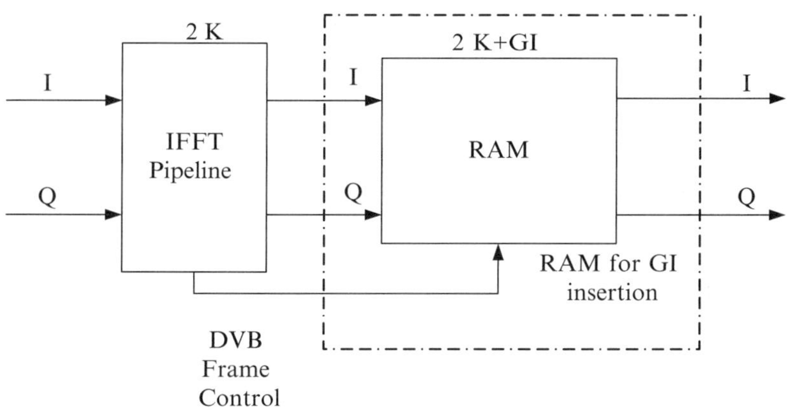

Fig. 2.11 Guard interval insertion structure

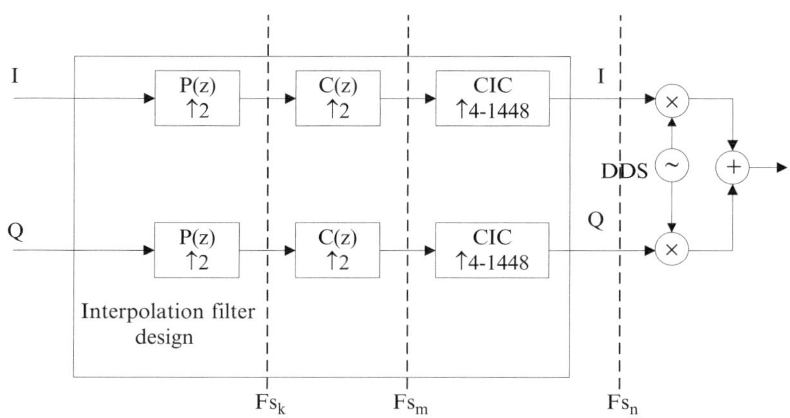

Fig. 2.12 Digital up-converter

higher sampling rate before being modulated onto a direct digital synthesized (DDS) carrier frequency. The DUC performs pulse shaping and modulation of an intermediate carrier frequency appropriate for driving a final analog up-converter. From the beginning of the interpolation filter, the complex input signals are passed through three stages of filtering, each of which performs a sampling rate change and associated low pass interpolation filtering. The three filtering stages are as follows. Firstly, pulse shaping FIR filter P(z) provides a sampling rate increase of 2 and performs transmitter Nyquist pulse shaping. Secondly, compensation FIR filter C(z) provides a sampling rate increase of 2 and is used to compensate for the passband distortion of the third stage cascaded integrator-comb (CIC) filter. The FIR filters, P(z) and C(z), are implemented using efficient multiplier-accumulator (MAC) blocks. The multipliers are realized using the embedded multipliers available in the Virtex-4.

Finally, the CIC interpolation filter structure implements sampling rate increases from 4 to 1448. The CIC filter design parameters permit the synthesis of a circuit with a fixed sampling rate increase, or one whose sampling rate increase is programmable in real time [14]. The complex data stream from the filtering stages is up-converted to an IF band by a mixing operation with a local oscillator generated by DDS in digital domain. The I and Q mixer outputs are combined to form the final DUC passband centered at 36 MHz. CIC filters are multirate filters used for realizing large sample rate changes in digital systems. The filter structure uses only addition operators, subtraction operators, and registers. The CIC filter supports interpolation factors from 4 to 1448. The CIC filter structure can be presented in Fig. 2.13.

Figure 2.13 shows the CIC filter structure for three stages and four times interpolation. As for the implementation for DVB transmitter, the configuration of the interpolation filter is P(z) with one time interpolation rate, C(z) with two times interpolation rate, and CIC with five times interpolation rate. Overall, the magnitude response of the interpolation filters is presented in Fig. 2.14.

The magnitude response is center at 36 MHz with passband from 31.2 to 40.8 MHz and stopband from 30 MHz to 42 MHz. The shoulder can be suppressed to 60 dB to meet the requirement of DVB spectrum masks.

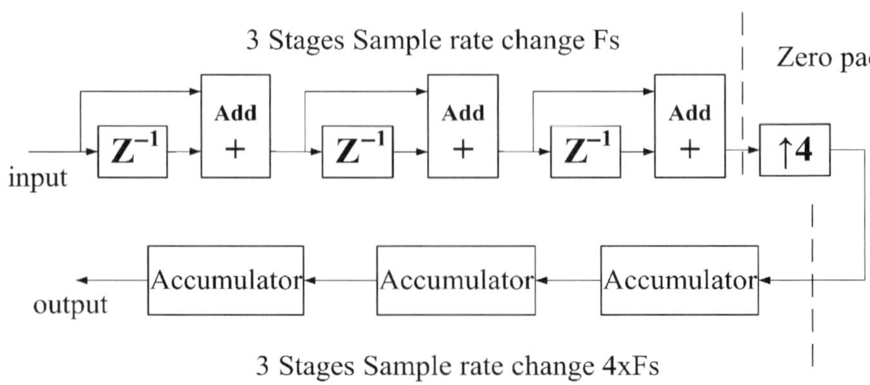

Fig. 2.13 CIC filter structure block diagram

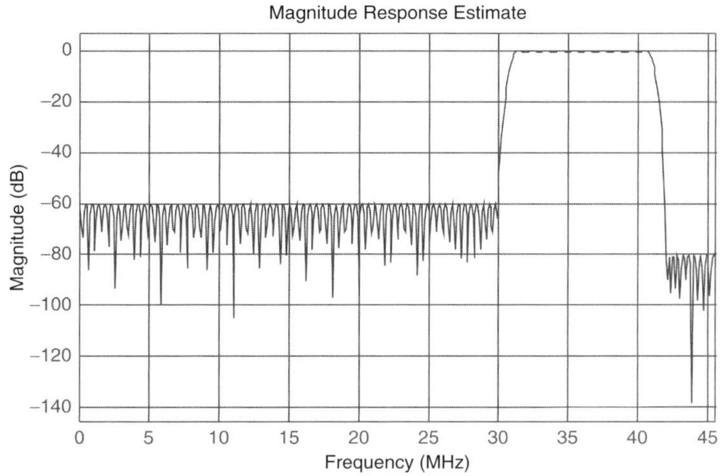

Fig. 2.14 Frequency response for CIC filter

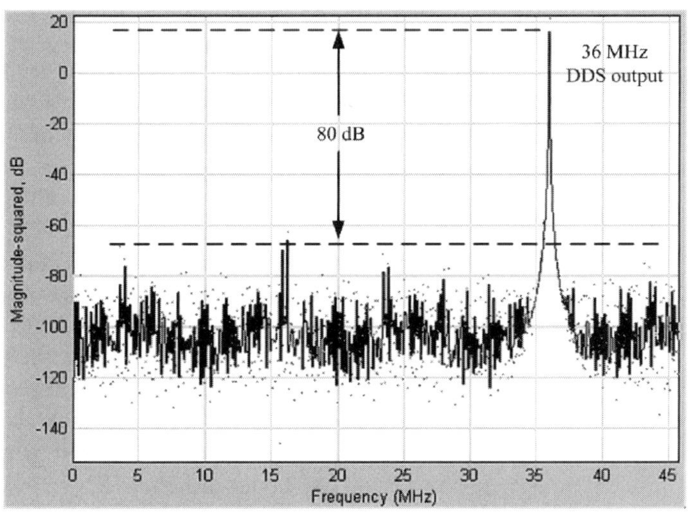

Fig. 2.15 Spectrum of DDS with Fc = 36 MHz, Fs = 91.4 MHz and SFDR = 80 dB

The DDS is realized using the embedded multipliers available in the Virtex-4 FPGAs. The DDS employs an error feed-forward circuit, which produces a high-quality local oscillator with up to 115 dB of spurious free dynamic range (SFDR). The SFDR of the DDS is a selectable parameter with three choices, 115, 80 or 60 dB, to provide the option for reduced utilization of FPGA resources. The implementation requires a single block memory and multipliers. The phase accumulator is fixed at 32 bits. Figure 2.15 provides the spectrum plot of the DDS with Fs 91.4 MHz, SFDR 80 dB, and center 36 MHz.

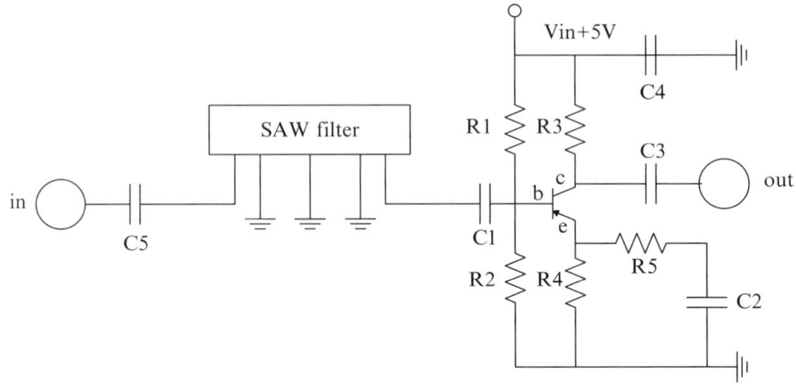

Fig. 2.16 SAW filter

2.3.2 Analog Front-End Circuit for DVB Transmitter

As shown in Fig. 2.7, the analog front-end circuit includes IF surface acoustic wave (SAW) filter, RF mixer, local oscillator, high power amplifier and channel filter. The IF SAW filter circuit is to remove unwanted mixer products from the IF signals constructed by the DUC (see Fig. 2.12). The SAW filter can be easily obtained off the shelf with a range of supported frequencies. However, there is a significant power loss in the SAW filter; hence there is a need to employ an amplifier circuit after the filter stage. Figure 2.16 shows a simple interfacing of SAW filter to a Class-A amplifier circuit.

The RF mixer (see Fig. 2.7) is to get RF band signal from IF band signal according to local oscillator. The high power amplifier is the UHF band low noise amplifier. The amplifier should be low noise with max 2 dB noise figure and at least have moderate gain over 20 dB. It also should be sufficiently linear to cope with full range of possible input signals. The selected component is RF low noise amplifier: ZRL-1150LN + (NF = 0.7 dB typical, BW: 650 MHz–1.4 GHz, gain at 730 MHz = 32 dB).

The channel filter should be required to remove out of band power, which could saturate subsequent amplifier. It probably is a cavity or helical filter. However, this kind of UHF filter has some loss, makes RX noise figure worse. The filter should be low loss (a few dB max), and have a good rejection at frequencies where significant out of band power is expected. The selected UHF filter in the design is 730 MHz band pass filter: VBFZ-780 + (2.5 dB points 710 MHz–850 MHz, loss at 726 MHz–734 MHz$\sim$1.8 dB).

2.4 Hardware Implementation for DVB Receiver

Figure 2.17 illustrates the hardware system diagram for DVB receiver. It includes analog front-end and digital system implementation. Because the DVB-H standard has been recently developed to provide video broadcasting services for hand-held

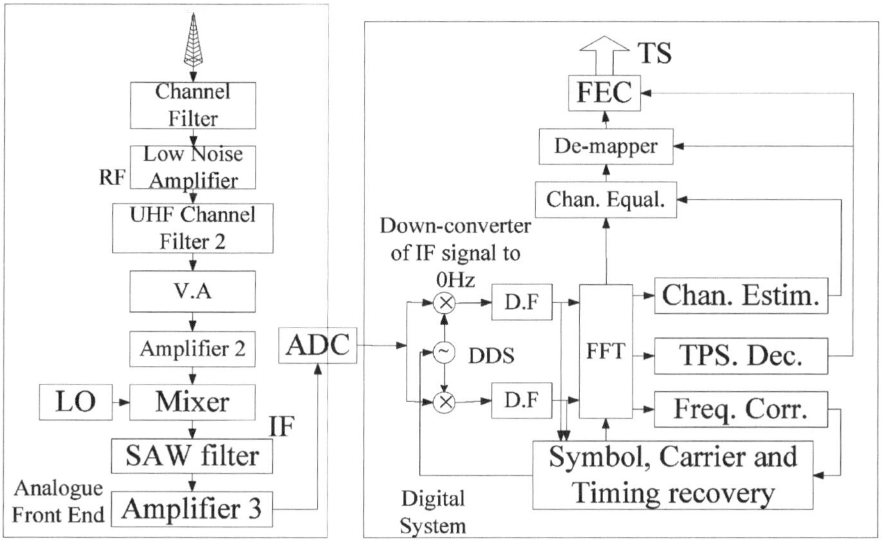

Fig. 2.17 Hardware design for DVB receiver

devices, the power consumption of a DVB-H baseband receiver must be strictly controlled in order to achieve long battery life. In the analog front-end for the receiver, it consists of several low power channel filters, amplifiers, attenuators and mixers. In the digital system for receiver, it comprised of an OFDM demodulator and FEC blocks for complete symbol, carrier and timing recovery, channel equalization and data demodulation and correction.

2.4.1 Analog Front-End Circuit for DVB Receiver

In Fig. 2.17, UHF channel filter is required to remove out of band power (e.g. mobile phone) which could saturate the subsequent amplifier [15]. The characteristics of a suitable UHF filter should be low loss (a few dB max), good rejection at frequencies where significant out of band power is expected. The selected UHF filter in the design is 730 MHz band pass filter: VBFZ-780 + (2.5 dB points 710 MHz–850 MHz, loss at 726 MHz–734 MHz$\sim$1.8 dB).

The low noise amplifier for UHF band should be low noise with max 2 dB noise figure and at least moderate gain over 20 dB. It also should be sufficiently linear to cope with full range of possible input signals. The selected component is RF low noise amplifier: ZRL-1150LN + (NF = 0.7 dB typical, BW: 650 MHz–1.4 GHz, gain at 730 MHz = 32 db). The UHF channel filter 2 should at least be able to remove power at image frequency with very low bandwidth.

The variable attenuator as shown V.A in Fig. 2.17 is required in order to allow the analog front-end to work in dynamic range of RF frequency, i.e. UHF and IF band. The selected component is ZX76–31R5-PN-S + (0–2400 MHz; 0.5 dB steps).

The amplifier 2 is used to obtain a moderate gain. In normal operation, output power level should not overload the mixer. The selected component is RF low noise amplifier: ZX60–6013E (NF = 3.3 dB typical, BW 20 MHz–6 GHz, gain at 730 MHz = 16 dB).

The mixer is required to extract IF from the UHF band signal by setting the appropriate local oscillator. SAW filter is required to remove unwanted mixer products from the received IF signals.

Amplifier 3 is the IF low noise amplifier. It is used to obtain more gain to meet the signal level requirement of ADC. The selected component is ZX60–6013E (NF = 3.3 dB typical, wide bandwidth start 20 MHz–6 GHz, gain at 30 MHz = 16 dB).

After the analog front-end, the DVB signal is converted down to a lower IF at 36 MHz. This is followed by ADC sampling at 91.4 MHz. Following the ADC, the data stream should be now available at data rate of 91.4 MSPS.

2.4.2 Digital Circuits System Design for DVB Receiver

The digital system design can be divided into six stages including DDC (digital down converter), symbol, carrier and timing recovery, OFDM demodulation, channel equalization, QAM demodulation and FEC for standard 4.98 Mb/s to 31.67 Mb/s data rates. The OFDM frame structure with a standard operational baseband frequency of 9.14 MHz is used as an important parameter to synchronize the symbol timing and carrier recovery.

2.4.2.1 Digital Down Converter

DDC is composed of DDS and several decimation filters as shown in Fig. 2.18. It is the reverse process of DUC (see Fig. 2.12).

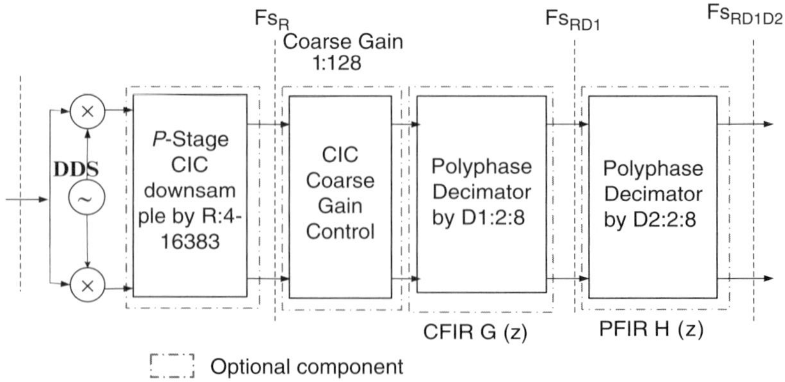

Fig. 2.18 Digital down converter design

The desired channel is translated to baseband using DDS comprising the two multipliers. DDS is the same component as in the DVB transmitter digital system design. The sample rate of the signal is then adjusted to match the DVB bandwidth. This is performed using a multistage multirate filter consisting of the CIC filter and the two polyphase decimators G(z) and H(z) as shown D.F in Fig. 2.17. In order to provide maximum flexibility in the datapath, each of the filters may be optionally inserted or excluded from the signal processing chain. Depending on a combination of master clock frequency, signal sample rate and decimation rates R, D1 and D2, the multipliers in the mixer, various components in the CIC filter and the CFIR and PFIR may be time division multiplexed to service both the inphase (I) and quadrature (Q) signals.

Furthermore, a course gain adjustment may be applied at the output of the CIC filter to compensate for the bandwidth reduction performed by the CIC decimator. The signal gain can be boosted by 42 dB. If the signal power across the input bandwidth is relatively flat, as is the case in most frequency division multiplexed (FDM) systems, then one would want to boost the signal power out of the CIC filter by a factor of $\sqrt{R}$, where R is the decimation rate of the CIC filter [14].

2.4.2.2 Symbol Carrier and Time Recovery Design

After getting the DVB baseband signal from DDC, the next important stage is to get symbol synchronization. There are two stages of symbol synchronization in the DVB receiver design. Firstly, the robust synchronization is searched by using guard interval correlation structure. After that fine symbol synchronization is performed making use of the scattered pilots.

As for the coarse symbol synchronization, autocorrelation based on guard interval is used to derive synchronization information [16]. Using autocorrelation, symbols are detected which exist in the signal several times and in the same way. For example, the received time domain samples are denoted by $r(k)$. The estimation of the starting position of the symbol, k_{est}, can be found in Eq. 2.3 as follows:

$$k_{\text{est}} = \arg\max_{k} \left| \sum_{i=0}^{L-1} r(k-i) r^*(k-i-N) \right| \tag{2.3}$$

where N is the number of the carriers and L is the size of guard interval in samples.

The system diagram of coarse symbol synchronization is illustrated in Fig. 2.19. To obtain the best performance, the length of the moving sum has to be same as the length of the guard interval, which is typically the same as the lengths of the correlating parts. The Max block will detect maximum correlation value position inside a certain window, whose length is typically $N+L$ samples. In an ideal case the maximum correlation values lie at $N+L$ samples apart from each other. The coarse synchronization result of DVB-T 2K mode is shown in Fig. 2.20.

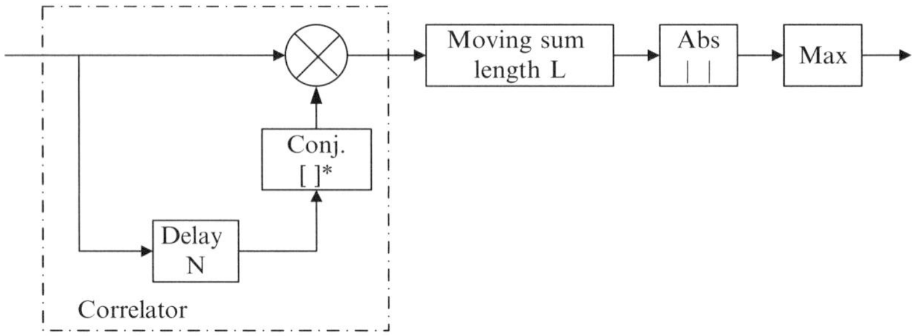

Fig. 2.19 Coarse symbol synchronization design

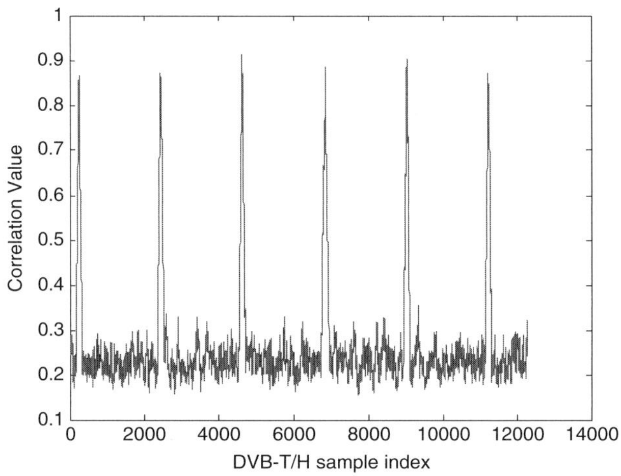

Fig. 2.20 Result of coarse symbol synchronization for DVB-T samples

From Fig. 2.20, six DVB-T symbols are detected by the coarse symbol synchronization. As for the fine symbol synchronization in DVB system the strategy is to use scattered pilots, which are also used for channel estimation and equalization. By collecting scattered pilots, the channel can be estimated in the frequency domain after post-FFT operation. After IFFT, the CIR (channel impulse response) can be estimated according to the scattered pilots. The received demodulated signal with incorrect FFT-window position can be described as in the following equation:

$$\widetilde{X}(n) = \frac{1}{N}\sum_{n=0}^{N-1} x(n)e^{j2\pi k\frac{\varepsilon}{N}}e^{-j2\pi k\frac{n}{N}} \tag{2.4}$$

where ε is the offset of the fine FFT-window in samples. Because $e^{j2\pi k\varepsilon/N}$ is constant for each k, this can be seen as a time shift of the impulse response. It is assumed that the offset is in the direction that causes no Inter Symbol Interference (ISI) [16]. The optimal FFT-window position can be found by indicating the direction and the amount of the CIR movement from the ideal place, which is known for the first

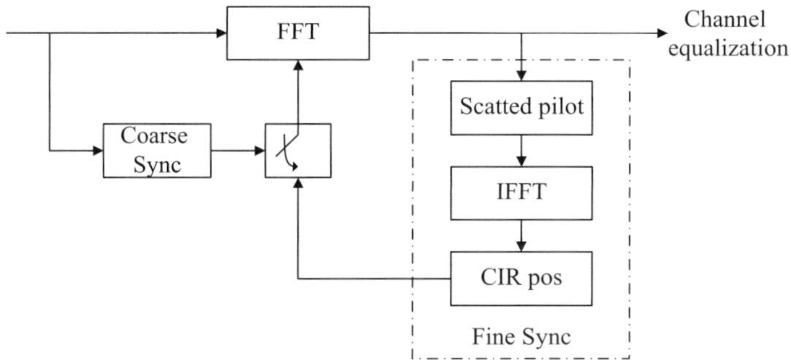

Fig. 2.21 Fine symbol synchronization design

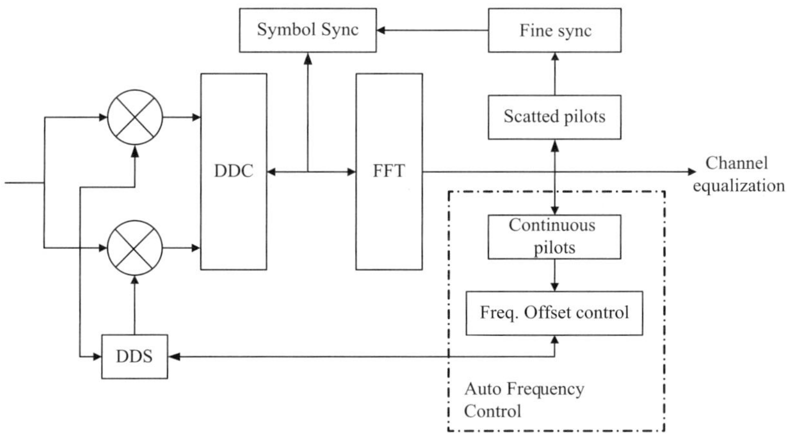

Fig. 2.22 Auto frequency control design

tap of CIR. The place of the first tap of CIR can be detected by using a threshold and detecting the parts where CIR values are over this threshold [17]. The hardware design is shown in Fig. 2.21.

2.4.2.3 Auto Frequency Control Design

Besides the symbol synchronization, the frequency synchronization is also necessary for the DVB receiver implementation [18]. If the receiver frequency differs from the transmitted frequency, all the constellation diagrams will rotate more or less quickly clockwise or anticlockwise. The system diagram is illustrated in Fig. 2.22.

It is necessary to measure the position of the continuous pilots in the constellation diagram. The phase difference is a directly controlled variable for the AFC and DDS. The DDS output frequency is changed until the phase difference becomes zero. The continuous pilots are located on the real axis, i.e. I (inphase) axis, at either

0° or 180° degrees, and have a defined amplitude. The continuous pilots are boosted by 3 dB compared with the average signal power and used for locking the receive frequency to the transmit frequency.

2.4.2.4 Channel Estimation Design

In DVB receivers, channel estimation is usually carried out in frequency domain using known symbols, referred to the pilots as shown in Fig. 2.23 [19].

After the synchronization process, the relationship between received signals $Y(n)$, transmitted signals $X(n)$, and the channel transfer function $H(n)$ can be formulated as

$$Y(n) = H(n) \cdot X(n) + W(n) \tag{2.5}$$

where $W(n)$ is the AWGN for the kth subcarrier after FFT. The received pilot signals are extracted from $Y(n)$. For the pilot subcarriers, the transmitted inform $X(n_p)$ is known to the receiver. Therefore, the channel coefficients at the pilot frequencies can be estimated simply using:

$$\widetilde{H}(n_p) = \frac{Y(n_p)}{X(n_p)} + W'(n_p) \tag{2.6}$$

where $W'(n_p)$ is the noise effect existing at the estimated channel coefficients and n_p is the subcarrier index corresponding to pilot locations. Channel estimation scheme in Eq. 2.6 is based on the least-squares (LS) algorithm. In order to estimate the channel coefficients at data subcarriers, the channel coefficients obtained at pilot frequencies are used for interpolation. After the interpolation, all channel coefficients $\widetilde{H}(n)$ can be obtained. The pilot signals are first extracted from the received signals after FFT, and the reference channel coefficients for the interpolation are

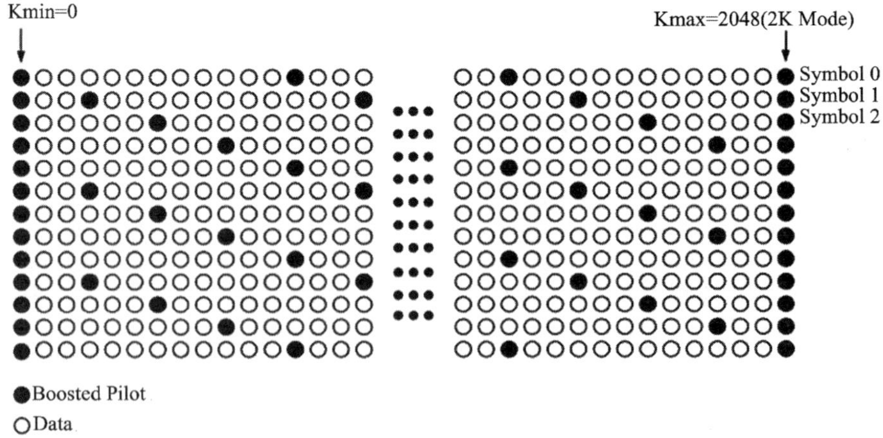

Fig. 2.23 Pilot arrangement in DVB systems

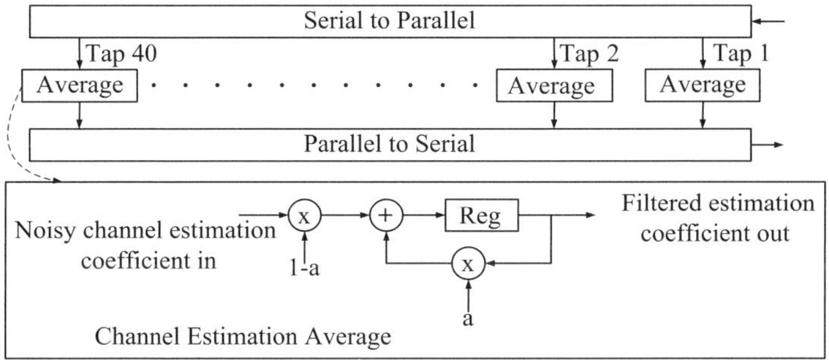

Fig. 2.24 Channel estimation average

estimated in frequency domain by comparing the pilot signals before and after transmission. Then, the channel responses of the subcarriers carrying data are estimated by interpolating neighboring pilot channel coefficients. In order to estimate the channel coefficients at data subcarriers, the piecewise-linear interpolation method is applied. In the linear interpolation, two consecutive pilot carriers are used to determine the channel response for data subcarriers. Assuming that the channel coefficient between $H(n_p)$ and $H(n_{p+1})$ is estimated, then the channel response is obtained by

$$\widetilde{H}(n) = \left(\frac{\widetilde{H}(n_{p+1}) - \widetilde{H}(n_p)}{n_{p+1} - n_p} \right) (n - n_p) + \widetilde{H}(n_p) \tag{2.7}$$

However, the LS estimator is very sensitive to noise because it does not consider noise effect in obtaining its solution. To compensate for the weakness of the LS estimator, auto-regressive averaging method based on pilot-block is illustrated in Fig. 2.24.

Since there is noise at the outputs of the FFT and IFFT circuits in the FPGA, it is necessary to take averages for the channel estimation. A simple first-order IIR (infinite impulse response) filter is used for averaging the channel estimation in which the function can be expressed as

$$\widetilde{E}(n) = (1 - \varsigma) \cdot (E(n)) + \varsigma \cdot (\widetilde{E}(n-1)) \tag{2.8}$$

where $\widetilde{E}$ is the averaged channel estimation coefficient and E is the new coming channel estimation coefficient, and ς is the averaging factor which is just less than 1.

2.4.3 Forward Error Coding Implementation

The FEC implementation in the receiver includes symbol/bit deinterleaver, Viterbi decoder, convolutional deinterleaver, and Reed-Solomon decoder as shown in Fig. 2.25. These functions are available in the library of Xilinx ISE FPGA software tool. Hence, the implementation is rather straightforward.

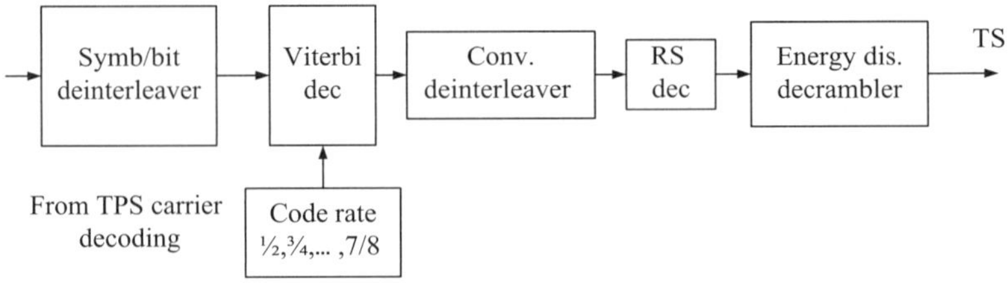

Fig. 2.25 FEC system diagram

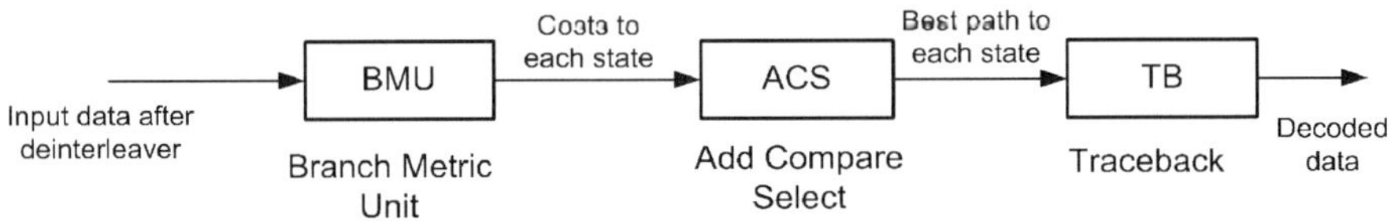

Fig. 2.26 Viterbi decoder block diagram

2.4.3.1 Viterbi Decoder

The demapped data passes from the demapper into the symbol and bit deinterleaver where they are resorted and fed into the Viterbi decoder. The Viterbi decoder decodes the convolutional coded bitstream by finding maximum likelihood path through trellis for convolutional code [20]. The basic decoder consists of three main blocks as shown in Fig. 2.26.

The first block is the branch-metric-unit (BMU). The incoming data can be hard coded with bit width 1 or soft coded with a configurable bit width which can be set to any value from 3 to 8. Hard coded data is decoded using the Hamming method, whereas the soft data is decoded using a Euclidean metric. In hard decoding, the demodulator makes a firm (or hard decision) on whether a one or zero is transmitted and provides no other information to the decoder on how reliable is the decision. In soft decoding, the demodulator provides the decoder with some side information together with the decision. The extra information provides the decoder with a measure of confidence for the decision. Soft coded data gives a significantly better BER performance compared with hard coded data. Soft decision offers approximately a 3 dB increase in coding gain over hard decision decoding.

The second block in the decoder is the add-compare-select-unit (ACS). This block selects the optimal path to each state in the Viterbi trellis. The ACS block uses the convolutional codes to extract the correct cost from the BMU for each branch in the trellis. The BER performance of the Viterbi algorithm varies with different convolutional code sets. The ACS module decodes for each state in the trellis. Thus, for a constraint length 7 decoder which has 64 states, there are 64 subblocks in the ACS block. If the hardware is implemented in serial mode, the amount of silicon required for each subblocks is only 2.5 slices.

The final block in the decoder is the traceback block (TB). The actual decoding of symbols in order to obtain the original data sent is accomplished by tracing the

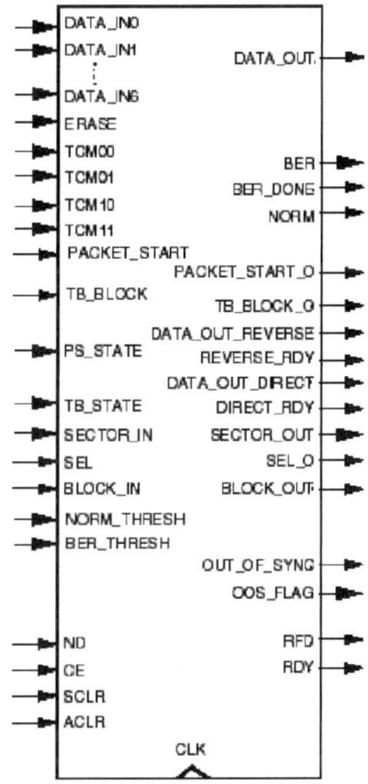

Fig. 2.27 Schematic symbol for Viterbi decoder in FPGA

maximum likelihood path through the trellis in a backward manner. Up to certain limit, a longer sequence of tracing through the trellis will result in a more accurate path. After a number of symbols equal to at least six times the constraint length, the decoded data is output. The traceback length is the number of trellis states processed before the decoder makes a decision on a bit. For the reduced latency option, the length of the traceback can be set to a multiple of 6 between 12 and 126. A best state option is to select the starting location for the traceback from the state with minimal cost.

Figure 2.27 illustrates Xilinx Viterbi decoder IP core. It provides two implementation methods which may vary the decoding operation: (i) a parallel implementation which gives fast data throughput at the expense of silicon area and (ii) a serial implementation which occupies a small area but requires a fixed number of clock cycles per decoded result.

2.4.3.2 Convolutional Deinterleaver Design

After the Viterbi decoder, the output data is sent to the convolutional deinterleaver. Figure 2.28 illustrates the operation of the Xilinx convolutional interleaver/deinterleaver IP core. The core operates as a series of delay line shift registers given the DIN input and DOUT output symbols. Both commutator arms start at branch 0 and advance to the next branch after the next rising clock edge. After the last branch has been reached, the commutator arms both rotate back to branch 0 and

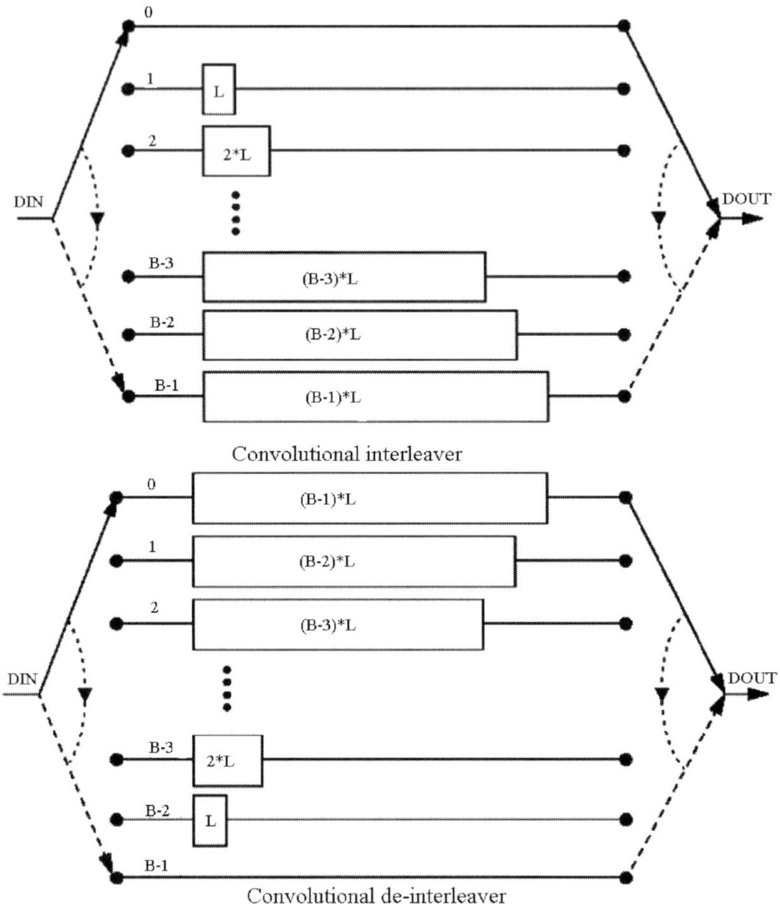

Fig. 2.28 Convolutional interleaver/deinterleaver

the process is repeated. Although branch 0 appears to be a zero-delay connection, there still is a delay of number of clock cycles between DIN and DOUT because of the fundamental latency of the FPGA core. Furthermore, the only difference between an interleaver and a deinterleaver is that branch 0 is the longest in the deinterleaver and the branch length is decremented by L rather than incremented and branch (B-1) has length 0.

2.4.3.3 Reed-Solomon Decoder Design

Thereafter the convolutional deinterleaver, the output data is sent to Reed-Solomon decoder. Figure 2.29 illustrates the schematic symbol of Reed-Solomon decoder obtained from the Xilinx IP core library. The Reed-Solomon decoder samples the n

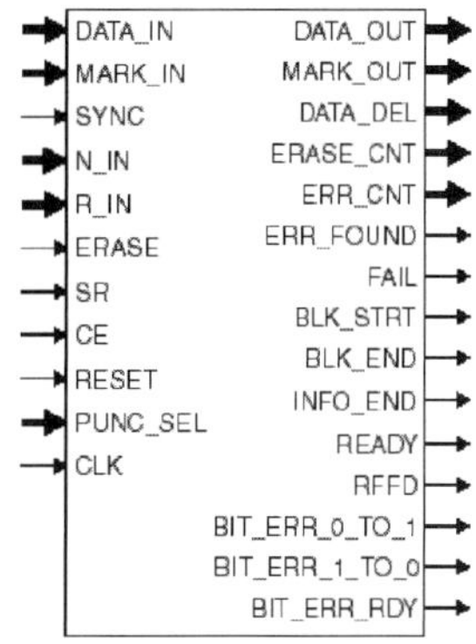

Fig. 2.29 Schematic symbol for RS decoder in FPGA

symbols on the DATA_IN port and attempts to correct errors. The corrected symbols are output on the DATA_OUT port after a fixed latency. The maximum number of symbol errors in a block is guaranteed to be corrected by the Reed-Solomon algorithm is $t = (n - k)/2$. The decoder can generally detect and indicate a failure to decode if the received block of symbols contains more than t errors.

2.5 Summary

This chapter discussed the fundamental functions of DVB physical layer which includes outer coder/decoder, inner coder/decoder, QAM mapping, frame adaptation, TPS insertion and OFDM structure. The hardware implementation of DVB transmitter/receiver that includes the digital system for transmit/receive baseband signal processing and the analog front-end for RF-to-baseband signal conversion or vice versa were then presented. This chapter covered a detailed discussion on DUC/DDC, carrier and timing recovery channel estimation and FEC, which are the functions implemented in the digital system. The analog front-end circuit design which includes SAW filter, mixer, low noise amplifier, and UHF/VHF channel filter was discussed. As a whole, the chapter intended to provide a good overall picture of both how the DVB system operates and how it can be implemented together with some design references which have not been previously covered by the existing literature.

References

1. Terrestrial digital television planning and implementation considerations (summer 2001), BNP005, third issue
2. Wu, Yiyan, Pliszka, E, Caron, B, Bouchard, P, Chouinard, G (June 2000) “Comparison of terrestrial DTV transmission systems: the ATSC 8-VSB, the DVB-T COFDM, and the ISDB BST-OFDM”, IEEE Trans. Broadcasting, Vol 46, Issue 2, pp 101–113

3. Kornfeld, M, May, G (March 2007) "DVB-H and IP datacast-broadcast to handheld devices", IEEE Trans. Broadcasting, Vol 53, Issue 1, part 2, pp 161–170
4. Rast, RM (Oct. 2005) "The dawn of digital TV", IEEE Spectrum, Vol 42, Issue 10, pp 26–31
5. European Telecommunications Standard Institute ETSI.York (July 1999) Digital Video Broadcasting (DVB); Framing Structure, Channel Coding and Modulation for Digital Terrestrial Television. EN 300 744 V1.2.1
6. European Telecommunications Standard Institute ETSI.York (Feb 2005) Digital Video Broadcasting (DVB); DVB-H Implementation Guidelines. TR 102 377 V1.1.1
7. Liang, Longfei, Yu, Songyu, Wang, Xingdong (Dec. 2002) "Implementation of a new MPEG-2 transport stream processor for digital television broadcasting", IEEE Trans. Broadcasting, Vol 48, Issue 4, pp 348–352
8. Zhang, Yue, Cosmas, J, Song, YongHua, Bard, M (Oct. 2006) "Future transmitter/receiver diversity schemes in broadcast wireless networks", IEEE Commun. Mag., Vol 44, Issue 10, pp 120–127
9. Zhang, Yue, Cosmas, J, Bard, M, Song, YH (Dec. 2006) "Diversity gain for DVB-H by using transmitter/receiver cyclic delay diversity", IEEE Trans. Broadcasting, Vol 52, Issue 4, pp 464–474
10. Zhang, Yue, Cosmas, J, Loo, KK, Bard, M, Bari, RD (March 2007) "Analysis of cyclic delay diversity on DVB-H systems over spatially correlated channel", IEEE Trans. Broadcasting, Vol 53, Issue 1, part 2, pp 247–255
11. Zou, William Y, Wu, Yiyan (March 1995) "COFDM: An overview", IEEE Trans. Broadcasting, Vol 41, Issue 1, pp 1–8
12. Proakis, JG, Manolakis, DG (1992) Digital Signal Processing Principles, Algorithms and Applications, Second Edition, Maxwell Macmillan International, New York
13. Hogenauer, EB (April 1981) "An economical class of digital filters for decimation and interpolation", IEEE Trans. Acoust, Speech Signal Processing, Vol 29, Issue 2, pp 155–162
14. Abu-Ai-Saud, WA, Stuber, GL (May 2003) "Modified CIC filter for sample rate conversion in software radio systems", IEEE Signal Processing Lett., Vol 10, Issue 5, pp 152–154
15. Amiot, S, Bassement, G, Daubenfeld, A, Fillatre, V, Maurice, E, Mercier, F, Mevel, T, Richard, Y, Tourret, J-R (March 2007) "A low power DVB-T/H zero-IF tuner IC design in 0.25 um BiCMOS technology for mobile TV reception", IEEE Trans. Broadcasting, Vol 53, Issue 1, pp 434–440
16. Beek, JJ, Sandell, M, Borjession, PO (1997) "ML estimation of time and frequency offset in OFDM systems", IEEE Trans. Signal Process, Vol 45, Issue 7, pp 1800–1805
17. Frescura, F, Pielmeier, S, Reali, G, Baruffa, G, Cacopardi, S (Sept. 1999) "DSP based OFDM demodulator and equalizer for professional DVB-T receivers", IEEE Trans. Broadcasting, Vol 45, Issue 3, pp 323–332
18. Li, Chih-Peng, Hu, Wei-Wen (June 2007) "Super-imposed training scheme for timing and frequency synchronization in OFDM systems", IEEE Trans. Broadcasting, Vol 53, Issue 2, pp 574–583
19. Speth, Michael, Stefan, AF, Fock, G, Meyr, H (April 2001) "Optimum receiver design for wireless broadband systems using OFDM-part II. A case study", IEEE Trans. Commun, Vol 49, Issue 4, pp 571–578
20. Viterbi, AJ (April 1967) "Error bounds for convolutional codes and an asymptotically optimum decoding algorithm", IEEE Trans. Information Theory, Vol IT-13, pp 260–269

Chapter 3
Digital Audio/Video Broadcasting and Digital Implementation of Analog TV

Daniel Iancu, Hua Ye, Joon-Hwa Chun, John Glossner, Andrei Iancu, and Mayan Moudgille

3.1 Introduction

After nearly a century of deployment, analog radio and television are being replaced by digital broadcasting systems. In digital systems, the amplitude and frequency modulation characteristics inherent in analog systems are replaced by digital modulation techniques resulting in more robust communication systems, reduced power consumption, significantly improved sound and picture quality, and much better spectrum efficiency.

Historically, digital communication systems were designed using Application Specific Integrated Circuit (ASIC) technology. In this technology multiple hardware blocks specific to certain functions are integrated together on the same silicon chip. As the size of the transistors shrink, Digital Signal Processors (DSPs) are becoming a more attractive solution for implementing digital communication systems. In this technology the communications functions are implemented in software. This is sometimes also referred to as Software Defined Radio (SDR). In contrast to ASIC technology, DSP technology allows multiple broadcasting systems to coexist on the same device. Multiple digital audio and TV standards can also coexist with traditional analog standards.

In this chapter we briefly present a software defined platform capable of executing multiple digital radio and television standards as well as analog radio and television standards in software only.

The chapter is structured as follows: Sect. 3.2 describes digital audio broadcasting followed by Sect. 3.3 dedicated to the terrestrial video broadcasting. A software implementation of DVB-H is described in Sect. 3.4. Analog television is briefly described in Sect. 3.5 followed by the NTSC standard software implementation in Sect. 3.6. A brief summary section ends the chapter.

D. Iancu (✉), H. Ye, J.-H. Chun, J. Glossner, A. Iancu, and M. Moudgille
Sandbridge Technologies Inc., Tarrytown, New York, USA
e-mail: diancu@sandbridgetech.com

F.-L. Luo (ed.), *Mobile Multimedia Broadcasting Standards: Technology and Practice*,
DOI: 10.1007/978-0-387-78263-8_3,

3.2 Digital Radio Broadcasting

Since its beginning in the 1980s digital radio broadcasting, despite its uncontestable advantages, suffered from lack of worldwide standard agreement. Different countries adopted different standards either on top of preexisting AM/FM services or in different frequency bands. The most common digital radio systems are:

- European Eureka 147 DAB (Digital Audio Broadcasting) described in ETSI 300 401 standard and coordinated by the WorldDMB (former World DAB) Forum [1]. DAB+ (ETSI TS 102 563 standard) [2] is an evolution of Eureka 147 DAB with no backward compatibility but new features such as Advanced Audio Coding (AAC+) and added error protection through Reed-Solomon coding and a virtual interleaver. The rest of the standard is unchanged.
- United States HD Radio is the registered trademark for the In-Band-On Channel (IBOC) offering digital audio on the top of existing AM/FM radio services.
- Japanese Integrated Services Digital Broadcasting-Terrestrial narrowband (ISDB-Tn) system provides CD quality digital radio.

In addition to terrestrial digital radio systems there are also satellite digital radio systems such as XM and SIRIUS radio offering a multitude of CD quality audio channels.

3.2.1 Eureka 147 DAB System

3.2.1.1 Transport Mechanism

DAB system supports three channels:

a. The Main Service Channel (MSC) carries audio and data service components. Multiple subchannels, individually encoded for error protection, form a time interleaved data channel. Each subchannel may carry multiple data and audio services. The MSC contains a sequence of Common Interleaved Frames (CIF). Each CIF is a data field containing 55,296 bits transmitted every 24 ms.
b. The Fast Information Channel (FIC) is made up of Fast Information Blocks (FIB) and carries Multiplex Configuration Information (MCI) and Service Information (SI) to the receiver. The Conditional Access (CA) management information and the Fast Information Data Channel (FIDC) can also be included in the FIC. The FIC is error protection encoded but not time interleaved.
c. The synchronization channel (SC) is used for receiver synchronization, channel state information (CSI), and transmitter identification.

Each channel receives data from multiple sources. Multiple channels form a frame as shown in Fig. 3.1. The DAB conceptual emission block diagram [1] is illustrated in Fig. 3.2.

The blocks illustrated in Fig. 3.2 are described in the next sections.

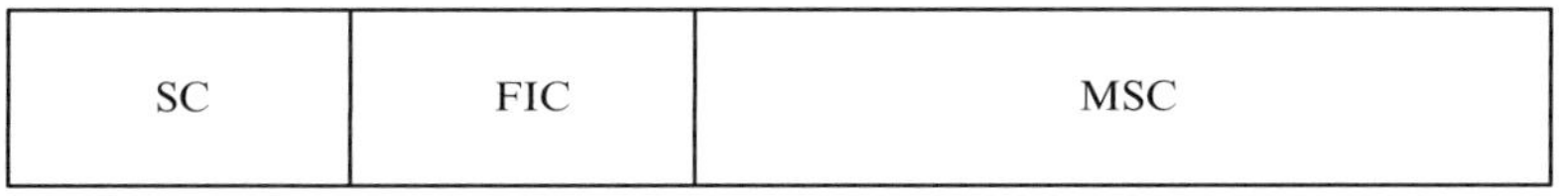

Fig. 3.1 Transmission frame structure

3.2.1.1.1 Audio Coding

The DAB system uses the MPEG Audio Layer II (MPEG-2) with two profiles: ISO/IEC 13818-3 [4] for a 24 kHz sampling rate and ISO/IEC 11172-3 [3] for 48 kHz sampling rate. The encoder input is a Pulse Code Modulated (PCM) signal at 24 or 48 kHz and the output is a compressed audio bit stream ranging from 8 to 384 kbps. The DAB+ system specifies MPEG 4 HE AAC v2 audio coding with sampling rates operated at 16 or 24 kHz with Spectral Band Replication (SBR) and 32 or 48 kHz if SBR is disabled. The maximum bit rate is specified to 192 kbps per subchannel.

Data features: The SI features include: service component language, service linking, date and time, program number, program type, regional identification, FM and AM services, frequency information, transmitter identification number, etc. Certain features may be redirected to the MSC and may be also prioritized due to limited FIC capacity.

Energy dispersal: In order to avoid patterns in the transmitted signal the data is randomized by a modulo-2 addition with the output of a Pseudo-Random Binary Sequence (PRBS) generator defined by the polynomial:

$$P(X) = X^9 + X^5 + 1$$

The randomizer is initialized with all 1's.

3.2.1.1.2 Error Correction Coding and Time Interleaving

The channel coding specified by DAB is based on a convolutional code with constraint length 7 defined by the following octal form generator polynomials: 133OCT, 171OCT, 145OCT, and 133OCT [1]. The block diagram of the encoder is shown in Fig. 3.3. The output of each convolutional encoder, for all subchannels of the MSC, is time interleaved. The convolutional encoded and time interleaved logical frames are combined into CIF. Each CIF contains 55,296 bits grouped in 864 Capacity Units (CU). The CFI configuration is signaled by the MCI carried in the FIC. In DAB+ the audio super frames are Reed-Solomon (RS) encoded and byte-wise virtual interleaved in order to increase the error immunity. The RS$(120,110,\ t=5)$ shortened code is derived from the systematic RS$(255,245,\ t=5)$ code in Galois Field GF(2^8), $\alpha^1 = 2$, and polynomial $P(X) = X^8 + X^4 + X^3 + X^2 + 1$ [2].

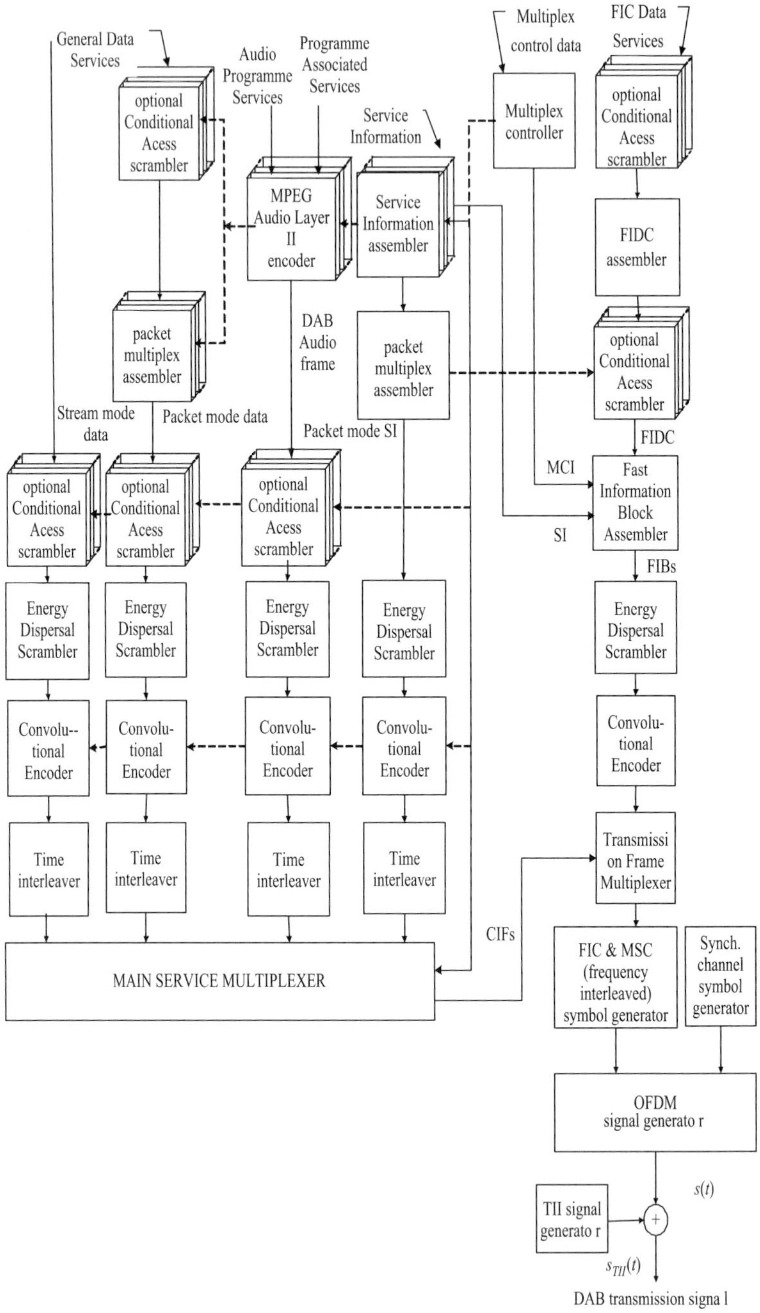

Fig. 3.2 Conceptual DAB emission block diagram

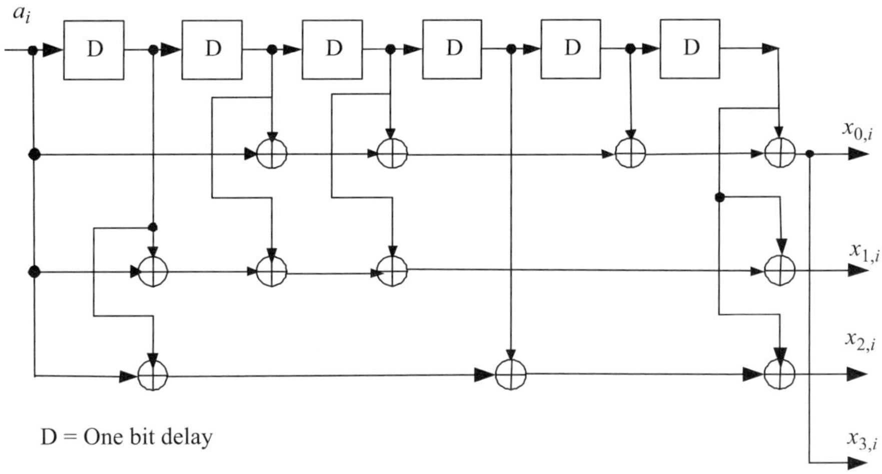

Fig. 3.3 Convolutional encoder

Table 3.1 Transmission modes parameters

Parameter	Mode I	Mode II	Mode III	Mode IV
Number of symbols per frame	76	76	153	76
Number of carriers	1536	384	192	768
Frame duration	96 ms	24 ms	24 ms	48 ms
Null symbol duration	1297 μs	324 μs	168 μs	648 μs
OFDM symbol duration	1246 μs	312 μs	156 μs	623 μs
Inverse of carrier spacing	1000 μs	250 μs	125 μs	500 μs
Guard interval	246 μs	62 μs	31 μs	123 μs

3.2.1.2 DAB Transmission Signal

The transmission frame is illustrated in Fig. 3.1. There are four transmission modes defined. The use of one particular mode depends on the network configuration and operating frequencies. Sequences of OFDM symbols are the building blocks of each frame. The four transmission modes are described in Table 3.1. The SC is used for receiver synchronization and tracking and consists of two OFDM symbols placed at the beginning of each frame. The first symbol is the NULL symbol while the second symbol is the phase reference symbol and constitutes the reference for the differential modulation of the next OFDM symbol. Details of these two symbols generation are described in [1].

The block diagram of the main signal generation is shown in Fig. 3.4. The block partitioner defines the parsing of the encoded FIBs into data blocks associated with the OFDM symbols and it is transmission dependent. Each carrier is QPSK modulated and then frequency interleaved before the differential modulator and multiplexing.

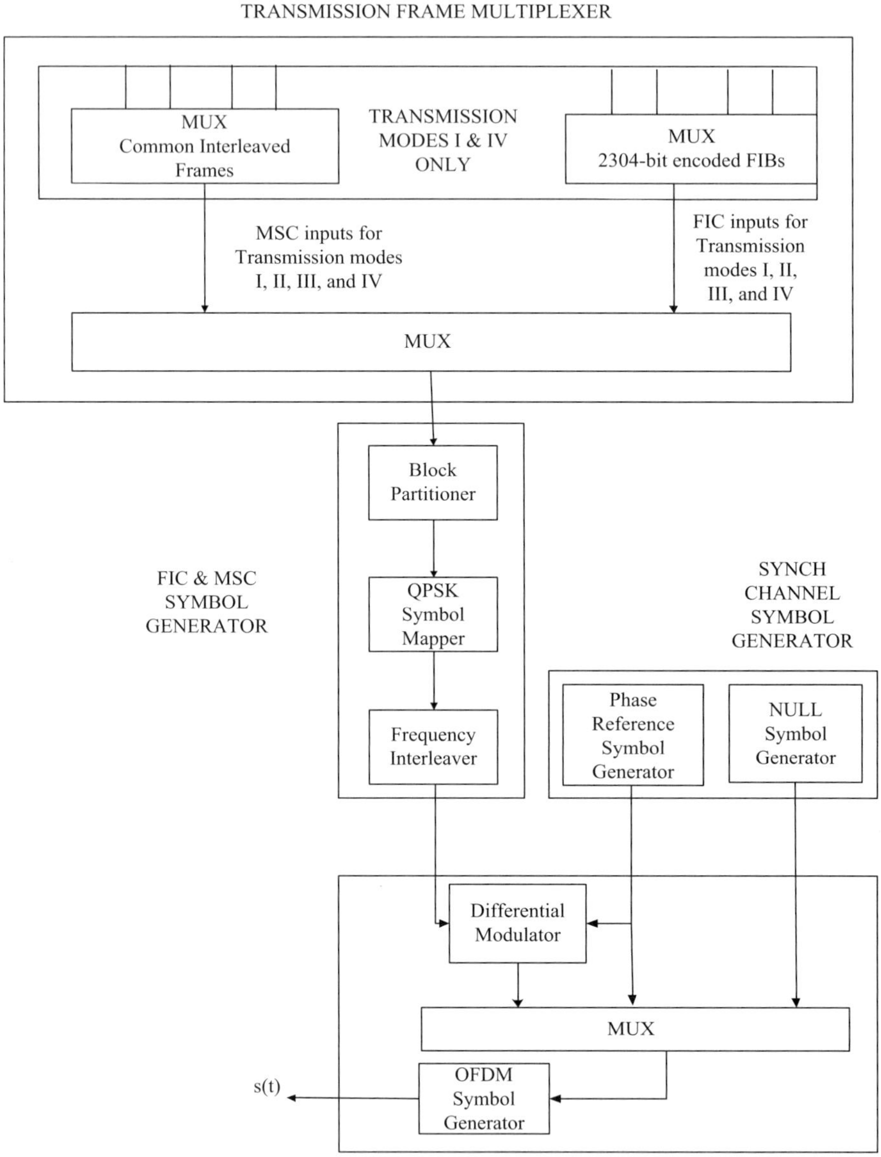

Fig. 3.4 Block diagram of the main signal generation

3.2.1.3 Radio Frequency Characteristics

There are four transmission modes with the following use:

a. Mode I for terrestrial and Single Frequency Networks (SFN)

b. Mode II and IV for the terrestrial local broadcasting in Bands I, II, III and IV and in the 1452 to 1492 MHz (L Band) and the satellite and hybrid satellite broadcasting in L-band
c. Mode III for terrestrial, satellite and hybrid satellite-terrestrial broadcasting below 3GHz

3.3 Digital Television

The first TV transmission was performed by John Logie Baird in 1926 (30 lines/frame, 5 frames/s). Conceptually, analog TV transmission has not changed since. The same AM/FM combination for image and sound lasted until the early 1990s when digital television became a reality. As in the case of digital radio, multiple competing standards for both terrestrial and satellite broadcasting with different technical characteristics exist today.

There are two main terrestrial video transmission standards: Digital Video Broadcasting (DVB) in Europe and the Advanced Television Systems Committee (ATSC) standards in the USA.

- The European terrestrial DTV standards is called DVB-T (ETS 300 744) [5].
- The USA DTV primary standard is ATSC DTV.
- The Japanese Digital Broadcasting Expert Group (DiBEG) has its own terrestrial DTV standards called (ISDB-T).
- There are also other standards such as Digital Media Broadcasting (DMB) and Digital Video Broadcasting – Handheld (DVB-H) [6] targeting handheld devices.
- Over the internet there is Internet Protocol TV using IP packets to carry the video data and the internet TV via the open Internet infrastructure.

In the next section we will describe the DVB-H standard [6].

3.3.1 DVB-T/H System Description

3.3.1.1 General Considerations

DVB is an OFDM system [9] directly compatible with MPEG-2 coded TV signals. It is designed to operate in the existing 6, 7, and 8 MHz VHF and UHF spectrum allocation for Analog TV transmission. There is also a new 5 MHz channel added for the DVB-H [6, 7].

There are two modes of operation defined for DVB-T: 2K and 8K. There are three modes of operation defined for DVB-H: 2K, 4K and 8K. The 8K mode is used for small, medium, and large SFN. It provides the lowest Doppler shift tolerance compared to the 4K and 2K modes but provides the highest bit rate transmission.

The 4K mode can be used for small and medium SFN, with less Doppler shift tolerance but still allowing for very high-speed reception. The 2K mode can be used

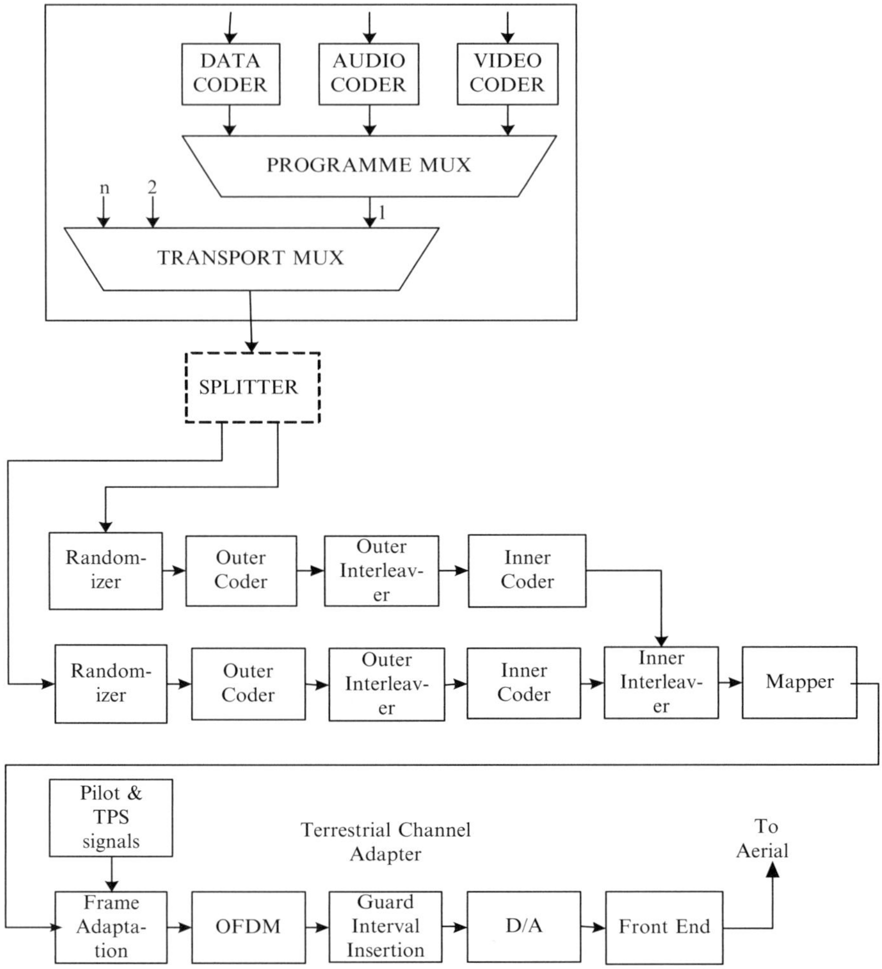

Fig. 3.5 Transmitter processing chain

for small SFN with limited range. It provides the best Doppler shift tolerance at high speed and high data rates. Thus, the 4K mode is a compromise between transmission rate, range, and Doppler shift.

The transmitter processing chain is illustrated in Fig. 3.5. As shown in Fig. 3.5, multiple programs, data, video and audio are multiplexed together into a transport stream. The system supports two hierarchical levels of channel coding and modulation, including uniform and multiresolution constellation. The hierarchical nature is restricted to channel coding and modulation only allowing the simulcast of a program service (or different programs) at different bit rates and transmission ruggedness. The MPEG frames are split into two streams – high and low priority streams. The two separate streams are separately randomized, passed through the outer coder, outer interleaver and inner coder and then inner-interleaved together. After mapping and frame adaptation the data is OFDM modulated and a guard is appended. The data is then converted from the digital to analog domain and passed to the front end to be transmitted. The main physical layer processing blocks are described in the following section.

3.3.1.2 Channel Coding and Modulation

The input stream is organized in 188 bytes MPEG-2 packets including one synch byte as illustrated in Fig. 3.6. The data is randomized using a PRBS generator with the polynomial $P(X) = 1 + X^{14} + X^{15}$.

The outer coding is performed using a Reed-Solomon RS(204, 188, $t = 8$) encoder, derived from the systematic RS(255, 239, $t = 8$) code with generator polynomial: $P(X) = X^8 + X^4 + X^3 + X^2 + 1$ and followed by convolutional byte-wise interleaving. The inner coding is performed by a 64 states convolutional encoder, shown in Fig. 3.7, with the polynomials: G1 = 171_{OCT} and G2 = 133_{OCT}. Following the convolutional encoder, the bit stream is bit-wise interleaved. More detailed information can be found in the DVB standard [5]. At the link layer DVB-H has a second RS encoder followed by virtual time interleaving for the purpose of improving the overall mobile system performance and increasing the tolerance to impulse noise.

3.3.1.3 Signal Constellation and Mapping

Data in each OFDM frame is modulated using QPSK, 16-QAM, 64-QAM, nonuniform 16-QAM, or nonuniform 64-QAM constellations. Details of the constellations for both hierarchical and nonhierarchical mapping are described in the standard [5].

3.3.1.4 OFDM Frame Structure

The DVB-T/H transmission is organized in frames. Every four frames form a superframe and each frame contains 68 OFDM symbols. Each OFDM symbol contains a fixed number of carriers depending on the operation mode. They carry data, system information, and pilots required for receiver synchronization and channel equalization. The system parameters for the 8 MHz channel are listed in Table 3.2.

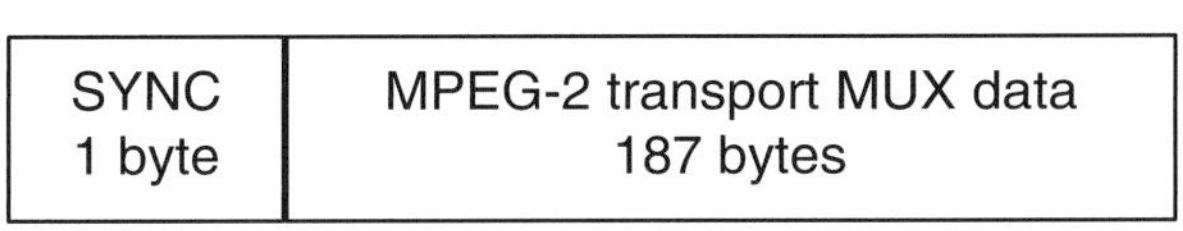

Fig. 3.6 MPEG-2 transport MUX packet

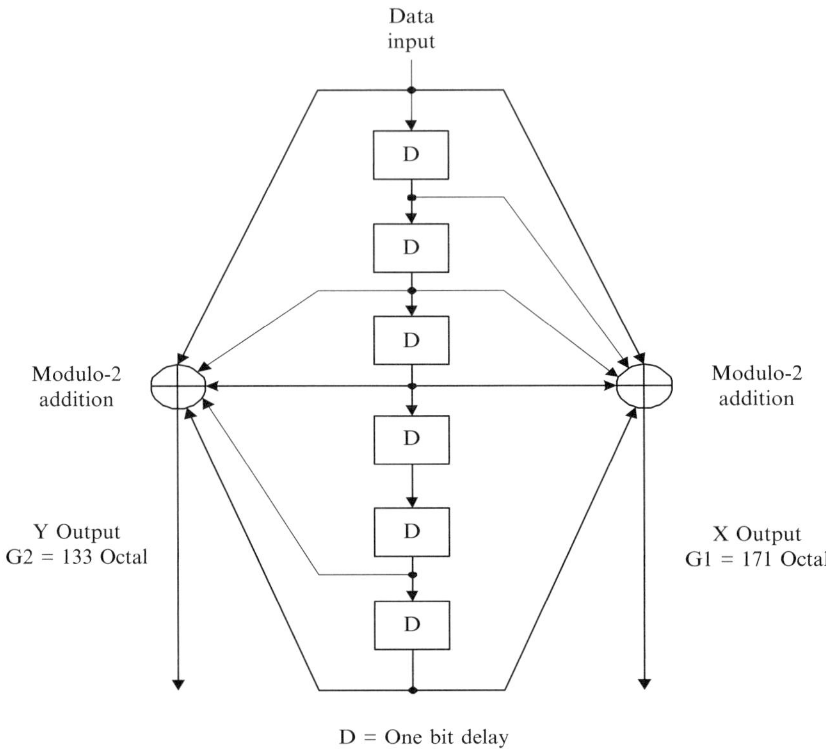

Fig. 3.7 DVB-T/H convolutional encoder

Table 3.2 System parameters for 8 MHz channel

Mode	2K	4K	8K
No. of carriers K	1705	3409	6817
K_{min}	0	0	0
K_{max}	1704	3408	6816
Carrier spacing (Hz)	1116	2232	4464
Freq. spacing between K_{min} and K_{max} (MHz)	7.61	7.61	7.61
Useful Symbol Duration (μs)	224	448	896
Guard ratio	1/4, 1/8, 1/16, 1/32	1/4, 1/8, 1/16, 1/32	1/4, 1/8, 1/16, 1/32

Certain carriers, at boosted power level, called pilots, carry reference signals known by the receiver. There are continual and scattered pilots. The continual pilot's positions coincide with the scattered pilot positions every four symbols with values derived from a PRBS generator.

The system information includes modulation, hierarchy information, code rate, guard interval, etc. System information is carried by the Transmission Parameter Signaling (TPS) carriers. The TPS carriers are DBPSK modulated and encoded using BCH (67, 53, $t = 2$) derived from the original BCH (127, 113, $t = 2$) code.

An OFDM transmitted signal for DVB-T/H can be described in mathematical format by the following formula:

$$s(t) = Re\left\{ e^{j2\pi f_c t} \sum_{m=0}^{\infty} \sum_{l=0}^{67} \sum_{k=K_{\min}}^{K_{\max}} c_{m,l,k} \times \Psi_{m,l,k}(t) \right\},$$

with

$$\Psi_{m,l,k}(t) = \begin{cases} e^{j2\pi k'/T_u(t-\Delta-l\times T_s-68\times m\times T_s)}, & (l+68m)T_s \leq t \leq (l+68m+1)T_s \\ 0, & \text{else} \end{cases}$$

where: k is the carrier number, l is the OFDM symbol number, m is the transmission frame number, K is the number of transmitted carriers, T_s is the symbol duration, f_c is the central frequency of the RF signal, k' is the carrier index relative to the central frequency, $k' = k - (K_{\max} - K_{\min})/2$, $c_{m,0,k}$ is the complex symbol for carrier k of the data symbol number 1 in the frame number m, $c_{m,1,k}$ is the complex symbol for carrier k of the data symbol number 2 in the frame number m, and so on. $c_{m,67,k}$ is the complex symbol for carrier k of the data symbol number 68 in the frame number m.

3.4 DVB-H Implementation on SB3011 DSP

Since DVB-H has a lower average data rate than the DVB-T, it can be easily implemented entirely in software and executed in real time in DSP. This section describes the software implementation of DVB-H on the SB3011 [10] DSP. For the interested readers more system implementation details are available in [13–16].

3.4.1 Receiver Main Processing Blocks

The DVB-H receiver processing chain is illustrated in Fig. 3.8. Briefly, in Fig. 3.8, the processing blocks perform the following functions:

The derotation function performs fine frequency correction and tracking through complex multiplication of the received samples with precalculated complex values. Preprocessing (PP) is required since the DVB-H receiver functions expect $f_c = 32/7$ MHz for the Low Intermediate Frequency (LIF) and $f_s = 4f_c$ for the sampling frequency. If the LIF frequency and sampling rate are not what the receiver expects, then preprocessing is needed to adjust the LIF and sampling rate. The Guard Length Detection (GLD) function is responsible for detecting one of the guard lengths employed by the transmitter. Once the guard length is detected, the coarse fractional offset is estimated and corrected in the Coarse Fractional Frequency Offset Estimation and Correction (CFFOEC) block. The Coarse Symbol Synchronization (CSS) and Integer Frequency Offset Estimation and Correction (IFOEC)

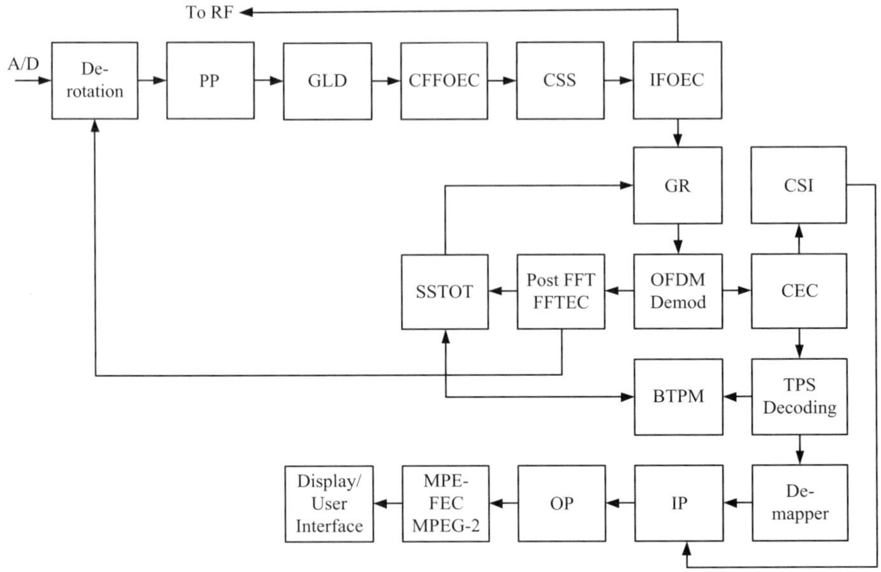

Fig. 3.8 DVB-H receiver block diagram

follow. After the Guard Removal (GR) Fine Frequency and Timing Estimation and Correction (FFTEC) take place during steady-state operation after the OFDM demodulation process. The Symbol Synchronization Tracking (SST) uses timing information from the Burst Timing Power Manager (BTPM) and provides the complex values to the de-rotation block. The large frequency correction is achieved by changing the RF chip PLL configuration. Channel Estimation and Correction (CEC) will also provide the CSI required by the Viterbi and TPS decoders in the Inner Processing (IP) and TPS decoding block, respectively. In our implementation, these processing blocks are executed completely in software. Both RS decoders, in the Outer Processing (OP) and MPE–FEC processing blocks, are also executed in software using Galois field multiplication vector instructions.

3.4.2 Main Software Functions

The main software functions used in the receiver are as follows:

1. ExtractRxSubcarriers extracts the received subcarriers and then calculates the Channel Transform Function (CTF) sample values on the scattered pilot locations for the current two OFDM symbols. The very first time when ExtractRxSubcarriers function is called there is no symbol synchronization. The OFDM starting symbol position is found through cross-correlation. The peak position indicates the starting position for an OFDM symbol. Research for the scattered pilot start-

ing position (the modular 12 scattered pilot index) for the current OFDM frame is also needed the first time when ExtractRxSubcarriers gets called. This is due to the fact that there may have been corrections for integer frequency offset; this will result in change of scattered pilot locations.

2. ChannelEstimation performs channel estimation and correction for the current OFDM symbol. Channel estimation makes use of scattered pilots. For each OFDM symbol, the scattered pilots are spaced 12 subcarriers apart. The first step in channel estimation is to estimate the CTF samples on the scattered pilot positions for the current OFDM symbol. The second step is to perform interpolation for three virtual pilot groups. The CTF estimates for the current symbol on the virtual pilots, spaced three subcarriers apart, are obtained through interpolation of the previous three symbols and future three symbols. During the third step, the remaining two CTF samples between each virtual pilot pair, for the current symbol, are estimated. Once CTF samples are estimated, the channel correction can be readily performed to get the corrected received subcarriers, further used for QAM demapping and TPS decoding. CTF estimates are also used to calculate the CSI, values acting as estimates of the reliability measure for each subcarrier. Currently the CSI is estimated as the amplitude of the CTF at each subcarrier frequency. Further, the CSI is used in the Viterbi decoder as a weighting factor for the path metrics calculation so that the input soft bits that are less reliable will contribute less to the path metrics accumulation. The CSI is also used as a weighting factor in TPS DBPSK demodulation and decoding to assure that the less reliable TPS carriers have less contribution in the estimation process [11].
3. FracOffsetTracking performs steady-state fractional carrier frequency offset tracking. It is based on estimating the average phase difference between two pilots located in two consecutive OFDM symbols at the same frequency. An averaging window of configurable length is used to smooth the estimated phase difference information.
4. SymbolSyncTracking performs steady-state symbol synchronization tracking. This is called per each OFDM symbol and it is based on estimating the averaged phase differences between the neighboring scattered pilot subcarriers. An averaging window of configurable length is used to smooth the estimated phase difference information.
5. SymbolSyncIntOffset first performs a coarse OFDM symbol synchronization, and then OFDM demodulation (4K FFT [11]) is performed to extract the subcarriers for pilot detection. This is done by calling PilotDet4k function. In the pilot detection routine, the first step is to search for scattered pilot locations. This can easily be done by exploiting the fact that for each OFDM symbol, the scattered pilots are located 12 subcarriers apart, and they are transmitted at a boosted power level compared to the other data bearing subcarriers. The second step is to search for continual pilots. The continual pilots are transmitted at fixed subcarrier locations. They are also transmitted at a boosted power level. The continual pilots are spaced at a multiple of three carriers from the scattered pilots. The search range for the continual pilot starting position determines the range of the integer carrier frequency offset that the receiver is designed to cover.

6. GuardLenCoarseFracOffset detects the guard length and performs coarse fractional frequency offset estimation. The first step is to generate a virtual complex sample sequence from the real input sample sequence. Then four cross-correlations are performed on the complex sample sequence. For each cross-correlation sequence, we estimate the corresponding distance between the two peak positions and then the four distances are compared with the four valid guard lengths. The guard length that is closest to any of the distances will be selected as the valid guard length. Once the guard length is detected, the phase difference between the guard period and the corresponding symbol period can readily be calculated and this will give a coarse estimate of the fractional carrier frequency offset.

3.4.3 Receiver State

From cold start to steady-state operation the receiver transitions through several states illustrated in Fig. 3.9.

STATE0 is the initial state right after power ON. The receiver will perform the following functions: Guard Length Detection (GLD) and Coarse Fractional Frequency Offset Detection and Correction (CFFODC). To improve the robustness of the detection algorithm in the presence of noise and fading, a majority detection of guard length is required. Coarse fractional frequency offset estimation is also estimated once the guard length is detected. The estimated coarse frequency offset will be corrected for before the receiver transitions to the next state, STATE1.

In STATE1, the receiver performs OFDM Initial Coarse Symbol Synchronization (ICSS), Continual/Scattered Pilots Detection (C/SPD) and Integer Frequency Offset Estimation Detection (IFOED) and correction. First, a cross-correlation is performed on the received sample sequence, assuming the guard length is known, which will produce a rough estimate of the OFDM symbol sampling position. Then, the OFDM symbol is extracted and demodulated (i.e. 2K FFT for 2K mode) to

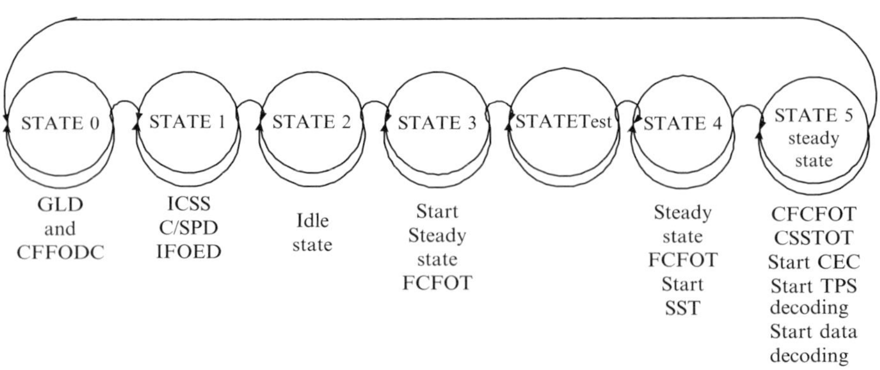

Fig. 3.9 Receiver state transitions

recover the subcarriers. Since continual pilots are located at fixed carrier locations, they are detected first, and then scattered pilot locations are detected next. As in the case of GLD, majority detection is also performed for pilot detections. Integer carrier frequency offset is also estimated in the process of continual pilot detection. The estimated integer carrier frequency offset will be corrected before the receiver transitions to the next state, STATE2.

STATE2 is an idle state to remove any transient effects after the frequency adjustments.

STATE3 starts steady-state fractional carrier frequency offset tracking (FCFOT). Continual pilots are used for the tracking. The receiver will stay in this state for some OFDM symbols to allow any residual carrier frequency offset tracking to converge.

STATETest is a special test state which is used to generate the ideal channel response for the SNR / BER performance testing. In this state, the transmitter will send special OFDM symbols where all the subcarriers are training pilots. At the receiver, the channel response can be calculated for all the subcarriers without the need for interpolation. This state is used only during testing and calibration.

A receiver in STATE4 will continue FCFOT. It will start OFDM Symbol Synchronization Tracking (SST). The receiver will stay in this state for a programmable number of OFDM symbols to allow any initial residual symbol sync offset to converge.

Receiver STATE5 is the final receiver state, the steady state. The receiver will continue continual fractional carrier frequency offset tracking (CFCFOT), and continual symbol synchronization timing offset tracking (CSSTOT) to keep the frequency and timing in sync with the transmitter. Also, channel estimation and correction is performed in this state. If the receiver looses synchronization, it will transition back to STATE0.

Among all the states, only STATE3, 4 and 5 have real-time processing requirements. The flow chart and the computational performance of STATE5 are illustrated in Fig. 3.10. Based upon the computational performance shown in Fig. 3.10 the thread allocation is as depicted in Fig. 3.11. The usage of the SB3011 is less than 50% for 14% duty cycle. The simulated performance for the test case described in [5] Annex A can be found in Table 3.3 [8].

3.5 NTSC Analog Television

3.5.1 NTSC Spectruma

The NTSC TV baseband spectrum is illustrated in Fig. 3.12. The total bandwidth is 6 MHz with Vestigial Side Band (VSB) on the lower side band and full side band on the upper band. The video signal is amplitude modulated on 4.2 MHz bandwidth while the audio signal is frequency modulated centering at $f_s = 4.5\,\text{MHz}$. The suppressed color subcarrier is centered at $f_{sc} = 3.579\,\text{MHz}$. The total number of lines is 525 with $f_h = f_s/286 = 15734.2657343\,\text{Hz}$ line frequency and $f_v = f_h/525 = 29.9700299701\,\text{Hz}$ is the frame frequency.

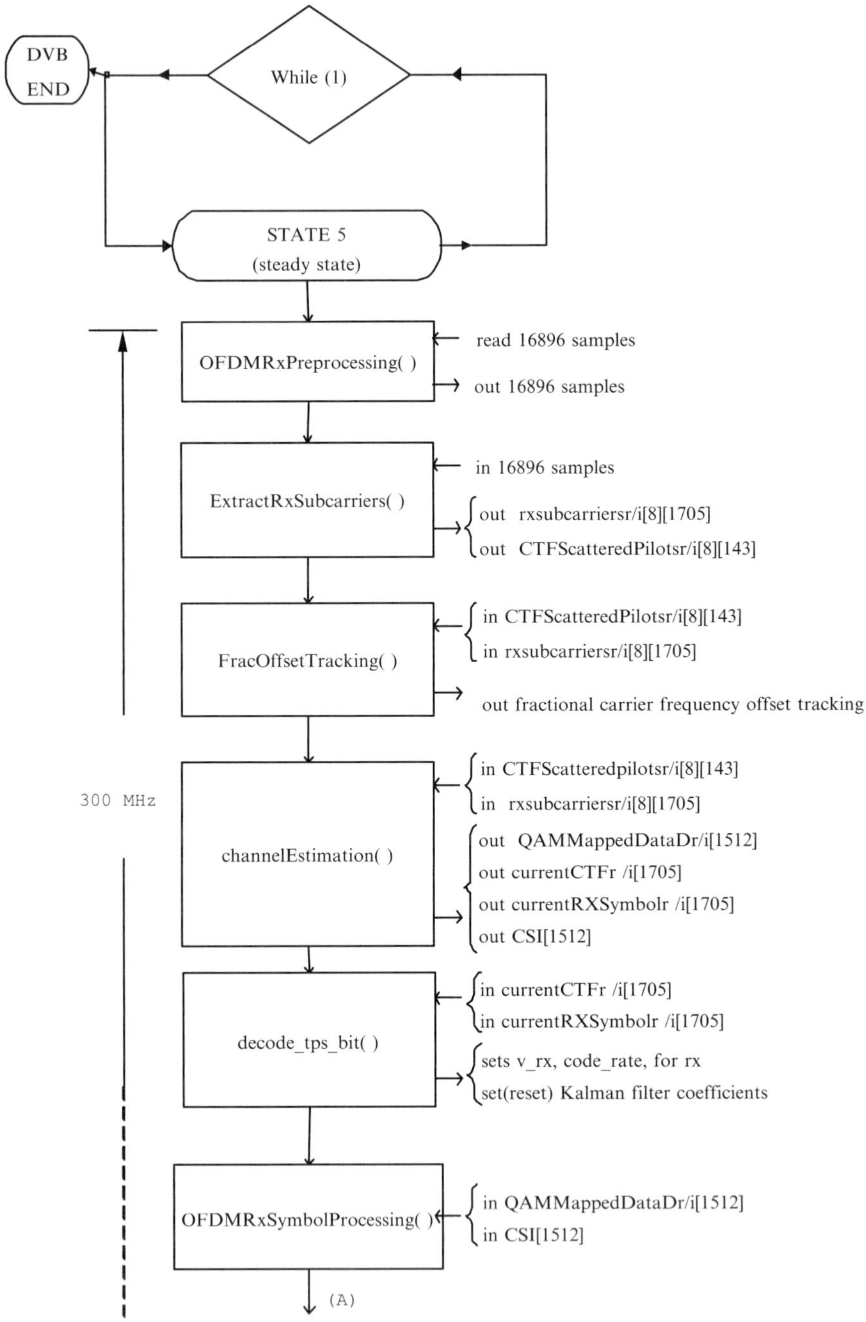

Fig. 3.10 STATE5 flow chart

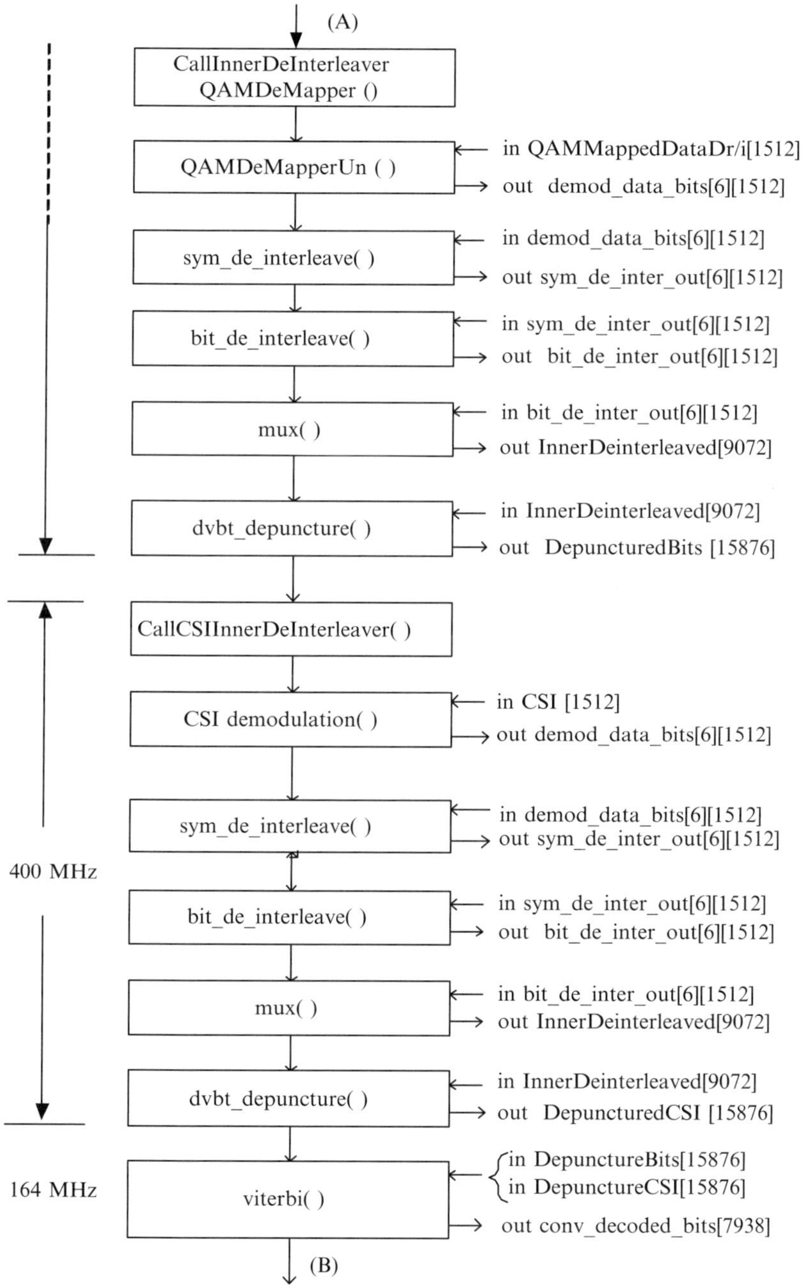

Fig. 3.10 *Continued*

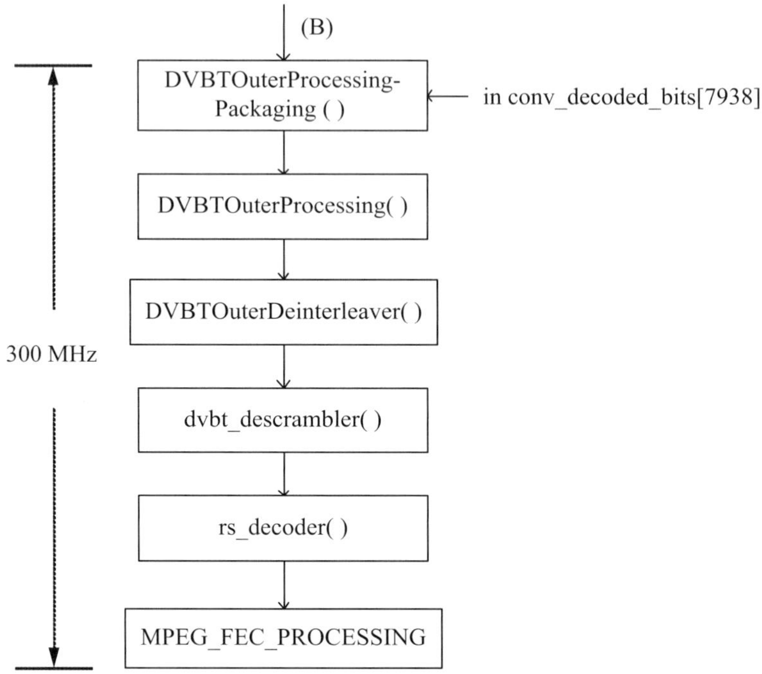

Fig. 3.10 *Continued*

3.5.2 Video Signal

For the NTSC system the luminance information (Y) as a function of the RGB signals (primary colors: Red, Green, and Blue) is a weighted sum of the primary colors:

$$Y = 0.299R + 0.587G + 0.114B$$

The transmitted color information has only two components $B - Y$ and $R - Y$. The component $G - R$ is not transmitted but can be easily recovered as

$$G - Y = -0.51(R - Y) - 0.19(B - Y)$$

The raw $B - Y$ and $R - Y$ signals are used to form the complementary color pairs Q and I as

$$Q = \cos(33^\circ)(R - Y) - \sin(33^\circ)(B - Y) = 0.21R - 0.52G + 0.31B$$
$$I = \cos(33^\circ)(B - Y) - \sin(33^\circ)(R - Y) = 0.60R - 0.28G - 0.32B$$

To avoid cross talk between Q and I samples the bandwidth of the Q samples is reduced to double sideband 0.5 MHz while the bandwidth of the I samples is reduced to single sideband 1.4 MHz. At the end, Q and I will be QAM modulated on to the color subcarrier.

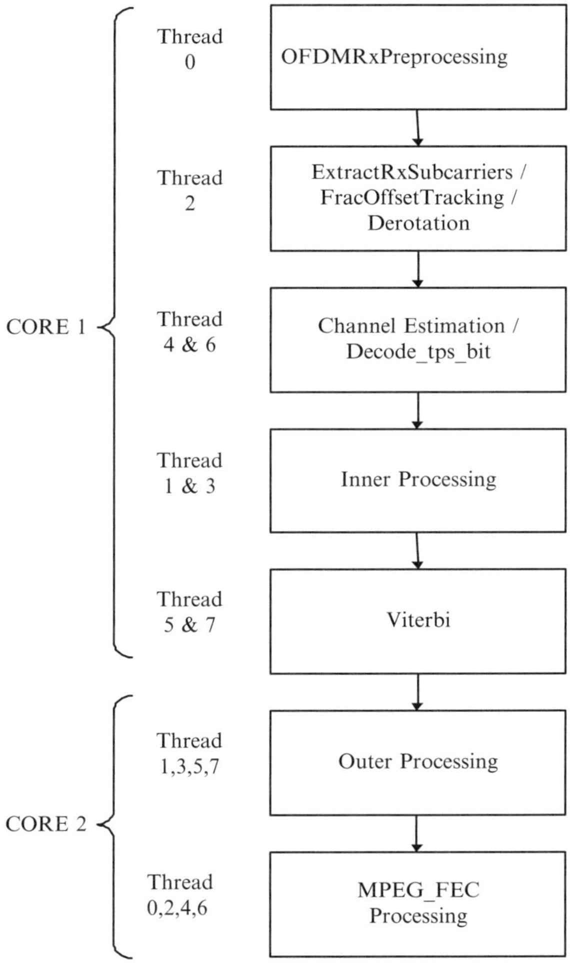

Fig. 3.11 Thread allocation

3.5.3 Synchronization and Timing

Each video frame consists of 2 color fields or 525 video lines [18, 19]. As shown in Fig. 3.13, the first 20 lines of color field I is the field-blanking period that carries the vertical synchronization pulses. The vertical synchronization pulses indicate the start of a new video frame. The color field II starts from the middle of line 263, carrying another 20 lines for the second field-blanking period indicating the middle of the video frame. The rest of lines in color field I and II are the displayable video lines. Figure 3.14 illustrates the detailed video line signal timing and level information. The various video line information must be constructed at predefined signal

Table 3.3 System parameters for 8 MHz channel

	Required C/N in dB for QEF after Reed-Solomon decoder	Required C/N in dB for QEF after Reed-Solomon decoder	Required C/N in dB for QEF after Reed-Solomon decoder
Modulation mode	Gaussian channel	Ricean channel	Rayleigh channel
QPSK rate 1/2 Simulated performance from Annex A in DVB-T	3.1 dB	3.6 dB	5.4 dB
QPSK rate 1/2 1 rx channel	2.61 dB	5.58 dB	4.87 dB
QPSK rate 1/2 2 rx channel	0.43 dB	2.93 dB	2.93 dB
QPSK rate 1/2 3 rx channel	−0.91 dB	1.59 dB	1.59 dB
QPSK rate 1/2 4 rx channel	−1.41 dB	0.83 dB	0.83 dB
16QAM rate 1/2 Simulated performance from Annex A in DVB-T	8.8 dB	9.6 dB	11.2 dB
16QAM rate 1/2 1 rx channel	8.47 dB	11.25 dB	11.01 dB
16QAM rate 1/2 2 rx channel	6.24 dB	8.14 dB	8.47 dB
16QAM rate 1/2 3 rx channel	4.61 dB	7.79 dB	6.65 dB
16QAM rate 1/2 4 rx channel	3.23 dB	6.03 dB	5.58 dB
64QAM rate 1/2 Simulated performance from Annex A in DVB-T	14.4 dB	14.7 dB	16.0 dB
64QAM rate 1/2 1 rx channel	13.90 dB	16.20 dB	16.26 dB
64QAM rate 1/2 2 rx channel	11.01 dB	13.45 dB	13.07 dB
64QAM rate 1/2 3 rx channel	9.83 dB	11.60 dB	11.71 dB
64QAM rate 1/2 4 rx channel	7.96 dB	11.25 dB	10.51 dB

levels/timing to allow the receiver to reconstruct the RGB signals correctly. Each displayable video line consists of the following two parts: The "Blanking Period" carrying "Front Porch," "Sync Tip or horizontal sync pulse" and "Back Porch" with $9+/-1$ cycles of color burst riding on it. The horizontal sync pulse can be used by the receiver for line synchronization and frequency/timing offset tracking purposes.

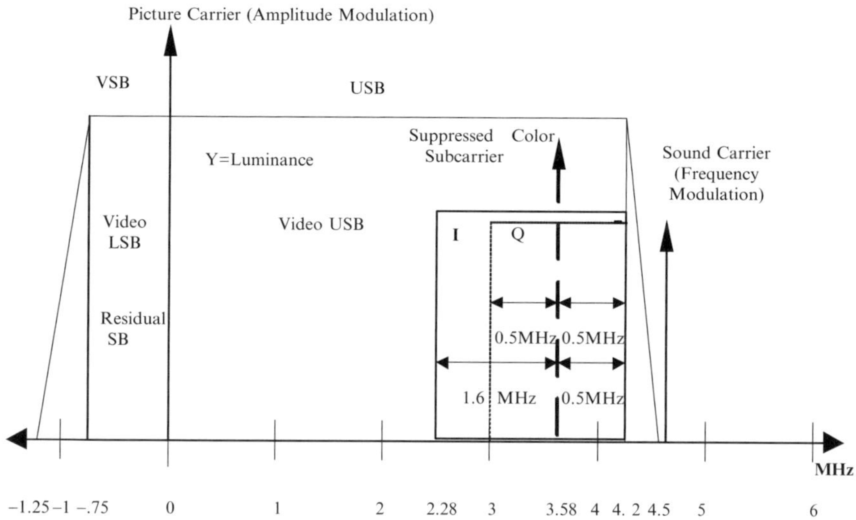

Fig. 3.12 Analog TV base band spectrum

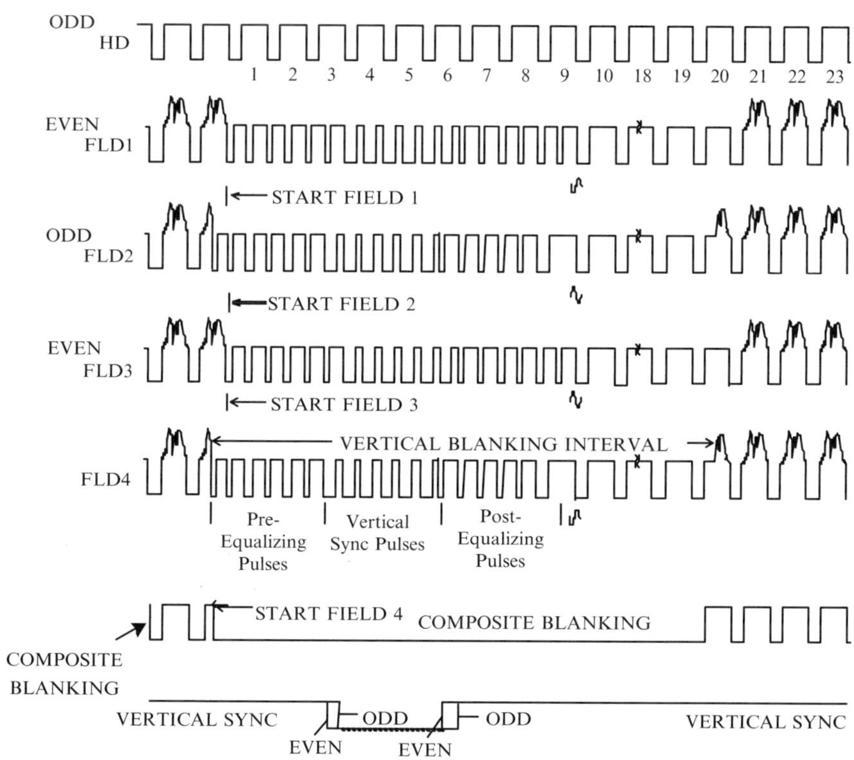

Fig. 3.13 Color Field Blanking Sequence for NTSC

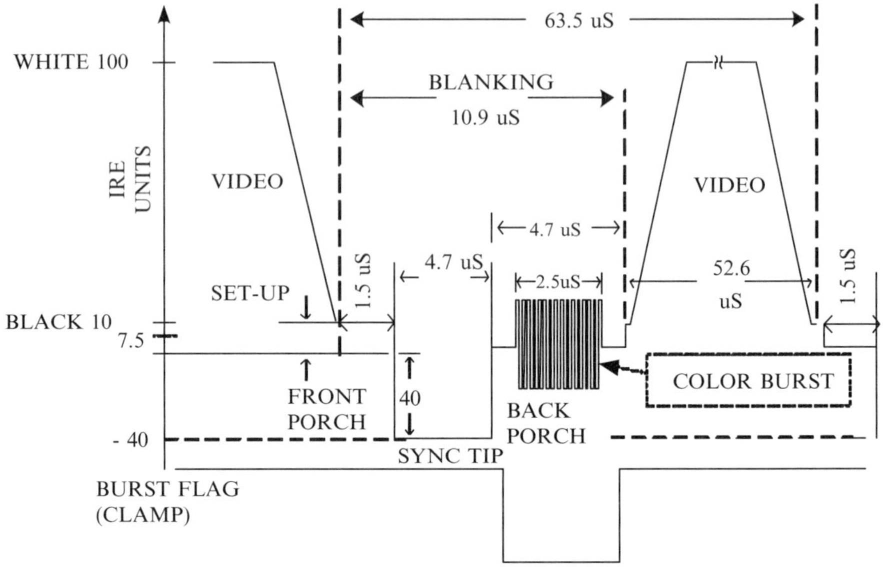

Fig. 3.14 Details of Video Line Timing/Levels Information

The color burst is a sine wave reference signal at the video carrier frequency. It is used to extract the phase and frequency information for decoding the I/Q modulated chroma signal. Following the back porch is the second part of the video line carrying the actual displayable video information (luminance and I/Q modulated chroma signal). The following summarizes the NTSC video encoding process:

a. Primary color components *R*, *G*, *B* are transformed into luminance/chrominance components Y, I, Q. The transform is linear:

$$[YIQ]^T = M * [RGB]^T,$$

where M is the 3×3 color space conversion matrix.
b. The luminance component Y is transmitted directly (in baseband).
c. The two chrominance components I and Q carry color information. They are QAM modulated at the color subcarrier frequency of 3.58 MHz. The video portion of baseband signal carries $Y + QAM(I,Q)$. This signal has the bandwidth of 4.2 MHz. For broadcasting, sound is added; it is FM modulated around 4.5 MHz. The overall baseband encoding is:

$$[YIQ]^T = M * [RGB]^T$$

$$Baseband = Y + QAM(I,Q) + FM(Sound)$$

3.6 Digital Implementation of the NTSC TV System

3.6.1 System Algorithms

Figure 3.15 shows the system block diagram of an analog TV receiver. The analog TV signal received is first down converted to baseband by the RF/IF analog front-end. The baseband composite video signal is then digitized and sent to the DSP. From this stage, the entire video processing is executed digitally in software.

Figure 3.16 shows the block diagram for the analog TV receiver digital signal processing, implemented on SB3011 DSP [12]. In the initial training mode, the horizontal synchronization and vertical synchronization sequences are detected and tracked; the input video composite signal is then adjusted to the proper DC and IRE (Institute of Radio Engineers) scaling levels. Once the receiver training is completed, the active video information can be extracted line-by-line. Y/C (luminance and color) separation and color I/Q demodulation are performed to reconstruct the

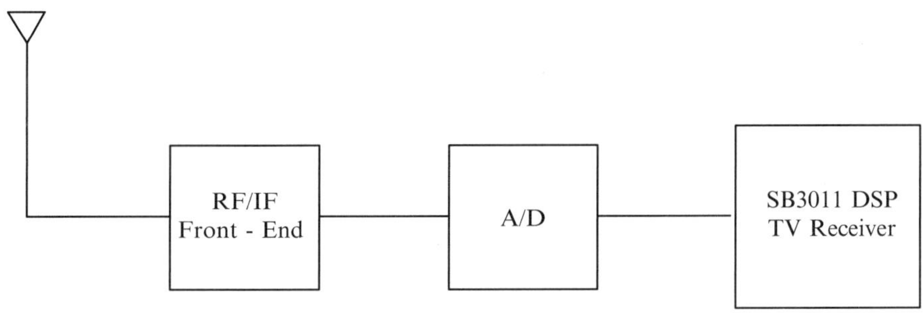

Fig. 3.15 System block diagram of analog TV receiver

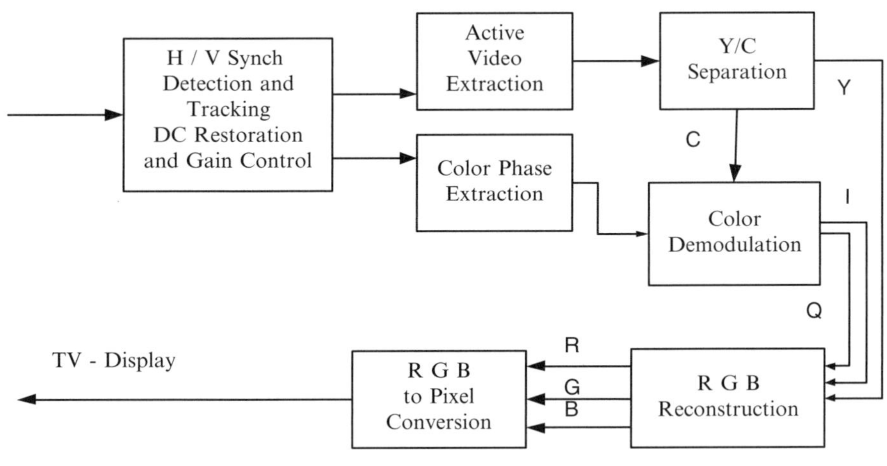

Fig. 3.16 Signal processing block diagram of NTSC receiver

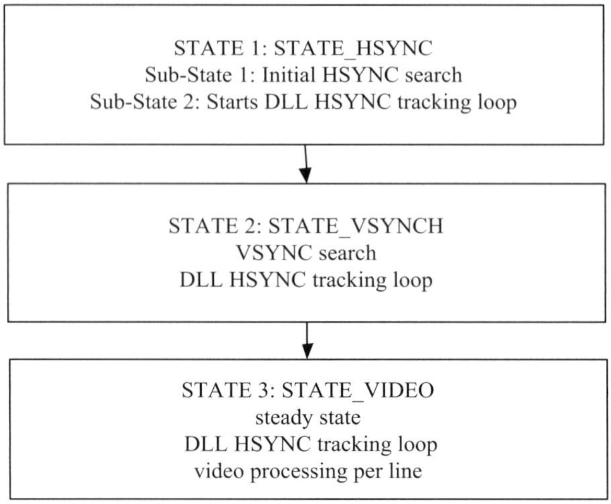

Fig. 3.17 NTSC receiver synchronization process

transmitted RGB signals. The RGB signals for the current video frame are then converted to pixels and displayed on the LCD screen.

Figure 3.17 illustrates the NTSC receiver synchronization process, the receiver synchronization achieves horizontal synchronization first, vertical synchronization next, and then video extraction and processing can be started.

In STATE 1 (HSYNC for horizontal synchronization), the receiver goes through two substates to achieve horizontal video line synchronization. The horizontal synchronization pulses are processed by a Delay Lock Loop (DLL) processing block for correcting any frequency offset that may exist between the transmitter and the receiver due to transmitter receiver frequency misalignment or sampling clock drift.

Substate 1 performs the initial blind search for HSYNC pattern on the sampled video composite signal r_{in}. The HSYNC pattern search is performed as following:

$$R_{rH}[i] = \sum_{k=0}^{N-1} r_{in}[k+i]$$

where $N = MF_s$, M is the line duration in microsecond, F_s is the sampling frequency in MHz, R_{rH} is the cross-correlation of the HSYNCH pattern with the sampled video complex signal at $i = 0, 1, L-1$, where L is the number of samples in a line. The HSYNC position is found at the sample position p for which R_{rH} reaches the maximum.

Substate 2 performs HSYNC Tracking. Once HSYNC signal is detected, the HSYNC needs to be continuously located for over a certain number of lines around the initially detected HSYNC sample position p:

$$R_{rH}[p+k] \geq 0.9 \times R_{rH}[p], \quad k = -2, -1, 0, 1, 2$$

If the above condition is not satisfied, the detected HSYNC is declared to be false and the receiver will fall back to the initial HSYNC detection substate 1. Starting from substate 2, a Delay Line LOOP (DLL) logic is activated to track the movement of the HSYNC position caused by a frequency offset between the transmitter and the receiver. The maximum of the HSYNC match results for $k = -2, -1, 0, 1, 2$ is found and the receiver buffer pointer is manipulated to adjust the maximum HSYNC match position to p. The purpose of the DLL tracking loop is to skip or repeat samples for each video line to track any frequency offset that may exist between the transmitter and the receiver. The DLL tracking loop should be enabled once the HSYNC is detected. The DLL tracking loop should be temporally disabled during the vertical synchronization period.

STATE 2(VSYNC for vertical synchronization): Since the receiver has achieved horizontal synchronization entering this state, the video line starting sampling position is known. As the VSYNC signals on lines 4, 5, and 6 of Fig. 3.13 are the repeated patterns, the VSYNC search is performed for the duration of one video frame, the maximum VSYNC position can be found as follows:

$$\max\left[\sum_{i=0}^{2} R_{rV}[i]\right]$$

where R_{rV} is the correlation of the HSINC pattern with the r_{in} samples. The summation is performed over three lines: $i = 0, 1, 2$.

Once vertical synchronization is achieved, VSYNC high and low pulse levels can be measured to adjust the r_{in} composite signal to the right IRE levels as shown in Fig. 3.14. The VSYNC and HSYNC pulse levels, blanking levels, and reference white levels are required to be adjusted to the correct IRE levels to properly recover the RGB signals. Assuming that the measured low pulse level for VSYNC is V_L, the measured blanking level for VSYNC is B_L, the DC level and scaling factor are calculated as follows:

$$g = \frac{40}{B_L - V_L}$$

$$DC = -g \times B_L$$

The video composite input from the A/D should be adjusted by g and DC as follows:

$$g \times r_{in} + DC$$

After the DC level and gain adjustment, the video composite signal samples will be at the proper IRE levels for further video signal video signal decoding as shown in Fig. 3.14.

Once the maximum VSYNC position is found, the line count that runs from 0 to 524 can be set accordingly to achieve vertical (frame) synchronization. Since HSYNC is tracked continuously by the DLL, vertical synchronization will be maintained automatically. The receiver will stay in this state until the line count reaches 0 indicating the start of a new video frame. At this point, the receiver transitions to the steady state to start video frame decoding.

STATE 3(VIDEO signal decoding) is the receiver steady state. It continues HSYNC and DLL tracking to maintain synchronization with the transmitter in the presence of frequency offset and timing offset. Video data extraction and decoding for reconstruction of RGB signals are performed line-by-line and frame-by-frame. There are a total of 484 displayable video lines for each video frame.

Figure 3.18 shows the flow chart for per line video processing. In each displayable video line, the receiver needs to first extract the color burst to reconstruct the I/Q demodulation phase and chroma demodulation frequency. Since the receiver maintains line synchronization by the HSYNC DLL tracking loop, one can safely assume the demodulation frequency to be the nominal chroma carrier frequency.

Two novel methods for extracting the chroma phase are illustrated [12]. The first algorithm extracts the chroma phase in time domain while the second algorithm extracts the chroma phase in frequency domain:

Time Domain Chroma Phase Extraction: In order to facilitate the estimation of the chroma phase, assume that the A/D sampling rate F_s is set to be M times the chroma carrier frequency: $F_s = Mf_{sc}$, there should be M samples per each color burst cycle, the phase change between the adjacent color burst samples is exactly $\frac{360}{M}$ degree. One can simply take any one color burst sample $x(k)$ as $\sin(k)$ and

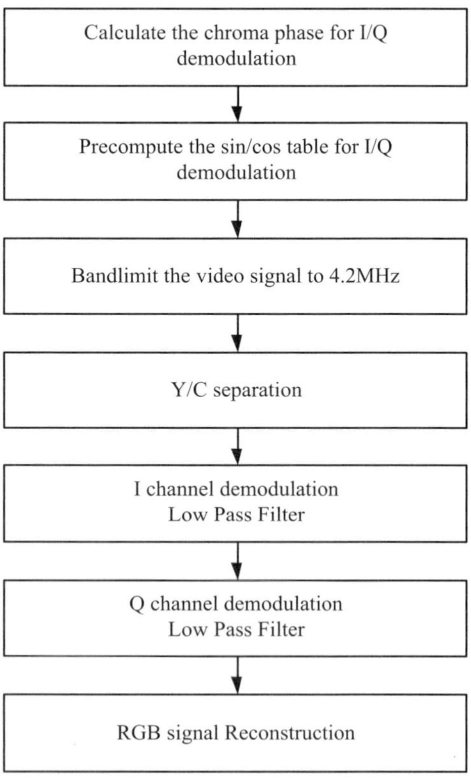

Fig. 3.18 Video processing per line

$x(k+1)$ as $\cos(k)$. The receiver can calculate the chroma phase at sampling position k by the following equation:

$$\phi[k] = \tan^{-1}(\sin(k)/\cos(k)) = \tan^{-1}(x(k)/x(k+1))$$

Frequency Domain Chroma Phase Extraction: First, N time domain color burst samples are extracted, a Fourier Series is computed for frequency domain index $k = \frac{N}{M}$ as follows:

$$X(k) = \sum_{n=0}^{N-1} x(n)e^{-j2\pi kn/N} \tag{3.1}$$

The chroma phase can then be estimated as

$$\phi[k] = \tan^{-1}\frac{Im[X(k)]}{Re[X(k)]}$$

Simulation results show that the chroma phase estimated from frequency domain is much more robust to noise disturbance than the chroma phase estimated from the time domain due to the fact that the previous equation is the phase Maximum Likelihood Estimation (MLE) [12, 17]. Once the chroma phase is detected, the I/Q demodulation phase for the current video line can be set as the following:

$$\phi_D[k] = \phi(k) + 180^\circ + 33^\circ$$

After the chroma phase is recovered, the rest of the video line processing will become quite straightforward: Y/C separation, I/Q demodulation, and RGB reconstruction.

3.6.2 Single Threaded Software Implementation on SB3011

The baseband signal coming from the tuner is oversampled. At each rising edge of the A/D clock a sample is stored within the processor memory. Each sample occupies 2 bytes of memory. The overall sequential algorithm for decoding video signal from baseband is as follows:

Sequential video decoding

1. $[gain, level] = analyze\_sync\_tip(input)$
2. $input = input * gain + level$
3. $video = LPF1(input)$
4. $Y = LPF2(video)$
5. $Sin = gen\_sin(get\_color\_burst(input))$
6. $I = LPF3((video - Y). * Sin)$
7. $Q = LPF3((video - Y). * Sin(2:))$
8. $[RGB]^T = clip(inverse(M) * [YIQ]^T)$
9. $pixels = B \ll 10 | G \ll 5 | R$

The *analyze_sync_tip*() function (step 1) computes the average of sync tip samples and calculates gain and level that normalize the top of the sync tip to be at 0 and the bottom of sync tip to be at IRE-40. Step 2 performs the normalization. The low-pass filter in step 3 separates the video signal from sound, the output is the baseband video composite signal: Y+QAM(I,Q). Step 4 separates Y (luminance) component using another low-pass filter. The LPF1 and LPF2 filters are implemented as FIR filters. Step 5 generates the demodulation sine and cosine tables starting from the extracted chroma phase as described in the previous sections. Steps 6 and 7 demodulate the chrominance components using the computed sine wave. The LPF3 filter is implemented as an FIR filter.

Step 8 performs the inverse color-space transform to obtain the primary color components. Colors are clipped to the range 0 to 31. The last step, Step 9 computes the bitmap that is displayable on the LCD screen. It packs the colors into lower 15 bits of a 16-bit pixel.

The above Sequential video decoding algorithms were implemented as ANSI C and profiled on the Sandblaster simulator. Video decoding takes 25516 cycles per video line. The data memory used in the implementation amounts to just under 20 kB. Thus it easily fits into the processor's local memory.

Step 1 in the algorithm takes little time. All other steps take time proportional to the number of samples decoded. Fortunately, steps 2 through 9 are "embarrassingly" parallel. So we may parallelize the algorithm with little overhead. If we split video decoding among T threads, then the time to decode a line is (computation + memory copy):

$$25516/T + 12004761.9$$

This gives us the lower bound on the number of threads required:

$$T \geq 7.16$$

Since T must be an integer, the lower bound on the number of threads is 8. Decoding every line for a quarter VGA display is overkill. We only have to decode every other line. This gives as a relaxed bound on T:

$$\begin{aligned} 25516/T + 1200 &\leq 2 * 4761.9 \\ T &\geq 3.06 \end{aligned}$$

This indicates that we may fit our decoder into four threads, even accounting for parallelization overhead.

3.6.3 Parallel Implementation

The parallel implementation is outlined below:

Parallel implementation:

DriverThread {

```
        Perform line / frame synchronization
        Forever {
            copy the next line from A2D buffer
            perform DLL tracking, adjust read position
            enqueue the line to video team
        }
    }
    VideoTeamThread (int thread_index){
        If(thread_index == 0)
            { dequeue next line; }
        barrier();
        run the partitioned
        barrier();
    }
```

The work is partitioned between one driver thread and T video threads. The driver thread copies data from the A2D buffer (in L2) into fast L1 memory. The Sandblaster RTOS includes an API that abstracts a circular buffer as an infinite stream with a movable read position. The read position can be advanced or retarded.

DLL tracking involves finding the start of the sync-tip and adjusting the read position of the A2D abstract stream. This is necessary to adjust for frequency offset between the A2D converter and the signal source. DLL tracking requires little computation. Afterwards the driver thread queues up the line for processing by video threads. The queue, in fact, implements double buffering between the driver and the video threads. The queue is implemented using a common circular buffer algorithm. The circular buffer has three slots. This data structure helps to implement concurrent producer/consumer computations. If there is only one producer and only one consumer, then no additional synchronization mechanism is required to maintain the queue in consistent state. After the initial barrier, each thread runs the steps of the sequential algorithm applied to quarter of the data. Additional barriers are necessary, due to overlaps required by the filters. We use light-weight spin barriers [5]. They have very low overhead of about 40 cycles.

We note that our current implementation can be scaled to full VGA by adding another team of four threads.

3.7 Summary

In this chapter we succinctly described the most popular digital audio and video broadcasting systems. Since DAB is much less computational intensive than DVB we described the DVB-H implementation as a more attractive example. The DAB implementation is similar to DVB-H but requires only a fraction of processor resources. The entire DVB-H execution requires only 50% of a SB3011 processor resources at the highest bit rates and modulation complexity. An additional 50% of processor capacity is still available to execute other functions [20–22]. When implemented on the recently announced SBX core, the expected utilization will drop to less than 10% of the capacity [23]. Analog TV is fading out. Therefore, a platform that can execute both analog and digital TV is an attractive solution and we have shown how easily can be implemented in software.

References

1. ETSI EN 300 401 V1.3.3. (2001–05) Radio Broadcasting Systems; Digital Audio Broadcasting (DAB) to Mobile; Portable and Fixed Receivers.
2. ETSI TS 102 563 V1.1.1 (2007–02) Digital Audio Broadcasting (DAB); Transport of Advanced Audio Coding (AAC) Audio.
3. ISO/IEC 11172-3 (March 1993) Information Technology-Coding of Moving Pictures and Associated Audio for Digital Storage Media at up to 1.5 Mbit/s – Part 3: Audio.
4. ISO/IEC 13818-3 (1998) Information Technology-Genetic Coding of Moving Picture and Associated Audio Information – Part 3: Audio.
5. ETSI EN 300 744 V.1.5.1 (2004–11) Digital Video Broadcasting (DVB): Framing Structures, Channel Coding, and Modulation for Digital Terrestrial Television.
6. Transmission Systems for Handheld Terminals (DVB-H), Draft DVB-H Standard, DVB document A081, June 2004.
7. ETSI EN 301 192 V.1.4.1 (2005–9) Digital Video Broadcasting (DVB): DVB Specification for Data Broadcasting.
8. D. Iancu, H. Ye, Y. Abdelilah, E. Surducan, John Glossner, "On the Performance of Multiple OFDM Receivers for DVB", Proceedings of the Joint IST Workshop on Mobile Future and Symposium on Trends in Communications (SympoTIC'04), Bratislava, Slovakia, pp. 1–4, October 24–26, 2004.
9. R. Van Nee, Ramjee Prasad, OFDM For Wireless Multimedia Communications, Artech House Publishers, 2000.
10. J. Glossner, D. Iancu, M. Moudgill, G. Nacer, S. Jintukar, S. Stanley, M. Schulte The Sandbridge SB3011 Platform. EURASIP Journal on Embedded Systems, Vol. 2007, 2007.
11. D. Iancu, Hua Ye, US20070183514, "Orthogonal Frequency Division Multiplexing Receiver", Hua Ye, Daniel Iancu US Patent: US7123669 "TPS Decoder in an Orthogonal Frequency Division Multiplexing Receiver".
12. Hua Ye, Daniel Iancu, US Patent US20070008434 "Digital Implementation of Analog TV Receiver".
13. H. Bolcskei, M. Borgmann, A.J. Paulraj, "Impact of the Propagation Environment on the Performance of Space-Frequency Coded MIMO-OFDM", Journal on Selected Areas in Communications, Vol. 21, No. 3, pp. 427–439, April 2003.
14. "Digital Video Broadcasting (DVB)", ETSI EN 300 744 V1.4.1 (2001–01).
15. Yeo Chee Bang, "Effect of Mutual Coupling in Small Array Antennas", MS thesis, Naval Postgraduate School, Monterey, CA, 2002.
16. E. Surducan, D. Iancu, J. Glossner, "Modified Printed Dipole Antennas for Wireless Multi-Band Communication Devices", URSI EMTS Int. Sym. on Electromagnetic Theory, pp. 161–163, Pisa, Italy, May 23–27, 2004.
17. Steven M. Kay Fundamentals of Statistical Signal Processing Estimation Theory, PTR Prentice-Hall, Inc., 1993.
18. Recommendation ITU – R BT.470-4 "Television Systems".
19. Online information site: http://www.ntsc-tv.com/.

20. J. Glossner, D. Iancu, J. Lu, E. Hokenek, M. Moudgill, "A Software Defined Communications Baseband Design", IEEE Communications Magazine, Vol. 41, No. 1, pp. 120–128, January 2003.
21. J. Glossner, D. Iancu, G. Nacer, S. Stanley, E. Hokenek, M. Moudgill, "Multiple Communication Protocols for Software Defined Radio", IEE Colloquium on DSP Enable Radio, pp. 227–236, September 22–23, 2003, ISIL, Livingston, Scotland.
22. Vladimir Kotlyar, Daniel Iancu, John Glossner, Hua Ye, Andrei Iancu, "Real-Time Software Implementation of NTSC Analog TV on Sandblaster SDR platform", WSR06 Workshop.
23. J. Glossner, M. Moudgill, D. Iancu, S. Jintukar, G. Nacer, M. Schulte The Sandblaster SBX 2.0 Architecture. 2007 Software Defined Radio Technical Conference, Denver, Colorado, 2007.

Chapter 4
Video Streaming over DVB-H

Mehdi Rezaei

4.1 Introduction to DVB-H System

Digital Video Broadcasting for Handheld terminals (DVB-H) is an ETSI specification for delivering broadcast services to battery-powered handheld receivers [4, 7]. DVB-H is mainly based on the DVB-T specification for digital terrestrial television.

Comparing to digital terrestrial television, handheld television is much more difficult from technical points of view. First, mobile terminals have very small antenna size comparing to standard television antennas. Consequently, handheld terminals need stronger signals than standard television. Second, mobile reception is expected everywhere, especially inside buildings and in vehicles. This demands extraordinary robustness for transmission signal. Third problem of handheld terminals is limited power supply which restricts consumption of TV content. The DVB-H mobile TV standard addresses these issues by adding a number of features to DVB-T standard. DVB-H adds functional changes in the link and physical layers while it is backward compatible with DVB-T. In the link layer, DVB-H has two new elements when compared to DVB-T: *Time Slicing* and *Multiprotocol Encapsulation Forward Error Correction* (MPE-FEC). In the physical layer, DVB-H also has two extensions to DVB-T: *DVB-H Signaling* and *4K-Mode OFDM*. More details about these extensions are presented in the sequel.

4.1.1 Time-Slicing

Services used in mobile handheld terminals require relatively low bit rates. The estimated maximum bit rate for streaming video using advanced compression technology like AVC/H.264 is in the order of a few hundred kilobits per second. A DVB

M. Rezaei
Tampere University of Technology, Finland
e-mail: mehdi.rezaei@ieee.org

F.-L. Luo (ed.), *Mobile Multimedia Broadcasting Standards: Technology and Practice*,
DOI: 10.1007/978-0-387-78263-8_4,

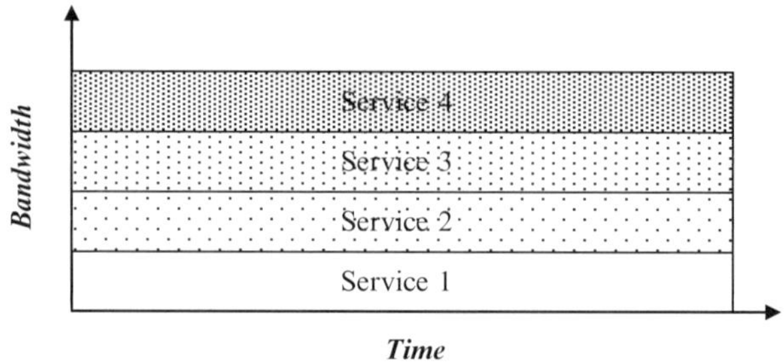

Fig. 4.1 Continuous transmission of parallel services in DVB-T channel

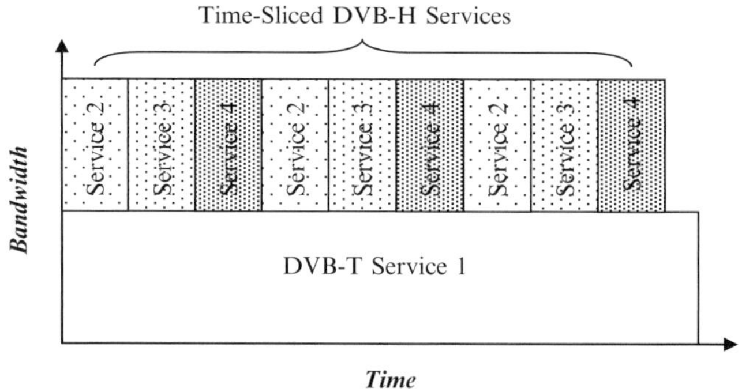

Fig. 4.2 Time-sliced DVB-H services in parallel and compatible with DVB-T services

transmission system usually provides a bit rate of 10 Mb/s or more. This provides a possibility to significantly reduce the average power consumption of a DVB-H receiver by introducing a transmission scheme based on time division multiplexing. This scheme is called time-slicing. To reduce the power consumption in handheld terminals, the service data is time-sliced and then sent through the channel as bursts at a significantly higher bit rate compared to the bit rate of the audio-visual service. Time-slicing enables a receiver to stay active only a small fraction of the time, while receiving bursts of a requested service. It significantly reduces the power consumption used for radio reception parts. Figures 4.1 and 4.2 compare the services bandwidth usage in typical DVB-T and DVB-H channels. Figure 4.2 also shows the backward compatibility of DVB-H with DVB-T system. Time-slicing also supports a quasi-optimum seamless handover by accomplishing the changing of the reception from one transport stream to another during the off-time between bursts.

4.1.2 MPE-FEC

DVB-H employs an additional forward error correction (FEC) to further improve mobile and indoor reception performance of DVB-T. In FEC a number of p extra symbols, called parity, are added to equal-length data symbols by an FEC code.

In case of error, an FEC code can ideally reconstruct any *p* corrupted data symbols, when the location of the error is known and the number of corrupted symbols is less than *p*/2. The *Reed-Solomon (RS)* FEC code [11] is a good example of an FEC code that is optionally used in DVB-H. Since RS code is a systematic code, that is, the source data remains unchanged after FEC encoding, MPE-FEC decoding is made optional for DVB-H receivers. MPE-FEC is computed over IP packets and encapsulated into MPE sections.

Time-slicing and FEC are performed in a network element called *IP encapsulator*. The IP encapsulator encapsulates IP packets into Multi-Protocol Encapsulation (MPE) sections which are further packetized into MPEG-2 *Transport Stream* packets. The optional FEC codes are also encapsulated into MPE-FEC sections. MPE sections are transmitted prior to MPE-FEC sections such that a receiver can just receive the unprotected data and ignore the protection data that follows.

The optional MPE-FEC improves the carrier-to-noise (C/N) ratio and *Doppler* performance in mobile channels and provides greater tolerance to impulse interference. The optional MPE-FEC can be combined with different unequal-error-protection (UEP) algorithms to provide different levels of protection for data according to their degrees of importance. See [8] as an example.

4.1.3 DVB-H Signaling

Two signaling bits are added to the *Transmitter Parameter Signaling* (TPS) bit that indicates the presence of DVB-H service and the possible use of MPE-FEC to enhance and speed up service discovery. A cell identifier is also carried in the TPS bit to support faster signal scan and frequency handover on mobile receivers.

4.1.4 4 K-Mode OFDM

Signal modulation in DVB-T transmission is based on COFDM (*Coded Orthogonal Frequency Division Multiplex*). It uses about 2000 (called 2 K-mode) or 8000 (called 8 K-mode) carrier frequencies on which transmitted streams are overlaid. In addition to 2 K-mode and 8 K-mode, an optional 4 K-mode (using about 4000 carries) has been adopted in DVB-H. The 4 K-mode provides a compromise solution between the two other modes. It allows for a doubling of the transmitter distance in *Single Frequency Networks* (SFN) compared to the 2 K-mode. In comparison to the 8 K-mode, the 4 K-mode is less susceptible to the effect of Doppler shifts in the case of mobile reception and it allows a higher speed. The 4 K-mode offers a new degree of flexibility for network planning.

For 2 K and 4 K-modes the in-depth interleavers increase the flexibility of the symbol interleaving, by decoupling the choice of the inner interleaver from the transmission mode. This allows a 2 K or 4 K signal to take benefit of the memory of

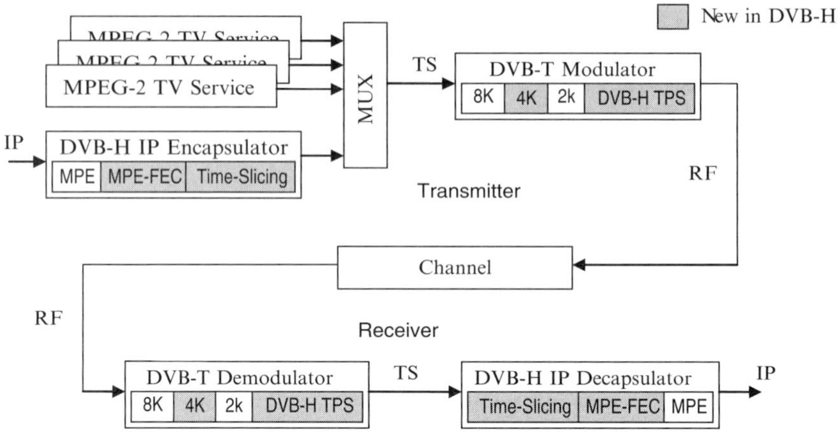

Fig. 4.3 An example of using a DVB-H system

the 8 K symbol interleaver to effectively increase, four times for 2 K and two times for 4 K, the symbol interleaver depth to improve reception in fading channels. This provides also an extra level of protection against impulse noise.

An example of using DVB-H for transmission of IP-services is depicted in Fig. 4.3. In this example, both traditional MPEG-2 services and time-sliced DVB-H services are carried over the same multiplex. The handheld terminal consumes IP services only. Note that 4 K-mode and the in-depth interleavers are not available in cases where the multiplex is shared between services intended for fixed DVB-T receivers and services for DVB-H devices.

4.1.5 Protocol Stack

The protocol stack for the DVB-H is presented in Fig. 4.4. The *Internet Protocol (IP)* packets are encapsulated into MPE sections and also the FEC codes are encapsulated into MPE-FEC sections. Then the MPE and MPE-FEC sections are packetized into MPEG-2 Transport Stream packets.

Using IP allows the coding process to be independent of the transport. This provides a number of advantages for handheld terminals including a variety of encoding methods. Time-slicing and MPE-FEC may be used with IPv6 and also IPv4. DVB has specified four methods for data broadcasting; *Data Piping*, *Data Streaming*, *MPE*, and *Data Carousel*. MPE is well suited to the delivery of streaming services as well as files and other data objects. DVB has specified IP address resolution on MPE, that is, INT table. In addition, MPE supports delivery of other protocols that provides more flexibility. Finally, the implementation of time-slicing on MPE is simple.

More details about the protocol stack for DVB-H are presented in Fig. 4.5. IP packets are encapsulated to MPE sections for transmission over DVB protocols in

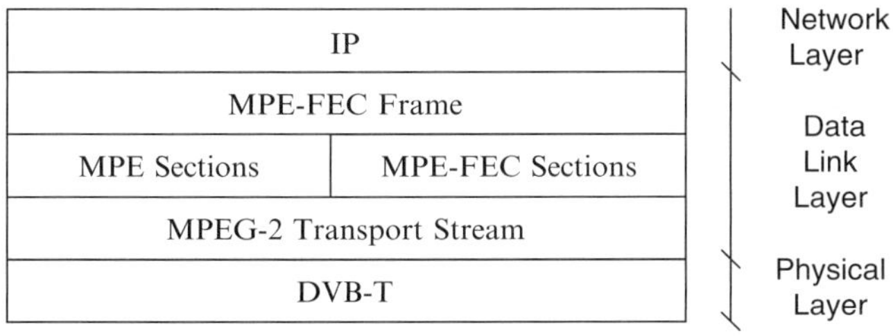

Fig. 4.4 DVB-H protocol stack

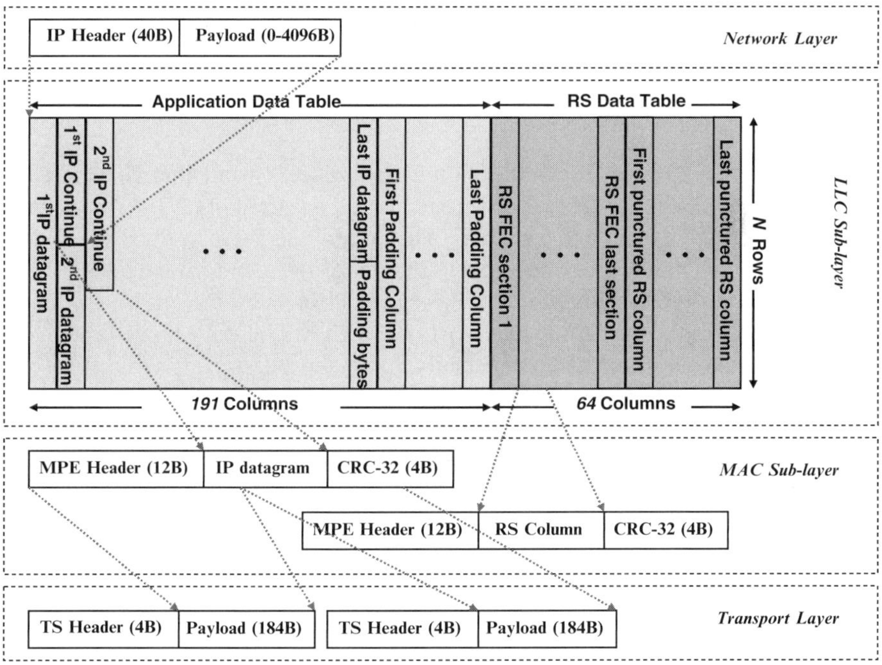

Fig. 4.5 A subset of the protocol structure of DVB-H

the *Medium Access* (MAC) sublayer. Each MPE section consists of a header and the IP datagram as payload. It also includes a 32-byte *Cyclic Redundancy Check* (CRC) for the verification of payload integrity. The MPE section header contains information that could be used by the receiver to identify the scope of the MAC address. The MPE sections can be logically arranged to application data tables in the *Logical Link Control* (LLC) sublayer, over which RS FEC codes are computed and MPE-FEC sections are created. The MPE and MPE-FEC sections are mapped onto MPEG-2 TS packets.

To create MPE-FEC, IP packets are filled into an ($N \times 191$) matrix where each cell of the matrix hosts one byte of information. According to the standard the number of rows in the matrix, i.e. N can be selected from the following values:

256, 512, 768, or 1024. The datagrams are filled into the matrix column wise. RS codes are computed for each row and concatenated such that the final size of the matrix is of size $(N \times 255)$. The $(N \times 191)$ part of the matrix is called the *Application Data Table* (ADT) and the adjacent $(N \times 64)$ part of the matrix is called the *RS Data Table* (RSDT). The ADT need not be completely filled. The unfilled part of the ADT is called padding. The padding allows rate control and preventing fragmentation of IP packet between two MPE-FEC frames. All the 64 columns of RSDT need not be transmitted, i.e. the RSDT may be punctured. This allows control of code rate. Further information on the MPE-FEC matrix construction can be obtained from [3].

4.2 DVB-H Broadcast System Design

Video streaming over DVB-H channel significantly differs from other video streaming applications. The time-sliced transmission scheme is used in DVB-H to increase the battery lifetime of handheld terminals. The end-to-end delay of DVB-H services has been increased due to the time-slicing scheme. A lower end-to-end delay is possible by more power consumption at the receiver. Moreover, the end-to-end delay is increased due to the variations that exist in the bit rate of the encoded media bit streams. Variable bit rate (VBR) especially for video bit streams is used to provide higher visual quality for broadcast services. For a given level of variation in bit rate, the end-to-end delay can be decreased by a higher transmission bandwidth that means a less utilization of transmission bandwidth. The bandwidth utilization can be improved by statistical multiplexing of VBR broadcast services. Therefore, the video quality, end-to-end delay, bandwidth utilization, and power consumption of DVB-H receiver are related parameters. These parameters should be considered simultaneously when a DVB-H broadcast system is designed. The rest of this chapter explains the relationships between these parameters in the system. Moreover, a number of research works related to video streaming over DVB-H are reviewed that try to improve the performance of DVB-H broadcast system.

4.3 Time-Slicing and Power Consumption

Service data in DVB-H is time-sliced and then sent into the channel as bursts at a significantly higher bit rate compared to the bit rate of the audio-visual service. This enables a receiver to stay active only a small fraction of the time, and thereafter decreases the consumed power for data reception. Figure 4.6 depicts the burst parameters in time-slicing scheme. *Burst Size* (B_B) refers to the number of *Network Layer* bits within a burst. *Burst Bit Rate* (R_B) is the bit rate used by a time-sliced elementary stream while transmitting a burst. *Constant Bit Rate* (R_C) is the average bit rate required by the elementary stream when not time-sliced. *Burst Duration*

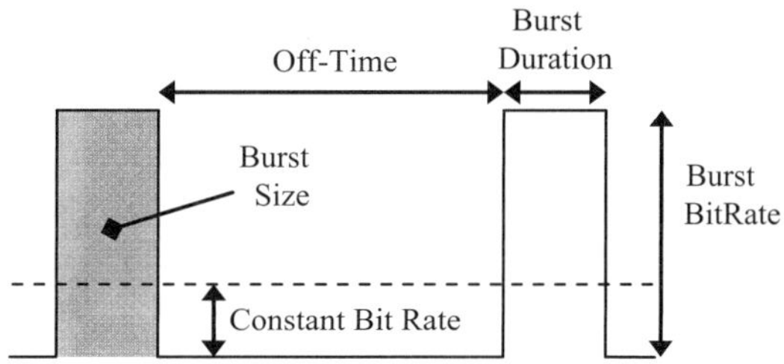

Fig. 4.6 Burst parameters

(T_B) is the time from the beginning to the end of the burst. *Off-time* (T_{off}) is the time between bursts. The percentage of power saving resulting from time-slicing can be expressed as

$$P_S = \frac{T_{off}}{T_{on} + T_{off}} \times 100, \tag{4.1}$$

where T_{on} denotes on-time or the time duration when the radio receiver is on. An extra time namely *Synchronization Time* (T_S) is required by a receiver to re-acquire lock onto the signal before the start of the reception of the next burst. Therefore the on-time can be expressed as

$$T_{on} = T_B + T_S. \tag{4.2}$$

The burst duration depends on the burst size and burst bit rate as

$$T_B = \frac{B_B}{R_B(1-h)}, \tag{4.3}$$

where h compensates the overhead caused by the transport packet and section headers. The constant bit rate or the average bit rate of service data can be calculated as

$$R_C = \frac{B_B}{T_{on} + T_{off}}. \tag{4.4}$$

Using Eqs. 4.1–4.4, for a given R_B and R_C, the percentage of power saving is derived as

$$P_S = \left(1 - R_C\left(\frac{1}{R_B(1-h)} + \frac{T_S}{B_B}\right)\right) \times 100. \tag{4.5}$$

This means a higher percentage of power saving is achieved for a lower service bit rate and by a higher channel bandwidth.

4.4 Channel Switching Delay

Channel switching delay or tune-in time in DVB-H refers to the time between the start of switching to a new channel and the start of the media rendering. Figure 4.7 shows a graphical demonstration for the channel switching delay in DVB-H application when switching from service 1 to service 2. The shown delay

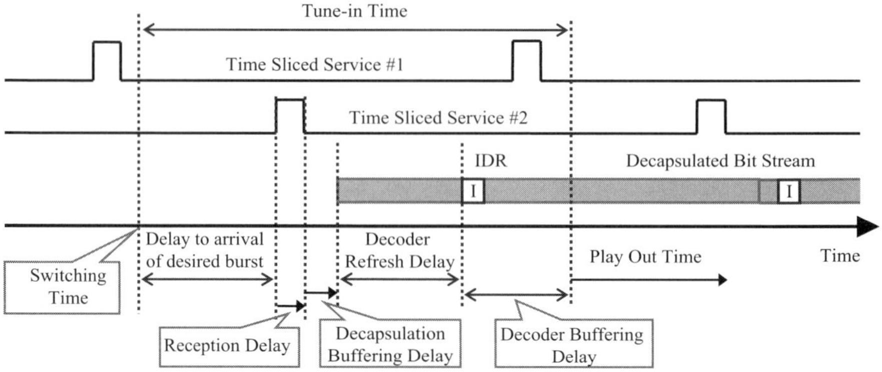

Fig. 4.7 Channel switching timing graphs (©2007 IEEE) (Rezaei et al., "Optimal Channel Changing Delay for Mobile TV over DVB-H," Portable 2007)

times in the graph are: *Arrival Delay* (delay to arrival of desired burst), *Reception Delay* (reception duration of desired burst), *Decapsulation Buffering Delay*, *Decoder Refresh Delay* (delay to the first random access point, e.g. an *Instantaneous Decoder Refresh (IDR)* picture in AVC/H.264 video coding standard or I-picture in other video coding standards), *Decoder Buffering Delay* (initial buffering period of *Coded Picture Buffer*). The decapsulation buffering delay includes two buffering delays for the *Multi-protocol Decapsulation Buffer (MDB)* and *RTP (Real Time Protocol) Decapsulation Buffer (RDB)* [2, 6]. The decapsulation buffering delay is required to compensate the variations that exist in the burst size and the decoder buffering delay is needed to compensate the variations that exist in the bit rate of media bit streams. Moreover, another delay is needed for synchronization between the associated streams (e.g. audio and video) of the streaming session which is not shown in the graph.

Channel switching delay is one of the significant factors in end-to-end delay of DVB-H system. In comparison with other video broadcast systems, DVB-H suffers from a higher channel switching delay that has been enlarged by the time-sliced transmission scheme. In the following sections the details of channel switching delay in conjunction with power consumption of DVB-H receiver and service quality are studied. Furthermore, a number of techniques proposed for the reduction of channel switching delay are reviewed.

4.5 Power Consumption and Channel Switching Delay

One of the significant factors in channel switching delay is the arrival delay or the time until the start of the desired time-slice. The arrival delay depends on the time-slicing parameters that define the percentage of power saving for DVB-H receiver. The lower the receiver power consumption, the higher is the arrival delay.

To find out the relationship between the power saving and channel switching delay, consider a simple case where neighboring channels have similar time-slicing parameters. When channel switching, the arrival delay is required until the start of the desired burst and the reception delay is required until the burst is completely received. The sum of these two delays or receiving delay in average is expected to be

$$D_R = \tfrac{1}{2}(T_{on} + T_{off}) + T_{on}. \tag{4.6}$$

Combining Eqs. 4.2– 4.4 with 4.6 yields

$$D_R = T_S + B_B\left(\frac{1}{2R_C} + \frac{1}{(1-h)R_B}\right). \tag{4.7}$$

Again by combining Eqs. 4.5 and 4.7, a closed form expression relating the percentage of power saving and the receiving delay is given as

$$P_S = \left(1 - \frac{1}{D_R - T_S}\left(\frac{T_S}{2} + \frac{R_C D_R}{R_B(1-h)}\right)\right) \times 100. \tag{4.8}$$

For the transmission of an elementary stream with a desired average bit rate of R_C, by a channel with bandwidth R_B, different operating points on the power saving curves (P_S, D_R) can be found from Eq. 4.8. Considering a typical values for h (0.04 or 4%) and T_S(200 ms), a set of power saving curves are depicted in Fig. 4.8. The power saving curves are computed for two different values of R_B (5and 10 Mb/s) and three different values of R_C (0.3, 0.4, 0.5 Mb/s). As graphs show the

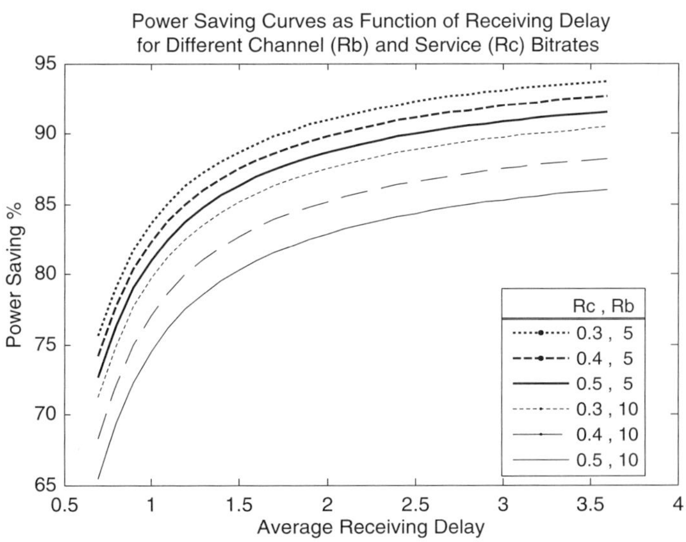

Fig. 4.8 Power saving as a function of receiving delay (©2007 IEEE) (Rezaei et al., "Optimal Channel Changing Delay for Mobile TV over DVB-H," Portable 2007)

percentage of power saving is an increasing function of the receiving delay. Moreover, the power saving curves drop dramatically when the on-time, including the synchronization time and the burst duration, becomes comparable with the off-time.

4.6 Video Quality and Channel Switching Delay

The decoder buffering delay and the decoder refresh delay are two major parts of the channel switching delay that are correlated to the video quality. The decoder buffering delay is required to compensate the variations of bit rate. For video streaming over DVB-H application the advantages of VBR video are exploited. For most video content, a VBR video can provide better visual quality and coding efficiency than a constant bit rate video. A higher quality and compression performance can be obtained by more variations in bit rate. The more the variations in bit rate, the higher is the initial buffering delay. The decoder buffering delay in DVB-H would be minimized if video bit streams are encoded with constant bit rates. However, by allocating a reasonable delay to the system, the advantages of VBR are utilized. In order to control the decoder buffering delay, the variations in the bit rate of bit streams should be controlled by a rate controller. The compression performance and quality of encoded VBR video also depend on the used rate control algorithm. Presented video rate control algorithms in [12–15, 19] are examples of rate controllers targeted for VBR video and streaming applications.

The decoder refresh delay means the required time until a media decoder is refreshed by a random access point to produce correct output frames. The video decoder refresh delay can be minimized if each time-slice is started with a random access picture such as an IDR picture in AVC/H.264 video coding standard. It should be remarked that if the decoder started decoding from an IDR picture that is not at the beginning of a time-slice immediately when the time-slice is received, the input buffer for decoding would drain before the arrival of the next time-slice and there would be a gap in video playback corresponding to the payout duration from the beginning of the time-slice to the first IDR picture.

In DVB-H, the content encoding and the encapsulation to time-slices are implemented independently and it is hard to align the exact location of IDR pictures to the boundaries of time-slices. The average of decoder refresh delay can be decreased by using frequent random access pictures in the bit streams. The problem is that a random access picture usually consumes a bit budget several times more than other pictures and using frequent random access pictures in the bit stream drops the compression efficiency and the video quality remarkably. Therefore, the minimum decoder refresh delay is constraint to compression performance. On the other hand, while at least one random access point in each time-slice is desired, the maximum value of decoder refresh delay is defined by the percentage power saving.

As discussed the video quality of DVB-H services depends on the frequency of random access pictures and also on the variations in bit rate of bit streams that are related to the decoder refresh delay and decoder buffering delay, respectively. Moreover, the quality of encoded video depends on the video content and the used

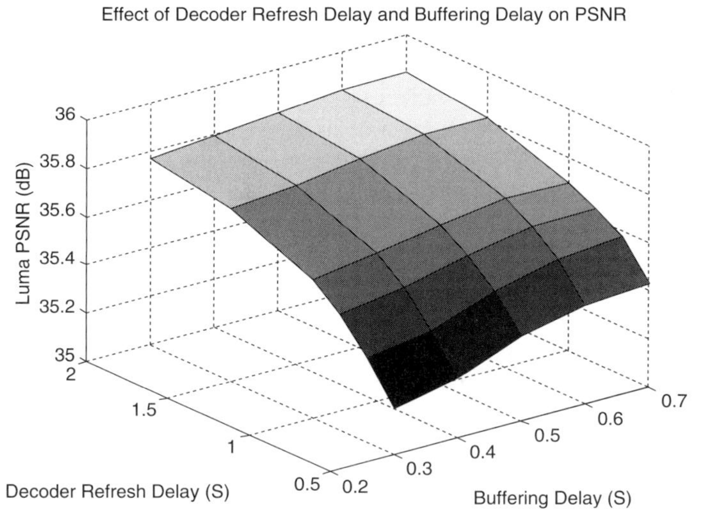

Fig. 4.9 Dependency of video quality in terms of PSNR to the decoder refresh delay and the decoder buffering delay (©2007 IEEE) (Rezaei et al., "Optimal Channel Changing Delay for Mobile TV over DVB-H," Portable 2007)

rate control algorithm. However, the functions between the video quality, decoder refresh delay, and decoder buffering delay can be found experimentally. Some experimental results have been presented by Rezaei et al. [23]. Figure 4.9 shows the video quality in terms of PSNR in average over a number of different video content as a function of the decoder refresh delay and decoder buffering delay. As results show the PSNR is an increasing nonlinear function of both the decoder refresh delay and decoder buffering delay. The PSNR drops dramatically where the delays become small and it saturates to a limit where the delays become very large. Note that the PSNR itself cannot present the visual quality of encoded video especially in VBR. The visual quality also depends on the variations in quality. A higher visual quality can be achieved by less variation in quality or more variations in bit rate. Therefore, although the PSNR is limited for large values of decoder buffering delay, the visual quality still can be improved by higher buffering delays.

4.7 Channel Switching Delay Reduction Methods

In previous sections the relationships between channel changing delay, power consumption, and video quality were explained. Compromising channel changing delay, power consumption and video quality an optimal operation point can be found for a DVB-H broadcast system. Moreover, some techniques have been proposed that decrease the channel switching delay of DVB-H services [16–18, 20, 21, 24]. The proposed techniques are implemented at the encoder and/or at the IP encapsulator. A number of techniques are presented in the sequel.

4.7.1 Video Splicing

A video encoding and splicing method is proposed by Rezaei et al. that can considerably decrease the decoder refresh delay as a major part of channel switching delay at the expense of some reasonable degradation in video quality [21]. The proposed method is implemented partially at the encoder and at the IP encapsulator. To explain the video splicing method first a conventional IP datacasting (IPDC) system over DVB-H is reviewed.

A simplified block diagram of a conventional IPDC system over DVB-H is depicted in Fig. 4.10. As shown, a content encoder receives a source signal in analog format, uncompressed digital format, compressed digital format, or any combination of these formats. The content encoder encodes the source signal into a coded media bit stream. Content encoder is typically capable of encoding more than one media type, such as audio and video. Alternatively, more than one content encoder may be required to code different media types of the source signal. Figure 4.10 illustrates the processing of one coded media bit stream of one media type.

The coded media bit stream is transferred to a server. Examples of the format used in transmission include an elementary self-contained bit stream format, a packet stream format, or one or more coded media bit streams encapsulated into a container file. The content encoder and the server may reside on the same physical device or may be included in separate devices. The content encoder and the server may operate with live real-time content, in which case the coded media bit stream may not be stored permanently, but rather buffered for small periods of time in the content encoder and/or in the server to smooth out the variations of processing delay, transfer delay, and coded media bit rate. The content encoder may also operate considerably earlier than when the bit stream is transmitted from the server. In such a case, the system may include a content database, which may reside on a separate device or on the same device as content encoder and/or server.

The server may be an IP multicast server using real-time media transport over RTP. The server is configured to encapsulate the coded media bit stream into RTP packets according to an RTP payload format. Although not shown in Fig. 4.10, the system may contain more than one server. The server is connected to an IP encapsulator. The connection between the server and IP network may be a fixed-line private network. As shown in Fig. 4.3, the system further includes at least one radio transmitter which is not essential for the operation of the proposed splicing system.

According to the proposed video splicing method, an additional stream consisting of refresh pictures only is encoded and transmitted to the IP encapsulator. The IP

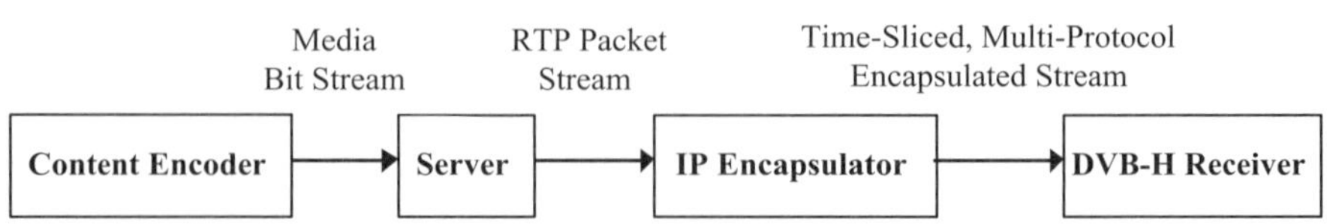

Fig. 4.10 Simplified block diagram of a conventional IPDC system

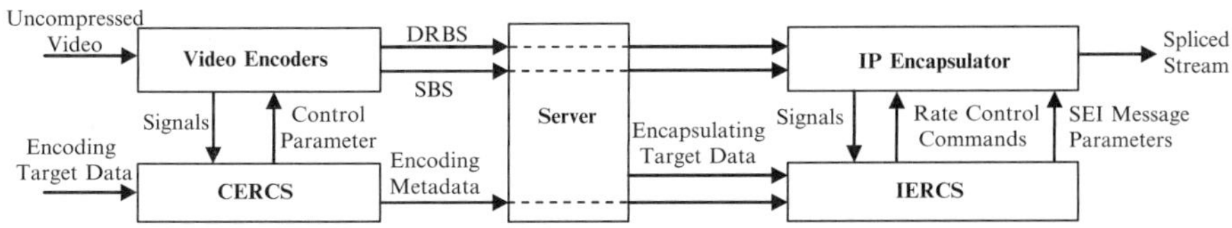

Fig. 4.11 Block diagram of proposed splicing and rate control method (©2007 IEEE) (Rezaei et al., "Tune-in Time Reduction in Video Streaming over DVB-H," Transactions on Broadcasting, Vol 53, March 2007)

encapsulator replaces some pictures in a normal bit stream with the refresh pictures according to time-slice boundaries in order to achieve the minimum decoder refresh delay.

A simplified block diagram of the proposed IPDC system for the implementation of video splicing is depicted in Fig. 4.11. At the content encoding level, one or two video encoders encode the video source to two encoded primary bit streams including a *Spliceable Bit Stream* (*SBS*) and a *Decoder Refresh Bit Stream* (*DRBS*) from the same source picture sequence. The SBS includes frequent spliceable pictures which are reference pictures constrained as follows: no picture prior to a spliceable picture, in decoding order, is referred to in the inter prediction process of any reference picture at or after the spliceable picture. Nonreference pictures after the spliceable picture may refer to pictures earlier to the spliceable picture in decoding order. These nonreference pictures cannot be correctly decoded if the decoding process starts from the spliceable picture, but can be safely omitted as they are not used as reference for any other pictures. The DRBS contains only intra or IDR pictures corresponding to spliceable pictures and with a picture quality similar to corresponding spliceable pictures. The DBRS and the SBS are transmitted from the server to the IP encapsulator. The IP encapsulator composes MPE-FEC frames, in which the first picture in decoding order is an IDR picture from the DBRS and the other pictures are from the SBS. The IDR pictures at the beginning of MPE-FEC frames minimize the tune-in time for newly joined recipients. No changes in the receiver operation are required in the proposed system.

Replacing an inter picture with an intrapicture in the SBS causes a mismatch in the pixel values of the reference pictures between the encoder and decoder. The reference mismatch propagates an error until the next IDR picture in the spliced stream. A technically elegant solution to avoid mismatch altogether would be to use SP and SI pictures of H.264/AVC, but they are only included in the *Extended Profile* of H.264/AVC [9]. The extended profile of H.264/AVC is not allowed in the current DVB-H standard [5]. According to experimental results the reference mismatch error is saturated to a limit and the overall degradation in quality is reasonably small. Furthermore, the mismatch error decreases if the spliceable pictures and corresponding IDR pictures are encoded with a higher quality than other pictures. In addition to the mismatch error, the proposed method involves two buffering related challenges. Firstly, the standard *HRD* (*Hypothetical Reference Decoder*) compliancy of the spliced stream is hard to be verified. Secondly, the initial delay for coded picture buffering in the HRD is hard to derive in the IP encapsulator.

According to the proposed splicing method the spliceable pictures and corresponding intra/IDR pictures in two primary streams should be encoded with similar qualities. In a similar quality an IDR frame can consume a bit budget from 5 to 10 times more than corresponding inter picture. Furthermore, similar quality for corresponding frames in two primary streams means that only the bit rate of one primary stream can be controlled. Consequently, there is no real short-term control on the bit rate of spliced streams and therefore it is hard to verify the HRD compliancy of spliced streams. Moreover, the encoding parameters cannot be controlled according to the results of splicing, because the encoding and splicing are performed independently and without any feedback link.

To solve the problem above, a comprehensive rate control system as depicted in Fig. 4.11 is proposed which is implemented in both the content encoder and the IP encapsulator. The content encoding rate control system (CERCS) controls the bit rate of two primary streams considering a fixed value for the frequency of IDR pictures in a desired spliced stream. However, the frequency of IDR picture in the spliced stream can have variations around an average value since the number of video pictures in MPE-FEC frames is not fixed. Moreover, in offline encoding, the IDR frequency which has been used for the rate control of primary streams at the content encoder may be different from the average IDR frequency of spliced stream. The IP encapsulating level rate control system (IERCS) implements another control to compensate the above variations and to provide HRD compliancy for the spliced bit stream. Furthermore, the H.264/AVC *Supplemental Enhancement Information* (*SEI*) message parameters related to buffering of the spliced bit stream can be provided by IERCS.

The CERCS controls the bit rate of primary streams according to encoding target data which is set by the user and also according to several signals which are extracted from the uncompressed and compressed video. The encoding target data includes target bit rate of spliced stream and average frequency of IDR pictures in the desired spliced stream. Furthermore, some encoding metadata as complementary information are provided by CERCS which are sent to the server and then IP encapsulator.

The IERCS controls the bit rate of spliced stream according to the encoding metadata, encapsulating target data defined by the server. The encapsulating target data includes target bit rate of spliced stream and IDR frequency of spliced stream. More details about the proposed rate control systems can be found in [17,18,20,21]. According to the proposed method the decoder refresh delay as a major part of tune-in time can be minimized to a very small value and even to zero.

4.7.2 Redundant Intrapicture Insertion

In AVC/H.264, an encoded picture can be either a *primary* encoded picture or a *redundant* encoded picture. A primary encoded picture is used to decode valid bit streams. The redundant encoded pictures are used to improve the robustness of encoded bit streams against transmission error. A redundant encoded picture is decoded only when the decoding of the primary encoded or decoded picture is corrupted.

Vadakital et al. used a redundant intrapicture insertion scheme to decrease the decoder refresh delay of tune-in time [24]. According to the proposed method a regular bit stream is encoded with a restriction that every sth picture, called *S picture*, is coded such that no picture succeeding the S picture, is inter-predicted from any picture preceding the S picture, all in decoding order. This constraint of S pictures causes a negligible drop in compression efficiency. For every S picture, an intra-coded redundant picture is also encoded with a lower quality compared to the S picture. The encoded bit stream is transmitted to the IP encapsulator. The IP encapsulator composes the MPE-FEC frames such that only the first intracoded redundant picture is placed in the ADT and all other redundant pictures in the bit stream are removed. The bit rate increases due to the redundant intracoded picture. This increase is, however, significantly lower compared to the bit rate increase that would result with coding periodic primary intrapictures at the frequency of S pictures.

The receivers can either start decoding from the beginning of the first GOP, or the first redundant intracoded picture, whichever comes earlier in decoding order. If the decoding is started from a redundant intracoded picture, there is a reference mismatch error in the reconstructed following pictures, compared to the case where the decoding started from an IDR picture preceding the current burst. However, the mismatch error only propagates until the next primary IDR picture occurs in the bit stream and is therefore not likely to be annoying. Assuming that the IP encapsulator makes no effort to align S pictures with the boundaries of bursts, the expected value of decoder refresh delay is derived as $\Delta_{\text{refresh}} = s/2f$, where f is the picture rate of coded pictures and S denotes the frequency of S pictures.

4.7.3 Time-Interleaved Simulcast

Simulcast is a fusion of "**simul**taneous broad**cast**". In the context of scalable multimedia, simulcast is a transmission mechanism, where multiple programs of the same source are coded in different qualities, and transmitted simultaneously. The receivers choose the most appropriate of the simulcasted programs according to their consumption capabilities. Vadakital et al. proposed time-interleaved simulcast methods that can decrease the channel switching delay of DVB-H services [24]. Consider a number of programs are simulcasted over DVB-H channels. Each program is encoded at different levels of qualities targeted for different classes of receivers. A receiver can decode and consume the bit streams of proper class. It is assumed that a receiver also can decode and consume all the bit streams targeted for receivers in lower classes. When channel switching, the receiver can start receiving and decoding of the earliest decodable bit stream of desired program, before receiving the targeted version, to decrease the receiving delay as a major part of tune-in time. To obtain the maximum performance, the time-sliced bursts of the simulcast programs of the same content are transmitted maximally apart from each other in the time line.

4.8 Video Encoding Methods for Broadcasting over DVB-H

In streaming applications usually VBR video bit streams are used. VBR video bit streams generally provide higher visual qualities and compression performances than constant bit rate video bit streams at the expense of more resources in terms of transmission bandwidth and delay [10, 15]. Encoding VBR video can be *open-loop* or *closed-loop*. In open-loop encoding, the video pictures are encoded with an almost constant quantization parameter (QP) to provide a relative constant quality for encoded video regardless of the coding complexity of video source. In close-loop encoding, the bit rate of encoded bit stream is controlled by a VBR rate controller according to feedback signals from encoding results and coding complexity of video source. The rate controller imposes some constraints on the degree of variability allowed in the bit rate. For example, the variations in bit rate are controlled such that the required buffering delay remains below a limit.

In video streaming over DVB-H, the closed-loop VBR encoding method is preferred for several reasons. First, for a given percentage of bandwidth utilization, unlike the open-loop encoded bit streams that need a relatively high buffering delay, the required buffering delay of constrained bit streams is limited. While the end-to-end delay is a serious bottleneck in DVB-H, the constrained bit streams are preferred to guarantee a limited end-to-end delay. Second, for a given transmission delay, the closed-loop encoded bit streams provide a higher bandwidth utilization than open-loop encoded bit streams. Finally, although the bandwidth utilization can be improved by statistical multiplexing of broadcast services, due to small number of DVB-H services that can be multiplexed to one DVB-T channel, the performance of statistical multiplexing is not as much as in other applications in which a large number of services are multiplexed. Therefore, the required resources in terms of bandwidth and delay should be controlled by the closed-loop encoding before multiplexing.

When multiple programs are broadcasted simultaneously, two main scenarios for the closed-loop encoding of the bit streams are possible: *independent video encoding* and *joint video encoding*. In first scenario each video source is encoded independently of the others using a separate rate controller. In the second scenario multiple sources are encoded simultaneously using a common rate controller. More details about the independent and joint video encoding are presented in the sequel.

4.8.1 Independent Video Encoding

In this scenario the video sources are encoded independently of each other and by separate encoders. The video encoders can be separated geographically. Figure 4.12 shows a block diagram for independent video encoding in DVB-H application. The encoded bit streams are sent to the servers and then to the IP encapsulator. Also the bit streams can be stored to be sent later on. As shown, each encoder utilizes an independent rate controller. VBR rate controllers are used to provide a higher level

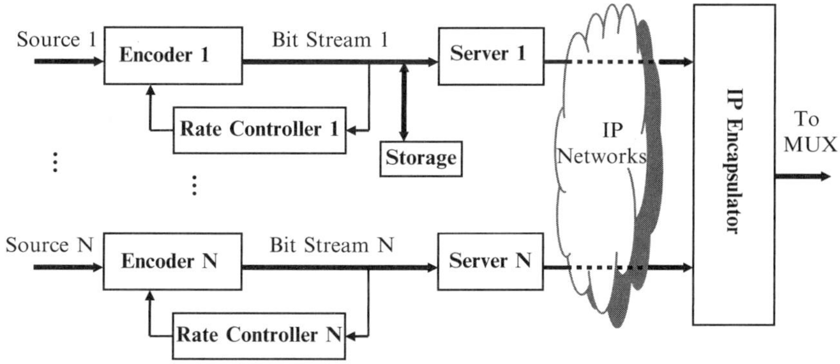

Fig. 4.12 A block diagram of independent video encoding in DVB-H

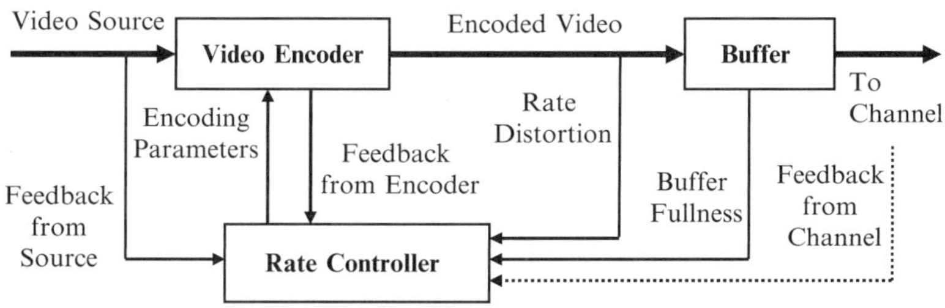

Fig. 4.13 A block diagram of independent close-loop video encoding

of quality for compressed video and also to guarantee a limited transmission delay with a level of bandwidth utilization.

Numerous rate control algorithms for VBR applications have been presented in the literature. See presented algorithms in [12–15, 19] as examples. A general block diagram for the rate control of independent encoded video bit stream is depicted in Fig. 4.13. As shown, a rate controller defines the encoding parameters for the encoder according to some feedback signals from the encoded video, encoder, video source, a real or virtual smoothing buffer, and sometimes from the channel. QP is the main encoding parameter that is used for the rate control. Frame rate, picture type, and picture resolution can also be used for the rate control. The rate and the distortion of encoded video are two parameters that are widely used as feedback signals by the rate controller. The level of allowed variations in bit rate is defined by the size of smoothing buffer while the occupancy of buffer is used as a feedback signal to the rate controller. In real-time applications the real buffer between encoder and transmission channel can be used for the rate control. In real-time applications in which the transmission channel is not stable, a feed back signal from the channel can also be used by the rate controller. For video streaming over DVB-H channels, a virtual buffer for the both real-time and offline encoding can be used. In this application, no feedback from the channel is required. In fact, no any other feedback from the IP encapsulator to encoder is used.

A fuzzy VBR video rate controller that is optimized for DVB-H application has been proposed in [19]. The proposed rate control algorithm has been optimized to provide VBR bit streams including frequent IDR pictures with a relative constant visual quality and low average buffering delay. Moreover, it has a small degree of computational complexity. The proposed rate controller controls the bit rate by adjusting the QP on picture base. It utilizes a relative small size virtual buffer, a fuzzy controller and several other tools to calculate the QP for different types of pictures. The fuzzy controller tries to minimize the fluctuations in QP and thereafter in the quality of encoded video while the buffering delay is limited. More details about the rate controller can be found in [19].

4.8.2 Joint Video Encoding

When a number of services are encoded and broadcasted simultaneously and the encoders are collocated geographically, it is possible to use a joint rate controller for all encoders instead of using independent rate controllers, one for each encoder. A joint rate controller can improve the average quality of encoded bit streams by sharing of resources among the bit streams. In fact, the joint rate controller imposes a type of statistical multiplexing on the encoded bit streams.

In video broadcast over DVB-H channels, the joint encoding can be utilized to improve the overall video quality and also to decrease the end-to-end delay of broadcast system. In joint encoding, a common bandwidth is shared between the bit streams according to their temporal complexities by a joint rate controller. Depending on the joint rate controller, the bandwidth can be shared for a long term such that the average bit rate of bit streams can be changed after long time or the bandwidth can be shared only for short term such that such that the average bit rate of bit streams is constant for long times. In comparison to independent encoding, in a similar buffering delay, the joint encoding with a long-term share of bandwidth can provide more constant qualities for encoded bit streams. On the other hand, when compared with independent encoding, in a similar quality, the joint encoding with a short-term share of bandwidth can provide a lower buffering delay.

The major problem of joint video encoding is how to allocate the available bit budget among the video sources that share the bandwidth. A number of joint rate control algorithms have been proposed in the literature for different applications [1, 22, 25, 26]. Two major approaches for the bit allocation in the joint rate control algorithms have been used including: *Preprocessing* and *Modeling* approaches. In the first approach, a preanalysis is performed on video sources to gather statistics about the coding complexity. The real coding process can operate based on the statistics obtained by the preanalysis. In the modeling approach, the attempt is to model the performance of video encoder and the coding complexity of video sources and then the allocated bits to video sources are controlled based on provided models while the models are updating during encoding. See the proposed methods in [1, 25, 26] as examples for the two approaches. The system presented in [1] consists of several preprocessors and video encoders. Each preprocessor analyzes a video source and derives picture statistics. Using these statistics, a joint rate controller calculates dynamically the bit rate for each encoder based on the relative complexities of the sources. Another bit allocation method for joint coding of multiple video sources is presented in [25]. In this method, the input video sources are divided into *Super GOPs* (a number of GOP) and *Super Frames* (a collection of frames, one from each program) and then, the bit budget is distributed hierarchically between the video super GOPs, super frames and frames according to their relative complexities while the encoder and decoders' buffers are prevented from overflowing and under flowing. Finally, using a rate-distortion model, a QP is calculated for each frame according to the allocated bits to the frame. A similar approach to that in [25] is presented in [26]. A new approach for joint rate control has been presented for DVB-H application in [22] that utilized a fuzzy controller. The fuzzy joint rate control algorithm distributes the bandwidth among the DVB-H services. The bandwidth is shared in short term. Provided results show that it decreases the buffering delay of DVB-H service with no change in the overall quality of encoded video. The proposed method can be used for joint encoding of multiple bit streams even with different qualities and bit rates.

A general block diagram of joint video rate control is depicted in Fig. 4.14. As shown a number of encoders utilize a joint rate controller simultaneously. A joint buffer is used for controlling variations in the bit rate of aggregated bit stream. The occupancy of buffer with some other feedback signals from the encoded bit streams is used by the joint rate controller to compute the QPs for encoders. Similar to the block diagram of independent rate control shown in Fig. 4.13, a joint rate controller can also use some feedback signals from the encoders and uncompressed video sources that are not shown in Fig. 4.14. The joint rate control algorithms proposed in [25, 26] are examples that can be fit to this block diagram.

When the joint encoded bit streams share the bandwidth only for short term, it is possible to have different bit rates and qualities for the bit streams. In this case, beside the joint rate controller and joint buffer, each encoder can also utilize an individual rate controller and buffer to control the long-term bit rate. A simplified

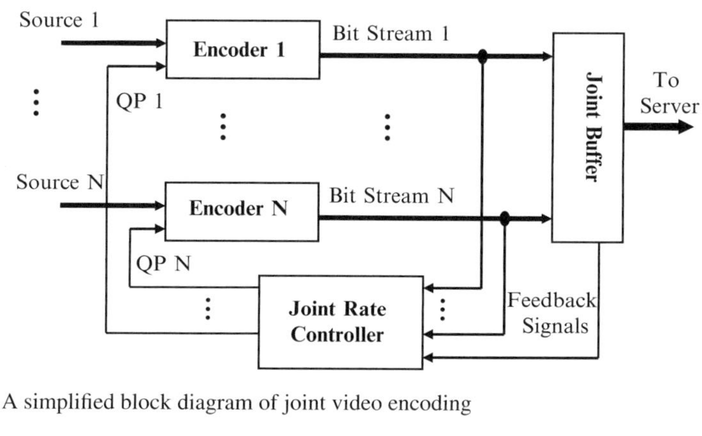

Fig. 4.14 A simplified block diagram of joint video encoding

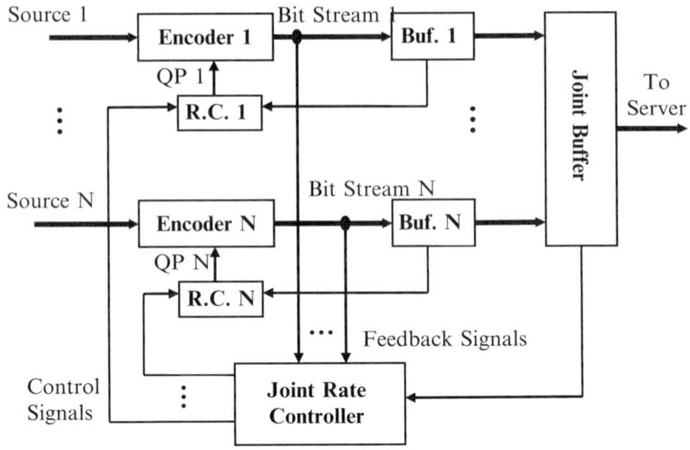

Fig. 4.15 A block diagram for joint video encoding with superficial bandwidth share

block diagram for this case is depicted in Fig. 4.15. According to this diagram the bit rate of each bit stream is mainly controlled by the individual rate controller (R.C. in the diagram) that uses a separate buffer. The joint rate controller provides control signals for the individual rate controllers according to feedback signals from the joint buffer and encoded bit streams. The control signals are used by the individual controllers for calculating QPs. The proposed fuzzy joint rate control algorithm in [22] is an example that operates according to this block diagram.

In DVB-H application, the joint encoded bit streams are sent to the server and then to the IP encapsulator. Different time-slicing and encapsulation scenarios for joint encoded bit streams are possible. As an option in DVB-H, multiple services can be encapsulated to one time-slice or burst. The aggregated bit stream resulted by joint encoding can be a perfect match for this option. However, the aggregated bit stream can be distributed over parallel time-slices. When the multiplexed services are encapsulated to one burst, which is completely received by receiver, channel switching delay between the multiplexed services can be decreased considerably at the expense of more power consumption for the receiver. In practice a combination of all encoding and encapsulation methods including open-loop encoding, closed-loop encoding, independent encoding, joint encoding, sharing of bandwidth in short term, sharing of bandwidth in long term, encapsulation of multiple service into one time-slice, and encapsulation to individual time-slices is possible.

4.9 Statistical Multiplexing and Time-Slicing

In video broadcasting over DVB-H the media sources are encoded to VBR bit streams to utilize the advantage of VBR. When VBR bit streams are used, utilizing statistical multiplexing would be beneficial for the transmission of bit streams.

In statistical multiplexing, a fixed bandwidth communication channel is virtually divided into several VBR channels. The resulting VBR channels are adapted to the instantaneous traffic demands of the bit streams that are transferred over the channels. Statistical multiplexing is used in many communication applications to improve the overall performance of communication channels. The performance of communication channel can be evaluated in terms of transmission delay, data drop rate, and bandwidth utilization.

The described joint video encoding methods in previous section are forms of statistical multiplexing that are implemented in conjunction with encoding. A new form of statistical multiplexing for DVB-H services that is implemented in conjunction with time-slicing is possible. In DVB-H, time-slicing and MPE-FEC are implemented by IP encapsulator. Moreover, a *Time Division Multiplexing* (*TDM*) is implemented by the IP encapsulator on the time-sliced services. The multiplexed services may fill a DVB-T transmission channel. The implementation of statistical multiplexing in IP encapsulator means statistical TDM.

Due to the time-sliced transmission scheme, the implementation of statistical multiplexing in DVB-H is significantly different from other applications and has some associated difficulties. According to the time-sliced transmission scheme used in DVB-H, during transmission of a burst data, a *Delta-T* or the time to the beginning of the next burst of the same service is signaled to the receiver in order to indicate to the receiver when to expect the next burst. In deterministic multiplexing the bandwidth is allocated to a number of services with fixed burst sizes and determined delta-t. Unlike deterministic multiplexing, in statistical multiplexing, the burst sizes and the duration of time-slices may vary over time according to the temporal bit rate of service bit streams. While in statistical multiplexing the time divisions should be proportional to the instantaneous bit rate of bit streams, the problem that arises due to the variation over time of the duration of time-slices is how to calculate the Delta-T for each service. When the data for the current burst is encapsulated, the time-slice boundaries of next burst of the same service are unknown. While a typical time-cycle can be about few seconds, even the estimation of the time-slice boundaries according to the variations in bit rate is difficult. However, any estimation error may provide even worse results for statistical multiplexing than deterministic multiplexing case. Shortly, a desired statistical multiplexer should know the required bandwidth for each service in the next time-cycle when the current time-cycle is time-sliced. This is possible with a long-time look ahead or by buffering of service data for a relatively long time before encapsulating. However, a long-time buffering imposes a long delay to the system that is in contradiction with the objectives of statistical multiplexing. Research on statistical multiplexing in DVB-H is ongoing.

4.10 Conclusions

DBV-H standard adds a number of features to the DVB-T specification for digital terrestrial television to provide broadcast services for handheld receivers. The available infrastructures for DVB-T can be used for DVB-H with few modifications.

DVB-H allows low power consumption for handheld receiver and seamless service handover by the time-sliced transmission scheme. Moreover, it improves the mobile reception quality by MPE-FEC. Finally, the 4 K-mode OFDM in DVB-H offers a new degree of flexibility for network planning. DVB-H standard is very flexible in different features and parameters. Duo to this flexibility there are many rooms for competition between a variety of players from broadcast and cellular operators to chip and equipment manufacturers. In this chapter, a number of key parameters and techniques related to the performance of DVB-H broadcast system were reviewed. To improve the performance of DVB-H broadcast systems, more research based on feedback from practical systems is required.

Acknowledgments I would like to thank my colleagues in Tampere University of Technology and Nokia Research Center especially Professor Moncef Gabbouj, Miska M. Hannuksela, Dr. Imed Bouazizi, and Vinod Kumar Malamal Vadakital for their help.

References

1. Boroczky L, Ngai AY, Westermann EF (2000) "Joint Rate Control with Look-ahead for Multi-program Video Coding," Circuits and Systems for Video Technology, IEEE Transactions, Vol 10, No 7, pp. 1159–1163.
2. Bouazizi I, Rezaei M (2006) Approaches for Smooth Buffering in Mobile TV over DVB-H. IEEE Broadcast Symposium, Washington, USA.
3. European Telecommunications Standards Institute (ETSI), "Digital video broadcasting (DVB); DVB Specification for Data Broadcasting," European Standard EN 301 192, version 1.4.1, Nov. 2004.
4. European Telecommunications Standards Institute (ETSI), "Digital Video Broadcasting (DVB): Transmission Systems for Handheld Terminals," ETSI standard, EN 302 304 V1.1.1., 2004.
5. European Telecommunications Standards Institute (ETSI), "Specification for the use of Video and Audio Coding in DVB Services Delivered Directly over IP Protocols DVB," ETSI standard, TS 102 005, Nov. 2005.
6. European Telecommunications Standards Institute (ETSI), "Digital Video Broadcasting (DVB); IP Datacast over DVB-H: Content Delivery Protocols," ETSI Standard, TS 102472 V1.2.1, 2006.
7. Faria G, Henriksson JA, Stare E, Talmola P (2006) DVB-H: "Digital Broadcast Services to Handheld Devices," IEEE Proceedings, Vol 94, No 1, pp. 194–209.
8. Hannuksela MM, Vadakital VKM, Jumisko-Pyykkö S (2007), "Comparison of Error Protection Methods for Audio-Video Broadcast over DVB-H," EURASIP Journal on Advances in Signal Processing, vol. 2007, Article ID 71801, 12 pages, doi:10.1155/2007/71801.
9. Karczewicz M, Kurceren R (2003) "The SP- and SI-frames Design for H.264/AVC," Circuits and Systems for Video Technology, IEEE Transactions, Vol 13, No 7.
10. Lakshman TV, Ortega A, Reibman AR (1998) "VBR Video: Tradeoffs and Potentials," IEEE Proceedings, Vol 86, No 5, pp. 952–973.
11. Reed IS, Solomon G (1960) "Polynomial Codes over Certain Finite Fields," SIAM Journal of Applied Mathematics, Vol 8, pp. 300–304.
12. Rezaei M, Gabbouj M (2005) Bit Allocation for Variable Bitrate Video. IEEE Workshop on Multimedia Signal Processing (MMSP), Shanghai, China.
13. Rezaei M, Wenger S, Gabbouj M (2005) Video Rate Control for Streaming and Local Recording Optimized for Mobile Devices. IEEE Symposium on Personal Indoor and Mobile Radio Communications (PIMRC), Berlin, Germany.

14. Rezaei M, Hannuksela MM, Gabbouj M (2006) Low-Complexity Fuzzy Video Rate Controller for Streaming. IEEE Conference on Acoustic, Speech and Signal Processing (ICASSP), Toulouse, France.
15. Rezaei M, Hannuksela MM, Gabbouj M (2006) Region Based Minimization of Quantization Error Propagation in Variable Bitrate Video. Picture Coding Symposium (PCS), Beijing, China.
16. Rezaei M, Hannuksela MM, Gabbouj M (2006) Video Splicing for Tune-in Time Reduction in IP Datacasting over DVB-H. IEEE Symposium on Broadband Multimedia Systems and Broadcasting, Las Vegas, USA.
17. Rezaei M, Hannuksela MM, Gabbouj M (2006) Video Encoding and Splicing for Tune-in Time Reduction in IP Datacasting (IPDC) over DVB-H. IEEE Conference on Multimedia & Expo (ICME), Toronto, Canada.
18. Rezaei M, Hannuksela MM, Gabbouj M (2006) Video Splicing and Fuzzy Rate Control in IP Multi-Protocol Encapsulator for Tune-in Time Reduction in IP Datacasting (IPDC) over DVB-H. IEEE Conference on Image Processing (ICIP), Atlanta, USA.
19. Rezaei M, Gabbouj M, Bouazizi I (2006) Delay Constrained Fuzzy Rate Control for Video Streaming over DVB-H. IEEE Conference on Intelligent Information Hiding and Multimedia Signal Processing (IIH-MSP), Pasadena, California, USA.
20. Rezaei M, Hannuksela MM, Vadakital VKM, Gabbouj M (2006) Spliced Video and Buffering Considerations for Tune-in Time Minimization in DVB-H for Mobile TV. IEEE Symposium on Personal Indoor and Mobile Radio Communications (PIMRC), Helsinki, Finland.
21. Rezaei M, Hannuksela MM, Gabbouj M (2007) "Tune-in Time Reduction in Video Streaming over DVB-H," IEEE Transactions on Broadcasting, Special Issue on Mobile Multimedia Broadcasting, Vol 53, No 1, Part 2, pp. 320–328.
22. Rezaei M, Bouazizi I, Gabbouj M (2007) Fuzzy Joint Encoding and Statistical Multiplexing of Multiple Video Sources with Independent Quality of Services for Streaming over DVB-H. IEEE Conference on Intelligent Information Hiding and Multimedia Signal Processing (IIH-MSP), Splendor Kaohsiung, Taiwan.
23. Rezaei M, Bouazizi I, Vadakital VKM, Gabbouj M (2007) Optimal Channel Changing Delay for Mobile TV over DVB-H. IEEE Conference on Portable Information Devices (PORTABLE), Orlando, USA.
24. Vadakital VKM, Hannuksela MM, Rezaei M, Gabbouj M (2006) Method for Unequal Error Protection in DVB-H for Mobile Television. IEEE Symposium on Personal Indoor and Mobile Radio Communications (PIMRC), Helsinki, Finland.
25. Vadakital VKM, Hannuksela MM, Gabbouj M (2007) Time-Interleaved Simulcast and Redundant Intra Picture Insertion for Reducing Tune-In Delay in DVB-H. Packet Video Workshop, Lausanne, Switzerland.
26. Wang L, Vincent A (1999) "Bit allocation and Constraints for Joint Coding of Multiple Video Programs," Circuits and Systems for Video Technology, IEEE Transactions, Vol 9, No 6, pp. 949–959.
27. Xiong H, Sun J, Yu S, Luo C, Zhou J (2004) "Design and Implementation of Multiplexing Rate Control in Broadband Access Network TV Transmission System," Consumer Electronics, IEEE Transactions, Vol 50, No 3, pp. 849–855.

Chapter 5
Overview of Mobile TV Standards and their CMOS Tuners

Iason Vassiliou, Nikos Haralabidis, and Kostis Vavelidis

5.1 Introduction

The upcoming switch to digital TV in most of the world has renewed interest in terrestrial tuner implementations. More importantly, digital video reception is emerging as the latest feature towards multimedia-enriched handheld devices. Mobile, battery operated devices require small size tuners that consume low power and are amenable to single-chip integration with the baseband demodulator.

In broadcast TV, the main standards expected to coexist worldwide are DVB-T (Digital Video Broadcast-Terrestrial) in Europe, ISDB-T (Integrated Services Digital Broadcasting-Terrestrial) in Japan based on OFDM modulation (Orthogonal Frequency-Division Multiplexing) and ATSC-DTV (Advanced Television Systems Committee-Digital TV) in the USA based on 8-VSB (Vestigial Sideband Modulation).

In mobile TV, several standards (summarized in Table 5.1) that will enable reception to cell-phones and other hand-held devices such as PDAs and portable music players compete globally. The most popular appears to be DVB-H (DVB-Handheld) [1], which is an extension of DVB-T. Its main features are:

(a) low-power reception, enabled by operation in bursts ("time-slicing") with a typical 1:10 on/off ratio at the expense of lower resolution and
(b) additional error-coding (FEC) for more robust Doppler performance, needed for mobile operation.

DVB-H is defined mainly for a portion of the UHF band, 470–750 MHz, since upper range is excluded as it may interfere with cell-phone operation in the 900 MHz GSM band in Europe. Deployment in the VHF III band (170–240 MHz) is also discussed, even though it will be hindered by the larger antenna size required for lower frequencies. A significant advantage of DVB-H deployment in VHF/UHF

I. Vassiliou (✉), N. Haralabidis, and K. Vavelidis
Broadcom Hellas
S.A., Poseidonos Ave. 75, 17455 Alimos, Athens, Greece
e-mail: iasonv@gmail.com, iason@broadcom.com

F.-L. Luo (ed.), *Mobile Multimedia Broadcasting Standards: Technology and Practice*,
DOI: 10.1007/978-0-387-78263-8_5,

Table 5.1 Overview of mobile TV standards

Standard	Modulation Features	BW (MHz)	Bands (MHz)
DVB-T/H (EU – USA)	COFDM 4–64 QAM $^1/_2$ to $^3/_4$ CR FFT: 2k,4k,8k Operation in bursts (time-slicing) for low power Additional FEC	6, 7, 8 5 6, 7, 8	470–862 174–230 1670–1675 1452–1492
T-DMB (Korea/EU)	COFDM DQPSK – $^1/_2$ CR	1.54	1452–1492 174–240
ISDB-T mobile (Japan)	COFDM 1–3 "segments" used (out of 13) 4–16 QAM	0.43–1.29	90–222 470–770
DMB-T (China)	4–64 COFDM (similar to DVB-H) VSB (similar to ATSC)	8	470–862 170–230
		18	2605–2655
MediaFLO (USA)	COFDM 4–16 QAM Operation in bursts for low power	6 (US) 5, 6, 7, 8 6	698–746 (US) 470–862 48–890 (TBD)
DVB-SH (EU)	Satellite, terrestrial OFDM, SC modes 4–16 QAM Turbo codes	1.7, 5, 6, 7, 8	2170–2200 2600

frequencies is that it will allow DVB-T broadcast equipment to be used for DVB-H, with potentially only a software or firmware upgrade. In addition, usage of the 1450–1490 MHz L-band is also discussed in Europe, while in the USA a 5 MHz channel has been allocated for DVB-H operation in the L-band (1670–1675 MHz).

T-DMB (Terrestrial Digital Multimedia Broadcasting) [2] is also OFDM-based and uses the DAB (Digital Audio Broadcasting) physical layer [3] for mobile TV operation. It is defined in the VHF III and 1450–1490 MHz bands and has already been deployed in Korea, while it will compete with DVB-H in Europe.

In Japan a lower bandwidth, mobile version of ISDB-T will be using 430 kHz channels in UHF. Finally, several more mobile TV standards are emerging worldwide, such as MediaFLO [4] and ATSC-mobile in the USA and DMB-T, CMMB (China Mobile Multimedia Broadcasting) in China.

As shown in Table 5.1, the above standards have several common characteristics, such as OFDM modulation and operation in the same or similar bands (VHF III, UHF IV and V, 1450–1490 MHz L-band and 1670–1675 MHz L-band). Since mobile TV subsystems will be battery operated and will need to fit in the crowded cellular phone case, implementations that are small in size and consume low power are required. In addition, multistandard capability is desirable since it will reduce cost for integrators and will allow users to receive mobile TV in their devices as they travel.

The rest of this chapter attempts to address the tuner design for such mobile devices in CMOS. First, an overview of typical tuner architectures used for broadcast TV is given. Subsequently, the specific challenges related to mobile TV standards are presented, with an emphasis to DVB-T/H radio specifications and their implications. The first implementation discussed is a DVB-H dual band tuner on 0.18 μm CMOS process [5], which occupies 9.7 mm^2. The tuner achieves a 4 dB noise figure (NF) at both UHF and L-band, eliminating the need for an external LNA. By using a fractional-N synthesizer both 470–862 MHz and 1.4–1.8 GHz bands are supported, while achieving an integrated phase noise of less than –41 dBc. Sixth-order low-pass filters support channel bandwidths from 4 to 10 MHz. By using this as an example, techniques enabling the design of low-power, high-integration tuners are discussed in more detail. The second implementation example is a triband, multistandard tuner implemented on a 65 nm digital CMOS process [6], which is based on the same architecture. To achieve multistandard operation and coverage of most popular mobile TV bands, it also supports the VHF III band (174–240 MHz) and features additional programmability in the baseband filters to support channel bandwidths from 0.2 to 8 MHz. Compared to the first implementation it achieves better performance at half the power consumption, while it occupies less than 7 mm^2. The description in the second design case is focused on the additional features added to achieve multistandard operation and the specific challenges related to the 65 nm CMOS process. For both tuners, detailed measurement results are presented, including system-level using a digital demodulator.

5.2 Mobile TV Challenges and Architecture Trade-Offs

5.2.1 Broadcast TV Tuners and Mobile TV

The main issue that traditional analog and digital TV tuners have to overcome is the coverage of a wide bandwidth spanning several octaves, typically from 48 to 862 MHz. The implications of wideband coverage are the harmonic conversion and the image conversion issues, shown in Fig. 5.1. When multiplying the TV signal with a local oscillator signal (LO) to convert it to DC or an intermediate frequency (IF) for further processing, interference at integer multiples of the LO can also be converted to DC or the same IF by multiplication with harmonics of the LO, thus corrupting the wanted signal. Similarly, when converting the wanted signal to a typical IF lower than the wanted signal carrier, interference at the image of the signal can also convert to the same IF. In narrowband wireless systems such interference is usually out of band and can be eliminated by a simple preselection filter.

To alleviate such issues in TV tuners, several different techniques are used. In the first solution, shown in Fig. 5.2, a "tracking" tunable filter that tracks the LO ensures that both image and harmonics are filtered before down-conversion. These filters are typically bulky and require high external control voltages to tune the variable tracking filter across the 48–862 MHz band.

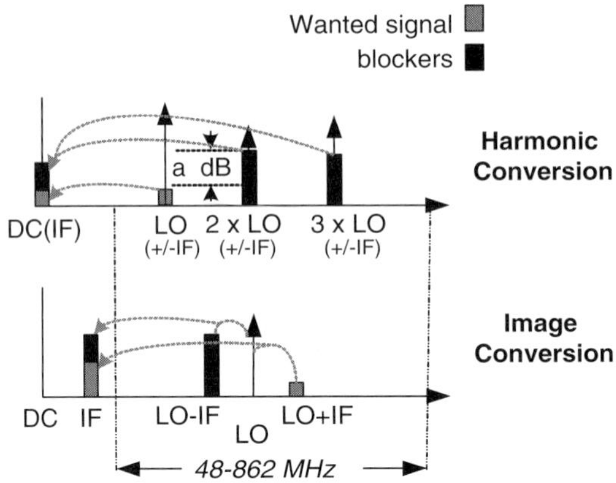

Fig. 5.1 Harmonic conversion and image conversion in wideband tuners

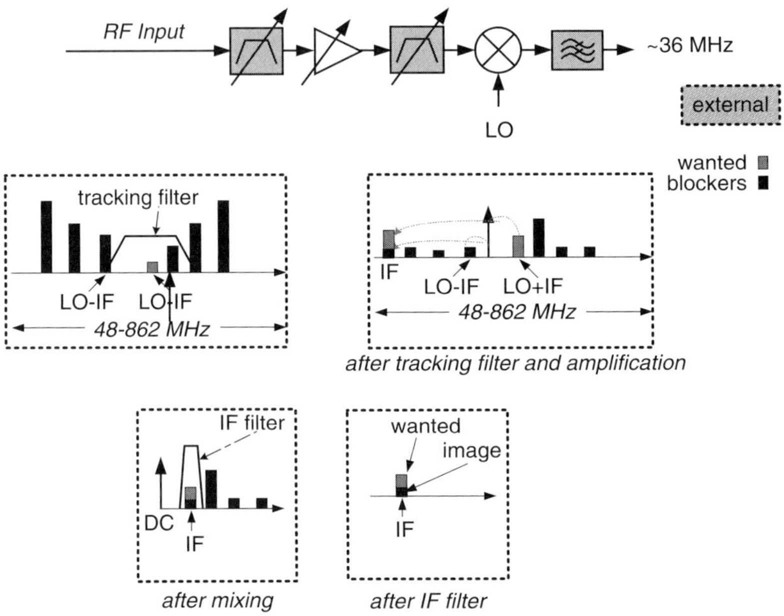

Fig. 5.2 Tracking filter architecture

Since such filters are hard to integrate on an IC, the up–down conversion technique shown in Fig. 5.3 is also employed. By converting to an IF which is higher than any signal in the TV band, it ensures that all harmonics of the LO and images fall out of band and can be eliminated by an off-chip SAW filter at a fixed high IF [7, 8]. Following that, conversion to a lower IF is employed and direct sampling is used to convert the signal to digital for further processing. The off-chip SAW filter can be eliminated by using on-chip image rejection [9].

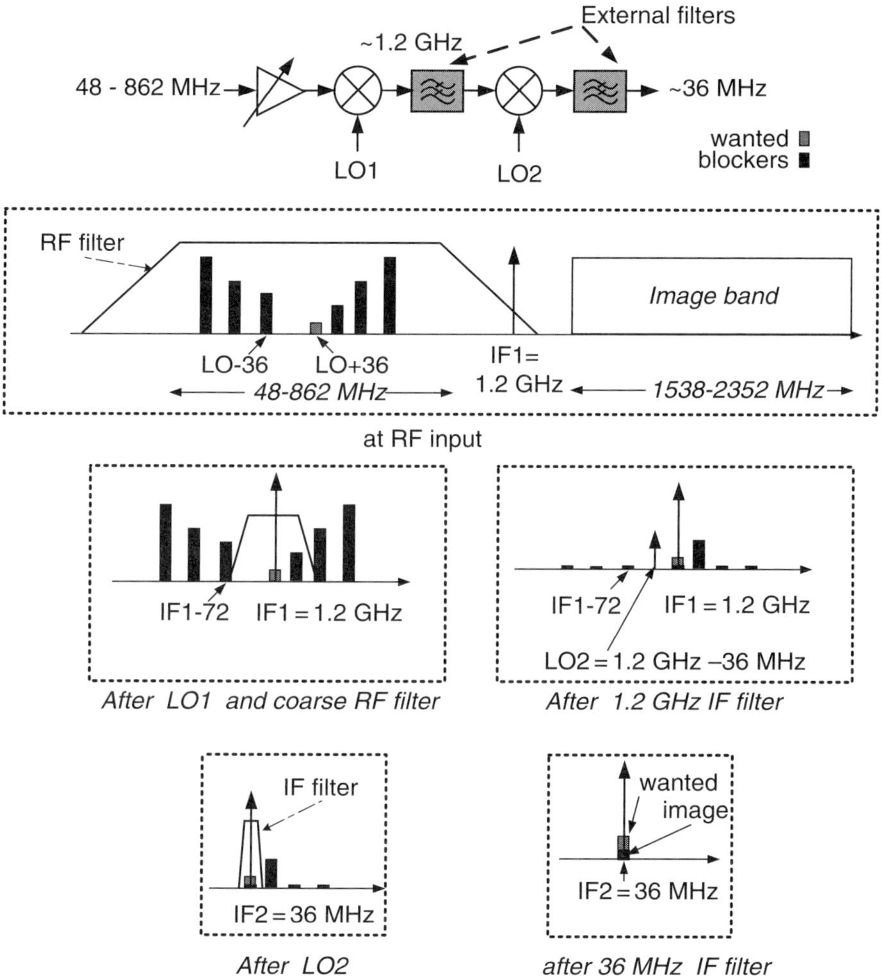

Fig. 5.3 "Up-down", dual IF conversion architecture

In [10] a system-in-package (SiP) uses external varicaps to implement an LC tracking filter integrated in a module. Additional image rejection is provided by an on-chip image reject scheme.

To avoid harmonic conversion, the cable TV tuner at [11] implements on-chip harmonic rejection mixers [12], which use several phases of the LO in order to cancel down-conversion of unwanted interference at harmonics of VHF frequencies.

The aforementioned solutions are unsuitable for mobile TV, since they either use bulky and expensive external components or consume high power (0.5–1 W).

To address the need for smaller form factor, lower cost and power consumption, direct conversion TV tuners covering the UHF band [13] or both the UHF and L-band [5, 14] have been implemented. In the case of UHF-only coverage, a 470–750 MHz preselection filter eliminates interference at harmonics of the LO. To cover the different RF bands, separate RF front-ends can be used with different preselection filters. Portable devices for UHF or L-band typically will use one of the two bands and, if both bands are used, separate antennas or a common antenna with an external, highly linear low-noise amplifier (LNA) followed by separate filters can be used. The solution of separate front-ends is also used in terrestrial TV tuners [15].

5.2.2 Mobile TV Standard Requirements

In mobile TV standards, OFDM modulation and stringent interference specifications pose significant challenges for the tuner design, which are aggravated when direct conversion is used.

The DVB-T/H radio specifications require a noise figure (NF) below 4 dB. In addition to the low NF, the high-order modulation used (up 64 QAM for DVB-T/H) requires an SNR in excess of 25 dB in mobile reception conditions [16]. In OFDM systems, when a high SNR is required other impairments such as I/Q mismatch, phase noise may affect the overall bit error-rate (BER) [17]. Such impairments can be viewed as additional "noise":

$$\mathrm{SNR}_{\mathrm{TOTAL}} = \frac{1}{\frac{1}{\mathrm{SNR}_{\mathrm{AWGN}}} + \left[\frac{1}{\mathrm{SNR}_{\mathrm{disto}}} + \frac{1}{\mathrm{SNR}_{\mathrm{interference}}} + \frac{1}{\mathrm{SNR}_{\mathrm{pnoise}}} + \cdots\right]} \tag{5.1}$$

To avoid degrading an SNR of 25 dB due to additive white Gaussian noise (AWGN), additional noise-like impairments should be at least 10 dB below the AWGN noise floor, which translates into an equivalent SNR of 35 dB. Using analytical calculations and simulations, this requirement can be translated to an *rms* in-band integrated phase noise error of less than 1° (–37 dBc) and I/Q mismatch of less than 1° and 0.1 dB.

Since the tuner is continuously tuned to a channel, I/Q mismatch can be corrected by "blind" adaptive calibration in the demodulator. However, designing a synthesizer that achieves low phase noise and at the same time has the resolution to tune at channel spacing required by DVB-T/H and T-DMB is a challenging task.

Both T-DMB and DVB-T/H standards define several reception conditions in the presence of other interfering TV channels, some of which have significant impact on the tuner specifications. Namely:

(i) S2 pattern in DVB-T/H (Fig. 5.4): digital TV interference 2 channels away, up to 40 dB stronger than the 64-QAM wanted signal. In addition to sharp channel-select filters, in direct conversion or low-IF architectures, this implies high second-order linearity. Calculation of the in-band interference from a modulated signal due to second-order distortion requires the computation of the convolution of the OFDM blocker with itself in the frequency domain. Alternatively, the result could

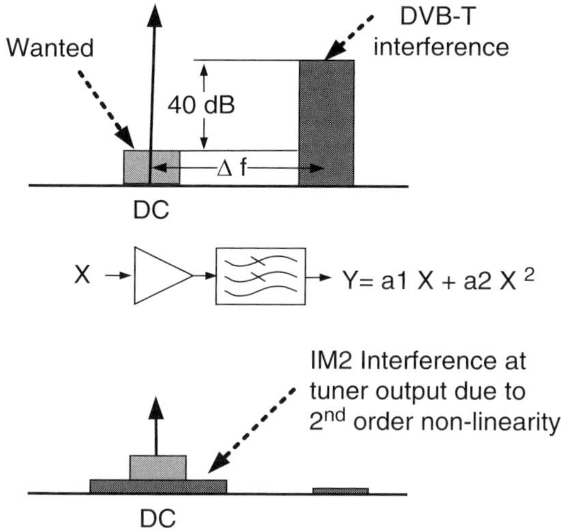

Fig. 5.4 S2 specification for DVB-H and second-order intermodulation

be approximated by using a two-tone signal as the interferer and doing a simple IM2 analysis:

$$\mathrm{IM2}_{\mathrm{input}}(\mathrm{dBm}) = 2 \cdot (P_{\mathrm{in,blocker}} - 3\,\mathrm{dB}) - \mathrm{IIP2}_{\mathrm{input}} \tag{5.2}$$

where $P_{\mathrm{in,blocker}}$ is the total power of the OFDM modulated blocker which is approximated by two tones at half the total power each. Since the standard defines the blocker at –28 dBm and the wanted 64-QAM signal at –68 dBm, assuming 20 dB SNR, which is typically required to achieve the wanted BER of 2×10^4, and allowing for additional 3 dB degradation due to other sources according to Eq. 5.1, using Eq. 5.2 it can be easily found that the required IIP2 at the input is higher than +29 dBm.

Taking into account other degradation sources, such as ADC quantization noise and phase noise, analysis based on Eq. 5.1 shows that the NF should be less 10 dB. It is important to set an NF specification at different input conditions, since the automatic gain control (AGC) loops of the system may set the tuner at different gain settings compared to sensitivity, where typically gain is maximum and NF is minimum.

To achieve this IIP2 performance, careful gain planning in the receive chain and accurately matched differential layout of mixers and filters is needed. In T-DMB, the 40 dB far-away blocking specification for digital interference results into a lower IIP2 requirement.

(ii) S1 pattern in DVB-T/H (Fig. 5.5): analog TV (PAL/SECAM) adjacent channel interference, up to 35 dB stronger than the wanted 64-QAM signal. As the carrier of the analog TV picture carrier (where the signal power is concentrated) is only 5.25 MHz away from the center of the wanted channel, this specification poses stringent analog filtering requirements. For an 8 MHz channel spacing, to avoid saturation of the ADC, at least 25 dB attenuation is needed at 5.25 MHz. In addition,

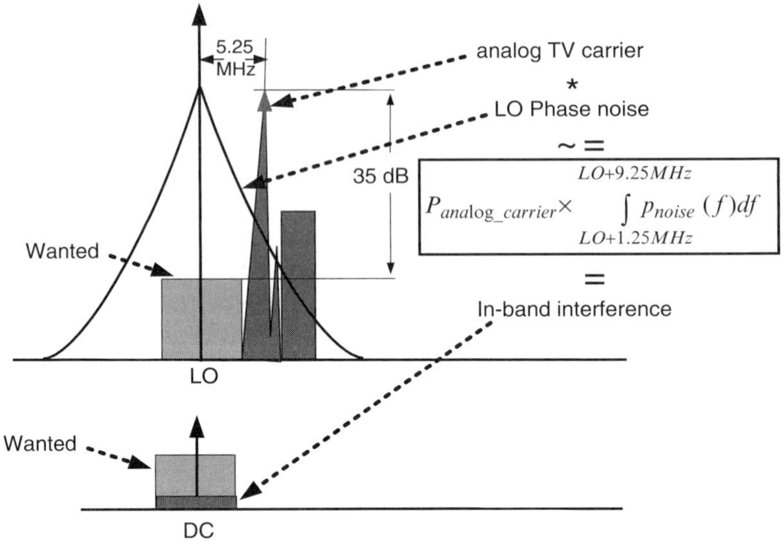

Fig. 5.5 S1 N + 1 specification for DVB-H and phase-noise reciprocal mixing

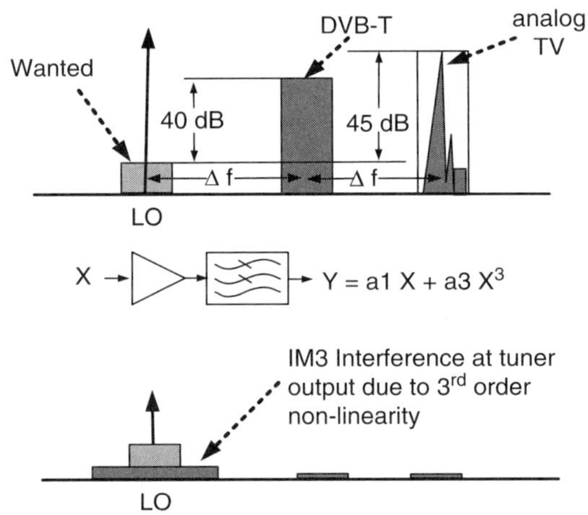

Fig. 5.6 L1 specification for DVB-H and third-order order intermodulation

reciprocal mixing of the LO phase noise with the picture carrier (represented as a convolution in the frequency domain with the approximation that all analog TV power is at the picture carrier – Fig. 5.5) will also result in in-band interference. To meet this specification with at least 3 dB margin, the integrated phase noise from 1.25 to 9.25 MHz offset from the LO should be less than –60 dBc.

(iii) L1 pattern in DVB-T/H (Fig. 5.6): interference from digital and analog TV, two and four channels away from the wanted channel. Since the interferers are defined to be 40 and 45 dB stronger than the wanted 16-QAM signal, taking into account other sources of SNR degradation, such as NF, and allowing for 3 dB margin as in the case of S2 specification, this translates into an IIP3 requirement of –5 dBm and an NF of less than 8 dB.

Additional challenges include the presence of GSM interference from a cellphone up-link in DVB-H, which requires high LNA linearity and RF filtering. In this worst-case scenario, the GSM up-link is transmitting a signal up to +33 dBm at 880 MHz, while the TV signal is received at 746 MHz, which is the highest TV channel at UHF for DVB-H receivers embedded in cellular phones. Assuming 15 dB loss between the transmit GSM and the receive UHF antennae, minimum sensitivity degradation is required while a +18 dBm blocker is present. This can only be met using a combination of sharp external RF filters, on-chip filtering, and high LNA linearity.

Given the requirements for low cost and low power, the above specifications call for careful design and innovative techniques. In addition, real-world environment conditions are often worse than what standards specify, thus it is prudent to aim for a significant implementation margin.

5.3 DVB-H Dual-Band Tuner

5.3.1 Design Considerations

The first design example is a dual-band tuner for DVB-H. Direct conversion was selected as the most efficient architecture to address the need for low cost and low power. Since the final target is implementation on SoC, using CMOS was unavoidable. To meet the stringent requirements imposed by the standard and enable a future, multistandard implementation that takes advantage of the processing power of a companion baseband demodulator, several techniques were used, both at circuit and architecture level.

First, was using an architecture that can be expanded to cover more bands and standards. Direct conversion allows addition of different front-ends, since it is not limited by the choice of a specific IF. In addition, a fractional-N synthesizer was used, which can operate from a wide range of external reference frequencies and produce almost arbitrarily accurate frequency synthesis to 50 Hz, within the range of 1.2–1.8 GHz. With appropriate divisions most TV bands and FM radio can be supported. Baseband filters are programmable to different channel bandwidths, 4–10 MHz. Since different protocols and modulation schemes may allow for different handling of DC offset, different modes of DC offset correction and a programmable DC offset notch were implemented.

Second, digital calibration was used whenever possible. Mostly digital oriented, CMOS processes can exhibit large variations in critical analog parameters across lots. However, they offer the possibility to integrate digital calibration mechanisms to overcome several IC process and architecture related imperfections, such as CMOS parameter variability, I/Q mismatch, and baseband filter tuning.

Third, most analog circuits were designed with programmable bias. Along with digital calibration, this may help overcome the occasionally poor process modeling in CMOS and avoid re-spins of the chip. In addition to that, a "smart" baseband demodulator can dynamically adjust power consumption based on detected signal conditions.

Fourth, as the tuner is eventually intended to operate in a noisy SoC environment, fully differential circuit topologies were used. Single-ended LNA topologies [14] may be attractive since they eliminate the need for external balanced-to-unbalanced transformers (baluns), but in fully integrated solutions digital noise, harmonics of the clocks of the baseband demodulator as well as crystal harmonics may couple to the LNA input degrade the overall SNR.

Fifth, on-chip voltage regulators were extensively used. This allows operation from a wide range of external voltages and at the same time protects sensitive RF circuits from noise coupled to the power supplies.

Finally, critical RF cells were separated as far from each other as possible and different supply and ground pins were used for sensitive circuits such as the synthesizer and the LNA.

5.3.2 Architecture

Figure 5.7 shows the block diagram of the dual-band tuner, in the context of a full system implementation. The chip is implemented on a 0.18 μm CMOS process and supports the 470–862 MHz UHF IV, V bands as well as the 1400–1800 MHz band using two separate front-ends.

The RF front-ends consist of variable gain LNAs and passive mixers. To achieve high linearity, which is necessary in order to meet the challenging blocking and intermodulation specifications for DVB-H/T in the UHF band, it was determined that it is necessary to reduce the RF gain in the presence of strong interference. To detect wideband interference within the UHF range, an on-chip logarithmic amplifier connected at the output of the UHF LNA provides a receive signal strength indicator (RSSI). The RSSI is digitized by the companion baseband demodulator, which reduces the RF gain when the total signal power exceeds a certain threshold, thus closing the RF AGC loop in the digital domain.

To achieve low in-band phase noise independent of the reference crystal and channel spacing, a ΣΔ fractional-N synthesizer is used for LO generation. Fractional synthesis (multiplication of a reference by a noninteger) also allows sharing virtually any crystal that the mobile phone RF transceiver may use, while maintaining fine tuning capability. Since the multiplication factor is noninteger, a large reference frequency can be used to produce the required LO, which has two effects: the multiplication ratio N is low and the loop bandwidth, which is usually limited for stability to less than one tenth of the reference frequency, can be made high. A large loop bandwidth helps reduce the integrated voltage-controlled oscillator (VCO) phase noise

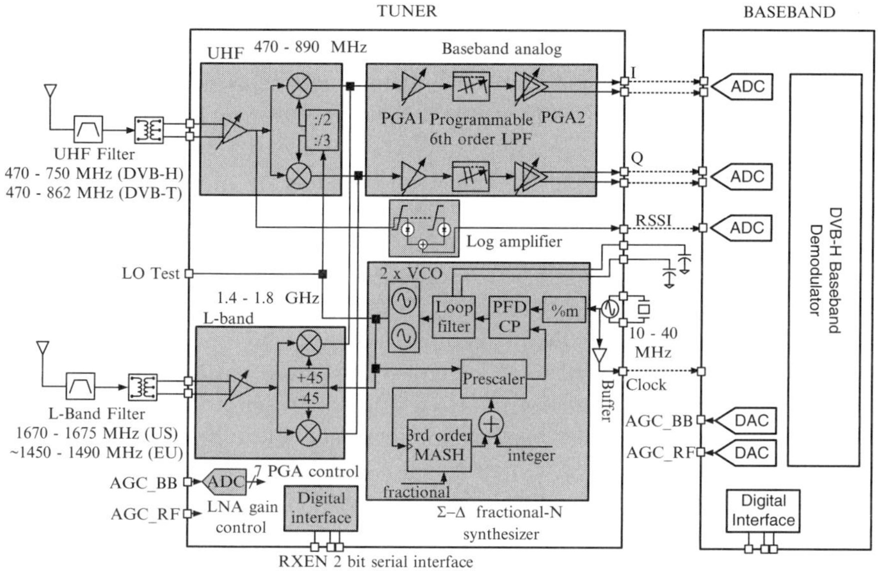

Fig. 5.7 Dual-band, direct conversion tuner block diagram

and a low N reduces the contribution of reference crystal, dividers and charge pump, thus improving integrated phase noise error as required by OFDM modulation.

UHF frequency range spans almost one octave (470–862 MHz), which could make a VCO implementation quite challenging. Large operating range requires very high VCO gain and/or a large number of switching components, which trades off with good phase noise performance. To reduce the required operating range, the range for the synthesizer was chosen to be 1.2–1.8 GHz. Quadrature LO for the UHF mixers is generated by dividing the VCO frequency by three [18] (470–600 MHz) or two (600–862 MHz). For the L-band the VCO frequency is used undivided and a first-order polyphase filter is used to generate quadrature LO. A polyphase filter uses passive RC components to produce a 90° phase shift at a specific frequency between its two outputs. Compared to other approaches that use a lower frequency and generate quadrature outputs after mixing the VCO with a divided version of itself, polyphase filters avoid active circuitry that may increase power consumption, degrade phase noise and generate spurious products. The main disadvantage of this solution, which is the phase and gain inaccuracy over a wide frequency range, can be eliminated by using digital calibration in the baseband modem.

Both RF front-ends converge into a common baseband path consisting of digitally programmable gain amplifiers (PGA) before and after the channel selection I/Q filters. To accommodate analog gain control, as required by typical baseband demodulators, the PGAs can also be controlled by an on-chip analog-to-digital converter (ADC). Gain control step is accurate to 0.5 dB, to minimize the SNR degradation from gain changes during data reception under fading conditions.

To reduce linearity requirements for the filters, a pole with programmable cutoff is used at the output of the first PGA to attenuate interference. DC-servo loops are implemented around gain stages, since in direct conversion a large amount of gain is implemented in baseband, thus small static or dynamic DC offsets may reduce the available dynamic range or even saturate the receiver output. Such DC offsets may occur even at the mixer outputs as a result of LO self-mixing. Based on the S1 and S2 adjacent channel specifications, a sixth-order low-pass, Chebyschev response was selected for the filter implementation. The filters have programmable response to meet different channel bandwidth requirements.

The tuner is controlled by an I2C interface. A receive enable pin is available to switch the tuner on and off for the DVB-H time-slicing mode. For a complete tuner implementation in either band, a minimum amount of external components is required: a balun, RF filter, a crystal, and a few capacitors. The crystal can be eliminated in embedded applications, since the tuner can operate from a wide range of crystal frequencies.

5.3.3 RF Front-End

The topology for the variable-gain LNA block used in both bands is shown in Fig. 5.8a. Instead of using techniques that achieve high bandwidth at the expense of NF, as the common-base LNA in [13], a degenerated differential cascode topology was chosen. Wideband input matching is achieved by using external series inductors and an on-chip switched capacitor bank Cb, which is adjusted according to the desired tuning frequency.

Continuous analog gain control is achieved by using the current steering technique, which reduces the LNA gain by steering current from the output load to a replica branch (M3C, M3D in Fig. 5.8a). It can be easily shown that the gain reduction is proportional to the current in the replica branch, which in turn is controlled by another replica scheme M3Rn, M3Rp, which regulates the current of devices M3C, M3D. A feedback loop generates voltage Vctrl so that the current through devices M3Rn, M3Rp is proportional to transconductance Gm, which is variable to enable programmability of the gain slope with respect to the external control voltage VAGC. Depending on programming, the LNA can achieve a gain control range of 20–30 dB. To achieve high linearity and low flicker noise, both RF paths use double-balanced current mode passive mixers (Fig. 5.8b), which is a technique commonly used in applications demanding high linearity.

To implement the wideband RSSI function, a logarithmic amplifier is used, tapping the signal at the UHF LNA output. The pseudo-logarithmic, piecewise linear function is implemented as a cascode of limiters with their outputs rectified and summed together.

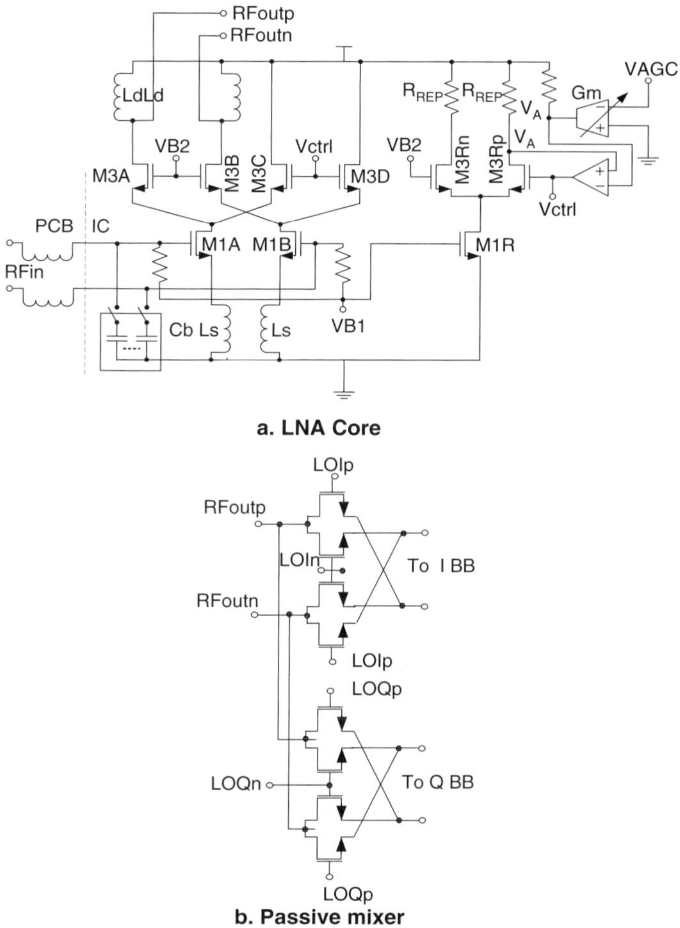

Fig. 5.8 Variable-gain RF Front with passive mixers

5.3.4 Analog Baseband

Figure 5.9 shows a detailed block diagram of the analog baseband section. Following the current-mode I/Q mixers, low-noise/high gain-bandwidth, current-to-voltage amplifiers are used (PGA1). Gain programmability is achieved by using switched feedback resistors. The second programmable gain block (PGA2) consists of three stages of resistive feedback amplifiers.

All gain stages use DC-servo loops. The output is integrated on a sampling capacitor using an active RC filter with a programmable pole. By using a transconductor that converts the integrated output voltage to a current, a correction is injected at the input of the gain stage that compensates DC offset. To enable fast initial acquisition of DC offset, a higher frequency pole can be programmed during a training phase, while during normal reception the pole can be programmed to a lower frequency to

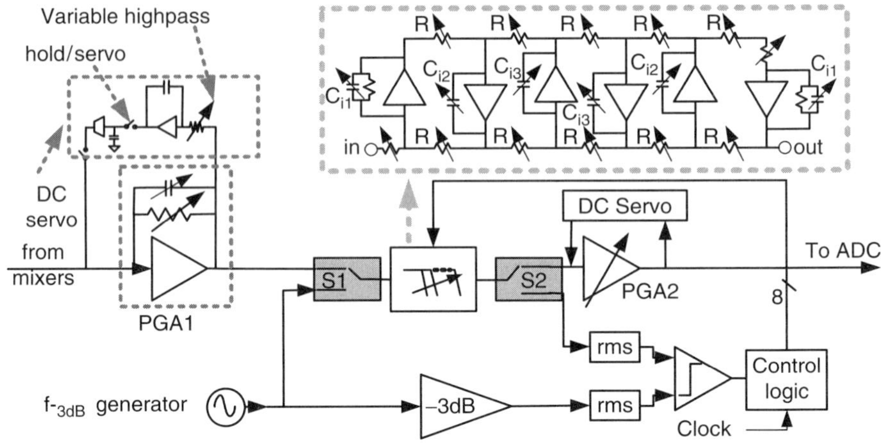

Fig. 5.9 Analog baseband simplified diagram (Q-path)

avoid corrupting the data. In DVB-H, the initial time before the "on" period in time-slicing can be used. A hold/servo switch enables the DC-servo loop to also operate in an open loop mode after initial acquisition of the DC offset, if required by the demodulator.

The sixth-order, Chebyschev filters are implemented using opamp-RC integrators in a leapfrog configuration. An on-chip auto-calibration loop shown in Fig. 5.9 is activated upon power-up and sets the Fo to a value from 2 to 5 MHz, supporting the different channel bandwidths of 5–8 MHz required by DVB-H. During auto-calibration, a tone at the appropriate cutoff frequency generated on-chip is applied at the Q-filter input and is compared with the filter output using an rms detector. A digitally controlled loop adjusts the filter bandwidth by varying binary weighted capacitors *Ci*, until the input attenuated by 3 dB equals the output. By accurately tuning the baseband filters we can ensure that S1 blocking specification is met under process variations. Filter matching, which can impact the tuner noise floor, is ensured by careful layout.

5.3.5 Frequency Synthesizer

A block diagram of the fractional-N phase lock loop (PLL) used to achieve low phase noise and fine frequency resolution is shown in Fig. 5.10. To minimize fractional spurs, the fractional part of the division ratio is generated by a 19-bit, third-order, MASH $\Sigma\Delta$ modulator. The fine granularity of the modulator output results in spurs below –75 dBc and a frequency resolution of 50 Hz at the VCO output. A fourth order filter is used in the loop to ensure proper attenuation of the quantization noise produced by the modulator.

The synthesizer employs a multimodulus prescaler, implemented as an array of cascoded 2/3 dividers. Several dividers are daisy-chained to achieve the required division ratio and range.

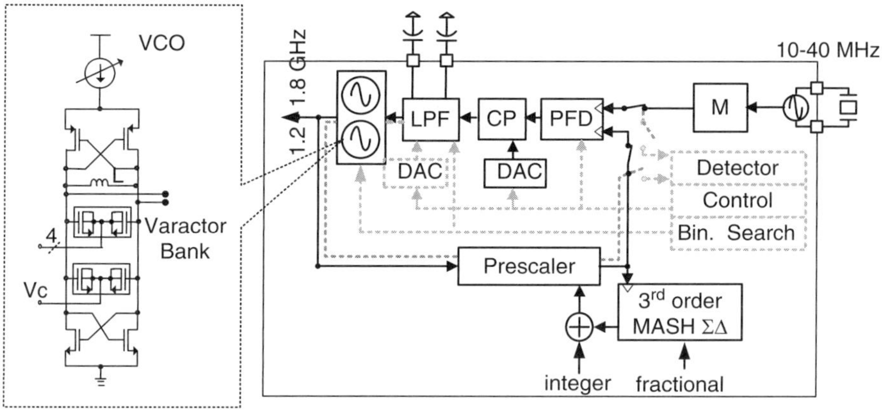

Fig. 5.10 ΣΔ Fractional-N PLL

In fractional-N PLLs, the prescaler divides with a different ratio N in each comparison cycle where the mean N value equals the desired exact fractional division of the VCO frequency. The N range is defined by the fractional part generation scheme employed. The modulator used produces a 3-bit stream that results in an N range of $N-3$ to $N+4$ around the integer value. Since the division range of any m-stage prescaler is 2^m to $2^{m+1}-1$, operating in an N range around, i.e., $N=64$ requires that for any $N<64$ the prescaler consists of five stages while for any $N \geq 64$ it consists of six stages. For this reason, a technique is used to switch in and out the appropriate divider stages. A timing scheme is employed to track the division number and engage/disengage the appropriate divider stages, so that a truly continuous-range prescaler is implemented, having no "dead" division range. This type of prescaler can fit in virtually any wireless system with any crystal frequency and alleviates the need to design an application-specific prescaler.

Two VCOs cover the range from 1.2 to 1.8 GHz using complementary cross-coupled pairs and MOS varactors in accumulation mode for tuning. Covering a large frequency range with a single varactor requires a high voltage-to-frequency gain which may degrade phase noise performance since any noise coupled at the sensitive VCO control input translates to phase noise at the VCO output. By using a switched varactor bank, the operating frequency range is split into discrete regions, each having a low VCO gain.

To automatically select the appropriate combination of varactors that bring the VCO in the desired region, a coarse tuning mechanism (shown by dotted lines in Fig. 5.10) is activated after a switch-channel command has been issued. Upon coarse tuning activation, the loop filter node is connected to a digital-to-analog converter (DAC), charging it to a desired DC level at mid-range and the charge pump output is in high-impedance mode. By using binary search, the tuning algorithm selects the varactor combination for which the frequency error between the PLL reference signal and the VCO divided signal is minimized. Once the appropriate varactor combination is selected, normal loop operation is resumed.

5.3.6 Measurement Results

The chip shown in Fig. 5.11 occupies 9.7 mm^2 and is encapsulated in a 6×6 mm MLF 40-pin package. An ultra-thin, chip-scale package of 3.5 mm $\times$ 3 mm is also available for SiP. The analog/RF performance is summarized in Table 5.2.

The tuner consumes 100–110 mA from a 2.7 V supply in continuous mode, which translates to less than 30 mW in a 1:10 time-slicing mode for DVB-H. NF is less than 4 dB at both L-band and the 470–750 MHz UHF range (Fig. 5.12a), used in GSM convergence terminals. Minimum and maximum gain is 3–83 dB and 6–86 dB for the L and UHF bands, respectively.

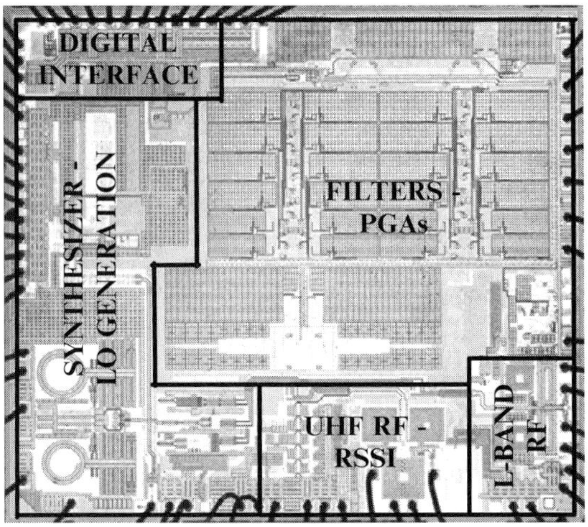

Fig. 5.11 Die micrograph of dual-band DVB-H tuner at 0.18 μm CMOS

Table 5.2 DVB-H tuner performance summary[a]

Parameter	UHF	L-Band
Frequency range (MHz)	470–862	1400–1800
Supported RF channel BW (MHz)	4–10	
Gain max/min/step (dB)	86/6/0.5	83/3/0.5
NF @ max. gain (dB) (balun included)	4	4
IIP3 ($N+2$, $N+4$) @ 0/6/10 dB attn. (dBm)	$-9/-3/-0.5$	$-5.5/-2.5/-2$
IIP2 ($N+2$) @ 0/10/15 dB attn. (dBm)	+21/30/34	+20/29/35
RX path attn. at 5.25/8/12 MHz (3.9 MHz BW) (dB)	27/50/75	
Integrated phase noise (100 Hz–4 MHz)	0.3° rms	0.5° rms
Phase noise @ 1 MHz offset (dBc/Hz)	−133	−127
Power cont. mode–2.7 V supply (mW)	295	280

[a]Measurements at the PCB connector including all balun and PCB losses.

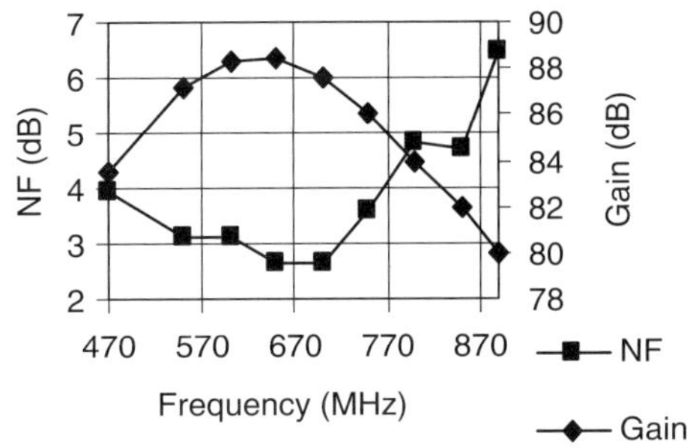

Fig. 5.12a NF and gain vs. frequency at UHF

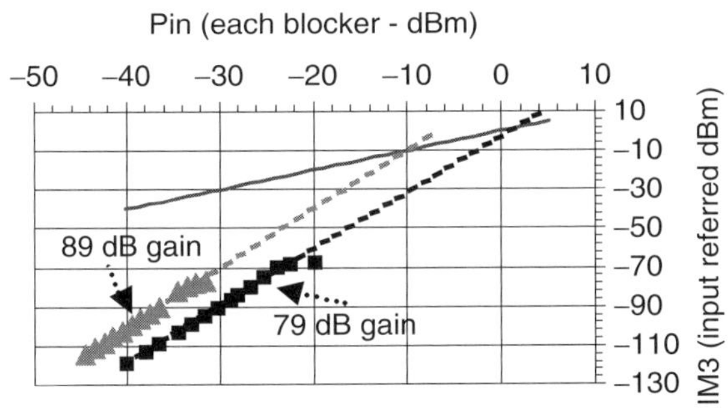

Fig. 5.12b IIP3 plots for maximum and maximum – 10 dB RF gain

Linearity was measured by applying tones emulating the worst-case interference defined by the standard. To measure IIP2 two tones were applied at 16 and 17 MHz offset from the carrier, emulating the modulated interference at two channels away (16 MHz). For IIP2, the tones applied were at 16 and 33 MHz away, emulating the 2 blocker scenario and two and four channels away from the carrier. An IIP2 of +30 dBm and IIP3 of –0.5 dBm are achieved at 10 dB below maximum gain, where the respective NF is 7 dB. Figure 5.12b shows an IIP3 plot at UHF channel 45 at maximum RF gain and at 10 dB backoff, which is the typical AGC loop target in the blocking conditions specified by MBRAI [16]. Since as discussed earlier, blocking and intermodulation specification compliance strongly depend on both linearity and NF, in Fig. 5.13. NF and gain are shown as a function of the RF AGC control voltage. Similarly, Fig. 5.14 shows the IIP2 and IIP3 at the same channel as a function of the RF AGC voltage. As expected, with the reduction of gain input referred IIP2 increases, since it only depends on the baseband IIP2 performance (LNA second-order distortion results in out-of-band products). In the case of IIP3, initially it decreases as the baseband third-order nonlinearity dominates but further reduction of the LNA gain causes nonlinear behavior. This can be explained by the current-steering variable gain LNA topology: reduction of the current in the main branch gradually reduces linearity and at some point the LNA IIP3 dominates the overall performance.

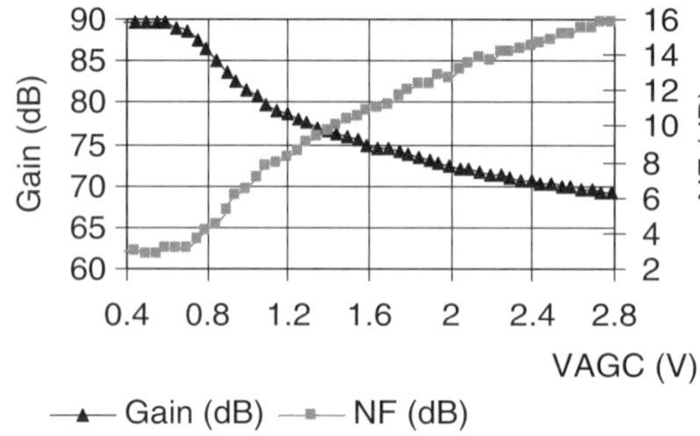

Fig. 5.13 NF and gain vs. RF AGC voltage at 666 MHz

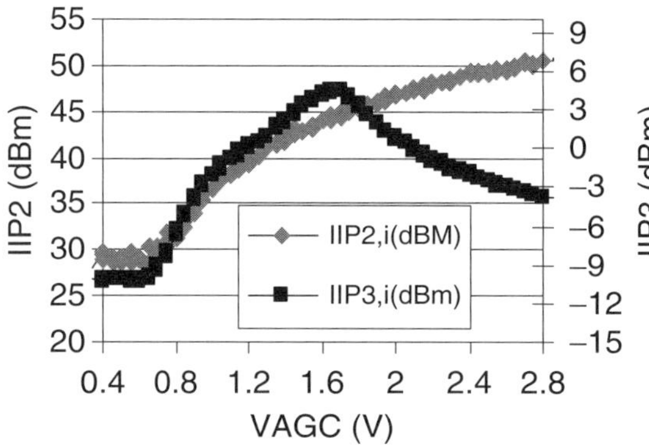

Fig. 5.14 IIP2 and IIP3 vs. RF AGC voltage at 666 MHz

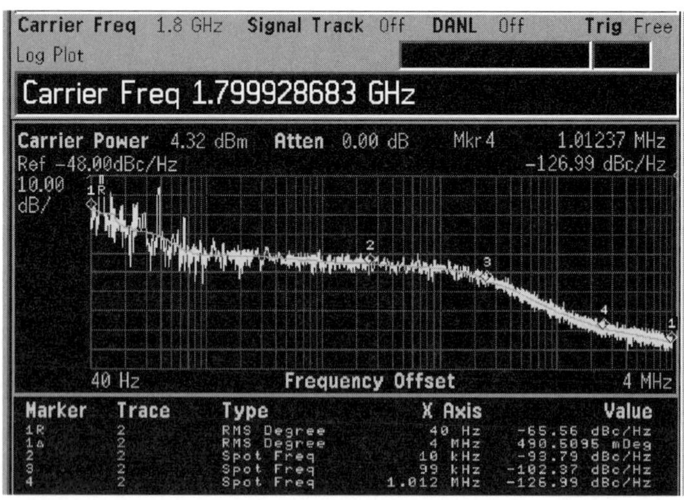

Fig. 5.15 Phase noise plot at 1.8 GHz (upper limit of PLL range)

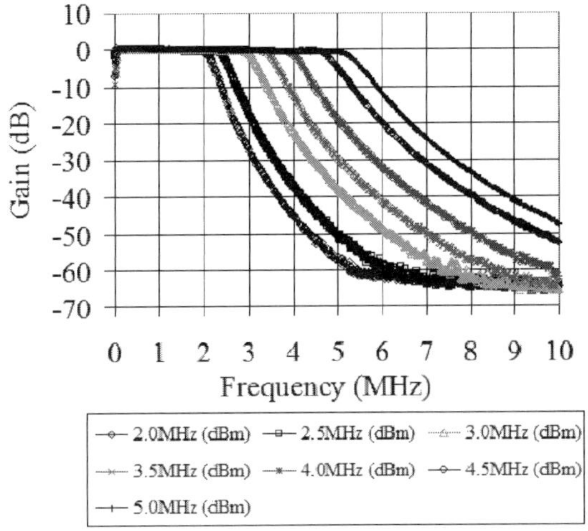

Fig. 5.16 Tunable BB Filter Response. The cutoff frequencies from the bottom curve to the upper curve are 2.0, 2.5, 3.0, 3.5, 4.0, 4.5 and 5.0 MHz, respectively

Figure 5.15 shows a measured phase noise plot at 1.8 GHz, (upper limit of operation of the PLL), where the VCO phase noise is highest. Integrated phase noise error within the signal bandwidth (100 Hz – 4 MHz), which is usually a significant limitation in OFDM radios, is less than 0.5°(−41 dBc) before division by 2 or 3 for UHF, thus having a negligible contribution to the overall system SNR. The phase noise at 1 MHz offset from the carrier is −127 dBc/Hz.

The tunable response of the baseband analog filters from 2 to 5 MHz cutoff is shown in Fig. 5.16. The filters were initially tuned using the autocalibration mechanism described earlier. The typical resolution of the programming step for the cutoff response was 200 kHz.

I/Q gain and phase mismatch for both the L-band and UHF was measured less than 0.1 dB and 3°. An SNR of higher than 30 dB was measured for an input signal of −70 to −25 dBm, limited by the digital demodulator accuracy.

The tuner has been validated with several DVB-T/H demodulators, showing compliance to the MBRAI blocking specifications [16]. Table 5.3 summarizes typical system measurements using a demodulator that achieves the required 2×10^{-4} BER DVB-T threshold for QPSK-1/2 mode with a minimum 4 dB SNR. Sensitivity achieved using the above criterion was –97.5 and −82.5 dBm for the QPSK (code rate 1/2) and 64-QAM(code rate 3/4) modes, respectively (vs. −96.6 and −81.3 dBm required by the standard), which is in-line with the measured NF for UHF.[1] Digital blocking specification S2 for 64-QAM-3/4 is exceeded by 3 dB, while L1 specification for 16-QAM-2/3 is also exceeded by 3 dB.

[1] The noise floor for a 7.61 MHz TV channel is −105.2 dBm at 27°C which corresponds to −97.2 dBm sensitivity, when SNR is 4 dB for QPSK1/2 and NF is 4 dB.

Table 5.3 System performance summary[a]

Parameter	Mode	UHF	L-Band
Sensitivity 2×10^{-4} BER (dBm)	QPSK-1/2	−97.5	−96
	16-QAM-1/2	−92	−91.5
	64-QAM-2/3	−83.5	−82.5
	64-QAM-3/4	−82.5	−81.5
Max. signal 2×10^{-4} BER (dBm)	64-QAM-3/4	−10	−15
	QPSK-1/2	−2	−5
MBRAI L1 immunity meas./spec (dB)	16-QAM-2/3	48/45	N/A
MBRAI L3 immunity meas./spec (dB)	16-QAM-2/3	44/40	N/A
MBRAI S1, N + 1 immunity meas./spec (dB)	64-QAM-3/4	41/35	N/A
MBRAI S2, N + 2 immunity meas./spec (dB)	64-QAM-3/4	43/40	N/A

[a] Measurements at the PCB connector including all balun and PCB losses.

Fig. 5.17 Snapshot from live DVB-T demo

The tuner was also tested in real-world conditions: Figure 5.17 shows a snapshot of high-quality live DVB-T video using a setup including the tuner and DVB-T demodulator in FPGA.

5.4 Migration to 65 nm Multistandard Tuner

Due to the complexity of OFDM demodulators and the memory required in mobile TV systems[2] multistandard SoCs have high digital gate counts. Therefore, to reduce chip area and cost, implementation in nanometer CMOS technologies is desirable. In this design example, as a first step towards a single-chip system, a multistandard

[2] Operation in bursts needs memory to store information received during a burst.

TV tuner was implemented on a 65 nm digital CMOS process [6]. The tuner is intended to support the most popular broadcast and mobile TV standards, such as DVB-T, DVB-H, T-DMB and ISDB-T fixed and mobile.[3]

5.4.1 Architecture

The tuner block diagram is shown in Fig. 5.18. The architecture was based on the direct-conversion DVB-H tuner discussed earlier [5], with the following main additional features and differences: (a) A VHF III band RF front-end was added to support T-DMB and potential deployment of DVB-T/H in this band. (b) The frequency plan remained the same, using a ΣΔ fractional-N synthesizer to support a range of 1.2–1.8 GHz with a single VCO. VHF LO generation was achieved by additional dividers: frequencies above 220 MHz use divide-by-6, while frequencies from 170

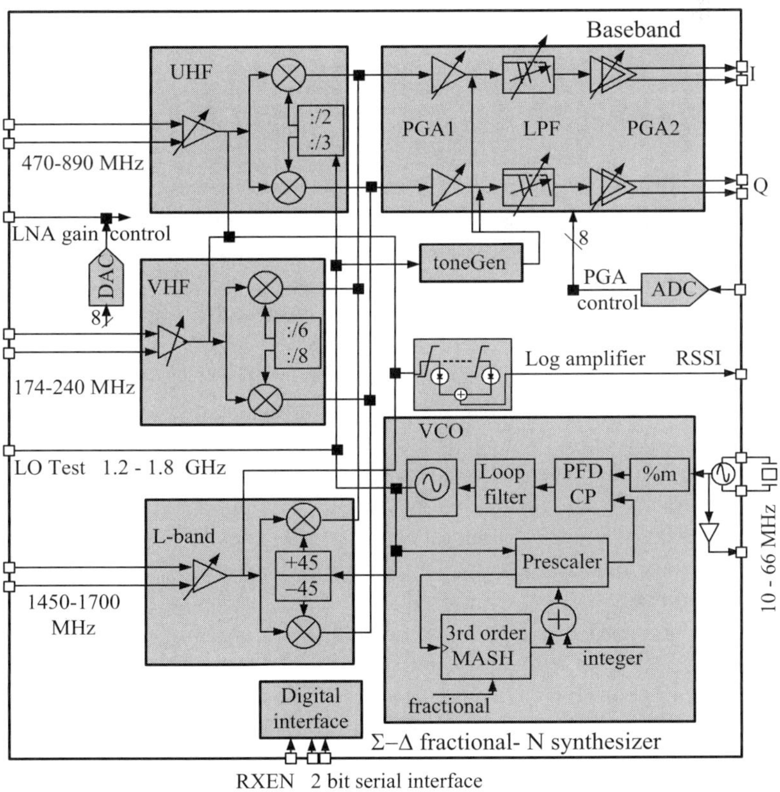

Fig. 5.18 Multistandard TV tuner block diagram

[3] Other standards mentioned in the introduction may also be supported, if their requirements are a subset of DVB-T/H, T-DMB, and ISDB-T.

to 220 MHz use divide-by-8. (c) A novel LNA topology was used supporting both single-ended and differential input operation depending on programming and external components. (d) The LO signal was used to fine-tune the LNA loads during an initial calibration cycle. (e) The baseband analog filters were designed with extended range, from 0.2 to 5 MHz cutoff. This was achieved by using the same filter topology with additional programmability in the passive components. In addition, the filter tuning step was reduced to 40 kHz in DVB-H mode (10 and 2 kHz in T-DMB and ISDB-T modes, respectively) (f) The $\Sigma\Delta$ fractional-N synthesizer was used to produce the reference signal for filter calibration, which allowed tuning to any cutoff frequency (limited by the filter programming step) within the supported range. Furthermore, each filter can be calibrated independently, thus achieving a guaranteed 0.5% matching in cutoff frequency between I/Q filters. (g) By using an on-chip DAC and an ADC, both analog and digital gain control are possible for RF and baseband. (h) The RSSI range was extended to support all three bands.

For the case of T-DMB and ISDB-T 1 or 3-segment, the direct conversion tuner can also support low-IF architectures, which are typical for these standards, by implementing a Weaver image-reject scheme in the digital domain.

5.4.2 Circuit Design

Typically in a system-on-a-chip (SoC) environment, it is preferable to use fully differential RF input stages to minimize digital noise coupling. However, some performance degradation from single-ended operation could be acceptable in applications sensitive to cost/area, where eliminating external baluns is important. For this reason, the UHF and L-band LNAs can support both differential and single-ended inputs by adjusting the external matching and appropriately programming internal switches. The new LNA topology is shown in Fig. 5.19 and is based on common-source, cascode topology with degeneration, using an on-chip switched capacitor bank at the input for wideband matching, as in [5].

In the single-ended mode, device M1B is switched off by S1 and RF switch S2 is turned on to adjust the primary coil inductance of the output transformer. The VHF LNA uses external output load inductors and can also operate as single-ended input. Switched capacitor banks (CL) are also used at the outputs of the LNAs to maintain optimal gain across the desired frequency ranges. Gain control is achieved by using the current-steering technique introduced in the previous generation DVB-H tuner. To control RF gain, either an external voltage or a digital word controlling an on-chip DAC can be used.

During a calibration cycle at the power-up of the device, the RSSI is used to fine-tune the LNA output resonance at different carrier frequencies. By using an RF switch, the LO signal is injected at the LNA output and an algorithm, which runs at the companion digital demodulator, finds the optimal combination of the switched capacitor bank that maximizes the RSSI indication. This is especially useful for the VHF LNA where narrow tuning is used to filter out potential interference at the third harmonic of the LO that can be down-converted by the third harmonic of the mixer. With this tuning the RF front-end is stabilized vs. process tolerances, power supply variations, and temperature.

5.4.3 Process Considerations at 65 nm

Continuously shrinking CMOS geometries have a direct benefit in silicon area and allow more complex systems to be integrated on a single die. However, this comes at the cost of increased design cycle and a multitude of issues involving complex device models, interconnect impedance estimation/extraction, and manufacturability concerns.

Analog and RF functions may suffer as standard circuit topologies are not directly transferable to nanometer scale digitally oriented processes. The combination of reduced power supply, high channel conductance, and low gate oxide breakdown voltage, along with the relatively high device threshold, leads to small output dynamic range in stacked-device topologies and difficulties in meeting linearity requirements in high power/gain RF regime. To alleviate those concerns accurate device modeling, detailed RF tuning and high-Q, inductors and capacitors were used. In the analog baseband path, the 2.5 nominal I/O supply was used along with a mix of both thin and thick oxide devices. A well-observed principle was to employ self-calibration techniques that fine-tune the receiver blocks upon start-up, like, i.e., channel select filters, thus eliminating any process parameter spread.

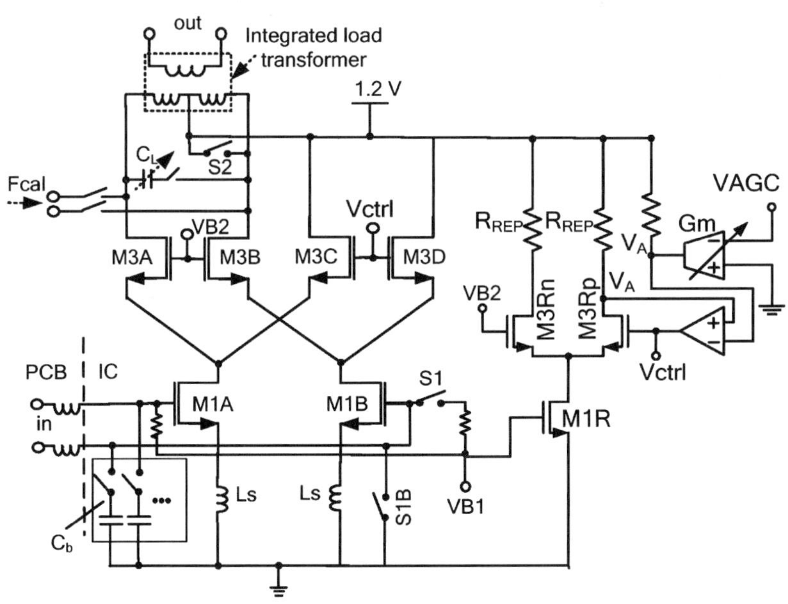

Fig. 5.19 Dual-mode single-ended/differential LNA schematic

To avoid performance degradation due to gate leakage, which is also significant in nanometer processes, standard design practices were used: currents and not voltages were used for bias, bias currents were kept above a certain threshold and used low mirror ratios, MOS capacitor usage was avoided and fringe metal capacitors were extensively used. Induced lattice stress and nonuniformity issues (e.g. relative size of active area, LOD, Well Proximity Effect, WPE) that affect matched-device performance requirements have been treated by proper device sizing, dummy insertion, accurate layout geometry modeling, and optimization of local floorplan. More intense aging phenomena (e.g. Hot Carrier Injection, HCI, Punchthrough, Negative Bias Temperature Instability, NBTI) are detrimental to analog circuits and should be taken into account during design and layout iterations. These phenomena were handled by setting/optimizing the per-device operating conditions in order not to stress it out of the combined safe operating limits.

Finally, efficient ESD protection, especially for RF devices, should handle both the requirements for low capacitance load and protecting very sensitive thin-oxide devices with small oxide breakdown voltage. To that end, all critical ESD paths were treated to present minimum impedance, provide fast clamping action with adequate current capacity, and ensure uniform turn-on of the protecting devices by proper layout. Increased per square metal resistivity and intermetal layer capacitance due to thinner dielectrics required accurate modeling of high bandwidth signal routing and balance between signal integrity concerns and die area consumed.

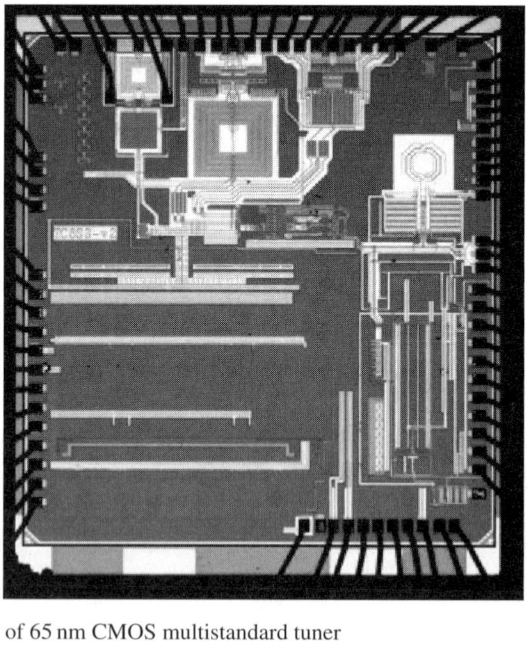

Fig. 5.20 Die photo of 65 nm CMOS multistandard tuner

Table 5.4 Multistandard TV tuner performance summary[a]

Parameter	VHF	UHF	L	MBRAI 2.0[b]
Noise figure (dB)	3	3[c]	2.8	
Gain max (dB)	98	96	96	
Gain min (dB)	−2	−4	4	
RF gain range (dB)	28	28	20	
Baseband low-pass filter Fo range (MHz)		0.1–7		
IIP2 Gain(Max RF-10 dB) (dBm)	36	40	35	
IIP3 Gain(Max RF) (dBm)	−4	−6	−12	
IIP3 Gain(Max RF-5 dB) (dBm)	−2	−3	−7	
Power DVB-T mode (mW)	133	138	135	
Power T-DMB mode (mW)	102	N/A	102	
Power ISDB-T 1-segment mode (mW)	N/A	93	N/A	
Phase noise integrated 1 kHz–4 MHz (°rms)	0.25°	0.5°	1.1°	
Phase noise 10/100/1000 kHz (dBc/Hz)	−102/ −107/ −135	−96/ −101/ −129	−90/ −95/ −123	
Sensitivity DVB-T 64-QAM 3/4 (dBm)	−83	−83	−83	−81.3
Sensitivity DVB-T 16-QAM 2/3 (dBm)	−90.5	−90.2	−90	
Sensitivity DVB-T QPSK 1/2 (dBm)	−98	−97.8	−98	−96.6
S2, $N+1$ DVB-T 64-QAM 3/4 (dB)	35	35	N/A	27
S2, $N+2$ DVB-T 64-QAM 3/4 (dB)	44	45	N/A	40
S1, $N+1$ PALG DVB-T 64-QAM 3/4 (dB)	40	40	N/A	35
S1, $N-1$ SECAM-L DVB-T 64-QAM 3/4 (dB)	35	34	N/A	30
S1, $N+2$ SECAM-L DVB-T 64-QAM 3/4 (dB)	47	48.5	N/A	43
L1 DVB-T 16-QAM 2/3 (dB)	44/49	44/49	N/A	40/45
L3 DVB-T 16-QAM 2/3 (dB)	45	46	N/A	40

[a] Measurements at the PCB connector including all balun and PCB losses.
[b] All system measurements using 8 MHz channel, GI = 1/8, 8 k FFT and 2×10^{-4} Vitterbi BER criterion for DVB-T.

5.4.4 Measurement Results

The chip implemented in a standard digital 65 nm CMOS occupies a total area of less than 7 mm^2. A die microphotograph is shown in Fig. 5.20 and typical measurement results are summarized in Table 5.4. Performance has been characterized in all bands and modes, verifying multistandard, multiband capability.

As shown in Figs. 5.21–5.23, NF is less than 3 dB for all three bands. Both single-ended and differential operation were verified, showing essentially the same performance in NF, linearity and related system-level measurements. A slight increase of NF of less than 0.5/0.2 dB at maximum gain was observed in ISDB-T/T-DMB modes, respectively, where lower signal bandwidth increases the contribution of flicker noise to the overall SNR and baseband opamp-RC filters use larger resistors.

Figure 5.24 shows the integrated phase error (after LO divider) and spot phase noise at 100 kHz (before LO divider) for the 474–858 MHz UHF range. Low integrated phase noise error and accurate filter matching help achieve an SNR in excess

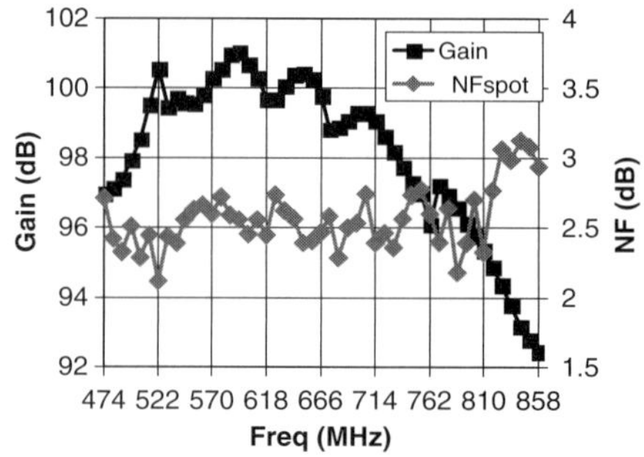

Fig. 5.21 UHF NF, gain vs. frequency

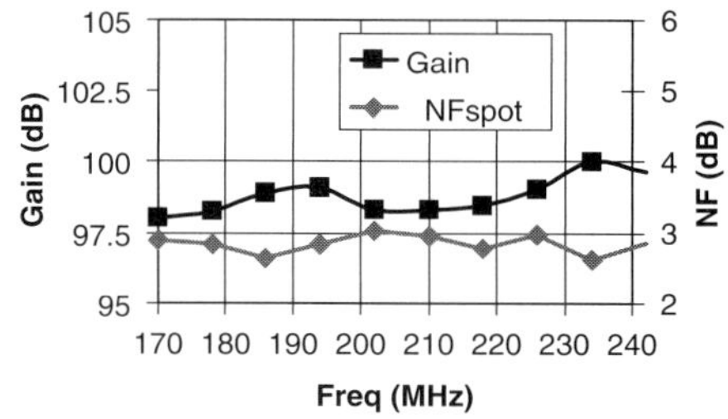

Fig. 5.22 VHF NF, gain vs. frequency

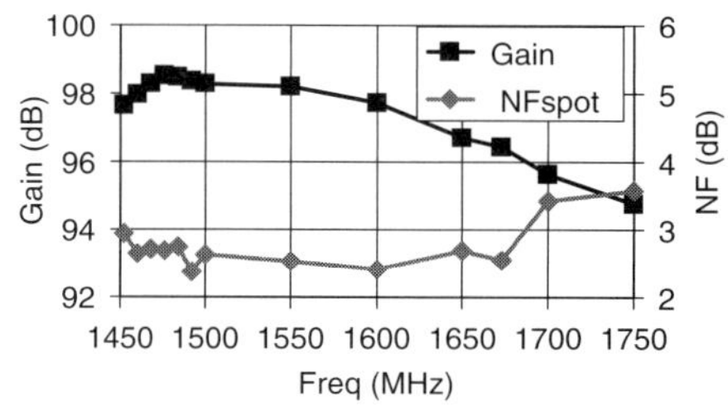

Fig. 5.23 L-band NF, gain vs. frequency

of 35 dB for a wide input signal range (Fig. 5.25), which allows robust mobile operation in conditions where high SNR is needed.

In continuous DVB-T mode, the tuner consumes less than 140 mW in all bands, split between the 1.2 and 2.5 V nominal supplies, while lower channel bandwidth and linearity requirements help reduce power consumption in 1-segment ISDB-T and T-DMB modes.

As indicated by sensitivity measurements for DVB-T/H (Fig. 5.26), the tuner performs better than state-of-the-art implementations at less than half the power consumption [5, 13, 14]. System-level measurements in DVB-T/H modes using the companion demodulator in FPGA indicate that good NF and linearity help exceed MBRAI specifications by a significant margin.

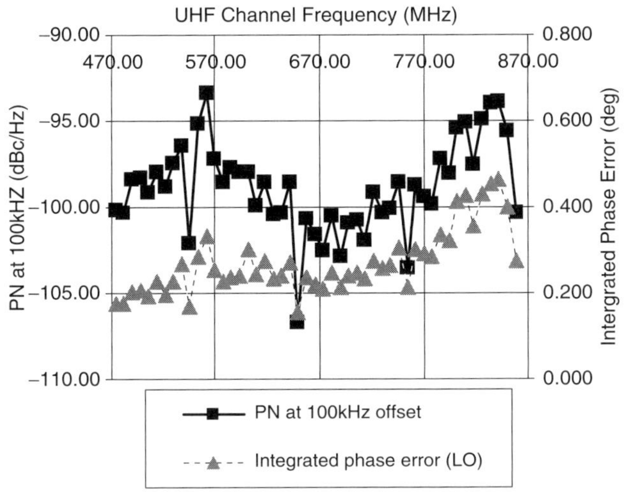

Fig. 5.24 Integrated phase error vs. frequency

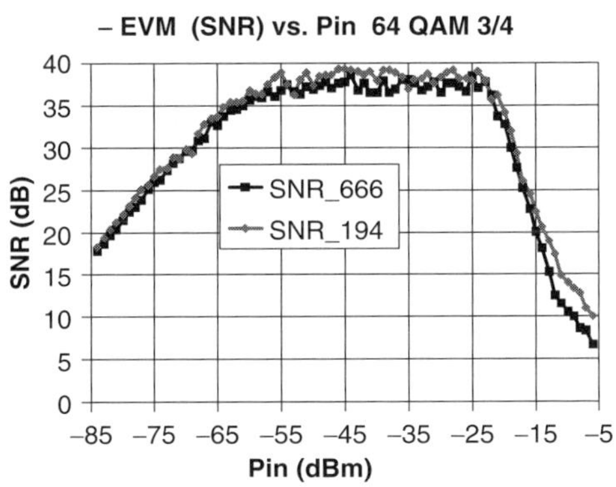

Fig. 5.25 SNR (calculated by DVB-H demodulator) vs. input power at UHF (Ch. 45 – lower curve), VHF (Ch. 9 – upper curve)

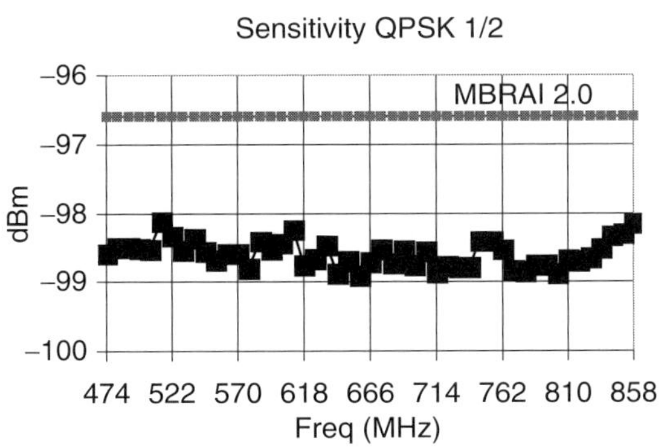

Fig. 5.26 Sensitivity in DVB-T QPSK-1/2 vs. channel at UHF, using the 2×10^{-4} Vitterbi BER error criterion

5.5 Conclusions

After reviewing typical broadcast TV architectures, mobile TV standards, and related specifications, this chapter presented CMOS mobile TV tuner implementation examples. The first was a direct-conversion DVB-H dual-band tuner implemented in 0.18 μm CMOS and the second used the same architecture adding the necessary features for multistandard, multiband support in an aggressive 65 nm digital CMOS. Both tuners take advantage of the digital processing power of CMOS to add calibration capabilities compensating for process and architecture-related imperfections.

Highlights include the implementation of a ΣΔ fractional-N synthesizer with continuous division ratio prescaler, programmable baseband filters with built-in cutoff frequency auto-calibration and RF front-ends with the capability to operate both as single-ended and differential. Compared to other implementations, these tuners achieve state-of-the-art performance while using CMOS process, which allows for cointegration on an SoC with a digital demodulator.

Acknowledgments The contributions of following people to the work described in this chapter are greatly acknowledged: Aris Kyranas, Yiannis Kokolakis, Stamatis Bouras, Ilias Bouras, Michalis Margaras, Kosmas Tsilipanos, George Kamoulakos, Charalambos Kapnistis, Spyros Kavadias, Nikos Kanakaris, Emmanouil Metaxakis, Christos Kokozidis, and Hamed Peyravi.

References

1. Faria G et al. (2006) DVB-H: Digital Broadcast Services to Handheld Devices. Proc. IEEE, vol. 94, no. 1, pp. 198–209, January 2006
2. European Telecommunications Standards Institute (ETSI) (2006) Digital Audio Broadcasting (DAB): DMB Video Service: User Application Specification, V1.1.1, ETSI TS 102 428, pp. 2005–06
3. International Electrotechnical Commission (IEC) (2003) International Standard 62104, Characteristics of DAB Receivers, IEC 2003–3
4. Chari MR et al. (2007) FLO Physical Layer: An Overview. IEEE J. Trans. Broadcasting, vol. 53, pp. 145–160, March 2007
5. Vassiliou I et al. (2006) A 0.18 μm CMOS, Dual-Band, Direct-Conversion DVB-H Receiver. Proc. IEEE Int. Solid-Stage Circuits Conf., pp. 606–607, February 2006
6. Vavelidis K et al. (2007) A 65 nm CMOS Multi-Standard, Multi-Band Mobile TV Tuner. European Solid-State Circuits Conf., pp. 424–427, September 2007
7. Connel L et al. (2002) A CMOS Broadband Tuner IC. Proc. IEEE Int. Solid-Stage Circuits Conf., pp. 400–401, February 2002
8. Dawkins M, Burdett AP, Cowley N (2003) A Single-Chip Tuner for DVB_T. IEEE J. Solid-State Circuits, vol. 38, pp. 1307–1317, August 2003
9. Heng CH et al. (2005) A CMOS TV Tuner/Demodulator IC With Digital Image Rejection. IEEE J. Solid-State Circuits, vol. 40, pp. 2525–2535, December 2005
10. Fillatre V et al. (2007) A SiP Tuner with Integrated LC Tracking Filter for both Cable and Terrestrial TV Reception. Proc. IEEE Int. Solid-Stage Circuits Conf., pp. 208–209, February 2007
11. Gupta M et al. (2007) A 48-to-860MHz CMOS Direct-Conversion TV Tuner. Proc. IEEE Int. Solid-Stage Circuits Conf., pp. 206–207, February 2007

12. Weldon J et al. (2001) A 1.75-GHz Highly Integrated Narrow-Band CMOS Transmitter with Harmonic-Rejection Mixers. J. Solid-State Circuits, vol. 36, pp. 2003–2015, December 2001
13. Antoine P et al. (2005) Direct-Conversion Receiver for DVB-H. IEEE J. Solid-State Circuits, vol. 40, pp. 2536–2546, December 2005
14. Womac M et al. (2006) Dual-Band Single-Ended-Input Direct Conversion DVB-H Receiver. Proc. IEEE Int. Solid-Stage Circuits Conf., pp. 610–611, February 2006
15. Infineon Technologies (2004) TUA6034, TUA6036 3-Band Digital TV Set-Top-Box Tuner IC. TAIFUN Version 2.5
16. European Information & Communications Technology Industry Association (EICTA) (2007) MBRAI 2.0 Mobile and Portable DVB-T/H Radio Access – Part 1: Interface specification
17. Van Nee R and Prasad R (2000) OFDM for Wireless Multimedia Communications. Artech House
18. Magoon R et al. (2002) A Single-Chip Quad-Band (850/900/1800/1900 MHz) Direct Conversion GSM/GPRS RF Transceiver with Integrated VCOs and Fractional-n Synthesizer. IEEE J. Solid-State Circuits, vol. 37, no. 12, pp. 1710–1720, December 2002

Chapter 6
Multimedia Broadcasting and Communications with WiMAX and Implementation for Its Downlink Physical Layer

Daniel Iancu, Joon-Hwa Chun, Hua Ye, Murugappan Senthilvelan, John Glossner, and Mayan Moudgill

6.1 Introduction

Currently, dramatic changes are affecting the way we communicate. New standards for both wired and wireless communications allow increasingly higher data rates at significantly lower costs. As internet access became a commodity, feature-rich communication and multimedia services such as Voice over Internet Protocol (VoIP), multiplayer online gaming, Internet Protocol Television (IPTV), and multi-mode teleconferencing are expanding the traditional voice and messaging services. With technology miniaturization and rapid reductions in integrated circuit size, in addition to wired and wireless communication convergence, we are also witnessing the convergence of fixed and mobile wireless communications. Contrary to multiplatform systems for multiservices, IP based Multimedia Systems (IMS) offer a single platform for a multitude of services resulting in increased revenues at lower operator expense. Mobile communications are playing an increasing role in the IMS services. A high level diagram of an IMS system is illustrated in Fig. 6.1.

The latest adoption of WiMAX (Worldwide Interoperability for Microwave Access) as part of the 3G standards by the International Telecommunications Union (ITU) assures compatibility and interoperability of 3GPP with broadband wireless access networks [28]. As a result, WiMAX may be part of the cellular systems as a next generation technology. Together they cover a significant portion of the broadband wireless access. WiMAX is a long-range, fixed, portable, and mobile wireless technology specified in the IEEE 802.16 standard. It provides high-throughput broadband connections similar to IEEE 802.11 wireless LAN systems but with a broader coverage range. Possible applications for WiMAX include: the "last mile" broadband connections, hotspot and cellular backhaul, and high-speed enterprise connectivity for business. The IEEE 802.16 standard defines a Media Access Control (MAC) layer that supports different physical layers and also defines the same

D. Iancu (✉), J.-H. Chun, H. Ye, M. Senthilvelan, J. Glossner, and M. Moudgill
Sandbridge Technologies Inc., Tarrytown, New York, USA
e-mail: diancu@sandbridgetech.com

F.-L. Luo (ed.), *Mobile Multimedia Broadcasting Standards: Technology and Practice*,
DOI: 10.1007/978-0-387-78263-8_6,

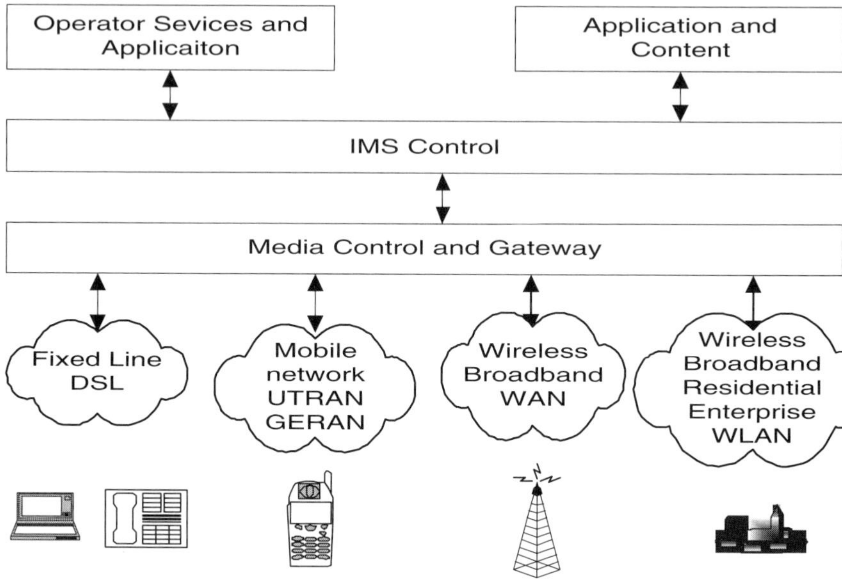

Fig. 6.1 IMS environment

Logical Layer Control (LLC) level 1 for different Local and Wide Area Networks (LAN and WAN), it opens up the possibility of bridging different communication networks together. The IEEE 802.16 standard has been revised to IEEE 802.16-2004 which forms the basis for fixed applications. Further enhancements in 2005 resulted in the finalized IEEE 802.16e-2005 which adds mobility support features to IEEE 802.16-2004 aiming at mobile applications. Fixed WiMAX is based on the IEEE 802.16-2004 OFDM PHY while mobile WiMAX is based on the IEEE 802.16e-2005 OFDMA PHY. In mobile WiMAX, various channel bandwidths ranging from 1.25 to 20 MHz are supported with constant OFDM subcarrier spacing. That means the number of OFDM subcarriers varies according to channel bandwidth. This concept is called Scalable OFDM and was introduced to 802.16e OFDMA mode. The Scalable OFDMA (SOFDMA) can reduce system complexity for small channel bandwidths and improves performance for wider channel bandwidths [2, 29]. On the contrary, fixed WiMAX uses a constant number of subcarriers regardless of various channel bandwidths raging from 1.5 to 28 MHz. Thus, for larger channel bandwidths, a larger subcarrier spacing is expected and vice versa for the smaller channel bandwidths. Since a software implementation of both fixed and mobile systems can coexist, throughout the chapter we will focus on two special cases: the 7 and 10 MHz bandwidth systems for fixed and mobile applications, respectively. The system profiles are shown in Table 6.1.

This chapter is structured as follows: In Sect. 6.2 we describe the DownLink (DL) physical layer features of both mobile and fixed WiMAX. The receiver algorithms focusing on timing and frequency synchronization are presented in Sect. 6.3.

Table 6.1 System Parameters in fixed and mobile WiMAX

	Fixed WiMAX	Mobile WiMAX
Channel bandwidth	7 MHz	10 MHz
Sampling frequency	8 MHz	11.2 MHz
Total number of subcarriers	256	1024
Number of used subcarriers	200	DL FUSC: 864
(DC subcarrier is not included)		DL PUSC: 840
Subcarrier spacing	31.25 kHz	10.9 kHz
Duration of cyclic prefix	8 μs	11.4 μs
OFDM symbol time	32 μs	102.8 μs

Section 6.4 gives an overview of the Sandbridge DSP architecture and describes the software implementation for both fixed and the mobile WiMAX systems on the Sandblaster processor. A brief summary in Sect. 6.5 ends the chapter.

6.2 Downlink Physical Layer of WiMAX

6.2.1 Mobile WiMAX

The DL physical layer functional blocks of the Mobile WiMAX are illustrated in Fig. 6.2. The system implementation begins with the randomizer that scrambles the data from the MAC layer in order to avoid the occurrence of long zero or one sequences. It is then followed by Forward Error Correction (FEC) coding, interleaving, and QAM symbol mapping block. After channel coding, the subchannelization block maps QAM modulated symbols onto appropriate subcarriers and constructs the OFDMA symbols in the frequency domain. The frequency domain symbols are transformed to the time domain by an Inverse Fast Fourier Transform (IFFT). The resulting time domain sequences are then filtered, to contain the spectrum, and converted into the analog domain by a Digital to Analog (D/A) converter. The following sections give brief description of the mobile WiMAX physical layer components and frame structure.

6.2.1.1 Randomization

Randomization is performed on each block of data prior to Forward Error Correction. The randomization consists of an XOR operation of the data to be randomized with a pseudo random sequence, generated by a Pseudo Random Binary Sequence (PRBS) generator. The PRBS generator is reinitialized for each FEC block, as described in the standard.

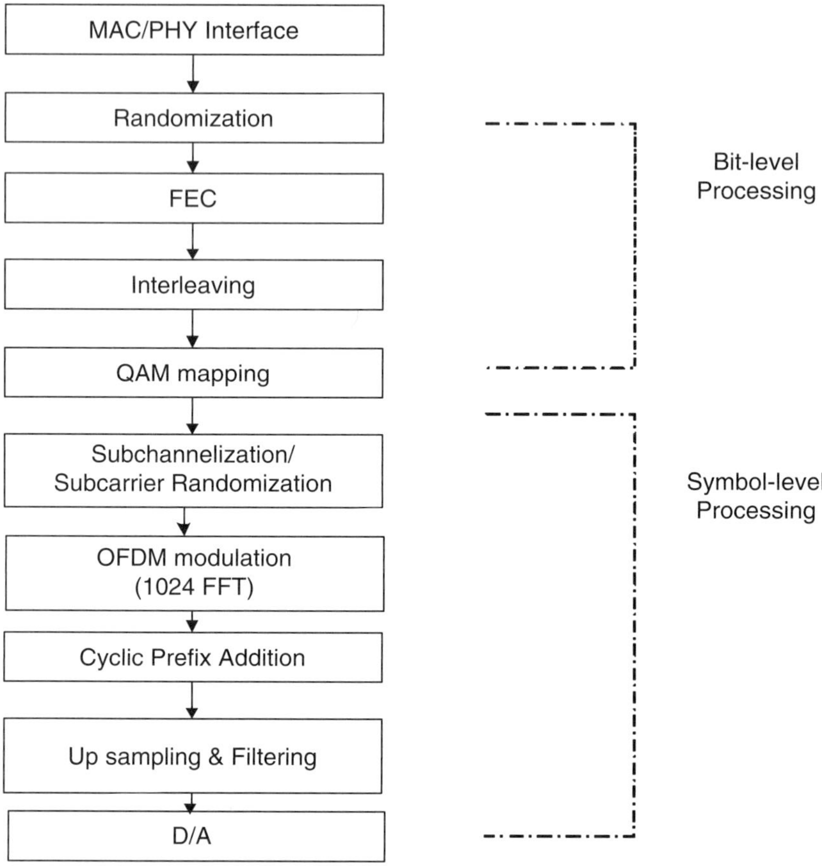

Fig. 6.2 DL physical layer processing blocks in mobile WiMAX

6.2.1.2 FEC

Mobile WiMAX supports four types of FEC encoders: Convolutional Coding (CC) with tail biting or zero tailing, Convolutional Turbo Coding (CTC), Block Turbo Coding (BTC), and Low Density Parity Check (LDPC) coding. Only CC with tail biting and CTC are mandatory whereas other FEC coding schemes are optional [26]. CC with zero tailing is a nonrecursive convolutional coding technique with a constraint length of 7 and native code rate of 1/2, in which a single byte filled with zero bits is appended to the input to the encoder. In this technique, zero tail bits flush out the memory of the encoder so that at the next encoding, the process can start with a state of all zeros. In tail biting CC, instead of padding zero bits, the last 6 bits of the data are used as flush bits meaning that the initial state of the encoder is data dependent. Compared to the optional zero tailing CC, this scheme will reduce the overhead by one byte thus improving the spectral efficiency at increased receiver complexity.

The CTC encoding supported in mobile WiMAX supports duo-binary turbo codes [5–7, 17, 30, 31] with a constituent Recursive Systematic Convolutional (RSC) encoder of a constraint length 4. The native code rate of the duo-binary turbo encoder is 1/3. As the term "duo-binary" implies, it encodes two consecutive bits simultaneously. One important feature in the WiMAX duo-binary turbo coder is that it uses circular constituent encoding, in which the ending state is the same as the starting state. It is the same tail biting concept as the above CC coding case, however, since the constituent encoder is recursive, the last few bits of the input sequence cannot be used for the starting state as in the case of CC with tail biting. To determine the circular state, data blocks must be encoded twice. In the first stage, the encoder is initialized to a state of all zeros. After the data block is encoded, the final state of encoder is used to calculate the circular state through an operator defined in [1]. The first constituent encoder encodes the data in the natural order while the second encoder encodes the interleaved data. It has been claimed that, compared to binary turbo coding used in 3GPP [4, 20, 25], the WiMAX duo-bit turbo coding has advantages of better convergence, less performance drop with max-log-map implementation, larger minimum distances, and less sensitivity to puncturing patterns [5, 6].

6.2.1.3 Channel Interleaving, QAM Modulation, and Subcarrier Randomization

The FEC encoded bits are then interleaved in two steps. The first step ensures that the adjacent coded bits are mapped onto nonadjacent subcarriers, which provides frequency diversity. The second step ensures that adjacent coded bits are mapped alternately onto more or less significant bits of the constellation to prevent long runs of low reliability bits [14]. The second step is necessary since in QAM modulated systems probability of error in the LSB is higher than that of MSB [2]. The details of interleaving can be found in [14]. The interleaved data is then mapped onto QAM symbols. According to the mobile WiMAX system profile [26], QPSK, 16-QAM, and 64-QAM modulations are supported. Each constellation needs to be scaled properly in order to have equal average power. The QAM modulated symbols are then further randomized by pseudo random sequence. This step is called subcarrier randomization. In addition to the encryption provided by FEC block randomization, the subcarrier randomization gives another physical layer encryption [2]. The initial seed of the subcarrier randomization sequence generator is determined by the cell ID and the segment ID.

6.2.1.4 Subchannelization and Subcarrier Mapping

The QAM modulated symbols are mapped in both the time and frequency domains in two steps. In the first step, the symbols are mapped onto logical subchannels, where a subchannel is a logical collection of subcarriers. In the second step, the subcarriers mapped in each logical subchannel are permuted and mapped onto physical subcarriers. Permutation and mapping of logical subcarrier is based on either distributed or adjacent mode. In distributed mode, the data subcarriers are distributed throughout the frequency band while in the adjacent mode, data subcarriers are mapped adjacent to each other. The distributed mode provides better frequency diversity and intercell interference averaging while the adjacent mode provides better frequency selective loading gain [2]. The distributed mode performs well in mobile application while the adjacent mode may be more suitable for stationary or low mobility environments [29]. Mobile WiMAX supports five permutation modes: Full Usage of SubChannels (FUSC), Partial Usage of SubChannels (PUSC), Optional FUSC (O-FUSC), Optional PUSC (O-PUSC), and Adaptive Modulation and Coding (AMC). Among these, AMC mode is based on adjacent permutation mode, and all others are based on distributed mode. The FUSC and O-FUSC modes are for the DL only. In FUSC, each subchannel is mapped onto physical subcarriers distributed throughout the entire physical channel. O-FUSC differs from FUSC in that pilots are allocated in different way and more data subcarriers are used. In DL PUSC, all the data subcarriers are partitioned into clusters, each contains 48 subcarriers over 2 OFDMA symbols. Clusters are then rearranged through renumbering and partitioned into 6 major groups and within each group, 48 subcarriers in each set are permuted and assigned to subchannels. Each of these groups can be allocated to a sector (or segment) of the serving cell. DL PUSC mode is mandatory for the first OFDMA symbols in both DL and UL subframes. In UL PUSC, grouping and permutation is based on a unit called tile structure. Each tile consists of 12 subcarriers over 3 OFDMA symbols. These tiles are consecutively partitioned into 6 groups. Six tiles from each group form a subchannel through UL PUSC permutation. In O-PUSC, size of tile and the way of allocating pilot subcarriers are different as compared to PUSC. As previously mentioned, the AMC mode is based on an adjacent permutation, i.e., all subcarriers forming a subchannel through bins are contiguous to each other.

6.2.1.5 Ranging

Ranging is the process which allows the Mobile Station (MS) to synchronize with Base Station (BS) at network entry. In the uplink since each MS may have different physical distance from a serving BS, this process is essential for synchronization. Once an MS initiates the ranging, a BS can estimate the link quality such as channel strength, carrier frequency offset, and timing offset between MS and BS. In steady-state mode, through ranging the MS and BS can maintain the reliability of the radio link. There are four types of ranging operations depending on the current connection state of the MS: initial ranging, periodic ranging, bandwidth request, and handover ranging. Through initial ranging, BS will recognize the presence of the MS and adjust the timing. During periodic ranging the BS will update the timing and carrier frequency offset. Bandwidth request ranging is used when the MS requests access to the shared spectrum and handover ranging for performing hand-off. Depending on the ranging mode, the ranging code is transmitted over two to four consecutive OFDMA symbols without phase discontinuity.

6.2.1.6 Hybrid Automatic Repeat Request (H-ARQ)

H-ARQ is a combination of FEC and ARQ. This technique may improve the overall error performance of the receiver [16]. During the H-ARQ procedure, the receiver combines the previously transmitted code block into the current block to reduce the bit error rate. The procedure stops when bits are decoded without error or a timeout is generated. In the IEEE 802.16e specification [14], two types of H-ARQ are supported: chase-combining H-ARQ and Incremental Redundancy (IR)-HARQ. In chase combining H-ARQ, a puncturing pattern of FEC coded bits remains the same from one transmission to the next. In IR-HARQ the puncturing pattern of FEC coded bits changes per each retransmission. It has been claimed that the IR H-ARQ performs better than the chase combining H-ARQ in terms of Bit Error Rate (BER) [2]. However, the IR H-ARQ usually requires larger buffer size. Currently, only chase combining H-ARQ with CTC is mandatory in the mobile WiMAX system profile [26].

6.2.1.7 Multiple Antenna Schemes in WiMAX

MIMO technologies may be used to increase data rates, cell coverage area, and link reliability [8]. MIMO technologies that include space time coding and spatial multiplexing with two, three, or four antennas have been adopted in the WAVE 2 mobile WiMAX profile [27]. Space time coding used in WiMAX provides the transmit diversity scheme originally proposed by Alamouti [1]. In this technique, a block of K symbols is space time encoded and mapped to the transmit antenna according to a coding matrix. The encoded symbols are then transmitted in K consecutive transmission periods from K transmission antennas. For example, for the case of the two transmit antennas, the following coding matrix is used:

$$A = \begin{bmatrix} S_1 & -S_2^* \\ S_2 & S_1^* \end{bmatrix},$$

where the symbols S_1 and S_2 represent two consecutive OFDM symbols, and $(\cdot)^*$ denotes complex conjugate. During a first symbol period, S_1 is transmitted from antenna 1 while S_2 from antenna 2. Next, during the second symbol period, $-S_2^*$ is transmitted from antenna 1 while S_1^* from antenna 2. Note that the two transmitted symbols from the two antennas are orthogonal, that is $AA^{\mathrm{H}} = (|S_1|^2 + |S_2|^2)I_{2\times 2}$, where $(\cdot)^{\mathrm{H}}$ denotes the Hermitian transpose and $I_{2\times 2}$ is the 2×2 identity matrix. With the orthogonal property of the coding matrix, the diversity combining technique can be simplified at the receiver. Assume one receive antenna and two transmit antennas. The channel response between the first transmit antenna and the receiver antenna is denoted by h_1 while the channel response between the second transmit antenna and the receiver antenna is denoted by h_2. Also, assume that those channel responses do not vary within two symbol periods. Under these system assumptions, order 2 diversity gain can be achieved. Let y_1 and y_2 denote the received symbols corresponding to two consecutive symbol periods.

Taking conjugate of y_2 results in

$$\begin{bmatrix} y_1 \\ y_2^* \end{bmatrix} = \begin{bmatrix} h_1 & h_2 \\ h_2^* & -h_1^* \end{bmatrix} \begin{bmatrix} s_1 \\ s_2 \end{bmatrix} + \begin{bmatrix} n_1 \\ n_2^* \end{bmatrix},$$

where n_1 and n_2^* represent Gaussian noise. The Least Squares Estimator (LSE) of the transmitted symbols is given by Kay [15]

$$\hat{s}_1 = c^{-1}(h_1^* y_1 + h_2 y_2^*) = c^{-1}(c \cdot s_1 + \tilde{n_1}),$$

$$\hat{s}_2 = c^{-1}(h_2^* y_1 - h_1 y_2^*) = c^{-1}(c \cdot s_2 + \tilde{n_2}),$$

where $c = (|h_1|^2 + |h_2|^2)$, $\tilde{n_1} = h_1^* n_1 + h_2 n_2^*$, and $\tilde{n_2} = h_2^* n_1 - h_1 n_2^*$.

In this decoding technique, the received symbols are weighted by the channel coefficients and they are linearly combined. Also, due to the orthogonal property of coding matrix, spatial interference can be successfully removed. When two receive antennas are used, this transmit diversity can be combined using Maximum Ratio Combing (MRC), which results in achieving array gain as well as diversity gain. The second MIMO technique adopted in WiMAX is the spatial multiplexing technique. For 2×2 (2 transmit antennas and 2 receive antennas) case, the spatial multiplexing technique is based on so called Matrix B:

$$B = \begin{bmatrix} S_1 \\ S_2 \end{bmatrix}.$$

According to the coding matrix, at each symbol period, two independent symbols are simultaneously transmitted from two transmit antennas and as a result this coding scheme does not provide diversity gain at the transmitter. However, order 2 diversity gain can still be achieved at the receiver depending on the decoding method that is used. Let h_{ij} denote the channel response between the transmit antenna j and the receive antenna i. Then, the received symbols per each symbol period can be expressed as

$$\begin{bmatrix} y_1 \\ y_2 \end{bmatrix} = \begin{bmatrix} h_{11} & h_{12} \\ h_{21} & h_{22} \end{bmatrix} \begin{bmatrix} s_1 \\ s_2 \end{bmatrix} + \begin{bmatrix} n_1 \\ n_2 \end{bmatrix}.$$

The optimum decoding technique for the symbols is based on the Maximum Likelihood (ML) estimation. However, since the complexity of ML estimator grows exponentially with the size of constellation, other suboptimal decoders (e.g. minimum mean square error decoder, sphere decoder, etc.) have been considered for practical implementation [8].

6.2.1.8 Mobile WiMAX Frame Structure

Figure 6.3 shows the frame structure for OFDMA TDD mode. Each frame begins with a DL preamble followed by a DL transmission period and then an UL transmission period. In the frame, the Transmit Transition Gap (TTG) shall be inserted between the downlink and uplink, and Receive Transition Gap (RTG) at the end of

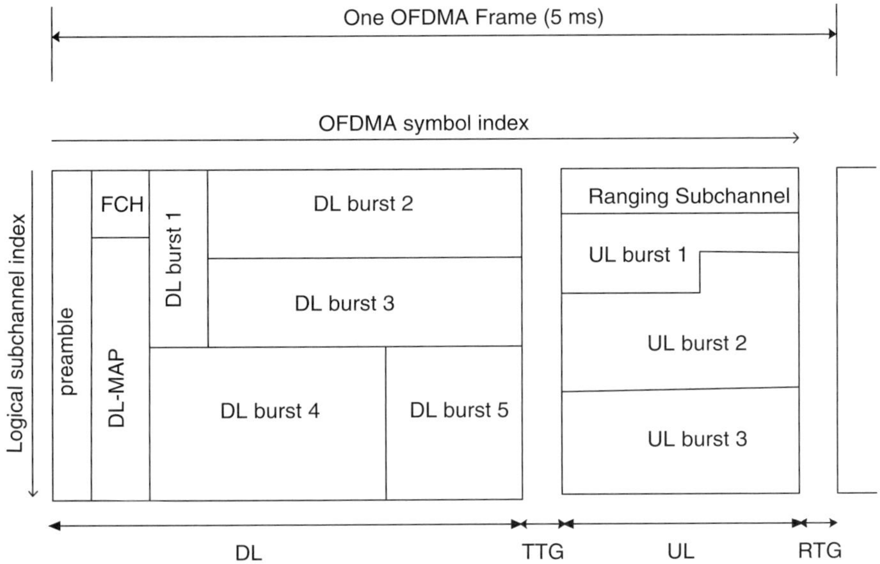

Fig. 6.3 Example of OFDMA TTD frame structure

each frame. In TDD mode, the TTG and RTG provide ramping time for the power amplifier and a guard period to account for the round trip delay. The BPSK modulated Pseudo-Random (PN) sequence that carries the information of the cell ID and segment ID is mapped onto subcarriers in the preamble. The preamble subcarriers are grouped into three segment carrier sets and their indices are given by

$$PreambleCarrierSet_n^k = n + 3k, \tag{6.1}$$

where n represents the segment index, $n = 0, 1, 2, k$ the subcarrier index $k = \{0, 1, K-1\}$, and K denotes the length of PN sequence. The next symbol after the preamble is the Frame Control Header (FCH). It contains 48 bits with code rate $1/2$ and a repetition of 4. The FCH symbol is QPSK modulated and mapped in PUSC format. The FCH contains DL Frame Prefix (DLFP) to specify the burst profile. The FCH is followed by DL-MAP and UL-MAP, which respectively specify the information of DL and UL bursts such as OFDMA symbol offset, subchannel offset, number of OFDMA symbols, number of subchannels, Connection IDentifier (CID), etc.

6.2.2 Fixed WiMAX

Figure 6.4 shows the DL transmitter baseband chain for the fixed WiMAX OFDM PHY. Functionally this chain is similar to the one for mobile WiMAX except that there is no DL subchannelization, and the mandatory FEC scheme is concatenated

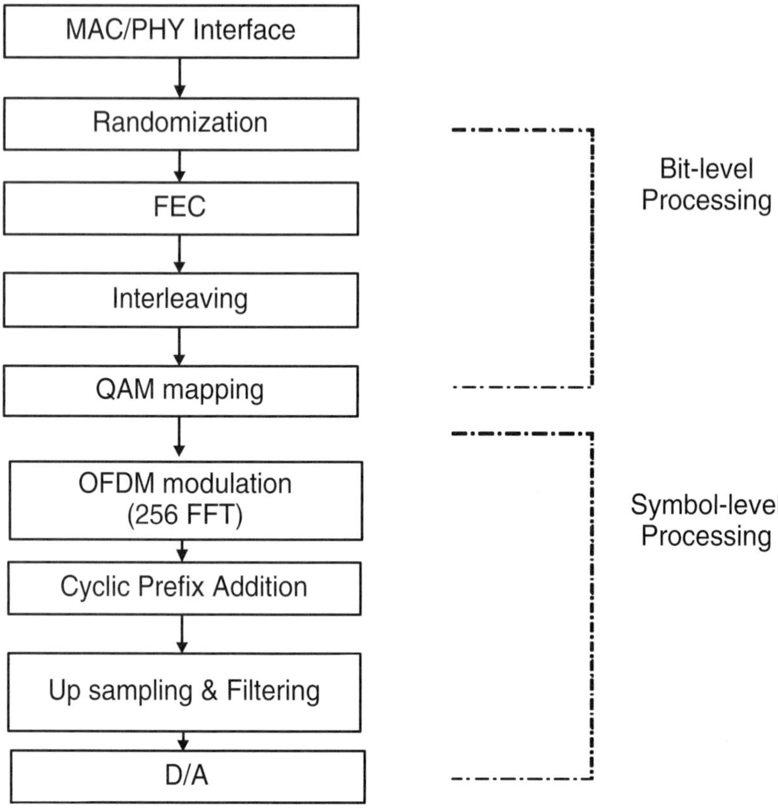

Fig. 6.4 DL physical layer processing blocks in fixed WiMAX

Reed Solomon Convolutional Code (RS-CC). RS-CC operates in such a way that randomized data is first encoded using an RS code and then the output from RS coder is again encoded using a rate $1/2$ convolutional encoder. The Reed Solomon code is derived from a systematic RS code ($N = 255$, $K = 239$, $T = 8$) where N represents the number of encoded bytes, K the number of input bytes, and T the number of data bytes which can be corrected [26]. CTC and BTC are optional FEC schemes for the fixed WiMAX.

6.2.2.1 Preamble and Frame Structure

The preamble in fixed WiMAX is called a long preamble and it consists of two consecutive OFDM symbols. The first symbol consists of four copies of 64 time samples preceded by a cyclic prefix. The second symbol is composed of two copies of 128 time samples following a cyclic prefix. The preamble is depicted in Fig. 6.5.

For the first symbol, QPSK modulated preambles are mapped onto every fourth subcarrier, and for the second symbol, they are mapped onto every even numbered

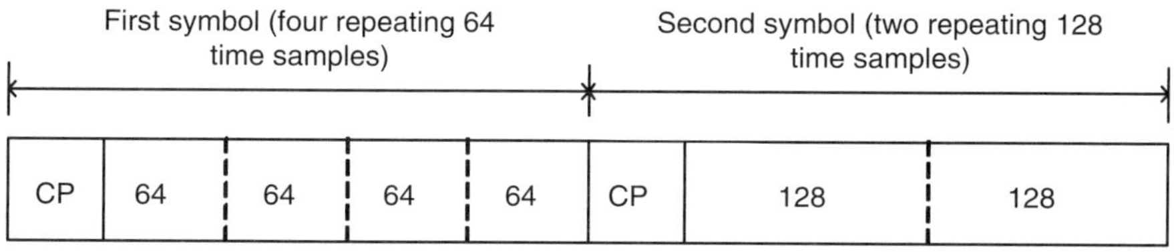

Fig. 6.5 DL long preamble

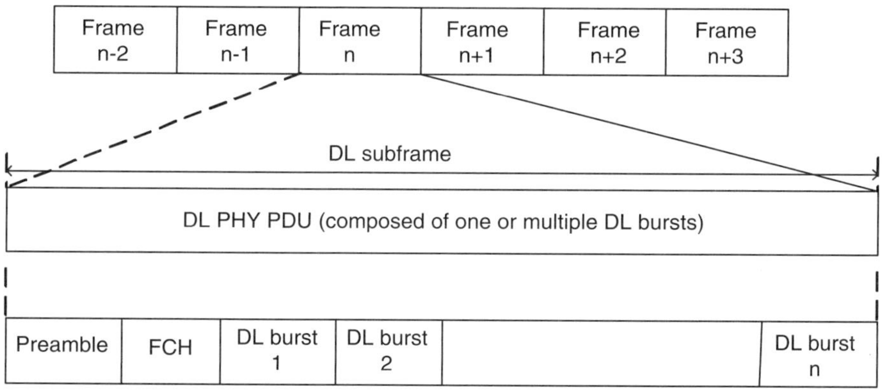

Fig. 6.6 Frame structure with FDD

subcarrier. These preambles can be used for assisting receiver functions such as timing and frequency synchronization and channel estimation. Preambles are transmitted with 3 dB more power than other DL symbols in a frame. The preamble is then followed by an OFDM symbol which carries the Frame Control Header (FCH). The FCH contains control information for the entire radio frame such as the frame length. The remaining bursts in the frame following FCH symbol are for the pay load.

Our fixed WiMAX design focuses on Frequency Division Duplex (FDD). Figure 6.6 shows the OFDM frame structure using FDD. A DL subframe consists of multiple DL bursts that carry one DL PHY Packet Data Unit (PDU).

6.3 DL Baseband Receiver

The standard does not specify the receiver implementation. The receiver implementation is open for competition. Figures 6.7 and 6.9 show the mobile and the fixed WiMAX receiver block diagrams, respectively. The Automatic Gain Control (AGC) block calculates the new value required to establish the appropriate control bits used to set the gain level for the two gain stages in the RF chip based on the signal energy measurements as follows:

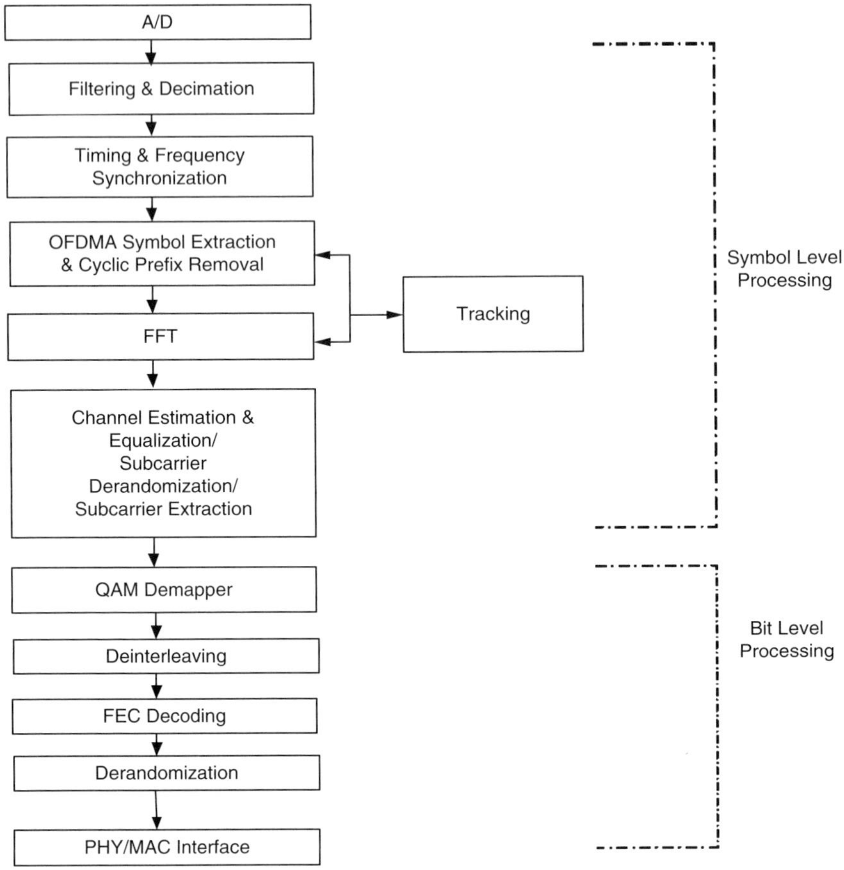

Fig. 6.7 Mobile WiMAX baseband receiver block diagram

$$E = \frac{\sum_{i=0}^{N-1} |r_i|^2}{N}, \tag{6.2}$$

where r_i are the input samples and N is the number of samples for energy calculation [15]. The AGC algorithm runs under coarse and fine setting modes. In the coarse setting mode, the AGC monitors the input energy E and once the incoming signal is detected, an initial AGC setting is calculated by comparing the measured energy E with a preset target energy level. The AGC coarse setting allows the Voltage controlled Gain Amplifier (VGA) to pull the input signal within the A/D's dynamic range in such a way that the Error Vector Magnitude (EVM) is minimized. Once the coarse setting is complete, the AGC gain will be kept constant while the receiver goes through a training process to achieve synchronization with the transmitter. The fine setting mode measures the energy E using the preamble symbols and compares the measured E to the preset target energy level. Based on the energy comparison results, the fine setting mode adjusts up or adjusts down the VGA gain setting.

6.3.1 Mobile WiMAX

The DL baseband receiver chain for mobile WiMAX is shown in Fig. 6.7. The receiver operations can be divided into startup and steady-state stage. During the startup stage, the MS receiver performs initial synchronization that includes timing and frequency synchronization and acquisition of network information in order to camp on a network. Once the receiver is synchronized with the BS, it transits to steady-state stage and starts performing the regular data processing. During the steady-state stage the receiver will also perform synchronization tracking and correction.

6.3.1.1 Timing and Frequency Synchronization

Timing synchronization consists of two steps: symbol boundary and frame boundary search. The symbol boundary can be estimated based on a correlation which utilizes repeated cyclic prefix samples. This approach takes advantage of the fact that the cyclic prefix is identical to the last part of the OFDM symbol. In this scheme, the estimate is an index at which the maximum correlation is observed within a correlation window.

$$\hat{i} = \arg\max_{i} \left| \sum_{v=0}^{G-1} y_{i+v} y^{*}_{i+v+N_{FFT}} \right|, \tag{6.3}$$

where y_i is the filtered and decimated samples, G is the size of cyclic prefix samples and N_{FFT} is the FFT size. This correlation based estimator is similar to Maximum Likelihood (ML) estimation. Beek et al. [3] applied a joint ML estimator for timing and carrier frequency offset estimation utilizing the characteristic of repetitive cyclic prefix. Similar work has been reported by Morelli and Mengali [19], Schmidl and Cox [22], and Moose [18] but instead of the cyclic prefix, a repetitive pilot structure in WLAN is used for estimation. In general, a carrier frequency offset Δf can be divided into integer multiples of subcarrier spacing, Δf_I , and a fraction of subcarrier spacing Δf_f:

$$\Delta f = \Delta f_I + \Delta f_f, \tag{6.4}$$

The fractional frequency offset will be estimated and compensated in the time domain while the integer frequency offset will be corrected in the frequency domain. ML estimation of the fractional frequency offset based on correlator outputs are well known [3, 21, 32] and given by

$$\Delta \hat{f}_f = \frac{1}{2\pi \times N_{FFT} \times T_s} \arg \left\{ \sum_{v=0}^{G-1} y_{\hat{i}+v} y^{*}_{\hat{i}+v+N_{FFT}} \right\} \tag{6.5}$$

where $\hat{i}$ is the estimate of symbol starting position expressed in Eq. 6.3 and T_s is the sampling time interval.

In mobile WiMAX systems since the first symbol in every radio frame is the preamble, searching for the frame boundary could be the same as searching for the preamble symbol. Once the frame staring position is estimated, the integer frequency offset can be estimated by detecting the maximal energy calculated per spectral shift of pilot tones that are distributed in OFDMA symbols next to the preamble. The technique is based on the fact that the power of the pilot tones is 2.5 dB higher than the average power level of data tones [14]. Compensation of the integer frequency offset is done in the frequency domain by circularly shifting the subcarriers according to the estimated integer shift. Next, the acquisition of the segment ID and cell ID based on the post-FFT preamble symbol y_{FFT} is performed. Considering the preamble pattern described in Eq. 6.1, the segment ID could be estimated based on the measured energy of subcarriers per each segment offset:

$$\hat{n} = \arg \max_{n \in \{0,1,2\}} \left\{ \sum_{k=0}^{K-1} \left| y_{FFT}(PreambleCarrierSet_n^k) \right|^2 \right\}, \tag{6.6}$$

where y_{FFT} denotes the post-FFT preamble symbol and the index variables n and k follow the description in Eq. 6.1. The cell ID search can be performed based on a lag-zero cross correlation of y_{FFT} and cell ID specific PN sequences. This procedure used in the startup stage is summarized in Fig. 6.8.

6.3.1.2 Channel Estimation and Soft QAM Demapper

Once the OFDMA symbol is converted from the time domain to the frequency domain, the amplitude and phase distortion caused by the wireless channel has to be estimated and equalized. The channel frequency response could be estimated using linear interpolation in time and frequency domains based on preamble and/or pilot subcarriers. There are a number of different algorithms for channel estimation. One can take advantage of the pilot subcarrier positions that are distributed differently from even to odd symbols in a certain subchannelization mode. In the frequency domain the equalized OFDMA symbols are QAM demapped into either hard or soft bits and transferred to the channel decoder. The hard bits have binary form, 1 or 0, while the soft bits are in a certain numeric range and carry bit reliability information (how likely the demapped bit is a 0 or 1). Soft bits are preferred to hard bits since it provides more information to the channel decoder thus improving the error correction capability. A soft demapper can be implemented based on the Log-Likelihood Ratio (LLR) of the QAM modulated bits. In [23, 24] a simplified suboptimal soft demapper is derived from an approximated log likelihood function. This simplified function demaps the QAM symbols onto soft bits with sign according to the partition the QAM symbol falls in. The magnitude is the measured distance of the received QAM symbols from the nearest partition boundary [24]. In soft demapping the sign of the bits is equivalent to hard bits and the magnitude provides the reliability information of the decision. QAM demodulated bits are then deinterleaved and decoded according to the reverse chain shown in Fig. 6.7.

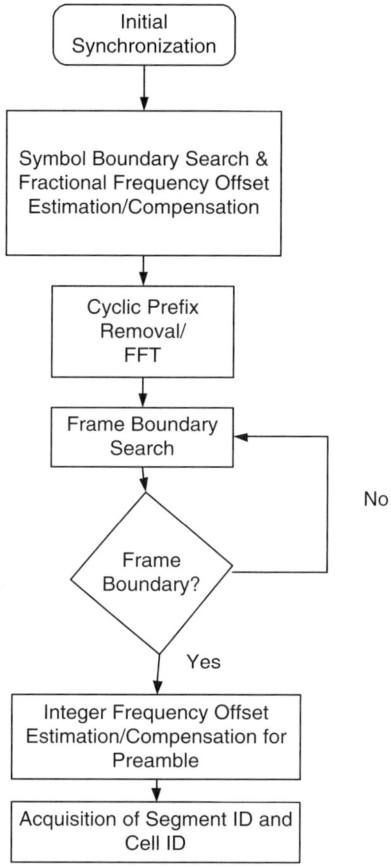

Fig. 6.8 Initial synchronization for mobile WiMAX receiver

6.3.2 Fixed WiMAX

Unlike mobile WiMAX, the fixed WiMAX preamble carries time domain periodicity that can be used to detect the symbol and frame boundary. The following equation is used to detect the preamble sequences:

$$\hat{i} = \arg\max_{i} \left| \sum_{v=0}^{127} y_{i+v} y^{*}_{i+v+128} \right| . \tag{6.7}$$

The coarse fractional carrier frequency offset is estimated as

$$\Delta\hat{f}_f = \frac{1}{2\pi \times 128 \times T_s} \arg \left\{ \sum_{v=0}^{127} y_{\hat{i}+v} y^{*}_{\hat{i}+v+128} \right\} . \tag{6.8}$$

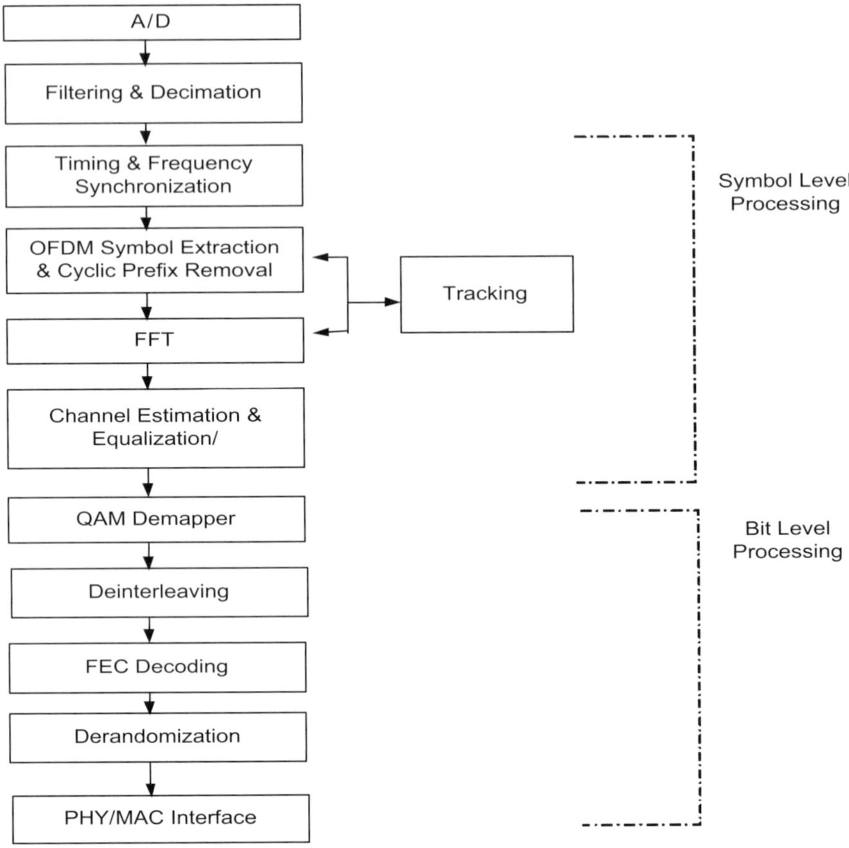

Fig. 6.9 Fixed WiMAX baseband receiver block diagram

Initial coarse symbol timing and frequency offset estimation are performed in the time domain. Fine estimation follows in the frequency domain. The fractional frequency offset is estimated in the time domain using the long preamble, while the integer frequency offset is estimated in the frequency domain. There are eight pilot signals inserted into each data-bearing OFDM symbol. These are used to perform post-FFT carrier frequency offset tracking, symbol synchronization tracking, and sampling rate offset tracking. The frequency and timing compensation is performed through a combination of Radio Frequency (RF) chip setting controlled by the base band processor and frequency derotation. For fixed WiMAX we assume that the channel is quasi-stationary within one frame duration so that channel estimation is performed only when receiving the long preamble symbol. The long preamble has 100 pilots spaced two subcarriers apart. A Least Squares Estimator (LSE) followed by linear interpolation can be used for the channel estimation.

6.4 Implementation of Downlink Baseband Receiver on SandBridge DSP

In this section we describe the implementation of both fixed and mobile WiMAX on the Sandbridge Technologies SDR platform. Both of these systems are executed in software and coexist on the same platform. After a brief overview of the Sandbridge processors [9, 10, 12] in Sect. 6.4.1, we will describe the detailed software implementation of the fixed WiMAX in Sect. 6.4.2 followed by a brief description of the mobile WiMAX implementation in Sect. 6.4.3.

6.4.1 Sandbridge SB3011 Platform

The Sandbridge SB3011 DSP processor is part of the Sandblaster 1.0 architecture and is composed of four DSP cores which are connected through a unidirectional ring network. Each core runs at a minimum of 600 MHz and has four execution units and two levels of on-chip memory. The execution units include a branch unit, an ALU (Arithmetic Logic Unit), an SIMD (Single Instruction Multiple Data) vector unit, a load/store unit. The on-chip memories are a 32 KB Level 1 (L1) instruction cache, a 64 KB L1 data cache, and a 256 KB Level 2 (L2) data memory. All data memory is noncached ensuring that there is no cache coherency issue in SB3011 [9]. While the L2 memory can be accessed by all other cores through the ring network, the L1 memory of a core is not shared by other cores to prevent pipeline stalls for L1 memory accesses. There are eight hardware threads supported in each DSP core. With multiple hardware threads, each core can execute eight independent instruction streams concurrently. These eight hardware threads share the core execution resources equally so that each thread can be viewed as a processor running at the speed of 75 MHz (= 600 MHz/8). This reduced clock cycle is termed a thread cycle and is used as a measurement for computational complexity and memory latencies. The SB3011 also supports data level parallelism with vector/SIMD operations. Using vector operations, four MAC (Multiply and Accumulate) operations can be performed within one thread cycle so that a peak performance of 2.4 billion MACs per core is achieved at 600 MHz operation. In total, the SB3011 provides for cores and 9.6 GMACs of DSP performance. In the Sandbridge platform (Fig. 6.10), there are DSP specific peripherals which are used to move RF (Radio Frequency) or TDM (Time Division Multiplexed) voice and data samples directly into the DSP's L2 memory. These are shown in the upper left corner of Fig. 6.10 and labeled PSD (Parallel Streaming Data) interfaces. An ARM9 core is also provided with its own peripherals which support multiple I/O processing streams through both an AHB (Advanced High Speed) and APB (Advanced Peripheral Bus). The platform also supports external memory up to 2 GB. Typical external memories used are FLASH, SRAM, DRAM, and SDRAM. Both the DSPs and the ARM can initiate external memory requests that are routed through an MMC (Multiport Memory Controller) unit either directly or through on-chip DMA (Direct Memory Access).

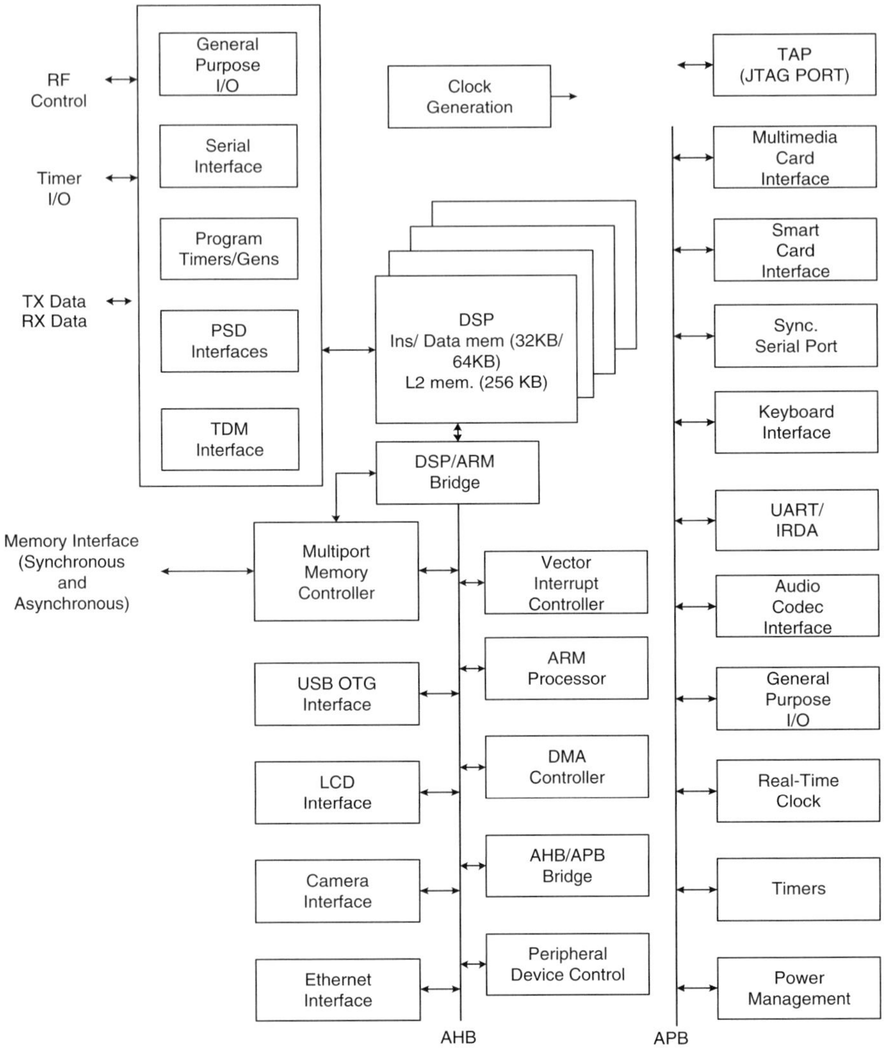

Fig. 6.10 Sandbridge SB3011 SDR platform

6.4.2 Software Implementation of Receiver Baseband Processing

A software implementation of the receiver typically follows the flow depicted in Figs. 6.7 and 6.9. Initially, in order to camp on a network, the MS should perform the initial synchronization that includes time/frequency synchronization and acquisition of network information. Once the communication link between BS and MS is established, the process of the MS receiver can move to steady-state mode in which filtering, tracking, cyclic prefix removal, FFT, channel estimation/equalization, and subcarrier extractions are performed in OFDM at the symbol level. All other back-end processing is performed at the bit level. Partitioning the processing workload onto multiple cores and multiple threads can be based on several factors. For some front end processing blocks such as filtering, it is critical that the processor is able to meet real-time requirements. Additionally, for bit level processing, meeting the throughput requirements of the service can be an important factor when considering total memory requirements. Partitioning can therefore be viewed as a constraint driven optimization. For example, let us consider the baseband filtering operation in mobile WiMAX. As the A/D converter generates samples with a sampling frequency of 22.4 MHz, i.e., two times oversampled data, I/Q DMA must move the sampled data to L2 memory which is associated with a specific core. As mentioned in the Sect. 6.4.1, since each thread effectively operates at 75 MHz, the samples generated from the A/D converter are delivered at the rate of about 3.35 thread cycles per sample. Assume the filer is implemented as a 16 tap FIR. Considering four MACs per cycle vector operations in the SB3011, it is expected that about four thread cycles are required for filtering I and Q samples. That is, two threads are required for filtering each I and Q data to meet real-time performance requirements. Partitioning threads to the processing of the rest of processing blocks can be done in a similar way. The real-time constraint for initial synchronization is more relaxed as compared to the steady-state mode. It is performed before network entry. However, schemes such as counting samples that passed during processing are needed to maintain synchronization. Threads can communicate with each other by either round-robin scheduling with a state machine or by using flags. In the following sections, the detailed software implementation for both fixed and mobile WiMAX receivers is discussed.

6.4.2.1 Fixed WiMAX FDD

One of the goals of WiMAX is to cover the "last mile." This can be accomplished only by employing the most robust communication profile – the 256 carrier full rate BPSK, FDD mode. In the section we describe in detail the software implementation of this mode on the SB3011 processor along with the computational performance complexity and power consumption results. Briefly, the implementation parameters are as follows:

- Symbol duration – 40 μs (3000 thread cycles @ 75 MHz)
- Samples per symbol – 640 samples (oversampled by 2)
- FIR in RX-16 tap, decimation by 2
- FIR in TX-20 tap, oversample by 2
- FFT size – 256 point
- Number of subcarriers – 256 (192 data subcarriers, 8 pilots, 56 unused subcarriers)
- Guard period – 64 samples (1/4 of 256)
- Viterbi decoding block length – 196 bits input and 88 bits output

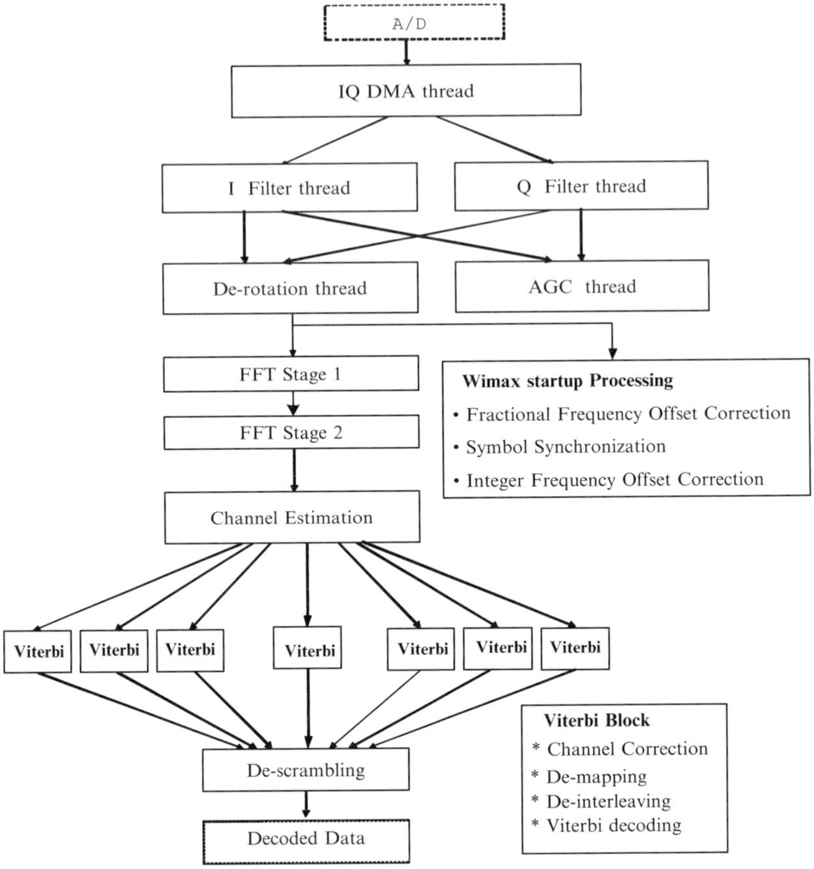

Fig. 6.11 Example of receiver pipeline

The receiver algorithms are implemented in a concurrent multithreaded pipeline. The pipeline consists of all the processing steps such as filtering, FFT, etc. To implement a pipeline on the Sandbridge processor, we first aggregate the processing chain steps into stages and second, we partition threads to the computations within a stage. The pipeline implementation is shown in Fig. 6.11.

Our implementation illustrates two methods for partitioning work to threads: either we partition a unit of work, i.e., OFDM symbol, across multiple threads, or we process multiple units of work concurrently. In general, we might have multiple units processing concurrently, with each unit being partitioned across a team of threads. Therefore, for each stage we have to specify the number of concurrent teams and the number of threads in each team. The partitioning of work within each team is dependent on particular computations. The receiver has two major modes of operation: startup and steady state. During the startup process the receiver goes through several states of a state machine until the steady state is reached. The states can be defined as follows: state 1: initial energy detection and initial AGC setting, state 2:

coarse carrier frequency offset estimation and correction, state 3: OFDM symbol synchronization via preamble sequence, state 4: integer frequency offset detection and correction, and state 5: steady-state processing. The first four states are not real time constrained. During non-real-time processing the synchronization is maintained through a CPU cycle clock accurate time keeping block. In the steady-state mode, the following functions are performed:

- IQDMA, data samples are transferred from A/D to L2 memory. One thread is responsible for this function.
- I/Q signal filtering, two threads are executing the filtering in parallel.
- Energy monitoring and AGC fine tuning, one thread is responsible.
- I/Q signal derotation, one thread. It performs the derotation based on frequency offset estimation.
- FFT stage 1. One thread performs the first FFT stage.
- FFT stage 2. One thread performs the second FFT stage.
- Channel estimation, one thread performs the channel estimation and also computes the frequency and timing offset tracking.
- Channel Correction, demapping, deinterleaving and Viterbi decoding are executed by seven threads.
- Finally, the descrambling is performed by one thread.

Overall, the receiver uses 16 threads, i.e., 2 cores. The receiver performance for 2.9 Mbps has been tested according to IEEE 802.16 specifications. The targeted receiver SNR was 3.0 dB when using BPSK modulation with $1/2$ rate convolutional coding. The entire software development has been initially performed using the Sandbridge simulator, sufficient for a complete ANSI C implementation of the entire physical layer processing. The laboratory tests confirmed 1.59 dB SNR when using 4-bit soft decoding. For the transmit chain the threads were allocated as follows:

- Symbol generation including the convolutional encoder and symbol packaging, one thread
- IFFT, two threads
- Guard generation stage, one thread
- FIR stage, four threads to filter and oversample the OFDM symbols

For the transmitter, the pipelining and multithreading were done automatically by the Sandbridge development tools. The entire FDD WiMAX chain requires just under three cores when both Rx and Tx are implemented. A sample of real-time code is depicted in Fig. 6.12.

The computational performance of the receiver is described in Table 6.2.

6.4.2.2 Mobile WiMAX

In the case of mobile WiMAX the required computational complexity is higher. Sandbridge Technologies has developed a new architecture that supports wireless data rates necessary for 3.5G and 4G systems. Building upon the Sandblaster 1.0 architecture, the fully object code compatible Sandblaster SBX 2.0 architecture

```
Void thread_function_name(){

        initialization code

        load all local/global variable values owned by the thread

        barrier() // wait for all other threads to reach this point

        exchange global variable values with other threads

        while(1){   // steady state processing loop

            while( wait for signal from previous stage) {
                // do NOTHING (or) keep checking for RESET flag
            };

            do the data processing

            signal the next stage

            signal previous stage that operation complete

    }
```

Fig. 6.12 Real-time sample code

Table 6.2 WiMAX fixed installation receiver performance

Fixed WiMAX Receiver execution stages	Execution cycles	MHz	Number of threads
I filter	1,961	49.03	1
Q filter	1,915	47.88	1
Derotation	2,429	60.73	1
AGC	447	11.18	1
FFT stage 1	1,979	49.48	1
FFT stage 2	2,382	59.55	1
Channel estimation	2,330	58.25	1
Channel correction and convolutional decoding	17,839	445.98	7
Descrambler	1,884	47.10	1
Total processing time required for one symbol if single threaded	33,166	829.15	15

extends support for high bit rate processing, MIMO-OFDM acceleration, wider vector execution, and code compression [11]. Architectural performance improvements range from 4x to more than 10x for a variety of signal processing applications while providing 100% object code compatibility with the Sandblaster 1.0 architecture. The multithreaded programming and pipeline technique used for implementing the fixed WiMAX receiver was applied to the mobile WiMAX receiver. After the startup procedure the receiver process reaches the steady-state mode in which functional blocks such as filtering/down sampling, FFT, channel estimation/equalization, subcarrier extraction, and channel decoding can be performed in pipelined fashion similar to the example shown in Fig. 6.11. Up to the subcarrier extraction stage, the received data can be processed per OFDM symbol period (102.8 μs). In the DL PUSC mode, since QAM modulated data are mapped onto subcarriers across two OFDMA symbols, subcarriers can be extracted per two OFDMA symbol periods (205.6 μs). In OFDMA mode, multiple bursts of data are mapped onto the same OFDMA symbol period. One could arrange buffers in such a way that each buffer holds data from each burst. It is not necessary to wait until the whole burst of data is written into the buffer. As soon as one code block amount of data is filled into a buffer, the data can be processed through the channel decoding stage and the decoded data can be executed in another data unit which is to be sent to MAC layer. The buffer can be reused for another code block of data. In H-ARQ mode, extra buffers are required for holding burst data. In chase combining H-ARQ, the maximum size of the buffer is about 23K soft bits per H-ARQ channel [26].

6.5 Summary

This chapter describes the downlink physical layers of both mobile and fixed WiMAX including the software implementation of the receive chains on the Sandblaster processors. Due to the flexibility provided by software implementation, both of these systems may coexist and execute as required. Throughout the chapter, two special cases, 7 and 10 MHz bandwidth systems for the fixed and mobile WiMAX applications, respectively, have been the focus. The entire software development has been initially performed using the Sandbridge simulator, sufficient for a complete ANSI C implementation of the entire physical layer processing. The 7 MHz bandwidth and 256 carrier mode executes in the SB3011 processor has been successfully field tested and demonstrated. A new core has been developed to implement a complete WAVE II profile mobile WiMAX. Overall, we have shown that software implementation may be used for high throughput communication systems.

Acknowledgments The authors wish to acknowledge the significant contributions of Gary Nacer, Sanjay Jinturkar, Vitaly Kalashnikov, Vasile Surducan, Emanoil Surducan, Stuart Stanley, Sitij Agrawal, Andrei Iancu, and Saurabh Lahoti. Without their significant ingenuity and effort this work would not have been possible.

References

1. S.M. Alamouti (1998) A simple transmit diversity techniques for wireless communications. IEEE Journal on Selected Areas in Communications, Vol. 16, pp. 1451.
2. J. Andrews, A. Ghosh, and R. Muhamed (2007) Fundamentals of WiMAX: Understanding Broadband Wireless Networking. Prentice Hall Communications Engineering and Emerging Technologies Series.

3. J. Beek, M. Sandell, and P.O. Borjesson (1997) Ml estimation of time and frequency offset in OFDM systems. IEEE Transactions on Signal Processing, Vol. 45, pp. 1800.
4. C. Berrou, A. Glavieux, and P. Thitimajshima (1993) Near Shannon limit error-correcting coding and decoding: Turbo-codes. IEEE International Conference on Communications, Vol. 2, pp. 1064–1070.
5. C. Berrou, M. Jezequel, C. Douillard, and S. Kerouedan (2001) The Advantages of Non-Binary Turbo Codes, Information Theory Workshop, pp. 61–63.
6. C. Berrou, M. Jezequel, C. Douillard, S. Kerouedan, and L.C. Canencia (2001) Duo-Binary Turbo Codes Associated with High-Order Modulations. ESA TTC '01, Noordwijk, The Netherlands.
7. C. Douillard, M. Jezequel, and C. Berror (2000) The Turbo codes Standards for DVB-RCS. Proceedings of the second International Symposium on Turbo Code.
8. D. Gesbert, M. Shafi, D. Shiu, P. J. Smith, and A. Naguib (2003) From theory to practice: an overview of MIMO space-time coded wireless systems. IEEE Journal on Selected Areas in Communications, Vol. 21, pp. 281.
9. J. Glossner and D. Iancu (2006) The Sandbridge SB3011 SDR Platform. Proceedings of the Symposium on Trends in Communications. Bratislava, Slovakia.
10. J. Glossner, D. Iancu, M. Moudgill, G. Nacer, S. Jintukar, S. Stanley, and M. Schulte (2007) The Sandbridge SB3011 Platform. EURASIP Journal on Embedded Systems, Vol. 2007.
11. J. Glossner, M. Moudgill, D. Iancu, S. Jintukar, G. Nacer, and M. Schulte (2007) The Sandblaster SBX 2.0 Architecture. 2007 Software Defined Radio Technical Conference, Denver, Colorado.
12. D. Iancu et al. (2006) Software Implementation of WiMAX on the Sandbridge SandBlaster Platform. Proceedings of the Sixth Workshop on Embedded Computer Systems: Architectures, Modeling, and Simulation.
13. IEEE 802.16 – 2004 Air Interface for Fixed and Mobile Broadband Wireless Access Systems.
14. IEEE 802.16e – 2005 Part 16: Air Interface for Fixed and Mobile Broadband Wireless Access Systems.
15. S. Kay (1993) Fundamentals of Statistical Signal Processing. Estimation Theory. Prentice Hall.
16. S. Lin and P. S. Yu (1982) A Hybrid ARQ Scheme with parity retransmission for error control of satellite channels. IEEE Transactions on Communications, Vol. 30.
17. Matthew C. Valenti, Shi Cheng, and Rohit Iyer Seshadri, Digital Video Broadcasting, *http://www.csee.wvu.edu/ mvalenti/documents/DVBChapter.pdf*.
18. P.H. Moose (1994) A technique for orthogonal frequency division multiplexing frequency offset correction. IEEE Transactions on Communications, Vol. 42, pp. 2908.
19. M. Morelli and U. Mengali (1999) An improved frequency offset estimator for OFDM applications. IEEE Communications Letters, Vol. 3, pp. 75.
20. W. Ryan. A Turbo Code Tutorial. http://www.ece.arizona.edu/ ryan/publications/turbo2c.pdf
21. M. Speth et al. (2001) Optimum receiver design for OFDM-based broadband transmission – Part 2: A Case Study. IEEE Transactions on Communications, Vol. 49.
22. T.M. Schmidl and D.C. Cox (1997) Robust frequency and timing synchronization for OFDM. IEEE Transactions on Communications, Vol. 45, pp. 1613.
23. K. Song and S.A. Mujtaba (2003) A low complexity space-frequency BICM MIMO-OFDM system for next-generation WLAN. GLOBECOM, Vol. 2, pp. 1109, 2003.
24. F. Tosato and P. Bisaglia (2002) Simplified soft-output demapper for binary interleaved COFDM with application to HIPERLAN/2. IEEE International Conference on Communications, Vol. 2, pp. 664–668.
25. M. Valenti and J. Sun (2001) The UMTS Turbo Code and an efficient decoder implementation suitable for software-defined radios. International Journal of Wireless Information Networks, Vol. 8.
26. WiMAX Forum Mobile System Profile V 1.2.0.
27. WiMAX Forum Mobile Radio Conformance Tests (MRCT) V0.9.1.
28. WiMAX and IMT-2000 (2007) WiMAX Forum White Paper V2.3

29. H. Yaghoobi (2004) Scalable OFDMA Physical layer in IEEE 802.16 WirelessMAN. Intel Technical Journal, Vol. 8, Issue 3.
30. Youssouf Ould-Cheikh-Mouhamedou (2005) On Distance Measurement Methods for Turbo Codes,. Ph.D. Thesis, McGill University.
31. C. Zhan et al. (2006) An efficient decoder scheme for double binary turbo codes. ICASSP, Vol. 4.
32. H. Zou, B. McNair, and B. Daneshrad (2001) An integrated OFDM receiver for high-speed mobile datacommunications. GLOBECOM, Vol. 5, pp. 3090.

Chapter 7
MediaFLO Technology: FLO Air Interface Overview

Qiang Gao, Murali Chari, An Chen, Fuyun Ling, and Kent Walker

7.1 Introduction

MediaFLO^{TM} is a mobile broadcast technology based on open and global standards. A key component of MediaFLO is the FLO^{TM} (Forward Link Only) air interface technology which has multiple published Telecommunications Industry Association (TIA) [13] specifications. FLO is also recognized by ITU-R as a recommended technology for mobile broadcasting and is in the approval process at the European Telecommunications Standards Institute (ETSI). Global standardization efforts are driven and supported by the FLO Forum [3], an industry consortia consisting of 90+ member companies throughout the global mobile broadcast value chain. MediaFLO technology has been launched commercially in the United States (USA) through the nationwide mobile broadcast network built by MediaFLO USA, Inc [10]. Verizon Wireless has deployed MediaFLO services in 50 U.S. markets, and AT&T expects to launch commercial services in early 2008 leveraging the MediaFLO USA network. In addition, MediaFLO technology is being trialed in major markets around the world.

Unlike other mobile broadcast technologies that have evolved from legacy systems, MediaFLO was designed from the ground up for the mobile environment. Consequently, it can deliver mobile broadcast services in a very efficient manner and offers unique advantages such as low receiver power consumption, fast channel switching time, robust reception in mobile fading channels, high spectral efficiency and efficient statistical multiplexing of service channels. MediaFLO technology achieves all of these advantages simultaneously without compromising one for another.

Figure 7.1 shows the MediaFLO protocol stack on the interface between the MediaFLO network and the MediaFLO device. The FLO Air Interface specification consists of Physical Layer, MAC (Medium Access Control) Layer, Control Layer

Q. Gao (✉), M. Chari, A. Chen, F. Ling, and K. Walker
Qualcomm Inc., San Diego, CA 92121, USA
e-mail: qgao@qualcomm.com

F.-L. Luo (ed.), *Mobile Multimedia Broadcasting Standards: Technology and Practice*,
DOI: 10.1007/978-0-387-78263-8_7,

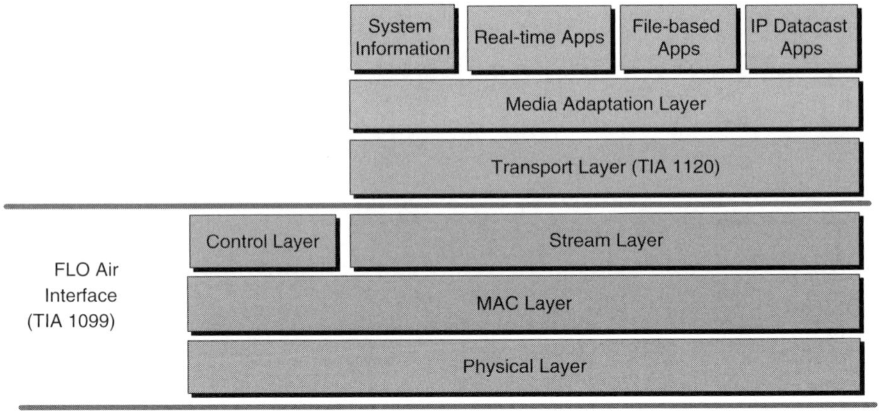

Fig. 7.1 MediaFLO protocol suite

and Stream Layer. The Transport Layer [5] forms a sequence of application service packets for a service component such as video and audio into fixed-size blocks or an octet stream every second for delivery over the Stream Layer. The Media Adaptation Layer provides adaptations that are specific to the class of content being transported, such as real-time streaming services, non-real-time services and IP (Internet Protocol) datacast packets.

This chapter gives an overview of the FLO Air Interface (standardized in TIA-1099 [4]). The rest of this chapter is organized as follows: Section 7.2 presents the overall layering architecture of the FLO Air Interface. Section 7.3 is devoted to the Physical Layer of the FLO system including Orthogonal Frequency Division Multiplexing (OFDM) modulation characteristics, interlace structure, superframe structure, Physical Layer subchannels, and waveform generation for data and OIS (Overhead Information Symbol) channels. The MAC Layer of the FLO system is addressed in Sect. 7.4 with coverage on data encapsulation, Reed-Solomon code and time-frequency resource allocation to services. Sections 7.5 and 7.6 deal with the Control Layer and Stream Layer, respectively. In Sect. 7.7, the FLO Air Interface handling scenarios at the FLO receiver including MLC (Multicast Logical Channel) reception and MLC switching are given.

7.2 FLO Air Interface Layering Architecture

The description on the FLO Air Interface in this section is based on the TIA-1099 standard published in Aug 2006 [4]. Future revisions of the standard may add new features and enhancements.

A MediaFLO service is an aggregation of one or more data components. Each data component of a service is called a *flow*. For example, a MediaFLO TV service may have a video flow, an audio flow, a subtitle flow and a signaling flow. Services are classified into two types based on their coverage: wide-area services and local-area services (Fig. 7.2). A local-area service is typically multicast for reception within a metropolitan area while a wide-area service is typically multicast in one or more metropolitan areas.

MediaFLO services and control information may be carried over one or more logical channels called MLCs. An MLC may be divided into a maximum of three logical subchannels called *streams*. Each application *flow* is mapped to a single stream in an MLC. The flows belonging to a single instance of a MediaFLO service may be sent over multiple MLCs within a single RF allocation, e.g. one 6 MHz channel. The MLCs are multiplexed to the FLO Physical Layer data channel. The relationship of flows, streams, MLCs and Physical Layer data channels is illustrated in Fig. 7.3. Note that although MLC is a MAC Layer concept, it is distinguishable at the Physical Layer. This is an important feature for mobility since it enables access to fraction of the Physical Layer bandwidth without demodulating the entire waveform.

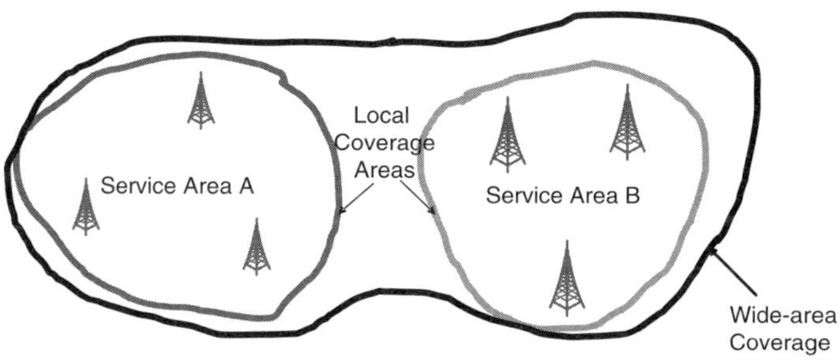

Fig. 7.2 Local-area and wide-area services

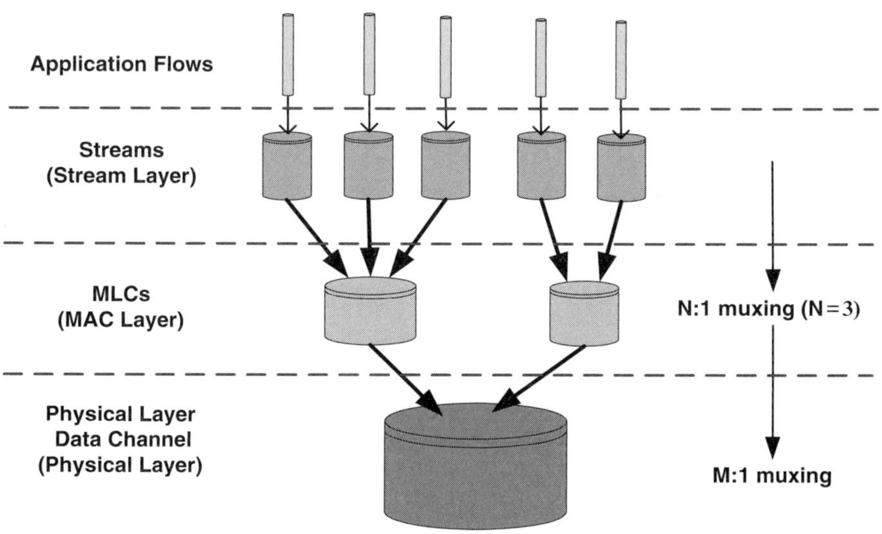

Fig. 7.3 Flows, streams, MLCs and Physical Layer data channels

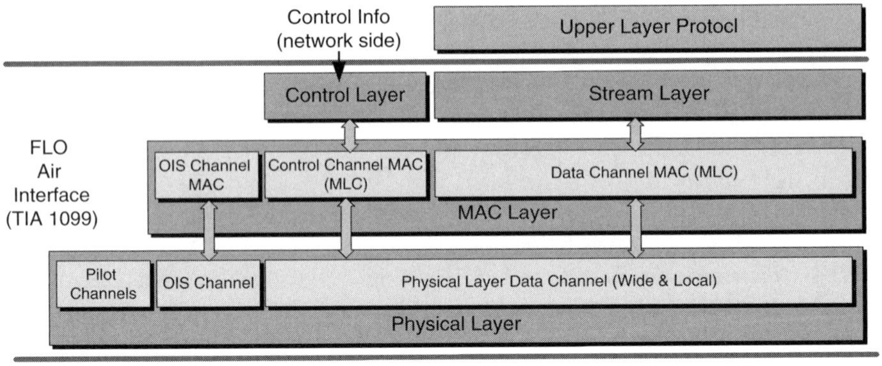

Fig. 7.4 FLO Air Interface layering architecture

Figure 7.4 depicts the FLO Air Interface layering architecture:

- *Physical Layer*: The Physical Layer provides the channel structure, frequency, modulation and encoding specification for the Forward Link.
- *MAC Layer*: The MAC Layer defines the procedures used to receive and transmit over the Physical Layer. It is responsible for resource allocation on the Physical Layer data channel The MAC Layer also multiplexes packets belonging to different media streams associated with the same MLC.
- *Stream Layer*: The Stream Layer binds Upper Layer flows to streams and MLCs.
- *Control Layer*: This layer is used by the network to disseminate control information to facilitate the device operation in the MediaFLO system. The device uses the Control Layer to maintain synchronization of its control information with that in the network.
- *Upper Layers*: The Upper Layer protocols provide multiple functions including encoding of multimedia content, controlling access to the multimedia content and formatting of control information.

7.3 Physical Layer

7.3.1 OFDM Modulation Characteristics

MediaFLO network deployment is typically based on the concept of SFN[1] (Single Frequency Network) [8]. In such a network, multiple transmitters transmit identical waveforms from time synchronized transmitters. The signals from these transmitters can be viewed by the receiver as multipath signals from the same source but with different propagation delays. The distance between transmitters in a mobile broadcast network can possibly be large ($20 \sim 30$ km), which can cause large delay spreads

[1] FLO also supports Multiple Frequency Network deployment.

Table 7.1 FLO OFDM parameters

Parameters	Value
RF channel bandwidth	6 MHz
Chip rate (FFT bandwidth)	5.55 MHz
Chip duration	0.18018 μs
Number of subcarriers (FFT size)	4,096
Subcarrier spacing	1.355 KHz
FFT interval (useful OFDM symbol interval)	738.02 μs (4,096 chips)
Cyclic prefix (flat guard interval)	92.25 μs (512 chips)
Window interval	3.06 μs (17 chips)
(Effective) OFDM symbol interval	833.33 μs (4,625 chips)
Number of guard subcarriers	96
Number of active subcarriers	4,000
Number of pilot subcarriers	500

up to or more than 100 μs (as opposed to 5–6 μs in cellular networks). OFDM is a form of multicarrier modulation that is especially suitable for the radio environment with large multipath delay spreads. OFDM receivers are also simple to implement as they do not require complicate equalizer or more specifically the equalization and demodulation are executed jointly. For these reasons, MediaFLO uses OFDM as the modulation technique.

The FLO Physical Layer supports transmission in radio frequency bands with RF channel bandwidths of 5, 6, 7 and 8 MHz. The description in this chapter assumes a 6 MHz RF channel bandwidth with the parameters specified in Table 7.1, unless otherwise stated. The system parameters for all the RF channel bandwidths supported by FLO are listed in the Appendix.

For an RF channel of 6 MHz, the FLO OFDM subcarriers span a bandwidth of 5.55 MHz, which is called the *chip rate.*[2] The chip rate is 92.5% of the allocated RF bandwidth in order to meet regulatory requirements for the transmit spectral mask. The frequency location of the 4,096 subcarriers at base band is given by

$$f_{SC}(i) = (i - 2048) \times (\Delta f)_{SC}, \quad i = 0, 1, \ldots, 4095$$

where i is the subcarrier index and $(\Delta f)_{SC}$ is the subcarrier spacing given by

$$(\Delta f)_{SC} = \frac{\text{chip rate}}{\text{number of subcarriers}} = \frac{5.55 \times 10^6}{4096} = 1.35498046875\,\text{kHz}$$

The 96 subcarriers with indices 0, ..., 47, 2,048, 4,049, ..., and 4,095 are not used and referred to as *guard subcarriers* (Fig. 7.5). The subcarrier 2,048 corresponds to DC. The 95 guard subcarriers on each side make it easier for the FLO receiver to cope with the adjacent channel interference. The remaining 4,000 subcarriers are referred to as *active subcarriers* that are modulated by data or pilot symbols.

[2] The chip rates for 5, 7 and 8 MHz RF channel bandwidths are 4.625, 6.475 and 7.4MHz, respectively.

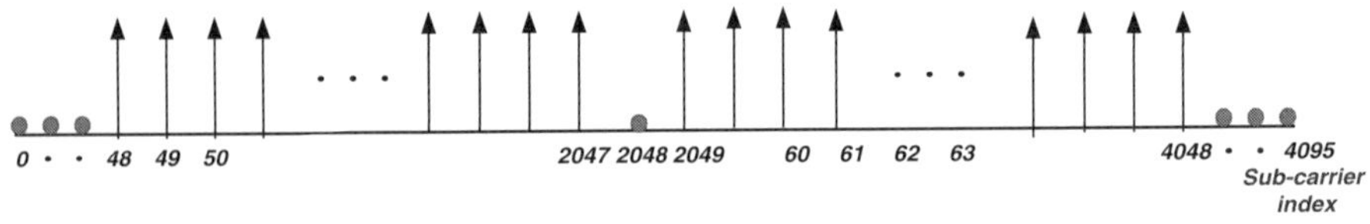

Fig. 7.5 Guard and active subcarriers

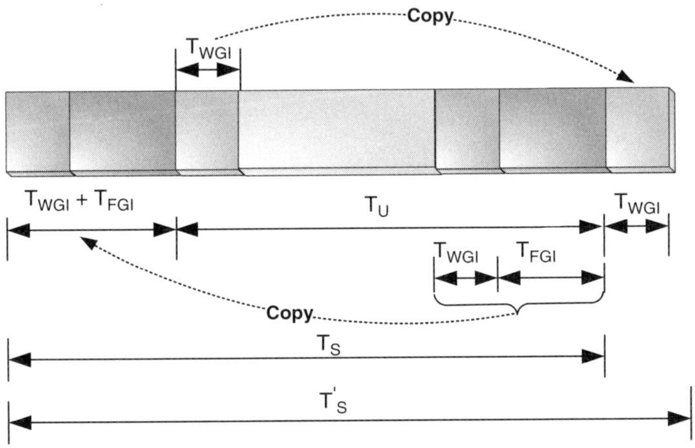

Fig. 7.6 OFDM symbol time-domain structure

The OFDM symbol time-domain structure is shown in Fig. 7.6. At base band, it consists of a number of time-domain samples called *OFDM chips*, which are transmitted at the chip rate 5.55 MHz (chip duration 0.18018 µs). The total OFDM symbol interval T_s' is comprised of four parts:

- T_U : The useful part with duration 4,096 chips. This is also duration of the FFT interval.
- T_{FGI} : The flat guard interval (also known as cyclic prefix) with duration 512 chips (1/8th of T_U). This duration of cyclic prefix allows for a multipath channel with delay spread up to $512 \times 0.18018 = 92.25\,\mu s$ which corresponds to 27.7 km path length difference without either inter-symbol interference or inter-carrier interference at the receiver [12].
- A raised cosine windowed interval of duration $T_{WGI} = 17$ chips is implemented on the beginning and end of each symbol.

The OFDM symbol is windowed by the following raised cosine function:

$$w(t) = \begin{cases} 0.5 + 0.5\cos(\pi + \pi t/T_{WGI}) & 0 \leq t \leq T_{WGI} \\ 1 & T_{WGI} < t < (T_{WGI} + T_{FGI} + T_U) \\ 0.5 + 0.5\cos(\pi + \pi(T_s' - t)/T_{WGI}) & (T_{WGI} + T_{FGI} + T_U) \leq t \leq (2T_{WGI} + T_{FGI} + T_U) \end{cases}$$

The purpose of the windowing function is to improve the spectral mask of the based-band OFDM signal by attenuating the side-bands. As shown in Fig. 7.7, two

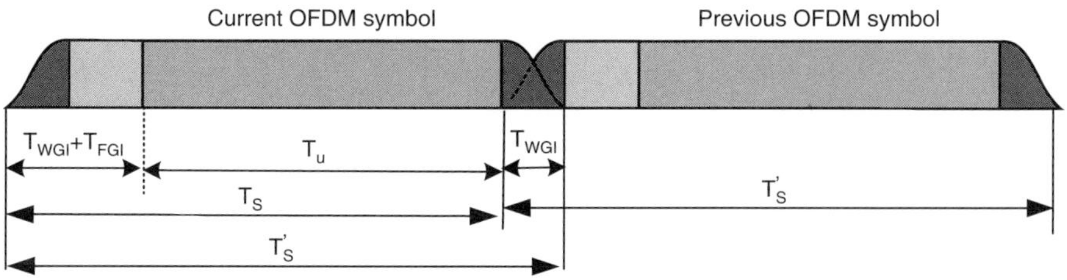

Fig. 7.7 Overlap of Windowed OFDM Symbols

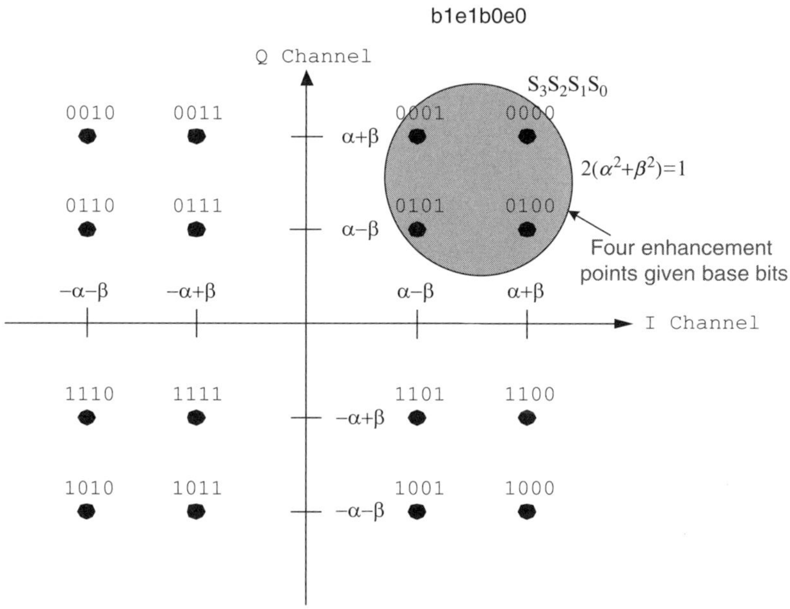

Fig. 7.8 Layered QPSK Signal Constellation

consecutive OFDM symbols overlap by T_{WGI}. Hence, the *effective OFDM symbol interval* is $T_s = T_{\text{WGI}} + T_{\text{FGI}} + T_{\text{U}} = 4625$ chips. The effective OFDM symbol interval is also referred to as the OFDM symbol interval. During an OFDM symbol interval, a modulation symbol is carried on each of the active subcarriers.

Each active subcarrier is modulated by QPSK, 16-QAM or Layered QPSK (16 states). If Layered QPSK is used, the data stream is divided into base layer and enhancement layer. Two base layer bits *b1b0* and two enhancement layer bits *e1e0* are combined to *b1e1b0e0* and mapped to a uniform or nonuniform 16-QAM signal constellation (Fig. 7.8). The two base layer bits *b1b0* determine the quadrant the signal point resides in and the two enhancement layer bits *e1e0* determine the exact location of the signal point within the quadrant selected by the two base layer bits. The enhancement bits can be successfully decoded only by users with higher SNR. In Fig. 7.8, the energy ratio between the base layer and the enhancement layer is α^2/β^2. It is equal to either 4.0 or 6.25 in FLO standard.

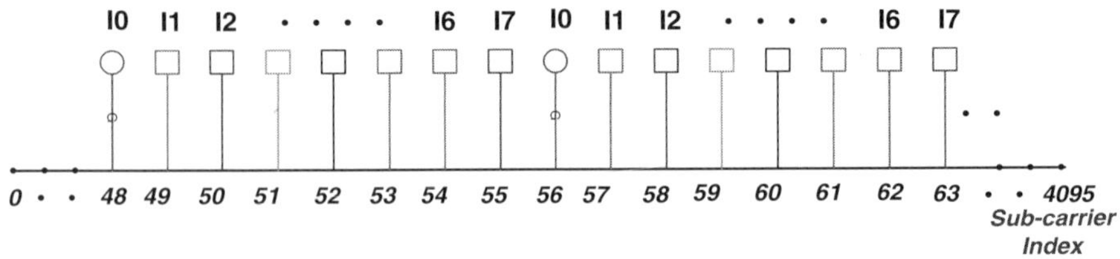

Fig. 7.9 FLO Interlace Structure

Layered modulation is suitable for applications that can produce base layer and enhancement layer bit streams associated with the same service. For example, a video application can use base layer bits to represent the basic details of the video and the enhancement layer bits to represent the fine details.

7.3.2 Interlace Structure

The 4,000 active subcarriers are divided into eight disjoint and equally sized groups called *interlaces*. The interlaces are indexed from 0 to 7. A subcarrier with index i belongs to the interlace with index equal to i *modulo 8*. Therefore, the 500 subcarriers each interlace has are evenly spaced and span the total FLO signal bandwidth (Fig. 7.9). An interlace is the minimum frequency allocation unit the system can allocate to an MLC within an OFDM symbol. It enables the frequency-division multiplexing of MLCs and allows for fine granularity of bandwidth allocation to MLCs. Since the subcarriers within an interlace span the total FLO signal bandwidth, there is no loss of frequency diversity within an interlace.

Slot is a concept closely related to interlace. A slot corresponds to a group of 500 constellation symbols. It is the smallest unit of bandwidth allocated to an MLC over an OFDM symbol. Each slot is mapped to one interlace (see Sect. 7.3.4). The eight slots are also indexed from 0 to 7. Note that the term *interlace* refers to a group of subcarriers, while the term *slot* refers to a group of constellation symbols and is defined for bandwidth allocation purposes.

7.3.3 Superframe Structure and Physical Layer Subchannels

7.3.3.1 Superframe Structure

The FLO transmitted signal is organized into superframes. Each superframe has duration 1 s, and consists of 1,200 OFDM symbols for a 6 MHz RF allocation. This is 200 symbols per MHz which is a common feature of all bandwidths. The OFDM symbols in a superframe are numbered 0 through 1,199. Figure 7.10 shows the FLO superframe structure. A superframe has four main portions:

- *TDM pilots*: The four OFDM symbols at the beginning of each super frame are TDM pilot 1 (TDM1), Wide-area Identification Channel (WIC), Local-area Identification Channel (LIC), and TDM pilot 2 (TDM2). TDM1 marks the beginning of a superframe. It can be used for superframe synchronization, initial frequency synchronization and coarse timing determination. The WIC and LIC symbols carry the WID (Wide-Area Differentiator) and LID (Local-Area Differentiator), respectively. Wide areas broadcasting the same wide-area service are allocated the same WID, and local areas broadcasting the same local-area services are allocated the same LID and WID. The TDM2 symbol is used for fine timing determination so the receiver can immediately start decoding the OIS symbols (see below). It can also be used to generate an initial estimate of the channel.
- *OIS*: The OIS portion has two sections: the Wide-Area OIS and the Local-Area OIS. Each section consists of five OFDM symbols. The Wide-Area OIS and the Local-Area OIS carry overhead information regarding the wide-area and local-area data channels (see below), respectively. The overhead information is the time-frequency allocation for each MLC in the current superframe (see Sect. 7.4.3).
- *Data frames*: The data portion has four data sections of equal duration, called *frames*. The frames are used to transmit MLCs. When an MLC is transmitted in a superframe the payload is divided into four equal bursts, with each burst transmitted in a *unique* frame. To support wide-area and local-area services (explained in Sect. 7.2), each data frame is further divided into two parts: the Wide-Area Data Channel for carrying wide-area service data and the Local-Area Data Channel for carrying the local-Area service data. The numbers of OFDM symbols for the Wide-Area and Local-Area data channels are specified by the information carried in the Wide-Area and Local-Area OIS channels, respectively.
- *Positioning Pilot Channel (PPC)/Reserved Symbols*: This portion is either the PPC or Reserved Symbols. The number of the OFDM symbols allocated to this portion is specified by the information carried in the OIS. The PPC symbols carry unique information for each transmitter and can be used for transmitter identification and/or the FLO receiver positioning based on the measured distances to transmitters. They will be completely defined in the future revisions of the FLO Air Interface standard.

There is an OFDM symbol, called the Transition Pilot Channel (TPC) symbols, on each side of every Wide- and Local-Area OIS or Wide- and Local-Area Data Channels. The TPC symbols are used to assist channel estimation at the boundary between the Wide- and Local-Area channels for demodulation of the OIS or data OFDM symbols adjacent to them. They also facilitate timing synchronization for the first Wide-Area or Local-Area MLC in each data frame.

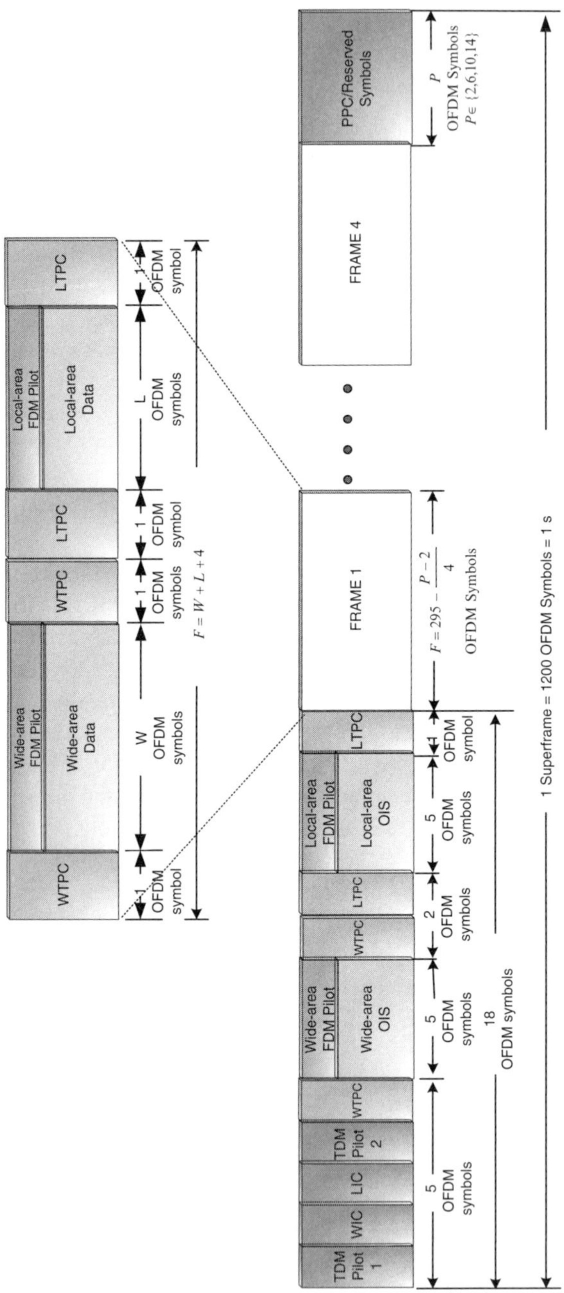

Fig. 7.10 FLO superframe structure

7.3.3.2 Impact of Superframe Duration

The FLO superframe duration was chosen to be 1 s based on the good trade-off among the following performance metrics:

- *Channel switching time:* A longer superframe duration leads to longer channel switching time. This is explained in Sect. 7.7.2.
- *Time diversity*: FLO enables data recovery for an MLC from data loss over the duration of up to one or two frames with the use of Reed-Solomon code as explained in Sect. 7.4.2, especially at code rate 8/16 or 12/16. The frame duration needs to be as large as the channel coherence time to allow de-correlation of the transmission in frames and provide time diversity gain. The Doppler frequency for an FLO receiver operating at 719 MHz with a velocity of 3 kmph is 2 Hz, which translates to a 212 ms coherence time. The coherence time is calculated by 0.423/Doppler frequency [6].
- *Statistical multiplexing gain*: the MediaFLO network dynamically allocates bandwidth (time-frequency resource allocation) to the multiple active MLCs on a per-superframe basis (see Sect. 7.4.3). The longer the superframe duration, the lower the standard deviation of the aggregate capacity allocation, and the greater the effective gain in capacity.
- *Receiver decoding buffer size and video/audio delay*: A longer superframe duration leads to a larger decoding buffer on the receiver and longer latency for the video and audio data.

7.3.3.3 Physical Layer Subchannels

The FLO Physical Layer Subchannels are shown in Fig. 7.11. The allocated subcarriers or interlaces, subcarrier modulation, error control coding and carried data on the subchannels are summarized in Table 7.2 (the scrambling PN sequences are generated by the method described in Sect. 7.3.4).

The data transmitted in each subchannel except for the TDM Pilot 1 is scrambled by a bit sequence that depends on the subchannel type, OFDM symbol index and slot index (see the Scrambling subsection in Sect. 7.3.4). Basically, all waveforms

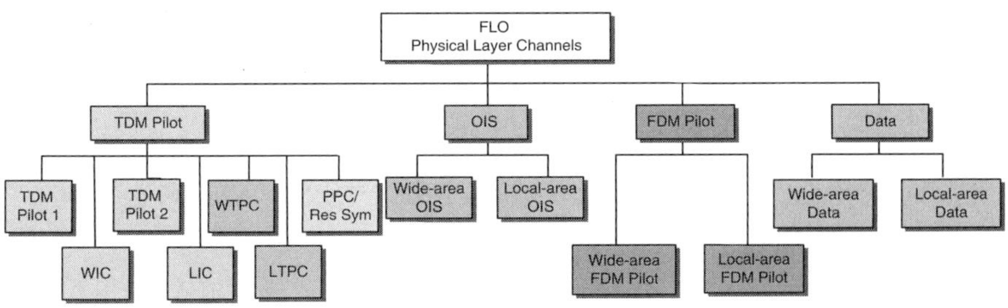

Fig. 7.11 FLO Physical Layer subchannels

Table 7.2 Summary of Physical Layer subchannels' transmission characteristics

Physical Layer subchannels	Allocated subcarriers or interlaces	Modulation and coding	Carried data
TDM1	One hundred and twenty-four evenly spaced subcarriers	QPSK	TDM Pilot 1 Information packet, a 248-bit fixed pattern from a PN-sequence
WIC	Interlace 0	QPSK	One thousand bits all-zero sequence scrambled by a PN sequence based on the WID assigned to the transmitter
LIC	Interlace 0	QPSK	One thousand bits all-zero sequence scrambled by a PN sequence based on the Wide-area Differentiator and the LID assigned to the transmitter
TDM2	Two thousand evenly spaced subcarriers (four interlaces)	QPSK	One thousand bits all-zero sequence scrambled by a PN sequence
WTPC/ LTPC	All eight interlaces	QPSK	One thousand bits all-zero sequence scrambled by a PN sequence on slot 0. A 1,000 bit fixed pattern (generated by an 11-tap linear feedback shift register) scrambled by a PN sequence on slot 1 to 7
FDM Pilot	One interlace (2 or 6)	QPSK	One thousand bits all-zero sequence scrambled by a PN sequence
OIS	Seven interlaces	QPSK, 1/5 Turbo code (Transmit Mode 5)	OIS overhead information
Physical Layer Data Channel	Seven interlaces	Modulation and coding on an interlace depends on the transmit mode of the MLC the interlace is allocated to	Service application data or control information

transmitted within a wide area are scrambled using the 4-bit WID corresponding to that area, and all waveforms transmitted within a local area are scrambled using a 4-bit LID, in conjunction with the WID, corresponding to that area. The WID and LID are transmitted in the WIC and LIC, respectively. The control information transmitted in the local control channel (see Sect. 7.5) also contains WID and LID for each of the neighboring local area. The purposes of the scrambling are explained in [2].

7.3.4 Waveform Generation for Data and OIS Channels

FLO services and control information are carried in logical channels called MLCs. An active MLC is allocated some slots in some OFDM symbols in a data frame and all active MLCs are multiplexed into the same data frame as described in Sect. 7.4.3.

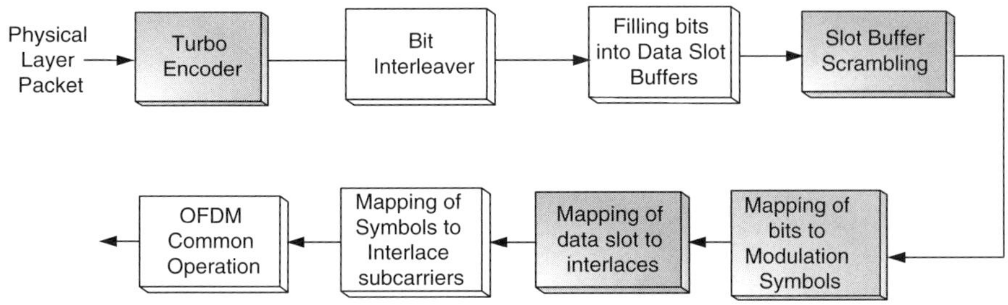

Fig. 7.12 Procedure to Generate Data Channel waveform contributed by an MLC

Table 7.3 FLO transmit modes and spectral efficiency

Transmit Mode	Modulation	Turbo code rate	Physical Layer data rate (Mbps)	Spectral efficiency (bps/Hz)
0	QPSK	1/3	2.8	0.47
1	QPSK	1/2	4.2	0.70
2	16-QAM	1/3	5.6	0.93
3	16-QAM	1/2	8.4	1.40
4	16-QAM	2/3	11.2	1.86
5[a]	QPSK	1/5	1.68	0.28
6	Layered QPSK with energy ratio 4	1/3	5.6	0.93
7	Layered QPSK with energy ratio 4	1/2	8.4	1.40
8	Layered QPSK with energy ratio 4	2/3	11.2	1.86
9	Layered QPSK with energy ratio 6.25	1/3	5.6	0.93
10	Layered QPSK with energy ratio 6.25	1/2	8.4	1.40
11	Layered QPSK with energy ratio 6.25	2/3	11.2	1.86

[a]This mode is used for the OIS channels only.

Figure 7.12 shows the procedure to generate the Physical Layer Data Channel waveform contributed by a particular MLC (OIS is a special case of data channels and uses the same procedure). The Turbo code rate and subcarrier modulation type for an MLC or OIS are specified by their transmit mode. The transmit modes defined in the FLO standard are shown in Table 7.3, along with the Physical Layer data rates,[3] and spectral efficiency for each transmit mode. Note that OIS always uses transmit mode 5. The transmit mode for the MLC mapped to a wide-area or local-area control channel (see Sect. 7.5) in a superframe is specified by the overhead information carried in the corresponding wide-area or local-area OIS channel in the same

[3] The Physical Layer data rates for all the RF channel bandwidths supported by FLO are listed in the Appendix.

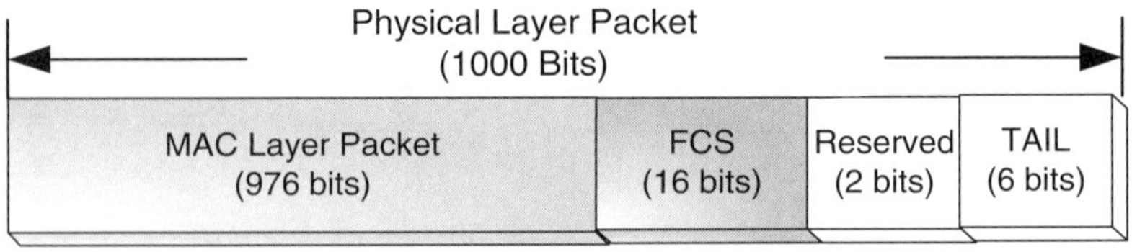

Fig. 7.13 FLO Physical Layer packet structure

superframe. The transmit modes for all other MLCs in the superframe are specified by the control information carried in the corresponding control channels in the same superframe.

7.3.4.1 Physical Layer Packet

The Physical Layer packet (PLP) is the unit of data transmission in the FLO Physical Layer. As shown in Fig. 7.13, a PLP has 1,000 bits in total and its payload size of 976 bits is equal to the MAC Layer packet size. The FCS (Frame Check Sequence) bits are 16 CRC bits calculated using the standard CRC-CCITT generator polynomial:

$$g(x) = x^{16} + x^{12} + x^5 + 1$$

The FCS bits are used by the receiver to determine if the PLP was correctly decoded or not.

7.3.4.2 Turbo Encoding

The PLPs are encoded by Turbo code derived from the Turbo codes defined in the CDMA2000 and 1x-EV-DO standards [1, 11]. In the FLO standard, the Turbo code is also referred as the inner code as compared to the Reed-Solomon code based outer code done on the MAC Layer. It is mainly used to exploit the frequency-diversity inherent in the channel. The FLO standard defines Turbo code rates of 1/3, 1/2 and 2/3. An MLC's Turbo code rate is specified by its transmit mode. The OIS uses code rate specified in transmit mode 5. The encoded PLPs are called Turbo Encoded Packets (TEP).

7.3.4.3 Bit Interleaving

The output bits of the Turbo encoder are bit-interleaved. The purpose of interleaving is to ensure that consecutive bits in the receiver's Turbo decoder input are transmitted on subcarriers that are sufficiently separated in the frequency domain such that they experience uncorrelated fading. This helps to disperse error bursts in the received PLP and increases the probability that the decoder will recover the encoded PLP as whole. Please refer to [4] for more details on the interleaver.

7.3.4.4 Filling Slot Buffers

Slots are allocated to MLCs at the MAC Layer. The TEPs of MLCs are transmitted in the slots allocated to the MLC. Slot allocations are done in such a way that multiple MLCs do not share the same slots within an OFDM symbol.

7.3.4.5 Scrambling

The bits in each allocated slot buffer are XORed sequentially with a scrambling bit sequence to randomize the bits prior to modulation.

As shown in Fig. 7.14, the scrambling bit sequence is generated according to the following procedure:

- For each slot at the start of every OFDM symbol, the state of the 20-tap linear feedback shift register with the generator sequence $h(D) = D^{20} + D^{17} + 1$ is initialized to $[\mathrm{d_3 d_2 d_1 d_0 c_3 c_2 c_1 c_0 b_0 a_{10} a_9 a_8 a_7 a_6 a_5 a_4 a_3 a_2 a_1 a_0}]$ that depends on the channel type and the OFDM symbol index as summarized in Table 7.4.
- The scrambling bit sequence is generated by a modulo-2 inner product of the state vector of the linear feedback shift register and a 20-bit mask associated with the slot index. The mask is chosen based on the slot index according to Table 7.5. The purpose of the mask is to divide the LFSR's state space into eight nonoverlapping segments so that each slot corresponds to a unique segment.

In general, the scrambling bit sequence depends on the channel type, OFDM symbol index and the slot index.

7.3.4.6 Slot to Interlace Mapping

The FLO system allocates time-frequency resources to FDM pilots, OIS and MLCs based on slots. The slots are mapped to interlaces on the Physical Layer. Figure 7.15 shows the mapping pattern that repeats after 14 consecutive OFDM symbols. The numbers in boxes are interlace indices. Slot 0 is always used for FDM pilot and mapped to interlace 6 and 2 alternatively on a per OFDM symbol basis.

The slot-to-interlace mapping for OIS and MLCs distributes their transmission over all interlaces in multiple OFDM symbols despite the fact that they are allocated a fixed set of slots within a superframe, which improves the performance on channels with periodic nulls. The mapping for the FDM pilots doubles the maximum channel multipath delay spread that can be estimated as explained earlier.

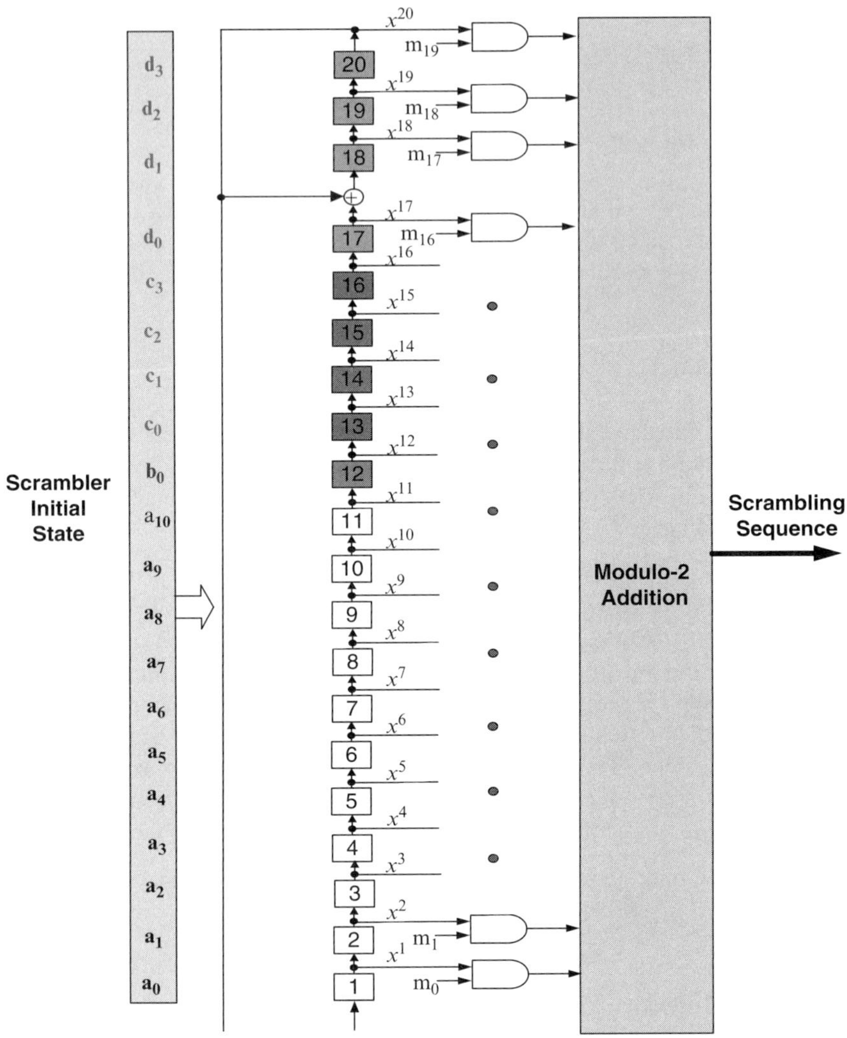

Fig. 7.14 Scrambling bit sequence generator

7.3.5 FDM Pilots

The FLO system uses one interlace as FDM pilots when the OIS or data frames are transmitted (Fig. 7.10). The FDM pilot is frequency division multiplexed with OIS or data channels (Fig. 7.16). The FDM pilot carries known modulation symbols and is used for channel estimation [7]. The subcarriers of the pilot interlace are modulated with QPSK symbols with the *same* energy as the MLC constellation symbols on the other interlaces.

Table 7.4 Initial state for the linear feedback shift register in the scrambling bit sequence generator

	D3d2d1d0	C3c2c1c0	b0	a10a9a8a7a6a5a4a3a2a1a0
TDM Pilot 2	WID	0000	1	Equal to OFDM symbol index number in a super-frame, which ranges from 0 through 1,199
WIC	WID	0000	1	
WTPC	WID	0000	1	
Wide-area FDM Pilot	WID	0000	1	
Wide OIS	WID	0000	1	
Wide Data	WID	0000	1	
LIC	WID	LID	1	
LTPC	WID	LID	1	
Local-area FDM Pilot	WID	LID	1	
Local OIS	WID	LID	1	
Local Data	WID	LID	1	
PPC/ Reserved OFDM Symbols	WID	LID	1	

Table 7.5 Masks associated with different slots

Slot Index	m_{19}	m_{18}	m_{17}	m_{16}	m_{15}	m_{14}	m_{13}	m_{12}	m_{11}	m_{10}	m_9	m_8	m_7	m_6	m_5	m_4	m_3	m_2	m_1	m_0
0	0	0	1	0	0	0	0	0	0	0	0	0	1	0	0	0	0	0	1	0
1	0	1	0	0	0	0	0	0	0	0	0	0	0	0	0	0	1	0	0	0
2	1	0	0	1	0	0	0	0	1	0	0	0	0	1	1	0	0	0	1	1
3	0	0	1	0	0	0	0	0	0	0	0	0	1	0	0	0	0	0	0	0
4	1	1	0	0	0	0	0	0	0	0	1	0	0	0	0	0	0	0	0	0
5	1	0	0	1	0	0	0	0	1	0	0	0	0	1	0	0	0	0	1	0
6	0	1	1	0	0	0	1	0	0	0	0	1	0	0	0	0	1	1	0	0
7	1	0	0	0	0	0	0	0	0	0	0	0	0	0	0	0	0	0	0	0

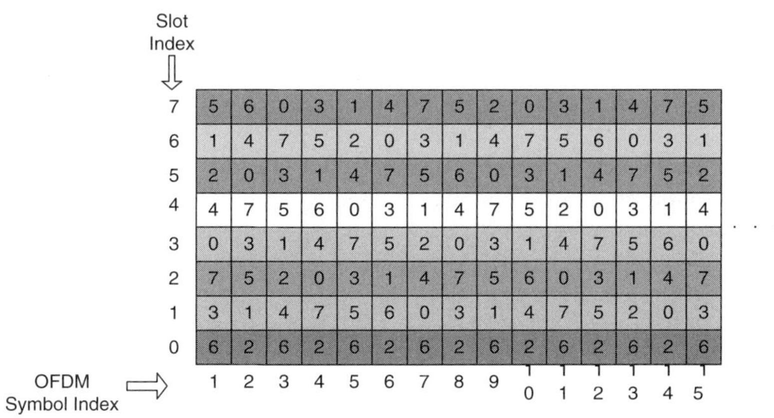

Fig. 7.15 Slot to Interlace Mapping

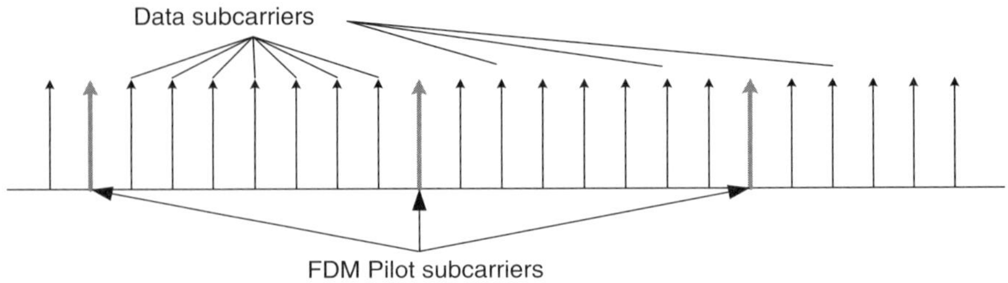

Fig. 7.16 Pilot and data subcarriers

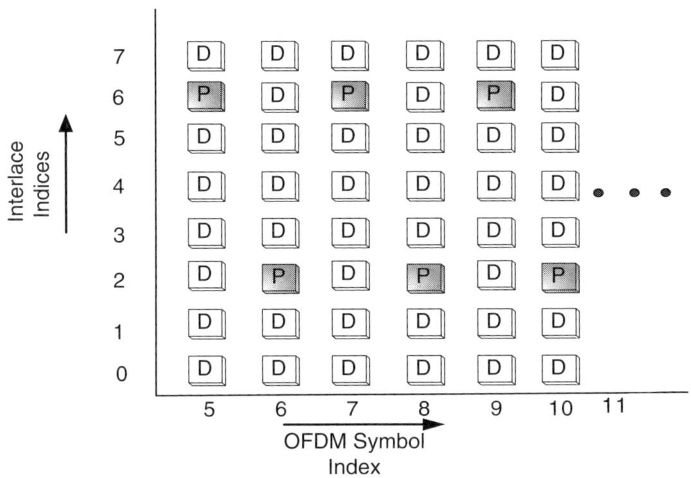

Fig. 7.17 FDM pilot interlaces

The FLO system allocates Slot 0 to the FDM pilot. The remaining seven slots are for OIS or data channels. As shown in Fig. 7.17, Slot 0 is mapped to interlace 2 on OFDM symbols with even indices and interlace 6 on OFDM symbols with odd indices. This allows the FLO receiver to use the pilot observations from two consecutive OFDM symbols to estimate channel with multipath delay spread up to two times the duration of the cyclic prefix (512 chips).[4] This allows graceful degradation of its performance for channels with multipath delay spreads greater than the cyclic prefix. Receivers in an OFDM system for which the channel multipath delay spread is not longer than the applied cyclic prefix do not experience inter-symbol interference or inter-carrier interference.

[4] The maximum channel multipath delay spread in chips that can be estimated in an OFDM system is limited by the number of FDM pilot subcarrier observations (FFT outputs corresponding to the pilot).

7.4 MAC Layer

The MAC Layer principally accomplishes the following:

- Defines the data encapsulation format and procedure for the data channels, control channels and OIS channels.
- Performs Reed-Solomon encoding and decoding.
- Allocates time-frequency resources to MLCs.

7.4.1 Data Encapsulation

7.4.1.1 Data Channel MAC

A data channel[5] is used to carry MediaFLO service data. It is mapped to MLCs. The content of a data channel for one superframe is encapsulated in an entity referred to as Data Channel MAC protocol capsule. Figure 7.18 illustrates how the Data Channel MAC protocol encapsulates Stream layer packets within a Data Channel MAC protocol capsule for an MLC configured for nonlayered mode of operation. Basically, the MAC Layer on the network side does the following:

- Concatenating Stream Layer packets in reverse order of stream ID.
- Adding a MAC Capsule Trailer to the Stream Layer Trailer in the stream 0 packet. The MAC Capsule Trailer has time-frequency allocation for the MLC in the *next* superframe in which the MLC will be present, which is also referred as "embedded OIS." Note the MAC Capsule Trailer is part of the Stream Layer Trailer.
- Add stuffing packets, if necessary, to make the MAC Protocol Capsule size to be equal to the next integer multiple of Reed-Solomon information block size (as described in Sect. 7.4.2).
- Perform Reed-Solomon encoding and generate parity packets according to Sect. 7.4.2.
- Fragment the MAC Protocol Capsule into MAC Layer packets and send them to the Physical Layer for transmission. A MAC Layer packet is 122 octets in size and forms the payload of one Physical Layer packet (PLP).

Note that the sizes of the Stream Layer packets are always an integer multiple of the MAC packet size (see Sect. 7.6).

The encapsulation for an MLC configured for layered mode is similar and shown in Fig. 7.19. The Base Layer and Enhance lLayer Data Channel MAC protocol capsules have the same size.

The information about the number of MAC Layer packets for each stream in a Wide- or Local-Area MLC in a superframe is transmitted in the Wide- or Local-Area OIS of the same superframe. The MAC Layer on the FLO receiver uses this information to demultiplex MAC Layer packets that are received from an MLC into streams.

[5] Note it should not be confused with the Physical Layer data channel.

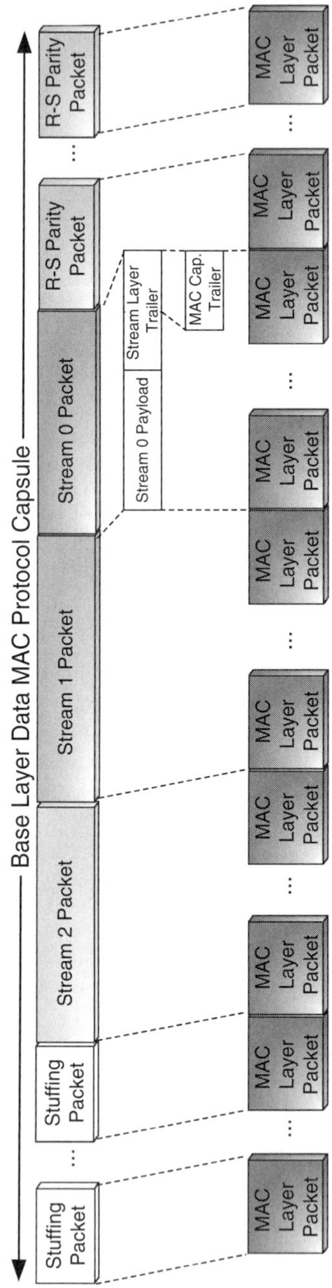

Fig. 7.18 Data Channel MAC protocol encapsulation for nonlayered transmit mode

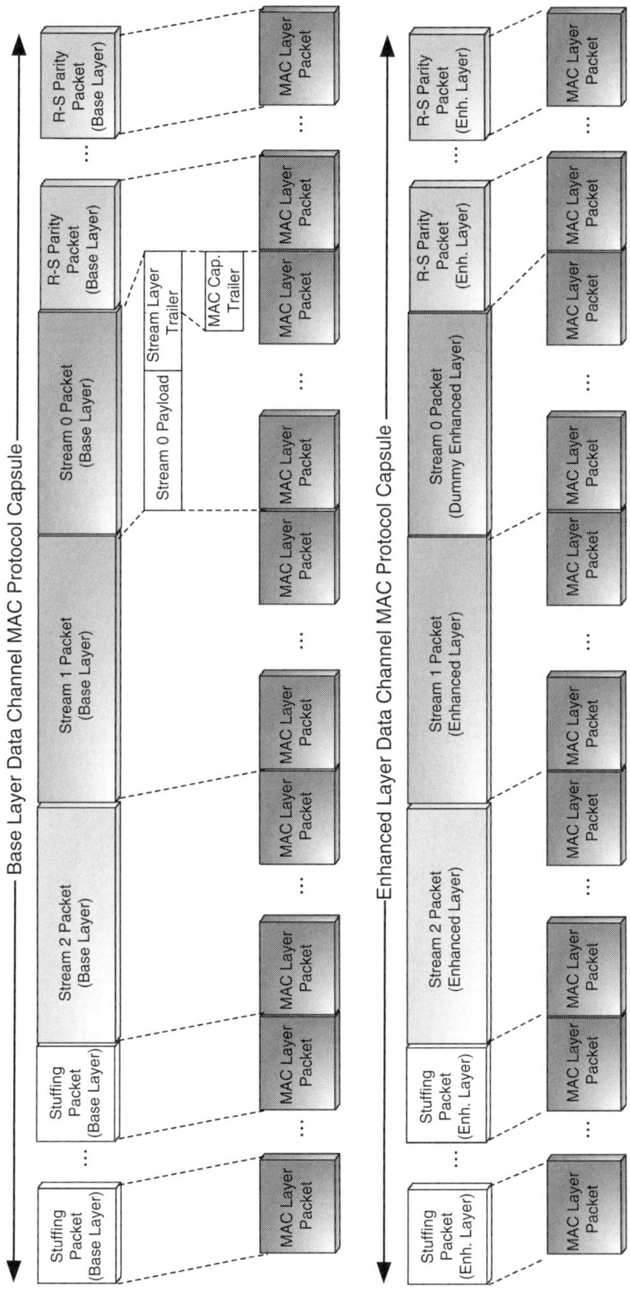

Fig. 7.19 Data Channel MAC protocol Encapsulation for layered transmit mode

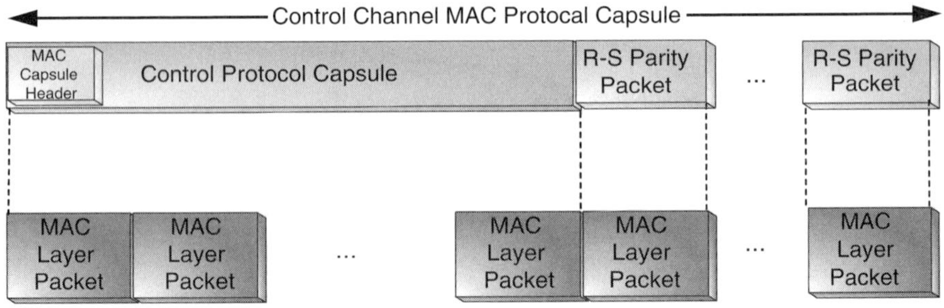

Fig. 7.20 Control Channel MAC protocol encapsulation

7.4.1.2 Control Channel MAC

Control channels are used to carry the overhead information (not carried by the OIS channel) that is needed by FLO receivers to acquire the FLO data channels. Like data channels, these are mapped to MLCs as well. The MLC mapped to a control channel is configured for nonlayered mode operation. The content of a control channel for one superframe is encapsulated in an entity referred to as the Control Channel MAC Protocol Capsule.

Figure 7.20 shows how the Control Channel MAC protocol encapsulates the Control Channel Capsule received from the Control Protocol within the Control Channel MAC Protocol Capsule. Basically, the MAC Layer on the network side does the following:

- Receives the Control Protocol Capsule from the Control Protocol.
- Puts the MAC Capsule Header in the Fill Field of the Control Protocol Capsule. Note that the Fill Field is reserved by the Control Protocol in the Control Protocol Capsule to accommodate the MAC Capsule Header (see Sect. 7.5.1). The Control Protocol Capsule size is always an integer multiple of the MAC Layer packet size.
- Performs Reed-Solomon encoding and generates parity packets according to Sect. 7.4.2.
- Fragments the Control Channel MAC Protocol capsule into MAC Layer packets and send them to the Physical Layer for transmission.

7.4.1.3 OIS Channel MAC

The OIS channels carry overhead information regarding the wide and local data channels such as the time-frequency allocation for each MLC in the current superframe. The overhead information is formatted as the System Parameters Message.

As shown in Fig. 7.21, the OIS Channel MAC Protocol on the network side constructs the System Parameters Message, then breaks in it into MAC Layer packets and passes it to the Physical Layer for transmission.

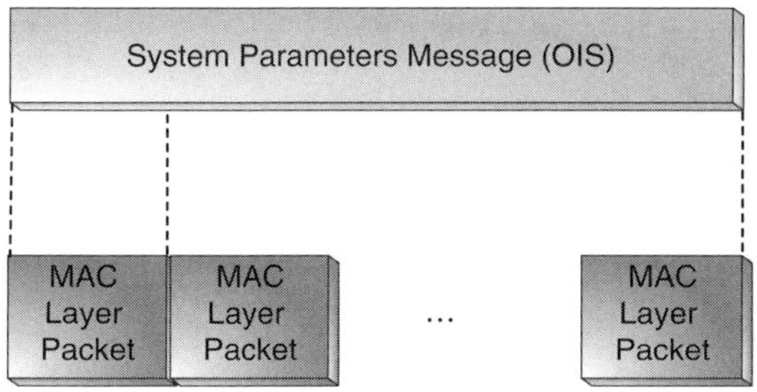

Fig. 7.21 OIS Channel MAC protocol encapsulation

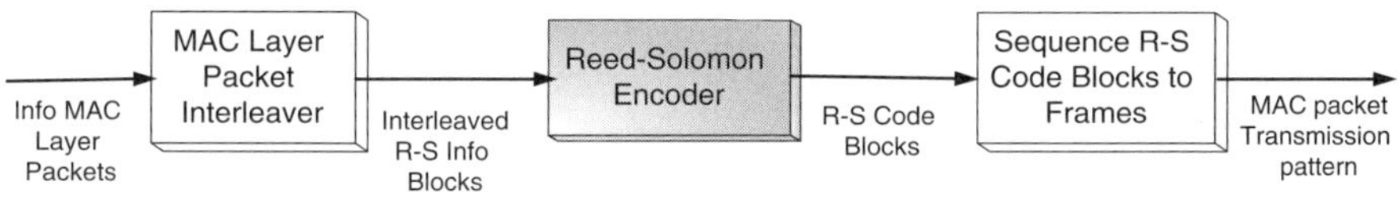

Fig. 7.22 Procedure to protect the Information MAC Layer packets

7.4.2 Reed-Solomon Code

The MAC Layer information packets in an MLC are protected by a (N, K, R) Reed-Solomon code, also referred to as the outer code in the FLO standard. The procedure is shown in Fig. 7.22: the MAC Layer Packet Interleaver receives the information MAC Layer packets formed for each superframe by the methods in Sect. 7.4.1 and generates interleaved Reed-Solomon information blocks that are passed to the Reed-Solomon Encoder. The Reed-Solomon Encoder generates code blocks from the received information blocks and the sequencer delivers the MAC Layer packets in the code blocks in sequence to the Physical Layer for transmission.

Figure 7.23 shows how the Reed-Solomon encodes an interleaved information block:

- Groups the bits in each of the K packets in the information block into 8-bit octets.
- Performs Reed-Solomon encoding on each column of K octets in the information block to generate a codeword of N octets with $R = N - K$ parity octets. The parity octets from all the codewords form $R = N - K$ parity MAC Layer packets. The K information MAC packets and R parity MAC packets form a Reed-Solomon code block.

Note that MLC transmissions in each superframe are always in integer multiples of Reed-Solomon code blocks.

The FLO standard defines the following Reed-Solomon code rates for MLCs: (16, 16, 0), (16, 14, 2), (16, 12, 4) and (16, 8, 8). The Reed-Solomon code rate for an MLC mapped to a control channel is specified in the overhead information (System Parameters Message) transmitted in the OIS while the code rates for all other MLCs are specified in the control information transmitted in the control channels. This is the same method as MLC transmit modes are specified.

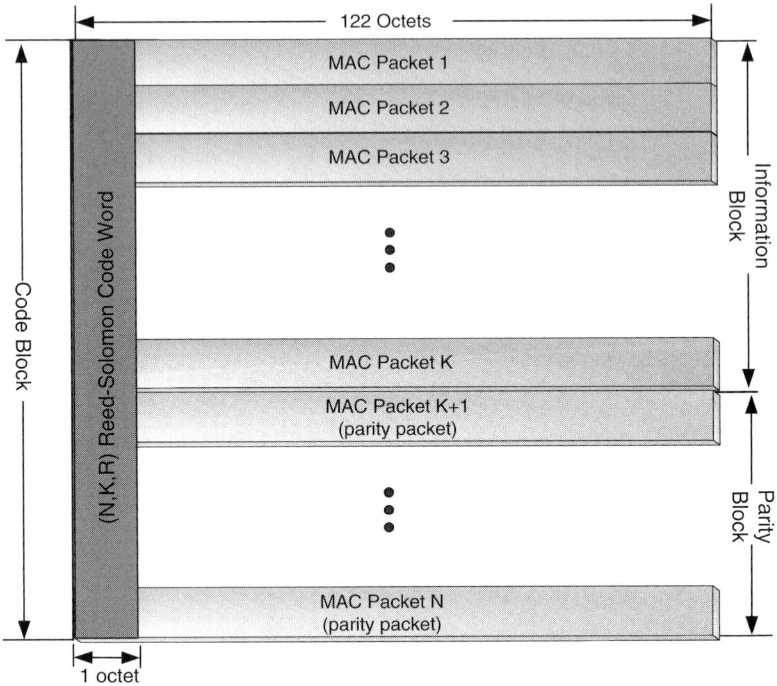

Fig. 7.23 Reed-Solomon code block in FLO

Figure 7.24 shows how the MAC Layer packets in a Reed-Solomon code block is sequenced to the four data frames within a superframe for transmission: the code block is split into four equal-size subblocks and each subblock is sent in a unique frame within the superframe. This process is repeated for every Reed-Solomon code block to be transmitted in a superframe. Since the sequencing process distributes the MAC packets in a Reed-Solomon code block evenly over the four data frames in a superframe, it increases the time-diversity gained across each code block.

The interleave and sequencer in Fig. 7.22 are designed in such a way that the MAC packet transmission pattern generated by the procedure for an MLC in a superframe has the characteristics as illustrated in Fig. 7.25. Each box represents a MAC layer packet. In this example, the MLC has two Reed-Solomon code blocks for the superframe):

- The MAC packets are transmitted in the same order as they are formed by the methods in Sect. 7.4.1. The number in each box indicates the corresponding packet's order.
- Due to the MAC Layer Packet interleaver, any two consecutive packets that belong to a code block and are in the same data frame are separated by $n-1$ packets belonging to other code blocks where n is the total number of the code blocks the MLC has in the superframe. This helps to disperse error bursts in each code block and increases the probability that the Reed-Solomon decoder can recover it. Please refer to [4] for more details on the interleaver.

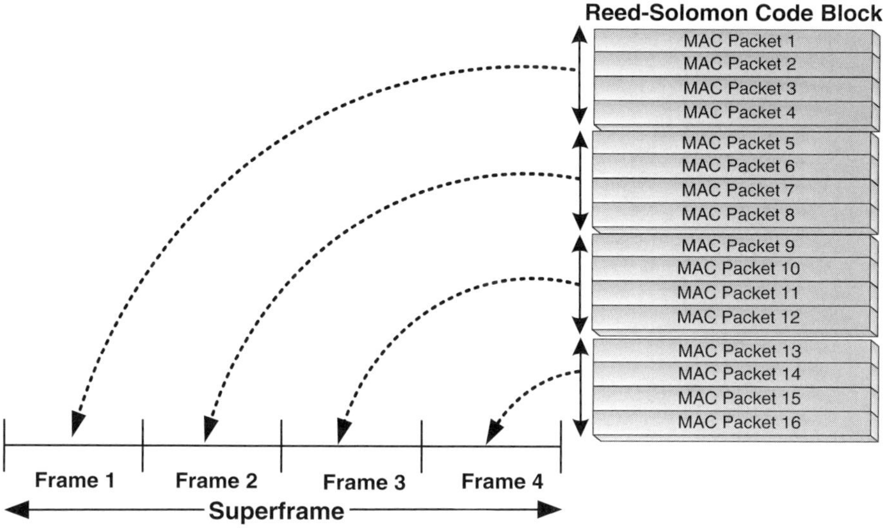

Fig. 7.24 Sequence MAC Layer packers in an R-S code block to data frames

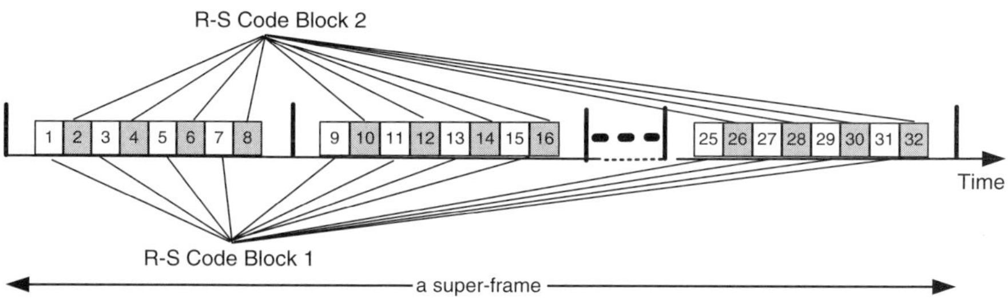

Fig. 7.25 Reed-Solomon code blocks sequenced to data frames in a super-frame

7.4.3 Time-Frequency Resource Allocation to MLCs

In the FLO system, the data carried in an MLC in a superframe is divided into four equal-size bursts and each burst is transmitted in a unique data frame in the superframe. The MAC Layer on the network allocates time-frequency resources in each data frame to the MLC. The allocation is in units of OFDM symbols on time domain and in units of slots on frequency time. The time-frequency allocation for an active MLC is identical in all the four data frames within a superframe and conveyed to the receiver by the following attributes for the MLC in the System Parameters Message transmitted in the OIS (see Fig. 7.26: an area with a unique color represents the time-frequency allocations for the corresponding MLC; the numbers in the box are stream index):

- *StartOffset*: the index of the first allocated OFDM symbol.
- *StartSlot*: the index of the lowest slot in the first allocated OFDM symbol.

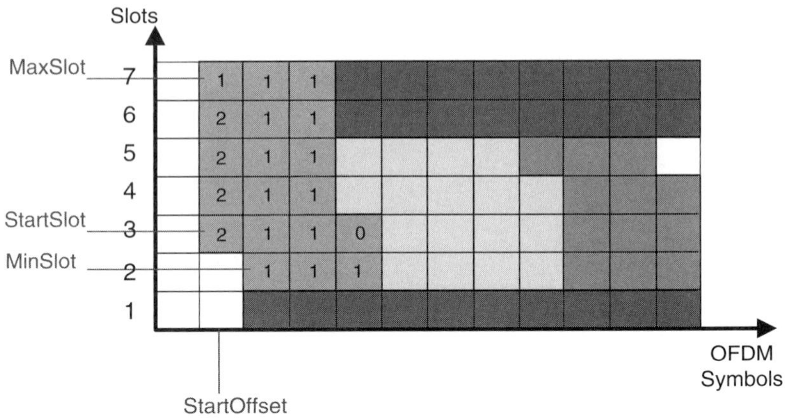

Fig. 7.26 Time-frequency allocation specification

- *MaxSlot*: the index of the highest slot in all the allocated OFDM symbols.
- *MinSlot*: the index of the lowest slot in the allocated OFDM symbols.
- *StreamLength*: it specifies the number of MAC Layer packets for each stream in the MLC. The total number of slots allocated to the MLC in the superframe can be determined from the total number of MAC Layer packets for the MLC in the superframe.

The area corresponding to the time-frequency allocation for an MLC in the time-frequency plot (Fig. 7.26) does not have to be a rectangular area. This flexibility offers the following advantages:

- Enables the FLO network to allocate bandwidth to an MLC that is very closely matched to what the MLC requires.
- Minimizes the number of slots that cannot be allocated due to constraint on the shapes of the allocated areas.

The time-frequency allocation is typically adjusted on a per-superframe basis according to the traffic variation. This mechanism allows the statistical multiplexing in an FLO Physical Layer Data Channel to be very efficient.

7.5 Control Layer

The control channel carries the control information that is required by FLO receivers to acquire and navigate the FLO data channels. The control information is carried in the following three types of control protocol messages:

- *Flow Description Message*: conveys information that maps upper layer flow ID to stream, MLC and RF Channel, and specifies the MLC parameters (e.g., transmit mode, Reed-Solomon code rate) and the stream parameters.

- *RF Channel Description Message*: conveys information about the RF Channels in use such as frequency and bandwidth, e.g. 5, 6, 7 or 8 MHz.
- *Neighbor List Description Message*: conveys information about the neighboring local-area e.g. RF Channels, WID, and LID. This information aids device transition to new local-areas.

The FLO system uses sequence numbers to indicate the versions of the control information. The sequence numbers are carried in the OIS that helps the FLO receiver determine if it has the latest control information. This allows the receiver to save power as the receiver will only decode the control channel when newer control information is present.

The control channel uses the Control Channel MAC Protocol to insert a special format MLC into the Physical Layer Data Channel (it does not use Stream Layer). The characteristics of the MLC mapped to the wide-area or local-area control channel in a superframe such as time-frequency allocation, transmit mode, Reed-Solomon code rates are carried in the System Parameters Message transmitted in the wide-area/local-area OIS channel in the same superframe.

7.5.1 Data Encapsulation

Each control protocol message is encapsulated in one or several Control Protocol Packets (CPP) for transmission. The CPP has the same size as the MAC Packet (122 octets) and there is one-to-one mapping from CPPs to Control Channel MAC Packets. The CPP structure is shown in Fig. 7.27.

The Control Protocol uses the CPPs to construct the Control Protocol Capsule. A Control Protocol Capsule is defined as a group of CPPs transmitted or received in a single superframe and its structure is shown in Fig. 7.28. The control information is

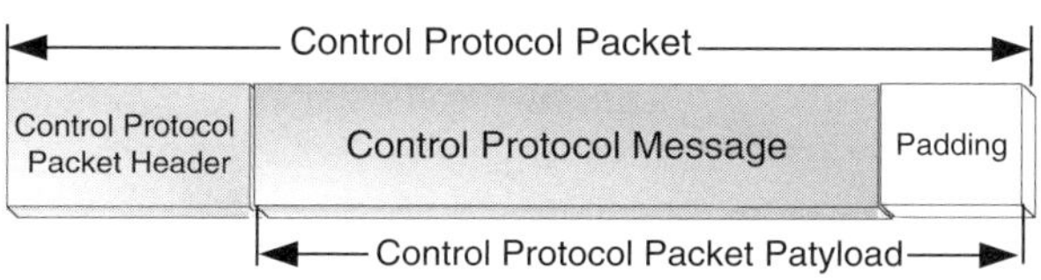

Fig. 7.27 Control Protocol Packet structure

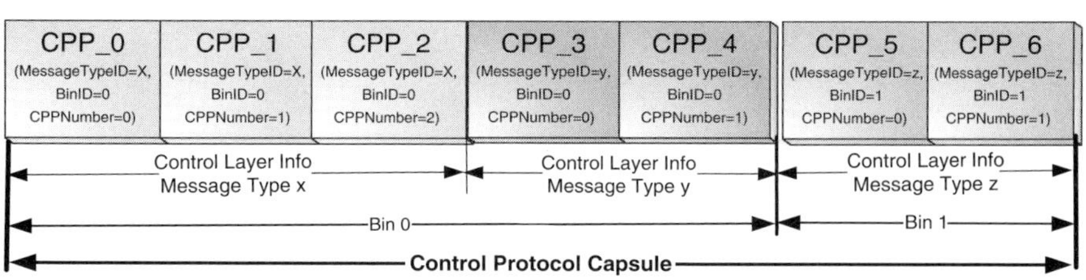

Fig. 7.28 Control Protocol Capsule Structure

organized into two logical groups called *bins* for efficient update purpose. Each bin has its own sequence number transmitted in the OIS to indicate its version, which allows the system to update the control information in a particular bin independently. Note the CPP header in the *first* CPP in a Control Protocol Capsule has a Fill Field where the MAC Layer will put the MAC Capsule Header (see Sect. 4.1).

Since the CPP has the same size as the MAC Packet, the Control Protocol Capsule size is always an integer multiple of the MAC Packet size. Each Control Protocol Capsule is carried in a Control Channel MAC Protocol Capsule (see Sect. 4.1).

7.6 Stream Layer

The Stream Layer maps the application flows to the streams and MLCs (the mapping is transmitted to the device in the Flow Description Message on the Control Channel). It receives data from application flows, constructs Stream Layer packets from the data and sends the packets to the Data Channel MAC protocol for transmission. The size of a Stream Layer packet is always an integer multiple of the MAC Layer packet size (122 octets).

The Stream Layer interface exposed to the upper layer supports two interface modes:

- *Transparent or Block Flow Mode* in which the Stream Protocol in the network receives a stream of 122-octet blocks from the Upper Layer and the peer protocol in the device delivers these fixed sized octet blocks to the Upper Layer. Each of the 122-octet blocks will be carried by a unique PLP. This mode is supported for streams 1 and 2.
- *Octet Flow Mode* in which the Stream Protocol in the network receives a stream of octets from the Upper Layer and the peer protocol in the device delivers a stream of octets to the Upper Layer. Stream Layer Packets constructed from the octet stream may have padding as the size of a Stream Layer Packet is always an integer multiple of the MAC Layer packet size. Stream 0 is operated in this mode.

7.7 FLO Air Interface Handling Scenarios at the Receiver

7.7.1 MLC Reception

The transmission for an MLC in a superframe consists of four equal-sized bursts with each burst transmitted in a unique frame. The time-frequency allocation for each MLC is identical for the four frames within a superframe and specified by the information transmitted in the OIS of the same superframe. The FLO receiver uses this information to locate the MLC in the superframe as shown in Fig. 7.29.

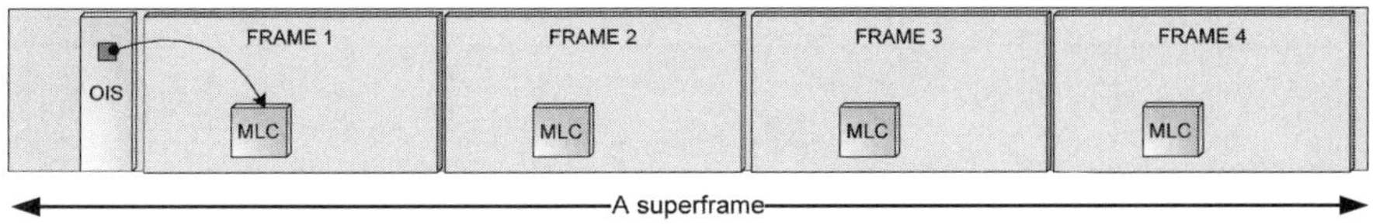

Fig. 7.29 Locating MLC Using OIS Information

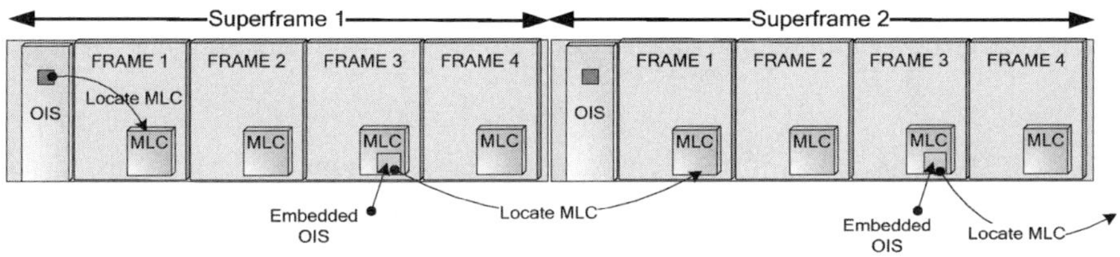

Fig. 7.30 Locating MLC Using Information in embedded OIS

The MAC protocol on the network copies the time-frequency allocation for an MLC in the *next* superframe to the *embedded OIS* (see Sect. 4.1). When the receiver needs to receive data in a specific MLC for a sustained period of time, it is more power efficient to utilize the embedded OIS. As shown in Fig. 7.30, the FLO receiver may use the embedded OIS received as part of the MLC payload in the current superframe to locate the MLC in the next superframe in which the MLC will be present, and avoid reading the OIS channel in every superframe.

The MLC reception on the FLO receiver can be very power efficient for the following reasons:

- The receiver only performs demodulation and decoding for the slots that are allocated to the MLC in the OFDM symbols that are allocated to the MLC. The FFT block in the receiver can be designed such that only the interlaces allocated to the MLC are demodulated.
- The embedded OIS allows decoding the OIS to be optional in superframes other than the first one in which the receiver starts the MLC reception.

7.7.2 Service Channel Switching

MLC switching on the FLO receiver is typically triggered by the user action like switching service channel on the receiver. A timeline for an FLO receiver switching from an old MLC to a new one is shown in Fig. 7.31:

1. The user selects a new service channel and the receiver is instructed to switch to the MLCs corresponding to the new service channel in the superframe 1.
2. The receiver decodes the System Parameters Message from OIS in superframe 2 and finds out it has the current control information based on the control information sequence number in the received System Parameters Message (see Sect. 7.5).

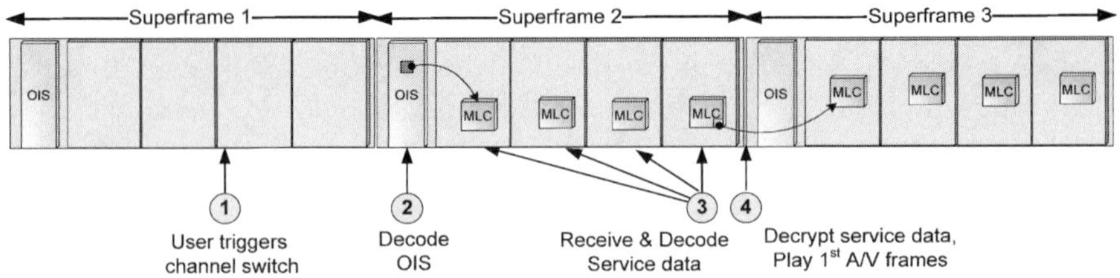

Fig. 7.31 Timeline for a device with current control channel data to acquire a real-time service

It then maps the flows in the new selected service channel to MLCs using the control information (Flow Description Message) and finds the locations of the MLCs in the superframe using information from the received System Parameters Message.

3. The device demodulates and decodes PLPs for the MLCs mapped to the flows in the new selected service from the four data frames in superframe 2, and Reed-Solomon decodes the PLPs to recover the service data after receiving the PLPs from the fourth data frame.
4. The service data is decrypted and the first video frame and the corresponding audio are played at the beginning of superframe 3. It is convenient to assume the first video frame is a frame like an Independent Decoder Refresh (IDR) frame that can be decoded independently.

7.8 Conclusions

MediaFLO is an open global technology standard that is purpose-built for mobile broadcast. This chapter gives an overview on the FLO Air Interface between a MediaFLO network and a device. The focus is on the efficient techniques that FLO is able to employ due to lack of constraint on backward compatibility with legacy technology. These techniques allow FLO to offer the following unique advantages:

- *Low receiver power consumption*: FLO makes the MLCs carrying service data distinguishable on the Physical Layer, which enables the FLO receiver to consume power only on receiving and decoding the desired data. The embedded OIS can allow the receiver reduced power consumption by eliminating the need for decoding OIS most of the time.
- *Fast service channel switching time*: The FLO superframe structure and one second superframe duration allows for fast service channel switching while maintaining high time diversity gains.
- *High spectral efficiency*: Both simulations and field tests have shown FLO capable of high spectral efficiency [2, 9]. FLO achieves this by offering significant time and frequency diversity gains and using Turbo code instead of convolutional code.

- *Efficient statistical multiplexing*: The variable bandwidth available for allocation to each MLC in an FLO system can be closely matched to service requirements. This bandwidth allocation can be adjusted dynamically on a per-superframe basis according to the traffic variation. This capability allows effective and efficient statistical multiplexing to be implemented across the FLO Physical Layer.

Appendix

Table 7.6 FLO System parameters for different RF channel bandwidths

Parameters	5 MHz	6 MHz	7 MHz	8 MHz
Chip rate (FFT bandwidth)	4.625 MHz	5.55 MHz	6.475 MHz	7.4 MHz
Chip duration	0.21622 μs	0.18018 μs	0.15444 μs	0.13514 μs
Number of subcarriers (FFT size)	4,096	4,096	4,096	4,096
Subcarrier spacing	1.129 kHz	1.355 kHz	1.581 kHz	1.807 kHz
FFT interval (useful OFDM symbol interval)	885.64 μs (4,096 chips)	738.02 μs (4,096 chips)	632.59 μs (4,096 chips)	553.53 μs (4,096 chips)
Cyclic prefix (flat guard interval)	110.70 μs (512 chips)	92.25 μs (512 chips)	79.07 μs (512 chips)	69.19 μs (512 chips)
Window interval	3.68 μs (17 chips)	3.06 μs (17 chips)	2.63 μs (17 chips)	2.30 μs (17 chips)
(Effective) OFDM symbol interval	1,000.00 μs (4,625 chips)	833.33 μs (4,625 chips)	714.29 μs (4,625 chips)	625.02 μs (4,625 chips)
Number of guard subcarriers	96	96	96	96
Number of active subcarriers	4,000	4,000	4,000	4,000
Number of pilot subcarriers	500	500	500	500
OFDM symbols per superframe	1,000	1,200	1,400	1,600

Table 7.7 FLO Physical Layer Data Rates for Different RF Channel Bandwidths

Transmit mode	Modulation	Turbo code rate	5 MHz Physical Layer data rate (Mbps)	6 MHz Physical Layer data rate (Mbps)	7 MHz Physical Layer data rate (Mbps)	8 MHz Physical Layer data rate (Mbps)
0	QPSK	1/3	2.33	2.80	3.27	3.73
1	QPSK	1/2	3.50	4.20	4.90	5.60
2	16-QAM	1/3	4.67	5.60	6.53	7.47
3	16-QAM	1/2	7.0	8.40	9.80	11.20
4	16-QAM	2/3	9.33	11.2	13.07	14.93
5[a]	QPSK	1/5	1.40	1.68	1.96	2.24
6	Layered QPSK with energy ratio 4	1/3	4.67	5.60	6.53	7.47
7	Layered QPSK with energy ratio 4	1/2	7.00	8.40	9.80	11.20
8	Layered QPSK with energy ratio 4	2/3	9.33	11.20	13.07	14.93
9	Layered QPSK with energy ratio 6.25	1/3	4.67	5.60	6.53	7.47
10	Layered QPSK with energy ratio 6.25	1/2	7.00	8.40	9.80	11.2
11	Layered QPSK with energy ratio 6.25	2/3	9.33	11.20	13.07	14.93

[a]This mode is used for the OIS channels only.

References

1. CDMA2000 High Rate Packet Data Air Interface Specification, 3GPP2 C.S2024-A v1.0, March 2004.
2. M. Chari, F. Ling, A. Mantravadi, et al, "FLO physical layer: an overview", *IEEE Trans. Broadcasting*, vol. 53, no. 1, pp. 145–160, 2007.
3. FLO Forum Web Site http://www.floforum.org/
4. Forward Link Only Air Interface Specification for Terrestrial Mobile Multimedia Multicast, TIA-1099, Aug. 2006.
5. Forward Link Only Transport Specification, TIA-1120.
6. W. C. Jakes, Ed., *Microwave Mobile Communications.* Piscataway, NJ: IEEE Press.
7. Y. Li, "Pilot symbol aided channel estimation for OFDM in wireless systems", *IEEE Trans. Vehicle Technology*, vol. 49, pp. 1207–1215, July 2000.
8. A. Mattson, "Single frequency networks in DTV", *IEEE Trans. Broadcasting*, vol. 51, no. 4, pp. 413–422, 2005.
9. "MediaFLO Field Test Report", QUALCOMM White Paper. Available at http://www.qualcomm.com/mediaflo
10. MediaFLO USA Web Site http://www.mediaflousa.com/
11. *Physical Layer Standard for cdma2000 Spread Spectrum System-Release C, 3GPP2 C.S2002-C v1.0, May 2002*
12. M. Speth, S. A. Fechtel, G. Fock, and H. Meyr, "Optimum receiver design for wireless broadband systems using OFDM. I", *IEEE Trans. Communications*, vol. 47, no. 11, 1999.
13. TIA Web Site http://www.tiaonline.org/

Part II
Baseband Processing in Mobile Multimedia Broadcasting

Chapter 8
DVB-H Link Layer

Onno Eerenberg, Arie Koppelaar, and Peter H.N. de With

8.1 Introduction

In the fall of 2004, the European Telecommunications Standards Institute (ETSI) approved the Digital Video Broadcast Handheld (DVB-H) standard [19], which is specifically tailored to battery-powered mobile reception and developed by the International Digital Video Broadcasting (DVB) Project [8]. The DVB-H standard, formerly known as DVB-X [20], is an extension to the DVB-T standard [17] with extra features added to the physical and link layer.

With respect to the DVB-T physical layer, the DVB-H physical layer is extended with Transmission Parameter Signalling (TPS) to, e.g., fasten service discovery, a 4K mode to trade off Doppler sensitivity versus echo sensitivity and an in-dept symbol interleaver to increase the robustness to, e.g., impulsive noise. The DVB-H link layer uses a Time-Division-Multiplex (TDM) broadcast technique, called time-slicing, to transmit a service, enabling power-efficient service reception. On top of this, the DVB-H link layer is foreseen with a second Forward Error Correction (FEC) layer, called MPE-FEC, which protects the received service against various reception impairments, e.g., Carrier-to-Noise Ratio (CNR) ratio, Doppler or impulsive noise influences. The DVB-H additional protection by the MPE-FEC provides a CNR improvement of 4–6 dB advantage to DVB-H transmission with respect to DVB-T and reduces the influence of the speed factor while receiving the signal [7]. According to the DVB-H standard, time-slicing is a mandatory feature, whereas the second FEC layer (MPE-FEC) is an optional feature.

O. Eerenberg (✉) and A. Koppelaar
NXP Semiconductors, Research High Tech Campus 32, 5656 AE Eindhoven, The Netherlands
e-mail: onno.eerenberg@nxp.com, arie.koppelaar@nxp.com

P.H.N. de With
University of Technology Eindhoven, Fac. EE, Postbus 513, 5600 MB Eindhoven, The Netherlands
e-mail: P.H.N.de.With@tue.nl

F.-L. Luo (ed.), *Mobile Multimedia Broadcasting Standards: Technology and Practice*, 223
DOI: 10.1007/978-0-387-78263-8_8,

Unlike the traditional DVB broadcast members DVB-C, DVB-S, or DVB-T, DVB-H is a datagram-based broadcast transmission standard because of its endurance to buffering, delays, and easy network integration. This allows transmission of data that pertains, e.g., to multimedia services or file-downloading services. The usage of the Internet Protocol (IP) allows the coding to be decoupled from the transport, opening the door to a number of features benefiting handheld mobile terminals including a variety of encoding techniques, which only require low power from a decoder. Therefore, IP is the Open System Interconnection (OSI) [24] Layer-3 protocol used in mobile handheld convergence terminals, creating an IP datagram stream which consists of IPv4 or IPv6 datagrams, each sharing the same IP source and destination address.

Whereas the traditional DVB broadcast members use the Packetized Elementary Stream (PES) container format [25] to encapsulate audiovisual access units, DVB-H uses Multi-Protocol Encapsulation (MPE) sections [18] to carry OSI Layer-3 datagrams or so-called Multi-Protocol Encapsulation Forward Error Correction (MPE-FEC) sections [18] to carry Reed-Solomon (RS) parities. Figure 8.1 visualizes the relation between the documents that describe the DVB-H standard [34]. DVB-H is based on the DVB-T standard. It is for this reason that, besides the new DVB-H system specification standard [19], the system is described by a number of existing documents, which have been amended with the appropriate DVB-H features. The standards depicted in Fig. 8.1 lack the description of the OSI Layers 3-7 protocols, because the definition of the layers above the IP layer is outside the scope of the DVB-H specification. To overcome this shortage, the DVB-H ad hoc group Convergence of Broadcast and Mobile Services (CBMS) developed the Internet Protocol Data Broadcast (IPDC) specification [45]. The purpose of the IPDC specification is to provide both the higher layer protocols for DVB-H that enable the construction of an end-to-end system and the integration

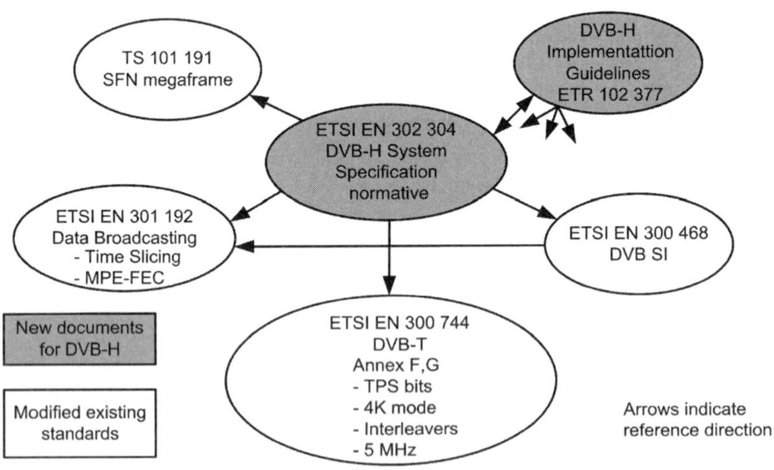

Fig. 8.1 The DVB-H family of standards

of a cellular communication system [12]. The IPDC over DVB-H standard complements the DVB-H standard, by defining OSI Layers 3-7 and influences some of the OSI Layer-2 (link layer) Program Specific Information (PSI) and Service Information (SI).

This chapter describes the aspects of an efficient and robust link layer. This link layer is an interface between the OSI Layer-1 (radio layer) operating in the physical domain and the OSI Layer-3 (network layer). The key words for this interface are *efficiency* and *robustness*. The efficiency aspects of the DVB-H link layer are multifold and are split into two categories. Category one is characterized in the sense that reliable and unreliable received data is both employed to maximize data recovery and reconstruction using erasure FEC decoding. Erasure information is scarcely assigned, leading to a higher flexibility in the usage of error correction. Category two is characterized with respect to the link layer implementation, optimizing on memory footprint, logic area, and cycle consumption. The robustness aspects of a DVB-H link layer are (1) that the link layer subsystem shall not collapse when incorrect data is processed and moreover (2) correctly received data shall always be transferred to the network layer, regardless of the outcome of the FEC decoding.

The sections of this chapter describe the various aspects that are related to the DVB-H link layer. Section 8.2 addresses the positioning of the link layer, a description of the signals processed, starting from the radio signals up to the IP layer, and the DVB-H link-layer features. In Sect. 8.3, the link-layer aspects related to the OSI layers are discussed. Section 8.4 presents the building blocks of an efficient and robust DVB-H link layer. Section 8.5 elaborates on the possible IP de-encapsulation methods required for an efficient link layer. Section 8.6 addresses the verification and validation of an efficient and robust DVB-H link layer. Finally, conclusions are presented in Sect. 8.7.

8.2 Features of the DVB-H Link Layer

This section is divided in to two parts. First, we depict the position of the link layer in a DVB-H terminal and briefly elaborate on the involved information signals and their definitions. Second, we introduce the DVB-H link-layer features, time-slicing aspects, and the MPE-FEC. Finally, the datagram and RS-parity data encapsulation are discussed in detail.

8.2.1 DVB-H Link Layer and Its Information Signals

Figure 8.2 indicates the position of the link layer in a basic DVB-H terminal. At the left-hand side of Fig. 8.2, the antenna signal enters the (silicon) tuner. The channel decoder and demodulator operate at the tuner output signals I and Q. The resulting MPEG-2 Transport Stream (TS) is processed by the link layer. The main

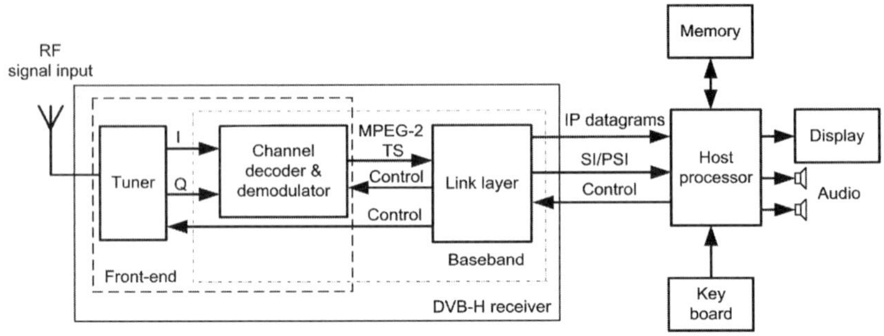

Fig. 8.2 Basic DVB-H terminal setup

responsibilities of the link layer are filtering of IP datagrams from a selected Elementary Stream, filtering of sections from SI and PSI, maintaining the synchronization with the time-sliced service, and front-end control of the receiver.[1] Because the IP datagram information, RS-parity data, and SI/PSI information are all transmitted using sections, a DVB-H link layer only needs to be able to handle section-based information, as this is the only MPEG-2 container format used.

An IP-based DVB-H service is associated to a so-called IP platform. In DVB-H, the SI/PSI information lists the available IP platforms and information to trace them. The SI/PSI information is processed by the middleware, resulting in a database that links an IP address to a particular Elementary Stream and the Transport Stream(s) that carry this Elementary Stream. A service IP address is obtained via the Electronic Service Guide (ESG), which is broadcasted using IP. An IP platform may contain more than one ESG. A special IP/port-number combination is used in every IP platform to transport a service that announces all ESGs to be found in that IP platform. This is the bootstrap ESG Service. The ESG consists of two essential types of information: user attraction and acquisition information. The majority of the ESG information is expressed as eXtended Markup Language (XML) fragments, but a part of the acquisition information are Session Description Protocol (SDP) files that the terminal needs to locate service streams and configure service consumption applications appropriately [13].

The link layer requires configuration, to extract an IP-based service from the received TS. Basically, this means setting the Packet IDentifier (PID) filter, setting up the MPE-FEC decoder (if used), indicating the maximum burst duration and initializing the possible IP filters to source and/or destination addresses. The configuration is done by the middleware.[2] For this, the middleware is invoked by the application requesting for a service with a particular IP address. This IP address is obtained by the application from the ESG. The middleware can be embedded in the DVB-H baseband but may also run on a host processor. Depending on the

[1] Front-end control by the link layer avoids time-slicing knowledge on the host, thereby simplifying the overall system design.

[2] We use the name middleware but for DVB-H this function is also known as Transport Stream Controller (TSC).

middleware system partitioning, the output of the DVB-H link layer contains either SI/PSI sections and IP datagrams as depicted in Fig. 8.2, or IP datagrams and the output of the middleware, e.g., a list of services for a particular IP platform.

8.2.1.1 Relation Between PSI/SI and a DVB Network

Figure 8.3 indicates the relation between DVB networks, Transport Streams, DVB services, and components after [14]. A DVB network is uniquely identified by a *network_id*. A DVB network consists of one or more Transport Streams, each carrying a multiplex and being transmitted by one or more DVB Radio Frequency (RF) signals. Information about a DVB network is available within the Network Information Table (NIT) *sub_table* (identified by *network_id*). The NIT lists all multiplexes and DVB RF-signals available within the DVB network. The NIT is carried within each DVB network. A single multiplex is a set of DVB services multiplexed together and transported by a TS. A multiplex and the corresponding TS are identified by *transport_stream_id* and *original_network_id*. The *Transport_stream_id* parameter is unique within the *original_network_id*. The *Original_network_id* parameter is the *network_id* of the DVB network generating the multiplex. Information about a particular multiplex is available within the Program Association Table (PAT) carried within the multiplex [25]. Each multiplex contains exactly one PAT, listing all DVB services available within the multiplex. A DVB service is a sequence of programme

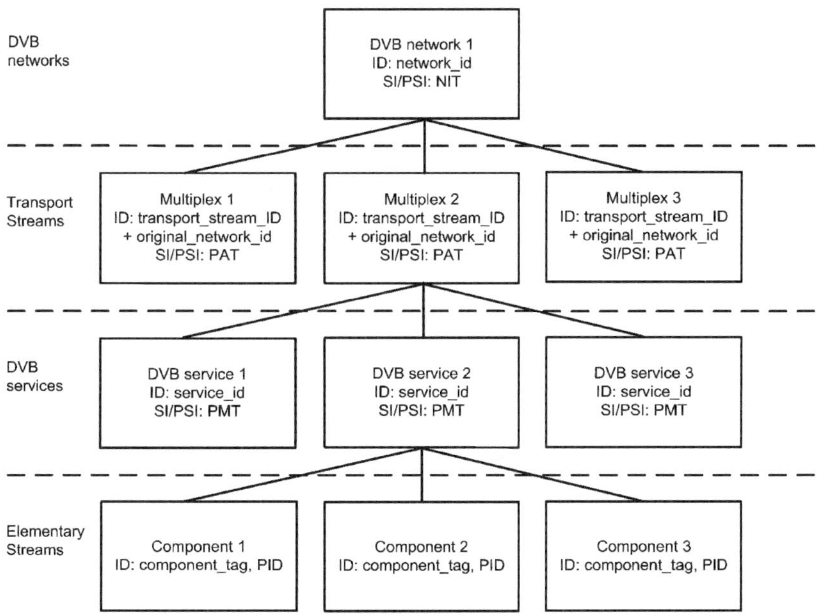

Fig. 8.3 Relation between DVB networks, Transport Streams, DVB services, and components

events, each of which groups together a set of components. Components can either have a seperate Elementary Stream, or share an Elementary Stream.[3] An Elementary Stream is a collection of Transport Stream packets sharing a common *PID* [25]. A DVB service is globally and uniquely identified by the triplet of *service_id*, *transport_stream_id*, and *original_network_id*. The *service_id* is unique at least within a multiplex. All the components of a DVB service are carried within a single multiplex. Information about a DVB service is available within the Program Map Table (PMT) *sub_table* (identified by *service_id*) [25], carried within the same multiplex. A component is identified by the *component_tag*, which is unique within a DVB service. The component is carried within an Elementary Stream, identified by the *PID*. The *PID* is unique within a TS. Mapping between a *component_tag* and the *PID* is signalled in the PMT. It is possible to have one Elementary Stream carrying a component of more than one DVB service.

8.2.1.2 Relation Between IP Platform, IP flow, and IP Stream

An IP platform is a set of IP flows managed by a service provider. The IP platform represents a harmonized IP-address space, that has no address collisions, and represents a set of IP/MAC streams and/or receiver devices. An IP platform may be available on multiple TSs, within one or multiple DVB networks. In such a case, each of the TSs may carry any subset of the IP flows of the IP platform. An IP platform is identified by the *platform_id*, which is carried by the IP/MAC Notification Table (INT) [18]. This table provides a flexible mechanism for carrying information about availability and location of IP/MAC streams within DVB networks. *Platform_id* values are divided into two ranges. One range consists of *platform_id* values that are globally unique. If such a *platform_id* value is signalled in two different DVB networks, it refers to the same IP platform. The second range consists of *platform_id* values, which are unique only within the scope of a DVB network. Data from an IP platform identified by such *platform_id* is not available in any other DVB network. Such an IP platform is globally and uniquely identified by the combination of both *platform_id* and *network_id* only. Each IP platform contains one or more IP flows, each mapped into one or more IP streams. An IP flow is identified within an IP platform by its source and destination addresses, and the involved port number. IP flows are IP platform specific, and two different IP flows on two different IP platforms may use the same source/destination addresses. Each IP flow may be mapped into one or more IP streams. An IP stream is a data stream delivering exactly one instance of an MPE-encoded IP flow. Figure 8.4 depicts the mapping of a single IP datagram on a single MPE section, which is mapped on one or more Transport Stream packets. An IP stream is identified by *transport_stream_id*, *original_network_id*, *service_id*, *component_tag*, and IP source/destination addresses (all together), which are further described hereafter.

---

[3] Note that in DVB-H, the definition of an Elementary Stream differs from the definition as used in MPEG-2, where an Elementary Stream refers to the basic information stream prior to TS packetization.

- *Transport_stream_id* and *original_network_id* together identify a single Transport Stream.
- *Service_id* and *component_tag* together identify a single Elementary Stream within a Transport Stream. The Elementary Stream consists of TS packets with the same *PID*.
- IP source/destination addresses identify a single IP stream within an Elementary Stream. This is required to differentiate between multiple IP streams carried within the Elementary Stream.

Figures 8.4 and 8.5 indicate the hierarchical relation and structure of IP platforms, IP flows and IP streams after [14]. Figure 8.4 depicts that a stream of MPE sections sharing the same MAC address, also indicated as MPE-section stream, is delivered within an Elementary Stream. At the top, it is shown that an IP flow consists of a number of IP datagrams. A single IP datagram is encapsulated in a single MPE section, resulting in an MPE-encoded IP flow, which is transported in an Elementary Stream. Figure 8.5 visualizes the hierarchical structure of an IP platform, IP flows, and IP streams. An IP platform consists of one or more IP flows, which can

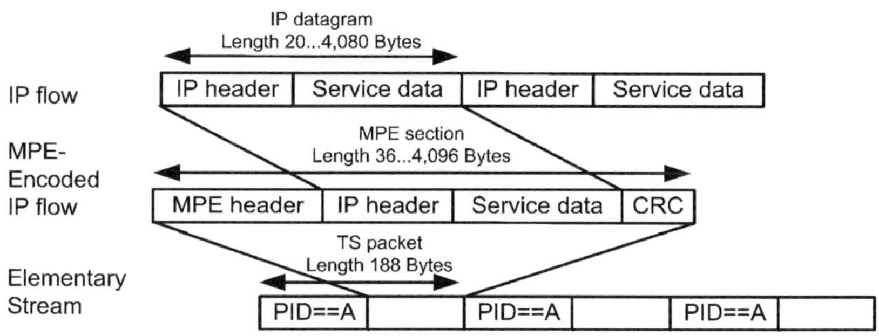

Fig. 8.4 Relation between IP stream, MPE-section stream, and Elementary Stream

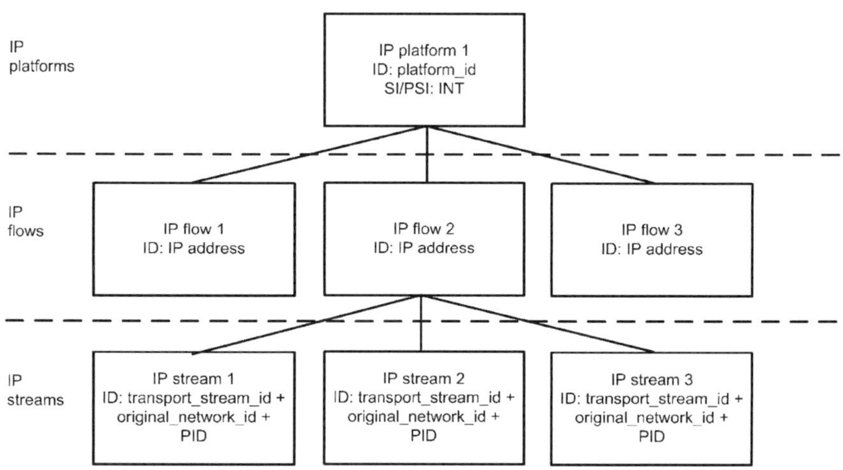

Fig. 8.5 Relation between IP platform, IP flow, and IP stream

be transmitted in one or more IP streams. As a result, the same data is available on one or more Transport Streams in one or more multiplexes and thus on different broadcast frequencies. Two different Elementary Streams, carrying IP streams which contain the same IP flow, may be located in different multiplexes and thus different broadcast frequencies. In such a case, an IPDC DVB-H receiver may accomplish a handover [38, 46] between these multiplexes, while receiving the IP flow. To optimize the bandwidth usage, the INT does not always announce the source address of an IP flow. In this case, IP stream differentiation is based on the destination address only. An IP stream is a single data stream delivering an IP flow. An IP stream is identified by associated parameters indicating its location, *network_id*, *original_network_id*, *transport_stream_id*, *service_id*, and *component_tag*. An IP flow represents the data content of a stream, while an IP stream represents an instantiation of such an IP flow. An IP flow belongs to exactly one IP platform. An IP stream may be announced by multiple INT *sub_tables*, eventually making it part of multiple IP platforms. This enables the transmission of a single service over multiple areas via multiple frequencies.

8.2.2 Time-Slicing

Although the DVB-T broadcast standard allows mobile reception, it is not optimized to support reception with mobile battery-powered terminals. Therefore, the DVB-H broadcast standard is equipped with a feature called time-slicing, in which services[4] are broadcasted in periodic bursts, see Fig. 8.6 after [18]. This feature is added to the DVB-H link layer to allow service distribution via an existing DVB-T network. Time-slicing enables the receiver front-end to be active for only a fraction of the time, receiving bursts of a requested service [35]. Major power consumers in, e.g., a Personal Digital Assistant (PDA) based terminal are the Liquid Crystal Display

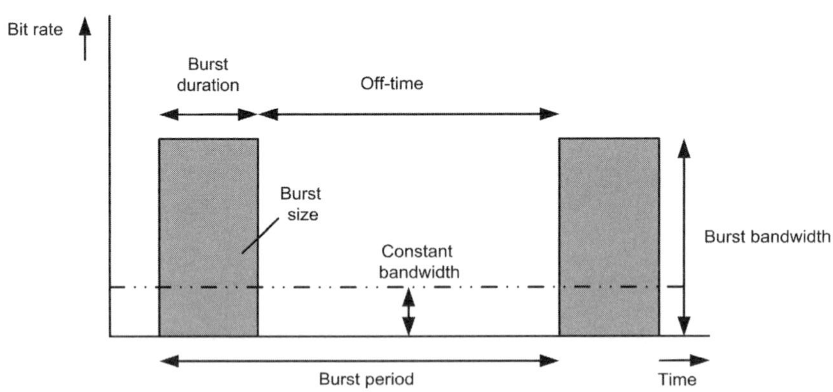

Fig. 8.6 A DVB-H service burst

[4] Note that the transmission of the Transport Stream is not interrupted.

(LCD) backlight and the front-end. In the early phase of DVB-H standardization, the continuous power consumption of such a front-end was estimated at 1 W [34]. As of today, the power consumption has been pushed back to around 150 mW for the silicon tuner and 200 mW for the baseband for continuous reception.[5] With current technology, the consumed DVB-H receiver power in off-time can be brought down to 3 mW. Depending on the values for the burst duration and burst period, power savings up to 95% can be achieved [42]. Although services in DVB-H are broadcasted in time-sliced mode, the SI/PSI information are non-time-sliced broadcasted. A time-sliced service can be sequential or parallel of nature and broadcasted in combination with a continuous broadcasted DVB service, see Fig. 8.7b. For example, in Fig. 8.7a, DVB-H Service 2 and DVB-H Service 3 represent parallel service bursts, whereas DVB-H Service 1 and DVB-H Service 2 form consecutive service bursts. Besides parallel services at Elementary Stream level, as indicated in Fig. 8.7a, parallel services can also share an Elementary Stream and differ in multicast address. Although the parallel services as indicated in Fig. 8.7a share the same burst start and end-time, the standard does not impose any restrictions on this behavior. Parallel services are beneficial for several reasons:

- Fast zapping time
- Simultaneous reception of the main services in combination with low-speed services (ESG, Alarms, ...)
- Local insertion of services
- Better optimization of bandwidth, e.g.
  - Maintain burst length (for impulsive noise / Doppler performance)
  - One service does not utilize the full bandwidth. Inserting two services in parallel results in a better bandwidth utilization

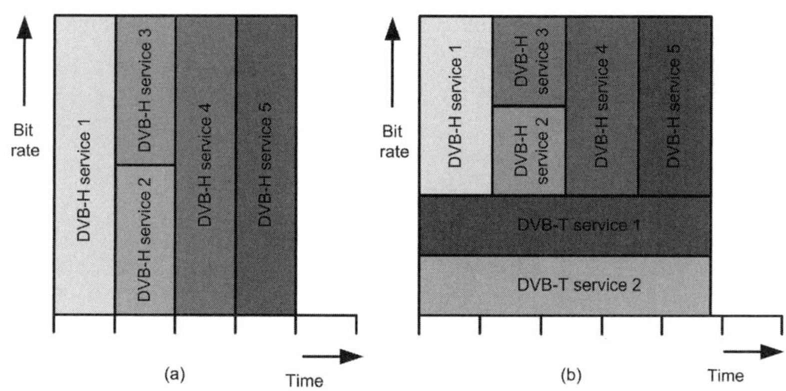

Fig. 8.7 DVB-H service broadcast methods. **(a)** Multiplex snapshot containing only DVB-H services, whereby service 2 and service 3 are broadcasted in parallel. **(b)** Multiplex snapshot containing a mixture of DVB-T and DVB-H services

[5] Note that the LCD backlight was and still is a major power consumer, but with state-of-the-art signal processing, the LCD backlight can be dynamically controlled, thereby reducing the backlight power consumption roughly with a factor 30%.

If parallel services are sharing an Elementary Stream, this may lead to the following complications. For example, one service may require all bandwidth in the Internet Protocol Encapsulator (IPE) leaving no bandwidth for the remaining services. Furthermore, it is difficult to re-encapsulate services in an existing Elementary Stream due to, e.g., sharing the MPE-FEC setting.

Similarly, if parallel services do not share the same Elementary Stream, some beneficial properties occur. First, one service will not influence the available bandwidth of other services. Second, IP data can be encapsulated locally due to the separate MPE-FEC setting. Third, some service types such as, e.g., ESG and alarm services can occur with higher periodicity, due to the disentanglement of the main service to which they were attached. Combining a continuous broadcasted DVB service and time-sliced services, see Fig. 8.7b, decreases the power saving due to a lower available data rate for time-sliced services [36]. When a DVB-H multiplex is combined with a regular statistical multiplexed information signal, the DVB-H burst character is not jeopardized [7]. A time-sliced broadcast technique allows the terminal to switch off the front-end for the off-time period, in which no service data is transmitted. Besides power reduction, time-slicing has also the advantage that the front-end can be reused during the off-times for peeking into neighboring cells to perform service discovery and enabling seamless horizontal handover for a specific service [37].[6]

Time-slicing information is carried by the *real_time_parameters*, which are available in each of MPE and MPE-FEC sections. The off-period starts after the end of a service burst. When due to a transmission error, the *frame_boundary* (see Sect. 8.2.5) is missed, the *max_burst_duration* of the *time_slice_fec_identifier_descriptor* (see Sect. 8.3.2) allows determination of the service burst end. This enables the activation of the power-down mode, provided that at least one MPE or MPE-FEC section has been correctly received. If the time-slicing information for a service burst is missed, the terminal's front-end must be active to detect the Elementary Stream containing the next service burst.

8.2.3 MPE-FEC Feature

To enhance the DVB-H receiver robustness under various reception conditions, an error-protection scheme is added to the DVB-H link layer. This scheme is called MPE-FEC, employing powerful channel coding on top of the DVB-T channel coding, offering extensive time interleaving and allowing service distribution via an existing DVB-T network. The FEC is based on the Reed-Solomon (RS) code [255,191,65] RS. The parity data is stored together with the OSI Layer-3 IP datagrams in a so-called MPE-FEC frame. DVB-H supports four different MPE-FEC frame sizes. The MPE-FEC frames can be constructed using either 256, 512, 768 or 1,024 rows. Figure 8.8 indicates an MPE-FEC frame of k rows, which consists of

[6] Note that there exists also vertical handover, which means service handover between two different standards, e.g., DVB-H and UMTS and is outside the scope of this chapter.

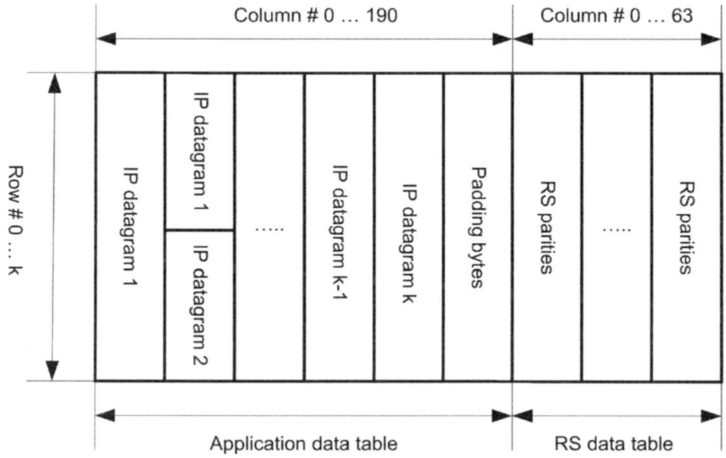

Fig. 8.8 MPE-FEC frame of k rows, one padding column, and 64 parity columns

an application data table and an RS-data table. Although the MPE-FEC frame has a two-dimensional structure, it can be considered of being two one-dimensional arrays that contain the corresponding data. Due to this one-dimensional structure, data of a single IP datagram can be split over two or more application data table columns.[7] Such a situation is depicted in Fig. 8.8, where the length of IP datagram 1 exceeds the number of MPE-FEC frame rows. The number of rows that construct an MPE-FEC frame is signalled in the *time_slice_fec_identifier_descriptor*, see Sect. 8.3.2. For the situation that the transmitted number of IP datagram bytes is less than the number of bytes corresponding to the product of application data table columns and MPE-FEC rows, zero-valued padding bytes fill up the remaining positions of the application data table. The number of fully padded columns is signalled in the MPE-FEC section header, see Sect. 8.2.4 and may vary per MPE-FEC burst. The [255,191,65] RS mother code allows to correct up to e erasures, if the erroneous positions are known, or t unknown errors according to the inequality

$$2t + e < 65. \tag{8.1}$$

Padding columns and subsequently not transmitting these padded columns can also be used to transmit less IP data making the code *stronger*, a process also known as shortening. Alternatively, transmission of less parity columns can also be applied making the code *weaker* than the mother code, a process also known as puncturing. Although MPE-FEC is part of the DVB-H standard, it is not obligatory. As the transmission of the application data table always precedes the transmission of the RS-data table, an MPE-FEC ignorant DVB-H receiver can skip the reception of RS data, reducing the power consumption by maximal 25%. For MPE-FEC capable

---

[7] Note that the row-wise calculated parity data is column-wise transmitted, using MPE-FEC sections.

DVB-H receivers, skipping the RS data of a service burst is beneficial when all IP datagrams of a particular service burst are received correctly. In general, the usage of MPE-FEC with code rate $\frac{3}{4}$ provides a spectacular improvement of the DVB-H coverage. A laboratory test showed a small variation of the CNR gain (i.e. about 3 dB) when the MPE-FEC coding rate varies from $\frac{7}{8}$ to a robust $\frac{1}{2}$ value. But for MPE-FEC code rate $\frac{7}{8}$, field trials were worse than in a laboratory environment, suggesting that the CNR gain is proportional to the MPE-FEC overhead/coding rate [7]. In the DVB-H standard, a quality measurement indicator MPE-FEC Frame Error Rate (MFER) is introduced, which is defined as the ratio between the number of defect received MPE-FEC frames divided by the total received MPE-FEC frames. Often, this indicator is used to determine the reception conditions that result in an MFER of 5%. Using an MFER of 5%, a Doppler performance of 120 Hz is achieved, corresponding to a speed of 160 km/h (100 mph) in the upper part of band V (800 MHz) or 640 km/h (400 mph) in the lower part of band III (200 MHz) [7].

8.2.4 DVB-H Data Encapsulation

DVB-H a is datagram-based broadcast standard. Unlike the traditional DVB broadcast members, which use the MPEG-2 system standard [25] for encapsulation of audiovisual access units into PES units, which are depicted at TS packets, DVB-H uses MPE for encapsulation of an OSI Layer-3 (Network) datagram (e.g., IP datagram) [18]. If MPE-FEC is used, RS data is encapsulated in MPE-FEC sections [18]. The mapping of the section into MPEG-2 Transport Stream packets is defined in the MPEG-2 Systems standard [25].

8.2.4.1 MPE Section

DVB-H is a broadcast transmission system for datagrams, which may be based on IP or other network layer datagrams [18, 24]. DVB has specified four methods for data broadcasting: data piping, data streaming, Multi-Protocol Encapsulation (MPE), and data carousel, which can all be used for delivering IP data. Data piping and data streaming are used so rarely that they are ignored in this context. Data carousels support delivery of files and other data objects, but are not suited for streaming services. Furthermore, implementing time-slicing on data carousels may be difficult. MPE is well suited for the delivery of streaming services as well as files and other data objects and it supports delivery of other protocols, while implementation of time-slicing on MPE leads to a low-cost solution. MPE sections are compliant to the *DSMCC_section* format for private data [26]. MPE sections provide means to handle either IP datagrams or LLC/SNAP datagrams [22, 23].

Table 8.1 presents the syntax of an MPE section. The *table_id* of an MPE section corresponds to the value "0x3E" [26] after [18]. The value for the *section_syntax_indicator* field is the complement of the *private_indicator* field. If the

Table 8.1 MPE section syntax

Syntax	Number of bits
datagram_section() {	
table_id	8
section_syntax_indicator	1
private_indicator	1
reserved	2
section_length	12
MAC_address_6	8
MAX_address_5	8
reserved	2
payload_scrambling_control	2
address_scrambling_control	2
LLC_SNAP_flag	1
current_next_indicator	1
section_number	8
last_section_number	8
MAC_address_4	8
MAC_address_3	8
MAC_address_2	8
MAC_address_1	8
If (LLC_SNAP_FLAG == '1') {	
LLC_SNAP()	
} else {	
for (j=0; j<N1; j++) {	
IP_datagram_data_byte	8
}	
}	
If (section_number == last_section_number) {	
for (j=0; j<N2; j++) {	
stuffing_byte	8
}	
}	
If (section_syntax_indicator == '0') {	
checksum	32
} else {	
CRC_32	32
}	
}	

section_syntax_indicator field is "1" (*private_indicator* is "0") the section uses a Cyclic Redundancy Checksum (CRC) to detect a transmission error. If the *section_syntax_indicator* field is "0" (*private_indicator* "1") the section uses a checksum to detect a transmission error. The *section_length* field indicates the number of bytes following this field including the CRC.[8] The maximum MPE section length is 4,096 bytes. With an MPE section overhead of 16 bytes, consisting of 12 bytes MPE-section header plus 4 bytes CRC, the maximum IP-datagram size becomes 4,080 bytes. Within the MPE-section header, a 6-byte field is allocated for the Medium Access Control (MAC) address. The length of the used MAC address is signalled in the *data_broadcast_descriptor*, see Sect. 8.3.2, which is inserted in the Service Description Table (SDT) or Event Information Table (EIT) [16]. The minimum MAC address length is 1 byte, leaving up to 5 bytes for other use. Four of these five MAC bytes, *MAC_address_1, ..., MAC_address_4*, are used to deliver the *real_time_parameters*, see Sect. 8.2.5, resulting in a 2-byte MAC field. In case of multicast IP streams, the MAC address is actually redundant data, as the MAC address is a function of the multicast group IP address. For all IP streams, the IP-datagram header following immediately the MPE-section header includes source and destination IP addresses, which uniquely identify the IP stream. A receiver can either ignore the MAC address entirely, filtering IP addresses only, or use the remaining MAC address bytes to differentiate IP streams within the Elementary Stream. Depending on the value of the *MAC_IP_mapping_flag* field, which is carried by the *data_broadcast_descriptor*, IP-to-MAC mapping is applied. As a result, the low-order IP bytes are mapped to *MAC_address_5* and *MAC_address_6*, as described for IPv4 [3] and for IPv6 [2]. The reserved fields are set to "1." The *payload_scrambling_control* field defines the scrambling mode of the section payload. This includes the payload starting after the *MAC_address_1* but excludes the checksum or *CRC_32* field. The applied scrambling method for the payload is user private. The *address_scrambling_control* field defines the scrambling mode of the MAC address. This field enables a dynamic change of the MAC addresses, the applied scrambling method for the MAC address is user private. If the *LLC_SNAP_flag* field is set to "1," the payload carries an LLC/SNAP-encapsulated datagram, following the *MAC_address_1* field. The *LLC_SNAP_flag* indicates the type of datagram conveyed. If this flag is set to "0," the section shall contain an IP datagram without LLC/SNAP information. If this flag is set to "1," the section shall contain an IP datagram preceded by LLC/SNAP fields. The *current_next_indicator* field is set to "1." The *section_number* field is set to "0," due to the fact that one section carries only a single IP datagram. The *last_section_number* shall be set to "0," again because there is only one section carrying a single IP datagram. The *MAC_address_1, ..., MAC_address_4* fields have obtained a new purpose in DVB-H and carry the *real_time_parameters*, see Sect. 8.2.5. The *CRC_32* field contains the calculated CRC according to [25].

8.2.4.2 MPE-FEC Section

The RS data of each MPE-FEC frame shall be delivered in MPE-FEC sections [18], using the syntax as depicted in Table 8.2 after [18]. The MPE-FEC sections are

[8] Note that the section length is always three bytes longer than indicated by the *section_length* field, due to the preceding generic part.

Table 8.2 MPE-FEC section syntax

Syntax	Number of bits
MPE-FEC_section() {	
table_id	8
section_syntax_indicator	1
private_indicator	1
reserved	2
section_length	12
padding_columns	8
reserved_for_future_use	8
reserved	2
reserved_for_future_use	5
current_next_indicator	1
section_number	8
last_section_number	8
real_time_parameters()	
for (j=0; j<N1; j++) {	
rs_data_byte	8
}	
CRC_32	32
}	

compliant to the *DSMCC_section* type, which are user private [26]. Each MPE-FEC section shall carry exactly one column of the corresponding RS-data table. The number of MPE-FEC sections used to carry RS data of an MPE-FEC frame shall not exceed the number of columns of the RS-data table. However, for the situation that puncturing is applied, the number of MPE-FEC sections indicated by *last_section_number* delivered may be less, indicating that not all RS data is transmitted. In the latter case, the RS decoder shall consider bytes within undelivered columns as unreliable. The number of delivered MPE-FEC sections is notified via the *last_section_number* field. The position of the delivered RS data in the RS-data table is indicated by the *section_number* field and the *real_time_parameters* field *address*. Section 0 carries the first (leftmost) column of the RS-data table, MPE-FEC section 1 carries the second column, and so on. The columns not delivered, e.g., due to puncturing, shall be the rightmost columns of an RS-data table. The *table_id* field equals the value "0x78." An MPE-FEC section only supports a CRC to be used for error detection, resulting in the *section_syntax_indicator* to be set to "1," forcing the *private_indicator field* to be "0" [26]. The two-bit reserved fields of the MPE-FEC section are set to "11." The *section_length* field indicates the number of remaining bytes in the section immediately following this field up to the end of the section, including the *CRC_32*. The number of *rs_data_bytes* carried in the MPE-FEC section shall be equal to the number of rows in the corresponding MPE-FEC frame, as indicated by the *frame_size* field of the *time_slice_fec_identifier_descriptor*, see Sect. 8.3.2. The amount of fully padded-padding columns in the application data table is denoted by the *padding_columns* field.[9] The *padding_columns* field value shall have a maximum value of 190 and may vary between successive frames.[10] When not used, the 8-bit *reserved_for_future_use* field is set to "0xFF." Similarly, when not used, the 5-bit *reserved_for_future_use* field is set to "0x1F." The *current_next_indicator* field is set to "1." The *section_number* field indicates the number of the section and corresponds to the RS-parity column in the RS-data table. The *section_number* field of the first section carrying RS data of an MPE-FEC frame is therefore set to "0x00." The *section_number* field shall be incremented with unity with each additional section carrying RS data of the concerned MPE-FEC frame. The *last_section_number* field indicates the number of the last section that is used to carry the RS data of the current MPE-FEC frame. The *rs_data_byte* field contains the RS-data bytes delivered. The *CRC_32* field contains the calculated CRC according to [25].

8.2.5 DVB-H Real Time Parameters

Table 8.3 indicates the *real_time_parameters* syntax, which are present in each MPE and MPE-FEC section after [18]. As mentioned in Sect. 8.2.4, the MPE-section syntax fields *MAC_address_1, ..., MAC_address_4* are replaced by the *real_time_parameters*. The interpretation of the *real_time_parameters* values depends on the broadcast mode, whether time-slicing and/or MPE-FEC are active or nonactive.

8.2.5.1 Real-Time Parameters: *delta_t*

In time-sliced transmission mode and regardless of the presence of MPE-FEC, the *delta_t* field indicates the time to the next time-slice burst within the Elementary

Table 8.3 DVB-H *real_time_parameters* syntax

Syntax	Number of bits
real_time_parameters () {	
delta_t	12
table_boundary	1
frame_boundary	1
address	18
}	

[9] Note that potential padding bytes after the last IP datagram for filling up a column, are not signalled.

[10] An MPE section should not be empty. As a result, the application data table shall contain at least one datagram per service burst, resulting in a maximum of 190 padding columns.

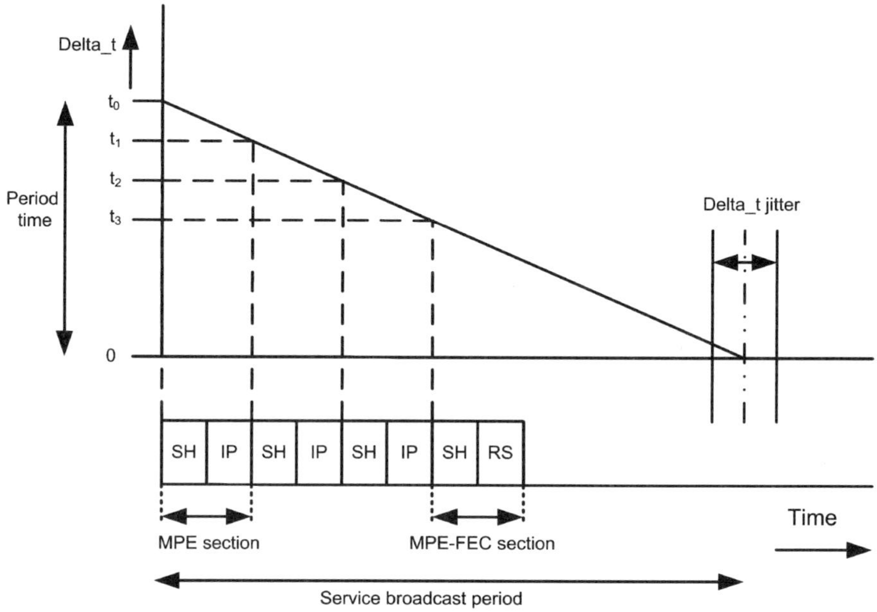

Fig. 8.9 *Delta_t* behavior in an MPE-encoded IP flow. The Section Header (SH) contains the *delta_t* value. The section payload is either an IP datagram (IP) or RS data (RS)

Stream. The time information is in all MPE sections and MPE-FEC sections[11] within a burst and the value may differ section by section, as indicated in Fig. 8.9. The resolution of the *delta_t* value is 10 ms. Value "0x00" is reserved to indicate that no further bursts will be transmitted within the Elementary Stream, i.e. indicating the end of a service. In such a case, all MPE sections and MPE-FEC sections within the burst shall have the same value in this field. The time indicated by *delta_t* shall exceed the end of the maximum burst duration, indicated by the *time_slice_fec_identifier_descriptor* field *max_burst_duration*, of the actual burst. This ensures that a decoder can always reliably distinguish two sequential bursts within an Elementary Stream. The basic objective of the *delta_t* method is to signal the time from the start of the currently received MPE (or MPE-FEC) section to the start of the next burst. To keep the *delta_t* insensitive to any constant delays within the transmission path, *delta_t* timing information is relative. The purpose of delivering *delta_t* in MPE and MPE-FEC section is to remove the need for synchronizing clocks between transmitter and receiver, as is the case for the DVB-T/C/S broadcast standards, which use the Program Clock Reference (PCR) [25]. The receiver has to provide sufficient accuracy for the duration of only one period, as the clock is restarted by each burst. As *delta_t* indicates the relative time rather than absolute one, the method is virtually unaffected by any constant delays within the transmission path. However, jitter on such delays does affect the accuracy of *delta_t*.

[11] Note that MPE-FEC may not be present because it is a nonobligatory feature.

This jitter is later referred to as *delta_t* jitter. If *delta_t* indicates the earliest possible time when the next burst may start, any *delta_t* jitter can be handled by decreasing the *delta_t* and thereby sacrificing the accuracy of the *delta_t*, thereby influencing the achieved power saving. Remultiplexing experiments indicate that the *delta_t* jitter remains in the vicinity of the allowed 10 ms [7].

For the situation that time-slicing is not used and MPE-FEC is applied, the *delta_t* field behaves like a cyclic MPE-FEC frame index within the Elementary Stream. The counter is incremented with unity for each subsequent MPE-FEC frame. If the maximum value "0xFFF" is reached, the counter value wraps back to zero. When large portions of data are lost, this parameter enables the identification to which MPE-FEC frame the actual received section belongs.

8.2.5.2 Real-Time Parameters: *table_boundary*

The *table_boundary* field is set to "1," when the current section is the last section of a table within the current MPE-FEC frame. If the section is an MPE section, this flag indicates the last section of application data table. For each MPE-FEC frame, exactly one MPE section with this flag set shall be transmitted. For each MPE-FEC frame for which any RS data is transmitted, exactly one MPE-FEC section with this flag set shall be transmitted. When MPE-FEC is not used on the Elementary Stream, this flag is reserved for future use, and set to "1."

8.2.5.3 Real-Time Parameters: *frame_boundary*

For the broadcast situation that time-slicing and MPE-FEC are active, the *frame_boundary* field when set to "1" denotes that the current section, which is either an MPE or MPE-FEC section, is the last section within the current burst. When the *frame_boundary* field is set in an MPE section, the burst does not contain MPE-FEC data.

8.2.5.4 Real-Time Parameters: *address*

For the situation that the *time_slice_fec_id* in the *time_slice_fec_identifier_descriptor* associated with the Elementary Stream is set to "0" and MPE-FEC is applied, the *address* field specifies the byte position in the corresponding MPE-FEC frame table that matches the first byte of the payload carried within the section. The first MPE section of a service burst has an *address* value corresponding to the value "0x00000." Consecutive sections carry *address* values in ascending order and can be calculated as follows. Let k be the MPE section index per burst, then the *address* value of section $k+1$ is calculated by adding the length of the IP datagram carried by section k with the *address* value of section k. The first MPE-FEC section after the MPE sections of a service burst starts again with the *address* field, reset to the value

"0x00000." Because an MPE-FEC section carries RS-parity data with a length that corresponds to the number of rows of a particular MPE-FEC frame, the *address* calculation for MPE-FEC section $k+1$ is reduced to adding the number of rows to the *address* value of MPE-FEC section k. If MPE-FEC is not used on the Elementary Stream, the *address* value is set to the fixed value "0x3FFFF."

8.3 OSI Layer Aspects Influencing the DVB-H Link Layer

This section presents the DVB-H broadcast stack and some aspects of the network and transport layer that influences a robust link-layer implementation. Furthermore, to properly de-encapsulate a received Elementary Stream, the link layer needs to be configured by the middleware, based on descriptors that are broadcasted via the SI. The second part of this section elaborates on the descriptors containing the configuration settings of the link layer.

8.3.1 DVB-H Broadcast Protocol Stack

DVB-H is an IP-datagram-based broadcast standard, which is visualized in Fig. 8.10, depicting the first four OSI layers of the DVB-H broadcast protocol stack [13]. From Fig. 8.10 it is clearly visible that the DVB-H link layer output has an IP and an SI/PSI section interface. The IP datagrams are offered to the IP stack on a host processor for further processing, whereas the DVB signalling information in the form of SI/PSI sections are requested by the middleware stack. This stack can be installed on either a host processor or embedded within the receiver baseband subsystem. Let us now highlight two aspects of the DVB-H broadcast stack that are relevant for a robust link-layer implementation. One of the aspects that characterize a robust link layer is the requirement that all correct IP datagrams, either correctly received or corrected by the link layer FEC, will be forwarded to the IP stack. A critical aspect of this requirement is the readout of IP datagrams from the MPE-FEC frame. The IP datagrams in DVB-H can have variable sizes up to 4,080 bytes, the maximum size that fits in an MPE section. It was mentioned in Sect. 8.2.3 that

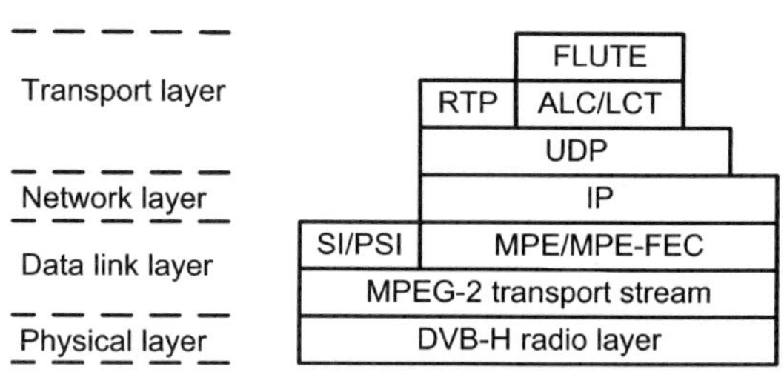

Fig. 8.10 DVB-H broadcast protocol stack

IP datagrams are stored in the application data table in a linear fashion. Due to a possible variance in IP-datagram length, it is necessary to determine the length field for each individual IP datagram in the MPE-FEC frame, which is located in the IP-datagram header. For the MPE-FEC frames that are correctly received or could be fully corrected by the link-layer FEC, IP datagram readout is possible due to the availability of a reliable IP-datagram length field. For the MPE-FEC frames that, even after applying the link layer FEC, still contain errors, the situation is more difficult and may result in the loss of all correct IP datagrams in that MPE-FEC frame. Hence, it is essential to have knowledge on the correctness of the IP-datagram length field. The protocol used by the network layer is based on either IPv4 or IPv6 [4, 39].[12] One of the points in which the IP network-layer protocols IPv4 and IPv6 differ from each other is the absence of an IP-header checksum in IPv6. The absence of an IP-header checksum results in an unreliable IP-datagram parsing in defect MPE-FEC frames, due to the fact that the length field may be corrupted. As a result, all correctly received IP datagrams may be lost, all depending on the location of the defect IP datagrams in the application data table. Figure 8.11 indicates the resulting loss of IP datagrams in the application data table, situated in column interval l, caused by a too large number of defect IP datagrams in column interval i, exceeding the FEC decoding capabilities. To overcome this protection shortage at the network layer (OSI Layer 3), the checksum of the transport layer (OSI Layer 4) can be used. The User Datagram Protocol (UDP) [40] uses a checksum that is calculated using a pseudo-header, information from the IP header, containing the IP-header source address, destination address, protocol field, and the UDP packet length. Because the IP version used for a service is fixed, the IP-datagram length

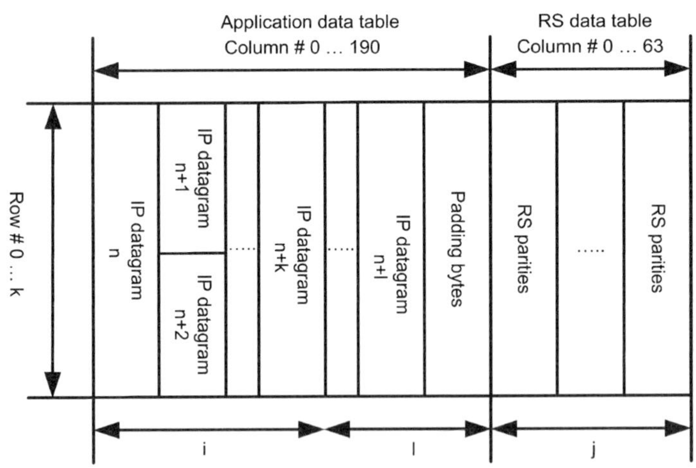

Fig. 8.11 Incorrect MPE-FEC frame, resulting in loss of IP datagrams for column interval l, for the situation that interval i contains more defect columns than the available parity columns j

[12] Note that it is prohibited to mix IPv4 and IPv6 in a single service.

can be calculated using the UDP length field.[13] The calculation of a UDP checksum requires the availability of all UDP data, which under certain conditions may be difficult, e.g., for the situation of a fragmented IP datagram, for example, such a situation occurs when a fragmented IP datagram is split over two time-sliced bursts. In a robust link-layer implementation, the applicability of the OSI Layer 3-4 checksum has to be considered to avoid unnecessary loss of IP datagrams. This aspect will be further discussed in Sect. 8.4.

8.3.2 DVB-H Service Information

Service discovery is performed by the receiver middleware, via interpretation of the received SI and PSI information. In order to receive a requested DVB-H service, the link layer must be configured. This section describes the *time_slice_fec_identifier_descriptor* and *data_broadcast_descriptor*, which contain essential information for configuring the DVB-H link layer.

8.3.2.1 Time_slice_fec_identifier_descriptor

A *time_slice_fec_identifier_descriptor* identifies whether time-slicing and/or MPE-FEC are used on an Elementary Stream, which is available for each time-sliced Elementary Stream describing the DVB-H specific link-layer settings. This descriptor can be transmitted by either the Network Information Table (NIT), the IP/MAC Notification Table (INT), or Conditional Access Table (CAT) [18]. An Elementary Stream can share a *time_slice_fec_iden-tifier_descriptor* with other Elementary Streams and a *time_slice_fec_identifier_descriptor* can override previous settings all depending on which *descriptor_loop* and which table the descriptor appears. Table 8.4 depicts the *time_slice_fec_identifier_descriptor* syntax after [18]. The *descriptor_tag* field is set to "0x77." The *descriptor_length* field specifies the number of bytes of the descriptor immediately following this field. The *time_slicing* field when set to "1" signals whether the referenced Elementary Stream is time-sliced. The value "0" indicates that time-slicing is not used. The *mpe_fec* field is a 2-bit field, which is further explained in Table 8.5. The *frame_size* field is a 3-bit field and provides information that a decoder may use to adapt its buffering usage. The exact interpretation depends on whether time-slicing and/or MPE-FEC are used. In case time-slicing is used (i.e. time-slicing is set to "1"), this 3-bit field indicates the maximum number of bits in section payloads within a time-slice burst for the Elementary Stream. For MPE sections, payload bits are counted over *ip_datagram_data_bytes* or *LLC_SNAP* field (whichever is supported), excluding any possible *stuffing_bytes*. For MPE-FEC sections, payload bits are counted over *rs_data_bytes*. When MPE-FEC is used (i.e. *mpe_fec* is set to "0x1"), this field

[13] The IP version used can be determined from the IP header of a correctly received IP datagram.

Table 8.4 Syntax *time_slice_FEC_identifier_descriptor*

Syntax	**Number of bits**
time_slice_fec_identifier_descriptor () {	
descriptor_tag	8
descriptor_length	8
time_slicing	1
mpe_fec	2
reserved_for_future_use	2
frame_size	3
max_burst_duration	8
max_average_rate	4
time_slice_fec_id	4
for(i=0; i<N; i++) {	
id_selector_byte	8
}	
}	

Table 8.5 MPE-FEC algorithm

Value	MPE-FEC	Algorithm
00	MPE-FEC not used	n/a
01	MPE-FEC used	[255,191,65] RS
01 to 11	Reserved for future use	Reserved for future use

Table 8.6 Size coding

Size	Max burst size	MPE-FEC frame rows
0x00	512 kbits	256
0x01	1024 kbits	512
0x02	1536 kbits	768
0x03	2048 kbits	1,024
0x04 to 0x07	Reserved for future use	Reserved for future use

indicates the exact number of rows in each MPE-FEC frame for the Elementary Stream. If both time-slicing and MPE-FEC are used for an Elementary Stream, both constraints (i.e. the maximum burst size and the number of rows) apply. If *time_slice_fec_id* is set to "0," the coding of the *frame_size* is according to Table 8.6. If *time_slice_fec_id* is set to any other value, coding of the *frame_size* is currently not defined. The *max_burst_duration* field is used to indicate the maximum burst duration in the Elementary Stream. A burst shall not start before $T1$ and shall end not later than at $T2$, where $T1$ is the time indicated by *delta_t* in the previous burst, and $T2$ is $T1 + MBD$, where MBD is the actually computed maximum burst duration in time (Fig. 8.12). If the *time_slice_fec_id* is set to "0," the computed value for MBD shall be between 20 ms and 5.12 s. The MBD parameter is computed according to

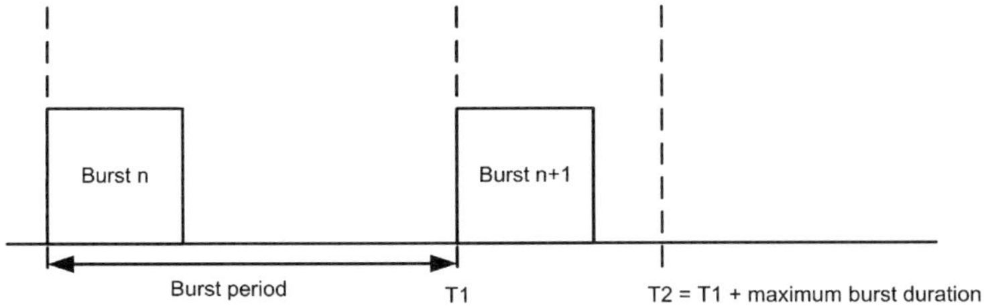

Fig. 8.12 Relation between service burst, *delta_t* and maximum burst duration

Table 8.7 Coding for max_average_rate

Bit rate	Description
0000	16 kbps
0001	32 kbps
0010	64 kbps
0011	128 kbps
0100	256 kbps
0101	512 kbps
0110	1,024 kbps
0111	2,048 kbps
1000 to 1111	Reserved for future use

the following formula: $MBD = (max\_burst\_duration + 1) \times 20ms$, with a resolution of 20 ms. If the *time_slice_fec_id* is set to any other value than "0," the coding of the *max_burst_duration* is currently not defined. When *time_slicing* is set to "0" (i.e. time-slicing not used), this field is reserved for future use and shall be set to "0xFF" when not used. The *max_average_rate* field is used to define the maximum average bit rate in the MPE-section payload measured within one time-slicing cycle or one MPE-FEC cycle, and it is derived by finding the maximum of

$$C_b = \frac{B_s}{T_c}, \tag{8.2}$$

where C_b denotes the actual calculated bit rate, B_s is the size of the current time-slicing burst or MPE-FEC frame in MPE-section payload bits and T_c is the time between the transport packet carrying the first byte of the first MPE section for the current burst (frame) and the transport packet carrying the first byte of the first MPE section for the next burst (frame) within the same Elementary Stream. Note that when MPE-FEC is used, the RS data is not included in B_s. If *time_slice_fec_id* is set to "0," the coding of the *max_average_rate* is according to Table 8.7. If *time_slice_fec_id* is set to any other value, coding of the *max_average_rate* is currently not defined. The *time_slice_fec_id* field identifies the usage of the following *id_selector_byte(s)*. Currently those bytes are not used, and this field shall be set to value "0x0," and *id_selector_byte(s)* shall not be present. Note that this field affects the coding of *frame_size*, *max_burst_duration*, and *max_average_rate* fields on the actual descriptor, and the *address* field of real-time parameters on the referred Elementary Stream. The definition of the *id_selector_byte(s)* of the *time_slice_fec_identifier_descriptor* will depend on the *time_slice_fec_id*.

Which *time_slice_fec_identifier_descriptor* applies to an Elementary Steam depends on the position within the SI information. When located in the first descriptor loop of the NIT, the descriptor applies to all Elementary Streams within all Transport Streams in the DVB network. If located in the second descriptor loop of NIT, the descriptor applies to all Elementary Streams within the referred Transport Stream, overriding any time-slicing or FEC information from the first descriptor loop. If located in the *platform_descriptor_loop* of an INT, the descriptor applies to all Elementary Streams referenced within the table, overriding any time-slicing or FEC information from the NIT. If located in the *target_descriptor_loop*, the descriptor applies to all Elementary Streams referenced within the current iteration of the *target_descriptor_loop* following the appearance of the descriptor, overriding any time-slicing or FEC information from the *platform_descriptor_loop* and in the NIT.[14] The descriptor may appear more than once, in which case each new occurrence overrides previous occurrence(s) [14].

8.3.2.2 Data_broadcast_descriptor

For each Elementary Stream carrying MPE-encapsulated IP stream(s), *SDT_actual* contains a *data_broadcast_descriptor* [18]. Table 8.8 indicates the *data_broadcast_descriptor* syntax after [16]. The *descriptor_tag* value is set to the value "0x64" [16].

Table 8.8 Syntax *data_broadcast_descriptor*

Syntax	Number of bits
data_broadcast_descriptor () {	
descriptor_tag	8
descriptor_length	8
data_broadcast_id	16
component_tag	8
selector_length	8
for(i=0; i<selector_length; i++) {	
selector_byte	8
}	4
ISO_639_language_code	24
text_length	8
for(i=0; i<text_length; i++) {	
selector_byte	8
}	
}	

[14] Note that the descriptor applies to Elementary Streams with *stream_type* "0x90."

The *descriptor_length* indicates the number of descriptor bytes following this field. The *data_broadcast_id* field is set to the value "0x0005," indicating that the *multiprotocol_encapsulation_info* structure is used. The *component_tag* is employed to label component streams with a unique tag value. The *component_tag* is unique within the DVB service and shall have the same value as a *component_tag* field of a *stream_identifier_descriptor* that may be present in the second descriptor loop of the PMT sub_table. The *selector_length* field is set to the value "0x02." There are two *selector_byte* fields, due to the value of the *selector_field*. The meaning of the *selector_byte* values is defined by the *multiprotocol_encapsulation_info* syntax, as depicted in Table 8.9 after [18]. The *MAC_address_range* field indicates the number of MAC address bytes that the service uses to differentiate the receivers according to Table 8.10. When the *MAC_IP_mapping_flag* field is set to "1," the service applies to the IP-to-MAC mapping as described for IPv4 multicast addresses [3] and for IPv6 multicast addresses [2]. If this flag is set to "0," the mapping of IP addresses to MAC addresses is done outside the standard [18]. The *alignment_indicator* field indicates the alignment that exists between the bytes of the datagram section and the Transport Stream bytes according to Table 8.11. The *alignment_indicator* is set to "1" according to [14]. The reserved field is set to the value "0x07." The

Table 8.9 Syntax for multiprotocol_encapsulation_info structure

Syntax	Number of bits
multiprotocol_encapsulation_info () {	
MAC_address_range	3
MAC_IP_mapping_flag	1
alignment_indicator	1
reserved	3
max_sections_per_datagram	8
}	

Table 8.10 Coding of the MAC_address_range

MAC_address_range	Valid MAC_address bytes
0x00	reserved
0x01	6
0x02	6, 5
0x03	6, 5, 4
0x04	6, 5, 4, 3
0x05	6, 5, 4, 3, 2
0x06	6, 5, 4, 3, 2, 1
0x07	Reserved

Table 8.11 Coding of the *alignment_indicator* field

Value	Alignment in bits
0	(8)-default
1	32

max_sections_per_datagram is set to the value "0x01," indicating that each IP datagram is carried within a single MPE section. The *component_tag* field is set to the value announced within the PMT sub_table of the DVB service for the referred component.

8.4 Efficient and Robust DVB-H Link Layer Implementation

This section will elaborate on the functional blocks together forming an efficient and robust link layer, using the definitions regarding efficiency and robustness as defined in Sect. 8.1.

8.4.1 DVB-H Link Layer Functional Blocks

Figure 8.13 depicts a block diagram of an efficient and robust DVB-H link layer, which is briefly described hereafter. The remaining subsections further elaborate on the individual functional blocks. The MPEG-2 demultiplexer is the first functional block that operates on the received Transport Stream, applying *PID* filters to select the requested Elementary Streams. After *PID* filtering, the Elementary Stream is either forwarded to the SI/PSI section filters or IP de-encapsulation filters. The queue manager forwards the correctly received SI/PSI sections to the host processor, using messages equipped with, e.g., locator information, indicating the queue from which the section originates. The reliable or unreliable de-encapsulated IP datagrams or IP-datagram fragments are temporally stored in the scratch buffer prior to final transferring them to the proper position in the MPE-FEC frame. During

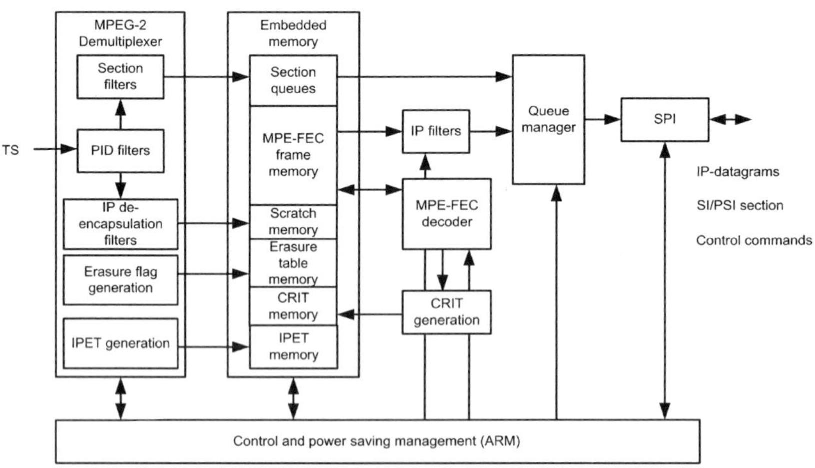

Fig. 8.13 Functional block diagram of an efficient and robust DVB-H link layer

IP de-encapsulation of the reliable and unreliable IP datagrams, locator information indicating the start position in the MPE-FEC frame of correctly received IP datagrams is stored in the Internet Protocol Entry Table (IPET). Furthermore, for each MPE-FEC symbol, corresponding erasure information is preserved in the erasure table. After receiving all data of a particular service burst, the [255,191,65] RS MPE-FEC decoding is applied for the case that not all IP datagrams were correctly received. For each row, the MPE-FEC decoding result is stored in the Corrected Row Index Table (CRIT). For the situation that all IP datagrams are correctly received, or are correctly available after applying MPE-FEC decoding, readout of the individual IP datagram is possible, using the traditional parsing method as discussed in Sect. 8.3.1. For the situation that not all IP datagrams were correctly received and the MPE-FEC decoder was not able to correct all MPE-FEC frame rows, readout of the correct IP datagrams is possible using the IPET and CRIT in combination with the erasure table. The IP datagram readout also applies filtering of IP datagrams on the basis of source and/or destination address(es). All filtered IP datagrams are forwarded to the host processor via a specific messaging technique, allowing the multiplexing of IP datagrams, SI/PSI section information, or system status information. To enable TDM-based broadcast reception, the link layer maintains service-reception synchronization and context control. Time-sliced service broadcasting allows the definition of two or more reception contexts, each of which can be regarded as a virtual receiver. In, e.g., the main reception context, the main service(s) are received. After main service reception, the receiver waits for the next service burst, based on the synchronization information (*delta_t*). During the off-time of the main context, an auxiliary context can be started. In the auxiliary context, different tuning parameters and filter settings for the available resources can be applied. The auxiliary context allows, e.g., to peek into neighboring channels. During such a peek operation, SI/PSI information can be collected as well as monitoring for *delta_t* values of the requested service. Based on the building blocks of Fig. 8.13, a data-flow chart is derived and depicted in Fig. 8.14. This data-flow chart is available for each context. The number of individual filters as indicated in Fig. 8.14, is derived in the corresponding subsections.

8.4.1.1 MPEG-2 Demultiplexer

At the left-hand side of Fig. 8.13, the received MPEG-2 TS enters the MPEG-2 demultiplexer. Elementary Streams, either SI/PSI, or those containing a DVB-H service, that are requested by the application, are filtered on the basis of their *PID* value. After *PID* filtering and processing of the other Transport Stream main header fields, the SI/PSI section information is routed through the SI/PSI section filters, where one or more SI/PSI filters can receive section data from the same *PID* filter. Although MPE encapsulation uses a section to encapsulate a single IP datagram, IP de-encapsulation differs somewhat from the traditional section filtering, due to the presence of the MPE-FEC. The IP de-encapsulation process is also responsible for the generation of erasure flags and the IPET entries. To allow proper erasure-flag

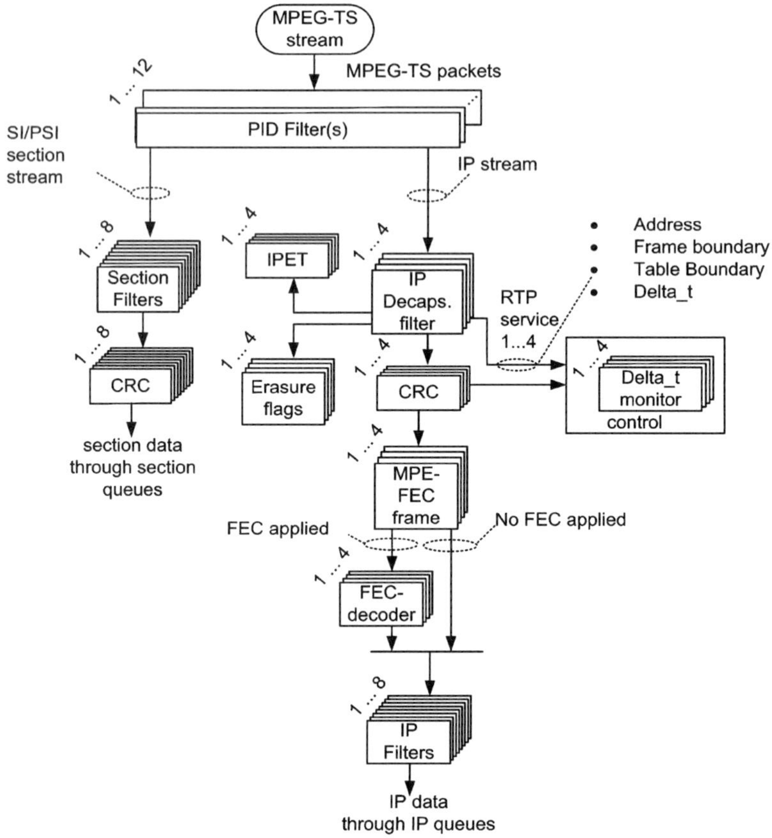

Fig. 8.14 DVB-H link layer data processing flow

generation, the TS main-header filtering (see Sect. 8.5) and the IP de-encapsulation filtering are jointly processed.

The number of available *PID* filters is equal to the sum of the maximum number of section filters and the maximum number of IP de-encapsulation filters.

The number of available section filters depends on the number of simultaneous section filters required by the middleware. From experiments, it can be concluded that the DVB-H middleware shows a satisfactory performance, when eight section filters are available.

The number of available IP de-encapsulation filters depends on the type of application. From an MPE-FEC frame size point-of-view, four IP de-encapsulation filters allow the reception of four parallel services, each with an MPE-FEC frame size of 256 rows. The sum of these four MPE-FEC frame sizes results in 1,024 rows, the maximum MPE-FEC frame size. The availability of four IP de-encapsulation filters allows the reception of parallel services with different MPE-FEC frame sizes, provided that the sum of the individual MPE-FEC frame rows for the various services is less than or equal to 1,024 rows.

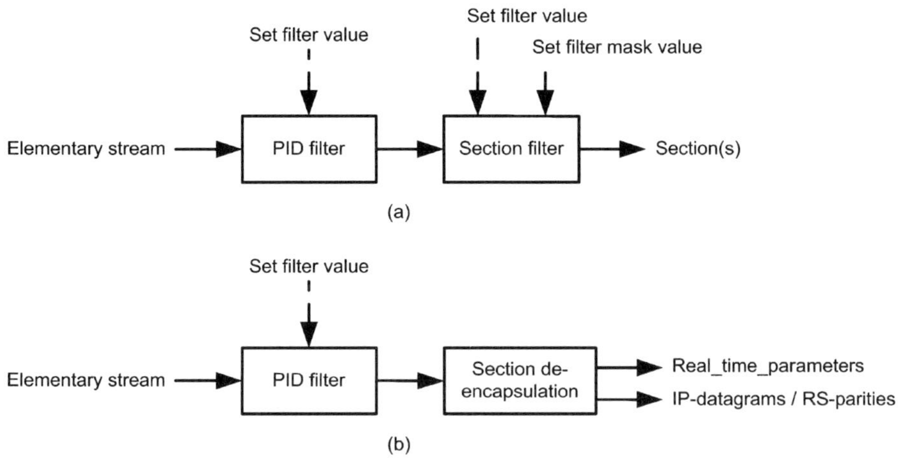

Fig. 8.15 Filter settings for a section-filter pair and IP de-encapsulation filter pair. **(a)** A section-filter pair consisting of a *PID* filter in combination with a section filter. **(b)** An IP-filter pair consisting of a *PID* filter in combination with an IP de-encapsulation filter

Figure 8.15a indicates the programming interface for a section-filter pair, consisting of a *PID* filter and a section filter. Figure 8.15b indicates the programming interface for an IP de-encapsulation filter and a *PID* filter, again forming a filter pair. An IP de-encapsulation filter requires no programming interface. The reason for this is twofold. First, the MPE header does not contain fields that require programmable filter operations. Although the standard allows for filtering on *MAC_address_5* and *MAC_address_6* fields, this filtering can be postponed until the actual IP datagrams are filtered from the MPE-FEC frame, using the IP source and/or destination address(es). On top of this, the distinction between MPE section and MPE-FEC section is achieved via the *table_id* of the corresponding section header, which is a fixed but different value for both MPE and MPE-FEC section type. Second, the payload of an MPE section is byte-aligned, see Sect. 8.3.2, resulting in the absence of padding bytes, avoiding notification of the IP de-encapsulation filter.

8.4.1.2 Section Queues

The total memory footprint for the section queues is the sum of the individual section-queue memory footprints. In the worst-case situation, all eight section filters operate on the same section data, which could be up to 4 kbyte in size, resulting in a total maximum section-queues size of 32 kbyte. Assigning 4-kbyte section-queue space rigid to each section filter will negatively influence the SI/PSI filtering performance. For the situation that a section queue contains a correctly received section, the queue manager needs to forward this data for further processing to a host processor. This is an interrupt-driven process and takes time to be executed. When the section queue contains a section, that is only a fragment of the 4 kbyte it can maximally store, the remaining space is not accessible by the section filter for further usage. As a result, the section filter is blocked for the duration that a section is forwarded to a host processor. On the average, the received sections will have a size that is much smaller than the maximum size of 4 kbyte. It is therefore advantageous to fragment the 32 kbyte section-queue memory into smaller fragments of, e.g., 256 bytes, allowing a better section-filter performance. The 256-byte fragments are used to construct a section queue that fits best to the received section. This section-queue buffer is allocated with aid of the *section_length* field, which is available in each section header. With some bookkeeping (e.g., section-queue number, memory start-address, and section length), the individual sections are traced in the section-queue memory space. This allows the queue manager to read the individual sections, create the corresponding message, send this to the host processor, and release the allocated memory-queue fragments. As long as there are free section-queue fragments, active section filters are prevented from being blocked, as long as the received sections fit within the available section-queue memory space, thereby increasing the section-filtering performance.

8.4.1.3 Erasure-Table Memory

The erasure table stores 2-bit erasure flags, see Sect. 8.5, generated by the IP de-encapsulation filter, for each MPE-FEC symbol that constructs the application data table. If each symbol of the application data table would have its own erasure flag, the memory footprint requires $2 \times 255 \times 1{,}024$ bits. Fortunately, the symbols of the application data table are transmitted via TS packets, which means that more than one symbol will have the same 2-bit erasure value. Let us assume that an IP datagram can be 32 bytes in size and transmitted in a single TS packet. An MPE-FEC frame of size 1,024 rows can hold 32 IP datagrams per column, resulting in 32 fragments of 32 bytes per MPE-FEC frame of size 1,024. We also assume that the RS data can be transmitted in a similar way as the IP datagrams. As a result, the erasure table will have $32 \times 255 = 8,160$ entries. Although the erasure-flag information requires a 2-bit storage, the erasure-transition coding requires 3-bit coding, see Table 8.12. Another 5 bits are required to indicate the position where a transition occurs within the 32 byte fragment. The total memory involved per erasure

Table 8.12 Coding of erasure type transition

Code	First stage	Second stage
000	OK	False
001	OK	Unknown
010	False	OK
011	False	Unknown
100	Unknown	OK
101	Unknown	False

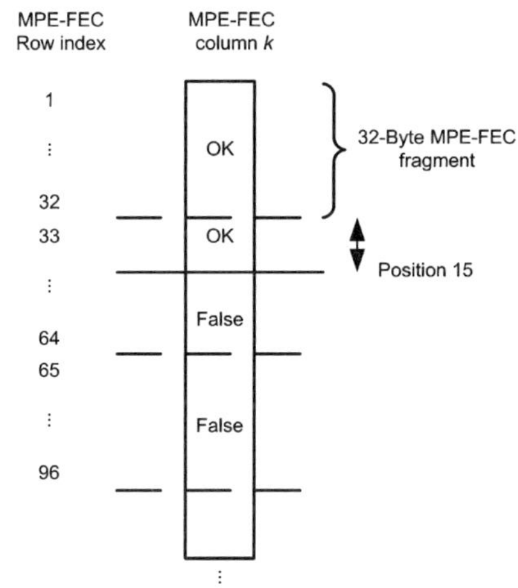

Fig. 8.16 Storage example of an erasure transition in the erasure memory

Table 8.13 Mapping of transition-type bits and transition-position bits forming an erasure entry

Bits	Name
7 : 5	Transition type
4 : 0	Transition position

entry is 1 byte, resulting in a total erasure table footprint of 8,160 bytes. Figure 8.16 indicates an example transition. The erasure-flag information in column *k* is OK for the first 46 MPE-FEC frame rows. When the erasure-type transition bits form the MSB bits of the erasure entry and the 5-bit transition position form the lower bits, see Table 8.13, then the first erasure entry for column *k* is "0x1F." The length value "0x1F" indicates that the transition lies outside this fragment. The second erasure entry will have an OK to False transition at position 15, resulting in an entry value of "0x0F."

8.4.1.4 Scratch Memory

This memory is used during the IP de-encapsulation and will be elaborated in Sect. 8.5. The size is determined by the maximum size of an IP datagram for MPE encapsulation, which is 4,080 bytes.

8.4.1.5 MPE-FEC Frame Memory

In principle, the MPE-FEC frame memory is equal to the product of the maximum MPE-FEC frame size, which is 1,024 rows and the sum of the application data table and RS-data table columns, which is 255 columns, resulting in total MPE-FEC

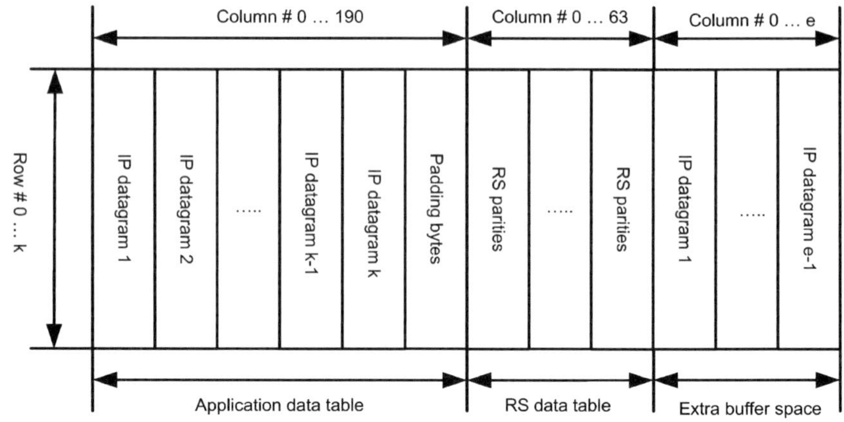

Fig. 8.17 MPE-FEC frame extended with extra columns allowing continuous reception of consecutive services

frame size of 255 kbyte. With this memory pool, various reception situations are supported as long as the sum of the individual MPE-FEC frame sizes are less than or equal to 1,024 rows. The MPE-FEC calculation time and IP datagram transfer time are neglected in the calculation, so is the data rate at which the service enters the link layer. These aspects have their influence on the reception of consecutive services. To allow the reception of two consecutive services with, e.g., an MPE-FEC frame size of 1,024 rows and MPE-FEC, extra memory is required to buffer the new incoming IP data, while the previous burst is still in process. Figure 8.17 indicates an MPE-FEC frame with an extra memory extension. Let us assume a situation, as depicted in Fig. 8.7, in which consecutive Services 4 and 5 both equipped with MPE-FEC are requested by the application. Let us consider the case that the application data table and RS data table are filled with data from Service 4. After finishing the Service-4 IP de-encapsulation process, the MPE-FEC is applied, provided that Service 4 was subject to errors. While the FEC is calculated, the Service-5 Elementary Stream simultaneously enters the MPEG-2 demultiplexer. To avoid Service-5 IP datagrams loss, extra MPE-FEC frame memory, columns 0...*e*, are added to the MPE-FEC frame. This extra memory space provides buffering to cover the time that the MPE-FEC calculation is performed. After finishing the MPE-FEC decoding of Service 4, the RS-data-table memory space is available for storing IP datagram or RS-parity data of Service 5. The columns that formed the application data table for Service 4 will become available somewhere during the reception of Service 5, but after transferring the Service-4 IP datagrams to the host. In this example, only two consecutive services are received. When the amount of consecutive services that is received is increased, MPE-FEC memory fragmentation occurs during the reception of those services. This fragmentation stops after reception of the last consecutive service burst.

8.4.1.6 IP Filter

The received Elementary Stream may contain more IP flows than actually required for the requested service. Such a situation occurs when services are multiplexed at IP level, as indicated in Sect. 8.2.2. An IP filter operates on IP source and/or destination address(es), allowing various filter operations. For the case that an Elementary Stream contains multiple IP flows, IP filtering lowers the amount of IP data that needs to be transferred to the host processor, reducing the transmission time of the IP data, lowering the power consumption. Because a service can be based on IPv4 or IPv6, two filter versions, one for each standard, would be required to perform the required source and/or destination address(es) filtering. Such a situation is avoided when the filtering is achieved by a filter which is IP-version agnostic. Such a filter consists of a filter value, filter-mask value, and an offset value, indicating the start position from where in the IP datagram, the filter value, and filter-mask value are to be applied. The offset is relative to the start of an IP datagram, as indicated in Fig. 8.18. In this way, only the application needs to be aware of the used IP version, calculating the proper filter value, filter-mask values, and offset value. The lengths of the filter value and filter-mask value are fixed to 40 bytes, allowing to operate on IPv4 and IPv6 datagram headers. With aid of the filter-mask value, a filter can be created that filters on only the source address, destination address, or both source and destination address(es) for both either IPv4 or IPv6. The number of IP filters is based on the number of available IP de-encapsulation filters and the number of IP flows per service. An audiovisual service results in two IP flows, one that contains the audio, while the other contains the video information. With a maximum of four IP de-encapsulation filters, minimally eight IP flows need to be handled, resulting in eight IP filters, which can be switched off for creating a bypass mode, resulting in all IP data to be delivered at the host processor.

8.4.1.7 Queue Manager

The queue manager is responsible for avoiding a queue from overflowing, by creating messages that are send-toward the host processor. Since the messages are sequentially transmitted over the external interface, a message header containing vital information regarding, e.g., data type, original queue number, is required to

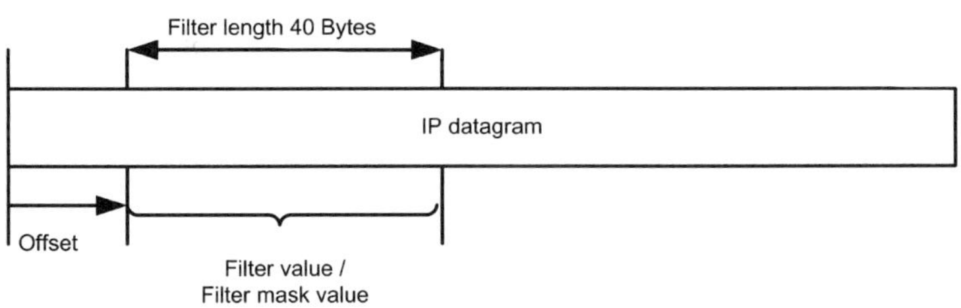

Fig. 8.18 An IP version agnostic IP filter method

allow proper routing of the message data at the host, e.g., invocation of the correct callback function for the middleware. A message can contain receiver-status information, or either one or more IP datagrams, or SI/PSI sections. Combining two or more IP datagrams, or SI/PSI sections into a single message, increases the transfer performance, avoiding unnecessary polling or even interrupt generation.

8.4.1.8 SPI Interface

A popular interface for connecting devices in, e.g., a mobile phone is the four-wire Serial Peripheral Interface (SPI). SPI requires a master to be present in the system providing the SPI clock. The value for the SPI clock depends on various aspect such as:

- Maximum link-layer input bit rate
- IP-datagram transfer time
- Receiver-software download time

The maximum link-layer input bit rate is determined by the transmission modulation settings, which is for an 8-MHz channel bandwidth equal to 31.6 Mb/s. Such an input bit rate results in an SPI clock of 32 MHz, if the amount of buffering is to be minimized. The modulation setting resulting in a channel throughput of 31.6 Mb/s will most probably not be used, due to poor mobile reception performance. Modulation settings that result in good or even excellent reception performance have a bit rate around the 14 Mb/s. As a result, the SPI clock will be at least 14 MHz, if the amount of buffering is to be minimized. The upper bound may be far above the 32 MHz, especially when the SPI bus is shared with other slaves or simply to reduce the receiver power consumption. A receiver can go to sleep mode,[15] consuming only a few milliWatts, see Sect. 8.2.2, as soon as all requested data has been fetched by the host processor. Finally, the SPI clock speed must be such that it allows to satisfy the requirements regarding the receiver-software download time. A guideline value for the software-image download time is in the range of 200 ms.

8.4.1.9 Link-Layer Control

The link-layer control has to deal with the following aspects:

- Front-end control
- Power-mode control
- Booting of the downloaded software image
- Command interpretation and control
- Context management
- Maintain service synchronization

[15] Note that the receiver front-end can be switched off, right after receiving the last data of a burst, or after the max burst duration if the service burst end is missed.

- Collect and calculate receiver condition
- Provide handover control

To avoid knowledge on the host concerning time-slicing, the link layer controls also the receiver front-end. Power consumption is reduced to a minimum, when the receiver is switched off completely. As a result, the receiver control will enable receiver booting, when the receiver module is activated. After booting, which may take around 10 ms, the link-layer control will notify the host that the receiver system is booted and ready for software download. Software download by the host avoids permanent storage inside the receiver module. Once the receiver is booted, the software image is downloaded and installed. An interrupt notifies the host that the receiver is ready to receive commands. Possible commands initiated by the host are, e.g., reset_link_layer, tune_to, setup_queue, start_queue, etc. The commands are equipped with a context parameter, indicating for which context, e.g., the filter settings apply. A context can be compared to a task in an Operating System (OS), where a task is executed based on its priority. A context in DVB-H has a priority but can be blocked by a context with an even higher priority. The context with the highest priority is the "main" context, delivering the service with the highest priority requested by the application. Other contexts are, e.g., peek context and handover context. The peek context allows to peek into neighboring channels and collect relevant information such as, e.g., SI/SPI, whereas the handover context prepares the receiver to switch to a new receiver setting. The link-layer control maintains service-reception synchronization for the service associated to the main context, using the transmitted *delta_t* value, and offers the receiver resources to the other contexts, e.g., the peek context, during the main-context off-time. At various stages in the receiver chain, signal-quality parameters, also known as Quality-of-Service (QoS) parameters, are available or can be calculated. Figure 8.19 indicates the high-level receiver building blocks and the associated signal-quality parameters. Such parameters are:

- Received Signal-Strength Indicator (RSSI), indicating the strength of the selected DVB-H signal, necessary for the handover algorithm
- Bit Error Count before Viterbi decoding (VBER), indicating the number of erroneous bits during the estimation period before Viterbi decoding
- Bit Error Count after Viterbi decoding (BER), indicating the number of erroneous bits during the estimation period after Viterbi decoding

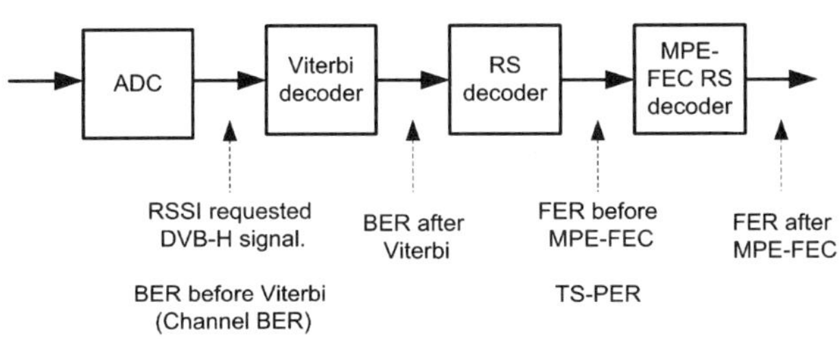

Fig. 8.19 Monitored signal-quality parameters

- Transport Stream Packet Error Count (TS-PER), indicating the number of erroneous TS-packets during the estimation period
- Frame Error Count before MPE-FEC decoding (FER)
- Frame Error Count after MPE-FEC decoding (MFER)

Horizontal handover is issued by the application, based on deterioration of the received QoS for the main context. The application configures a handover context and request for a handover. The actual handover is in the hand of the link-layer control, again avoiding the host processor from being aware of any form of time-slicing. For the situation that more than one service is received, the handover is performed for the service with the highest priority. This priority is set by the application and is, e.g., part of the setup_queue command.

8.4.2 IP Datagram Recovery in the DVB-H Link Layer

The DVB-H block diagram as depicted in Fig. 8.13 provides means to guarantee that correctly received IP datagrams will be forwarded to the host under all circumstances, realizing the required robustness as defined in Sect. 8.1. Furthermore, there are means to determine the location of corrected MPE-FEC frame rows, for MPE-FEC frames that remain defect after FEC decoding, which enable the location of corrected IP datagram bytes. The robustness is obtained via the usage of two tables. The first table is the Internet Protocol Entry Table (IPET) [5], which stores the start address of a correctly received IP datagram in the MPE-FEC frame. The second table is the Corrected Row Index Table (CRIT), which stores for each MPE-FEC frame row the result of the FEC decoder. Both tables are elaborated in the remaining part of this subsection.

Figure 8.20 depicts the results of the IPET and CRIT concept. Due to the IP de-encapsulation concepts as presented in Sect. 8.5, the receiver can correct up to a TS-packet error rate of 10%. For higher TS-packet error rates, the MPE-FEC frame becomes uncorrectable. With the aid of IPET and CRIT, considerable amount of IP datagrams can be retrieved from the defect MPE-FEC frame. Although the audiovisual information is corrupted, smart error-concealment techniques can camouflage this loss up to a certain extent. The IP datagram recovery due to the CRIT highly depends on the IP datagram size and on the MPE-FEC frame size. For IP datagram sizes around 128 bytes and MPE-FEC frame size of 1,024 rows, 4% of the total amount of IP datagrams contained by a defect MPE-FEC frame can be recovered. For IP datagram sizes larger than 512 bytes and MPE-FEC frame size of 1,024 rows, the recovery due the CRIT is below 1%.

8.4.2.1 Internet Protocol Entry Table

During reception of a service burst, the IP de-encapsulated IP datagrams are stored at predetermined positions in the MPE-FEC frame. The IP datagram start-position

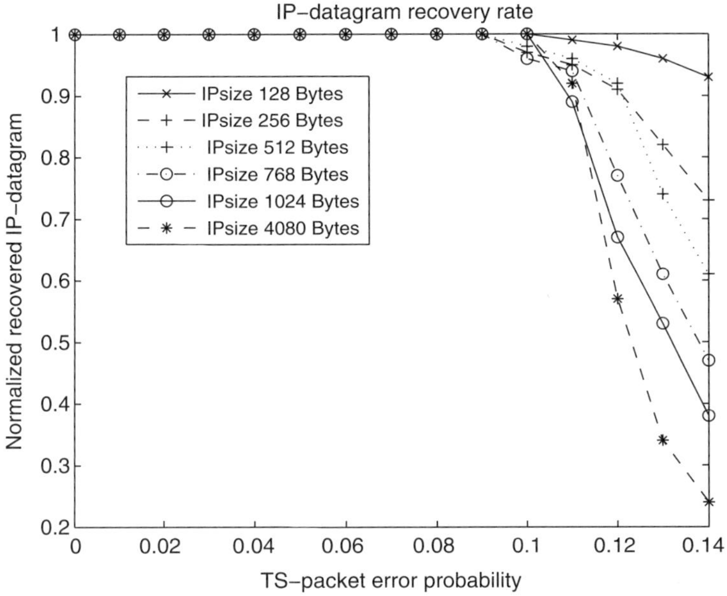

Fig. 8.20 Normalized IP datagram recovery in defect MPE-FEC frames for various IP datagram sizes and varying TS-packet error rate

in the MPE-FEC frame is for each IP datagram indicated by the broadcasted *real_time_parameters* field *address*, which is part of the MPE section header. When after IP de-encapsulation, the MPE section is signalled reliably by means of the CRC code, the *address* field value points to a correctly received IP datagram. The DVB-H standard does not provide means to store this *address* field value, resulting in the loss of this locator information after receiving the service burst. When the *address* field value is stored as locator information in the IPET, each correctly received IP datagram can be retrieved from the MPE-FEC frame. Discontinuities in the IPET-stored locator information indicate the positions of erroneously received IP datagram(s). The IPET locator information solves the intrinsic problem of locating the start positions of IP datagrams in the MPE-FEC frames, which is caused by linked-list way the IP datagrams need to be retrieved from the MPE-FEC frame. The received IP datagrams are stored in a linear fashion in the application data table, as mentioned in Sect. 8.3.1. IP datagram retrieval requires inspection of each IP header to determine its length, in order to retrieve the IP datagram and automatically locating the start position of the next IP datagram. This parsing method collapses as soon as an IP header length field is corrupted, or an IP datagram is missing. The IPET guarantees retrieval of correctly received IP datagrams, regardless of the outcome of the link-layer FEC. Besides information on the correct IP datagrams, IPET also provides information about unreliable or missing IP datagrams. Figure 8.21 visualizes two data application tables. The data application table of Fig. 8.21a contains only correctly received IP datagrams, whereas Fig. 8.21b holds three correct and

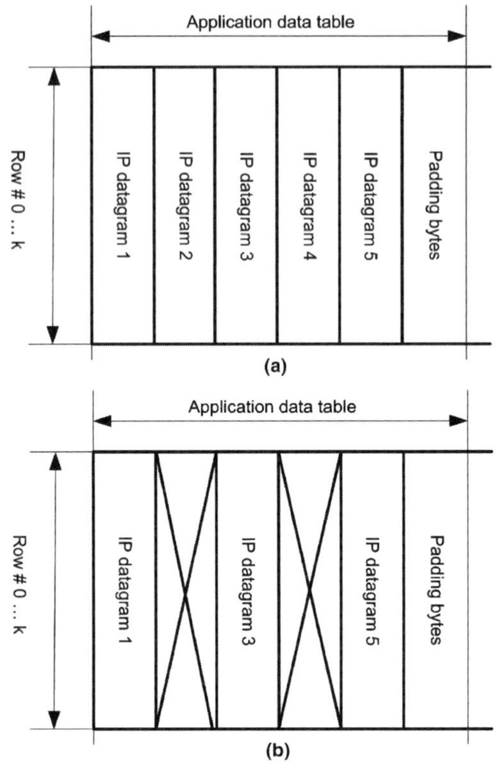

Fig. 8.21 Data application table of an MPE-FEC frame with size $k = 1,024$ rows, containing five IP datagrams. **(a)** Data application table contains only reliable IP datagrams. **(b)** Data application table contains reliable and unreliable IP datagrams

two defect received IP datagrams. Let the IPET locator information corresponding to Fig. 8.21a be $\{0; 1,024; 2,048; 3,072; 4,096\}$ and the IPET locator information corresponding to Fig. 8.21b be equal to $\{0; 2,048; 4,096\}$. The IPET of an application data table that is constructed of only correctly received IP datagrams shall not contain discontinuities. A discontinuity in the IPET occurs when the stored locator information incremented with the IP datagram length of the IP datagram indicated by the locator information does not correspond to the next IPET entry. The IPET corresponding to Fig. 8.21a does not contain discontinuities, because each IPET entry incremented with the length of the IP datagram to which it refers, which is in this example 1,024 bytes, corresponds to the next IPET entry. The IPET associated to Fig. 8.21b contains two discontinuities, indicating that two or more IP datagrams have been received incorrectly.[16] The IPET memory footprint depends on the IP

[16] In this particular example, the IP datagrams have a fixed length of 1,024 bytes. In practice, IP datagrams will have a variable length, resulting in an unknown number of IP datagrams per IPET discontinuity.

datagram size and the locator size. The locator size is directly derived from the *address* field size, which requires 18 bits. The IP datagram size depends on various aspects. On the one hand, there is a dependency on the Maximum Transmission Unit (MTU) size for Ethernet, resulting in IP datagrams with a maximum size of 1,500 bytes. On the other hand, it is influenced by the MTU for the MPE, which is 4,080 bytes. Finally, the third aspect is the MPE-FEC performance, which depends strongly on the IP datagram size [44]. According to [44], optimal IP datagram sizes have a range between 1,024 bytes and 2,048 bytes. The IPET memory footprint for an MPE-FEC frame size of 1,024 rows results in roughly 1,024 bytes. This memory footprint size is based on an IP datagram size of 1,024 bytes, using 3 bytes per IPET locator, although 573 bytes would be sufficient. In practice, H.264 encoded video will be transmitted using IP datagrams that are in the range of 1,024–1,500 bytes and AAC+ audio will use IP datagrams that are in the range of 512–1,024 bytes. Taking into account that video requires more bandwidth, the small-sized IP datagrams occur less often. As a result, the remaining IPET address space is sufficient for handling typical broadcast situation, with the mix of audio and video packets. Depending on the result of the FEC decoding, all erroneous positions in the MPE-FEC frame are repaired or some of the erroneous positions are still defective. The possible erroneous positions correspond to the regions covered by the IPET discontinuity positions. Data reliability can be determined using the OSI Layer 3-4 checksum, as discussed in Sect. 8.3, but this introduces some drawbacks. These drawbacks are avoided when using the CRIT, which is elaborated in the next subsection.

8.4.2.2 Corrected Row Index Table

The CRIT is a table giving the FEC decoding result for each MPE-FEC frame row. After FEC decoding there are two situations. In the first situation, all MPE-FEC frame rows are corrected. As a result, the CRIT entries all indicate a successful FEC decoding operation. In the second situation, not all MPE-FEC frame rows are corrected, so that not all CRIT entries indicate a successful FEC decoding operation. The CRIT is a table storing a 1-bit flag to signal the FEC decoding result, leading to a memory footprint of 128 bytes for an MPE-FEC frame size of 1,024 rows. For the case that an MPE-FEC frame still contains errors even after FEC decoding, the combination CRIT, IPET, and erasure-flag information allows to determine whether an erroneously received IP datagram is corrected. The advantage of this concept is that there is no need to process the data using higher OSI layer protocols, resulting in a uniform recovery approach, regardless of the used network and transmission protocol. Figure 8.22 visualizes the CRIT and erasure-flag processing, for a fragment of column k, indicated by the IPET as a discontinuity region. In Fig. 8.22, the CRIT positions indicated as "0" correspond to MPE-FEC frame rows that are correct, whereas the positions indicated by a "1" mean that the MPE-FEC frame row could not be corrected. The positions in the erasure table indicate the reliability of the symbol position in the MPE-FEC frame (for more details, see Sect. 8.5). The 2-bit erasure flags allow to differentiate between four reliability situations.

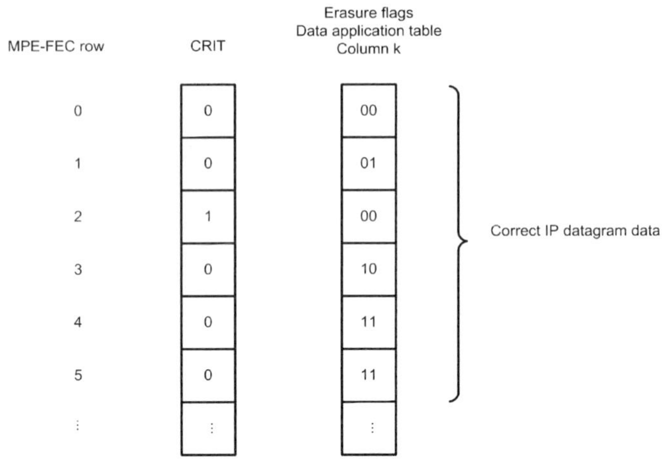

Fig. 8.22 IP datagram recovery using CRIT, IPET, and erasure table

The situation "00" indicates that the symbol position is reliable, whereas all other situations indicate that the symbol position is unreliable. For the situation that the FEC decoder could not correct an MPE-FEC row, the CRIT signals this by means of a "1." Although a particular MPE-FEC row could not be corrected, this does not mean that all symbols of that row are incorrect. By combining the IPET, CRIT, and erasure-flag information, an incorrectly received IP datagram may be recovered from the MPE-FEC frame. The mechanism works as follows. A discontinuity in IPET indicates a region that is unreliable, but this region can contain symbols with a corresponding erasure flag set to "00," signalling that the symbol is correct. The discontinuity region consists of a number of rows, which are either correct or could be corrected by the FEC decoder, or remain defective after FEC. When the CRIT and the erasure-flag information are combined for each defect row of a discontinuity region, each symbol of that region is correct when the CRIT is "1," and the erasure flag is "00," allowing to retrieve a correct IP datagram from a defective MPE-FEC frame. Moreover, the symbols that have an erasure flag value equal to "01" or "10" could also be correct, see Sect. 8.5. For the situation that an incorrectly received IP datagram has symbols constructing the IP header that are correct, but the payload has erasure flags "01" or "10" and the corresponding CRIT has value "1," this IP datagram can be reliably retrieved from the MPE-FEC frame, but it may still be defect. Retrieval of such an IP datagram is advantageous, because the erasure flags may not be correct, leading to another recovered IP datagram. Moreover, if the discontinuity region is larger than the length of this IP datagram, this means that there is at least another incorrectly received IP datagram. This *hidden* IP datagram would otherwise not be visible and definitely be lost. Transmitting possible incorrect IP datagrams will be detected by the IP-stack checksum calculation, preventing them from ending in the application.

8.5 IP De-encapsulation and MPE-FEC Decoding for DVB-H Link Layer

The DVB-H link layer IP de-encapsulation processing has the objective to retrieve in an efficient way one or more IP flows from an Elementary Stream. These efficiency aspects are visible as follows. First, the link layer IP de-encapsulation techniques use received data in such a way that under deteriorating channel conditions, the QoS can be maximized. Second, another efficiency aspect is that link layer processing results in a power and memory friendly way, hence the implementation is also efficient.

Let us briefly explain the IP de-encapsulation process once more. One or more Elementary Streams form a service, see Sect. 8.2.1 for a more elaborate description about the relation between IP streams, IP flows, and Transport Streams. Deriving the requested IP flows from the received Elementary Stream(s) requires a filter-chain as depicted in Fig. 8.14. The filter chain settings are derived from the received SI/PSI information, which is processed by the receiver middleware. The middleware configures the receiver such that the requested IP flows are extracted from the received Elementary Stream(s). For the link layer, the configuration is such that the requested Elementary Stream is *PID* filtered, resulting in MPE and MPE-FEC sections. These sections are forwarded to the IP de-encapsulation filters, which are elaborated extensively thereafter. Furthermore, MPE-FEC decoding is discussed, emphasizing the erasure-flag generation and the usage of the erasure flags in the MPE-FEC decoding. Within a Transport Stream (TS), an Elementary Stream can be uniquely identified by the *PID*, which is a 13-bit field in the TS packet header, for TS packet main header syntax details see Table 8.14. The *PID* filter, as shown in Fig. 8.15b, is a programmable filter that blocks TS packets, which *PID* of which the *PID* values do not match the filter criteria. Every output of the *PID* filter is a sequence of TS packets that have an identical *PID* value, i.e., these packets form an Elementary Stream, see Fig. 8.4. The payload of TS packets belonging to a DVB-H compliant Elementary Stream contains two types of sections: MPE sections and MPE-FEC sections. Each MPE section contains a single IP datagram and an MPE-FEC section carries a single column, having parity data for the MPE-FEC decoding. The *section_length* field and *address* field, indicating the start position of the section payload in the MPE-FEC frame, are present in the header of the MPE sections, see Tables 8.1 and 8.3. The MPE-FEC frame *address* is required if FEC is applied on the received IP data. For this situation, the IP datagrams have to be stored column-wise in the MPE-FEC frame. For checking the integrity of the received data (IP datagram or Reed-Solomon (RS) parity data), sections are provided with a 32-bit Cyclic Reduncancy Check (CRC). The DVB-H Implementation Guidelines [10] suggest to use CRC failures for discarding received data. Accordingly, the corresponding positions in the MPE-FEC frame should be declared as erasures in order to facilitate simpler RS decoding. As indicated earlier, the MPE-FEC sections contain the RS parity data for the extra layer of FEC, which is the FEC Reed-Solomon decoder. Each MPE-FEC section contains one MPE-FEC frame column with parity data. In the MPE-FEC section header, the *section_number* can be used for determining in

Table 8.14 Syntax MPEG-2 Transport Stream main header

Syntax	Number of bits
transport_packet() {	
sync_byte	8
transport_error_indicator	1
payload_unit_start_indicator	1
transport_priority	1
PID	13
transport_scrambling_control	2
adaptation_field_control	2
continuity_counter	4
if (adaptation_field_control == '10' || adaptation_field_control == '11') {	
adaptation_field()	
}	
if(adaptation_field_control == '01' || adaptation_field_control == '11') {	
for (i = 0; i < N; i++) {	
data_byte	8
}	
}	
}	

which column of the MPE-FEC frame the parity data should be placed. For storing the parity data in the MPE-FEC frame, one could also use the *address* field of the *real_time_parameters*, because both fields indicate the start position in the MPE-FEC frame. After receiving the IP datagrams and the RS-parity data, RS decoding is applied when erroneous IP datagrams were received.

8.5.1 Reconstructing the MPE-FEC Frame Under Erroneous Conditions

The IP de-encapsulation method as described in the Implementation Guidelines [10] operates at section level. This means that after calculation of the section CRC, the section payload is either accepted or discarded and erased. In the sequel, the term *Section level IP de-encapsulation* refers to the de-encapsulation method as described in the DVB-H Implementation Guidelines. The discarding of whole sections leads to large erased fragments in the MPE-FEC frame and complicates the reconstruction of the MPE-FEC frame. Therefore, it is advantageous to operate at TS packet level instead of the section level. In literature, several publications [21, 30, 31, 33] support the concept of TS level de-encapsulation. The following facts explain why TS-level IP de-encapsulation has to be preferred over section-level IP de-encapsulation.

A first fact is that not all fragments of a corrupt section have to be erroneous. Discrimination between correct and incorrect fragments is achieved by observing the TS packet field *transport_error_indicator* (TEI), for syntax details see Table 8.14. The TEI flag is a 1-bit flag in the TS packet header that signals whether the RS decoder in the physical layer ([204,188,17] RS) was able to correct the TS packet.

A second fact is that a corrupt TS packet (TEI=1) contains at least 9 byte errors per 204-byte code word. Hence, in the best case, only 9 byte errors occur, and quite some data of the 184-byte payload is useful and actually employed for restoring the original MPE-FEC frame. Both simulations and measurements with an experimental receiver [43] show that under typical channel conditions, often less than 50-Byte errors occur in a corrupted TS packet. In Fig. 8.23 relative distributions

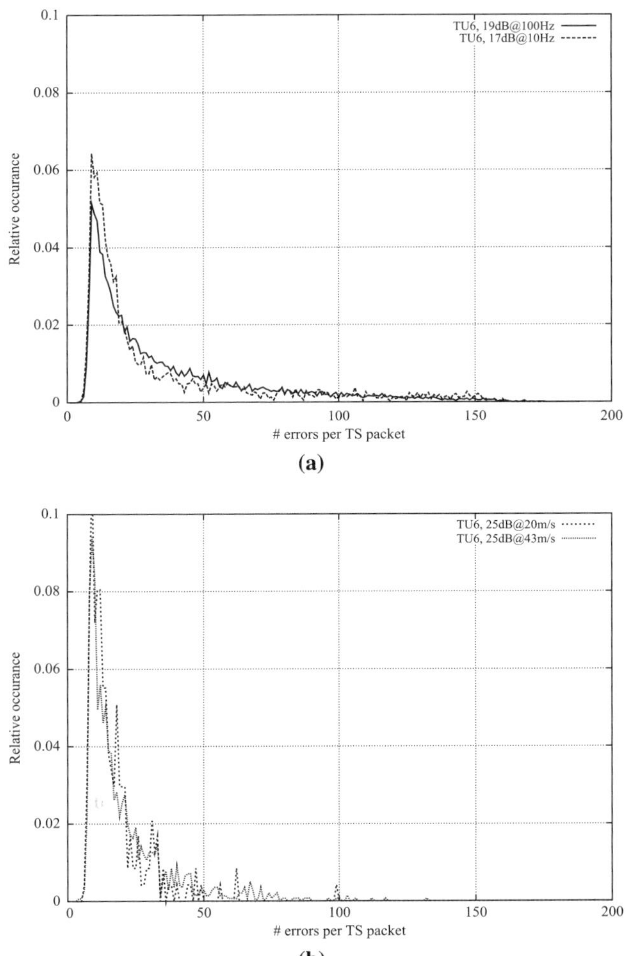

Fig. 8.23 Average number of defect bytes in an erroneous TS packet. (a) Simulated number of byte errors per TS packet. (b) Measured number of byte errors per TS packet.

of byte errors per corrupt TS packet are shown. Fig. 8.23a indicates distributions obtained with computer simulations and Fig. 8.23b depicts distributions obtained with measurements on an engineering sample of a DVB-H receiver. Although the distributions show a high similarity, the measured distributions are somewhat more concentrated to a lower number of errors per corrupt TS packet, an effect that is due to differences in the channel demodulator algorithms, but also from the different channel conditions.

A third fact is that a consistency check on the TS-packet-header *PID* value and *continuity_counter* value, see Table 8.14, is applied to detect whether the payload of a corrupt TS packet is useful. The consistency check involves a comparison of the received *continuity_counter* value of the current TS packet and the expected *continuity_counter* value. The expected *continuity_counter* value is equal to the previous *continuity_counter* value incremented with unity, provided that the *PID* value of the current TS packet is identical to the *PID* value of the previous TS packet. Since not all payload is corrupted, the stored fragment is equipped with an erasure flag of type "unknown," see Table'8.12 for possible erasure types. Because not all data is corrupted, the previously mentioned erasure flag is handled as a medium-priority erasure flag.

A fourth fact is that the MPE-FEC decoder can correct up to 64 erasures per row. Differentiating in erasure priority can prevent the decoder from being overloaded with erasures. For example, high-priority erasures should always be used by the RS decoder, and medium- and low-priority erasures only when there is sufficient correction capacity available within the boundary of the $2t + e < 65$ correction capacity of the FEC decoder. With 2-bit erasure flags, we can distinguish between missed or discarded fragments (high priority), fragments with some errors (low and medium priority) and fragments that have probably no errors (no priority). The use of multi-level erasure information is also reported in literature [21, 31].

Summarizing, TS packet-level IP de-encapsulation comprises:

- If (TEI=0), place the payload of the TS packet at the appropriate position in the MPE-FEC frame and assign "no priority" erasure information to the corresponding positions.
- If (TEI=1) and the TS packet header is consistent, put the TS-packet payload at the appropriate position in the MPE-FEC frame and assign "medium priority" erasure information to the corresponding positions.
- If (TEI=1) and the TS packet header is not consistent, discard payload and assign "high priority" erasure information to the corresponding positions in the MPE-FEC frame.
- Perform error and erasure decoding of the MPE-FEC frame with ordered granting of erasure information.

The first three techniques are related to extracting IP datagrams, fragments from a TS packet and placement of these fragments at the proper location in the MPE-FEC frame, combined with the erasure assignment, i.e., so-called IP de-encapsulation. In the next paragraph, TS-packet-level IP de-encapsulation is described.

8.5.1.1 TS-Packet-Level IP De-encapsulation

TS-packet-level IP de-encapsulation is based upon a concept in which fragments of IP datagrams are first collected in a kind of scratch memory, prior to a possible fragment placement of the partially reconstructed IP datagram in the MPE-FEC frame. In Fig. 8.24, the concept for TS-packet-level IP de-encapsulation is depicted. The purpose of the following concept is to linearly fill a temporary buffer (scratch memory) with both reliable and unreliable payload of the received TS packets. This filling process can fail, leading to data gaps in the linear address space which hinders proper placement of scratch data in the MPE-FEC frame. As a result, the MPE-FEC frame filling procedure starts at the beginning or at the end of the scratch memory, to maximize the capturing of useful data. The maximum IP datagram size in DVB-H context is 4,080 bytes. A TS packet can contain at most 184 bytes of payload. A fragment is defined as the part of an IP datagram that is contained in one TS packet. The scratch memory containing the fragments of an IP datagram is called fragment memory (scratch memory). Assuming that the fragments are efficiently encapsulated in a TS packet, a maximum-sized IP datagram is distributed over $N = \lceil 4,080/184 \rceil = 23$ fragments. From a received TS packet, the IP fragment is placed in the fragment memory. Using the *continuity_counter* in the TS packet header, one can determine (to a certain degree) the position of the fragment (i.e., the fragment pointer) in the fragment memory. The *continuity_counter* is also used for detecting missed fragments. Note that the *continuity_counter* is a 4-bit counter, so its range is actually too small for a unique identification of fragments of a maximum-sized IP datagram. However, according to [44], the optimal IP datagram size is between

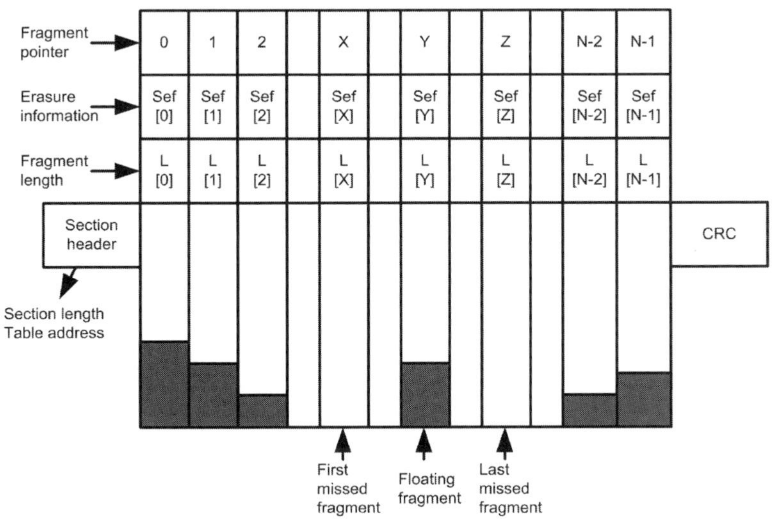

Fig. 8.24 TS level IP de-encapsulation using a fragment memory

the 1,024 bytes and 2,048 bytes. The number of TS packets required to transmit IP datagrams lying in this range fits within the *continuity_counter* range, whose maximum value is 15.

Fragments can vary in length, due to a multiplexing technique called stuffing. Stuffing is a technique for properly aligning data entities like sections, in several TS packets by introducing stuffing bytes (bytes with value "0xFF") in the payload of a TS packet. In case of private sections, two mechanisms can be used for stuffing:

- The first mechanism is based upon adaptation fields. In case of adaptation field stuffing, the stuffing is preceding the actual payload. If *adaptation_field* is used for stuffing, this is signalled in the TS header by the *adaptation_field_control* parameter.
- Another mechanism for stuffing occurs at section level, and thus also applies to MPE- and MPE-FEC sections. In this case, stuffing takes place between the last byte of a section and a new section starting in the next TS packet with a *pointer_field* having value zero. At the decoder, this kind of stuffing can be detected by comparing the section length with the number of extracted bytes. Hence, if the number of already extracted bytes is equal to the section length and the *payload_unit_start_indicator* does not signal the start of a new section, then the remaining bytes of the TS packet payload should be stuffing bytes, thus "0xFF."

As shown in Fig. 8.24, several types of fragments can be distinguished. All these fragment types have to be properly placed in the MPE-FEC frame. For this purpose, the MPE-FEC table address for every individual fragment is determined. The fragments that are received right after the section header can be placed in the MPE-FEC frame, starting from the table *address* that is signalled in the section header. The addresses of succeeding fragments can be determined, either from extrapolating the *address* in the section header together with the section length, or from backward extrapolation, based on the *address* signalled in the next section header. Floating fragments are fragments that are preceded and succeeded by one or more missed fragments. If all fragments (except for the first and maybe the last) have an equal length, one can use address interpolation for the floating fragments.

Since the fragments of IP datagrams originate from different TS packets, they can have different erasure priority information. Fragments that originate from TS packets with TEI = 0 are error-free and will obtain erasure information with the lowest priority level. Usable fragments from TS packets with TEI = 1 will obtain erasure information with medium-priority level. Missed fragments are definitely erroneous, resulting in high-priority level erasure information. Finally, it may happen that all fragments seem to be error-free (TEI = 0 for all TS packets), but that due to mis-correction in the channel decoder, one or more TS packets are mis-corrected. This is detected with the section CRC. The corresponding IP datagram should be assigned with low-priority level erasure information. In Table 8.15, the 4-level erasure assignment is summarized.

Table 8.15 Assignment of 4-level priority erasure's

level	TEI	Other conditions	Symbol state	scope
00	0	CRC=0	Error free	Section
01	0	CRC=1	Mis-corrected TS packet(s)	Section
10	1	'Readable' TS packet	>8 Errors in TS packet	TS packet
11	1	Missed fragment	False	TS packet

8.5.1.2 MPE-FEC Decoding

After the best-effort filling of the MPE-FEC frame, as described in the previous section, MPE-FEC decoding can be started. As stated earlier, error and erasure decoding is applied. The MPE-FEC code is a [255,191,65] Reed-Solomon code, with which one can correct up to t errors and e erasures, provided that $2t+e<d$, where $d=65$. The assignment of 4-level erasure information is incompatible with a conventional error and erasure RS decoder, in which only two levels of erasure information (reliable versus unreliable) can be handled. Moreover, by assigning erasure information to large data fragments, the risk exists that the number of erasures exceeds the erasure-correction capacity of the RS decoder. For this purpose, the erasure information is rearranged to 2-level erasure information in descending order of priority, a process called "granting of erasure information." Figure 8.25 show the flow-chart for granting erasure information (transforming 4-level into 2-level erasure information). The variables N_i stand for the number of erasures of level i, where a high level corresponds to a high-priority erasure and a low level to a low-priority erasure. The granting procedure shown in Fig. 8.25 implements a method in which erasures of lower priority class are granted only if the number of total erasures does not exceed the erasure correction capacity $d-1$ or a preselected maximum $p_{\max}$. For the situation that the correcting capacity is not exceeded, MPE-FEC decoding can be applied. The MPE-FEC decoding result is notified to the CRIT in the case that an MPE-FEC frame row could not be corrected, see Sect. 8.4.2. In [31], a similar algorithm for transforming 3-level erasure information to 2-level erasure information is described.

8.5.1.3 Performance of TS-Level IP De-encapsulation

In a simulation, the performance of TS-level IP de-encapsulation and decoding techniques is validated on a Mobile Channel (TU6) [29]. The modulation parameters are: 8K FFT, Guard Interval $\frac{1}{4}$, 16-QAM nonhierarchical and Convolutional Code with R=2/3. The simulated receiver does not use advanced Doppler compensation techniques in the channel demodulator.

In Fig. 8.26, the required average CNR for achieving an MFER of 5% is shown as function of the Doppler frequency on the TU6 channel. The gain in CNR of the TS-level IP de-encapsulation method compared to the section-level IP de-encapsulation

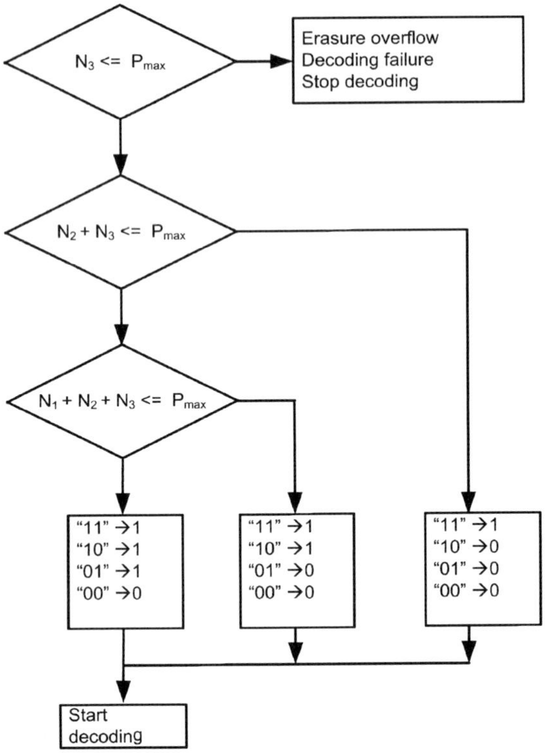

Fig. 8.25 Transformation of multilevel erasure information to bi-level erasure information

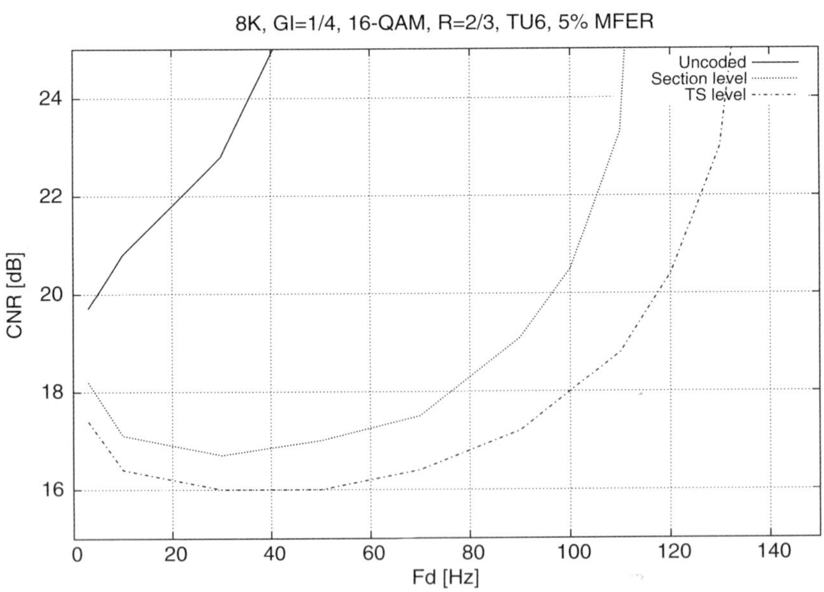

Fig. 8.26 Required CNR for achieving MFER 5%

Table 8.16 Performance figures at TU6 channel for (M)FER 5%

	$CNR_{\min}$ [dB]	$f_{d,3\text{dB}}$ [Hz]	$f_{d,\max}$ [Hz]
Uncoded	20.8	35	–
Section level	17.1	97	125
TS level	16.4	114	141

method ranges from less than 1 dB at 10 Hz to more than 1 dB at 70 Hz. For higher Doppler frequencies, the CNR gain is even higher. The TS-level method also results in a larger Doppler robustness. At an average CNR of 20 dB, we find a difference of approximately 15 Hz between the two curves. The simulation results for $f_d \leq 40$ Hz show that with increasing Doppler frequency, the required average CNR decreases. This phenomenon can be explained by the fact that due to increasing Doppler frequency, more time-interleaving is realized. Time-interleaving is beneficial in the case of rapid changes in channel conditions (e.g., due to motion), so that the duration of bad and good reception conditions is mixed. The advantage of the time-interleaving can be traded off with the limitations of channel-estimation algorithms. The performance criteria $CNR_{\min}$ and $f_{d,3\text{dB}}$ are often used as key performance figures. $CNR_{\min}$ is the required average CNR for achieving the desired Frame Error Rate (FER) (e.g., 5%) at TU6 with 10 Hz Doppler. The parameter $f_{d,3\text{dB}}$ is the maximum Doppler that is allowed for achieving the desired (M)FER with a CNR that is equal to $CNR_{\min}$ + 3 dB. In Table 8.16, these key performance figures are listed. Another performance criterion is $f_{d,\max}$, which is the maximum Doppler that is allowed for achieving the desired FER with a large CNR approaching infinity.

8.5.2 Techniques in the Link Layer for Extra Power Saving

In the DVB-H Implementation Guidelines [10] it is mentioned that in case the application data table is received without errors, MPE-FEC decoding is not applied. Hence, the reception of the RS data table is not necessary anymore. As a consequence, the receiver can be put into sleep mode and wait for the reception of the next burst. At good reception conditions (e.g., error-free reception), a power-saving ratio of up to 64/255 $\approx$ 25% can be realized, in case an [255,191,65] RS code is used. The number 64 in the previous fraction corresponds to the number of columns in the RS data table and 255 in the fraction is the total number of columns in an MPE-FEC frame. However, the requirement of error-free reception of the application table is a requirement that is difficult to satisfy. In practice significant power-savings are realized only if the channel conditions are rather good.

A more promising power-saving method (see also [1, 32]) is based upon the observation that by using an $[n,k,d]$ RS code and applying erasure decoding, one can correct $d-1$ erasures. This means that if k correct symbols are received, decoding of the corresponding word and reconstructing the code word are possible. This observation may be applied to a DVB-H receiver, where MPE-FEC decoding is then applied after reception of k correct columns, while simultaneously switching off the receiver front-end (tuner and channel demodulator). When k columns are received correctly, the RS decoder can reconstruct the whole MPE-FEC frame by erasing the columns that still had to be received (this is a kind of virtual puncturing). For example, when in the application data table one column is corrupt, only a single correct RS data column is required for reconstruction of an MPE-FEC frame. The state of received columns can be determined by monitoring the section CRC and TS packet headers (*transport_error_indicator* and *continuity_counter*). The decision to switch off the receiver front-end can be further refined. The requirement of receiving at least k correct columns can be transformed into a more relaxed requirement of having (at least) k correct symbols per row. This demands somewhat more book-keeping, but gives possibly an earlier switch-off time. In order to anticipate on some undetected errors, it is required to have $k+m$ correct columns or at least $k+m$ correct symbols per row. For example, with $m = 2$, a single undetected error can be corrected, in addition to the k erasures per row. Summarizing, two criteria for early switch-off of the receiver front-end can be distinguished [32]:

- **Column criterion.** Switch off front-end when $k+m$ correct columns $(m \geq 0)$ are received.
- **Row criterion.** Switch off front-end when all rows have $k+m$ or more correct symbols.

The performance of the extra power-saving methods has been simulated. The channel is modelled by having a TS packet error probability P_{TS}. It is assumed that in a corrupt TS packet all bytes are erroneous such that the payload cannot be used. Furthermore, TS packet errors occur independently from each other. In Sect. 8.5.1.3, it is shown that TS-level IP de-encapsulation has a positive effect on the MFER after MPE-FEC decoding, see Fig. 8.26, as it operates with the lowest CNR. Therefore, it can be expected that the power-saving performance also will improve.

In the simulation, 1,000 MPE-FEC frames are generated and TS packets are corrupted randomly (i.i.d.) using an error probability P_{TS}. From these 1,000 MPE-FEC frames, two histograms are made registering the number of required columns fulfilling each of the two criteria. The histogram is used for calculating the expected average power-saving, which is depicted in Fig. 8.27. The proposed extra power-saving method outperforms the "error-free" method significantly. In case of a TS-packet error probability of $P_{\mathrm{TS}} = 0.01$, no power-saving is realized with the "error-free" method, while the proposed methods achieve a power-saving 15% or more. As expected, the method based upon the row criterion outperforms the method based upon the column criterion. Moreover, applying the TS-level IP de-encapsulation method implicitly improves the power-saving performance. Furthermore, this power-saving improvement from TS-level IP de-encapsulation is larger for the row-based method than for the column-based method. The TS-level IP de-encapsulation method combined with the row-based power-saving method realizes a power-saving ratio of more than 10% over a wide range of TS-packet error probabilities.

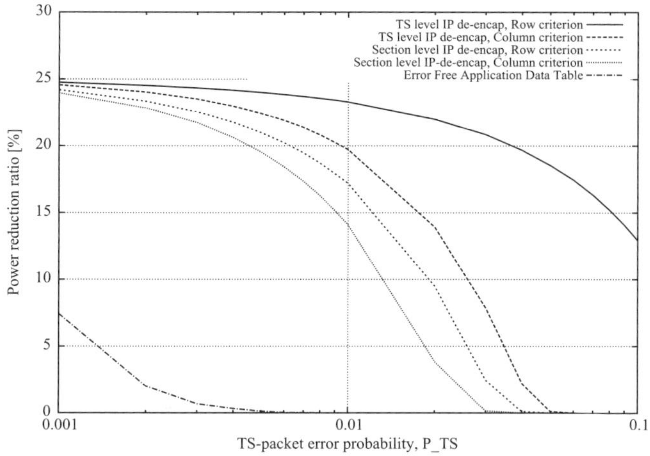

Fig. 8.27 Simulated power-saving

8.6 Verification/Validation of a Robust DVB-H Link Layer

One of the key functions in a DVB-H receiver is the link layer. In order to differentiate on performance, manufacturers can choose between basic implementation, e.g., according to the Implementation Guidelines [10], or advanced link-layer implementation concepts as discussed in Sect. 8.5, resulting in modest or excellent performance. This section presents the DVB-H generator and analyzer concept required for validation and verification of a *robust* DVB-H link layer, as depicted in Fig. 8.13, capable of generating and analyzing extreme signal conditions [6]. The generator and analyzer system aspects are based on the link-layer architecture to be validated. Five functional blocks, MPEG-2 demultiplexer, MPE-FEC decoding, IP-filters, queue management, and link-layer control, are distinguished in the robust DVB-H link-layer architecture, see Fig. 8.13. Four of the five blocks are directly influenced by the incoming TS, whereas the MPE-FEC decoder can only be indirectly influenced. The link-layer functional blocks impose the following system requirements on the DVB-H TS generator:

- **MPEG-2 demultiplexer:** The generator must be able to generate error-free and error-prone TSs, thereby exploring the syntax variations according to [19, 25].
- **MPE-FEC decoding:** This block operates on the MPE-FEC frame, using erasure information derived by the MPEG-2 demultiplexer block.
- **IP filter:** The generator must be able to create error-free or error-prone IPv4 or IPv6 traffic, with uniform or nonuniform destination addresses.
- **Queue management:** The generator must be able to generate error-free or error-prone SI/PSI traffic for average signal conditions up to worst-case timing conditions, maximum section sizes and parallel or consecutive IP-based services based on different MPE-FEC dimensions.

- **Link-layer control:** The generator must be able to generate time-sliced and dispersed elementary streams, either with or without FEC.
- **Reference data:** The generator must deliver the reference data for the analyzer to perform SI/PSI and IP-data comparisons.

The functional blocks of the link layer, see Fig. 8.13, impose the following system requirements on the DVB-H analyzer:

- **Erasure information:** The analyzer must have the knowledge about erasure-location information and its usage by the FEC decoder to determine the FEC decoder result.
- **MPE-FEC frame:** The analyzer must have knowledge about the transmitted MPE-FEC frame.
- **IP-filter settings:** The analyzer must have knowledge of the IP-filter settings to determine which IP datagrams are expected.
- **SI/PSI sections:** The analyzer must have knowledge on all SI/PSI sections, their position in time, the SI/PSI-filter settings, and the DVB-H receiver reception mode.

The DVB-H TS-generator system aspects result in a DVB-H TS-generator concept, in which the error-prone TS generation aspects dominate the DVB-H generator architecture. Verification of the robust link layer as depicted in Fig. 8.13 is to a large extent determined by the MPEG-2 demultiplexer, because of its ability to handle both reliable and unreliable TS packets. The MPEG-2 demultiplexer requires syntax variation to verify robustness, reliable and unreliable data to fill the erasure table, MPE-FEC frame, and IPET. Besides the MPE-FEC decoder, the remaining functional link-layer blocks operate on reliable data, which simplifies the verification.

Let us now discuss the DVB-H generator system aspects. The first system aspect is the syntax variation. For DVB-H, syntax variation is limited to the presence or absence of the *adaptation_field* and the occurrence of a section header split. Basically, a TS header can be succeeded by an *adaptation_field*, see Table 8.14, where the usage of an *adaptation_field* is selected at the IP encapsulator. In DVB-H, the presence of the *adaptation_field* allows stuffing at TS level, instead of section level. A section-header split can occur when the first bytes of a section, which by default are part of the section header, are placed in the payload of Transport Stream packet N and the remaining section header bytes are placed in the payload of Transport Stream packet $N+1$. Due to bandwidth optimization, all TS payload bytes are used for the transmission of section data, resulting in a possibility of a section header split.

To avoid the development of a link-layer reference model that produces the data reference set, the DVB-H generator is defined such that it provides data from which the reference set can be derived in a low-cost way. This allows the generation of error-free and error-prone streams. As a result, the DVB-H generator will apply a concept called error back-annotation, enabling the injection of errors anywhere in the stream-generation process. Error back-annotation indicates which MPE-FEC frame data is reliable and which data is unreliable. The error back-annotation takes

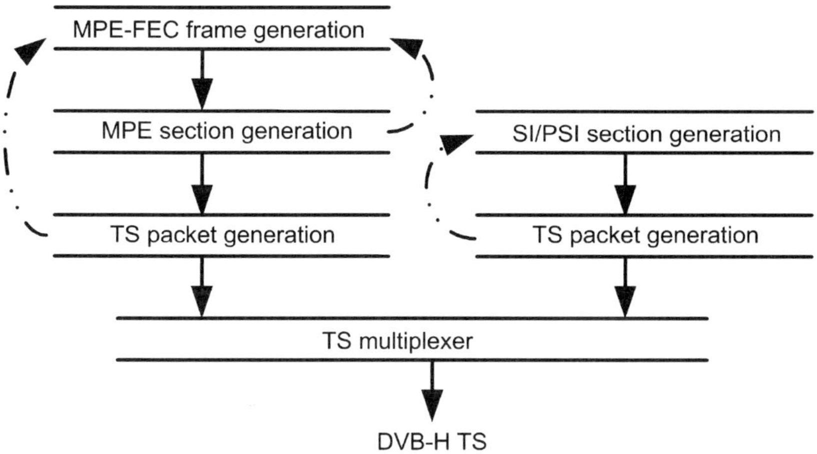

Fig. 8.28 DVB-H generator concept with error back-annotation

into account the IP de-encapsulation scheme of the link layer under test. Error back-annotation is invoked when errors are injected at MPE-section or TS-packet level, or when error expansion occurs. When error back-annotation is applied on SI/PSI data, erroneous sections will not be part of the SI/PSI reference set. Figure 8.28 indicates the stream generation process, where the dashed arrows indicate the error back-annotation.

8.6.1 DVB-H Stream Generator

Let us now discuss an error-prone stream generator allowing an analyzer to verify the link-layer result in an elegant way. At the upper-left side of Fig. 8.28, an MPE-FEC frame is generated. This MPE-FEC frame can be equipped with or without errors. The byte values at erroneous positions in the MPE-FEC frame are not modified, but equipped with a label that is used during the MPE encapsulation and TS packatization process to modify the byte value and finally cause the TS packet-header TEI flag to be set to "1." Error back-annotation will occur for the following situations:

- Injection of errors in the MPE-FEC frame
- Injection of an error at MPE-section level
- Injection of an error at TS-packet level

For the situation that a single error is injected in the MPE-FEC frame, error expansion can occur. Error expansion is caused by the fact that when the erroneous MPE-FEC frame byte is MPE encapsulated and TS packetized, that particular byte will most probably be part of a TS packet containing more than one byte payload. The TS packet that carries the erroneous payload byte(s) shall have the TEI flag set to "1," indicating that the TS packet is unreliable.[17] For the situation that the unreliable TS packet contains more than one byte payload, all other payload bytes are signalled as unreliable via the error back-annotation concept. As a result, a single injected error in the MPE-FEC frame causes error expansion in column direction, possibly affecting preceding or succeeding column data. Note that although the other payload bytes are signalled unreliable, they may still be correct, all depending on what the type of error causing the error back-annotation.

For the case that an error is injected at MPE-section level or TS-packet level, error back-annotation is applied taking into account the IP de-encapsulation process of the link layer that is to be tested. For example, if a link layer is implemented using the de-encapsulation method as described in the DVB-H Implementation Guidelines [10], a whole section is discarded when erroneous. Using the error back-annotation concept, all data of the IP datagram encapsulated in that particular MPE section is hard erased. The advantage of using the error back-annotation is that with little effort, various IP de-encapsulation schemes can be validated, without the need to modify the IP de-encapsulation analyzer. Furthermore, the DVB-H generator and associated analyzer allow simulation of various IP de-encapsulation concepts, without the need for an actual implementation.

The back-annotated MPE-FEC frame is the result of error-prone IP encapsulation. The SI/SPI-generation process can be implemented in a simpler way, due to the lack of an additional FEC layer. At the right-hand side of Fig. 8.28, the SI/PSI-generation process is depicted. So when the TS packet containing SI/PSI-section data is corrupted, this is notified to the SI/SPI generator. As a result, the SI/PSI section that is corrupted is not available in the reference set. Each section that is transmitted correctly is equipped with an absolute time-stamp and stored in the reference set. The inserted sections meet the requirements as indicated in [14], involving the time distance between succeeding sections and their corresponding bit rates.

The SI/PSI sections also allow to characterize the receiver wake-up time and power-down time, because of the attached absolute time-stamp. Such a measurement requires a test stream in which sections with a size smaller or equal to the TS-packet payload are inserted in each TS packet. For the situation that the receiver has started too early, sections outside the service burst are received and thereby appear at the output. Similar for the power-down mode, if sections are available in the link-layer output that are transmitted after the end of a service burst, the receiver is active for an unnecessary long period, thereby negatively influencing the power consumption. For this test method, the constraints implied in [14] are not applicable.

8.6.2 DVB-H Stream Analyzer

The analyzer verifies the link-layer-forwarded IP datagrams and SI/PSI sections. This requires the following parameters and data:

[17] Note that in a normal receiver situation, the TEI flag is set by the channel decoder if more than 8 byte errors occur.

- Error back-annotated MPE-FEC frame
- IP filter settings
- Rules for erasure granting
- SI/PSI reference set
- Reception mode, continuous or time-sliced
- Time-slice duty-cycle, and service start-time

The error back-annotated MPE-FEC frame contains all transmitted IP datagrams in reliable format and the corresponding labels of the fragments that will be unreliably received. For the situation that no IP filters are set, all transmitted IP datagrams will be expected, provided that the number of erasures per MPE-FEC row is less than 65. Simply counting the number of erasures (medium and high priority) per row in the error back-annotated MPE-FEC frame indicates which rows can be corrected and thus which IP datagrams can be expected.

When IP postfiltering is applied, the analyzer needs to have the filter values and filter-mask values to reduce the number of expected IP datagrams from the error back-annotated MPE-FEC frame.

For the situation that the number involved erasures exceeds the erasure correction capacity of the RS decoder, erasure granting can be applied. If applied, the DVB-H analyzer needs to be aware of this, and have knowledge of the applied rules.

The section reference set contains the time-stamped correctly transmitted sections. The analyzer uses this reference set to determine which section is to be delivered by the link layer under test. It therefore uses the link-layer reception mode and if operated in time-sliced mode, also the time-slice duty-cycle and the service start-time. Furthermore, it employs an off-set to determine the location of the first service burst in the TS, to extract the sections from the reference set using the absolute time-stamp value.

8.7 Conclusions

When designed for robustness and performance, the DVH-H link layer becomes a quite complex subsystem. The features time-slicing and MPE-FEC added to the DVB-H link layer result in a performance gain when compared to DVB-T. On the one hand, time-slicing reduces the power consumption, while on the other hand, it allows the creation of a virtual receiver, which positively influences the overall system cost. Such a virtual receiver allows, e.g., peeking into neighboring channels during the off-time of the received service, monitoring for possible candidate services, thereby avoiding the need for a second front-end. However, a virtual front-end is essential, due to the cell-based character of a DVB-H network, where for nomadic usage of a DVB-H receiver, horizontal handover is frequently applied without discontinuities in the reception.

The significant reception improvement gained by MPE-FEC can only be achieved when the correct and incorrect received service data is handled properly and the corresponding erasure flags are carefully assigned. At the expense of 12% extra memory on top of the 255 kbytes required for the MPE-FEC frame of maximum size, a robust and efficient DVB-H link layer is achieved. Such a link layer allows the reception of consecutive service reception for services with maximum MPE-FEC frame size and provides graceful degradation of the received audiovisual information under varying channel conditions. Furthermore, applying intelligent erasure monitoring either on a column or a row basis allows an extra 10% power reduction due to early front-end switch-off.

Fields of future research may be based using the higher OSI layers to enable improved IP de-encapsulation. This is based on the recognition that the OSI layers contain a considerable amount of redundancy. Another field that will provide valuable understanding is the transport-packet burst length. This burst length will strongly depend on the modulation settings and the channel characteristics and will directly influence the IP de-encapsulation performance, perhaps making certain techniques obsolete.

Acknowledgments We wish to acknowledge the NXP Research management for offering us the time and means to write this chapter. In addition, we would like to thank A. Stuivenwold (NXP BL-PES) for putting us on the DVB-H development team. Finally, we would like to thank all the colleagues from various NXP departments involved in the DVB-H project from the cities of Caen, Dresden, Eindhoven, and Nijmegen for their contribution.

References

1. Diego Balaguer de E, Fitzek FHP, Olsen O, Gade M (2005) Performance evaluation of power saving strategies for DVB-H services using adaptive MPE-FEC decoding. IEEE 16th International Symposium on Personal, Indoor and Mobile Radio Communications, Vol. 4, pp 2221–2226
2. Crawford M (1998) Transmission of IPv6 packets over ethernet networks RFC 2464
3. Deering S (1989) Host extensions for IP multicating RFC 1112
4. Deering S, Hinden R (1998) Internet protocol version 6 (IPv6) specification RFC 2460
5. Eerenberg O, Koppelaar A, Stuivenwold AM, With de PHN (2006) IP recovery in the DVB-H link layer for TV on mobile. International Conference on Consumer Electronics, pp 411–412
6. Eerenberg O, Wendrich P, Orsouw EHW, With de PHN (2007) Efficient validation/verification of a robust DVB-H link layer. International Conference on Consumer Electronics, pp 1–2
7. DVB-H validation task force: final report (2005)
8. Digital Video Broadcasting Project, www.dvb.org (2005)
9. European Telecommunications Standards Institute (ETSI) (2001) Digital video broadcasting (DVB); Allocation of Service Information (SI) Codes for DVB systems European Standard TR 101 162 v1.2.1
10. European Telecommunications Standards Institute (ETSI) (2007) Digital video broadcasting (DVB); DVB-H Implementation Guidelines European Standard draft TR 102 377 v1.3.1
11. European Telecommunications Standards Institute (ETSI) (2005) Digital video broadcasting (DVB); Transmission to Handheld Terminals (DVB-H); Validation Task Force Report European Standard TR 102 401 v1.1.1
12. European Telecommunications Standards Institute (ETSI) (2005) Digital video broadcasting (DVB); IP Datacast over DVB-H: Set of Specifications for Phase 1 European Standard draft TR 102 468 v1.1.1

13. European Telecommunications Standards Institute (ETSI) (2006) Digital video broadcasting (DVB); IP Datacast over DVB-H: Architecture European Standard TR 102 469 v1.1.1
14. European Telecommunications Standards Institute (ETSI) (2006) Digital video broadcasting (DVB); IP Datacast over DVB-H: Program Specific Information (PSI)/Service Information (SI)European Standard TS 102 470 v1.1.1
15. European Telecommunications Standards Institute (ETSI) (2006) Digital video broadcasting (DVB); IP Datacast over DVB-H: Electronic Service Guide (ESG) European Standard TS 102 471 v1.2.1
16. European Telecommunications Standards Institute (ETSI) (2006) Digital video broadcasting (DVB); Specification for Service Information (SI) in DVB systems European Standard EN 300 468 v1.7.1
17. European Telecommunications Standards Institute (ETSI) (2004) Digital video broadcasting (DVB); Framing structure, channel coding and modulation for digital terrestrial television European Standard EN 300 744 v1.5.1
18. European Telecommunications Standards Institute (ETSI) (2004) Digital video broadcasting (DVB); DVB specification for data broadcasting European Standard EN 301 192 v1.4.1
19. European Telecommunications Standards Institute (ETSI) (2004) Digital video broadcasting (DVB); Transmission system for handheld terminals European Standard EN 302 304 v1.1.1
20. Henriksson J (2003) "DXB-X" DVB-Scene magazine, p 7
21. Himmanen H, Hazmi A, Paavola j (2006) Comparison of DVB-H link layer FEC decoding strategies in a mobile fading channel. IEEE 17th International Symposium on Personal, p 1–5
22. International Organization for Standardization and the International Electrotechnical Commission (ISO/IEC) (2001) Information technology – Telecommunications and information exchange between systems – Local and metropolitan area networks – Specific requirements – Part 1: Overview of Local Area Network Standards International Standard 8802-1
23. International Organization for Standardization and the International Electrotechnical Commission (ISO/IEC) (1998) Information technology – Telecommunications and information exchange between systems – Local and metropolitan area networks – Specific requirements – Part 2: Logical link control International Standard 8802-2
24. International Organization for Standardization and the International Electrotechnical Commission (ISO/IEC) (1996) Information technology – Open Systems Interconnection – Basic Reference Model: The Basic Model International Standard 7498-1
25. International Organization for Standardization and the International Electrotechnical Commission (ISO/IEC) (2000) Information technology – Generic coding of moving pictures and associated audio information: Systems International Standard 13818-1
26. International Organization for Standardization and the International Electrotechnical Commission (ISO/IEC) (1998) Information technology – Generic coding of moving pictures and associated audio information – Part 6: Extensions for DSM-CC International Standard 13818-6
27. International Organization for Standardization and the International Electrotechnical Commission (ISO/IEC) (2001) Information technology – Coding of audio-visual objects – Part 3: Audio International Standard 14496-3
28. International Organization for Standardization and the International Electrotechnical Commission (ISO/IEC) (2003) Information technology – Coding of audio-visual objects – Part 10: Advanced video coding International Standard 14496-10
29. International Organization for Standardization and the International Electrotechnical Commission (ISO/IEC)(2002) Mobile and portable DVB-T/H Radio access – Part 1: Interface Specification International Standard ISO/IEC62002-1
30. Jokela T, Paavola J, Himmanen H, Ipatov V (2006) Performance analysis of different Reed-Solomon erasure decoding strategies at the DVB-H link layer. IEEE 17th International Symposium on Personal, pp 1–5
31. Joki H, Paavola J (2006) A novel algorithm for decapsulation and decoding of DVB-H link layer forward error correction. IEEE International Conference on Communications, Vol. 11, pp 5283–5288
32. Koppelaar AGC, Eerenberg O (2006) Power-saving by adaptive RS decoding in a DVB-H receiver. Proceedings of the 27th Symposium on Information Theory in the Benelux, pp 9–16

33. Koppelaar AGC, Eerenberg O, Tolhuizen LMGM, Aue V (2006) Restoration of IP-datagrams in the DVB-H link-layer for TV on mobile. International Conference on Consumer Electronics, pp 409–410
34. Kornfeld M , May G (2007) DVB-H and IP datacastBroadcast to handheld devices. IEEE Transactions on Broadcasting, Vol. 53 No. 1, pp 161–170
35. Loncaric S, Grgic S, Zovko-Cihlar B (2006) Minimizing power consumption of the DVB-H receiver. 48th International Symposium ELMAR-focused on Multimedia Signal Processing and Communications, pp 309–313
36. May G (2004) The IP datacast system overview and mobility sspects. IEEE International Symposium on Consumer Electronics, pp 509–514
37. May G (2005) Loss-free handover for IP datacast over DVB-H networks. Proceedings of the Ninth International Symposium on Consumer Electronics, pp 203–208
38. Milosavljevic ZD (2007) A varactor-tuned DVB-H antenna. International Workshop on Antenna Technology: Small and Smart Antennas Metamaterials and Applications, pp 124–127
39. Postel J (1981) Internet protocol – DARPA internet program protocol specification RFC 791. USC/Information Sciences Institute
40. Postel J (1980) User datagram protocol RFC 768. USC/Information Sciences Institute
41. Schulzrinne H, Fokus G.M.D, Casner S, Frederick R, Jacobson V (1996) RTP: A Transport Protocol for Real-Time Applications RFC1889
42. Television on a Handheld Receiver, www.digitag.org (2005) Ver. 1.1
43. Trauschke E (2007) Untersuchungen von Algorithmen zum Reed-Solomon Erasure-Decoding von MPE-FEC Daten nach dem DVB-H Standard unter Berücksichtigung der Eigenschaften des Mobilfunkkanals. Studienarbeit TU Dresden
44. Vadakital VKM, Hannuksela MM, Razaei M, Gabbouj M (2006) Optimal IP packet size for efficient data transmission in DVB-H. Proceedings of the 7th Nordic Signal Processing Symposium, pp 82–85
45. www.ipdc-forum.org (2005)
46. Yang X, Vre J, Owens TJ (2006) A survey of handover algorithms in DVB-H. IEEE Communications Survey 4th Quarter 2006, Vol. 8, No. 4, pp 16–29

Chapter 9
Transport and Time-Slicing Mechanisms in Multistandards for Mobile Broadcasting

Georgios Gardikis

9.1 Introduction

In a multimedia broadcasting system, the transport mechanism performs all the necessary adaptation and multiplexing functions in order to adapt the IP or the raw, non-IP multimedia streams to the final baseband format that will be sent to the modulator. In this sense, it can be actually considered as a separate layer in the broadcasting system layer stack (Fig. 9.1). However, we avoid calling it as the "transport layer" in order not to confuse it with the transport layer of the OSI (Open Systems Interconnection) model, which has a slightly different hierarchical position.

Some fundamental functions that the transport mechanism includes are:

- Packetization of contiguous non-IP multimedia streams
- Encapsulation and framing of IP data
- Separation of the entire TDM (Time Division Multiplex) into "logical channels" so as to convey discrete services
- Time-slicing organization and signalling, for energy conservation at the receiver
- Transport-Level FEC (Forward Error Correction), for extra per-service FEC protection in addition to the coding inserted at the modulator

The following sections provide an overview of the transport mechanisms which are used by the main contemporary multimedia broadcasting systems. The rest of this chapter is organized into as follows. Section 9.2 is devoted to the MPEG-2 Transport Stream (TS) employed in most broadcasting standards including DVB-H (Digital Video Broadcasting for Handheld mainly developed in Europe), T-DMB (Korean's Terrestrial Digital Multimedia Broadcasting), A-VSB (Advanced Vestigial Sideband, developed in the USA), DMB-T/H (Digital Terrestrial Multimedia Broadcast, being implemented in China), ISDB-T (Japanese Integrated Services

G. Gardikis
NCSR "Demokritos", Institute of Informatics and Telecommunications, Patriarchou Gregoriou & Neapoleos str., Aghia Paraskevi Athens 153 10, Greece
e-mail: gardikis@iit.demokritos.gr

F.-L. Luo (ed.), *Mobile Multimedia Broadcasting Standards: Technology and Practice*, DOI: 10.1007/978-0-387-78263-8_9,

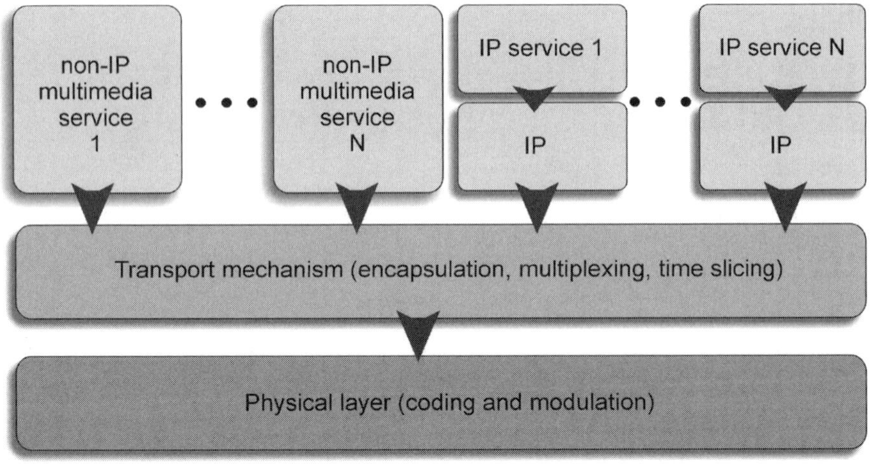

Fig. 9.1 Layered hierarchy of a multimedia broadcasting system

Digital Broadcasting) and CDMB (China's Digital Multimedia Broadcasting). In Sects.9.3 and 9.4, the transport mechanisms in DAB-IP (Digital Audio Broadcasting – IP System) and Media FLO (Media Forward-Link-Only system developed by Qualcomm, Inc) are presented. Some conclusions and further discussion are included in the last section.

9.2 The MPEG-2 Transport Stream (DVB-H, T-DMB, A-VSB, DMB-T/H, ISDB-T, CDMB)

Most multimedia broadcasting systems today utilize the MPEG-2 TS (TS), as defined in [1] as the transport mechanism. The MPEG-2 TS provides a fixed-packet-size statistical TDM multiplexing service for heterogenous multimedia and data streams, along with service and timing information. Originally adopted by the DVB (Digital Video Broadcasting) and ATSC (Advanced Television Systems Committee) family of standards as the baseband signal format, it was inherited by most contemporary broadcasting systems [2].

9.2.1 Structure and Framing

The MPEG-2 TS Statistical Multiplex is a TDM sequence of fixed-sized *Transport Packets (TPs)* of 188 bytes which can convey multiple constant-rate or variable-rate streams such as:

- Non-IP MPEG-2 and MPEG-4 audiovisual streams
- IP data streams

- Service information
- Conditional access signalling
- Timing information

The MPEG-2 Systems specification defines the format of the baseband TS to be sent to the broadcast modulator (Fig. 9.2). From that point on, the FEC encoding and modulation procedures are defined by the corresponding physical-layer specifications of the broadcasting standard (e.g. DVB-H, ISDB-T, etc.)

As aforementioned, each Transport Packet has a constant size of 188 bytes, out of which the first four constitute the Header (Fig. 9.3).

The upper-layer useful data (multimedia streams, IP data) is framed and fragmented under the procedures which will be described in the following sections, and are loaded into the Payload field of the Transport Packet. If the upper-layer data block is not an integer multiple of the TP payload (184 bytes), then the last TP has to contain some null bytes. In this case, the Adaptation Field is used.

All TPs containing data belonging to a specific service or subservice (e.g. the audio component of a multimedia service or the subtitles) are assigned a certain

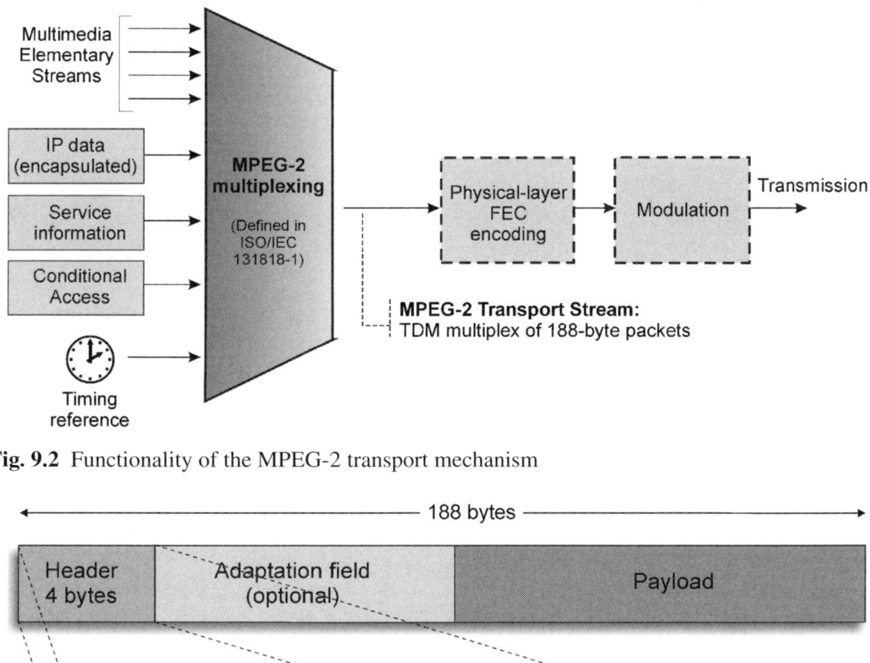

Fig. 9.2 Functionality of the MPEG-2 transport mechanism

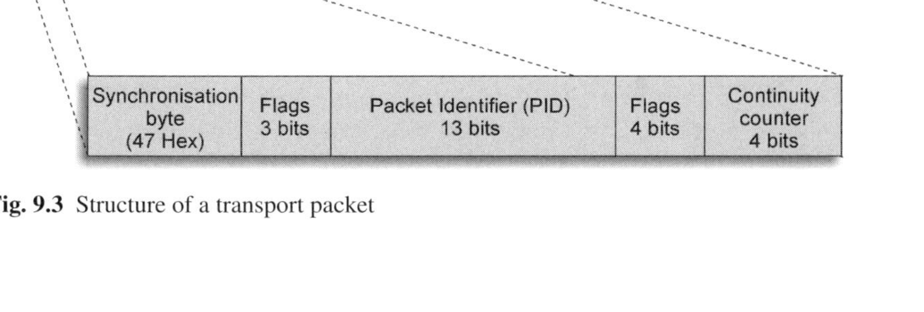

Fig. 9.3 Structure of a transport packet

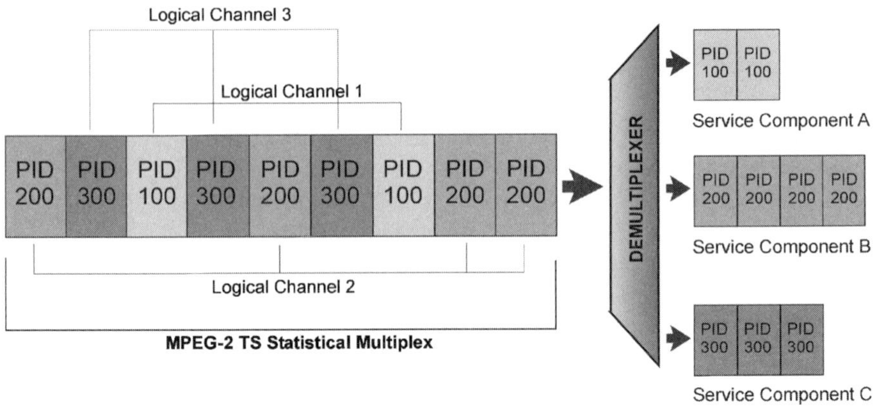

Fig. 9.4 Division of the MPEG-2 transport stream in logical channels

Packet Identifier (PID), which is a 13-bit value. For example, regarding a specific video broadcast, the following values could be assigned:

- Program information – PID 100
- Video component – PID 101
- Audio component – PID 102
- Subtitles – PID 103
- Associated data (teletext) – PID 104

The PID value assists the demultiplexer to easily isolate packets within the MPEG-2 TDM stream, which carry information for a specific service. In this sense, the TS is divided into statistically multiplexed "logical channels", each corresponding to a specific PID value (Fig. 9.4).

The use of relatively small and constant-sized packets makes the processing and error-protection of the multiplexed stream by the Modulator easier. For example, most DVB and DVB-based broadcasting systems protect each Transport Packet with a block Reed-Solomon code by adding 16 bytes of coding overhead. The TS format has been designed especially for transmission in erroneous conditions, such as mobile propagation.

The following sections describe in brief the procedure under which IP and non-IP service streams are framed and fragmented in order to be distributed into the payload of Transport Packets.

9.2.2 The A-VSB Enhancement

The A-VSB (Advanced Vestigial Sideband) system uses the MPEG-2 TS, as described above. However, in order to provide improved dynamic and mobile reception, it inserts some extra bytes inside some TS packets, right after the TS header, in the place of the Adaptation Field, when this is present:

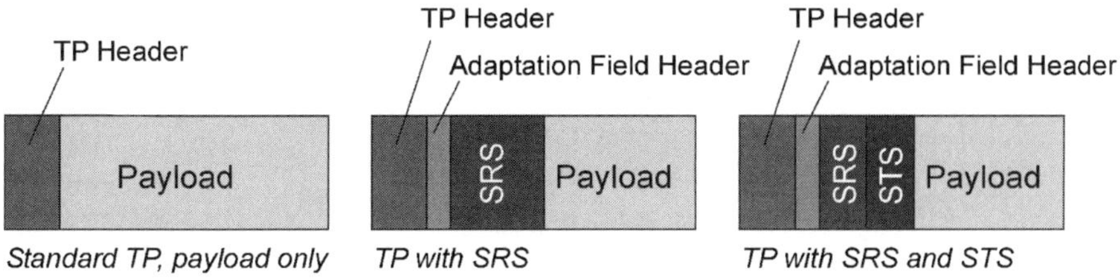

Fig. 9.5 Insertion of STS and SRS bytes by the A-VSB system

- The *Supplementary Reference Sequence (SRS)* is a training sequence which assists the receiver to adapt itself in the current channel conditions by properly training its equalizer.
- The *Scalable Turbo Stream (STS)* feature employs extra FEC redundancy to further protect the stream at a customized code rate (1/2 or 1/4). This feature is quite similar to the MPE-FEC technique to be described in Sect. 9.2.5, but this time Turbo coding is used instead of Reed-Solomon.

The insertion of the SRS and STS overhead in the place of the MPEG-2 Adaptation Field achieves full backward compatibility to older systems which do not recognize this overhead. The use of the Adaptation Field header leads legacy VSB receivers to just bypass these extra bytes and process the rest of the payload normally (Fig. 9.5).

9.2.3 Mapping of Non-IP Multimedia Streams on the MPEG-2 TS

The processing of an audiovisual stream by a source encoder (MPEG-2, MPEG-4 etc.) normally produces two bitstreams, containing the compressed digital video and audio data, respectively. These continuous bitstreams, named *Elementary Streams (ESs)*, have to be fragmented and framed before being loaded into the payload of the TPs. This is a two-stage procedure described in [1]. First, the ESs are fragmented into packets of non-constant-size (usually of a few KBs), in order to form a *Packetized Elementary Stream (PES)*. A PES preamble is added, as shown in Fig. 9.6.

It is usual (but not always mandatory) that the start and end of the payload (data bytes) of a PES packet are aligned to logical points within the ES. For example, one PES packet may carry one video frame.

Then, each PES packet is distributed into the payload of a sequence of Transport Packets, having the same PID. The last TP of this sequence may have some bytes left unused. In this case, the Adaptation Field is used, as aforementioned.

The receiver/decoder performs the inverse operation, by following these steps:

- Location of Transport Packets having a certain PID
- Extraction of their payload
- Re-construction of the PES packets
- De-encapsulation and re-construction of the audio and video ESs
- Decoding and presentation

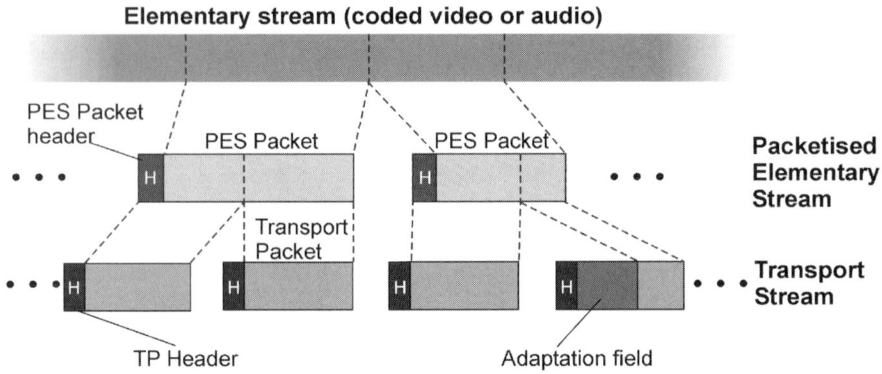

Fig. 9.6 Fragmentation of a non-IP audiovisual stream into a sequence of transport packets

9.2.4 MPE Encapsulation of IP Streams into the MPEG-2 TS

Initially intended to convey MPEG-2 encoded audio and video ESs, the MPEG-2 TS was eventually used also for the transport of IP traffic, with the adaptation method introduced in [3] and named as *Multiprotocol Encapsulation (MPE)*. The adoption of MPE accented the role of broadcasting platforms as access networks for IP-based broadband data and multimedia services. Broadcasters have the potential to use a part of the capacity of the broadcast channel to include unicast or multicast IP traffic along with the audiovisual streams. In certain cases, e.g. in DVB-H, all multimedia and data traffic are conveyed over IP, so that no non-IP streams are included.

In order that IP datagrams can be conveyed over a MPEG-2 TS, a fragmentation (and reassembly) procedure is required. The first stage of the encapsulation process is the framing of the datagram with the encapsulation protocol header and the CRC (Cyclic Redundancy Check) or checksum.

ETSI dealt with the IP-to-MPEG-2 encapsulation issue by standardizing four methods in [3], namely Data Piping, Asynchronous Data Streaming, Synchronous Data Streaming and MPE. The latter was proposed as the most suitable method for conveying IP datagrams and was soon adopted by all IP-to-MPEG-2 Gateways.

In MPE, IP datagrams are encapsulated within blocks called *Datagram Sections*. The framing process includes the addition of the Datagram Section header at the beginning (before the IP datagram), and a 4-byte CRC or checksum at the end calculated over the entire section. Among others, the headers contain fields declaring the length of the section, the scrambling status, and the MAC-layer (Medium Access Control) address of the receiver. Several MPEG-defined fields are also present, as shown in Fig. 9.7.

Once formed, the Datagram Section is split into several fragments and transferred to the payload of the MPEG-2 Transport Packets.If its length is not an integer mul-

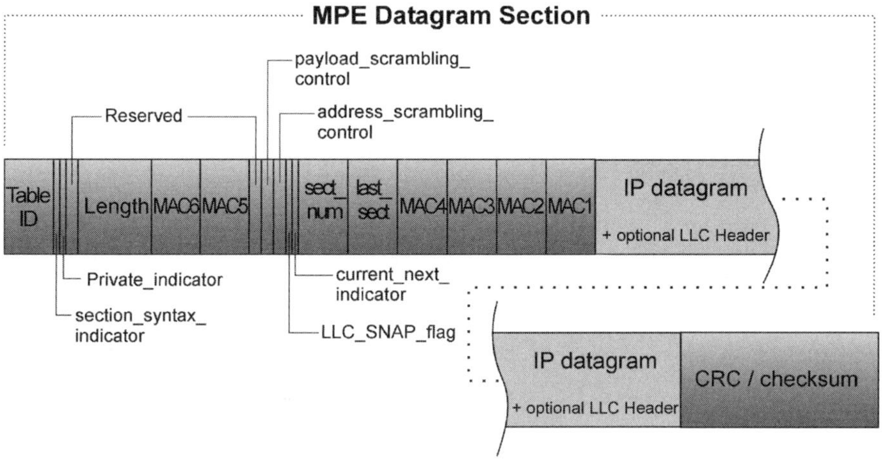

Fig. 9.7 MPE encapsulation of an IP datagram within a datagram section

tiple of the TP payload (184 bytes),there are two options: either padding the rest of the last TP with stuffing bytes ("padding") or beginning a new Datagram Section in the remaining area ("packing").

9.2.5 ULE Encapsulation of IP Streams into the MPEG-2 TS

Unidirectional Lightweight Encapsulation (ULE) [4] was designed with the aim of making the encapsulation process as lightweight as possible [5]. It follows the approach of "data piping," i.e. directly mapping the datagram into the TP payload, adding only a small header, thus forming a *ULE Sub-Network Data Unit (ULE SNDU)*, in proportion to the MPE Datagram Section. ULE header contains just a Length field which declares the length of the SNDU, and a Type field which has the same functionality as that of Ethernet, i.e. it declares the type of the payload. Thanks to the Type field, ULE provides native support for newer network protocols, such as IPv6 and MPLS.

The ULE header can also include a 6-byte destination address corresponding to the receiver's Network Point of Attachment (NPA), which is used to uniquely identify a receiver and may correspond to the receiver's MAC. Finally, a CRC-32 tail is appended (as in MPE) to ensure proper reception and synchronization. Figure 9.8 shows the structure of the ULE SNDU. The framing has become as lightweight as possible, retaining only the necessary fields for proper de-encapsulation and forwarding of the IP datagram.

After framing, the ULE SNDU is mapped to the payload of MPEG-2 Transport Packets. In the case that the SNDU length is not an integer multiple of the TP payload, one of the aforementioned techniques of Padding and Packing can be employed.

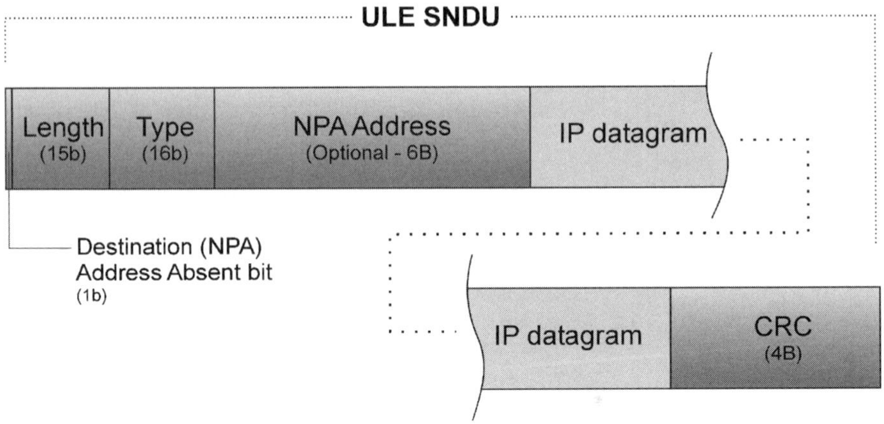

Fig. 9.8 ULE encapsulation of an IP datagram within a ULE SNDU

9.2.6 DVB-H Extensions to MPE: MPE-FEC and Time Slicing

As aforementioned, DVB-H relies on IP/MPE for the transport of both data and audiovisual streams [6]. Furthermore, it adopts two additional features in order to better support mobile reception, namely *Time Slicing* and *MPE-FEC*. Both techniques have been standardized in [3].

Time slicing. A basic issue in handheld operation is the limited battery time. This issue is of particular importance in terrestrial mobile reception, where the receiver/demodulator/demultiplexing/decapsulation chain consumes typically 1 W. The time-slicing feature of DVB-H aims at reducing the average power consumption by allowing the terminal to know when to expect data and to switch off the receiving chain when not needed (i.e. when the transmitted data is of no interest to the specific receiver). At the broadcaster, prior to encapsulation, IP data belonging to a certain stream is organized in TDM bursts. During encapsulation, each IP section is tagged with a "delta-t" value, which informs the receiver about the time interval until the next burst. This information allows the receiver to switch off until the next burst of data arrives (Fig. 9.9). Practically, the duration of one burst is in the range of several hundred milliseconds whereas the powersave time may amount to several seconds. A typical power saving up to 90% is expected, whereas this figure depends on the number and the bit rate of the IP services that the terminal is "listening" to [7].

MPE-FEC. As a transmission system, DVB-H includes two layers of error-protection coding, namely a transport-level Reed-Solomon and an inner convolutional coder. These methods protect the TS as a whole and have been proven to be very effective. DVB-H introduces an additional FEC layer, prior to encapsulation, which can be applied on a per-stream basis. The MPE-FEC method organizes the IP datagrams in a table, column-by-column, and then protects each row of the table with a Reed-Solomon overhead, as shown in Fig. 9.10. IP datagrams are then separately encapsulated and transmitted from the FEC data. The latter can be discarded

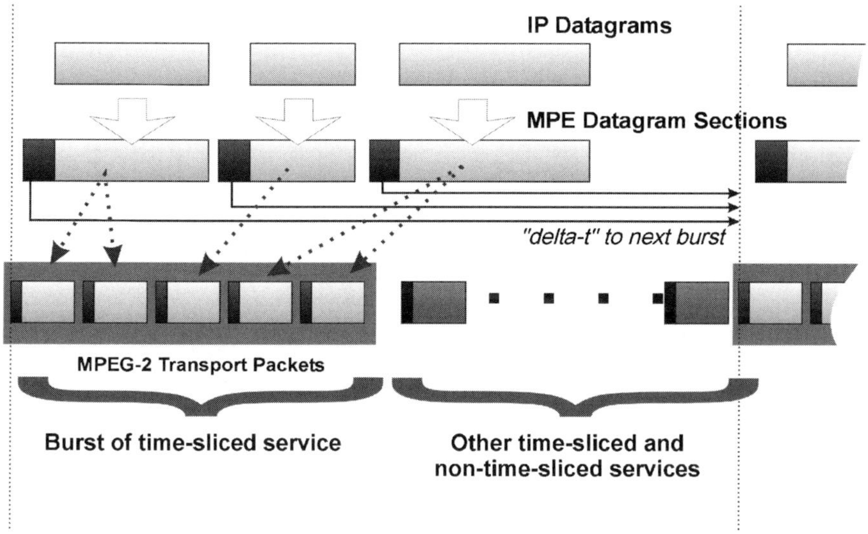

Fig. 9.9 The principle of MPE time slicing used in DVB-H

by FEC-ignorant receivers, thus making the method backwards compatible. MPE-FEC allows the broadcaster to apply a different level or protection on each broadcast IP service, depending either on the importance on the service and/or on the reception conditions of the terminal(s) to which the service is targeted.

9.3 The DAB-IP Transport Mechanism

Unlike most broadcasting systems which utilize the MPEG-2 TS format, DAB uses a completely different transport mechanism [8], more suitable for lower data rates. A DAB multiplex comprises of three main virtual channels multiplexed:

- The *Main Service Channel (MSC)* carries the content itself (audio and data)
- The *Fast Information Channel (FIC)* is mostly used for signalling information that needs to be quickly delivered
- The *Synchronization* channel is internally used (transparent to the user) within the transmission system for basic demodulator functions (synchronization, frequency control, channel estimation, etc.)

Here, we will examine particularly the MSC, which carries the useful information.

The MSC comprises of data blocks named *Common Interleaved Frames* (CIFs), sized 55,296 bits (6,912 bytes).

The data carried over the MSC is divided into "subchannels," similar to the MPEG-2 logical channels. Each service or service component, e.g., an audio stream or a data session, is carried over a dedicated subchannel.

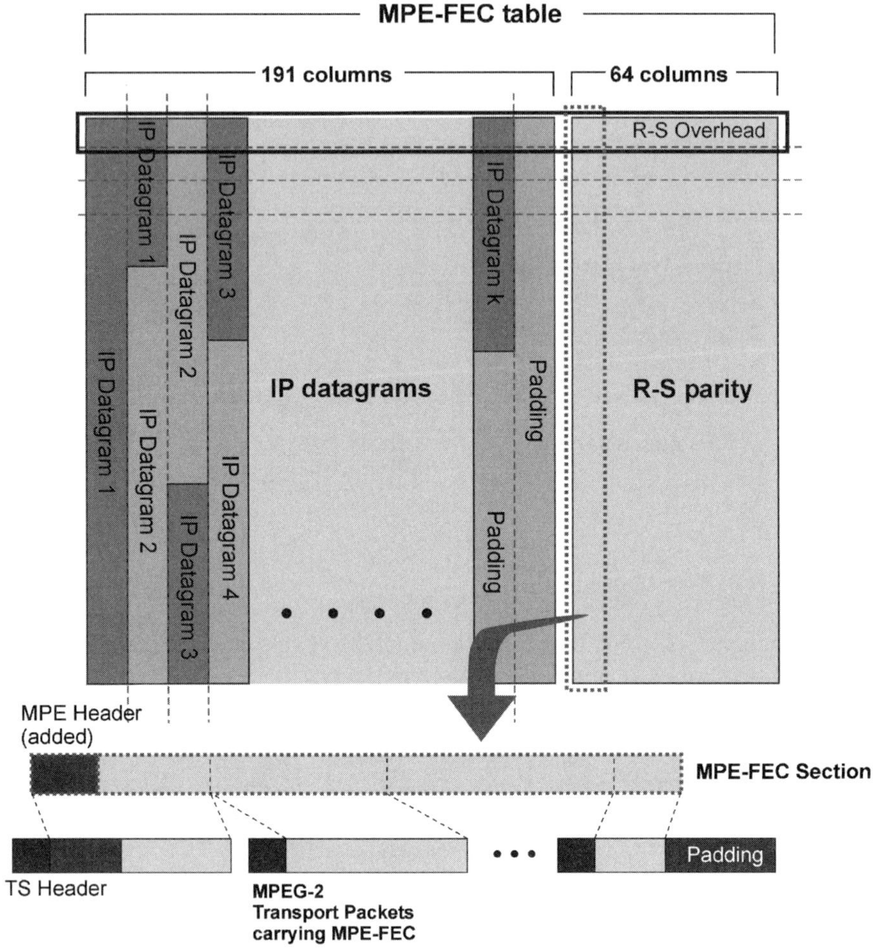

Fig. 9.10 IP data protection via MPE-FEC

DAB-IP employs a specific encapsulation procedure for carrying IP multimedia streams over a DAB MSC. First, IP packets are framed to form a *MSC data group*, as shown in Fig. 9.11. The payload of a Data Group is 8,191 bytes at maximum.

Then, the Data Group is fragmented into several *Packets*, as shown in Fig. 9.12. Packets are of specific size: they can be 24, 48, 72 or 96 bytes long, having a 3-byte header, a 2-byte CRC tail, and a payload of 19, 43, 67, and 91 bytes long, respectively.

Afterwards, Reed-Solomon FEC may be applied to a sequence of packets to add customizable transport-layer error protection in a way very similar to the MPE-FEC technique described in Sect. 9.2.6. in the case of MPE encapsulation. As with MPE-FEC, several packets are organized in an FEC frame consisting of a 188-by-12-byte *Application Data Table* and a 16-by-12-byte *RS Data Table* (Fig. 9.13).

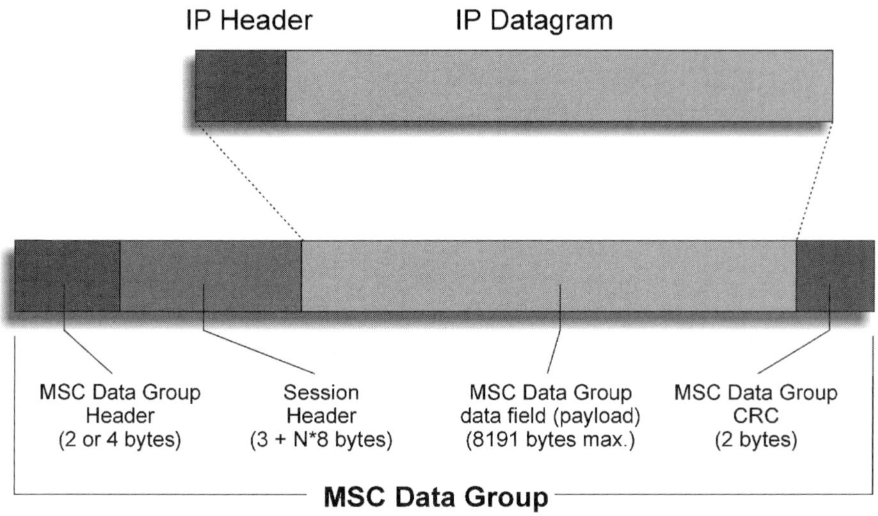

Fig. 9.11 DAB-IP Encapsulation of an IP multimedia stream in MSC data groups

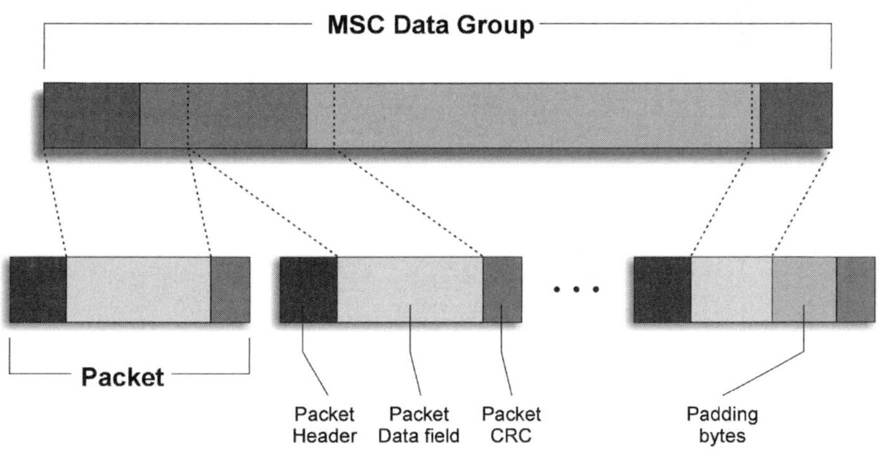

Fig. 9.12 Fragmentation of the data group into packets

The Application Data Table contains the packets to be protected, organized in a column-by-column basis. Depending on its length, each packet exactly fits in 2, 4, 6 or 8 columns. RS codewords are calculated over each row. Then, the packets are transmitted unaltered, followed by special FEC packets containing the RS overhead, as calculated via the aforementioned procedure.

In this way, the DAB-IP broadcaster can select the FEC rate, i.e. the signal robustness of each service against impairments inserted by poor propagation conditions. The receiver performs the inverse procedure, i.e. receives data and FEC packets, reorganizes the Application and FEC Data Table, and via R-S decoding, corrects any bit errors inserted during transmission.

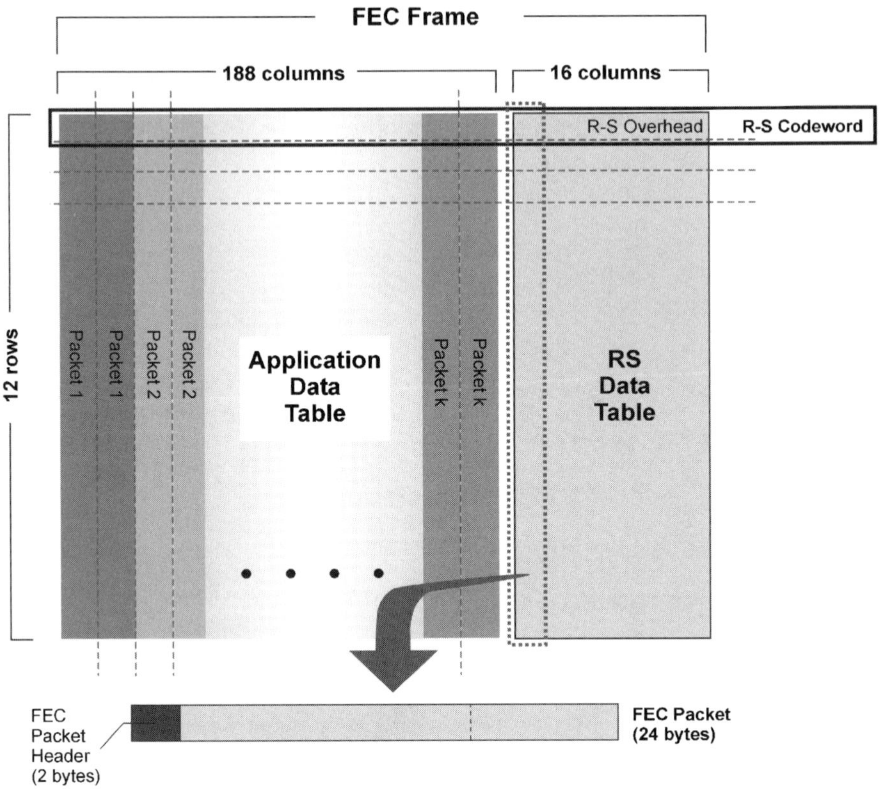

Fig. 9.13 Addition of FEC using the application data table structure

9.4 The MediaFLO Transport Mechanism

In contrast to DVB-related technologies, Qualcomm's MediaFLO designed a multiplexing format "from scratch" [9], instead of adopting the MPEG-2 TS. In FLO, useful information can come either as IP flows or non-IP, raw multimedia streams. Packetization is performed in constant-sized *MAC layer packets*, each carrying 976 bits (122 bytes) of information.

In the way that in an MPEG-2 stream, discrete streams are carried over separate logical channels, services within a MediaFLO multiplex are conveyed over *Multicast Logical Channels (MLCs),* each consisting of a flow of MAC layer packets. Each MLC carries a separate service or service component. It is possible, for instance, that the audio and video information of a multimedia service are carried over two separate MLCs, in the way that in an MPEG-2 multiplex, video and audio were transported having two discrete PIDs.

Data flows in an MLC can be either constant- or variable-rate. In the latter case, the Statistical Multiplexing process makes optimal use of the bandwidth, enhancing the system with the corresponding multiplexing gain. Also, each MLC can have its own coding rate and modulation constellation at physical layer. This feature gives the broadcaster the ability to control the propagation robustness of each service, and control in this way the extent of the radio coverage.

Furthermore, in order to minimize power consumption at the receiver, MediaFLO employs a Time-Slicing mechanism, like DVB-H, although a bit more sophisticated. An MLC containing a service occupies not only a certain time slot, but also a frequency slot, i.e. a portion of the OFDM (Orthogonal Frequency Division Multiplexing) symbol named an "interlace." As it is described in the corresponding chapter, each MLC is spread over frequency and time at a pattern known to the receiver. After receiving a block of data, the receiver is aware of the frequency- and time-slot to come, which will contain the next packet. In the mean time, it can switch off the corresponding parts of the receive chain, thus conserving energy [10].

9.5 Conclusions

The sections of this chapter presented the transport and time-slicing mechanisms which are used in contemporary mobile multimedia broadcasting systems. The MPEG-2 Transport Stream and the mechanisms which have been developed around it are still dominant and have been adopted in most cases. However, some recently standardized systems (i.e. DAB-IP and MediaFLO) have developed their own transport mechanism, more oriented to mobile use and to the handling of IP traffic. MPEG-2 TS is a widely adopted format and it is backed by numerous industrial players; therefore it is quite easy and cheap to implement its support in broadcasting platforms and receivers. However, in certain cases, it lacks efficiency and flexibility, and this has led to the design of new transport and time-slicing formats "from scratch." It can be foreseen that future broadcasting systems will also attempt to redesign their own transport layer, which will be more tailored to mobile use and will better exploit the benefits of the underlying physical layer.

References

1. International Organisation for Standardisation (2000) ISO/IEC 13818–1, Information Technology – Generic Coding of Moving Pictures and Associated Audio Information (MPEG-2) Part 1: Systems, Second Edition
2. Sarginson P (1996) MPEG2: Overview of the Systems Layer, BBC Research and Development Report, BBC RD 1996/2, February 1996
3. European Telecommunications Standards Institute (ETSI) (2004) Digital Video Broadcasting (DVB): DVB Specification for Data Broadcasting, European Standard EN 301 192, v.1.4.1
4. Fairhurst G, Collini-Nocker B (2005) Unidirectional Lightweight Encapsulation (ULE) for Transmission of IP Datagrams over an MPEG-2 Transport Stream (TS), IETF RFC 4326, [Online] Available: http://www.ietf.org/rfc/rfc4326.txt
5. Xilouris G, Gardikis G, Koumaras H, Kourtis A (2006) Unidirectional Lightweight Encapsulation: Performance Evaluation and Application Perspectives, IEEE Transactions on Broadcasting, 52 (3), pp. 374–380

6. Gardikis G, Xilouris G, Skianis H, Kourtis A (2008) Broadband Multimedia on the Move with DVB-H, Multimedia Tools and Applications, 36 (1–2), pp. 133–144
7. Gardikis G, Kokkinis H, Kormentzas G (2007) Evaluation of the DVB-H Data Link Layer, Proc. European Wireless '07, Paris, France
8. European Telecommunications Standards Institute (ETSI) (2006), ETSI, Radio Broadcasting Systems; Digital Audio Broadcasting (DAB) to Mobile, Portable and Fixed Receivers, European Standard EN 300 401 v.1.4.1
9. Telecommunications Industry Association (2007) Forward Link Only Transport Specification, TIA-1120, Rev. 07
10. Qualcomm Inc. (2007) FLO Technology Overview, [Online] Available: http://www.qualcomm.com/common/documents/brochures/tech_overview.pdf

Chapter 10
Basic Modulation Schemes in Digital Mobile Multimedia Broadcasting Systems

Qinghua Han and Chan Ho Ham

10.1 Introduction

Modulation is the mapping–superimposing process where the baseband bit stream message information is mapped to the radio carrier. Modulation and demodulation are paired with modulation being carried out at the transmitter, and the demodulation being carried out at the receiver.

Most of the latest digital mobile multimedia broadcasting (DMMB) standards such as Digital Video Broadcasting, Handheld (DVB-H) and Terrestrial Digital Multimedia Broadcasting (T-DMB) [1–10] are using the orthogonal frequency-division multiplexing (OFDM) based techniques with various inner modulation schemes while the Advanced Television Systems Committee (ATSC) [11] is adopting the vestigial sideband (VSB) modulation approach. The selection of digital modulation schemes used in the digital mobile multimedia broadcasting system (DMMBS) has almost the same requirements as other digital communication systems. The consideration may include, but is not limited to the following: Nyquist theoretical minimum bandwidth, Shannon limit (Shannon–Hartley capacity), low bit error rate (BER), high power and bandwidth efficiency, high power efficiency, low out of band radiation, low sensitivity to multipath fading, constant envelope, low cost, ease of implementation, flexible and compatible to existing system and easy for future upgrade, and the regulation. Some of the above requirements conflict each others. The final decision always reflects a trade-off, a balance of these factors. For certain application, some factors may be paid more attention than others.

In analog broadcasting and communication systems, the basic parameter is average signal power to average noise power ratio *SNR* or S/N, which is also called as the signal to noise ratio. For digital systems, the signal to noise ratio can be expressed in the form

Q. Han (✉) and C.H. Ham
Florida Space Institute/Univ. of Central Florida, MS: FSI, Rm 1130, Bldg M6–306, Kennedy Space Center, FL 32899, USA
e-mail: q_h_han@yahoo.com

F.-L. Luo (ed.), *Mobile Multimedia Broadcasting Standards: Technology and Practice*,
DOI: 10.1007/978-0-387-78263-8_10,

$$SNR = \frac{E_b}{N_0} = \frac{ST_b}{N/W} = \frac{S/R_b}{N/W} = \frac{S}{N} * \frac{R_b}{W} = \frac{S}{N} * \eta_s \tag{10.1}$$

where E_b is bit energy, W is bandwidth, S is signal power, T_b is bit time, R_b is bit rate, N_0 is noise power spectral density, N is noise power, and η_s is spectrum efficiency.

The noise source can arise from a number of contributory factors in addition to the receiver thermal noise. This includes, but is not limited to: intermodulation performance, local oscillator phase noise, channel filter response, and quadrature imbalance, and their effects will sum in an RMS (root mean square) manner. Spectrum efficiency η_s (also called as bandwidth efficiency η_B) is one of the key attributes of the DMMB communication systems, which almost always operate in a crowded and valuable licensed radio frequency spectrum. Spectrum efficiency is used to measure in units of bits per second (bps) per bandwidth.

$$\eta_s = \frac{R_b}{W} = \frac{1}{WT_b} \text{Log}_2(\text{M}) \text{ bps/Hz} \tag{10.2}$$

The capacity of a band-limited additive white Gaussian noise (AWGN) channel is governed by Shannon–Hartley theorem

$$\begin{aligned} C &= W \,\text{Log}_2\left(1 + \frac{\text{average signal power}}{\text{average noise power}}\right) \\ &= \frac{\text{average signal power}}{\text{average noise power} * \ln 2} (W \to \infty) \end{aligned} \tag{10.3}$$

It is possible to transmit information over a channel at a rate R with an arbitrarily small error probability, if $R <= C$. The rate limit is set by signal power (S), bandwidth (W) and noise power (N).

Nyquist theoretical stated that the theoretical minimum bandwidth needed for the based band transmission of R_s symbols per second without ISI is $R_s/2\,\text{Hz}$. For M-ary system, relationship between bit rate R_b and symbol rate R_s is

$$R_s = R_b / \log_2 M \tag{10.4}$$

The portable and mobile DMMBS needs to minimize the battery requirement; normally a nonlinear power amplifier is used. Nonlinear amplifier may degrade the bit error rate performance of some modulation schemes. Spectrum shaping and predistortion can be performed prior to up conversion and nonlinear application. Power efficiency, η_p, is a measure of how much received power is needed to achieve a specified BER.

In digital modulation, the modulation is implemented in digital form and named as "Shifting & Keying." The basic digital modulation schemes are Amplitude Shift Keying (ASK), Frequency Shift Keying (FSK), and Phase Shift Keying (PSK). These systems represent the transmitted data signal by varying the amplitude, frequency or phase of a certain frequency carrier, respectively. There is another important type of modulation, hybrid modulation, which modulates more than one parameter. If the parameters are amplitude and phase then the modulation is the Quadrature Amplitude Modulation (QAM). The QAM is also called as QASK (Quadrature Amplitude Shift Keying) or APSK (Amplitude and Phase Shift Keying).

The inner modulation schemes used in OFDM based DMMBS are very similar to the modulation schemes in other single carrier systems, which has been discussed thoroughly in many books [12–14]. Various digital modulation techniques have been used as modulation schemes in different DMMB standards [1–11]. The rest of the chapter will give introduction to various basic modulation schemes: BPSK (Binary Phase Shift Keying), QPSK (Quadrature Phase Shift Keying), $\pi/4$ DQPSK, 8PSK, 16QAM, 64QAM, 256QAM, 8VSB, and 16VSB.

10.2 Phase Modulation (PM)

For a sinusoidal carrier with frequency (f_c), the modulated signal can be expressed as

$$s(t) = A(t)\cos(2\pi f_c t + \theta(t)) \tag{10.5}$$

where $A(t)$ is amplitude and $\theta(t)$ is the phase. For phase modulation, the amplitude, $A(t)$, and carrier frequency are keeping constant. The information bit stream modulates the phase of the carrier. PSK modulation is a commonly used technique. It has constant envelop, high frequency efficiency, insensitive to channel fluctuations. It was the most popular modulation scheme during past decade. An interesting result is that the bit error probability of QPSK is same as BPSK, given by Eq. 10.6, and has twice the spectral efficiency with identical energy efficiency.

$$P_{b,\mathrm{BPSK}} = P_{b,\mathrm{QPSK}} = Q\left(\sqrt{\frac{2E_b}{N_0}}\right) \tag{10.6}$$

The theoretical symbol error probability of coherently detected MPSK in an AWGN channel is given by

$$P_s(M) = erfc\left(\sqrt{\frac{E_s}{N_0}}\sin\left(\frac{\pi}{M}\right)\right) \approx 2Q\left(\sqrt{\frac{2E_s}{N_0}}\sin\left(\frac{\pi}{M}\right)\right) \tag{10.7}$$

where *erfc* is the complementary error function, $Q(x)$ is the Q-function, E_s/N_o is the ratio of energy in a symbol to noise power spectral density, and M is the number of symbols or the size of the symbol set. The bit error probability, P_b, can be expressed in terms of upper and lower limits of a symbol error probability, P_s, function as

$$\frac{P_s(M)}{\log_2 M} \le P_b \le \frac{M/2\ P_s(M)}{M-1} \tag{10.8}$$

If $P_b \ll 1$, then

$$P_b \approx \frac{P_s(M)}{\log_2 M} \tag{10.9}$$

10.2.1 BPSK

The simplest digital phase modulation is BPSK. The phase item, $\theta(t)$ in Eq. 10.10, has only two values, one representing bit 0, with $\theta(t) = \pi$, and the other one representing bit 1, with $\theta(t) = 0$.

$$S(t) = \sqrt{\frac{2E_b}{T_b}}\cos(2\pi f_c t + \theta(t)) \tag{10.10}$$

where E_b is the energy per bit, and T_b is the bit period. It is well known that a phase change of π radians is equivalent to multiplying a "−1." One possible BPSK modulator can be a simple level translator, changing logic levels to $+/-1$, with a multiplier, as given in Fig. 10.1. An example of the pulse and waveform and the constellation of BPSK are shown in Fig. 10.2 and Fig. 10.3, respectively.

The spectrum of a rectangular pulse spans infinite frequency. In DMMS the transmitted signal must be restricted to a certain bandwidth. When rectangular pulses are passed through a low-pass filter, the pulses will spread into succeeding pulses and cause inter-symbol interference (ISI). The pulse-shaping filter is used to reduce ISI and spectral spreading. Nyquist invented a class of filters which have zero inter-symbol interference at the sampling times. The raised cosine filter is one of the most commonly used pulse-shaping filters in the digital communication system. With raised cosine filtering, when the receiver makes its decision at the middle of each pulse interval, the ripples from adjacent pulses are crossing through zero. Therefore, they do not introduce errors within the decision making process. The raised cosine filter transfer function is given by the following equation:.

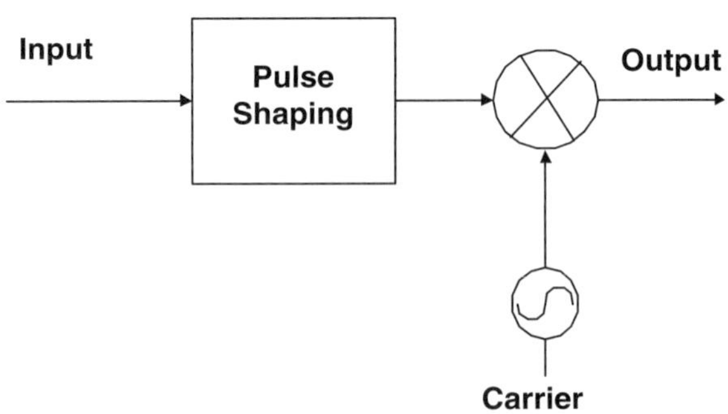

Fig. 10.1 BPSK modulator block diagram

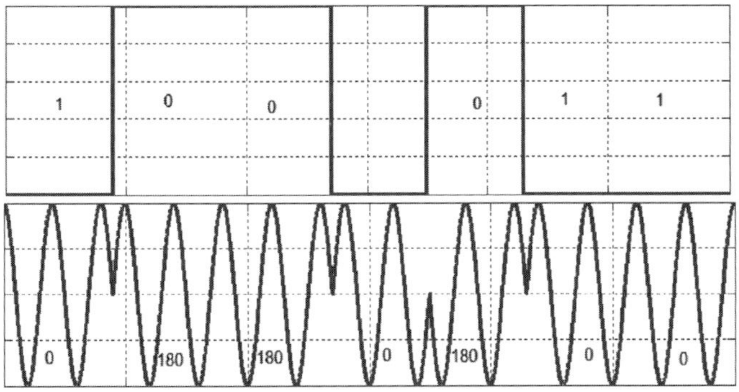

Fig. 10.2 Pulses and waveform of BPSK

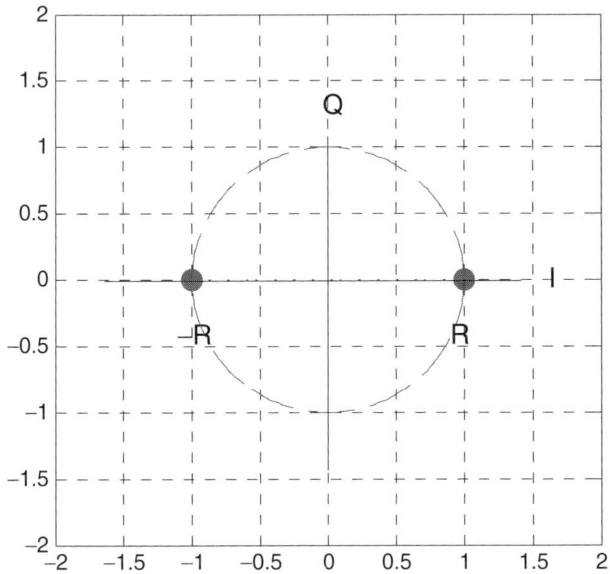

Fig. 10.3 BPSK signal constellation diagram

$$H(\omega) = \begin{cases} \tau & \tau \le \omega \le \dfrac{\pi(1-\alpha)}{\tau} \\ \tau\left\{\cos^2\left[\dfrac{\tau\left(\omega - \dfrac{\pi(1-\alpha)}{\tau}\right)}{4\alpha}\right]\right\} & \dfrac{\pi(1-\alpha)}{\tau} \le \omega \le \dfrac{\pi(1+\alpha)}{\tau} \\ 0 & \omega > \dfrac{\pi(1+\alpha)}{\tau} \end{cases} \tag{10.11}$$

where τ is the pulse period, ω is the radian frequency, and α is the roll off factor, which is the parameter used to adjust the frequency response of the pulse shaping filter ($0 \leq \alpha \leq 1$). The time domain impulse response of the raised cosine filter is given by

$$h(t) = \left\{ \frac{\sin\left(\frac{\pi t}{\tau}\right)}{\frac{\pi t}{\tau}} \right\} \left\{ \frac{\cos\left(\frac{\pi \alpha t}{\tau}\right)}{1 - \left(\frac{2\alpha t}{\tau}\right)^2} \right\} \tag{10.12}$$

with $h(0) = 1$ and $h\left(\pm\frac{\tau}{2\alpha}\right) = \frac{\alpha}{2}\sin\left(\frac{\pi}{2\alpha}\right)$

$$H(\omega) = \begin{cases} \sqrt{\tau} & \tau \leq \omega \leq \frac{\pi(1-\alpha)}{\tau} \\ \sqrt{\tau}\left\{\cos\left[\frac{\tau(\omega - c)}{4\alpha}\right]\right\} & c \leq \omega \leq \frac{\pi(1+\alpha)}{\tau} \\ 0 & \omega > \frac{\pi(1+\alpha)}{\tau} \end{cases} \tag{10.13}$$

The pulse shaping may be implemented as the product of two identical responses, one at the transmitter and the other at the receiver. In such cases, the response becomes a square-root raised cosine response. The square-root raised cosine response is given by Eq. 10.13. And the time domain impulse response of the square root raised cosine filter is given by the following equation:

$$h(t) = \frac{4\alpha}{\pi\sqrt{\tau}} \left\{ \frac{\cos\left(\frac{\pi(1+\alpha)t}{\tau}\right) + \frac{\tau}{4\alpha t}\sin\left(\frac{\pi(1-\alpha)t}{\tau}\right)}{1 - \left(\frac{4\alpha t}{\tau}\right)^2} \right\} \tag{10.14}$$

BPSK uses coherent demodulation. The block diagram of the coherent demodulator, which consists of a frequency recovery, phase recovery, timing recovery, and decision components, is shown in Fig. 10.4.

The commonly used close loop carrier recovery methods are squaring method and Costas loop. In the squaring method, the input signal is squared; the double frequency component is used to control a PLL to produce a signal at the carrier frequency which is used to demodulate the input BPSK signal.

The Costas loop, shown in Fig. 10.5, consists of the Arm Filters, the Loop Filter, Phase Detectors, and a Voltage-Controlled Oscillator (VCO). The Costas loop is able to obtain the phase and frequency information of the modulated carrier and achieve phase tracking, acquisition and synchronization to this extracted carrier while demodulating and extracting the data contained in the received signal. It provides not only suppressed-carrier tracking but also demodulates the received signal. Costas loop can be used for BPSK, QPSK, 8-PSK, and OQPSK.

With the high-performance digital signal processors, the open-loop phase and frequency compensation was used in some system recently.

Symbol Timing Synchronization is another key step for successfully recovering the received information. While there are a number of common approaches, three of

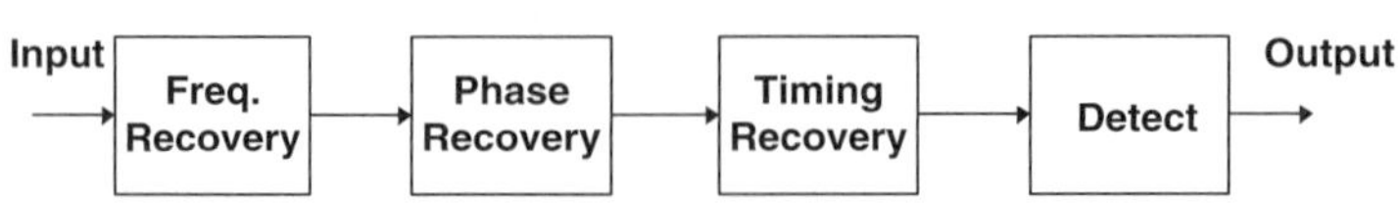

Fig. 10.4 The block diagram of the coherent demodulator

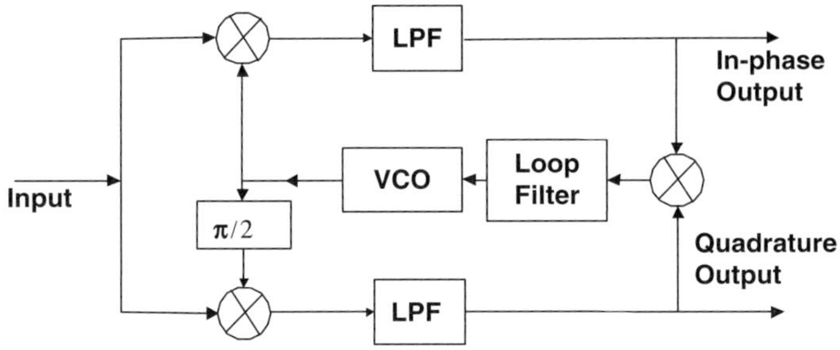

Fig. 10.5 Basic diagram of Costas demodulator

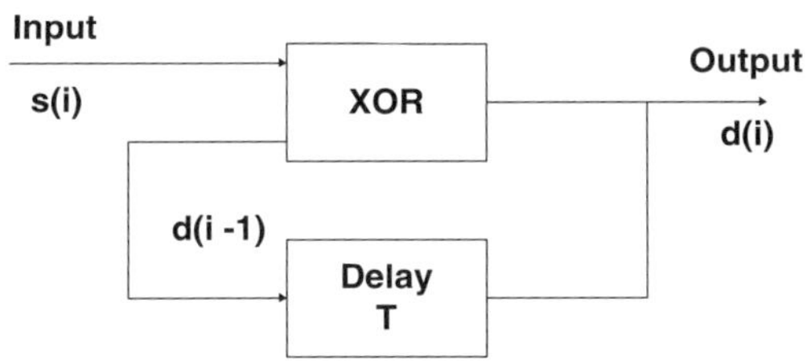

Fig. 10.6 Basic diagram of differential encoder

them are introduced here. The Early–Late Gate, which approximates the Maximum Likelihood detector locating the zero point of the approximate derivative of the eye. The data-transition tracking loop (DTTL), which finds the zero crossing point. The Gardner method is another means of timing recovery and bit estimating method. The result of an integration of the sampled bits becomes the input to an interpolator. The filtered output of this interpolator becomes the bit estimate. The interpolator may be a "Farrow" cubic interpolator.

The carrier and timing recovery can be implemented jointly. The joint estimation may perform remarkably better than individual estimation of the parameters in some case.

Both the squaring and Costas loop methods have the problem of 180° ambiguity. There are two commonly used methods to overcome this problem. One way is adding the known sequence, and the other way is using differential encode method. Demodulation can be classified into techniques that use coherent and noncoherent methods. Coherent demodulators generate a local carrier, which is phase synchronized with the transmitter. The differential detection does not require coherent detection. In DSPK system, the input data stream $s(i)$ is differentially encoded to $d(i)$ prior to modulation. A circuit for this is given in Fig. 10.6.

The DPSK scheme has the advantage of simplifying the receiver complexity. However, in an AWGN channel, they require up to 3 dB more transmit power to achieve the same bit error rate performance. DPSK has a difference with 1–2 dB comparing with PBSK. DQPSK is with around 2–3 dB. For larger constellation the difference is around 3 dB.

10.2.2 QPSK

The simple BPSK can be expanded to an M-ary scheme. We can obtain M-ary PSK by grouping bits together and choosing the phase modulation accordingly. With $M = 4$, the bits can be grouped into pairs, called "dibits" or "symbol," and the resulting signal is known as QPSK. There are two symbols for every sample of QPSK. One symbol mapping of dibits to phase angle is shown in Table 10.1 and the constellation is given by Fig. 10.7. The high-order modulation schemes in DMMB have used the Gray code labeling. Gray code labeling minimizes the effect of noise on the bit error rate by having labels between any two nearest neighbors differ by exactly one bit, which is an optimum manner for assigning bits to constellation points.

QPSK signal can be expressed in the equation

$$S(t) = \sqrt{\frac{2E_s}{T_s}} \cos\left(2\pi f_c t + \frac{\pi}{2}(i-1)\right), \quad 0 \le t \le T_s, \quad i = 1, 2, 3, 4 \qquad (10.15)$$

The QPSK may be viewed as the combination of two orthogonal BPSK signals. The bandwidth of each BPSK signal is same as the symbol rate, which is half the bit rate. As stated before, the bandwidth efficiency of QPSK is twice that of BPSK.

Table 10.1 QPSK mapping

Dibit $A_k B_k$	$\theta(t)$
00	$\pi/2$
01	$-\pi/2$
10	π
11	0

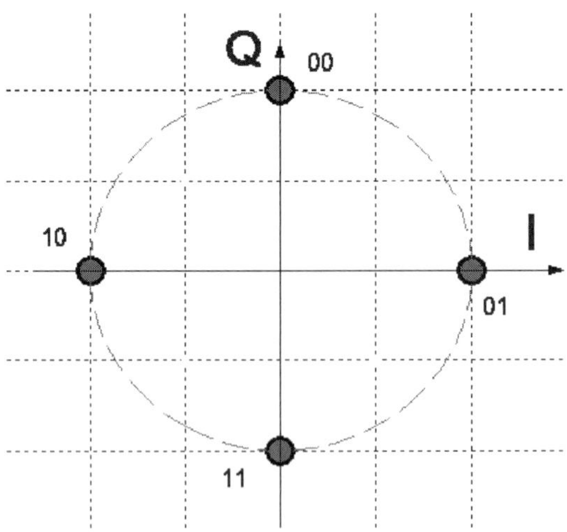

Fig. 10.7 QPSK constellation

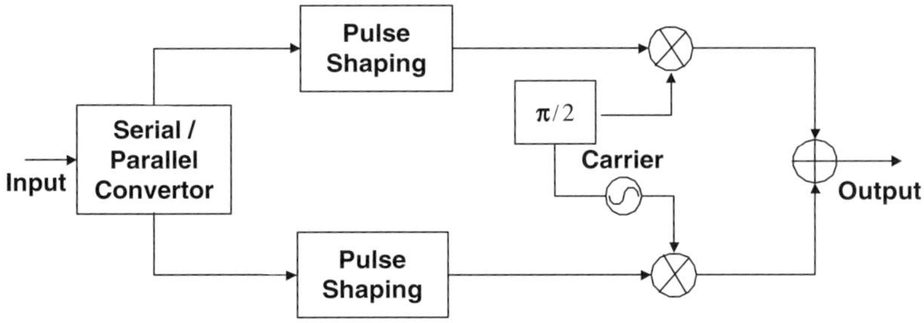

Fig. 10.8 QPSK modulator

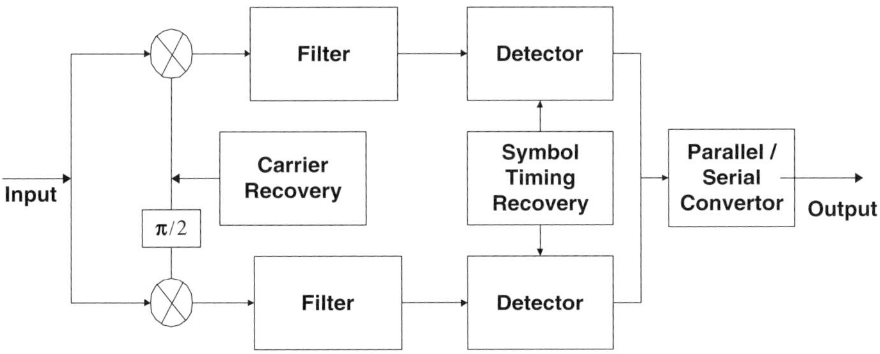

Fig. 10.9 QPSK demodulator

The block diagrams of typical QPSK modulator and demodulator are shown in Figs. 10.8 and 10.9, respectively.

QPSK has constant envelope amplitude and widely separated phase state, in turn, has a high degree of robust against channel noise and interference. It is very mature and well studied, widely used modulation scheme.

10.2.3 π/4 D-QPSK

$\pi/4$ D-QPSK ($\pi/4$ Differential-Quadrature Phase Shift Keying) may be viewed as $\pi/4$ QPSK combined with coherent and differential detection. It is similar to DQPSK with $k\pi/4$ representing the message mapping phases. The "D" in DPSK means the coding is a differential coding. It encodes the difference between the current input with the delayed output. The symbol mapping of $\pi/4$ DQPSK is given in Table 10.2. The recovery of phases can be done without the need knowing the exact phase of the carrier. Therefore, noncoherence detection can be carried out.

The phase increment, $\Delta\theta(t)$, is used to derive the phase angle $\theta(t)$

$$\theta_k(t) = \theta_{k-1}(t) + \Delta\theta_k(t) \tag{10.16}$$

Table 10.2 π/4 DQPSK phase angle mapping

Dibit A_kB_k	$\Delta\theta(t)$
00	$-3\pi/4$
01	$3\pi/4$
10	$-\pi/4$
11	$\pi/4$

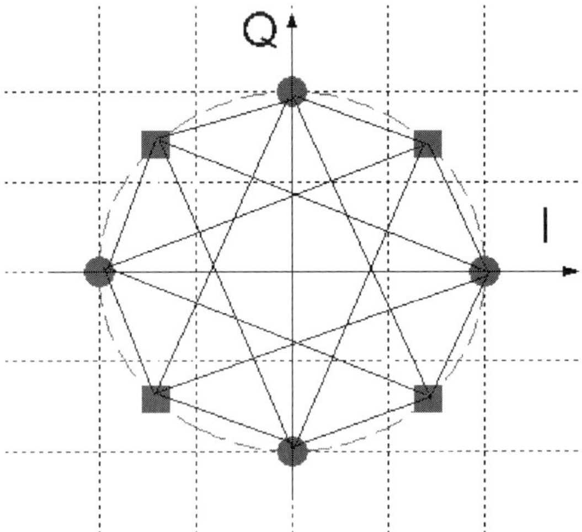

Fig. 10.10 $\pi/4$-DQPSK constellation

$\theta_k(t)$ is used to shift the phase of the carrier according to

$$S(t) = \sqrt{\frac{2E_s}{T_s}}\cos(2\pi f_c t + \theta_k(t)) \tag{10.17}$$

If the initial phase shift at time $t = 0$ is zero, the possible phase angles of $\pi/4$-DQPSK are $\{0, \pi/4, \pi/2, 3\pi/4, \pi, 5\pi/4, 3\pi/2, 7\pi/4\}$. The $\pi/4$-DQPSK has an eight-point constellation. The $\pi/4$-DQPSK constellation consists of two four-point subsets that are offset by $\pi/4$ radians from each other. The $\pi/4$-DQPSK constellation of Fig. 10.10 shows one of these subsets in circle and the other in square. The modulator symbol mapping may be implemented with combining Tables 10.2 and 10.3

The $\pi/4$D-QPSK has an advantage over offset QPSK in which it can be differentially detected. A stream of identical 1's or 0's will always produce a phase change. Among the many different demodulation approaches, the baseband differential receiver is the easiest to implement in a sampled signal environment for $\pi/4$-DQPSK. $\pi/4$-DQPSK performs about 2–3 dB worse than QPSK on an AWGN channel. However, on fading channels, $\pi/4$D-QPSK may perform better than QPSK.

Table 10.3 π/4 DQPSK mapping

$\theta(t)$	Real	Image
0	1	0
π/4	0.707	0.707
π/2	0	1
3π/4	−0.707	0.707
π	−1	0
5π/4	−0.707	−0.707
3π/2	0	−1
7π/4	0.707	−0.707

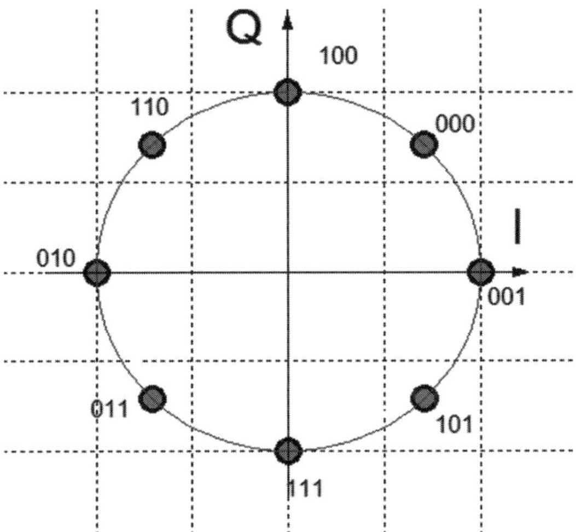

Fig. 10.11 The 8PSK constellation

10.2.4 8PSK

8PSK has virtually constant envelope modulations and can be used in nonlinear amplifier even driven near saturation. 8PSK signal can be expressed in the equation

$$S(t) = \sqrt{\frac{2E_s}{T_s}} \cos\left(2\pi f_c t + \frac{\pi}{2}(i-1)\right), \quad 0 \le t \le T_s,\ i = 1, \ldots, 8 \tag{10.18}$$

The 8PSK constellation is given in Fig. 10.11.

Even if the constellation location is the same as that of 8-PSK, $\pi/4$-DQPSK and 8-PSK are different forms of PSK. The greater the number of data locations in the modulated signal, the smaller the minimum distant. The minimum distant ($d_{\min}$) is the smallest distance between any two pints in the constellation. The smaller the minimum distant ($d_{\min}$), the smaller the allowable error caused by noise and intermodulation before a data location incurs a neighboring location. 8PSK system is less tolerant to such errors than a BPSK and QPSK based system. This effect is compounded when considering even higher modulation such as 16QAM.

10.3 QAM

The QAM is also called as QASK and APSK, which can be viewed as a combined amplitude and phase modulation, or as a two-dimensional modulation.

The general form of QAM signal is given by

$$s(t) = I_i \cos(2\pi f_0 t) - Q_i \sin(2\pi f_0 t) = A_i \cos(2\pi f_0 t + \phi_i) \tag{10.19}$$

where I_i and Q_i are the real and imaginary (in phase and quardrature) modulation components, the pair of (I_i, Q_i) forms the mapping signal point, the constellation. A_i is amplitude, and ϕ_i is the phase. The basic design goal for a QAM is to balance the high minimum distance, d_{min}, between the constellation points and the low average power. There are various QAM constellations. The most commonly used are type I – the inner ring and outer ring have same number of constellation points, type II – having more constellation points on the outer ring than on the inner ring, and type III the square constellation system. The QAM does not have a constant amplitude constellation.

Basically, type II has a 3 dB performance improvement than type I, and type III has a little better performance comparing with type II. If differential encoding is used, the type I and type II constellations may not need carrier recovery. The OFDM signal can never have constant amplitude.

10.3.1 Square QAM

16-QAM is the most commonly used QAM scheme. Its constellation is shown in Fig. 10.12.

If the minimum squared Euclidean distance for this 16 QAM signal constellation is d^2, the average energy is

$$E = \frac{1}{16}\left(4 * \frac{2}{4}d^2 + 8 * \frac{10}{4}d^2 + 4 * \frac{18}{4}d^2\right) = 2.5d^2 \tag{10.20}$$

Since there are 4 bits per signal point, the band efficiency is 4 bps/Hz

The typical QAM modulator block diagrams are shown in Fig. 10.13, and an example of I and Q pulses for 16 QAM is shown in Fig. 10.14.

The typical QAM demodulator block diagrams are shown in Fig. 10.15.

The carrier recovery is not required for noncoherent detection system. The clock recovery is required for all the modulation schemes. The common carrier recovery schemes are time N carrier recovery and decision directed carrier recovery. The block diagram of time N carrier recovery is shown in Fig. 10.16. The Nth power loops is similar to the times 2 or the squared carrier recovery used for binary schemes. The timing recovery and carrier recovery used in QAM schemes are very similar to the timing and carrier recovery scheme for BPSK and QPSK at previous sections.

The Square QAM constellations may be arranged uniformly or nonuniformly. Two examples of 16 QAM nonuniform constellations are given in Fig. 10.17.

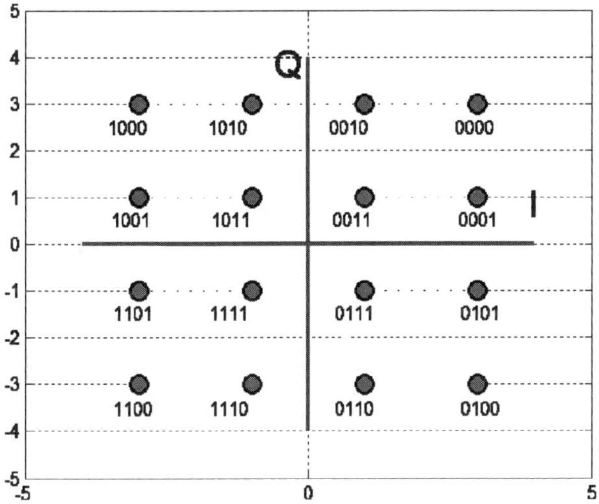

Fig. 10.12 16 QAM constellation

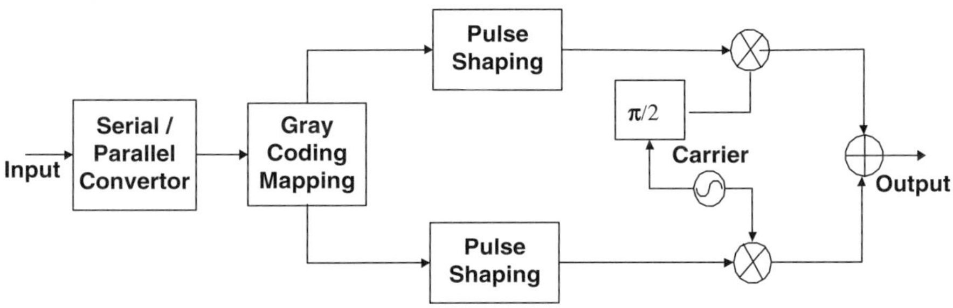

Fig. 10.13 Typical QAM modulator block diagrams

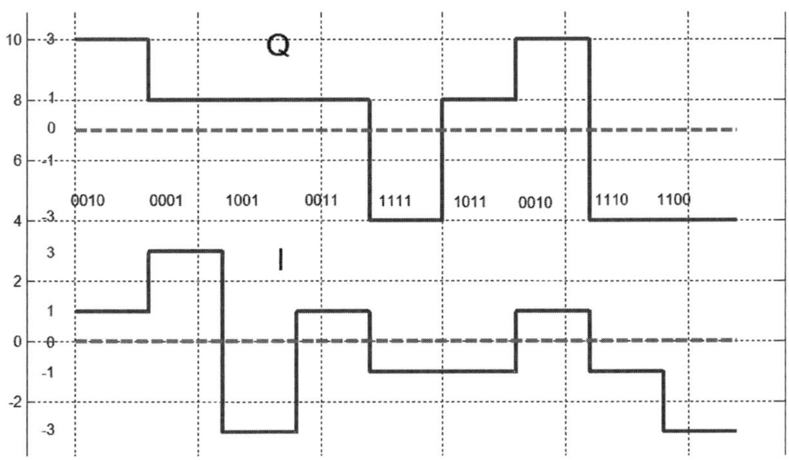

Fig. 10.14 I, Q Pulses waveform of 16QAM

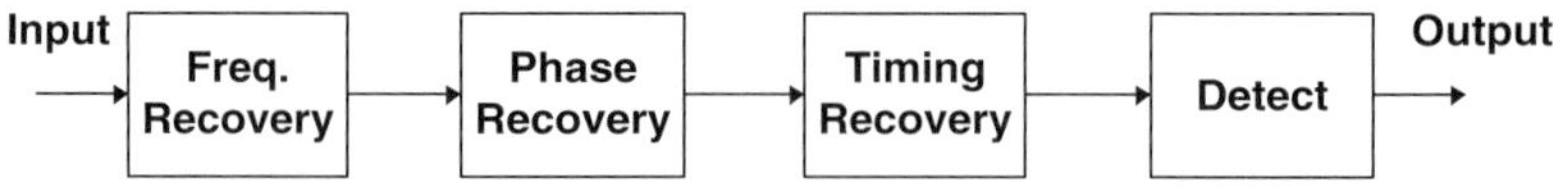

Fig. 10.15 The typical QAM demodulator block diagrams

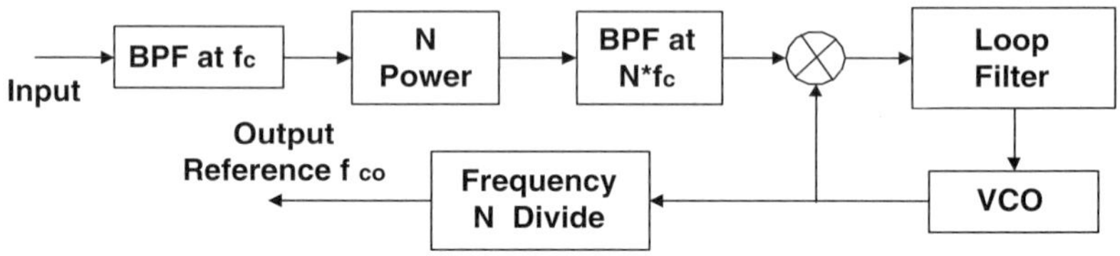

Fig. 10.16 The time *N* carrier recovery block diagram

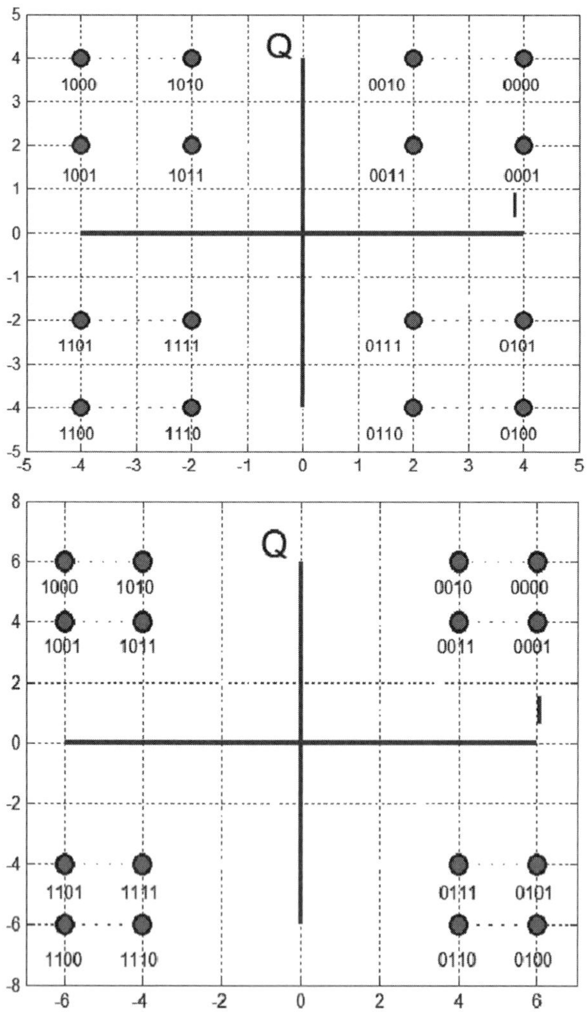

Fig. 10.17 16QAM nonuniform constellations

The above uniform and nonuniform mappings can be extended into higher order constellations such as 32-QAM, 64-QAM, 128-QAM, or 256-QAM. The constellations of the high-order QAM are not given here due to the paper length.

10.3.2 Start QAM

One of the commonly used QAM schemes is type II QAM, which has more constellation points on the outer ring than on the inner ring. Figures 10.19 and 10.18 show the examples of start 16 QAM and 32 QAM, respectively.

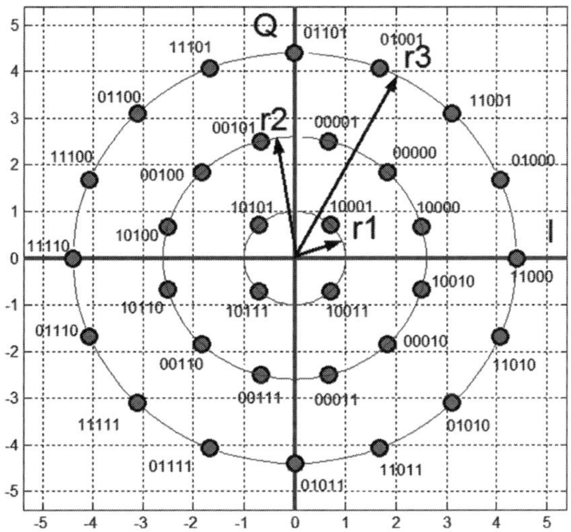

Fig. 10.18 Start 32 QAM constellation

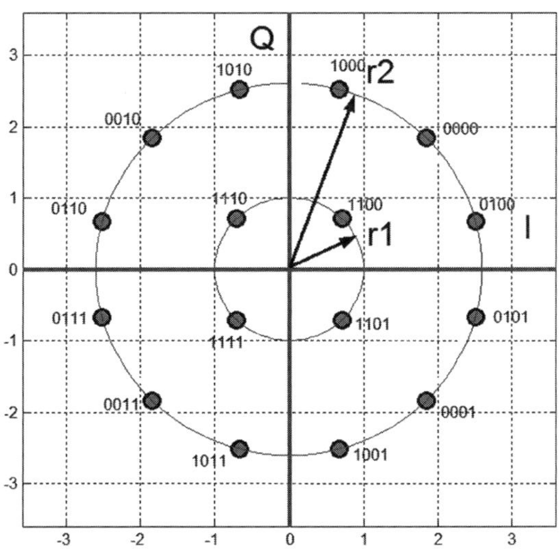

Fig. 10.19 Start 16 QAM constellation

Basically, type II has the similar performance as type III has, and type II constellations may not need carrier recovery. But, they need the adoption of advanced predistortion methods to minimize the effect of power amplifier nonlinearity. The 16APSK and 32APSK constellations have been optimized to operate over a nonlinear power amplifier by placing the points on circles. And, their performances on a linear channel are comparable with those of type III 16 QAM and type III 32 QAM, respectively.

The important parameter of start QAM is the ratio between the outer circle radius and the inner circle radius ($\gamma = r_2/r_1$), which is called constellation radius ratio. The modulation schemes in DMMS use the optimum constellation radius ratio.

10.3.3 Hierarchical Modulation

Hierarchical modulation provides a way for the new system to be backwardly compatible with existing receivers and the ability of making trade-off in the bit rate versus receiving robustness.

In hierarchical modulation, two separate data streams, the "High Priority" (HP) and "Low Priority" (LP) stream, are modulated into a single stream. Receivers with "good" reception conditions or "new" receiver can receive both streams, while those with poorer reception conditions or "old" receiver may only receive the "High Priority" stream. Typically, the LP stream is of higher bit rate, but lower robustness than the HP one.

Using a 16 QAM/QPSK hierarchical system as an example, the constellation has a QPSK stream buried within the 16 QAM stream. The two most significant bits (MSB) would be used as HP stream for the robust service, while the remaining two bits would contain the LP stream. A good quality reception/receiver allows receiver to resolve the entire 16 QAM constellation. In the case the receiver may only be able to resolve the HP portions of the constellation, a QPSK constellation is deduced.

The "hierarchical modulation" can be used with both type II and type III QAM schemes.

10.3.4 VSB

The VSB used in ATSC system is a digital version of conventional analog VSB modulation. The VSB signal's spectrum consists of one sideband and a vestige of the second sideband. The VSB modulation is easy to be understudied in frequency domain with block diagram. Figure 10.20 shows the generation of VSB signal.

If a real-valued baseband message signal $A_{\mathrm{m}}(t)$ is multiplied with a carrier, the resulting signal $S(t)$ is a double sideband (DSB) as given by the following equation:

$$S(t) = A_{\mathrm{m}}(t) * \cos(2\pi f_{\mathrm{c}} t) \tag{10.21}$$

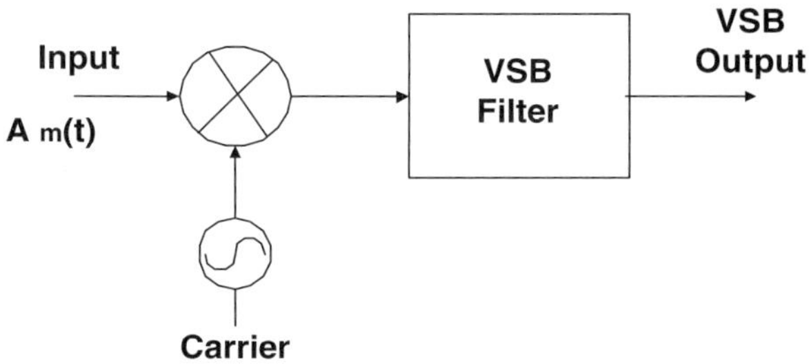

Fig. 10.20 The block diagram of VSB generator (modulator)

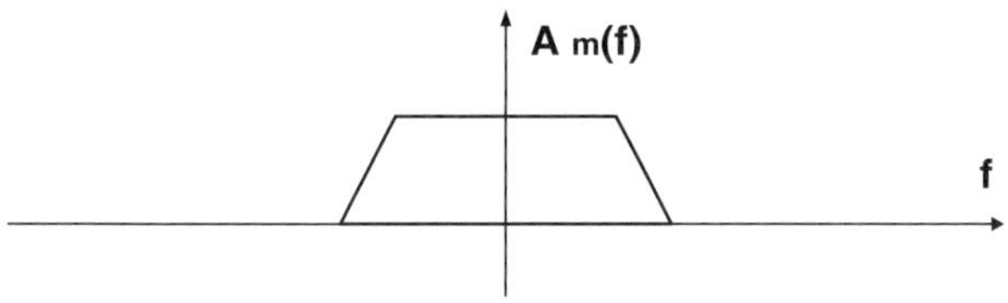

Fig. 10.21 Message spectrum

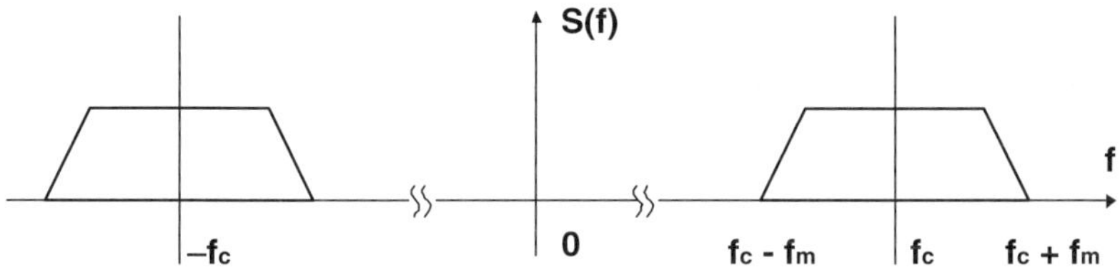

Fig. 10.22 The spectrum of DSB signal

The representations of $A_{\mathrm{m}}(t)$ and $S(t)$ in frequency domain are shown in Figs. 10.21 and 10.22, respectively.

Because $A_{\mathrm{m}}(t)$ is a real-valued signal, $A_{\mathrm{m}}(f)$, the spectrum of $A_m(t)$, has symmetry character with $A_{\mathrm{m}}(f) = A^*_{\mathrm{m}}(f)$. The spectrum of $S(t)$ can be shown to be

$$S(f) = \tfrac{1}{2}[A_{\mathrm{m}}(f+f_{\mathrm{c}}) + A_{\mathrm{m}}(f-f_{\mathrm{c}})] \tag{10.22}$$

The modulated signal can be divided into two parts, above and below the carrier frequency f_{c}. The part between f_{c} and $f_{\mathrm{c}} - f_{\mathrm{m}}$ is called the lower sideband. The part between f_{c} and $f_{\mathrm{c}} + f_{\mathrm{m}}$ is called upper sideband, where f_{m} is bandwidth of the message signal $A_{\mathrm{m}}(t)$. The VSB signal can be generated with the VSB filter, which allows one sideband to be passed completely together with a part of another sideband. The VSB filter transfer function should have odd symmetry about f_c and the relative level of $1/2$ at this frequency.

The ATSC VSB system has two modes. The terrestrial broadcast mode, 8-VSB, transmits data at a rate of 19.4 Mbps in a 6 MHz channel with eight signal levels of $(-7,-5,-3,-1,1,3,5,7)$.The high data rate mode, 16-VSB, provides the data rate of 38.8 Mbps in a 6 MHz channel with 16 signal levels of $(-15, -13, -11, -9, \ldots, 13, 15)$. The VSB system used in ATSC system has almost same performance as the DVB QAM rival.

10.4 Conclusions

This chapter introduced the basic modulation schemes widely used in DMMB standards [1–11]. These schemes are the fundamental building blocks of various DMMBS. These topics have been covered comprehensively by many books [12–14] and various standards from different point of view. The content in this chapter is to provide readers a synopsis of the modulation schemes used in DMMBS that would be a brief introduction and reference for further study on various DMMB standards.

References

1. European Telecommunications Standards Institute (ETSI) TR 101 198: Digital Video Broadcasting (DVB); Implementation of binary phase shift keying (BPSK) modulation in DVB satellite transmission systems
2. European Telecommunications Standards Institute (ETSI) (2004) Digital video broadcasting (DVB); transmission system for handheld terminals (DVB-H). ETSI EN 302 304 V1.4.1
3. European Telecommunications Standards Institute (ETSI) TR 102 376: Digital Video Broadcasting (DVB); User guidelines for the second generation system for Broadcasting, Interactive Services, News Gathering and other broadband satellite applications (DVB-S2)
4. Murali R. Chari, Fuyun Ling, Ashok Mantravadi, Raghuraman Krishnamoorthi, Rajiv Vijayan, G. Kent Walker, and Rob Chandhok, "FLO physical layer: an overview," IEEE Transactions on Broadcasting, Vol. 53, No. 1, 145–160, 2007
5. "FLOTECHNOLOGY OVERVIEW," QUALCOMM White Paper, 2006 [Online]. Available: http://www.qualcomm.com/mediaflo
6. U. Ladebusch and C. A. Liss, "Terrestrial DVB (DVB-T): a broadcast technology for stationary portable and mobile use," Proceedings of the IEEE, Vol. 94, No. 1, 183–193, 2006.
7. European Telecommunications Standards Institute (ETSI) EN 302 307 V1.1.2 (2006–06): Digital Video Broadcasting (DVB); Second generation framing structure, channel coding and modulation systems for Broadcasting, Interactive Services, News Gathering and other broadband satellite applications
8. 2007 world wireless standards, RF design
9. European Telecommunications Standards Institute (ETSI) TS 102 428 V1.1.1 (2005–06): Digital Audio Broadcasting (DAB); DMB video service; User Application Specification
10. European Telecommunications Standards Institute (ETSI) EN 300 401 V1.4.1 (2006–06): Radio Broadcasting Systems; Digital Audio Broadcasting (DAB) to mobile, portable and fixed receivers
11. ATSC Digital Television Standard, ATSC Standard A/53, September 16, 1995
12. John G. Proakis, Digital communications, McGraw Hill Higher Education; 4th edition, 2000
13. Bernard Sklar, Digital communications: fundamentals and applications, Prentice Hall PTR; 2nd edition, 2001
14. William Webb, Lajos Hanzo, Modern quadrature amplitude modulation principles and applications for fixed and wireless communications, London: Pentech; New York: IEEE Press, 1994

Chapter 11
Error Control for Broadcasting and Multicasting: An Overview

Ivan V. Bajić

11.1 Introduction

The term *error control* refers to the set of actions that may be taken by the transmitter, receiver, an intermediate node, or any combination of these devices, to reduce the effect of errors introduced by the communication channel. In this chapter we will survey some of the common error control schemes for video multicast, where a video clip is being transmitted to a group of users through either a wireless or a wired packet-based network. Multicast presents special challenges for error control; different users generally see different channels and experience different errors. The goal of error control in this case is to minimize video quality degradation due to errors for all users.

A simple illustration of video multicast is shown in Fig. 11.1: a video server sends a video stream to multiple users. Each arrow in the figure indicates a communication link, or a group of links; a sequence of links creates the communication channel between the server and the corresponding user. Different links can employ different communication infrastructure. Hence, when considering the channel between the server and a particular user, a part of that channel may be an Ethernet link, another part may be an optical link in a Wide Area Network, yet another part may be a wireless link in a cellular network. Physical-layer issues are different for each of these links, and physical-layer error control is tailored to each link. In most cases, a block of data (called a *frame*, or a *packet*, as we will call it here to avoid confusion with video frames) that is received and processed at the physical layer will be verified at the data link layer through a Cyclic Redundancy Check (CRC) [29]. The packet will only be fed to the upper layers if passes the CRC test. Hence, regardless of the error processes at the physical layer, higher layers usually see only packet losses. In addition to the packet losses caused by CRC failures, in packet-switched networks, loss may occur due to other reasons, such as congestion. For real-time data such

I.V. Bajić
School of Engineering Science, Simon Fraser University, Burnaby, BC V5A 1S6, Canada
e-mail: ibajic@sfu.ca

F.-L. Luo (ed.), *Mobile Multimedia Broadcasting Standards: Technology and Practice*,
DOI: 10.1007/978-0-387-78263-8_11,

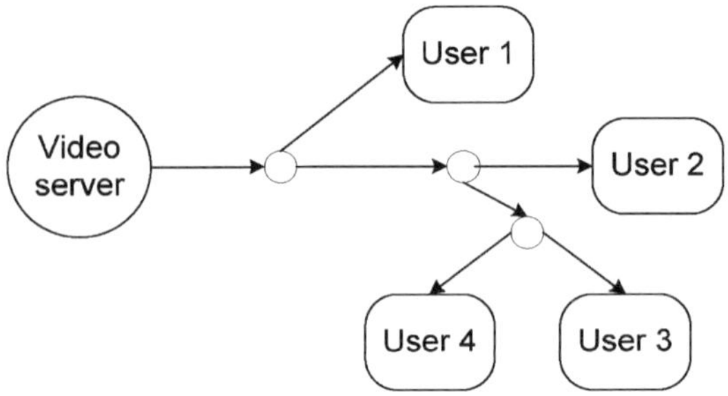

Fig. 11.1 Illustration of video multicast

as video, long delays may also cause effective packet loss, since in most cases late packets are no better than the lost ones. Our focus in this chapter will be on error control techniques that apply to communication layers above the data link layer, where packet loss is the name of the game.

Traditional error control schemes for video transmission over packet-lossy channels include packet retransmission, Forward Error Correction (FEC) and error concealment. These methods can also be combined to cover a wider range of communication scenarios and to improve system performance. In general, application requirements and infrastructure dictate the type of error control scheme to be used. For example, interactive applications such as videoconferencing require low end-to-end delay and may therefore favor FEC and error concealment over retransmission, although several packet retransmissions may be possible within the typical conversational delay allowance of 150–400 ms [36]. On the other hand, streaming applications have a much higher delay allowance of several seconds or even tens of seconds, which gives the system designer greater flexibility in choosing the error control scheme. In the case of streaming, retransmission-based strategies perform very well. In the next few sections we will review several traditional error control schemes and describe the benefits and challenges that arise in their application to video multicast.

11.2 Forward Error Correction

FEC was the earliest error control technique used in digital communications. Since Hamming's seminal paper [7] which described the design of Hamming codes, many other codes were developed. Of particular importance to the problem of packet loss are the codes that handle erasures, the so-called *erasure-correcting codes*. Among these, Reed-Solomon (RS) codes have been the most popular, owing to their good performance with short codeword lengths.

For the purposes of this section, a *symbol* is a block of q bits representing a basic unit of error control. An (n, k) RS code takes k data symbols and produces n code symbols, where $n = 2^q - 1 \geq k$. In a *systematic* RS code, the first k symbols of an

n-symbol codeword are the same as the original k data symbols. Systematic codes are preferred in many applications since the data symbols appear directly as a part of the codeword, which can greatly simplify subsequent processing. One may think of a systematic encoder as adding n–k *parity* symbols to a group of k data symbols to produce the n-symbol codeword, as shown in Fig. 11.2.

11.2.1 Equal Loss Protection

An (n, k) RS code can correct up to $n-k$ erasures; in other words, the RS decoder will be able to fill in the missing symbols as long as it knows the locations in the codeword from which they have been erased, and there are no more than $n-k$ missing symbols in a codeword. To use RS coding for packet loss recovery, one can employ a scheme illustrated in Fig. 11.3 [23], sometimes called *interlaced* RS coding. In Fig. 11.3, each packet is split into a sequence of q-bit symbols. RS encoding is then applied column-wise across the symbols from k different packets to create $n-k$ parity symbols, which are then combined to create parity packets. If no more than $n-k$ packets out of this group of n packets get lost, RS decoder at the receiver will be able to fill in the missing symbols from each RS codeword, thereby recre-

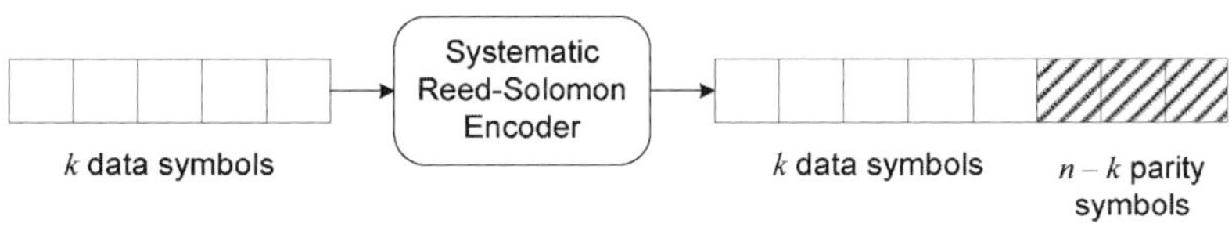

Fig. 11.2 Illustration of systematic RS encoding

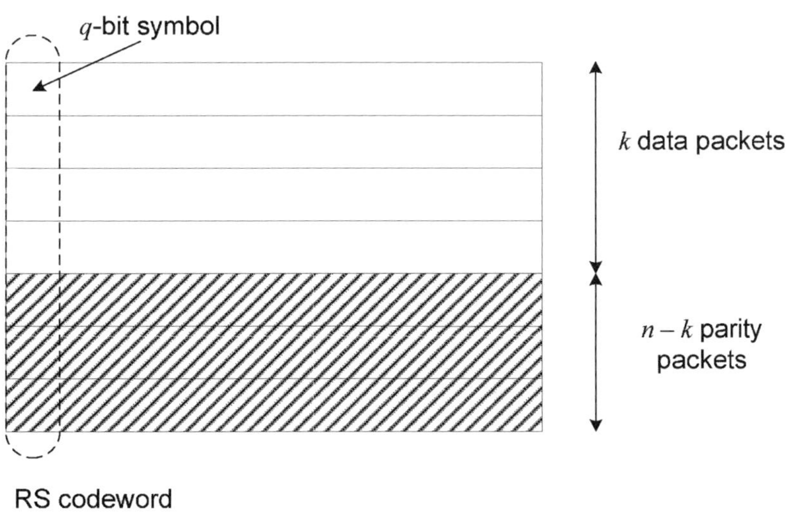

Fig. 11.3 RS coding for packet loss protection

ating the missing packets. Note that in the figure, the length of each data packet is the same. In practice, if data packets have different lengths, shorter packets can be padded by zeros up to the length of the longest data packet in the group. This may lead to substantial overhead if the lengths of data packets vary significantly.

If the packet loss probability is p, and if the losses are independent, the residual loss probability P_L of data packets after RS decoding is given by

$$P_{\mathrm{L}} = \sum_{l=n-k+1}^{n} \frac{l}{n} \binom{n}{l} p^{l}(1-p)^{n-l}. \tag{11.1}$$

For example, if $p = 0.1$ and we use the $(7,4)$ RS code, then $P_{\mathrm{L}} \approx 0.0016$. In this case, only four out of every seven packets carry the data in this scheme, while the other three carry redundant parity symbols. So FEC can significantly reduce the residual packet loss probability at the expense of reducing the effective rate of information transfer. Note that in Fig. 11.3, all information in the data packets is protected equally; each symbol from each of the data packets can be recovered if no more than $n-k$ packets are lost. Such a scheme is sometimes called *equal loss protection* (ELP). However, not all video data is equally important for the reconstructed video quality. For example, the loss of a piece of data from an I-frame in a MPEG-4 stream will affect many subsequent frames, whereas the loss of a piece of data from a B-frame will only affect that same frame. Later in this section we will discuss some error control schemes that take this unequal importance of data into account to provide *unequal loss protection* (ULP).

If one knows the packet loss probability p and the desired residual packet loss probability P_{L}, then n and k can be chosen to satisfy Eq. 11.1. Note that for proper FEC design it is important to have reasonably accurate *channel state information* (CSI), in this case packet loss probability p. In reality, channels are time-varying, so CSI needs to be updated on the fly. If there is a mismatch between the perceived CSI at the channel encoder and the actual channel realization, the encoder may end up generating too many or too few parity packets.

Another important aspect of using FEC is the delay introduced in the system. If a data packet is lost, the decoder usually needs to wait for the last packet in the block of n packets before recovering the missing data packet. In addition to this buffering delay, there is also a processing delay associated with computing the missing symbols from the available ones. Nonetheless, in terms of delay, FEC compares favorably to other error control schemes described later in this chapter, which makes it attractive for conversational services such as videoconferencing. Efficient implementations of FEC can be found in [15, 27].

11.2.1.1 Extensions to Multicast

A simple extension of FEC to video multicast can be done in the following way. The server estimates the loss rate p_i for each user $i = 1, 2, \ldots, N$, then takes $p = \max\{p_i\}$ and chooses n and k for the desired maximum residual loss rate P_{L} based on Eq. 11.1.

If the estimates of individual users' loss rates are accurate (and if the channel behaves as predicted), the user with the maximum loss rate p will end up having the residual loss rate P_L, while other users will have lower residual loss rates. Note that each user receives the same set of parity packets, but it uses them to correct its own losses, which may vary from user to user.

There are at least two problems that the above video multicast scheme would face in practice. First, in a large enough group of users, it is likely that users' access bandwidths will differ and/or that their viewing preference in terms of video resolution, frame rate, or quality will differ. In such a scenario, sending the same data and parity packets to each user would seem inappropriate. We will revisit this issue in the next section. Second, in order to estimate the loss rate p_i of each user, the server needs to receive feedback from each user. As the number of users increases, this may lead to the so-called *feedback implosion* problem.

11.2.1.2 Feedback Implosion

A number of solutions have been proposed to deal with feedback implosion in multicast. We will mention two representative ones. In Reliable Multicast Transport Protocol (RMTP) [24], users are organized into *local regions*, with each region having one *designated receiver* (DR). Users do not send feedback to the server, but to the DR in their region. The DR then sends feedback to the server. This approach effectively aggregates the feedback in each region and reduces the number of feedback messages the server has to deal with. Figure 11.4 depicts this approach in a video multicast.

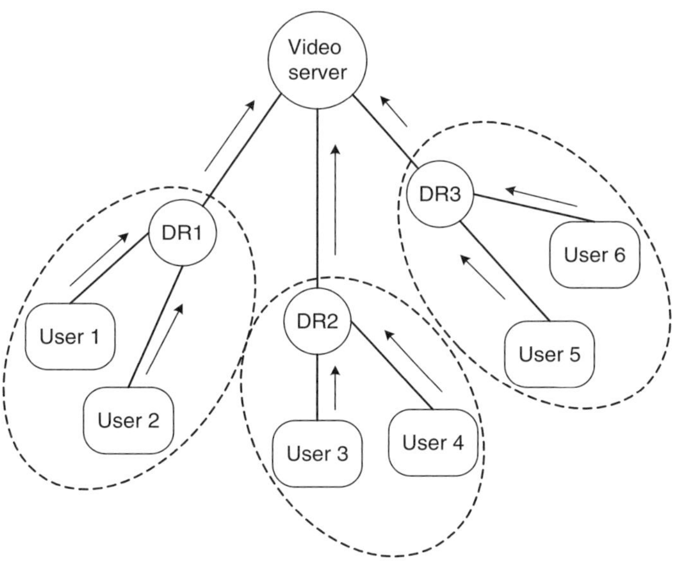

Fig. 11.4 Illustration of RMTP-like feedback aggregation

The DRs collect statistics about the users in their region, and inform the server about the highest loss rate in their region. From this information the server can determine the highest loss rate p among all users, which can be used in Eq. 11.1 to determine n and k. The drawback of this method is the need for DRs. If some users act as DRs, they suffer increased workload compared to other users in their region, without necessarily obtaining any benefit for this extra effort. If DRs are provided as dedicated devices independent of the users, then there is an additional infrastructure cost associated with this type of multicast. In either case, defining local regions and DRs requires some careful planning in order to be effective.

Another method to deal with feedback implosion problem is scalable feedback control [3]. Here, the server does not receive feedback from all users, but probes a randomly chosen group of users to try to infer channel conditions along the multicast tree. Probing is done in epochs. In each epoch, the server and all users generate a random 16-bit key. The server then multicasts a polling packet with its 16-bit key and a number specifying how many bits in the key are significant. If a user's key matches the server's key in the significant bits, that user will send feedback to the server. If the server does not receive any response within a specified timeout period, meaning that no user's key matched its own, it will reduce the number of significant bits in the key and resend the polling packet. Note that the probability that the keys randomly generated by the server and the users match depends on the number of users. Since the number of significant bits in the key will decrease in subsequent epochs if there is no response, the probability of a match will increase. In fact, it is shown in [3] that the relationship between the number of users N and the expected epoch $E[First]$ in which the first response arrives is

$$N \approx \exp\left(16.25 - E[First]/1.4\right). \qquad (11.2)$$

Hence, the server can get an estimate of the number of users in the multicast tree although it does not receive feedback from all of them. Depending on the type of information included in the feedback, the server can also obtain an estimate of the maximum probability of error, longest Round Trip Time (RTT), etc.

11.2.2 Unequal Loss Protection

As mentioned previously, not all video data is equally important in terms of the effect of its loss on the decoded video quality. For example, standard video coders such as MPEG-2/4 or H.261/3/4 code different frames in different ways. I-frames are coded as still images (intra-mode), P-frames are coded via prediction from previous I- or P-frames, while B-frames are coded via prediction from previous and future I- or P-frames. The loss of a piece of data from an I-frame will affect all future P- and B-frames until the next I-frame. On the other hand, the loss of a part of a B-frame will only affect that same frame. This suggests that I-frame data should be protected more than B-frame data. This idea has been developed in a scheme called priority encoding transmission (PET) [1]. PET relies on a user-defined "priority function" to specify the priorities of different pieces of data, and to choose protection for each piece of data in accordance with its priority. When applied to MPEG video, PET protects I-frames more than P-frames, which are in turn protected more than B-frames.

Instead of requiring a user to provide the "priority function," one can use the increase in video distortion caused by the loss of a piece of video data to determine how that piece of data should be protected. The most popular video distortion measure is the mean squared error (MSE), computed for each $M \times N$ frame as the average squared difference between its original version f_o and decoded version f_d,

$$\mathrm{MSE} = \frac{1}{MN} \sum_{x=1}^{M} \sum_{y=1}^{N} (f_o(x,y) - f_d(x,y))^2. \tag{11.3}$$

A related measure is the Peak Signal-to-Noise Ratio (PSNR) computed in decibels (dB) as

$$\mathrm{PSNR} = 10 \log_{10} \frac{255^2}{\mathrm{MSE}}. \tag{11.4}$$

Both measures can be averaged over a group of frames.

The influence of the loss of a piece of video data on MSE or PSNR depends on whether other pieces of data have been lost as well. For example, the loss of a piece of data from a P-frame may affect MSE differently depending on whether the previous I- or P-frame has been affected by loss. This suggests that the measure of importance of a piece of data needs to be determined by averaging over many loss realizations, which may be computationally intensive. Several methods have been proposed for accomplishing this task, including Recursive Optimal per-Pixel Estimate of decoder distortion (ROPE) [37] and Multi-Decoder Distortion Estimation (MDDE) [30].

A very flexible framework for video transmission can be created by applying ULP to layered or scalable video bit streams, where some parts of the bit stream provide refinement to video resolution, frame rate, or quality, obtained from other parts of the bit stream. Scalable video bit streams can provide a variety of resolutions, frame rates, and qualities to satisfy a diverse set of users, which makes them ideal for video multicast. Scalable video coding has been popular in the research community for some time now [8, 16, 33]. In the context of video coding standards, a simple form of frame-rate scalability existed in MPEG-1 [10] through B-frames which could be removed from the bit stream without affecting decoding of other parts of the stream. All three forms of scalability were incorporated into MPEG-2 [11], although high-performance quality scalability was only introduced in the form of Fine-Granularity Scalability (FGS) [34] as an extension of MPEG-4 [12]. At the time of the writing of this chapter, a full-featured high-performance scalable video coding standard [9] is being finalized as an extension to H.264/AVC [14].

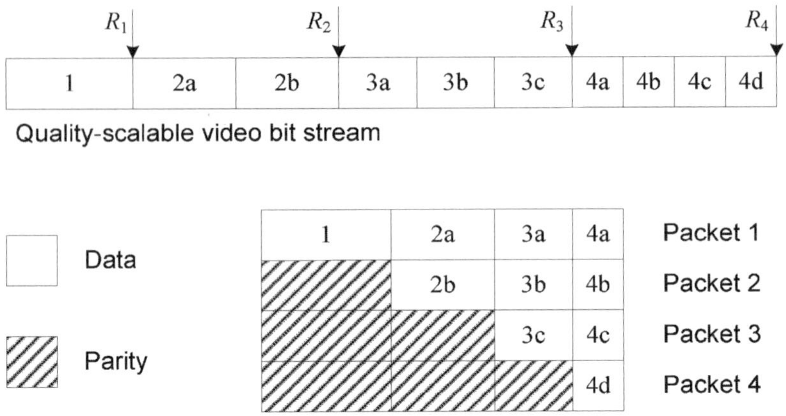

Fig. 11.5 Illustration of MD-FEC

As a representative of the methods for applying ULP to scalable video bit streams we mention multiple description coding through FEC (MD-FEC) [26]. The scheme is illustrated in Fig. 11.5 for a quality-scalable video bit stream. At the top, the scalable bit stream corresponding to one Group of Pictures (GOP) is split into four sections, with section k being further split into k subsections. Then the data from section k is protected by an (n, k) RS code. Note that $n = 4$ in Fig. 11.5. The $(4, k)$ RS code can be obtained from the $(7, k)$ RS code by removing three parity symbols, a process called *puncturing*. Data and parity symbols are then combined into four packets as shown at the bottom of the figure. The scheme has the following property: if k out of n packets are received, then the first k sections can be reconstructed by RS decoding, so the video decoder will be able to decode at least R_k bits.

An important issue with the scheme like this is how to split the bit stream into sections. In other words, for the case of N packets (i.e., N sections), how to choose $R_1, R_2, \ldots, R_N$. The approach taken in [26], as well as most other works on this topic, is to choose the R's in Fig. 11.5 that give the minimum expected distortion at the receiver under some channel model. Let $\mathbf{R} = (R_1, R_2, \ldots, R_N)$ be the vector of R's, and let D be the distortion at the decoder (for example, MSE in Eq. 11.4). Distortion depends on $\mathbf{R}$ as well as the packet loss realization. If the decoder decodes k sections when k packets are received, the expected distortion is

$$E[D(\mathbf{R})] = \sum_{k=0}^{N} P(n,k) D(R_k), \tag{11.5}$$

where $P(n,k)$ is the probability that exactly k out of n packets are received under some channel model, and $R_0 = 0$. The goal is to find $\mathbf{R}$ that minimizes (11.5) under certain constraints, for example, keeping the total bit rate below a given value.

11.2.2.1 Extensions to Multicast

Extending ULP to video multicast faces some of the same challenges described previously in the section on ELP. Since different users see different channels, it is important for the server to be aware of the channel conditions of different users. Methods for dealing with feedback implosion need to be considered in this context too. In addition, the flexibility provided by a combination of scalable video and ULP gives rise to numerous possibilities for designing an error control scheme.

One class of methods for video multicast with ULP keeps the server in charge of error control. In [20], MD-FEC is extended to the case of multiple users. The extension is not straightforward; however, since different users observe different values of $P(n,k)$ in Eq. 11.5, they will experience different expected distortions. It might not be possible to minimize expected distortions for all users simultaneously, so some kind of aggregate measure of distortion for all users is needed. Designing for the worst-case scenario is one option, but the approach taken in [20] is based on the *maximal regret* criterion. In this context, the optimal $\mathbf{R}$ is the one that minimizes the maximal difference that any user would observe between the expected distortion given by the chosen $\mathbf{R}$ and the expected distortion that user would see in a unicast scenario. It is also possible to choose other criteria for optimizing ULP for multiple users, for example, optimizing for the average user (which involves averaging expected distortion over all users) or sequential optimization starting with low-bandwidth users [4].

The second class of proposed methods places a large part of the responsibility for error control on the users themselves. In receiver-driven layered multicast (RLM) [21], the server sends different layers of scalable video to different multicast groups. Users join and leave groups depending on the channel conditions they see at a particular time. The more layers a user receives, the higher the quality/resolution/frame rate of the video it can decode. If the channel condition of a particular user worsens, that user can leave some of the groups it subscribes to. In turn, this will cause the IP routers upstream to stop forwarding the corresponding video layers to that user, which will hopefully clear the congestion and improve channel conditions. Server load in this scheme is relatively low, since there is no feedback sent to the server. However, the overhead traffic associated with users joining and leaving groups may well outweigh the savings made by eliminating feedback to the server. This is because join/leave traffic is generated under all channel conditions (join when the channel is good, leave when it is poor), while feedback traffic with negative acknowledgements (to be discussed in more detail in the next section) is generated only when the channel conditions are poor.

The RLM idea was further developed in [32] by layering the parity packets that protect different layers of scalable video. The server sends different layers of video and parity packets to different multicast groups. It is up to the users to subscribe to the appropriate amount of parity for their channel condition. In this scheme, parity packets are not sent in the same time interval as the video data they protect, which reduces the chance that the parity and the data it protects are both lost. Hence, time-diversity effect is achieved, but at the expense of some additional delay.

11.3 Retransmission-Based Error Control

Retransmission-based error control in the form of Automatic Repeat reQuest (ARQ) is probably the most widely used error control mechanism in the Internet, since it is an integral part of the Transport Control Protocol (TCP). In ARQ, when the receiver becomes aware that a packet has been lost, it requests the retransmission of that packet from the server. This can be done by either sending a negative acknowledgement (NACK) message for the lost packet back to the server, or by not sending a positive acknowledgement (ACK) message for the lost packet.

ARQ can be much more efficient than FEC [28], because it avoids explicit channel modeling and prediction about how the channel is going to behave in the future. Instead, the channel is simply observed. Error control action takes place only after the loss actually happens. But this efficient reactive strategy of ARQ comes at a price of additional delay. When the receiver becomes aware of the loss, some time is needed for its feedback message to reach the server, and additional time is needed for the retransmitted packet to reach the receiver. In total, one RTT will elapse before the loss is rectified. Additional time will be needed if the retransmitted packet is lost as well. This additional delay may limit the use of retransmission-based error control for some video applications. Each piece of video data has its playout deadline. It makes no sense to resend a lost piece of data if it cannot be received in time to be decoded and displayed. However, in noninteractive applications like video streaming, buffering time of several seconds at the receiver is common. During this time period several retransmissions can easily take place, which makes retransmissions a viable error control strategy for video streaming. In fact, retransmissions form an integral part of error control employed in industry standards for video streaming, such as Windows Media and RealVideo.

11.3.1 Retransmissions for Unicast

Retransmission-based error control in the form of ARQ was originally developed for unicast data transmission. As a result, not all ARQ schemes are suitable for video transmission. The three basic types of ARQ are the *stop-and-wait ARQ*, the *go-back-N ARQ*, and the *selective-repeat ARQ* [17].

Stop-and-wait ARQ is the simplest of the three ARQ schemes. The server sends a packet and waits for the response from the receiver. If the response is positive (ACK), the server sends the next packet. If the response is negative (NACK), the previous packet is retransmitted. The timing diagram of this scheme is shown in Fig. 11.6 for the case when packet 2 is lost. Retransmitted packet on the sender side is indicated in bold type. Observe that the time between two successive transmissions (which is one RTT) is idle, while the server waits for feedback from the receiver. This significantly reduces the throughput efficiency of stop-and-wait ARQ [17] and severely limits its applicability to video streaming.

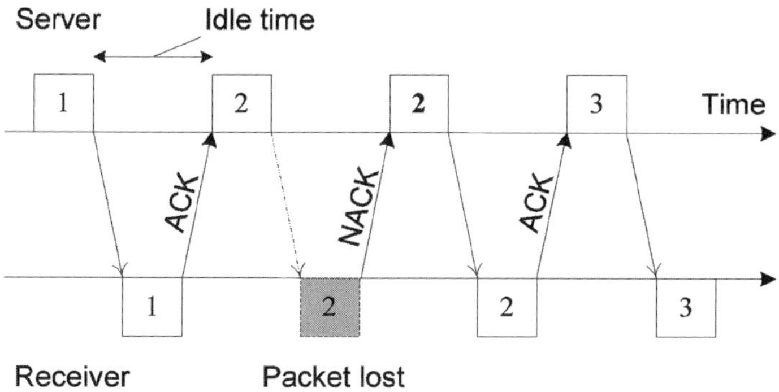

Fig. 11.6 Stop-and-wait ARQ

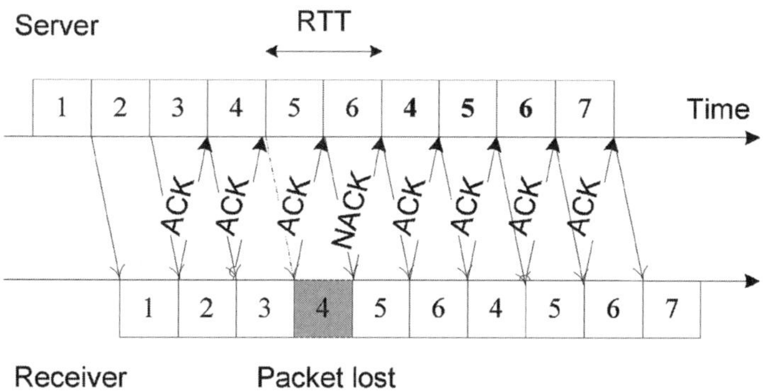

Fig. 11.7 Go-back-*N* ARQ for $N = 3$

Go-back-*N* ARQ eliminates idle time by allowing the server to send the data continuously, without waiting for feedback from the receiver. It takes one RTT for feedback to arrive. During this time, the server has transmitted N–1 packets. If the feedback message is an ACK, the server will simply send the next packet in the sequence. However, if the feedback message is a NACK, the server will "rewind" to the packet that was lost, and start resending all packets in sequence from that packet onwards. The timing diagram of go-back-*N* ARQ for $N = 3$ is shown in Fig. 11.7. Note that the loss of packet 4 has triggered retransmission of packets 4, 5, and 6. Go-back-*N* ARQ has higher throughput efficiency than stop-and-wait ARQ, but it is still inefficient – even packets that were not lost are retransmitted. This inefficiency is corrected in selective-repeat ARQ, described next.

Selective-repeat ARQ operates exactly as we would intuitively want a retransmission scheme to operate. The server continuously sends packets to the receiver. If a NACK is received for a particular packet, only that packet is retransmitted, and transmission of other packets resumes. The timing diagram is shown in Fig. 11.8. This is the most efficient of the three ARQ schemes, but the price for its efficiency is implementation complexity. The server needs to keep the last *N* packets in a buffer

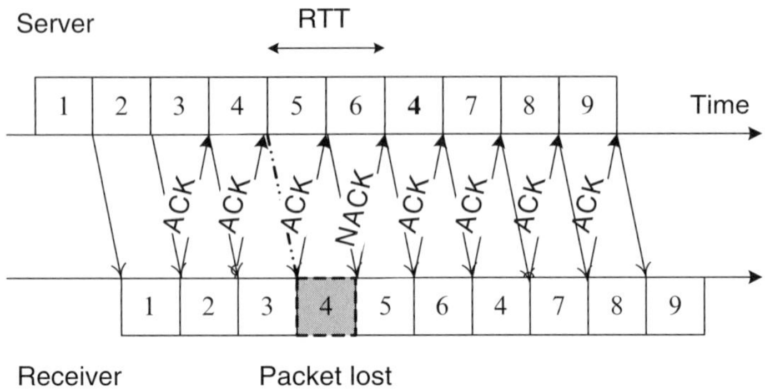

Fig. 11.8 Selective-repeat ARQ

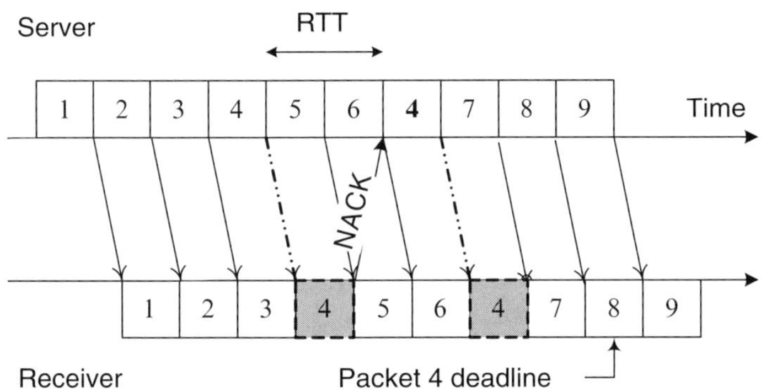

Fig. 11.9 Delay constrained retransmission with ACK suppression

(as in go-back-N ARQ), where N depends on the RTT. But in selective-repeat ARQ, the receiver may also need to keep an N-packet buffer and manage packet reordering if the data in the packets needs to be processed sequentially. Despite this added complexity, selective-repeat ARQ seems the best match to video streaming of all three ARQ schemes.

As mentioned previously, in real-time video streaming, a video packet is useful to the receiver if it is received in time to be decoded and displayed. There is no point in requesting a retransmission of a packet that would not be received on time. *Delay-constrained retransmission* is a version of selective-repeat ARQ, where the receiver sends a NACK only when the retransmitted version of the lost packet has a chance of being received on time, based on its RTT estimate. In addition, ACK messages are typically suppressed in delay-constrained retransmission to minimize the amount of feedback traffic. A timing diagram for delay-constrained retransmission is shown in Fig. 11.9. In this example, packet 4 is lost on the first transmission. The receiver estimates that the retransmitted copy of packet 4 will be received before the deadline, so it sends a NACK back to the server. However, the retransmitted copy gets lost as well. This time the receiver estimates that the retransmitted copy will not make it on time, so it does not send a NACK. In [19], such a scheme was used in real Internet video streaming experiments, with receiver buffers of 2.7 s. The authors report that 94% of recoverable packets (those packets whose loss was discovered at least one RTT before the deadline, so that their retransmission could be requested) were recovered by this simple retransmission scheme, which illustrates the effectiveness of delay-constrained retransmission for video streaming.

11.3.2 Retransmissions for Multicast

In a multicast scenario, different receivers may lose different packets. As the number of receivers grows, it becomes increasingly more probable that *each* packet is lost by *at least one* receiver. For example, if there are N receivers, and if each loses packets independently with probability p, then the probability P that a packet is lost by at least one receiver is

$$P = 1 - (1 - p)^N. \tag{11.6}$$

This packet loss probability is shown against the number of users in Fig. 11.10 for $p = 0.01, 0.02, 0.05$, and 0.10. As shown in the figure, this probability may become high even for a relatively small number of receivers if they lose packets independently. Therefore, multicast retransmission schemes must be carefully engineered. Feedback implosion could be a problem in the case of multicast retransmission, and previously described methods for handling feedback implosion can be used here as well.

In RMTP [24], receivers are organized into regions as shown in Fig. 11.4, each region having its DR. If a packet is lost between the server and a DR, the DR requests retransmission from the server. But if the packet is lost in the local region,

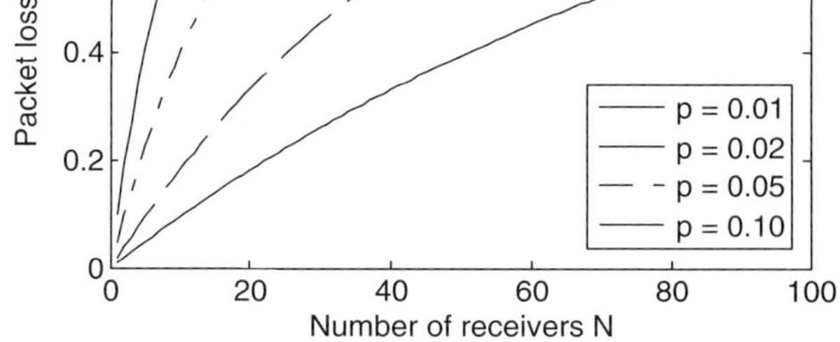

Fig. 11.10 Probability that a packet is lost by at least one of *N* receivers

the receiver requests retransmission from the DR. This kind of "local recovery" significantly reduces the effective packet loss probability, since both the number of DRs and the number of receivers in any given region are much smaller than the total number of receivers. Additional benefits include reduced RTT for retransmission and localization of feedback traffic.

In [25], a comprehensive study of retransmission-based error control for real-time multicast was performed. The authors consider a number of receivers with independent losses but different loss probabilities. They study several delay-constrained error control schemes based on retransmissions:

- Immediate Unicast Retransmission (IUR), where the server retransmits the lost packet only to the receiver who requested it immediately upon receiving a NACK;
- Immediate Multicast Retransmission (IMR), where the server multicasts to all users the packet requested by any user immediately upon receiving a NACK;
- Wait-for-All Multicast (WAM), where the server waits for all receivers (based on its estimate of largest RTT) to possibly send a NACK, and then multicasts the packet if at least one NACK is received.

These three schemes were compared in terms of four parameters: residual loss probability, the average time needed to deliver a packet correctly, the average number of retransmissions, and the average cost of retransmitting a packet (which is higher for multicast than for unicast because of packet duplication across different branches of the multicast tree). Although the simulations performed in [25] involved a relatively small number of users (up to 10), the results were very interesting and perhaps somewhat surprising. First, the choice of the error control scheme has most effect on receivers that are close to the server and have poor channel conditions, whereas those that are further away or have good channel conditions are less affected by it. Second, when the number of receivers is small, it makes little difference whether retransmitted packets are unicast or multicast. As the number of receivers increases, multicasting retransmitted packets leads to lower residual loss probability and reduces the time needed for packet delivery, but increases the cost to the network. Finally, under very general conditions, the best of the three schemes in terms of the four parameters mentioned above is IMR. These results indicate that retransmissions can be a viable option for a simple and effective error control scheme for video multicast.

11.4 Combining FEC and Retransmissions

FEC and retransmissions can be combined into a powerful error control framework known as *hybrid ARQ/FEC*. These schemes tend to have higher reliability than FEC alone, and higher throughput than ARQ alone [17]. Any of the three ARQ schemes described previously can be combined with FEC. There are two basic types of hybrid FEC/ARQ schemes: type-I and type-II.

11.4.1 Type-I Hybrid ARQ/FEC

In a type-I hybrid ARQ/FEC, the FEC subscheme works within the ARQ scheme. The goal of FEC here is to correct frequent error patterns [17], and let ARQ handle the infrequent ones. In the context of packet loss, if the loss probability is not very high, "frequent loss pattern" simply means a relatively small number of lost packets within a larger block of packets. A type-I scheme for packet loss recovery may consist, for example, of a selective-retransmission ARQ scheme coupled with an RS (n,k) code with relatively small n–k. For every k information packets, the server generates n–k parity packets and sends the entire block of n packets to the user. If no more than n–k packets are lost, RS decoder can recover any missing information packets, so no retransmission is needed. If more than n–k packets are lost, not all information packets can be recovered. The user then sends NACKs for the missing information packets, and the server retransmits those packets. An alternative is to send one NACK for the whole block of n packets, in which case the server retransmits all n packets, both information and parity.

To design such a scheme, one can first choose the probability P_R with which the retransmission is to be triggered. Retransmissions occur whenever the RS decoder cannot recover the missing information packets using the available parity packets. For an (n,k) RS code, this happens when there are more than n–k losses in a block of n packets. The probability of this event is given by Eq. 11.7 for independent losses with probability p. Equation 11.7 can be used to choose n and k for the RS code for the desired probability of retransmission P_R.

$$P_R = \sum_{l=n-k+1}^{n} \binom{n}{l} p^l (1-p)^{n-l}. \tag{11.7}$$

11.4.2 Type-II Hybrid ARQ/FEC

The main difference between type-I and type-II hybrid ARQ/FEC algorithms for packet loss recovery is that upon NACK reception, type-I method retransmits data packets, while type-II method transmits parity packets. A type-II method for packet loss recovery based on selective-repeat ARQ and RS coding might work as follows. For every k information packets, the server generates n–k parity packets. The k information packets are sent to the user, followed by m parity packets, where $m < n$–k. Note that k information packets and m parity packets form an $(m+k,k)$ RS code, punctured from the RS (n,k) code. If all information packets can be recovered, meaning that no more than m packets out of $m+k$ packets are lost, no retransmission is needed. Otherwise, the user sends a NACK message indicating how many *additional* parity packets would be needed to recover all k information packets. The server then sends this number of additional parity packets from the remaining n–k–m parity packets. If some of these parity packets get lost, another NACK message is generated, additional parity packets are sent, and so on. For this reason, type-II methods are sometimes called *incremental redundancy*.

The intuition behind sending m parity packets along with the k information packets is that if the loss is small, m parity packets will be sufficient to correct it, and no additional parity packets will be needed. This reduces the time needed to deliver information packets to the user and reduces feedback traffic. However, if no loss happens in a particular block, the corresponding parity packets are useless and simply reduce the throughput. If the loss process is bursty (the usual case in practice), then packet losses are clustered. Each block of packets either ends up with no loss, in which case m parity packets serve no purpose, or a relatively large burst loss, in which case m parity packets are likely not sufficient to correct the loss, and additional parity packets are needed. Hence, for bursty losses, one may simply set $m = 0$, i.e. send no parity packets on initial transmission.

11.4.3 Extensions to Multicast

A single parity packet can recover the loss of different packets at different receivers. This makes type-II methods very suitable for multicast [22]. In a type-II multicast scheme, assuming $m = 0$, the server multicasts k information packets to all receivers. Each receiver (or a selected group of receivers) reports the number of lost packets. The maximum number (among all receivers) of lost packets determines how many parity packets the server should multicast to correct the losses. This number of parity packets (if correctly received) is sufficient for each user to correct its own losses. The following simple example [2] illustrates why type-II hybrid ARQ/FEC is more efficient than pure ARQ for multicast. Assume there are three users and the server multicasts $k = 6$ packets to them. Let the packet loss realization be as indicated in the table below, where "1" stands for correctly received packet and "0" for lost packet. In this case, type-II scheme would send two parity packets to correct all losses at each receiver, while the pure ARQ scheme would need to retransmit four packets (2, 3, 4, and 6).

Packet	1	2	3	4	5	6
User 1	1	1	0	1	1	1
User 2	1	1	1	0	1	1
User 3	1	0	1	1	1	0

In [18], several state-of-the-art centralized and distributed error control schemes are evaluated. In the centralized schemes, the server is in charge of error control, while feedback implosion is handled via a method similar to scalable feedback control [3] described previously. Distributed schemes are similar to RMTP where a DR provides error control for each local region. The results in [18] show that the distributed scheme employing type-II hybrid ARQ/FEC in local regions is the best in terms of bandwidth utilization and the average time needed to deliver the packets to the receivers. However, as the size k of the block of information packets upon which error control is exercised increases, the centralized scheme based on a type-II method catches up with the distributed scheme based on a type-II method. How large the k can be depends on several factors. For video multicast, the crucial factor is the receiver buffer size; k should be chosen small enough to allow at least one, and possibly several rounds of additional parity packets to arrive while the information packets from the same block are waiting in the buffer prior to decoding and display.

11.5 Application-Layer Error Control

In the context of this chapter, application-layer error control schemes rely on video encoder and decoder to handle the loss of video data. Both FEC and retransmission-based error control (as well as their combinations) can also be employed at the application layer, but they can also run at the transport layer, perhaps in conjunction with the application in a cross-layer framework. Application-layer techniques are attractive because they do not require much support from the underlying communication infrastructure. They are also necessary because error control methods at lower layers typically cannot completely eliminate packet loss. In this section we will describe some of the methods employed at the encoder and decoder to handle the loss of compressed video data. The discussion follows an excellent review of this topic in [36], where further details can be found. We will pay special attention to the application of these methods to video multicast.

11.5.1 Encoder Side: Error-Resilient Coding

Compressed bit streams are very sensitive to bit errors and losses, mainly due to the use of Variable Length Coding (VLC). When a bit is flipped or erased from a compressed bit stream, VLC decoder loses synchronization with the VLC encoder. Without special provisions, the decoder may take a long time to resynchronize with the encoder. Popular methods for dealing with the issue of VLC synchronization are the insertion of resynchronization markers, and the use of Reversible Variable Length Codes [31]. The combination of these two schemes allows the VLC decoder to decode the bit stream up to the piece that was lost both in the forward direction and in the backward direction from the next available resynchronization marker. Both methods have been adopted in H.263 [13] and MPEG-4 [12].

Bit-level synchronization problems arise when any kind of compressed bit stream is subject to errors or losses. But for compressed video, additional problems exist at frame-level due to predictive coding. When a part of a video frame gets lost, error concealment (discussed below) is typically applied at the decoder, but the reconstructed frame is not an exact match to the frame produced by the encoder. This not only affects the visual quality of the damaged frame, but may also affect future frames that use the damaged frame as a reference for predictive coding. An effective way to stop such error propagation is to insert an occasional I-frame into the stream. Unfortunately, this causes significant local increase in the coded video bit rate, since I-frames consume a lot more bits than P- or B-frames. To spread this rate increase over many frames, one may code some blocks of each frame in the intra-mode, instead of using occasional I-frames. For example, the previously mentioned ROPE method [37] selects the coding mode (inter/intra) for each block by considering both its cost in terms of coded bits, and the distortion its loss may cause at the decoder under a given channel model.

In the multicast scenario with many users, each with its own channel, the coding mode that is best for one user may not be the best for another user. Users with good channels will prefer inter-mode coding due to its higher compression efficiency, while the users with bad channels will prefer intra-mode coding due to its error resilience. Similar to the case of ULP for multicast, one has to aggregate the effect of different users' channels when deciding the coding mode for a particular block. While it might be possible to extend ROPE to the case of multiple users, a more natural approach for mode decision in multicast seems to be provided by MDDE [30], since this method explicitly considers multiple decoders.

11.5.2 Decoder Side: Error Concealment

Error concealment refers to the process of estimating missing pieces of video data from the available ones, in order to improve the visual quality of the video that has been damaged by packet loss. A number of error control algorithms have been proposed. All these methods are based on the same principle – exploiting the residual correlation (or, more precisely, statistical dependence) in the coded video signal to fill in the missing pieces. Despite the same underlying principle, the details of different methods differ. One reason for this is that for good performance, error concealment needs to be tailored to the particular codec used in the system. Methods suitable for standard codecs employing motion-compensated prediction and block-based DCT coding will not, in general, be appropriate for subband/wavelet video codecs. A good overview of various error concealment algorithms for standard video codecs can be found in [36]. For the purpose of this chapter, we will illustrate error concealment on a simple example that captures several key ideas in error concealment for motion-compensated predictive codecs.

Figure 11.11 shows a frame from which a macroblock (MB) has been lost, indicated in gray. For simplicity, we assume all pixels surrounding this MB are available. *Spatial* error concealment methods try to interpolate the pixels of the missing MB from the surrounding available pixels. On the other hand, *temporal* methods try to fill in the missing pixels from the previous frame, for example, by using the motion vector of a neighboring MB. *Spatio-temporal* error concealment methods fill in the missing MB by looking for a block in one of the previous or future frames that best fits into the neighborhood of the missing MB. One way to do this is to extract the boundary of the missing MB, shown as a dashed hollow rectangle in the figure, and

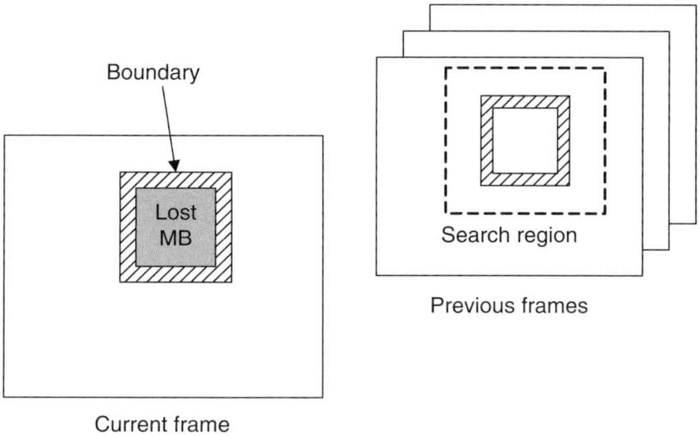

Fig. 11.11 Illustration of error concealment

use it as a search template in previous or future frames. Where the best match is found, the corresponding block enclosed by the boundary is copied into the location of the missing MB.

Of all error control techniques described in this chapter, error concealment is easiest to scale to a very large number of users in multicast, because it is fully distributed and requires no feedback. Each user tries to recover from its own losses by using whatever data it has available. Further, error concealment can be scaled to the user's processing power. For example, set-top box receivers and desktop workstations may employ sophisticated and computationally intensive error concealment algorithms, while battery-powered mobile users might settle for simpler methods.

11.5.3 Reference Picture Selection

Reference Picture Selection (RPS) [5] is an error control scheme that relies on the interaction between the video encoder and decoder. When a frame gets damaged by packet loss, error concealment is typically applied at the decoder. However, the concealed frame is not the exact replica of the transmitted frame. The error caused by this mismatch propagates to subsequent frames due to predictive coding employed at the encoder, creating the mismatch between subsequent decoded frames and the corresponding encoded frames. In order to stop this error propagation, the decoder can inform the encoder about the loss. The encoder then alters its prediction structure to stop error propagation.

One form of RPS is illustrated in Fig. 11.12, which shows the transmission of frames 1 through 9. Prediction used at the encoder is indicated by the curved arrows in the figure. Normally, the most recent frame is used as the reference frame for motion-compensated prediction. When frame 4 gets damaged by packet loss, the decoder sends a NACK message to the encoder. This informs the encoder that frame

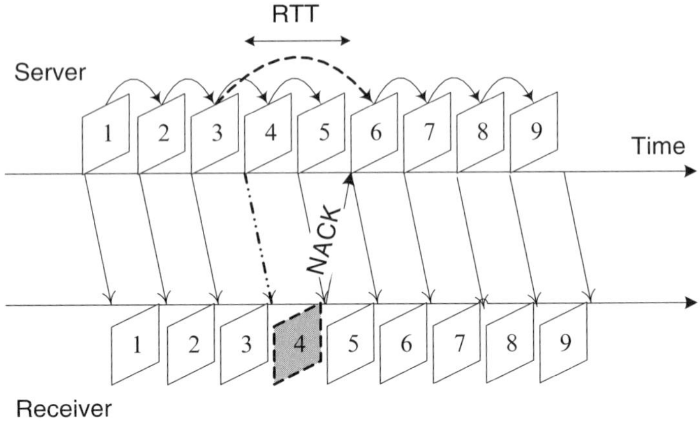

Fig. 11.12 Illustration of reference picture selection

4 has been damaged, and that subsequent frames are subject to error propagation. The encoder then chooses the most recent frame known to be received correctly (frame 3 in the figure) as a reference for encoding the next frame, which is frame 6. RTT determines the amount of time needed for prediction structure to be altered, and consequently determines the number of frames that will be affected by error propagation.

It is commonly thought that RPS is not suitable for multicast. One reason is the feedback implosion problem mentioned earlier. However, as we have seen, several methods have been developed to deal with feedback implosion. Another, perhaps more serious problem, is that in a large multicast group, it may be hard to find a frame that is correctly received by all receivers, which could be used as a reference frame for prediction. This argument is certainly valid if the loss process is memoryless and independent from user to user. However, if the losses are bursty and correlated among different users, there may be long stretches of time where no loss occurs, which could provide a number of potential reference frames for all users. RPS has been shown to have excellent performance in unicast [6]; perhaps it deserves to be studied more carefully in the multicast scenario as well.

11.6 Conclusions

Video multicast error control presents a set of challenging problems for the communication system designer. Heterogeneity of application requirements and communications infrastructure makes it virtually impossible to create a "one-fits-all" solution. Interactive applications, such as multipoint videoconferencing, favor low-delay error control, while streaming applications allow for a greater variety of error control schemes to be employed. Error control can be simplified if the receivers' capabilities are uniform and known to the system designer, as is the case in IPTV, where the service provider supplies set-top box receivers to the subscribers. On the other hand, Internet video applications are likely to serve a highly diverse set of users, ranging from mobile battery-powered devices to powerful workstations. This makes it difficult to design an error control scheme that is able to provide suitable video quality for all users.

In this chapter, we have presented an overview of error control methods that were traditionally developed for unicast (one-to-one) applications, and discussed their extensions to the multicast (one-to-many) scenario. Adapting an error control scheme from unicast to multicast can be straightforward in some cases (e.g., error concealment) and fairly involved in other cases (e.g., retransmission-based control). In addition to the user heterogeneity mentioned above, a major problem that arises in multicast communications is the amount of feedback information that the server needs to process. We have described several methods for dealing with such feedback implosion. We have also looked at some application-layer error control schemes that were developed specifically for video communications. Such schemes are attractive, since they generally require less support from the underlying communication infrastructure, and some of them, like error concealment, show excellent scalability to a very large number of users.

With an increasing number of services and a rapidly growing customer base, video multicast applications are receiving attention in both industrial and academic research communities. Some of the most promising directions for further development of video multicast error control are in the area of cross-layer design [35]. Combining error control mechanisms at different layers in a framework that optimizes both resource usage and video quality can provide efficient solutions and improved performance compared to the more traditional error control schemes [2]. Beyond the traditional engineering work, adoption and deployment of communication services depend heavily on the associated business issues as well. Market size, customers' habits and expectations, content rights management, and other regulatory constraints will all have an effect on whether or not a particular service and its associated error control scheme get deployed. Development of business models to go along with the quality of service provided by a given error control scheme is another interesting research area that has received relatively little attention thus far. There seem to be plenty of opportunities in multicast error control, for researchers and entrepreneurs alike.

References

1. Albanese A, Blömer J, Edmonds J, Luby M, Sudan M (1996) Priority encoding transmission. IEEE Transactions on Information Theory, vol. 42, no. 6, pp. 1737–1744
2. Bajić IV (2007) Efficient cross-layer error control for wireless video multicast. IEEE Transactions on Broadcasting, vol. 53, no. 1, pp. 276–285
3. Bolot JC, Turletti T, Wakeman I (1994) Scalable feedback control for multicast video distribution in the Internet. Proceedings of ACM SIGCOMM, pp. 58–67

4. Chou PA, Wang HJ, Padmanabhan VN (2003) Layered multiple description coding. Proc. 13th International Packet Video Workshop. Available online at: http://www.polytech.univ-nantes.fr/pv2003/papers/pv/html/main/home.htm
5. Fukunaga S, Nakai T, Inoue H (1996) Error resilient video coding by dynamic replacement of reference pictures. Proceedings of IEEE Globecom, vol. 3, pp. 1503–1508
6. Girod B, Färber N (1999) Feedback-based error control for mobile video transmission. Proceedings of the IEEE, vol. 87, no. 10, pp. 1707–1723
7. Hamming RW (1950) Error detecting and error correcting codes. Bell System Technical Journal, vol. 29, no. 2, pp. 147–160
8. Hsiang ST, Woods JW (2001) Embedded video coding using invertible motion compensated 3-D subband/wavelet filter bank. Signal Processing: Image Communication, vol. 16, no. 8, pp. 705–724
9. Huang HC, Peng WH, Chiang T, Hang HM (2007) Advances in the scalable amendment of H.264/AVC. IEEE Communications Magazine, vol. 45, no. 1, pp. 68–76
10. ISO/IEC IS 11172 (1993) Information technology – coding of moving pictures and associated audio for digital storage media at up to about 1.5 mbit/s (MPEG-1)
11. ISO/IEC IS 13818–2 (1995) Information technology – generic coding of moving pictures and associated audio information: Video (MPEG-2 Video)
12. ISO/IEC IS 14496–2 (1999) Information technology – coding of audio-visual objects – part 2: Visual (MPEG-4 Video)
13. ITU-T Recommendation H.263 (1998) Video coding for low bit rate communication.
14. ITU-T Recommendation H.264 (2005) Advanced video coding for generic audiovisual services.
15. Karn P (1997) C++ class library for Galois field arithmetic and algebra, with RS encoder/decoder. [Online] http://www.ka9q.net/code/fec/rs32. tar.gz
16. Kim BJ, Xiong Z, Pearlman WA (2000) Low bit-rate scalable video coding with 3-D set partitioning in hierarchical trees (3-D SPIHT). IEEE Transactions on Circuits and Systems for Video Technology, vol. 10, no. 8, pp. 1374–1387
17. Lin S, Costello Jr. D (2004), Error control coding, Second Edition. Pearson Prentice Hall, Upper Saddle River, NJ
18. Lacher MS, Nonnenmacher J, Biersack EW (2000) Performance comparison of centralized versus distributed error recovery for reliable multicast. IEEE/ACM Transactions on Networking, vol. 8, no. 2, pp. 224–238
19. Loguinov D, Radha H (2002) End-to-end Internet video traffic dynamics: statistical study and analysis. Proceedings of IEEE Infocom, vol. 2, pp. 723–732
20. Majumdar A, Sachs DG, Kozintsev IV, Ramchandran K, Yeung MM (2002) Multicast and unicast real-time video streaming over wireless LANs. IEEE Transactions on Circuits and Systems for Video Technology, vol. 12, no. 6, pp. 524–534
21. McCanne S, Jacobson V, Vetterli M (1996) Receiver-driven layered multicast. Proceedings of ACM SIGCOMM, pp. 117–130
22. Nonnenmacher J, Biersack EW, Towsley D (1998) Parity-based loss recovery for reliable multicast transmission. IEEE/ACM Transactions on Networking, vol. 6, no. 4, pp. 349–361
23. Parthasarathy V, Modestino JW, Vastola KS (1999) Reliable transmission of high-quality video over ATM networks. IEEE Transactions on Image Processing, vol. 8, no. 3, pp. 361–374
24. Paul S, Sabnani KS, Lin JCH, Bhattacharyya S (1997) Reliable Multicast Transport Protocol (RMTP). IEEE Journal on Selected Areas of Communications, vol. 15, no. 3, pp. 407–421
25. Pejhan S, Schwartz M, Anastassiou D (1996) Error control using retransmission schemes in multicast transport protocols for real-time media. IEEE/ACM Transactions on Networking, vol. 4, no. 3, pp. 413–427
26. Puri R, Lee KW, Ramchandran K, Bharghavan V (2001) An integrated source transcoding and congestion control paradigm for video streaming in the Internet. IEEE Transactions on Multimedia, vol. 3, no. 1, pp. 18–32
27. Rizzo L (1997) Effective erasure codes for reliable computer communication protocols. ACM Computer Communication Review, vol. 27, no. 2, pp. 24–36

28. Spragins JD, Hammond JL, Pawlikowski K (1991) Telecommunications: protocols and design. Addison-Wesley, Reading, MA
29. Stallings W (2004) Computer networking with Internet protocols and technology. Pearson Prentice Hall, Upper Saddle River, NJ
30. Stockhammer T, Hannuksela MM, Wiegand T (2003) H.246/AVC in wireless environments. IEEE Transactions on Circuits and Systems for Video Technology, vol. 13, no. 7, pp. 657–673
31. Takishima N, Wada M, Murakami H (1995) Reversible variable length codes. IEEE Transactions on Communications, vol. 43, no. 2, pp. 158–162
32. Tan W, Zakhor A (2001) Video multicast using layered FEC and scalable compression. IEEE Transactions on Circuits and Systems for Video Technology, vol. 11, no. 3, pp. 373–386
33. Taubman D, Zakhor A (1994) Multirate 3-D subband coding of video. IEEE Transactions on Image Processing, vol. 3, no. 5, pp. 572–588
34. van der Schaar M, Radha H (2001) A hybrid temporal-SNR fine-granular scalability for Internet video. IEEE Transactions on Circuits and Systems for Video Technology, vol. 11, no. 3, pp. 318–331
35. van der Schaar M, Shankar NS (2005) Cross-layer wireless multimedia transmission: Challenges, principles, and new paradigms. IEEE Wireless Communications Magazine, vol. 12, no. 4, pp. 50–58
36. Wang Y, Ostermann J, Zhang Y-Q (2001) Video processing and communications. Prentice-Hall, Upper Saddle River, NJ
37. Zhang R, Regunathan SL, Rose K (2000) Video coding with optimal inter/intra-mode switching for packet loss resilience. IEEE Journal on Selected Areas of Communications, vol. 18, no. 6, pp. 966–976

Chapter 12
Convolutional, Turbo, and Turbo-Like Codes for Digital Multimedia Broadcasting: Theory and Applications

Fred Daneshgaran, Massimiliano Laddomada, and Marina Mondin

12.1 Chapter Summary

The theory of channel coding is a well-established technical subject dating back to the seminal work by Shannon, who paved the way to what is nowadays recognized as Information Theory. It embraces both theory and a variety of practical applications in Digital Multimedia Broadcasting.

In this chapter we provide a review of the basic theory and results in this subject, with the emphasis on those channel coding architectures that deal with convolutional codes as well as low-density parity-check codes. The rest of the chapter is organized as follows. In Sect. 12.1.1 we briefly present the Shannon's view of a typical digital communication system, while in Sect. 12.1.2 we discuss the basic channel models employed as benchmarks for system performance evaluation. The theory of convolutional codes is presented in Sect. 12.2. In Sect. 12.2.1 we focus on a variety of alternative techniques for representing convolutional encoders. In Sect. 12.2.2 the focus is on the transfer function of convolutional codes needed for successive code performance evaluation. The decoding of convolutional codes is addressed in Sect. 12.2.3; both Maximum-Likelihood (ML) and Maximum A-Posteriori (MAP) decoding algorithms are presented and investigated in connection to a sample convolutional code. While maximum-likelihood decoding accomplished with the Viterbi algorithm, represents the de facto standard of many second-generation digital communications standards, MAP decoding is nowadays employed for the decoding

F. Daneshgaran (✉)
California State University, Los Angeles, CA, USA
e-mail: fdanesh@calstatela.edu

M. Laddomada (✉)
Texas A&M University-Texarkana, TX, USA
e-mail: mladdomada@tamut.edu

M. Mondin (✉)
DELEN-Politecnico di Torino
e-mail: marina.mondin@polito.it

F.-L. Luo (ed.), *Mobile Multimedia Broadcasting Standards: Technology and Practice*,
DOI: 10.1007/978-0-387-78263-8_12,

of turbo codes. The structural properties of convolutional codes are presented in Sect. 12.2.4.

Section 12.2.5 focuses on the performance evaluation of convolutional codes, while Sect. 12.2.6 presents a technique for obtaining high-rate codes starting from low-rate convolutional codes.

Section 12.3 deals with capacity-achieving channel codes, namely turbo and turbo-like codes. The design of turbo codes is addressed in Sect. 12.3.1 along with the key applications in digital multimedia broadcasting. The class of low-density parity-check codes is presented in Sect. 12.3.2.

This is the main focus of this chapter and for further investigations the reader is referred to some classical references in the subject. Finally, Sect. 12.4 outlines the conclusions.

12.1.1 Introduction

Channel coding dates back to the seminal work by Claude Shannon [1] who paved the way to the Information Theory Age.
To quote Shannon [1]:

The fundamental problem of communication is that of reproducing at one point either exactly or approximately a message selected at another point. Frequently the messages have meaning;... The significant aspect is that the actual message is one selected from a set of possible messages. The system must be designed to operate for each possible selection, not just the one which will actually be chosen since this is unknown at the time of design...

In Shannon's view, a typical communication system can be simply summarized as depicted in Fig. 12.1.

The main aim of a communication system is that of conveying a message **u** produced by a source to an intended user. The source can emit information in analog or digital form. Examples of source messages are a sequence of letters belonging to a file, the human voice or any other kind of desired information. By analog-to-digital conversion, analog signals can be converted to a digital form; so, there is no loss of generality in considering the message produced by the source as a sequence of bits $u_1,\ldots,u_i,\ldots,u_k$, with $u_i \in \{0,1\}$, $\forall i = 1,\ldots,k$, which constitute the information sequence **u**.

The transmitter is constituted by an encoder whose main purpose is to convert the information sequence **u** in a form suitable for transmission over the channel. Without loss of generality, the encoder can be interpreted as a mapper which associates a

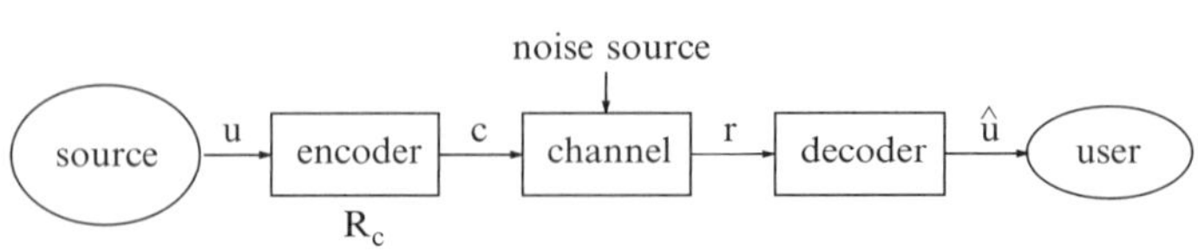

Fig. 12.1 Shannon's model of a general communication system

unique block **c** of n bits to every k information bits **u** at its input. The mapping operation is characterized by a positive number, R_c, defined as the ratio k/n. If $R_c < 1$, then redundancy bits are added to the information sequence **u** in order to overcome the effects of the noise introduced by the channel. If $R_c = 1$ (i.e., $k = n$), no encoding is performed at the transmitter and the information sequence is transmitted as it is towards the destination.

The transmitted sequence $\mathbf{c} = c_1, \ldots, c_i, \ldots, c_n$ is altered by the transmission channel, generating the received sequence $\mathbf{r} = r_1, \ldots, r_i, \ldots, r_n$, which is then passed on to the decoder block, which essentially operates the inverse mapping operation performed by the encoder at the transmitter. The output of the decoder is an estimate $\widehat{\mathbf{u}}$ of the information sequence **u**. Whenever $\mathbf{u} \neq \widehat{\mathbf{u}}$, at least one error has occurred in the information sequence during transmission over the channel. In this respect, the goal of the pair encoder/decoder is to introduce controlled redundancy in such a way that we ideally have $\mathbf{u} = \widehat{\mathbf{u}}$. Depending on the application, some errors could be tolerated on the transmitted sequence; this allows one to substantially reduce the redundancy introduced by the encoder at the transmitter side.

In connection with Shannon's model of a digital communication system, the role of the encoder is essentially to counteract the effects of the transmission channel on the information sequence **u**. As stated by Shannon [1], the objective is to reproduce the information sequence **u** at the receiver with a vanishingly small probability of error. Provided that the encoder rate R_c is below the channel capacity C, the price to be paid for guaranteeing a low probability of error is the decoder complexity.

12.1.2 Basic Channel Models

Different models are typically used for the communication channel shown in Fig. 12.1. In case of absence of memory and finite cardinality of the input and output samples, the discrete memoryless channel model is used. In connection with the block diagram shown in Fig. 12.2, a discrete memoryless channel [2] is a probabilistic system in which the output symbols y_i belonging to the set $Y = \{y_1, \ldots, y_O\}$ (O is the cardinality of the set Y), depend probabilistically on its input symbols x_j belonging to the set $X = \{x_1, \ldots, x_I\}$ (I is the cardinality of the set X) as specified

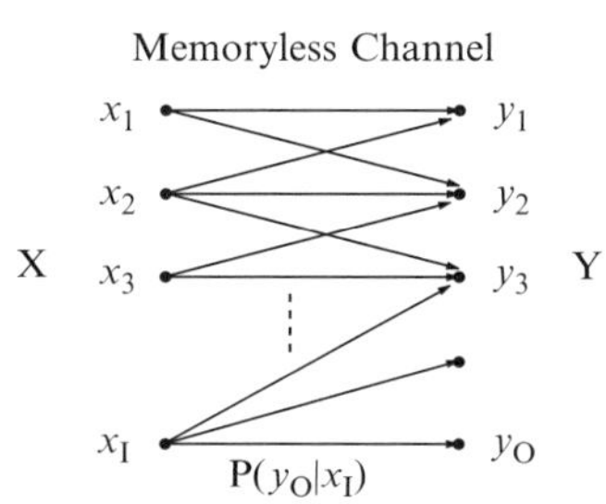

Fig. 12.2 Model of a memoryless probabilistic channel

by the probability transition matrix **M**, which specifies the conditional distributions related to the output symbols given each input symbol.

The matrix **M** is specified as follows:

$$\mathbf{M} = \begin{pmatrix} P(y_1|x_1) & P(y_2|x_1) & \dots & P(y_O|x_1) \\ P(y_1|x_2) & P(y_2|x_2) & \dots & P(y_O|x_2) \\ \vdots & \vdots & \vdots & \vdots \\ P(y_1|x_I) & P(y_2|x_I) & \dots & P(y_O|x_I) \end{pmatrix}. \tag{12.1}$$

Given a distribution $p(X)$ of the channel input symbols, the capacity C of the discrete memoryless channel is defined as [2]

$$C = \max_{p(X)} I(Y;X), \tag{12.2}$$

whereby $I(Y;X)$ is the mutual information between the two random variables X and Y, defined as:

$$I(Y;X) = \sum_{x_i \in X} \sum_{y_j \in Y} p(x_i, y_j) \log_2 \left(\frac{p(x_i, y_j)}{p(x_i) p(y_j)} \right).$$

12.1.2.1 Binary Symmetric Channel

The Binary Symmetric Channel (BSC) is a simple model of a communication channel conveying information in the form of bits. It easily stems from the general memoryless channel model with the matrix **M** defined in (12.1) upon setting $I = O = 2$, $x_1 = y_1 = 0$, $x_2 = y_2 = 1$. The reference model, depicted in Fig. 12.3, accepts at the input a bit c_i per unit of time, and transmits such a bit correctly with a probability $1 - p$, or erroneously with probability p independently from the previously transferred bits.

The probability p is defined as the probability to receive the bit 0 (i.e., $r_i = 0$), given that the input bit is $c_i = 1$, or, by symmetry, the probability to transmit $c_i = 0$ and receive $r_i = 1$. We have therefore:

$$\begin{aligned} p &= P(r_i = 0|c_i = 1) = P(r_i = 1|c_i = 0), \\ 1 - p &= P(r_i = 0|c_i = 0) = P(r_i = 1|c_i = 1). \end{aligned} \tag{12.3}$$

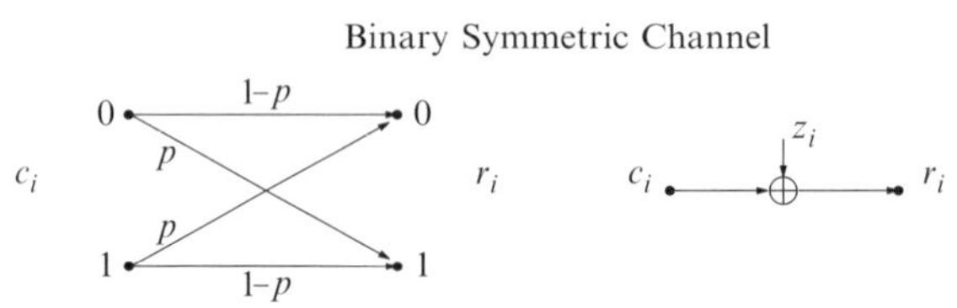

Fig. 12.3 The Binary Symmetric Channel model characterized by the transition probability p

Upon employing the rule of the addition over GF(2) (essentially Exclusive-Or of the logic variables),

$$0+0=0,\ 1+1=0,\ 0+1=1+0=1,$$

the output bits r_i can be represented as follows:

$$r_i = c_i \oplus z_i,\ \forall i,$$

whereby $z_i \in \{0,1\}$ depending upon the occurrence of a bit error during transmission, and

$$P(z_i) = \begin{cases} P(z_i = 1) = p \\ P(z_i = 0) = 1-p \end{cases}.$$

With this setup, the binary random variable z_i is distributed as an independent and identically distributed (i.i.d.) Bernoulli random variable with probability p [3].

The transmission of a data vector $\mathbf{c} = (c_1,\ldots,c_i,\ldots,c_n)$, with $c_i \in \{0,1\}$, $\forall i = 1,\ldots,n$ (n is the encoder block length), over a BSC is governed by the conditional probability

$$P(\mathbf{r}|\mathbf{c}) = P(r_1,\ldots,r_n|c_1,\ldots,c_n) = \prod_{i=1}^{n} P(r_i|c_i), \tag{12.4}$$

which easily stems from the i.i.d. assumption. Each conditional probability $P(r_i|c_i)$ is defined as in (12.3).

The channel capacity of the BSC is derived as [2]:

$$C_{\mathrm{BSC}} = 1 + p\log_2 p + (1-p)\log_2(1-p) \text{ bits per channel use.}$$

12.1.2.2 Memoryless AWGN channel

The memoryless Additive White Gaussian Noise (AWGN) channel is used when the transmission is impaired by the effect of the noise process generated in the electronic components constituting communications system receivers and possibly from a combined effect of independent noise processes with underlying Gaussian nature. The reference model of a transceiver system conveying information over an AWGN channel is depicted in Fig. 12.4.

As discussed earlier with reference to the BSC model, the information sequence $\mathbf{u}$ is mapped by a channel encoder with rate R_c in the encoded sequence $\mathbf{c}$. Each bit belonging to $\mathbf{c} = c_1,\ldots,c_n$ is BPSK modulated in order to form the real sequence

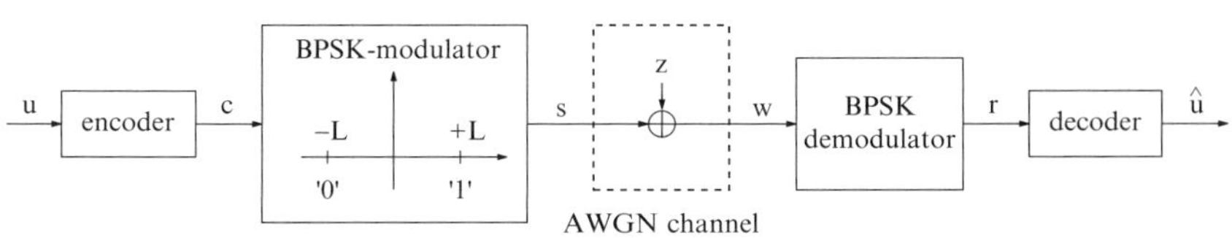

Fig. 12.4 Model of a coded transceiver system for the memoryless AWGN channel

$\mathbf{s} = s_1, \ldots, s_n$ with $s_i \in \{\pm L\}$. The signal amplitude L is defined as $\sqrt{R_c \cdot E_b}$, whereby E_b is the energy per information bit [4]. Each real symbol s_i is independently sent over the AWGN channel, producing the output symbol,

$$w_i = s_i + z_i,$$

whereby z_i is a zero-mean Gaussian random variable with variance $\sigma_Z^2 = N_o/2$. Given a transmitted symbol s_i, the conditional probability density function $f(w_i|s_i)$ is defined as

$$f(w_i|s_i) = \frac{1}{\sqrt{2\pi}\sigma_Z} \mathrm{e}^{-\frac{(w_i-s_i)^2}{2\sigma_Z^2}},$$

since $w_i = s_i + z_i$ and s_i is given. The transmission of a data vector $\mathbf{c} = (c_1, \ldots, c_i, \ldots, c_n)$, with $c_i \in \{0,1\}$, $\forall i = 1, \ldots, n$, over the AWGN channel is governed by the conditional probability:

$$\begin{aligned} f(\mathbf{w}|\mathbf{s}) &= f(w_1, \ldots, w_n|s_1, \ldots, s_n) = \prod_{i=1}^{n} f(w_i|s_i) \\ &= \prod_{i=1}^{n} \frac{1}{\sqrt{2\pi}\sigma_Z} \mathrm{e}^{-\frac{1}{2\sigma_n^2}(w_i-s_i)^2} \\ &= \left(\sqrt{2\pi}\sigma_Z\right)^{-n} \mathrm{e}^{-\frac{1}{2\sigma_Z^2}\|\mathbf{w}-\mathbf{s}\|^2}, \end{aligned} \tag{12.5}$$

which easily stems from the memoryless property of the AWGN channel, and upon using the Euclidean distance defined as,

$$\|\mathbf{w}-\mathbf{s}\|^2 = \sum_{i=1}^{n} (w_i - s_i)^2.$$

The channel capacity of the discrete-time memoryless AWGN channel accepting input symbols s with variance σ_S^2, is defined as [2]:

$$C_{\mathrm{AWGN}} = \frac{1}{2}\log_2\left(1 + \frac{\sigma_S^2}{\sigma_Z^2}\right) \text{ bits per channel use.}$$

The capacity C_{AWGN} takes on a different expression for bandlimited analog signals. Let us assume that the signals at the input of the AWGN channel are strictly bandlimited to the frequency interval $[-B_S, +B_S]$ Hz. Upon resorting to the sampling theorem [4], transmitted signals can be equivalently represented by employing at least $2B_S$ samples per second, whereby each sample is characterized by a variance σ_S^2 which also corresponds to the signal power P_S. On the other hand, noise power σ_Z^2 can be evaluated upon noting that the noise power spectral density is equal to $N_o/2$ over the signal bandwidth $[-B_S, +B_S]$. Therefore, we have $\sigma_Z^2 = \frac{N_o}{2} 2B_S = N_o \cdot B_S$. With this setup, the capacity C_{AWGN} can be rewritten as:

$$C_{\mathrm{AWGN}} = \frac{1}{2}\log_2\left(1 + \frac{P_S}{N_o \cdot B_S}\right) \text{ bits per channel use.}$$

Upon sampling analog signals with a sampling frequency as low as $2B_s$, signal samples are sent over the channel with a rate equal to $2B_s$ samples per second. Therefore, the capacity of the AWGN channel expressed in bits per second corresponds to

$$C_{\text{AWGN}} = B_s \log_2\left(1 + \frac{P_s}{N_o \cdot B_s}\right) \text{ bps.}$$

It is possible to demonstrate [4] that the BSC channel model depicted in Fig. 12.3 is equivalent to the system in Fig. 12.4 by setting,

$$p = \frac{1}{2}\text{erfc}\sqrt{\left(R_c \frac{E_b}{N_o}\right)}, \tag{12.6}$$

whereby E_b/N_o is the bit energy to one-side noise spectral density ratio.

12.2 Convolutional Codes

The block diagram of a general $k \times K$-stage convolutional encoder is shown in Fig. 12.5. A convolutional encoder is characterized by the following parameters:

- k denotes the number of bits belonging to the input sequence $\mathbf{u}$ which enters the $k \times K$-stage shift registers.
- n denotes the size in bits of the output sequence $\mathbf{c}$; the ratio $R_c = k/n$ is called encoder rate.
- K is the encoder constraint length, and it identifies the number of k-bits shift registers constituting the convolutional encoder.

As opposed to block codes, a convolutional code is operated by an encoder with memory. In other words, each sequence $\mathbf{c}$ depends on the input sequence $\mathbf{u}$ along with the $(K-1) \times k$ previous input sequences stored in the shift registers depicted

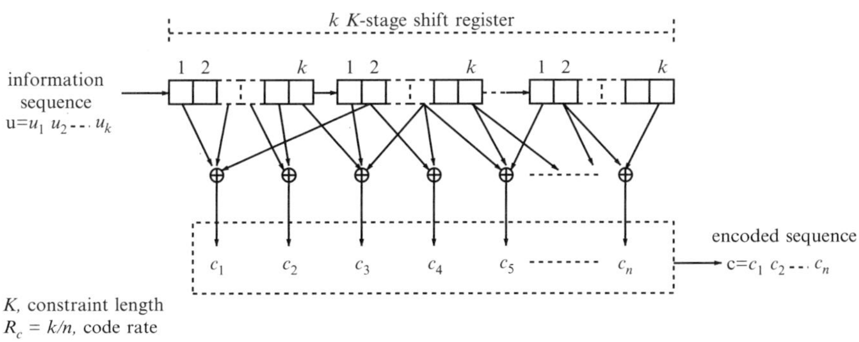

Fig. 12.5 Block diagram of a convolutional encoder of rate $R_c = k/n$ and constraint length K encoding k input bits, $u_1, \ldots, u_k$, at a time. The encoded sequence $\mathbf{c}$ is composed of n bits $c_1, \ldots, c_n$

in Fig. 12.5. In this respect, the constraint length K identifies the number of k-bits shifts over which an input bit u_j influences the output sequence $\mathbf{c}$, while $(K-1)\times k$ is the number of memory bits.

The input binary sequence $\mathbf{u} = u_1,\ldots,u_k$ is encoded in blocks of k bits at the time, which enter each shift register cell one at the time, while the output sequence $\mathbf{c}$ is generated once the k input bits have entered the first k shift register cells. Each output bit c_i, $\forall i = 1,\ldots,n$, is obtained by applying the modulo-2 summation of the bits contained in the shift register cells whose content is added in the ith modulo-2 adder (represented as $\oplus$ in Fig. 12.5).

To be specific, consider the convolutional encoder in Fig. 12.6 described by the following parameters $k = 1, n = 2$, $K = 3$, and memory $(K-1) = 2$. With respect to the general architecture of a convolutional encoder shown in Fig. 12.5, the block diagram depicted in Fig. 12.6 employs $K-1$ memory registers, each one of size $k = 1$. It is in fact only necessary to store in a memory cell the bits remaining from the previous step $u_2,\ldots,u_k$, while the current bit u_1 is available at the encoder input.

The operations of the convolutional encoder in Fig. 12.6 is as follows. At each bit time, a new bit u_1 appears at the encoder input, while the output bits c_1 and c_2 are evaluated according to the new bit u_1 along with the previous two input bits stored in the shift register cells s_1 and s_2. Before a new bit u_1 appears at the encoder input, the cell contents are right-shifted by one position, while the bit u_1 enters the memory cell s_1 upon a right shift of s_1 in s_2. These operations are then repeated for each new input bit u_1.

Convolutional encoding generally starts and ends in the zero state, since the memory content s_1, s_2 of the convolutional encoder is usually cleared at the beginning of the encoding operations, while an appropriate number of zero bits are appended to the encoded sequence. For the convolutional encoder depicted in Fig. 12.6, two zero bits are appended to the useful information sequence in order to force the memory content s_1, s_2 to zero, so that $s_1 = 0, s_2 = 0$. Of course, this technique slightly reduces the encoder rate below k/n. Since the block size of a convolutional code is not fixed like in the case of block codes, the rate penalty can be contained upon choosing a sufficiently long input sequence. As an example,

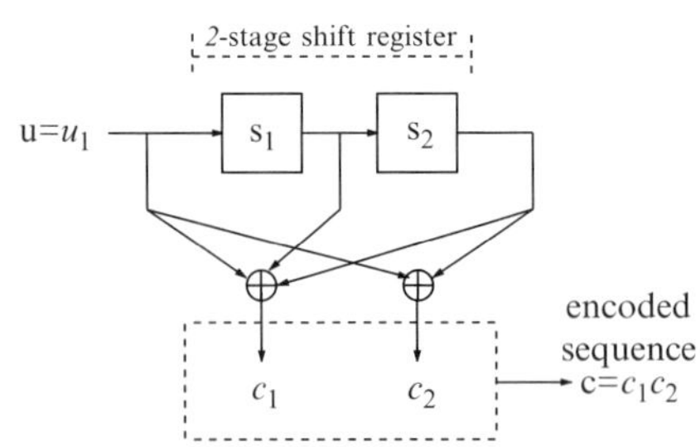

u_1	s_1	s_2	c_1	c_2
0	0	0	0	0
1	0	0	1	1
0	1	0	1	0
1	1	0	0	1
0	0	1	1	1
1	0	1	0	0
0	1	1	0	1
1	1	1	1	0

Fig. 12.6 Block diagram of a convolutional encoder of rate $R_c = k/n = 1/2$ and constraint length $K = 3$, encoding $k = 1$ input bits at a time. The encoded sequence $\mathbf{c}$ is composed of $n = 2$ bits

if the input sequence contains 100 bits, the actual rate of the encoder in Fig. 12.6 becomes:

$$R_c = \frac{k * 100}{n * (100+2)} = \frac{100}{2 * (100+2)} = \frac{100}{204} \approx \frac{k}{n} = \frac{1}{2}.$$

12.2.1 Representation of Convolutional Encoders

The encoding function accomplished by convolutional encoders can be described in a variety of methods. In connection with the encoder in Fig. 12.6, a compact notation consists of specifying the generators $\mathbf{g}_i$ associated to each bit c_i belonging to the output sequence $\mathbf{c}$. The ith generator $\mathbf{g}_i$ contains K bits (the code constraint length), where a 1 in the jth position identifies the presence of a connection between the ith code bit and the jth position in the shift register, whereby positions are numbered from left to right starting from the input bit.

The encoder depicted in Fig. 12.6 has the following generators $\mathbf{g}_1$ and $\mathbf{g}_2$:

$$\mathbf{g}_1 = [1\ 1\ 1] \Rightarrow c_1,$$
$$\mathbf{g}_2 = [1\ 0\ 1] \Rightarrow c_2.$$

A somewhat simpler method of describing the behavior of an encoder to one input bit at the time is described in the table shown in Fig. 12.6 for the considered code. The first column represents the possible values of the input bit, while the second and third columns represent the memory content of the shift register encoding the specified convolutional code. Depending on the memory content as specified by s_1, s_2, and on the input bit u_1, the output sequence $\mathbf{c} = c_1, c_2$ takes on the values highlighted in the two rightmost columns. Both c_1 and c_2 can be readily evaluated upon using the modulo-2 additions of the bits u_1, s_1, s_2 involved in the respective parity-check equations specified by the two generators $\mathbf{g}_1$ and $\mathbf{g}_2$.

An alternative notation for describing the function accomplished by convolutional encoders relies on polynomials with coefficients specified in an appropriate field. Since throughout this chapter the focus is on binary sequences, each coefficient of the polynomial can take values in the field GF(2). With this setup, any bit sequence $u_0, u_1, \ldots, u_i, \ldots$ is associated with the polynomial in the indeterminate D

$$\begin{aligned} u(D) &= u_0 + u_1 D + u_2 D^2 + \ldots \\ &= \sum_{i=0}^{+\infty} u_i \cdot D^i, \ u_i \in \mathrm{GF}(2), \ \forall i = 0, 1, \ldots . \end{aligned} \tag{12.7}$$

Upon representing the generators $\mathbf{g}_i$ in polynomial notations, each output sequence c_i can be evaluated by a polynomial multiplication over the same field GF(2) as follows:

$$c_i(D) = u(D) g_i(D).$$

Consider the convolutional encoder in Fig. 12.6, and assume the following input sequence:

$$\mathbf{u} = 1\,0\,0\,1\,1.$$

The polynomial associated with $\mathbf{u}$ is $u(D) = 1 + D^3 + D^4$, while the generator polynomials are $g_1(D) = 1 + D + D^2$ and $g_2(D) = 1 + D^2$. With this setup, the two polynomials associated with the two sequences of bits $\mathbf{c}_1$ and $\mathbf{c}_2$ can be easily evaluated as follows:

$$\begin{aligned} c_1(D) = u(D)g_1(D) &= (1 + D^3 + D^4)(1 + D + D^2) \\ &= 1 + D + D^2 + D^3 + D^6 \\ c_2(D) = u(D)g_2(D) &= (1 + D^3 + D^4)(1 + D^2) \\ &= 1 + D^2 + D^3 + D^4 + D^5 + D^6. \end{aligned}$$

In the most general case of an encoder with rate k/n, generalization of the previous results follow from defining a polynomial row vector for the input sequences

$$\mathbf{U} = [u_1(D)\; u_2(D) \ldots\; u_k(D)],$$

and a $k \times n$ polynomial generator matrix $\mathbf{G}(D)$

$$\mathbf{G}(D) = \begin{bmatrix} g_{1,1}(D) & g_{1,2}(D) & \ldots & g_{1,n}(D) \\ g_{2,1}(D) & g_{2,1}(D) & \ldots & g_{2,n}(D) \\ \vdots & \vdots & \vdots & \vdots \\ g_{k,1}(D) & g_{k,1}(D) & \ldots & g_{k,n}(D) \end{bmatrix}, \tag{12.8}$$

whereby the polynomial entry $g_{i,j}(D)$ represents the generator polynomial specifying the connections between the jth output bit c_j and the ith input polynomial $u_i(D)$, $\forall i \in \{1,\ldots,k\}, j \in \{1,\ldots,n\}$. The polynomials associated with the output sequence $\mathbf{c}$ can be obtained through a vector–matrix multiplication:

$$\mathbf{C}(D) = [c_1(D)\; c_2(D) \ldots\; c_n(D)] = \mathbf{U} \times \mathbf{G}(D).$$

In connection with the encoder in Fig. 12.6, $\mathbf{G}(D)$ is the following 1×2 polynomial matrix

$$\mathbf{G}(D) = \left[1 + D + D^2\;\; 1 + D^2\right],$$

while the output sequence associated with the input polynomial $u(D) = 1 + D^3 + D^4$ can be evaluated as follows:

$$\begin{aligned} \mathbf{C}(D) &= \mathbf{u}(D) \times \mathbf{G}(D) = u(D)\left[1 + D + D^2\;\; 1 + D^2\right] \\ &= \left[u(D)(1 + D + D^2)\;\; u(D)(1 + D^2)\right] = [c_1(D)\; c_2(D)]. \end{aligned} \tag{12.9}$$

The previous examples show that the convolutional encoder maps each input sequence $\mathbf{u}$ into a unique output sequence $\mathbf{c}$, also called codeword, through a semi-infinite mapping:

$$\mathbf{c} = G(\mathbf{u}).$$

The polynomial representation of both sequences and polynomial matrix is very useful for dealing with both recursive and systematic convolutional encoders.

An encoder is said to be *recursive* (as opposed to *not recursive*) when it contains at least one feedback connection (like in Fig. 12.7a and c). In this case, the polynomial generator matrix $\mathbf{G}(D)$ contains rational functions.

An encoder, whether recursive or not, is said to be *systematic* if an identity submatrix can be identified in its polynomial generator matrix $\mathbf{G}(D)$ by appropriate permutations of the rows and/or columns of $\mathbf{G}(D)$.
As an example, consider the three convolutional encoders depicted in Fig. 12.7. The polynomial generator matrices associated with these three encoders are as follows:

$$\begin{aligned}
\mathbf{G}^{(a)}(D) &= \left[1,\ \frac{1+D+D^2+D^3}{1+D^2+D^3},\ \frac{1+D+D^3}{1+D^2+D^3}\right], \\
\mathbf{G}^{(b)}(D) &= \left[1,\ 1+D+D^2\right], \\
\mathbf{G}^{(c)}(D) &= \left[1,\ \frac{1+D+D^2}{1+D^2}\right].
\end{aligned} \tag{12.10}$$

Owing to the previous definitions, the three encoders are systematic; the first and the third encoders are also recursive. For clarity, the encoder in Fig. 12.7a also shows the polynomial terms associated to the coefficients of the generators. Note that the encoder represented in Fig. 12.6 is neither systematic nor recursive.

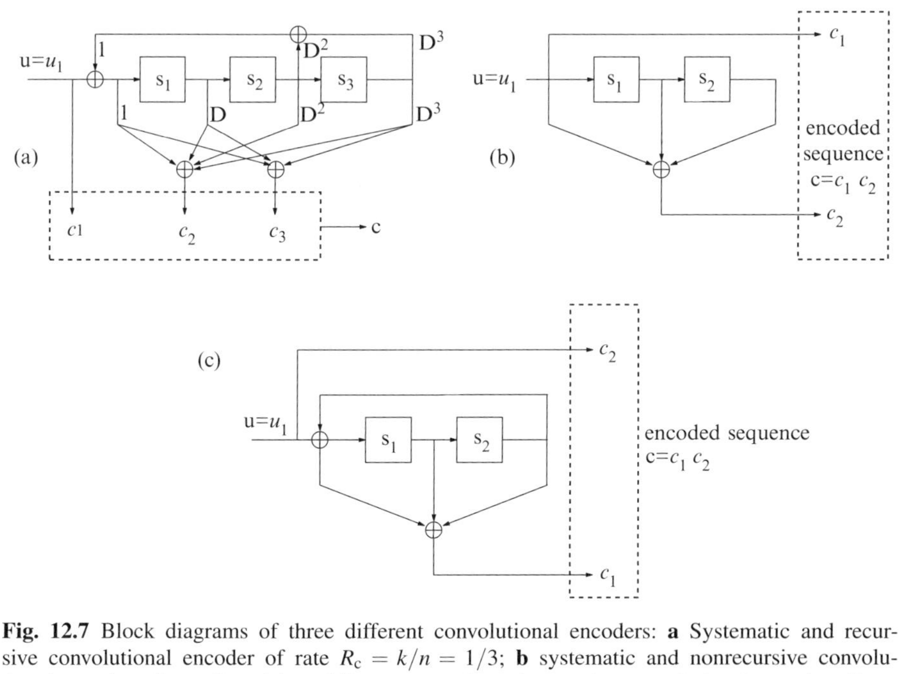

Fig. 12.7 Block diagrams of three different convolutional encoders: **a** Systematic and recursive convolutional encoder of rate $R_c = k/n = 1/3$; **b** systematic and nonrecursive convolutional encoder of rate $R_c = k/n = 1/2$; **c** systematic and recursive convolutional encoder of rate $R_c = k/n = 1/2$

Consider again the convolutional encoder in Fig. 12.6. As already noted in connection with the table employed for evaluating the output sequence, a convolutional encoder is nothing but a finite state machine, whereby the output sequence depends on both the current state of the encoder, as exemplified by the content of the cells s_1, s_2, and the input bit u_1. The evolution of this state machine can be described by the so-called state diagram, a meaningful representation widely employed in connection with convolution decoding. For a general rate $R_c = k/n$, constraint length K convolutional encoder, a state diagram has $2^{k(K-1)}$ states representing the contents of the $k \times (K-1)$ rightmost shift register cells (the memory cells). In each state, there are 2^k departing arrows identifying the k input bits yielding the underlined transition between states, and 2^k incoming arrows indicating the state from which transition originates from. Each arrow is associated with the k input bits producing the state transition and the n bits constituting the output sequence $\mathbf{c}$ separated by a vertical line.

As a reference example, the state diagram associated with the convolutional encoder depicted in Fig. 12.6 is shown in the left plot of Fig. 12.8. For the sake of simplifying the notation, for $k = 1$ convolutional codes, transitions originating from an input bit of one are represented as dashed arrows, while the ones associated with the input bit of zero are identified by a continuous arrow.

An alternative representation of the relationship between states relies on the one-stage trellis description shown in Fig. 12.8. In this scheme the transitions between states due to the current k-bits input sequence are shown. The $2^{k(K-1)}$ states are listed in each column with the appropriate transition arrows departing from each state. As for the state diagram, each arrow is associated with the k input bits producing the state transition and the n bits constituting the output sequence $\mathbf{c}$ separated by a vertical line. The trellis diagram represents the unfolding of the state transition diagram in time and it is very useful for visualizing the transitions occurring between states corresponding to a sequence of N consecutive input bits. As an example, the

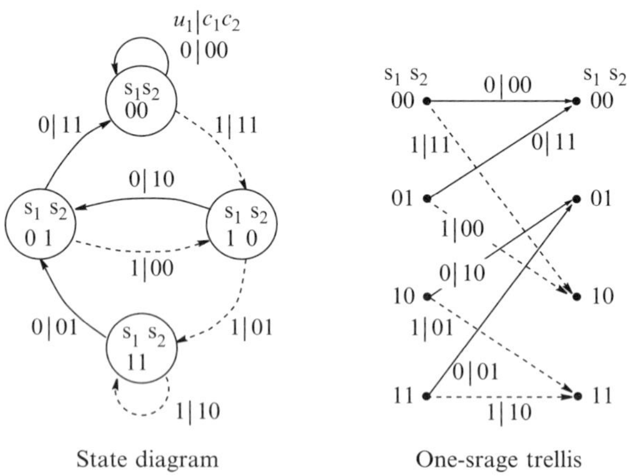

Fig. 12.8 State diagram (*left plot*) and one-stage trellis (*right plot*) associated with the convolutional encoder in Fig. 12.6 described by the generator matrix $\mathbf{G}(D) = \left[1 + D + D^2 \; 1 + D^2\right]$

trellis diagram related to the convolutional encoder in Fig. 12.6 is shown in Fig. 12.9 for a sequence of seven consecutive input bits, whereby any trellis stage is associated with a single input bit. Under the hypothesis that the encoder starts from the zero state at time t_1, there are two possible transitions occurring in response to an input bit. If $u_1 = 0$ the encoder remains in the zero state, while it transits to the state $s_1, s_2 = 1, 0$ if $u_1 = 1$.
The output sequence **c** along with the sequence of states traversed in accordance to the input sequence $\mathbf{u} = 1\ 1\ 0\ 1\ 0\ 0$ is, as highlighted in Fig. 12.10, $\mathbf{c} =$ 11 01 01 00 10 11.

An output sequence which departs at some time instant from the zero state and merges back into it after some state transitions throughout the trellis is called a codeword. As an example, the output sequence $\mathbf{c} =$ 11 01 01 00 10 11 identified in Fig. 12.10 is a codeword with Hamming weight 7 since it departs from the zero state at the time instant t_1 and merges back into it at time t_7. Such a codeword is generated by an input sequence with Hamming weight 3.

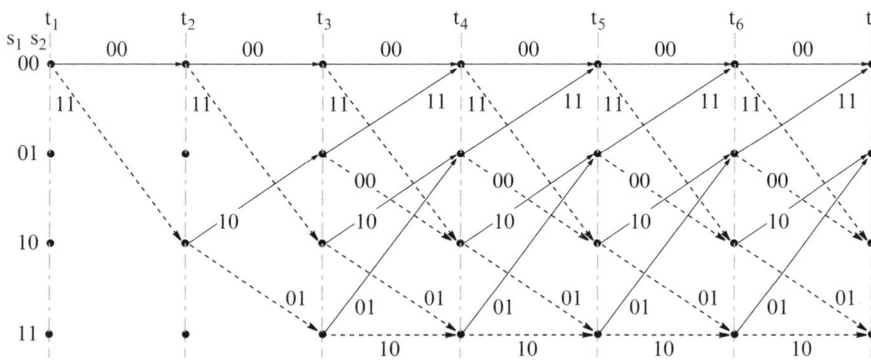

Fig. 12.9 Trellis diagram of the convolutional encoder of rate $R_c = k/n = 1/2$ and constraint length $K = 3$ described in Fig. 12.6, related to a sequence of six consecutive input bits

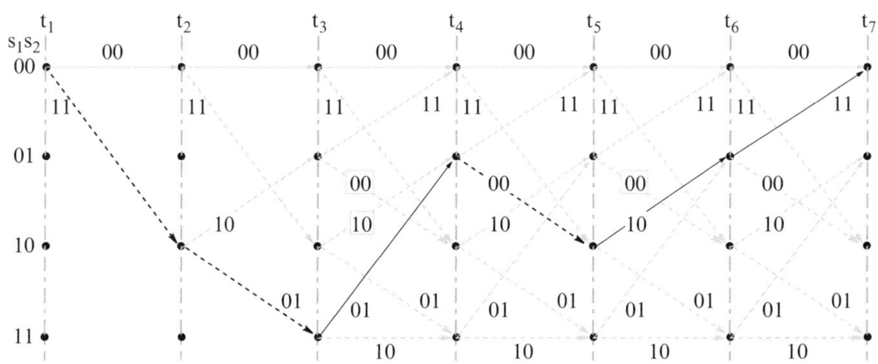

Fig. 12.10 Trellis path due to the input sequence of bits $\mathbf{u} = 1\ 1\ 0\ 1\ 0\ 0$. The output sequence is $\mathbf{c} =$ 11 01 01 00 10 11

A convolutional code is a linear code. For linear codes, the set of distances between any pair of codewords belonging to the code corresponds to the set of distances of any codeword from the zero sequence, which also corresponds to the Hamming weight of the considered codeword. Error correction capabilities of convolutional codes are strictly related to the code distance spectrum, that is the enumeration of the Hamming weights of the codewords of the code along with the Hamming weights of the input sequences which generated such codewords.

With reference to the code trellis depicted in Fig. 12.9, the distance spectrum can be obtained with the following steps:

- List the sequences which depart at the time instant t_1 from the zero state and which merge back into it at a certain future time instant, along with the input sequences producing such codewords.
- Evaluate the Hamming weights of such codewords.
- Create a 3-column table whereby any row contains increasing values of the Hamming weights of the codewords. Then, for each Hamming weight, list the number of codewords with the same Hamming weight, say d_c, along with the cumulative Hamming weights of all input sequences generating the codewords with the common weight d_c.

The minimum value of d_c other than zero is called the free distance d_f of the convolutional code. Note that d_f corresponds to the minimum distance of block codes.

Owing to the linearity of the code, the Hamming distances of the codewords from any other reference codeword is the same as the one obtained by evaluating the distances of all the codewords from the zero codewords. For this reason, the zero codeword is usually taken as a reference codeword in evaluating the code performance.

12.2.2 Transfer Function of Convolutional Codes

Distance properties of convolutional codes can also be obtained by resorting to the state diagram of the code. As a reference example for the derivations which follow, consider the state diagram, shown in Fig. 12.8, associated with the convolutional encoder in Fig. 12.6 described by the generator matrix $\mathbf{G}(D) = \left[1+D+D^2 \; 1+D^2\right]$. Since the self-loop on the zero state does not contribute to the distance spectrum, it can be eliminated. Furthermore, since codewords are paths along the trellis which depart from and merge back into the zero state, the zero state in Fig. 12.8 is split into two new zero states which represent, respectively, the starting and the final edge of a codeword. Upon enumerating each state s_1, s_2 with the labels S_i, the state diagram depicted in Fig. 12.8 can be equivalently represented as in Fig. 12.11, whereby each state transition is associated with a polynomial on the three dummy variables I, N, and W, which enumerate, respectively, the Hamming weight of the input sequence, the number of transitions, and the Hamming weight of the output sequence. In fact, for a given path in the trellis,

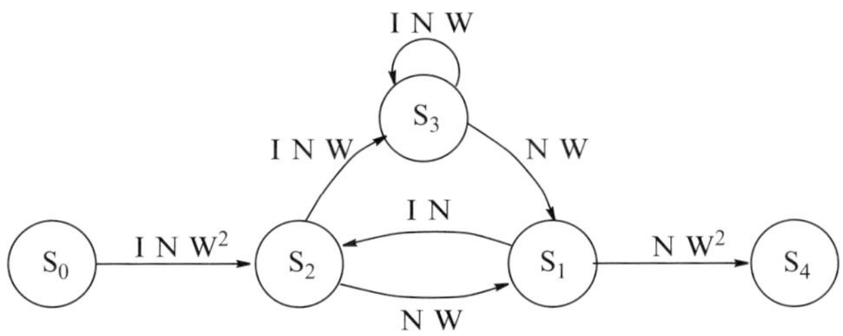

Fig. 12.11 State diagram associated with the convolutional encoder in Fig. 12.6 described by the generator matrix $\mathbf{G}(D) = \left[1+D+D^2 \; 1+D^2\right]$

the exponent of the variable I identifies the Hamming weight of the corresponding input sequence, the exponent of N is related to the number of transitions between states for the underlined path and the exponent of W indicates the Hamming weight of the corresponding output sequence.

The state diagram shown in Fig. 12.11 is the starting point for deducing the so-called transfer function of the underlying convolutional encoder, defined as:

$$T(I,N,W) = \frac{S_4}{S_0}. \tag{12.11}$$

The evaluation of $T(I,N,W)$ relies on the following four state equations:

$$\begin{aligned} S_1 &= N \cdot W \cdot S_3 + N \cdot W \cdot S_2, \\ S_2 &= I \cdot N \cdot W^2 \cdot S_0 + I \cdot N \cdot S_1, \\ S_3 &= I \cdot N \cdot W \cdot S_3 + I \cdot N \cdot W \cdot S_2, \\ S_4 &= N \cdot W^2 \cdot S_1, \end{aligned} \tag{12.12}$$

associated with the state diagram in Fig. 12.11. Next line of pursuit consists in finding a relation of the form $S_4 = f(S_1)$.

From the third equation in (12.12)

$$S_3 = \frac{INW}{1-INW} S_2.$$

Upon substituting the previous equation in the first equation in (12.12), the relation $S_1 = f(S_2)$ noted below easily follows:

$$S_1 = \left[\frac{IN^2W^2}{1-INW} + WN\right] S_2. \tag{12.13}$$

The second relation in (12.12) yields:

$$S_0 = \frac{S_2 - INS_1}{INW^2}. \tag{12.14}$$

Upon substituting (12.14) and (12.13) in (12.11), the transfer function simplifies to:

$$\begin{aligned} T(I,N,W) &= \frac{N \cdot W^2 \cdot S_1}{S_0} = \frac{IN^2W^4\left[WN + \frac{IN^2W^2}{1-INW}\right]}{1 - IN\left[WN + \frac{IN^2W^2}{1-INW}\right]} \\ &= \frac{\left[IN^3W^5 + I^2N^4W^6\left[1 + INW + \cdots\right]\right]}{\left[1 - IN\left[WN + \frac{IN^2W^2}{1-INW}\right]\right]}. \end{aligned} \tag{12.15}$$

Finally, the relation:

$$\frac{1}{1-x} = \sum_{i=0}^{+\infty} x^i = 1 + x + x^2 + \cdots,$$

allows us to rewrite (12.15) as an infinite sum of polynomial terms:

$$T(I,N,W) = IN^3W^5 + I^2N^4W^6 + \cdots. \tag{12.16}$$

The code transfer function in the form stated in (12.16) yields useful information about the distance properties of the code. In particular, the term IN^3W^5 in $T(I,N,W)$ indicates that there is only one output sequence with Hamming weight 5 (which corresponds to the code free distance, d_f) due to an input sequence with Hamming weight 1, obtained by traversing three consecutive states of the encoder (other than the starting zero state S_0), as shown by the term N^3. With a similar reasoning, the term $I^2N^4W^6$ indicates the presence of a codeword with weight 6 obtained by traversing 4 states with an input sequence with weight 2. The factor N is important when one is interested in a sequence constituted by a finite number of bits, say M. In this case, the transfer function can be truncated over the polynomial term containing terms in N of order at most M.

On the other hand, when one is only interested in the Hamming weights of the codewords, the dummy variables I and N can be set to 1, yielding a transfer function depending on the variable W:

$$T(W) = \sum_{i=d_\mathrm{f}}^{+\infty} m_i W^i,$$

whereby m_i is the number of codewords with weight i.

In the most general case, the transfer function of a convolutional encoder can be written as

$$T(I,N,W) = \sum_{r,p,t} T_{r,p,t} I^r N^p W^t, \tag{12.17}$$

whereby $T_{r,p,t}$ denotes the number of codewords with Hamming weight t corresponding to information sequences with weight r, and whose length is equal to p. Such a transfer function is also called *codeword Input–Output Weight Enumerating Function* (IOWEF) [5].

When the lengths of the codewords are not of concern, the IOWEF can be rewritten as

$$T(I,W) = T(I,N,W)\,|_{N=1} = \sum_{r,t} T_{r,t} I^r W^t, \tag{12.18}$$

where the term

$$T_{r,t} = \sum_p T_{r,p,t}$$

corresponds to the number of codewords with Hamming weight t due to information sequences with weight r.

12.2.3 Decoding of Convolutional Codes

Consider the transceiver depicted in Fig. 12.1 with a memoryless transmission channel that affects each transmitted bit independently from the others.
The decoding of convolutional codes consists in deciding in favor of a particular transmitted sequence $\mathbf{u}$ based on the knowledge of the received, possibly corrupted, sequence $\mathbf{r}$. To date, a variety of decoding algorithms for convolutional codes have been proposed.

When all the bit sequences at the input of the memoryless channel are equally likely, the decoding algorithm which minimize the probability of error consists in choosing the codeword $\mathbf{c}$ maximizing the conditional probability $P(\mathbf{r}|\mathbf{c})$ over all the possible codewords $\mathbf{c}_i$ belonging to the code [6, 7]:

$$\mathbf{c} = \arg\max_{\forall \mathbf{c}_i} P(\mathbf{r}|\mathbf{c}_i). \tag{12.19}$$

This decoding strategy is also called maximum-likelihood decoding.

12.2.3.1 Maximum-Likelihood Decoding: the Viterbi Algorithm over Memoryless Channels

Consider the transceiver depicted in Fig. 12.1 and assume that during transmission the sequence is corrupted due to the BSC shown in Fig. 12.3 with probability p. Moreover, assume that the received sequence $\mathbf{r}$ differs from the transmitted one $\mathbf{c}_j$ in $d_{\mathrm{H}}(\mathbf{r},\mathbf{c}_j)$ bits of known positions.

Let us assume that the receiver employs the maximum-likelihood strategy in (12.19) by comparing the received sequence to all the codewords belonging to the convolutional code space. Due to the memoryless property of the channel, the probability that any codeword $\mathbf{c}_j$ is corrupted in $d_{\mathrm{H}}(\mathbf{r},\mathbf{c}_j)$ known positions can be evaluated using (12.4) as follows

$$P(\mathbf{r}|\mathbf{c}_j) = p^{d_{\mathrm{H}}(\mathbf{r},\mathbf{c}_j)}(1-p)^{n-d_{\mathrm{H}}(\mathbf{r},\mathbf{c}_j)} = (1-p)^n \cdot \left(\frac{1-p}{p}\right)^{d_{\mathrm{H}}(\mathbf{r},\mathbf{c}_j)},$$

whereby n is the codeword length. Since $p < \frac{1}{2}$, we have $(1-p)/p < 1$. Therefore, over all the set of codewords $\mathbf{c}_j$, probability $P(\mathbf{r}|\mathbf{c}_j)$ gets maximized when $d_H(\mathbf{r}, \mathbf{c}_j)$ is minimum. In other words, the maximum-likelihood algorithm in (12.19) identifies the codeword $\mathbf{c}$ closest to the received sequence $\mathbf{r}$ in terms of Hamming distance. Then, because of the unique mapping between each information sequence $\mathbf{u}$ and the codewords $\mathbf{c}_i$ belonging to the code, it is a simple task to evaluate the information sequence $\mathbf{u}$.

The Viterbi algorithm [6] finds the codewords closest to the received sequence $\mathbf{r}$ in terms of Hamming distance (or other distance metrics). Is is essentially a maximum-likelihood decoder which reduces the computational burden due to the need of finding the correct sequence in a set containing many candidate sequences, by exploiting the particular structure of the code trellis. The effectiveness of the Viterbi algorithm relies on the fact that at each stage of the trellis, only the distances between the received coded symbols associated with that stage and the paths entering each state in the trellis need to be evaluated.

As a reference example, consider the transceiver depicted in Fig. 12.1 and assume that during transmission the sequence is corrupted due to the BSC shown in Fig. 12.3 with $p = 1/6$ (this means that on average 2 bits every 12 transmitted bits are corrupted during transmission). Suppose that initially the encoder is in the zero state and that the source produces the sequence $\mathbf{u} = 1\ 1\ 0\ 1$, which is then encoded by the rate 1/2 convolutional encoder shown in Fig. 12.6. The codeword associated with $\mathbf{u}$ is the sequence $\mathbf{c} = 11\ 01\ 01\ 00$. Suppose that the received sequence $\mathbf{r}$ contains one erroneous bit, underlined in the sequence $\mathbf{r} = 11\ 01\ \underline{1}1\ 00$. Note that since the free distance of this convolutional code is 5, a number of $t = \left\lfloor \frac{d_f - 1}{2} \right\rfloor = 2$ bit errors can be corrected.

Considering the trellis diagram in Fig. 12.9 extended over a number of stages equal to the number of transmitted bits, i.e. four stages, and assuming that the trellis starts from the zero state, the Viterbi algorithm searches for the sequence closest (in terms of Hamming distance) to $\mathbf{r}$ in the set of all sequences departing from the zero state at time t_1 and extending over six stages.

The simplified trellis highlighting the metrics evaluated in each trellis stage is depicted in Fig. 12.12, where t_l indicates the time instant at which the transition

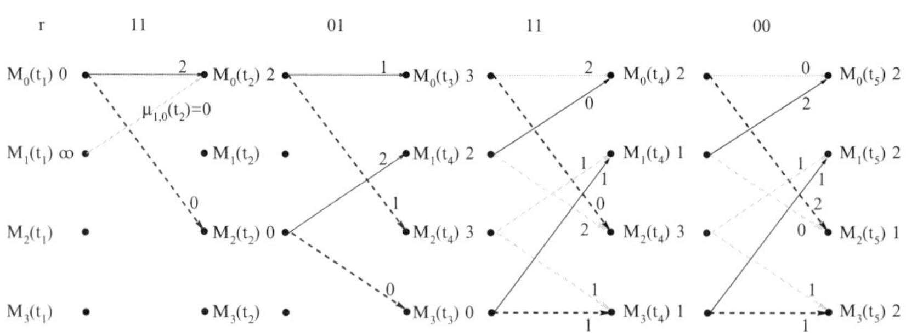

Fig. 12.12 Trellis diagram of the convolutional encoder in Fig. 12.6 described by the generator matrix $\mathbf{G}(D) = \left[1 + D + D^2,\ 1 + D^2\right]$

occurs. At the time instant t_1, the code symbols 11 are received, while the encoder is in the zero state. Each state is associated with a metric $M_i(t_1)$ with $i \in \{0,\ldots,2^{k(K-1)} = 2^2 - 1\}$ in the initial decoding step. Since the encoder starts in the zero state, $M_0(t_1) = 0$ while the other metrics can be set to ∞, meaning that these states are never reached at time t_1.

Given the underlying code symbols, from the zero state only two transitions are allowed, namely $S_0 \rightarrow S_0$ and $S_0 \rightarrow S_2$. The code symbols associated with the $S_0 \rightarrow S_0$ transition are 00, while the ones related to the state transition $S_0 \rightarrow S_2$ are 11. For each allowed transition $S_j \rightarrow S_i$, the path metric $\mu_{j,i}(t_2)$ is evaluated as the Hamming distance between the received code symbols and the ones associated to the transition $S_j \rightarrow S_i$.
The state metrics $M_i(t_2)$, $\forall i = 0,\ldots,3$ at the time instant t_2 are each evaluated as the state metrics $M_i(t_1)$ in the previous time instant plus the minimum path metrics $\mu_{j,i}(t_2)$ associated with all the transitions converging to the underlying state i

$$M_i(t_2) = \min_j \left\{ M_j(t_1) + \mu_{j,i}(t_2) \right\}, \quad \forall i = 0,\ldots,3,$$

whereby the index j takes on the state indexes which converge in to state i in t_2. As an example, the metric $M_0(t_2)$ associated with the zero state at the time instant t_2 in Fig. 12.12 can be evaluated as follows:

$$M_0(t_2) = \min\{M_0(t_1) + \mu_{0,0}(t_2), M_1(t_1) + \mu_{1,0}(t_2)\} = \min\{0+2, \infty+0\} = 2.$$

Let us assume to be in the time instant t_3 whereby the code symbols 11 are received, while the accumulated path metrics associated with each state are, as noted in Fig. 12.12, $M_0(t_3) = 3, M_1(t_3) = 2, M_2(t_3) = 3, M_3(t_3) = 0$. Upon evaluating the Hamming distances between the received code symbols 11 and all the transitions between states, the following accumulated state metrics at the time instant t_4 will result:

$$\begin{aligned}
M_0(t_4) &= \min\{M_0(t_3) + \mu_{0,0}(t_4), M_1(t_3) + \mu_{1,0}(t_4)\} = \min\{5,2\} = 2,\\
M_1(t_4) &= \min\{M_2(t_3) + \mu_{2,1}(t_4), M_3(t_3) + \mu_{3,1}(t_4)\} = \min\{4,1\} = 1,\\
M_2(t_4) &= \min\{M_0(t_3) + \mu_{0,2}(t_4), M_1(t_3) + \mu_{1,2}(t_4)\} = \min\{3,4\} = 3,\\
M_3(t_4) &= \min\{M_2(t_3) + \mu_{2,3}(t_4), M_3(t_3) + \mu_{3,3}(t_4)\} = \min\{4,1\} = 1.
\end{aligned}$$

Upon proceeding in this way for each trellis stage while retaining only the paths yielding the minimum accumulated Hamming distance between the received sequence and the paths allowed within the trellis, at the time instant t_5 the state accumulating the minimum metric is S_2. Traveling the trellis from the zero state at the time instant t_1 and retaining the path accumulating the minimum Hamming distance at each trellis stage, the decoded sequence 11 01 01 00 (see Fig. 12.13) is obtained by reading the code symbols associated with the chosen transitions. Finally, the estimated information sequence $\widehat{\mathbf{u}} = 1\ 1\ 0\ 1$ is obtained upon reading the input bits producing the state transitions yielding the decoded sequence 11 01 01 00.

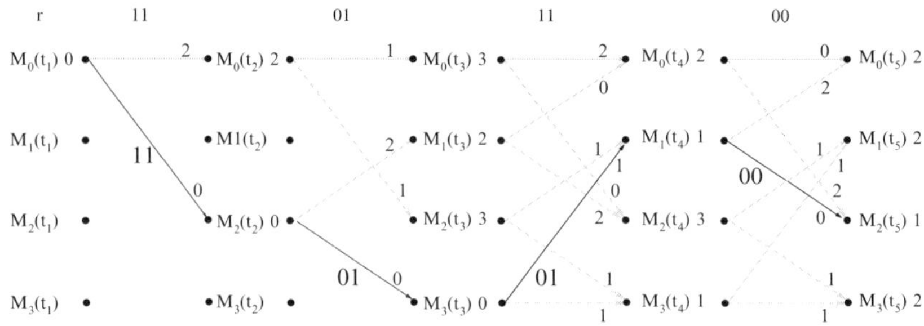

Fig. 12.13 Decoded sequence 11 01 01 00 in the trellis diagram of the convolutional encoder in Fig. 12.6 described by the generator matrix $\mathbf{G}(D) = \left[1+D+D^2,\ 1+D^2\right]$

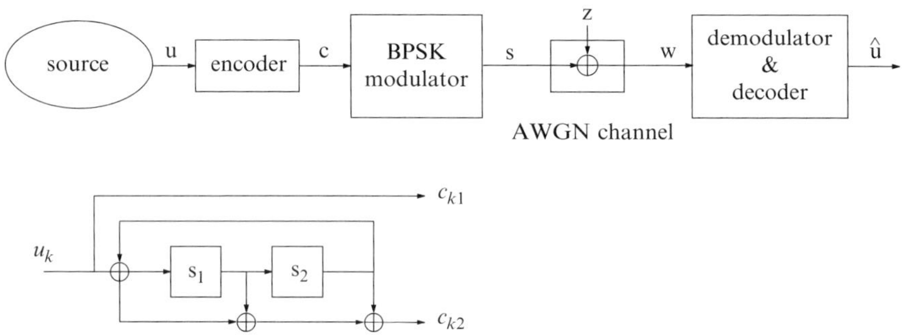

Fig. 12.14 Model of a coded transceiver system for the memoryless AWGN channel

Viterbi decoding for AWGN channels is accomplished in a similar fashion as discussed in connection to BSC channels, with the only difference that the distances between symbols are Euclidean distances, as discussed in Sect. 12.1.2.

12.2.3.2 Maximum A - Posteriori Probability Decoding: The BCJR Algorithm

The Viterbi algorithm, described in Sect. 12.2.3.1, is a maximum-likelihood decoding algorithm which minimizes the word error probability of the given convolutional codes.

An optimal algorithm which minimizes the bit error probability is the Bahl–Cocke–Jelinek–Raviv (BCJR) algorithm proposed in [8], whose main application is in connection with the iterative decoding of concatenated convolutional codes. As a reference scheme for the derivations which follow, consider the architecture shown in Fig. 12.14, and assume a rate $R_c = 1/2$ systematic convolutional encoder with constraint length K and memory $\nu = 2$.

The information sequence $\mathbf{u}$ (with length K_I) is encoded by producing a coded sequence $\mathbf{c}$ of length N. Each source bit u_k is encoded in the coded symbols c_{k1}, c_{k2}, whereby $c_{k1} = u_k, \forall k$. After BPSK modulation, the sequence $\mathbf{s}$ is transmitted over a memoryless AWGN channel which corrupts each modulated symbol $s_i \in \mathbf{s}, \forall i = 1,\ldots,N$ with a zero-mean Gaussian random variable with variance $\sigma_z^2 = N_o/2$. The received sequence $\mathbf{w}$ is passed to the channel decoder.

With this setup, the BCJR algorithm [8] evaluates the A-Posteriori Probabilities (APPs) of the decoded bits $\widehat{\mathbf{u}}$ by employing the joint probability $\Lambda_k^b(m)$ defined as follows [9]

$$\Lambda_k^b(m) = P\{u_k = b, S_k = m|\mathbf{w}\}, \tag{12.20}$$

whereby u_k is the kth decoded bit, $b \in \{0,1\}$, S_k is the state in the trellis at the time k and $\mathbf{w}$ is the received sequence. Given $\Lambda_k^b(m)$, the APP related to the kth decoded bit can be written as follows:

$$P\{u_k = b|\mathbf{w}\} = \sum_m \Lambda_k^b(m), \ \forall b \in \{0,1\}. \tag{12.21}$$

Employing (12.21), the following Log-Likelihood Ratio (LLR) associated with the kth decoded bit can be obtained:

$$L(\widehat{u}_k) = \log\left(\frac{P\{u_k = 1|\mathbf{w}\}}{P\{u_k = 0|\mathbf{w}\}}\right) = \log\left(\frac{\sum_m \Lambda_k^1(m)}{\sum_m \Lambda_k^0(m)}\right). \tag{12.22}$$

The decoder takes a decision on $\widehat{u}_k$ based on the sign of $L(\widehat{u}_k)$:

$$\widehat{u}_k = \begin{cases} 1, & \text{if } L(\widehat{u}_k) \geq 0 \\ 0, & \text{if } L(\widehat{u}_k) < 0, \end{cases} \tag{12.23}$$

which is a compact way to implement the decision MAP rule

$$\frac{P\{u_k = 1|\mathbf{w}\}}{P\{u_k = 0|\mathbf{w}\}} \gtrless 1.$$

Let us analyze the expression of the probabilities $\Lambda_k^0(m)$ and $\Lambda_k^1(m)$ in (12.22). Let I be the length in bits of the binary information sequence $\mathbf{u}$, and assume that the trellis starts and ends in the zero state. Bit u_k corresponds to the transition between the $(k-1)$th and the kth stage in the trellis. Let i and j be the indexes identifying the states in the trellis stages $(k-1)$ and k, respectively. With this setup, (12.22) simplifies to [10]

$$L(\widehat{u}_k) = \log\left(\frac{\sum_{(i,j),u_k=+1} P\{i,j,\mathbf{w}\}}{\sum_{(i,j),u_k=-1} P\{i,j,\mathbf{w}\}}\right). \tag{12.24}$$

The summations in (12.24) are over the transitions in the trellis between stages $(k-1)$ and k associated with bits $u_k = +1$ and $u_k = -1$, respectively. For a memoryless channel like the considered AWGN channel, such joint probabilities can be evaluated as follows [8]

$$P\{i,j,\mathbf{w}\} = P\{i,\mathbf{w}_{t<k}\} \cdot P\{j|i\} P\{\mathbf{w}_k|i,j\} \cdot P\{\mathbf{w}_{t>k}|j\},$$

whereby $\mathbf{w}_{t<k}$ identifies the received sequence from the first stage up to the trellis stage $(k-1)$, $\mathbf{w}_{t>k}$ denotes the received sequence from the $(k+1)$th stage up to the end of the trellis, and $\mathbf{w}_k$ denotes the received symbols at the kth stage of the trellis. Upon setting

$$\begin{aligned} P\{i, \mathbf{w}_{t<k}\} &= \alpha_{k-1}(i), \\ P\{j|i\}P\{\mathbf{w}_k|i,j\} &= \gamma_k(i,j), \\ P\{\mathbf{w}_{t>k}|j\} &= \beta_k(j), \end{aligned}$$

the forward recursion corresponds to the evaluation of $\alpha_k(j)$:

$$\begin{aligned} \alpha_k(j) &= \textstyle\sum_i \gamma_k(i,j)\alpha_{k-1}(i), \\ \alpha_0(0) &= 1, \\ \alpha_0(j) &= 0, j \in \{1,\ldots,2^{k(K-1)}-1\}, \end{aligned} \tag{12.25}$$

while the backward recursion leads to the evaluation of $\beta_{k-1}(i)$:

$$\begin{aligned} \beta_{k-1}(i) &= \textstyle\sum_j \gamma_k(i,j)\beta_k(j), \\ \beta_N(0) &= 1, \\ \beta_N(i) &= 0, i \in \{1,\ldots,2^{k(K-1)}-1\}, \end{aligned} \tag{12.26}$$

whereby

$$\gamma_k(i,j) = P\{\mathbf{w}_k|u_k\}P\{u_k\} \tag{12.27}$$

if there exists a transition in the trellis between the states indexed by i and j.

Note that the initial conditions $\alpha_0(0) = 1$ and $\beta_N(0) = 1$ refer to the fact that at the beginning and at the end of the codeword, the trellis is in the zero state. The latter condition is guaranteed by appending an appropriate bit sequence, denoted as *terminating pattern*, at the end of the information sequence.

The probability $P\{u_k\}$, called a-priori probability of the bit u_k, can be evaluated upon resorting to log-likelihood formulation of the problem

$$P\{u_k = \pm 1\} = \frac{\mathrm{e}^{\pm L(u_k)}}{1 + \mathrm{e}^{\pm L(u_k)}},$$

which can also be rewritten as [10]:

$$P\{u_k = \pm 1\} = A_k \mathrm{e}^{\frac{L(u_k)u_k}{2}}.$$

As far as the evaluation of $P\{\mathbf{w}_k|u_k\}$ is concerned, for systematic convolutional codes the following relation holds:

$$P\{\mathbf{w}_k|u_k\} = B_k \cdot \mathrm{e}^{\frac{1}{2}L_c w_{k1} u_k + \frac{1}{2}\sum_{p=2}^{n} L_c w_{kp} c_{kp}}.$$

For a rate $1/2$ systematic encoder, the previous equation simplifies to:

$$P\{\mathbf{w}_k|u_k\} = B_k \cdot \mathrm{e}^{\frac{1}{2}L_c w_{k1} u_k + \frac{1}{2}L_c w_{k2} c_{k2}}.$$

The evaluation of (12.27) can be accomplished by using:

$$\begin{aligned}\gamma_k(i,j) = P\{\mathbf{w}_k|u_k\}P\{u_k\} &= B_k \cdot \mathrm{e}^{\frac{1}{2}L_c w_{k1}u_k + \frac{1}{2}L_c w_{k2}c_{k2}} A_k \mathrm{e}^{\frac{L(u_k)u_k}{2}} \\ &= A_k B_k \mathrm{e}^{\frac{1}{2}u_k(L(u_k)+L_c w_{k1})}\gamma_k^e(i,j),\end{aligned}$$

whereby

$$\gamma_k^e(i,j) = \mathrm{e}^{\frac{1}{2}L_c w_{k2}c_{k2}}.$$

Upon noting that terms A_k and B_k are equal for each transition in the trellis between the stages $(k-1)$ and k, it is straightforward to note that such terms will cancel out in the ratio in (12.24). Furthermore, the exponential function $\mathrm{e}^{\frac{1}{2}u_k(L(u_k)+L_c w_{k1})}$ is common to the terms involved in the summations noted in (12.24); therefore, (12.24) can be simplified as follows:

$$L(\widehat{u}_k) = L(u_k) + L_c w_{k1} + \log\left(\frac{\sum_{(i,j),u_k=+1}\gamma_k^e(i,j)\alpha_{k-1}(j)\beta_k(i)}{\sum_{(i,j),u_k=-1}\gamma_k^e(i,j)\alpha_{k-1}(j)\beta_k(i)}\right). \quad (12.28)$$

In summary, the MAP algorithm can be implemented by the following steps:

- Initialize $\alpha_0(j)$ and $\beta_N(i)$ with the initial conditions noted in (12.25) and (12.26), respectively
- Compute the forward recursion $\alpha_k(j)$ in (12.25) for any $k = 1,2,\ldots,K_I$
- Compute the backward recursion $\beta_k(i)$ in (12.26) for any $k = K_I, K_I-1,\ldots,2$
- For any $k = 1,2,\ldots,K_I$, compute $L(\widehat{u}_k)$ in (12.28)
- For any $k = 1,2,\ldots,K_I$, make a hard decision on $L(\widehat{u}_k)$ following the MAP rule discussed in (12.23)

12.2.4 Structural Properties of Convolutional Codes

In this section, we will present several properties of convolutional encoders.

12.2.4.1 The Concept of Equivalence Among Encoders

Two convolutional encoders are said to be *equivalent* when they generate the same code, i.e., the same set of output codewords. Equivalence between encoders turns out to be quite useful for generating minimal encoders, i.e., encoders featuring the minimum number of memory cells among the set of their equivalent encoders. The mathematical theory of equivalent encoders has been described by Forney in [11].

As a reference example, the encoders shown in Fig. 12.6 and 12.7c are equivalent in the following sense; upon erasing the input bit along with the state labels in the state diagram shown in Fig. 12.8 for the convolutional encoder in Fig. 12.6, we would obtain the same state diagram as the one related to the recursive and systematic convolutional encoder depicted in Fig. 12.7c. In this respect, the two convolutional encoders are equivalent. The only difference between them is the mapping between input sequences and output codewords, as established by the specific convolutional encoder adopted. Dividing the generator matrix $[1+D+D^2, 1+D^2]$ associated with the encoder in Fig. 12.6 by $1+D^2$, the following generator matrix

$$\left[\frac{1+D+D^2}{1+D^2}, 1\right]$$

is easily obtained, which also corresponds to the generator matrix of the recursive and systematic encoder in Fig. 12.7c.

Note that, while equivalent encoders yield quite similar performance with maximum-likelihood Viterbi decoding, they do behave differently when embedded as constituent encoders in both parallel and serially concatenated convolutional codes. In other words, the mapping between input and output sequences due to a convolutional encoder does matter in concatenated convolutional codes, as it will be shown in a successive section.

12.2.4.2 Catastrophic Encoder

Given an encoder characterized by the generator matrix $G(D)$, such an encoder is catastrophic if there exist an input sequence $\mathbf{u}(D)$, with $w_\mathrm{H}(\mathbf{u}) = \infty$, for which the output codeword $\mathbf{c}(D) = \mathbf{u}(D) \times \mathbf{G}(D)$ has finite Hamming weight, that is $w_\mathrm{H}(\mathbf{c}) < \infty$.

A sufficient condition for a convolutional encoder to be catastrophic can be deduced by the encoder state diagram. In particular, the presence of self-loops in any of the encoder states, generated by nonzero input sequences, producing a zero output sequence is a sufficient condition for the encoder to be catastrophic. As an example, consider the encoder with generator matrix given by [5]

$$\mathbf{G}(D) = \left[1+D\,,\ 1+D^2\right].$$

The codeword associated with the information sequence

$$\mathbf{U}(D) = \frac{1}{1+D} = \sum_{i=0}^{\infty} D^i$$

is

$$\mathbf{C}(D) = \mathbf{U}(D) \times \mathbf{G}(D) = [1\,,\ 1+D],$$

which is a sequence with Hamming weight equal to 3, even though the input sequence has an infinite weight. If three errors during the transmission over a BSC flip these three bits, a maximum-likelihood decoder would decode the received sequence as an all-zero sequence, implying an infinite number of decoding errors. For this reason, catastrophic encoders should be avoided since they cause catastrophic error propagation.

In [5] several properties needed for avoiding catastrophic encoders are presented. Catastrophic encoders are not minimal, therefore they require a number of memory cells greater than the one required by an equivalent minimal encoder. Furthermore, systematic encoders are never catastrophic since the input sequence always belongs to the output codeword.

12.2.5 Performance of Convolutional Codes with Maximum-Likelihood Decoding

As already discussed earlier, a maximum-likelihood decoder finds the codeword **c** closest in an appropriate sense (Hamming or Euclidean distance) to the received sequence **w**.

Performance of Convolutional Codes (CCs) depends on the distance properties of the code other than on the employed decoding algorithm. Owing to the code linearity along with the uniform error property of convolutional codes, distance properties can be evaluated by considering the all-zero sequence as reference transmitted sequence [5]. By doing so, upper bounds on the bit error probability can be evaluated upon resorting to the transfer function $T(I,W) = T(I,N,W)|_{N=1}$ of convolutional codes presented in Sect. 12.2.2.

Assuming that all zero sequence is transmitted, the word error probability of a convolutional code can be upper bounded by adding the probabilities associated with all paths diverging from the zero state and merging back into it at some later stage in the trellis [5]. As far as hard decoding is concerned, the transmission channel is a BSC characterized by the transition probability p (i.e., the receiver makes a hard one-zero decision about what was transmitted before passing this data to the decoder). The word error probability can be evaluated as follows

$$P_{\mathrm{w}}(e) \leq T(W)\Big|_{W=2\sqrt{p-p^2}},$$

where $T(W) = T(I,N,W)|_{I=N=1}$. For small values of p, the dominant term in $T(W)\Big|_{W=2\sqrt{p-p^2}}$ is the one corresponding to the code free distance, and the word error probability can be approximated as follows

$$P_{\mathrm{w}}(e) \approx m_{d_{\mathrm{f}}} \cdot \left(2\sqrt{p-p^2}\right)^{d_{\mathrm{f}}}, \tag{12.29}$$

whereby $m_{d_{\mathrm{f}}}$ is the number of codewords with Hamming weight equal to the free distance d_{f}.

As far as the bit error probability $P_{\mathrm{b}}(e)$ is concerned, the following upper bound holds [5]:

$$P_{\mathrm{b}}(e) \leq \frac{1}{k}\left[\frac{\partial T(I,W)}{\partial I}\right]\Big|_{I=1,W=2\sqrt{p-p^2}}. \tag{12.30}$$

The upper bounds on both word and bit error probabilities presented above can be easily extended to the AWGN channel with BPSK modulation. Upon substituting the relation (12.6) in (12.29) and (12.30), the word and bit error probabilities for BPSK modulation and AWGN channels easily follow.

As far as soft-decision Viterbi decoding is concerned (i.e., the receiver quantizes the received statistic and passes this to the decoder rather than making a hard one-zero decision, which may be viewed as an one-bit quantization of the received statistic), the bit error probability (also denoted Bit Error Rate or BER) of a convolutional code of rate $R_c = k/n$ with BPSK or QPSK modulation in Additive White Gaussian Noise, can be well upper bounded by the following expression [4]:

$$P_b \leq \frac{1}{k} \sum_{d=d_f}^{\infty} w_d Q\left(\sqrt{2\frac{E_b}{N_o} R_c d}\right), \tag{12.31}$$

in which d_f is the minimum nonzero Hamming distance of the CC, w_d is the cumulative Hamming weight associated with all the paths that diverge from the correct path in the trellis of the code, and re-emerge with it later and are at Hamming distance d from the correct path, and finally $Q(.)$ is the Gaussian integral function, defined as $Q(t_o) = \frac{1}{\sqrt{2\pi}} \cdot \int_{t_o}^{\infty} e^{-\frac{t^2}{2}} dt$.

12.2.5.1 Best Convolutional Codes

There is now a large body of literature addressing the design of good convolutional codes for maximum-likelihood decoding. Owing to the definition of the bound in (12.31), a classic approach to the design of good convolutional encoders of given rate R_c and constraint length K consists of finding the generator matrix that yields a code whose distance spectrum has the property of having the maximum minimum distance d_f. A better approach is to obtain the distance spectra of the codes and to select the one which minimizes the BER upper bound in (12.31) based on the first few terms of the distance spectra. The optimum code which leads to the best distance spectrum, may be selected as the best convolutional code, provided that it is not catastrophic.

The best known codes are listed in a number of books and papers. Among the best references in this topic, we cite [5]. The generator matrices of the best known rate $1/2$ convolutional encoders are presented in Table 12.1. The notation is the same adopted in (12.8). Parameters noted in Table 12.1 are as follows: K is the constraint length, d_f the free distance, while the generator matrix is identified by $G(D) = [g_1(D), g_2(D)]$.

Simulated bit error probabilities ($P_b(e)$) obtained with unquantized Viterbi decoding applied to the convolutional encoders summarized in Table 12.1, are depicted in Fig. 12.15. Probability P_b is derived against the signal-to-noise ratio E_b/N_o per information bit, and it refers to the transmission of BPSK modulated data over the continuous-output AWGN channel discussed in Sect. 12.1.2.
For reference, the figure also shows the bit error rate performance for an uncoded transmission along with the coding gain γ guaranteed by each encoder with respect

Table 12.1 Best rate 1/2 convolutional codes with generator matrix $G(D)$

K	d_{f}	$g_1(D)$	$g_2(D)$
3	5	$1+D^2$	$1+D+D^2$
4	6	$1+D+D^3$	$1+D+D^2+D^3$
5	7	$1+D+D^2+D^4$	$1+D^3+D^4$
6	8	$1+D+D^3+D^5$	$1+D+D^2+D^3+D^4+D^5$
7	10	$1+D+D^2+D^3+D^6$	$1+D^2+D^3+D^5+D^6$

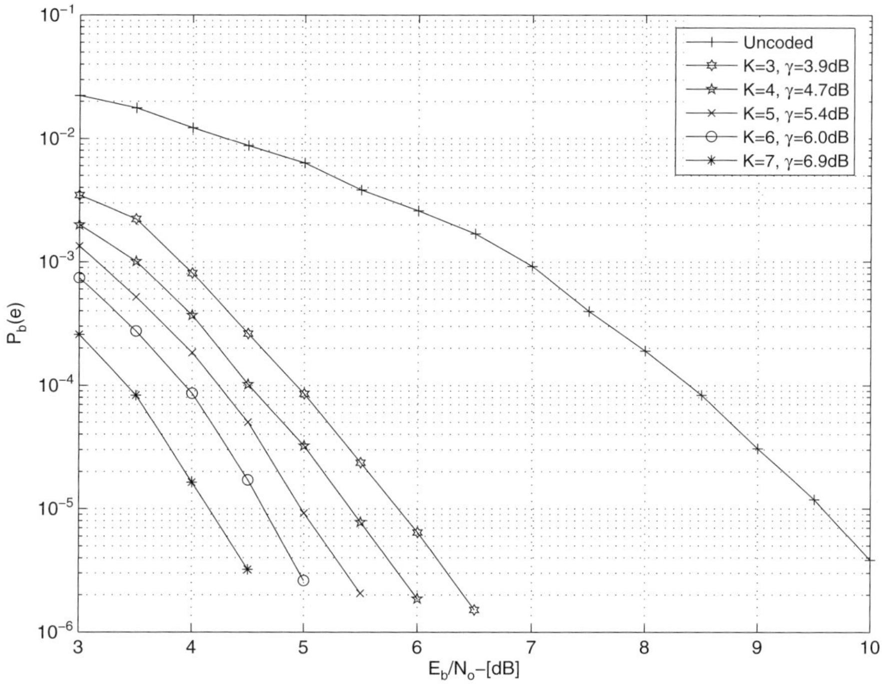

Fig. 12.15 Simulation results (bit error probability $P_b(e)$) with the Viterbi algorithm for the optimal rate 1/2 convolutional encoder shown in Table 12.1

to an encoded transmission at high signal-to-noise ratio. Such a coding gain is defined as follows [5]:

$$\gamma = 10 \times \log_{10}(R_c \times d_f) .$$

12.2.6 Punctured Convolutional Codes

For applications requiring high spectral efficiency, there is often a need for high rate codes that satisfy the system requirements in terms of the required Bit Error Rate (BER) or Frame Error Rate (FER) at a target Signal-to-Noise Ratio (SNR).

Furthermore, variable rate coding that can offer the possibility of providing different degrees of protection to the data is often required in system design. To this end, high rate punctured Convolutional Codes (CCs) are among the most commonly used codes for Forward Error Correction (FEC). Puncturing is the most commonly used technique to obtain high-rate CC since the trellis complexity of the overall code is the same as the lower rate mother code whose output is punctured [4]. Using puncturing, changing the code rate is equivalent to only changing the transmitted symbols and has no impact on the trellis structure of the code used at the receiver for decoding. Since the main complexity of a channel codec resides at the decoder, this solution offers a very flexible and cost effective method of supporting variable rate coding.

In 1979, Cain et al. in [12] suggested the use of a punctured code obtained from a mother code of lower rate, say $1/n$, for obtaining higher rate codes with simplified Viterbi decoding. Another approach to the design of punctured convolutional codes led to the development of the so called Rate-Compatible Punctured Convolutional (RCPC) codes introduced by Hagenauer in 1988 [13]. RCPC codes were essentially introduced because in some applications such as the transmission of compressed digital speech signals, information bits may require different degrees of error protection. In this scenario, the most important bits are transmitted with more redundancy by using the same convolutional code with variable redundancy, i.e., a code whose puncturing patterns as a function of increasing rate are subsets of each other.

A classic approach to the design of good punctured codes is analogous to the technique used for finding the best convolutional mother encoders. In other words, the technique consists of finding the puncturing pattern that yields a code whose distance spectrum has the property of having the maximum minimum distance d_{f}. A better approach is to obtain the distance spectra of the punctured codes and to select the one which minimizes the BER upper bound based on the first few terms of the distance spectra. The optimum code, which leads to the best distance spectrum, may be selected as the best punctured code provided that it is not catastrophic. We recall that the code is catastrophic if a finite number of channel errors can cause an infinite number of decoding errors. In terms of the state transition graph of a CC, this condition requires that this graph should not possess zero-weight cycles other than the self-loop associated with the zero state.

Another issue to consider is whether a recursive code remains recursive after puncturing. Indeed, when an Infinite Impulse Response (IIR) convolutional code is punctured, the resultant encoder is not necessarily recursive [14]. If the punctured CC has to be used in a parallel concatenated scheme, or as an inner code in a serially concatenated code, it must be recursive in order for the interleaver to yield a coding gain [15, 16]. This problem is addressed in [14], whereby a mathematical theory is developed for obtaining the closed-form representation of a punctured convolutional encoder generator.

Let us review some mathematical notation for punctured CC. Puncturing is obtained by regularly deleting some output bits of a mother code with rate $1/n$. As a result of puncturing, the trellis of the punctured code becomes periodically time varying. Consider a rate $1/2$ systematic mother code, identified by a 1×2 generator matrix $G(D) = \left[1, \frac{g_1(D)}{g_0(D)}\right]$ composed of two polynomials $g_1(D)$

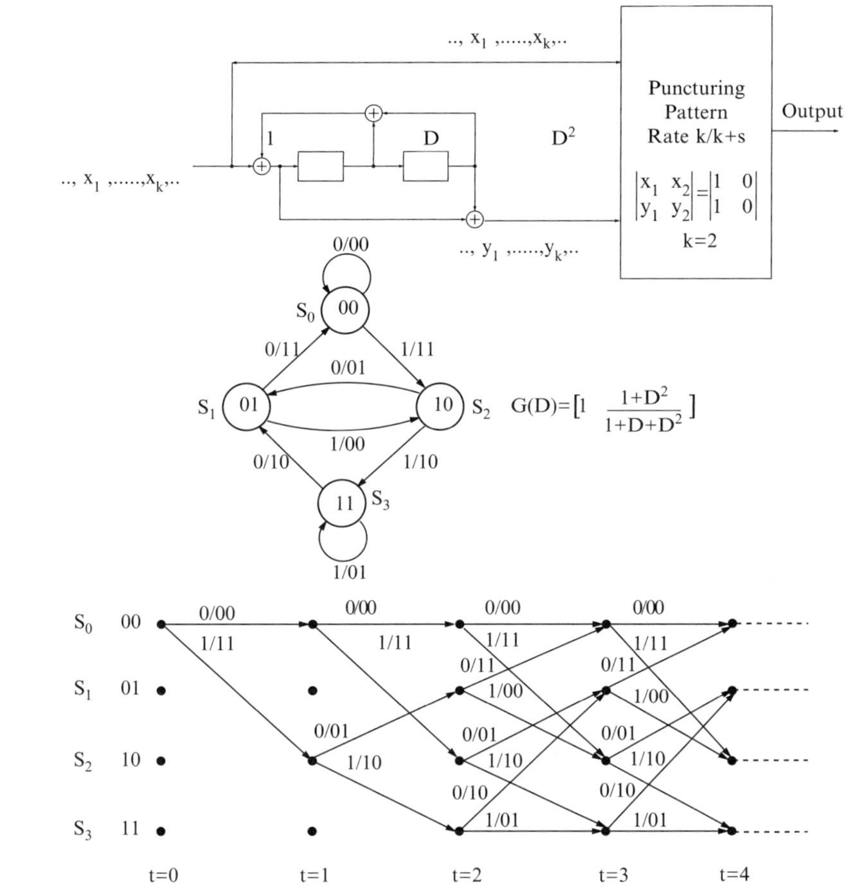

Fig. 12.16 Example of a four-state rate 1/2 mother CC and the puncturing mechanism

and $g_0(D)$ specifying the connections of the finite state encoder. In this formula, $g_i(D) = g_o^i + g_1^i D + \cdots + g_\nu^i D^\nu$, where $i = 0, 1$, $g_l^i \in \{0, 1\}$, and $l = 0, \ldots, \nu$. The variable ν specifies the memory of the convolutional code (the code constraint length is $K = 1 + \nu$). The form of the generator matrix $G(D)$ expresses the fact that the considered encoders are recursive (i.e., we have ratio of polynomials). As an example, Fig. 12.16 shows a 4-state, rate $1/2$, recursive systematic encoder with generator polynomials $g_1(D) = 1 + D^2$ and $g_0(D) = 1 + D + D^2$ with $\nu = 2$. The code symbols which are punctured are indicated through zeros in a suitable PP matrix, usually indicated as a $2 \times k$ matrix (for a rate-$1/2$ mother encoder) or as a sequence of length $2k$, $< x_1, y_1, x_2, y_2, \ldots, x_k, y_k >$, whereby x_i is the systematic bit and y_i the corresponding parity bit output from the encoder. To get a code rate $k/(k+s)$, input bits are grouped in blocks of k elements, and from the output of the encoder $2k - (k+s)$ bits are deleted. As an example, if the desired rate is $k/k+s = 3/5$, then $k = 3$ and $s = 2$.

Figure 12.16 shows the case in which a 4-state, rate-$1/2$ mother code is punctured in order to obtain a rate $2/3$ code. The trellis shown in the figure is the time-invariant trellis of the rate-$1/2$ mother code. The punctured code is obtained by deleting the first bit of every two input bits leading to the puncturing pattern $< x_1, y_1, x_2, y_2 >= 1101_2$.

The design of good puncturing patterns to be used in connection to maximum-likelihood (ML) decoding has been addressed in a number of papers. Tables of good puncturing patterns for various constituent code rates and constraint lengths can be found in [17–27].

12.3 Capacity-Achieving Codes

The theory of channel coding is a well-established technical area that had its origins in the seminal work of Shannon [1]. Theory as well as practical channel coding solutions are widely documented in a number of papers published in the past 50 years. However, most channel codes known so far have performance still far from the ultimate theoretical limits presented by Shannon (specially for short block lengths). In 1993, Berrou et al. [28] proposed a rate 1/2 concatenated architecture employing two equal recursive and systematic convolutional codes, capable of approaching the capacity limit of the AWGN channel within 0.7 dB at a target BER equal to 10^{-5}. Soon after this key discovery in coding theory, another class of block codes, namely, the Low-Density Parity-Check (LDPC) codes, actually discovered by Gallager in 1962 in connection with his PhD thesis [29], was proven to be capacity achieving by MacKay [30].

This section presents both turbo codes and LDPC codes by highlighting the key design techniques at the very basis of these capacity-achieving codes.

12.3.1 Parallel Concatenated Convolutional Codes (Turbo Codes)

Parallel Concatenated Convolutional Codes (PCCCs) or turbo codes are a powerful class of forward error correcting codes with a suboptimal iterative decoder that have a remarkable performance near the capacity limit [28, 31, 32]. There is now a large body of literature that addresses virtually every facet of the design, iterative decoding, application, and implementation of turbo codes.

An example two component PCCC composed of two identical Recursive Systematic Convolutional (RSC) codes coupled by an interleaver is depicted in Fig. 12.17. The puncturing matrix, which is optional, allows one to vary the PCCC rate by puncturing an appropriate set of coded bits. Since the constituent codes of the PCCCs at the beginning of the data block are usually in the all-zero state, they should also be properly terminated at the end of the data block so that the starting and terminal states of the encoder are known. The resulting terminated code resembles a large

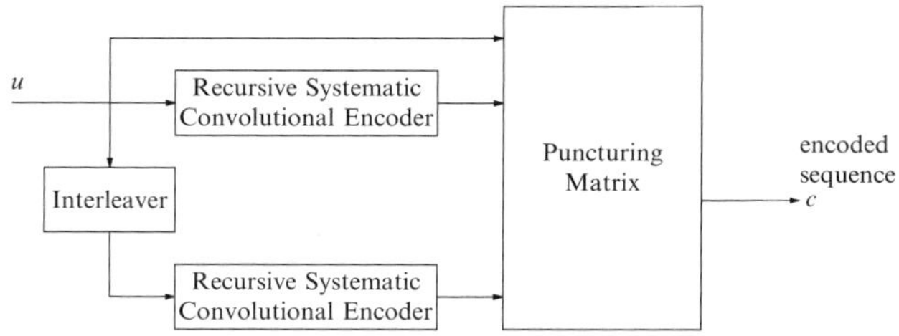

Fig. 12.17 Example of a rate 1/3 parallel concatenated convolutional code employing two identical RSC codes

block code. For trellis termination of the RSC constituent codes, various techniques have been proposed in literature so far. Perhaps, the most used is [32].

The BER and FER of a PCCC employing iterative soft decoding as a function of the Signal-to-Noise Ratio (SNR) is characterized by two regions. At low values of SNR the interleaving gain of the code dominates and the impact of the interleaver is a reduction of the codeword multiplicities that results in a rapid reduction of the BER and FER with increasing SNR. Soon however, an error floor region is reached marked by a change in the rate of decay of BER and FER with increasing SNR. In the error floor region, the asymptotic performance of the code is dominated by its distance spectrum and the code performance resembles that of a large block code. Hence, union bounding techniques can be used to provide a reasonably accurate estimate of the code's performance. In reality, performance of an iterative soft decoder can approach that of a ML decoder but can never reach it. Experimental evidence suggests [15, 33] that the performance of iterative soft decoders approaches that of Maximum-Likelihood (ML) decoding, essentially characterized by the asymptotic BER of the code obtained using bounding techniques, to within $\sim$0.5 dB. Obviously it is impossible to know precisely what the performance of the ML decoder would be since it is not possible to practically perform ML decoding of most turbo codes.

12.3.1.1 Design of Turbo Codes

The design of turbo codes have essentially been conducted as a two phase process. In the first phase, the constituent RSC codes in the configuration are optimized based on the use of a uniform interleaver [31], while in the second phase the interleaver itself can be optimized (or selected based on some criterion) for a given pair of constituent codes. The interleaver in the parallel concatenation has been proved to be essential for guaranteeing low error floors especially at high SNR values.

12.3.1.2 Design of Constituent Encoders in PCCCs

The constituent convolutional encoders in PCCCs must be recursive and systematic in order for the interleaver to yield a gain [16]. Furthermore, in PCCCs the dominant error events yielding the lowest terms of the distance spectra are due to input patterns with weight-2, especially for large interleaver sizes. In fact, it is known that the performance of the PCCC with large interleavers [34] for moderate-to-high SNR can be expressed as:

$$P_b \approx \frac{1}{N} Q\left(\sqrt{2\frac{E_b}{N_o} R_c d_2^{TC}}\right), \tag{12.32}$$

where R_c is the code rate of the PCCC, N is the interleaver length, d_2^{TC} is the effective free distance of the PCCC, i.e., the minimum weight of the PCCC codewords due to weight-2 error events, and $Q(.)$ is the Gaussian integral function, defined as

$$Q(t_o) = \frac{1}{\sqrt{2\pi}} \cdot \int_{t_o}^{\infty} e^{-\frac{t^2}{2}} dt.$$

For this reason, a good criteria for obtaining both good mother encoders as well as Puncturing Patterns (PPs) in a PCCC consists of maximizing the effective distance d_2^{TC}. Given a pair of RSC codes in the parallel concatenation, d_2^{TC} can be defined as follows [34]:

$$d_2^{TC} = 2 + \sum_{j=1}^{2} d_{j,2}^{p},$$

whereby $d_{j,2}^{p}$ is the minimum parity-weight of the codewords at the output of the jth RSC code in the PCCC, while the constant 2 is due to the fact that the RSC encoders are systematic. The notation above is shown in the PCCC sample depicted in Fig. 12.18.

We note that the maximum effective distance d_2 achievable with a recursive systematic rate 1/2 encoder with generator matrix $G(D) = \left[1, \frac{g_1(D)}{g_o(D)}\right]$ can be obtained from [34]:

$$d_2 \leq 4 + 2^{\nu-1}, \tag{12.33}$$

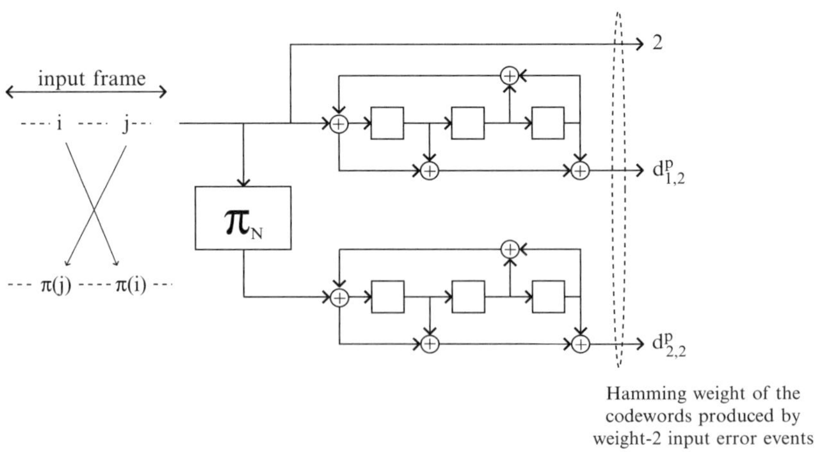

Fig. 12.18 Example of a rate 1/3 parallel concatenated convolutional code employing two identical RSC codes with polynomial generators $G(D) = \left[1, \frac{1+D+D^3}{1+D^2+D^3}\right]$

where ν is the code memory (the constraint length of the RSC code is $\nu+1$). In particular, equality is achieved when the denominator polynomial $g_o(D)$ is primitive and under two additional conditions which require that $g_o(d) \neq g_1(D)$ and that $\deg[g_i(D)] \leq \nu$ for $i=0,1$. The search for good mother encoders having highest d_2 can be conducted by considering the above conditions on the polynomials associated with $G(D)$.

In connection with the use of punctured encoders in a PCCC, the possible puncturing strategies can be different. Due to the complexity of finding good puncturing patterns for the overall PCCC codewords [35], a viable solution is to use punctured constituent encoders. For example, in reference to the general scheme of a PCCC, rate $k/k+1$ recursive systematic encoder can be used as the upper encoder of the PCCC, whereas the lower encoder can be punctured so that the systematic bits and some of its parity bits are completely eliminated in order to achieve the desired rate for the overall PCCC.

In summary, for the design of both systematic mother encoders and PPs a suitable objective function consists in the maximization of the effective distance d_2. Tables of good convolutional codes as well as puncturing patterns to be used for construction of PCCCs have been presented in [36–39].

12.3.1.3 Interleaver Design Techniques for PCCCs

The interleaver in the parallel concatenation has the effect of permuting the block of N bits at the input in such a way as to counteract the presence of error events yielding the lowest weight PCCC codewords.

So far, a wide variety of interleaver design techniques have been proposed in the literature (see for instance [40–58]). Among the many proposed techniques, there are a few that on the average have resulted in good turbo codes.

- Distance Spectrum Optimized (DSO) interleavers obtained via the application of the Interleaver Growth Algorithm (IGA) [45] and their delay constrained variants [40]. The resulting interleavers induce a quasirandom mapping but are obviously deterministically obtained and often lead to excellent results.
- S-Random (SR) or spread interleavers [43] and their variants [44] (the New S-Random (NSR), and Dithered-Diagonal (DD) interleavers) are a *class* of permutations which on the average result in much better performance than purely random interleavers [33]. We have emphasized the term "class" since the method of their construction does not lead to a unique permutation. Hence, a selection criterion must afterward be applied to select a member of the class with desirable properties.
- Interleavers designed based on a criterion of minimization of correlation coefficients of the extrinsic information and noise in iterative soft decoding [41]. We denote this class of interleavers as the Extrinsic Information Correlation Optimized (EICO) interleavers.

As a reference scheme for the techniques presented in the following, consider the PCCC shown in Fig. 12.18, whereby the block labeled Π_N identifies an interleaver of size N, while the input frame represents a sequence of information bits of length N at the input of both the upper RSC encoder and the employed permutation.

Let us discuss the main techniques for the design of the interleavers summarized above.

12.3.1.4 Design of S-Random Interleavers

The S-random interleaver [43] is a random permutation constructed as follows: each randomly selected integer is compared to the s previously selected integers. If the current selection equals any of the previous s selections within a distance of $\pm s$, the current selection is rejected. The process is repeated until a permutation of desired length is constructed. Note that there is no guarantee that the resulting construction will be successful. The spread of such an interleaver is defined as follows. Consider a window of size $W \leq s$. Let this window slide over the permutation vector and find the absolute minimum difference between the elements falling within the window, for all allowed window sizes and all window positions. The resulting number is defined as the spread of the permutation. Briefly, the goal of the spread interleaver is to separate out as much as possible the input bits that are close to each other, at the output of the interleaver. In particular, the spread S_o is defined as

$$
\begin{aligned}
S'(s,i,j) &= |\pi(i) - \pi(j)|, \ \forall\, |i-j| \leq s, \\
S_o &= \min_{i,j} S'(s,i,j).
\end{aligned}
\tag{12.34}
$$

It can be shown that for an interleaver of length N, the maximum achievable spread is $\lfloor \sqrt{N} \rfloor$.

12.3.1.5 Design of New S-Random Interleavers

The new S-random interleaver [44] uses the same concept as the SR permutation but with the recognition that for large separation between the input bits to the interleaver, the interleaver output bits need not have high separation (i.e., obviously there is a tradeoff between separation at the input and the output of the interleaver). The goal is to maximize the sum of the separation at the input and the output. The resulting spread, which is the minimum of the sum of the separation between input and output bits within a sliding window, is used as the new definition of spread. In particular, the new spread S_{new} is defined as

$$
\begin{aligned}
S_n'(i,j) &= |\pi(i) - \pi(j)| + |i-j| \\
S_{\text{new}} &= \min_{i,j} S_n'(i,j).
\end{aligned}
\tag{12.35}
$$

Upon using this technique, it can be shown that for an interleaver of length N, the maximum achievable spread is $\lfloor \sqrt{2N} \rfloor$.

12.3.1.6 Design of Dithered-Diagonal Interleavers

The dithered-diagonal interleaver [44] has the same goal of maximizing the spread based on the new definition, but uses a completely different approach for construction. The design is based on the following steps. In the first step, a rectangular grid of points with a spacing of $\sqrt{N}$ is generated. The extent of this grid needs to be a little larger than the enclosure used for defining the plot of the permutation points. In the second step, an entire row or column of points, but not both, is dithered by adding a random offset in the range $[-0.5\sqrt{N}, 0.5\sqrt{N})$. This is repeated for every row or column by adding offsets independently. In the third step, the grid is rotated by $45°$. The fourth step is to enclose N of these points inside a square with sides of length N. The coordinates of these N points represent the real read and write addresses of the interleaver. These real coordinate points are sorted and the sort order is used to define the interleaver.

12.3.1.7 Design of EICO Interleavers

The design of EICO interleavers is based on the correlation properties of the extrinsic information exchanged by the two SISO decoders employed in the decoding of PCCCs. In [41] the correlation coefficient of the extrinsic information and noise is modeled as an exponentially decaying function of the separation between the symbols.

12.3.1.8 Design of Interleavers Based on the Iterative Growth Algorithm

This section summarizes the main results of [45] on the IGA. For the sake of brevity, we shall keep the description of the IGA to the minimum, and invite the interested reader to review [45] for many theoretical details that are omitted here. The basic understanding of the IGA presented below requires representation of the permutations using their transposition vectors introduced in [45]. It is the transposition vector of a permutation and not directly the permutation itself that is iteratively grown to the desired size.

Transposition Vector of a Permutation

Consider an indexed set of elements $x_1, x_2, x_3, \ldots, x_N$. A given interleaver performs a particular permutation of this set of elements. The permutation π acts on the indices of the elements. Henceforth, the notation $\pi(i) = j$ is used to mean that the jth input symbol is carried to the ith position at the output. It is a basic result in group theory [59] that any permutation π on a set of elements S can be written as a product of disjoint cycles, and S may be divided into *disjoint subsets* such that each cycle operates on a different subset. A cycle of length two is called a *transposition*. It is easy to verify that any finite cycle can be written as a product of transpositions. Hence, we conclude that transpositions represent the elementary constituents of *any permutation*.

The Finite State Permuter (FSP) introduced in [45], is a realization of an interleaver in the form of a sliding window transposition box of fixed length equal to the delay of the permutation it effectuates on its input sequence, and with the property that the transposition performed at a given time slot is responsible for the generation of the output at the same time slot. The operation of the FSP can be understood by thinking of the sliding window as a queue. To generate any possible sequence of outputs, it is sufficient to exchange the head of the queue with the element that is to be ejected at that time slot.

Any permutation on a finite set of elements can be represented using a *unique transposition vector* associated with its FSP realization. As an example consider the permutation

$$\pi_1 = \begin{pmatrix} 1\ 2\ 3\ 4\ 5 \\ 4\ 3\ 1\ 2\ 5 \end{pmatrix}.$$

Consider the queue model of the FSP and assume that data enters from left to right. Let us label the transpositions to be performed sequentially with the head of the queue to generate the desired outputs, using positive integers. Let the integer 1 denote the case whereby no transposition is performed with the head of the queue and the element at the head of the queue is simply ejected. Then the transposition vector (to be read from left to right) that fully defines permutation π_1 is $T_\pi = (4,2,2,1,1)$. Take the binary sequence 10110 labeled from left to right. Permutation π_1 maps this sequence to 00111. Consider a new transposition vector $T_{\pi_2} = (3, T_\pi) = (3,4,2,2,1,1)$ associated with the permutation

$$\pi_2 = \begin{pmatrix} 1\ 2\ 3\ 4\ 5\ 6 \\ 3\ 5\ 4\ 2\ 1\ 6 \end{pmatrix}.$$

Note that π_2 looks quite different from π_1, yet its output corresponding to the *shifted* sequence 010110 is 001110 which is a shifted version of the output generated by π_1. In essence, the descriptions of π_1 and π_2 using the transposition vectors preserves information about the *prefix symbol substitution* property of these two permutations on error patterns whereby the first transposition exchanges a zero with a zero, or a one with a one.

Any permutation on N elements uniquely defines a transposition vector of size N. Conversely, any transposition vector of size N, defines a unique permutation on N elements. Note that when synthesizing a permutation using the transposition vector T, the kth element of vector T, when scanned from right to left, can only assume values in the set $\{1,2,\ldots,k\}$. In the interleaver design algorithm presented in [45], the interleaver is grown based on the minimization of a cost function related to the asymptotic BER (or FER) performance of the overall code. Hence, we shall first introduce this cost function, and then present the interleaver growth algorithm.

Cost Function and the Interleaver Growth Algorithm

Let E be a terminating error pattern of length L_{E} and Hamming weight $W_{\mathrm{H}}(E)$ associated with the RSC code in a PCCC configuration. By a terminating error pattern we mean a binary sequence at the input of the RSC encoder that corresponds to a path in the trellis of the code diverging and reemerging with the all-zero path. As an example, $E = 1001$ is one such sequence for the turbo code presented in Fig. 12.17. Let the interleaver length be $(M+1)$, and let $E_{n,(M+1)}$ denote the nth phase of error pattern E for $1 \leq n \leq (M - L_{\mathrm{E}} + 2)$ that can appear within the interleaver span of length $(M+1)$. For instance, for $(M+1) = 10$, $E_{1,10} = 0000001001, E_{2,10} = 0000010010, \ldots, E_{7,10} = 1001000000$. Then all the phases of E represent terminating *error events* prior to permutation and all have equal probability of occurring within the interleaver span.

Let T_{M+1} denote a transposition vector of length $(M+1)$. Let the elementary cost function for the interleaver design based on just one phase of one error event be denoted by $f(T_{M+1}, E_{n,(M+1)})$, where the notation used implies that $f(.,.)$ is a function of the transposition vector T_{M+1} and error pattern phase $E_{n,(M+1)}$, both of which are vectors whose size grows with M. We use the notation $\hat{E}_{n,(M+1)} = T_{M+1}E_{n,(M+1)}$ to mean that the transposition vector T_{M+1} permutes the components of $E_{n,(M+1)}$ generating an error pattern $\hat{E}_{n,(M+1)}$ with the same weight. Let $Y(\hat{E}_{n,(M+1)})$ denote the RSC encoder output sequence corresponding to the input $\hat{E}_{n,(M+1)}$, and set $d(\hat{E}_{n,(M+1)}) = W_{\mathrm{H}}(Y(\hat{E}_{n,(M+1)}))$, where $W_{\mathrm{H}}(\cdot)$ is the Hamming weight of the argument. Here, $d(\cdot)$ corresponds to the Hamming distance from the all-zero path of the path in the trellis of the RSC code corresponding to the input $\hat{E}_{n,(M+1)}$. We similarly define $d(E_{n,(M+1)}) = W_{\mathrm{H}}(Y(E_{n,(M+1)}))$ as the Hamming weight of the output of the RSC encoder receiving the unpermuted input. A suitable elementary cost function directly related to the upperbound on pairwise error event probability of the code is:

$$f\left(T_{M+1}, E_{n,(M+1)}\right) = \frac{W_{\mathrm{H}}(E)}{2(M+1)} \times \operatorname{erfc}\left(\sqrt{\frac{R_{\mathrm{c}}E_{\mathrm{b}}}{N_o}H(d_t)}\right) \tag{12.36}$$

where $R_{\mathrm{c}}E_{\mathrm{b}}/N_o$ is a constant related to the code rate R_{c} and the Signal-to-Noise Ratio (SNR) E_{b}/N_o, and $H(d_t)$ is defined as:

$$H(d_t) = \left(d(\hat{E}_{n,(M+1)}) + d(E_{n,(M+1)}) + W_{\mathrm{H}}(E)\right).$$

This elementary cost function can be used for optimization of the asymptotic BER of the code. If the goal is optimization of the asymptotic FER, we can simply drop the factor $W_{\mathrm{H}}(E)/(M+1)$ from the above expression.

We assume that the overall cost function is additive in the phases of our error patterns. For a PCCC containing two identical RSC encoders and one interleaver [32], we define the total cost function as the sum of the cost function associated with the forward permutation denoted $C_{\mathrm{f}}(T_{M+1})$, and the cost function associated with the inverse permutation denoted $C_{\mathrm{r}}(T_{M+1})$. Hence, the global cost function using P error patterns is

$$\hat{C}(T_{M+1}) = \sum_{j=1}^{P} \sum_{n=1}^{M-L_{E^j}+2} \left[f\left(T_{M+1}, E^j_{n,(M+1)}\right) + f\left(T^{-1}_{M+1}, E^j_{n,(M+1)}\right)\right]. \quad (12.37)$$

The expression in (12.36) clearly resembles the upperbound to the asymptotic BER of the code. The operational steps of the IGA, which aims at iteratively minimizing this cost function using a greedy search paradigm, are:

1. Set the iteration index which corresponds to the initial size of the interleaver to N_1. Initialize the recursion by exhaustively solving for

$$T^o_{N_1} = \arg\min_{T_{N_1}} \hat{C}(T_{N_1}), \quad (12.38)$$

   for a manageable value of N_1. Set $T^*_{N_1} = T^o_{N_1}$
2. Let $T_{N_1+1} = (k, T^*_{N_1})$ and find k_{N_1} from

$$k_{N_1} = \arg\min_{k} \hat{C}(k, T^*_{N_1}), \quad (12.39)$$

   and form $T^*_{N_1+1} = (k_{N_1}, T^*_{N_1})$
3. Set $N_1 = N_1 + 1$ and iterate step-2 to grow the transposition vector to the desired size $N_{\max}$

We note the following: (1) using IGA one can neither establish the local nor global strict optimality of the solution at a given recursion step. To establish local optimality, one needs to define a notion of perturbation about a point in an appropriate-sized space. If one takes the definition of a local perturbation of a permutation of length N to be allowing the first element of the transposition vector to assume any of the possible values, while the rest of the transposition is fixed, then IGA is by definition locally optimal. This notion of local perturbation may indeed be justified in light of the prefix symbol substitution property referred to above. Under this notion of local perturbation, IGA is essentially a steepest decent type algorithm; (2) In many optimization algorithms using the greedy approach, the input vector is often of a fixed dimension. In IGA, the dimensionality of the input vectors (i.e., the transposition vectors) actually increases from one recursion to the next. Hence, the search is over an ever growing input space of higher and higher dimensions; (3) The core theoretical justification for the use of IGA are Theorems 3.1 and 3.2 of [45]. These theorems demonstrate that the sequence of cost functions with respect to randomization of the first transposition with the rest of the transposition intact is asymptotically a martingale [60], meaning that for a sufficiently large interleaver length, even randomly growing the transposition vector one element at a time is a fair process. IGA does much better in that it does not randomly pick the transposition value at a given recursion step, indeed the transposition that gives the best value of the cost function at that step is selected. This leads to a deterministic process where on the average the value of the cost function should decrease. One cannot say the resulting process is a supermartingale since there is no randomness involved anymore, but the minimum of the cost function at each step beats the average unless the variance is zero and hence it is anticipated that IGA should perform better than a pure martingale. This is as good as having a supermartingale. Indeed, simulation results exhibit an almost monotonic and rapid decrease in the value of the cost function as a function of the recursion step.

To conclude the section, we note the following comparison of the performance of the noted interleaver design techniques based on their general features and simulation studies:

1. The DSO interleavers obtained using IGA in general result in PCCCs with better distance spectra compared to the other design techniques reported in the literature [46]. Furthermore, once the algorithm is initialized, the IGA results in a unique solution, unlike the other design techniques referred to above which generally lead to a class of interleavers to choose from. The IGA by virtue of its recursive nature (i.e., an optimized interleaver of length $(N+1)$ is constructed from one of length N) leads to *implicitly prunable* optimized interleavers.
2. The interleavers designed using IGA are *tailored to the specific RSC codes* used in the construction of the PCCC. This is in sharp contrast to most other interleaver design techniques that are not per se tailored to the RSC codes of the construction, if not through a trial-and-test process.
3. IGA can just as easily be applied for optimization of the FER performance of the code rather than the BER performance (it suffices to change the cost function). This difference is important since the cost functions are fundamentally different resulting in different optimized solutions. An interleaver optimized under the BER criterion at a target SNR is generally not optimal under the FER criterion and vice versa.
4. The DSO interleavers lead to turbo codes with better *asymptotic performance*. In this sense, they are not constrained by the number of iterations performed in iterative decoding. However, at sufficiently high SNR, the DSO interleavers outperform the other interleavers noted above for the same number of iterations due to their superior distance properties (i.e., the relative performance of different techniques for a fixed number of iterations is obviously dependent on the operating point).
5. The best interleavers of the class of S-random interleavers have an overall performance that is reasonably good both in terms of their asymptotic performance (but not as good as the DSO interleavers) and in terms of their performance for a fixed number of decoding iterations.

A recent performance comparison of the DSO (with and without delay constraint) and EICO interleavers was conducted in [42] where it is noted that the two interleavers have very similar performance in the pre-error floor region for a fixed number of iterations. However, the DSO interleavers ultimately have a superior performance by virtue of their lower error floor.

The design of prunable interleavers for turbo codes employing variants of the interleaver growth algorithm has been addressed in [61–63], while [64] proposed a method combining a search for good component codes with interleaver design for optimal distance spectrum designed turbo codes.

The decoding algorithm for PCCCs relies on the BCJR algorithm presented in Sect. 12.2.3. The description of the decoder for PCCCs is beyond the scope of this chapter. Nevertheless, many papers in literature addressed the functionalities of such algorithm. The interested reader is referred to [65] for the necessary details.

A comprehensive reference covering any facet of the design of turbo codes is the special issue recently appeared in the Proceedings of the IEEE [66].

To conclude this section, we present a set of simulation results of the rate-$1/3$ PCCCs employing the four and eight state constituent codes described by the following generator matrices:

$$G_2(D) = \left[1, \frac{1+D^2}{1+D+D^2}\right], \tag{12.40}$$

$$G_3(D) = \left[1, \frac{1+D+D^3}{1+D^2+D^3}\right]. \tag{12.41}$$

The number of iterations of the BCJR algorithm is equal to 12. Frame Error Rate (FER) and Bit Error Rate (BER) of these codes are reported in Fig. 12.19 for interleaver lengths equal to 40, 80, 160, and 320 in order to highlight the performance improvement available with PCCCs as a function of the interleaver size N.

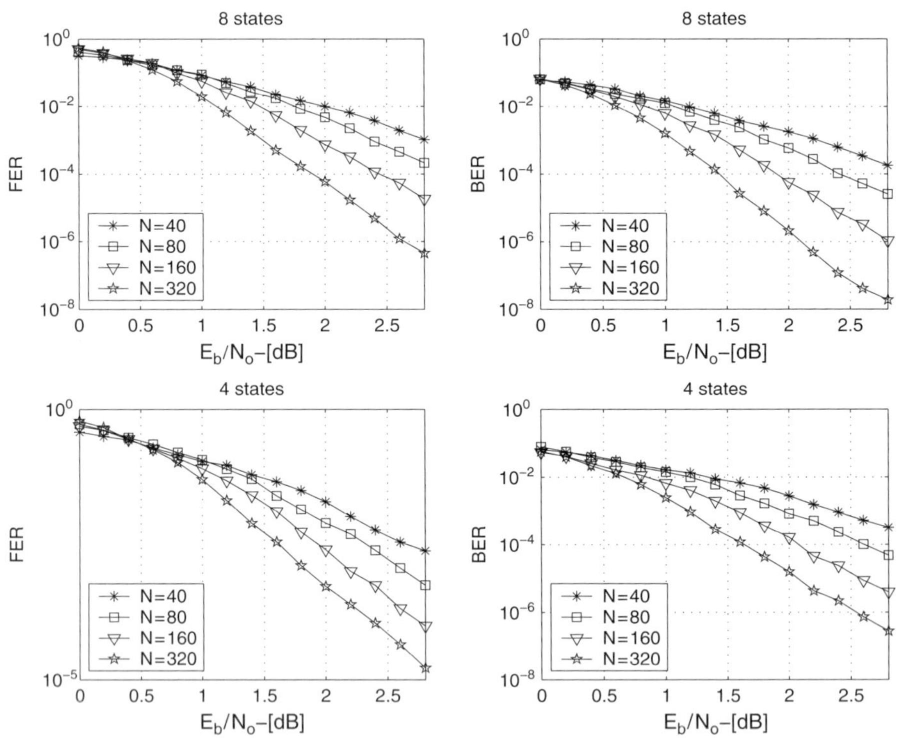

Fig. 12.19 Simulated FER and BER for the four-state and eight-state turbo codes employing the implicitly prunable optimized interleavers obtained using the modified IGA proposed in [61]. The interleaver length is denoted by N

12.3.1.9 Applications in Digital Broadcasting

The explosive growth of multimedia applications over the Internet and the ever-increasing users demands over commercial terrestrial digital multimedia broadcasting all over the world, calls for efficient physical and cross-layer techniques able to mitigate the potential problems limiting broadband services over wireless networks. In this scenario, mobile multimedia is expected to be one of the key services of future wireless mobile networks. Meanwhile, recent advances in digital communications have paved the way to a variety of standards aimed at providing multimedia services over terrestrial broadband networks. In this section, the focus is on the channel coding architectures foreseen in some key digital telecommunications standards.

Convolutional codes have been in use for decades as the standard for forward error correction in digital communication since the discovery of the Viterbi algorithm [6]. In a variety of applications, such codes are embedded in a concatenation with Reed-Solomon codes. However, the performance of this codes are still quite far from the ultimate limit presented by Shannon [1]. Soon after the discovery of turbo codes as well as LDPC codes, many standards have been renewed in order to embed such powerful codes.

The FEC architecture employed in the Digital Terrestrial Television Broadcasting (DTTB) standard in China [67], ratified in 2006, considers a concatenation between an outer BCH (762,752) code with an inner LDPC code presenting three different rates, as depicted in Fig. 12.20. The LDPC code blocks are noted in the figure with the notation (n,k), whereby n is the codeword size and k is the information block size.

The DTTB standard supports single-carrier as well as multi-carrier transmission, and it will guarantee high-definition broadcasting services.

The Terrestrial Digital Multimedia Broadcasting (T-DMB) system in Korea [68] guarantees CD quality audio broadcasting as well as data services devoted to fixed and mobile applications. Though mainly based on the European Digital Audio Broadcasting (DAB) system Eureka-147 DAB [69], the channel encoder employed in the Korean T-DMB system, is composed of a Reed–Solomon ($n = 204$ bytes, $k = 188$ bytes) code separated from a punctured convolutional encoder by a convolutional interleaver with depth $I = 12$. This scheme is needed for assuring high level of protection to the mobile reception of compressed video signals.

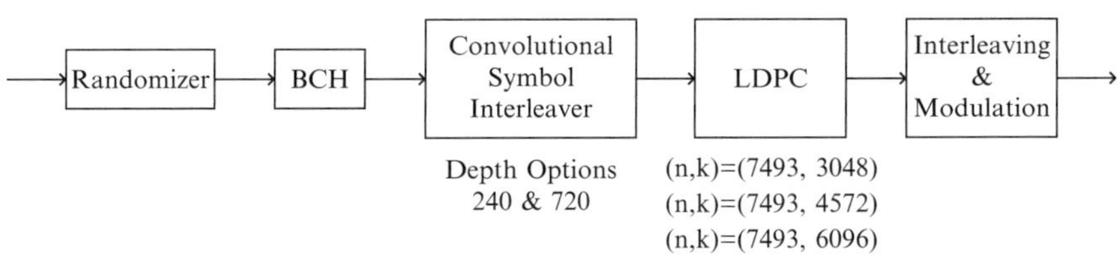

Fig. 12.20 FEC architecture employed in the DTTB standard in China

The serial concatenation between a Reed–Solomon code and a turbo code has been employed at the physical layer of the Forward Link Only (FLO) Air Interface [70]. The FLO Air Interface belongs to the MediaFLO system by QUALCOMM, and it has been proposed as an alternative mobile multicast technology to mobile devices employing TV.

The Reed–Solomon code has the following block sizes $(n = 16, k = 8, 12, 14, 16)$ bytes, whereby $k = 16$ is used to signify the fact that the block code is not used. The turbo code employed as inner code in the serial concatenation is constituted by the RSC codes with the following matrix generator:

$$G(D) = \left[1, \frac{1+D+D^3}{1+D^2+D^3}\right]. \tag{12.42}$$

Turbo codes have been adopted as the reference FEC scheme in a number of third generation standards. For instance, the turbo code embedding RSC encoders with generator matrix represented in (12.42) has been employed in W-CDMA, cdma2000, and TD-SCDMA [71]. Code performance in terms of both BER and FER has been presented in [71].

12.3.2 Low-Density Parity-Check Codes

Low-Density Parity-Check (LDPC) codes were discovered by Gallager [29] in 1962 but did not receive much attention at the time essentially for complexity reasons. In [72], Tanner resurrected LDPCs generalizing them through the introduction of the so-called Tanner graph. Only recently in 1999, Mackay [30] demonstrated that LDPCs when optimally decoded are capable of achieving information rates up to the Shannon limit.

As any linear block code, a generic binary LDPC is described in terms of its $m \times n$ parity-check matrix H whereby the rows of H span the null space of the codeword space. In case H is full-rank, it represents a linear code with rate $R = k/n$ and $m = n - k$, where m corresponds to the number of parity checks performed by the code.

An LDPC code is characterized by a parity-check matrix with a low density of 1s. When both the number n_c of 1s in any column of H and the number n_r of 1s in any row of H, is constant, then the LDPC is said to be regular. In this case, it follows that $n_r = n_c \cdot n/(n-k)$. Otherwise, the LDPC is said to be irregular. Not that, by virtue of the low density of 1s, it is $n_c \ll m$ and $n_r \ll n$. A rate R LDPC presents a number of 1s in each column equal to $n_c = (1-R)n_r$. For $R = 1/2$, in [73] it is shown that good irregular LDPC codes have an average number of 1s in each column equal to $n_c \approx 3.3$; this value assures a very low encoding complexity as well as decoding complexity, while guaranteeing very low BER for block sizes on the order of $n = 1000$ bits.

A popular irregular LDPC parity-check matrix, which has a very efficient encoding algorithm, is called staircase LDPC [74]. Each codeword $\mathbf{c}$ is composed of a systematic part $\mathbf{u}$, and a parity part $\mathbf{p_u}$ which together form $\mathbf{c} = [\mathbf{u}, \mathbf{p_u}]$. With this setup and given the parity check matrix $H^{n-k,n}$ of the LDPC code, it is possible to decompose $H^{n-k,n}$ as follows:

$$H^{n-k,n} = (H^u, H^{p_u}), \tag{12.43}$$

whereby H^u is a $(n-k) \times (k)$ matrix specifying the source bits participating in check equations and H^{p_u} is a $(n-k) \times (n-k)$ matrix of the form:

$$H^{p_u} = \begin{pmatrix} 1 & 0 & \dots & 0 & 0 \\ 1 & 1 & 0 & \dots & 0 \\ 0 & 1 & 1 & 0 & \dots \\ \dots & \dots & \dots & \dots & \dots \\ 0 & \dots & 0 & 1 & 1 \end{pmatrix}. \tag{12.44}$$

The choice of this structure for H has been mainly motivated by the fact that aside from being systematic, LDPC encoding has a complexity which is linear in the codeword length n [75]. In particular, with this structure, the encoding operation is as follows:

$$p_{u_i} = \begin{cases} \left[\sum_{j=1}^{k} u_j \cdot H^u_{i,j}\right] \pmod 2, & i = 1 \\ \left[p_{u_{i-1}} + \sum_{j=1}^{k} u_j \cdot H^u_{i,j}\right] \pmod 2, & i = 2, \dots, n-k, \end{cases} \tag{12.45}$$

where $H^u_{i,j}$ represents the element (i, j) of the matrix H^u, and u_j is the jth bit of the source sequence $\mathbf{u}$.

By virtue of the efficient encoding complexity, this class of LDPCs presents a small implementation area, while the codes have error-rate performance very close to the Shannon limit for a wide range of code rates ($R_c = 1/4$–$9/10$), and for various modulation schemes such as BPSK, QPSK, 8-PSK, 16-amplitude PSK (16-APSK), and 32-APSK [76]. For this reason, such LDPC codes have been standardized for next-generation digital video broadcasting DVB-RS2. Moreover, LDPC codes have been recently included in a number of standards, such as IEEE802.16, IEEE802.20, and IEEE 802.2.

An LDPC can also be described in terms of a bipartite graph [72], useful for visualizing the exchange of extrinsic information between nodes with the belief propagation decoder, i.e., a graph $T(H)$ with two distinct vertex classes V_1 and V_2, the set of bit and check nodes respectively, that satisfy $T(H) = V_1 \cup V_2$, $V_1 \cap V_2 = \emptyset$ and each edge joins a vertex of V_1 to a vertex of V_2 (see Fig. 12.21). With this setup, the element $h_{i,j}$ at the ith row and the jth column in the parity-check matrix H, represents an edge in $T(H)$ connecting the jth element in V_1 to the ith element in V_2. Many papers in the literature have attempted to determine the properties of the Tanner graphs associated with "good" LDPC codes. From these studies, the basic properties whose consideration in a design of LDPC codes can generally guarantee good error-performance with the suboptimal belief propagation decoder can be summarized as follows:

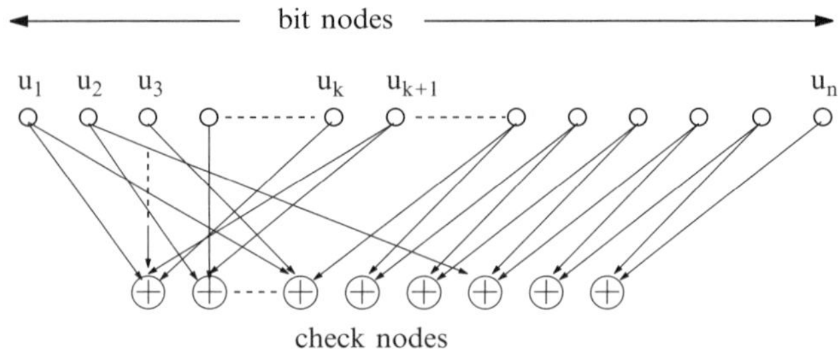

Fig. 12.21 Tanner graph of a low-density parity-check code

1. *Girth g*. The girth is the length of the shortest cycle in the Tanner graph $T(H)$ of the code. For good LDPCs, the girth of the Tanner graph of the code should be maximized. The reason for this is that the neighbors of a generic node in $T(H)$ are independent. That is, they locally form a tree-like structure only up to the $(g/2)$th iteration of the belief propagation algorithm for decoding.
2. *Minimum distance*. It corresponds to the smallest number of columns of H that add up to the zero vector. In the context of iterative decoding, the minimum distance of the code does not generally play as crucial a role as it does in classical coding theory. This is specially true for structured LDPCs for which it has been demonstrated that the weakness of such codes are largely due to near-codewords than to low weight codewords.
3. *Stopping set [77]*. A stopping set is a subset S of the set of the bit node V_1 in $T(H)$ such that all the neighbors[1] of S are connected to S at least twice. The reason for which stopping sets are problematic under iterative decoding on Binary Erasure Channel (BEC) has been demonstrated in [77], where the authors have shown that the residual erasure bits after belief propagation iterative decoding equals to the maximal size stopping set in $T(H)$. Clearly, this problem is strictly related to BECs. However, similar problems can arise in transmission over Additive White Gaussian Noise (AWGN) channel when the reliability of some bit nodes are so small that they can effectively be treated as erased bits. Furthermore, in [78], the authors have clarified the relationship between cycles, stopping sets and dependent columns of the parity-check matrix H of a generic LDPC, explicitly demonstrating that if a code has a minimum distance d_m, it must have a stopping set with d_m bit nodes. Thus, code design strategies avoiding all stopping sets in $T(H)$ with a number of bit nodes less than or equal to t also ensures that $d_m > t$.
4. *Near-codewords*. A (w,v) near-codeword of an LDPC defined through its parity-check matrix H is a vector $\mathbf{x}$ with Hamming weight w whose syndrome $\mathbf{z(x)} = H \cdot x$ has a Hamming weight equal to v. As pointed out in [79], the relatively low weights of these words makes them quite probable, while the fact that the syndrome $\mathbf{z}$ has a small weight indicates that a small number of checks in the Tanner graph of the code is affected by these error events. In fact, as opposed to what happens in the case of stopping sets, in a typical (w,v) near-codeword,

[1] The neighbors of bit nodes are the check nodes, i.e., the nodes belonging to set V_2.

there are v check nodes that are connected to the bits in the near-codeword only once. Typically, during the suboptimal iterative decoding, the decoder terminates in a noncodeword whose weight differs only by a small amount from the true codeword. Furthermore, in a typical decoding situation it often turns out that some near-codewords combine to yield a true codeword. In connection with the behavior of iterative decoding, we refer to [80], whereby the authors have shown that iterative decoding decodes to the pseudosignal defined as the signal that has the highest correlation with the channel output. The set of pseudosignals corresponds to pseudocodewords, only a vanishingly small number of which corresponds to codewords. At the same time, some pseudocodewords cause decoding errors.

5. *Trapping sets*: In [81] Richardson generalized the definition of both stopping sets and near-codewords associated with a specific iterative decoding algorithm when operating with a defined decoder input space. In particular, the notion of trapping sets were introduced in [81]. A (a,b) trapping set T_{r} is a set of a bit nodes and b odd degree check bit neighbors in the subgraph induced by T_{r}, to which an iterative decoding algorithm erroneously converges when the decoding fails after a predefined number of iterations.

To date, there are three main classes of LDPC constructions, that can be summarized as follows:

- ***Density evolution optimized LDPCs.*** Based on some analytic properties of the probability density functions of the Log-Likelihood Ratios (LLRs) associated with both the bit and check nodes in the Tanner graph during iterative decoding, in [82–84] the authors are able to find the threshold value for a randomly constructed irregular LDPC code that identifies the boundary of the error-free region as the block length approach infinity. Subsequently, both bit and check node degrees of the codes are optimized through a linear programming approach in such a way as to guarantee an early converging LDPC. Performance results in terms of Bit Error Rate (BER) have shown that these codes approach the Shannon limit within 0.045 dB at BER$= 10^{-6}$. We note in passing that early converging codes usually have high error-floors [73] (i.e., frame error rates as large as $10^{-5} - 10^{-6}$) due to the fact that optimization of the code in the very-low-to-moderate signal-to-noise ratios usually does not take into account minimum distance codewords. This issue should come as no surprise. Indeed, this is a behavior already observed for parallel concatenated convolutional codes [85] designed with the density evolution technique.
- ***Combinatorial and algebraically designed LDPCs ([85–101]).*** The approach is aimed at designing well-structured LDPC codes based on cyclic difference families, affine configurations, Margulis–Ramanujan algebraic designs, and pseudorandom constructions. One advantage of this class of LDPC codes over randomly constructed LDPCs resides in the fact that usually it is possible to obtain LDPCs with predefined girth and/or minimum distance properties, depending on the particular construction chosen. For instance some of the codes designed using the Margulis–Ramanujan construction (e.g., the Ramanujan

(17,5) code with block-length $n = 4{,}896$) achieve an almost optimal girth among all possible codes in a specific class. However, studies conducted on the sub-optimal iterative belief propagation decoder showed weakness in structured codes due to the near-codewords [79], generalized later by Richardson in [81] to trapping-sets as stated above. This problem affects the error-floor performance of the designed LDPC codes in the sense that the error-floor region of the code is constrained by near-codewords instead of low weight codewords.

- ***Graph-theoretically designed LDPCs ([102–104]).*** This class of codes make use of graph–theoretical properties to design good LDPCs. In [102], the authors proposed a code design technique based on finding a set of paths in a connected graph which satisfies the constraint that any two paths in the set are either disjoint or cross each other at one and only one vertex. In [103], the authors proposed a general method for constructing Tanner graphs with large girth by progressively inserting edges between bit and check nodes in the Tanner graph of the code (the so-called Progressive Edge Growth algorithm or PEG). In [104], the authors in addition to noting that irregular LDPCs perform better than regular ones, provide efficient methods of finding good irregular structures for use with iterative decoding algorithms.

Other literature in connection with design of LDPC codes includes [105–107]. In [108], the authors proposed an algorithm for the estimation of the minimum distance of LDPC codes. The use of LDPC codes has also been examined in connection with bandwidth-efficient modulation schemes (see [109], [110], and rate compatible punctured LDPC codes [111]).

12.3.2.1 A Decoding Algorithm for LDPCs

To date many algorithms have been proposed for the decoding of LDPC codes. Perhaps, the most important algorithm for soft-decision decoding is the belief propagation algorithm, known as sum–product algorithm [5]. In the following, a belief propagation decoder is discussed in connection with the use of LDPC codes over a binary-inputs AWGN channel, whereby binary data are BPSK modulated.

Consider a (n,k)-LDPC identified by the matrix $H^{(n-k,n)}$ as expressed in (12.43). Note that we only make reference to maximum rank matrix H since the particular structure assumed for H ensures this. In particular, the double diagonal on the parity side of the H matrix always guarantees that the rank of H is equal to the number of its rows, $n-k$.

It is well known that the parity check matrix H can be described by a bipartite graph with two types of nodes as depicted in Fig. 12.21; n bit-nodes corresponding to the LDPC code bits, and $n-k$ check-nodes corresponding to the parity checks as expressed by the rows of the matrix H. Let $B(m)$ denote the set of bit-nodes connected to the mth check-node, and $C(n)$ denote the set of check-nodes adjacent to the nth bit-node. With this setup, $B(m)$ corresponds to the set of positions of the 1s in the mth row of H, while $C(n)$ is the set of positions of the 1s in the nth column of H. In addition, let us use the notation $C(n) \setminus m$ and $B(m) \setminus n$ to mean the sets $C(n)$ and

$B(m)$ in which the mth check-node and the nth bit-node, respectively, are excluded. Furthermore, let us identify with $\lambda_{n,m}(u_n)$ the log-likelihood of the message that the nth bit-node sends to the mth check-node, that is, the LLR of the probability that nth bit-node is 1 or 0 based on all checks involving the nth bit except the mth check, and with $\Lambda_{m,n}(u_n)$ the log-likelihood of the message that the mth check-node sends to the nth bit-node, that is, the LLR of the probability that the nth bit-node is 1 or 0 based on all the bit-nodes checked by the mth check except the information coming from the nth bit-node. With this setup, we have the following steps of the sum–product algorithm over AWGN channels for BPSK modulation:

Initialization step: Each bit-node is assigned an a-posteriori LLR composed of the sum of the a-posteriori probabilities $L_c(\mathbf{r})$ evaluated from the sufficient statistic from the matched filter as follows,

$$L_c(r_{1,j}) = \log\left(\frac{P(u_{1,j}=1|r_{1,j})}{P(u_{1,j}=0|r_{1,j})}\right) = \frac{2}{\sigma_z^2} r_{1,j},\ \forall j = 1,\ldots,n, \tag{12.46}$$

plus an extrinsic LLR added only to the systematic bit nodes, i.e., to the bit nodes $u_{1,j},\ \forall j = 1,\ldots,k$. In (12.46), σ_z^2 is the noise variance at the matched filter output due to the AWGN channel. In summary, for any position (m,n) such that $H_{m,n} = 1$, set:

$$\lambda_{n,m}(u_n) = L_c(r_{1,j}),\ \forall\ j = 1,\ldots,n, \tag{12.47}$$

and

$$\Lambda_{m,n}(u_n) = 0. \tag{12.48}$$

(1) *Check-node update.* For each $m = 1,\ldots,n-k$, and for each $n \in B(m)$, compute:

$$\Lambda_{m,n}(u_{1,n}) = 2\tanh^{-1}\left(\prod_{p\in B(m)\backslash n} \tanh\left(\frac{\lambda_{p,m}(u_{1,p})}{2}\right)\right). \tag{12.49}$$

(2) *Bit-node update.* For each $t = 1,\ldots,n$, and for each $m \in C(t)$, compute:

$$\lambda_{t,m}(u_{1,t}) = L_c(r_{1,t}) + \sum_{p\in C(t)\backslash m} \Lambda_{p,n}(u_{1,t}). \tag{12.50}$$

(3) *Decision:* For each bit node $u_{1,t}$ with $t = 1,\ldots,n$, compute:

$$\lambda_t(u_{1,t}) = \begin{cases} L_c(r_{1,t}) + \\ \sum_{p\in C(t)} \Lambda_{p,n}(u_{1,t}) & t = 1,\ldots,k \\ L_c(r_{1,t}) + \sum_{p\in C(t)} \Lambda_{p,n}(u_{1,t}) & t = k+1,\ldots,n \end{cases} \tag{12.51}$$

and quantize the results such that $u_{1,t} = 0$ if $\lambda_t(u_{1,t}) < 0$ and $u_{1,t} = 1$ otherwise.

If $H \cdot \mathbf{u_1}^{\mathbf{T}} = \mathbf{0}$, then halt the algorithm and output $u_{1,t},\ t = 1,\ldots,k$ as the estimate of the transmitted source bits, $\mathbf{u_1}$, corresponding to the first source. Otherwise, if the number of iterations is less than a predefined maximum number, iterate the process starting from step (1).

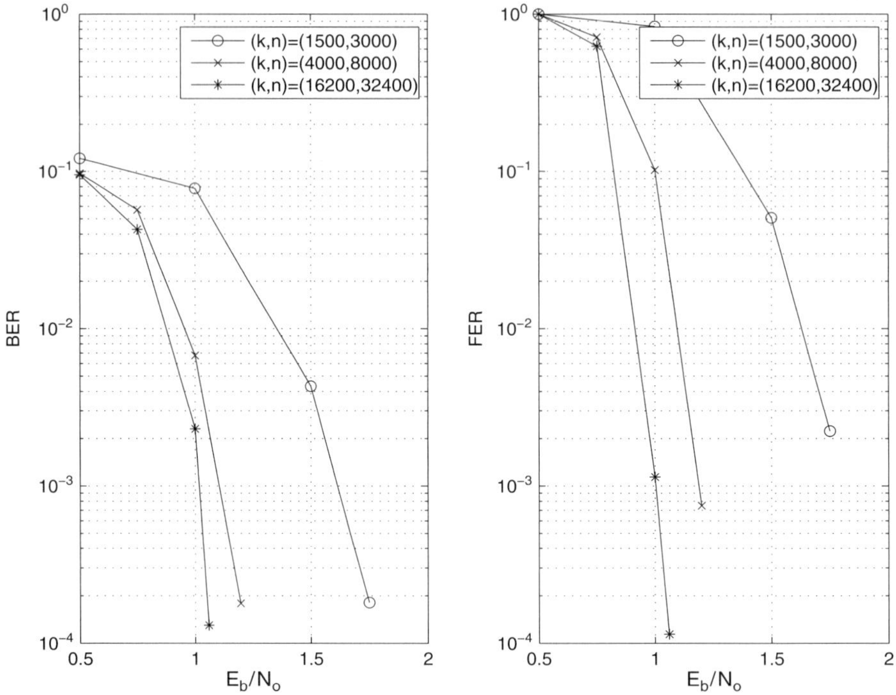

Fig. 12.22 Simulated FER and BER employing 80 iterations of the belief propagation decoder for irregular, rate 1/2 LDPC codes

To conclude this section, we present a set of simulation results of some rate-$1/2$, irregular LDPCs with different block sizes (n,k). The parity check matrix, designed with the technique proposed in [103], is compliant with the structure in (12.43) with an average bit node degree equal to 3.3, while the number of iterations of the LDPC decoding algorithm discussed above was 80. BER as well as FER performance of such codes is noted in Fig. 12.22.

12.4 Discussion and Conclusions

In this chapter we have provided an overview of different channel coding solutions for the next-generation digital multimedia broadcasting systems. The first part of the chapter started from the concepts at the very basis of Information Theory, and, then, it focused on the different facets of the design of convolutional codes. This class of codes represents the starting point for obtaining more efficient coding architectures, such as turbo codes. Each section provided basic technical information, current research activities on the subject, and applications as well as challenges of the aforementioned coding architectures.

The second part of the chapter focused on capacity-achieving channel codes, namely parallel concatenated convolutional codes and low-density parity-check codes. A common characteristic of these two recent coding architectures consists in the fact that they can be iteratively decoded using alternating exchanges of soft-decision information. Some degree of similarity between the decoders for these two capacity-achieving error-control techniques do exist. Nevertheless, low-density parity-check codes have superior BER performance for code lengths greater than, say, 1000 bits. BER performance as close as 0.045 dB from the AWGN channel capacity has been presented in the technical literature for very long block lengths (on the order of 10^6 bits). On the other hand, turbo codes are constructed with known constituent convolutional encoders with a relatively low complexity, and they show a very good BER performance, with respect to LDPC codes, for block lengths lower than 1000 bits. In connection to the decoding algorithms of these codes, many papers in the literature have been aimed at the problem of reducing the computational complexity of the iterative decoders. As a consequence of their superior performance, both turbo codes and LDPC codes have been standardized in a number of future digital multimedia broadcasting systems, such as DVB-RCS, DVB-RS2, MediaFlo, UMTS to cite but a few.

To probe further, the book [5] represents a classical reference on the subject of channel coding. Moreover, some special issues have appeared in the past few years in the technical literature focusing on recent advances on channel coding solutions. As a comprehensive reference on iteratively decodable channel codes, we suggest the special issue recently appeared on the Proceedings of the IEEE [66].

References

1. Shannon CE (1948) A mathematical theory of communication. The Bell System Technical Journal, 27:379–423
2. Cover TM, Thomas JA (1991) Elements of Information Theory, Wiley, 1st ed. New York, NY
3. Papoulis A, Pillai SU (2002) Probability, Random Variables and Stochastic Processes. Mc-Graw Hill, New York, NY
4. Proakis JG (2001) Digital Communications, 4th ed. McGraw Hill, New York, NY
5. Lin S, Costello DJ (2004) Error Control Coding, Prentice Hall, 2nd ed. Upper Saddle river, NJ
6. Viterbi A (1967) Error bounds for convolutional codes and an asymptotically optimum decoding algorithm. IEEE Transactions on Information Theory, 13(2):260–269
7. Viterbi A (1971) Convolutional codes and their performance in communication systems. IEEE Transactions on Communications, 19(5):751–772
8. Bahl L, Cocke J, Jelinek F, Raviv J (1974) Optimal decoding of linear codes for minimizing symbol error rate. IEEE Transactions on Information Theory, 20(2):284–287
9. Sklar B (1997) A primer on turbo code concepts. IEEE Communications Magazine, 35(12):94–102
10. Hagenauer J, Offer E, Papke L (1996) Iterative decoding of binary block and convolutional codes. IEEE Transactions on Information Theory, 42(2):429–445
11. Forney Jr. GD (1970) Convolutional codes I: Algebraic structure. IEEE Transactions on Information Theory, 16(6):720–738

12. Cain JB, Clark G, Geist JM (1979) Punctured convolutional codes of rate $\frac{n-1}{n}$ and simplified maximum likelihood decoding. IEEE Transactions on Communications, 25(1):97–100
13. Hagenauer J (1988) Rate-compatible punctured convolutional codes (RCPC codes) and their applications. IEEE Transactions on Communications, 36(4):389–400
14. Shen B, Patapoutian A, McEwen PA (2001) Punctured recursive convolutional encoders and their applications in turbo-codes. IEEE Transactions on Information Theory, 47(6):2300–2320
15. Benedetto S, Divsalar D, Montorsi G, Pollara F (1998) Serial concatenation of interleaved codes: Performance analysis, design, and iterative decoding. IEEE Transaction on Information Theory, 44(3):909-926
16. Benedetto S, Montorsi G (1996) Design of parallel concatenated convolutional codes. IEEE Transactions on Communications, 44(5):591–600
17. Daut DG, Modestino JW, Wismer LD (1982) New short constraint length convolutional code constructions for selected rational rates. IEEE Transaction on Information Theory, 28(5): 794–800
18. Kim M (1997) On systematic punctured convolutional codes. IEEE Transactions on Communications, 45(2):133–139
19. Yasuda Y, Kashiki K, Hirata Y (1984) High-rate punctured convolutional codes for soft decision viterbi decoding. IEEE Transactions on Communications, 32(3):315–319
20. Haccoun D, Begin G (1989) High-rate punctured convolutional codes for Viterbi and sequential decoding. IEEE Transactions on Communications, 37(11):1113–1125
21. Begin G, Haccoun D (1989) High-rate punctured convolutional codes: Structure properties and construction technique. IEEE Transactions on Communications, 37(12):1381–1385
22. Begin G, Haccoun D, Paquin C (1990) Further results on high-rate punctured convolutional codes for Viterbi and sequential decoding. IEEE Transactions on Communications, 38(11):1922–1928
23. Hole KJ (1988) New short constraint length rate $\frac{N-1}{N}$ punctured convolutional codes for soft-decision Viterbi decoding. IEEE Transactions on Information Theory, 34(5):1079–1081
24. Lee PJ (1985) New short constraint length, rate 1/N convolutional codes which minimize the required SNR for given desired bit error rates. IEEE Transactions on Communications, 33(2):171–177
25. Lee PJ (1986) Further results on rate 1/N convolutional code constructions with minimum required SNR criterion. IEEE Transactions on Communications, 34(4):395–399
26. Lee PJ (1988) Constructions of rate $\frac{n-1}{n}$ punctured convolutional codes with minimum required SNR criterion. IEEE Transactions on Communications, 36(10):1171–1174
27. Chang J, Hwang D, Lin M (1997) Some extended results on the search for good convolutional codes. IEEE Transactions on Information Theory, 43(5):1682–1697
28. Berrou C, Glavieux A, Thitimajshima P (1993) Near Shannon limit error-correcting coding and decoding: Turbo-codes. Proc. 1993 IEEE Int. Conf. on Comm., Geneva, Switzerland, 1064–1070
29. Gallager RG (1962) Low-density parity-check codes. IRE Transactions on Information Theory, 8(1):21–28
30. Mackay DJC (1999) Good error-correcting codes based on very sparse matrices. IEEE Transactions on Information Theory, 45(2):399–431
31. Benedetto S, Montorsi G (1996) Unveiling turbo codes: Some results on parallel concatenated coding schemes. IEEE Transactions on Information Theory, 42(2):409–428
32. Divsalar D, Pollara F (1995) Turbo codes for PCS applications. Proc. 1995 IEEE Int. Conf. on Comm., Seattle, WA, 54–59
33. Garello R, Pierleoni P, Benedetto S (2001) Computing the free distance of turbo codes and serially concatenated codes with interleavers: Algorithms and applications. IEEE Journal on Selected Areas in Communications, 19(5):800–812
34. Divsalar D, Pollara F (1995) On the design of turbo codes. JPL TDA Progress Report 42–123
35. Kousa MA, Mugaibel AH (2002) Puncturing effects on turbo-codes. IEE Proceedings of Communications, 149(3):132–138

36. Daneshgaran F, Laddomada M, Mondin M (2004) An extensive search for good punctured rate k/k+1 recursive convolutional codes for serially concatenated convolutional codes. IEEE Transactions on Information Theory, 50(1):208–218
37. Daneshgaran F, Laddomada M, Mondin M (2004) High-rate recursive convolutional codes for concatenated channel codes. IEEE Transactions on Communications, 52(11):1846–1850
38. Benedetto S, Garello R, Montorsi G (1998) A search for good convolutional codes to be used in the construction of turbo codes. IEEE Transactions on Communications, 46(9):1101–1105
39. Laddomada M, Scanavino B (2006) Some extended results on the design of punctured serially concatenated convolutional codes. JCOMSS – Journal of Communication Software and Systems, 3(2):235–244
40. Campanella M, Garbo G, Mangione S (2000) Simple method for limiting delay of optimized interleavers for turbo codes. IEE Electronics Letters, 36(14):1216–1217
41. Hokfelt J, Edfors O, Maseng T (2001) A turbo code interleaver design criterion based on the performance of iterative decoding. IEEE Communication Letters, 5(2):52–54
42. Garbo G, Mangione S (2001) Some results on delay constrained interleavers for turbo codes. Proc. SoftCOM 2001, Workshop on Channel Coding Techniques, 17–24
43. Dolinar S, Divsalar D (1995) Weight distribution of turbo codes using random and nonrandom permutations. JPL TDA progress report 42–122
44. Crozier S (2000) New high-spread high-distance interleavers for turbo codes. Proc. 20th Biennial Symposium on Communications, 3–7
45. Daneshgaran F, Mondin M (1999) Design of interleavers for turbo codes: Iterative interleaver growth algorithms of polynomial complexity. IEEE Transactions on Information Theory, 45(6):1845–1859
46. Daneshgaran F, Mondin M (2002) Optimized turbo codes for delay constrained applications. IEEE Transaction on Information Theory, 48(1):293–305
47. Nemati MA, Jamali HR, Kwon HM (2001) Interleaver and puncturing in turbo codes. Proc. VTC 2001, 149–153
48. Wang MZ, Sheikh A, Qi F (1999) Interleaver design for short turbo codes. Proc. IEEE Globecom 1999, 1b:894–898
49. Said F, Aghvami AH, Chambers WG (1999) Improving random interleaver for turbo codes. IEE Electronics Letters, 35(25):2194–2195
50. Fragouli C, Wesel RD (1999) Semi-random interleaver design criteria. Proc. IEEE Globecom 1999, 5:2352–2356
51. Hokfelt J, Edfors O, Maseng T (1999) Interleaver structures for turbo codes with reduced storage memory requirement. Proc. 50th VTC 1999, 3:1585–1589
52. Herzberg H (1997) Multilevel turbo coding with a short latency. Proc. 1997 IEEE Int. Sympo. on Inform. Theory, Ulm, Germany
53. Hokfelt J, Maseng T (1997) Methodical interleaver design for turbo codes. Proc. Intern. Sympos. on Turbo Codes & Related Topics, 212–215
54. Robertson P (1994) Illuminating the structure of code and decoder of parallel concatenated recursive systematic (turbo) codes. Proc. 1994 IEEE Global Commun. Conf, 1298–1303
55. Takeshita O, Costello D Jr. (1998) New classes of algebraic interleavers for turbo-codes. Proc. 1998 Int. Sympo. on Inform. Theory, MIT, Cambridge, MA
56. Andrews K, Heegard C, Kozen D (1998) Interleaver design methods for turbo codes. Proc. 1998 Int. Sympo. on Inform. Theory, MIT, Cambridge, MA
57. Crozier S, Guinand P (2001) High-performance low-memory interleaver banks for turbo-codes. Proc. IEEE 54th Vehicular Technology Conference, VTC Fall, USA
58. Crozier S, Guinand P (2002) Distance bounds and the design of high-distance interleavers for turbo-codes. Proc. 21st Biennial Symposium on Communications
59. Marshall Hall Jr. (1976) The Theory of Groups. Chelsea, 2nd ed. New York, NY
60. Meyer PA (1972) Martingales and Stochastic Integrals I. Springer, Berlin
61. Daneshgaran F, Laddomada M (2004) Optimized prunable single-cycle interleavers for turbo codes. IEEE Transactions on Communications, 52(6):899–909
62. Daneshgaran F, Laddomada M (2005) Reduced complexity interleaver growth algorithm for turbo codes. IEEE Transactions on Wireless Communications, 4(3):954–964

63. Laddomada M, Scanavino B (2006) A cost-function based technique for design of good prunable interleavers for turbo codes. IEEE Transactions on Wireless Communications, 5(8):1953–1958
64. Jinhong Y, Vucetic B, Wen Feng (1999) Combined turbo codes and interleaver design. IEEE Transactions on Communications, 47(4):484–487
65. Vucetic B, Yuan J (2000) Turbo codes: Principles and applications. Kluwer, USA
66. Berrou C, Hagenauer J, Luise M, Schlegel C, Vandendrope (2007) Special issue on turbo-information processing: Algorithms, implementations and applications. IEEE Proceedings of the IEEE, 95(6): 1146–1149
67. Zhang W et al. (2007) An introduction of the Chinese DTTB standard and analysis of the PM595 working modes. IEEE Transactions on Broadcasting, 53(1):8–13
68. Bae B, Kim W, Ahn C, Lee S-I, Sohng K-I (2006) Development of a T-DMB extended WIPI platform for interactive mobile broadcasting services. IEEE Transactions on Consumer Electronics, 52(4):1167–1172
69. ETSI EN 300 401 v1.3.3 (2001) Radio broadcasting systems; digital audio broadcasting (DAB) to mobile, portable and fixed receivers
70. Fuyun Ling MR, Mantravadi A, Krishnamoorthi R, Vijayan R, Walker GK, Chandhok R (2007) FLO physical layer: An overview. IEEE Transactions on Broadcasting, 53(1): 145–160
71. Gracie K, Hamon M-H (2007) Turbo and turbo-like codes: Principles and applications in telecommunications. Proceedings of the IEEE, 95(6):1228–1254
72. Tanner RM (1981) A recursive approach to low complexity codes. IEEE Transactions on Information Theory, 27(5):533–547
73. Richardson T, Urbanke R (2003) The renaissance of Gallager's low-density parity-check codes. IEEE Communications Magazine, 41(8):126–131
74. Yang M, Ryan WE, Li Y (2004) Design of efficiently encodable moderate length high-rate irregular LDPC Codes. IEEE Transactions on Communications, 52(4):564–571
75. Richardson T, Urbanke R (2001) Efficient encoding of low-density parity-check codes. IEEE Transactions on Information Theory, 47(2):638–656
76. Eroz M, Sun F-W, Lee L-N (2006) An innovative low-density parity-check code design with near-Shannon-limit performance and simple implementation. IEEE Transactions on Communications, 54(1):13–17
77. Di C, Proietti D, Telatar E, Richardson T, Urbanke R (2002) Finite length analysis of low-density parity-check codes on the binary erasure channel. IEEE Transactions on Information Theory, 48(6):1570–1579
78. Tian T, Jones C, Villasenor JD, Wesel RD (2003) Construction of irregular LDPC codes with low error floors. IEEE ICC'03, 5:3125–3129
79. MacKay DJC, Postol MS (2003) Weakness of Margulis and Ramanujan-Margulis low-density parity-check codes. In: Electronic Notes in Theoretical Computer Science. Elsevier, Amsterdam
80. Frey BJ, Koetter R, Vardy A (2001) Signal-space characterization of iterative decoding. IEEE Transactions on Information Theory, 47(2):766–781
81. Richardson T (2003) Error floors of LDPC codes. Proc. of 41st Annual Allerton Conf. on Comm., Control, and Comp.
82. Richardson TJ, Shokrollahi MA, Urbanke RL (2000) Design of provably good low-density parity check codes. IEEE International Symposium on Information Theory, 199
83. Chung S, Forney GD, Richardson TJ, Urbanke R (2001) On the design of low-density parity-check codes within 0.0045dB of the Shannon limit. IEEE Communications Letters, 5(2): 58–60
84. Richardson TJ, Shokrollahi MA, Urbanke RL (2001) Design of capacity-approaching irregular low-density parity-check codes. IEEE Transactions on Information Theory, 47(2):619–637
85. ten Brink S (2001) Convergence behavior of iteratively decoded parallel concatenated codes. IEEE Transactions on Communications, 49(10):1727–1737

86. Vasic B, Milenkovic O (2004) Combinatorial constructions of low-density parity-check codes for iterative decoding. IEEE Transactions on Information Theory, 50(6):1156–1176
87. Johnson SJ, Weller SR (2003) Resolvable 2-designs for regular low-density parity-check codes. IEEE Transactions on Communications, 51(9):1413–1419
88. Ammar B, Honary B, Kou Y, Lin S (2002) Construction of low density parity check codes: A combinatoric design approach. IEEE International Symposium on Information Theory, 311
89. Kou Y, Lin S, Fossorier MPC (2000) Low density parity check codes: Construction based on finite geometries. IEEE Global Telecommunications Conference
90. Fossorier MPC (2004) Quasy-cyclic low-density parity-check codes from circulant permutation matrices. IEEE Transactions on Information Theory, 50(8):1788–1793
91. Bond JW, Hui S, Schmidt H (2000) Constructing low-density parity-check codes. EUROCOMM 2000. IEEE/AFCEA Information Systems for Enhanced Public Safety and Security
92. Sankaranarayanan S, Vasic B, Kurtas EMA (2003) A systematic construction of irregular low-density parity-check codes from combinatorial designs. IEEE International Symposium on Information Theory
93. Johnson SJ, Weller SR (2001) Construction of low-density parity-check codes from Kirkman triple systems. IEEE Global Telecommunication Conference, 2:970–974
94. Bond JW, Hui S, Schmidt H (1999) Constructing low-density parity-check codes with circulant matrices. Information Theory and Networking Workshop
95. Rosenthal J, Vontobel PO (2000) Constructions of LDPC codes using Ramanujan graphs and ideas from Margulis. Proc. of 38th Allerton Conference on Communication, Control and Computing
96. Margulis GA (1982) Explicit constructions of graphs without short cycles and low-density codes. Combinatorica, 2(1):71–78
97. Echard R, Shih-Chun Chang (2001) The π-rotation low-density parity check codes. IEEE Global Telecommunications Conference
98. Prabhakar A, Narayanan K (2002) Pseudorandom construction of low-density parity-check codes using linear congruential sequences. IEEE Transactions on Communications, 50(9):1389–1396
99. Ammar B, Honary B, Kou Y, Xu J, Lin S (2004) Construction of low-density parity-check codes based on balanced incomplete block designs. IEEE Transactions on Information Theory, 50(6):1257–1268
100. Tang H, Xu J, Kou Y, Lin S, Abdel-Ghaffar K (2004) On algebraic construction of Gallager and circulant low-density parity-check codes. IEEE Transactions on Information Theory, 50(6):1269–1279
101. Chen L, Xu J, Djurdjevic I, Lin S (2004) Near-Shannon-limit quasi-cyclic low-density parity-check codes. IEEE Transactions on Communications, 52(7):1038–1042
102. Djurdjevic I, Lin S, Abdel-Ghaffar K (2003) Graph–theoretic construction of low-density parity-check codes. IEEE Communications Letters, 7(4):171–173
103. Hu X-Y, Eleftheriou E, Arnold DM (2001) Progressive edge-growth Tanner graphs. IEEE Global Telecommunications Conference
104. Luby MG, Mitzenmacher M, Shokrollahi MA, Spielman DA (2001) Improved low-density parity-check codes using irregular graphs. IEEE Transactions on Information Theory, 47(2):585–598
105. Echard R, Chang S-C (2002) Deterministic π-rotation low-density parity check codes. Electronics Letters, 38(10):464–465
106. Mackay DJC, Wilson ST, Davey MC (1999) Comparison of construction of irregular Gallager codes. IEEE Transactions on Communications, 47(10):1449–1454
107. Yang M, Ryan WE, Li Y (2004) Design of efficiently encodable moderate-length high-rate irregular LDPC codes. IEEE Transactions on Communications, 52(4):564–571
108. Daneshgaran F, Laddomada M, Mondin M (2006) An algorithm for the computation of the minimum distance of LDPC codes. ETT-European Transactions on Telecommunications, 17(1):57–62

109. Hou J, Siegel PH, Milstein LB, Pfister HD (2003) Capacity-approaching bandwidth-efficient coded modulation schemes based on low-density parity-check codes. IEEE Transactions on Information Theory, 49(9):2141–2155
110. Eleftheriou E, Olcer S (2002) Low-density parity-check codes for digital subscriber lines. Proc. of IEEE ICC, 3:1752-1757
111. Ha J, McLaughlin SW (2003) Optimal puncturing distributions for rate-compatible low-density parity-check codes. IEEE International Symposium on Information Theory

Chapter 13
ASIC Design for Broadcasting Baseband Processing: Practices and Architectures

Peijun Shan

13.1 Introduction

A multimedia broadcasting receiver typically consists of RF (radio frequency) and analog receiving circuitry, modulated signal processing, digital demodulation, channel decoding, link-layer control and processing, source decoding (media decompression), media processing (such as sampling rate conversion, error concealment, and filtering), and application functionalities. Digital custom hardware, in form of an ASIC (Application Specific Integrated Circuit), is the suitable solution to implementing physical-layer baseband processing, including filtering, synchronization, demodulation, channel error control decoding, etc., in order to meet the requirements of speed, power consumption, and cost. Custom hardware is also suitable for source decoding and media processing to enhance performance and relax the requirements on firmware and processor hardware.

Chip power consumption is critical for battery life of handheld terminal devices. System complexity of consumer electronic devices has been increasing in a much larger pace than advancement of battery technology. This continuing trend is applying growing pressure on low power implementation. This chapter will start with a description of a general ASIC design flow, followed by major issues and trade-offs in ASIC design for high date-rate digital processing, especially low-power design options.

Digital filtering is one of the most commonly seen functions in digital multimedia communications systems. With various application scenarios and system requirements, a digital filter could be implemented in many different architecture styles. Sometimes it is quite challenging to design a filter for a given semiconductor process, system performance, and product cost factors. Digital filtering architecture style variations will be examined, and the concepts and approaches discussed can also be applied to implementing other data processing functions.

P. Shan
Avnera Corperation, Beaverton, Oregon, USA
e-mail: pshan@ieee.org

F.-L. Luo (ed.), *Mobile Multimedia Broadcasting Standards: Technology and Practice*,
DOI: 10.1007/978-0-387-78263-8_13,

Another major part of digital broadcasting receiver is the error control decoder. Forward Error Correction (FEC) coding techniques including algebraic codes, convolutional codes, and their combinations are used in digital broadcasting. Galois Field (GF) arithmetic operations are the basic building blocks in implementing encoders and decoders for algebraic codes such as the widely used Reed-Solomon codes. High-speed ASIC implementation of GF operations will be explored, including efficiently multiplying two GF numbers in logic gates.

Complex number mathematics is an essential part of communication modems, especially when considering OFDM, phase modulation, QAM (Quadrature Amplitude Modulation), and even more classic single-side-band and analytic signals. For example, in an OFDM receiver the complex-valued data path starts from mixer, in form of in-phase (I) and quadrature (Q) channels, and is processed through analog filtering, amplification, A/D conversion, digital filtering, AGC logic, time and frequency synchronization, invert FFT, channel equalization, to the constellation de-mapping decision. Some skills for efficiently implementing complex number operations in ASIC design will be covered.

The rest of this chapter is organized as follows. Section 13.2 presents an overview of present ASIC design engineering flow and practices. Section 13.3 is devoted to low-power design, including motivations, principles, and common techniques from ASIC design point of view. The following sections are on architecture side. Using an FIR filter as an example, Sect. 13.4 covers architecture choices for data processing in the context of power and speed. Section 13.5 addresses ASIC implementation architectures and logics for basic GF arithmetic operations used in block code-based channel coding. Efficient implementation architectures for complex functions are described in Sect. 13.6. Conclusions and further discussions are given in Sect. 13.7.

13.2 ASIC Design Overview

More and more chips being designed today are more accurately classified as SOCs (System-on-Chip) rather than traditional ASICs. SOC design overall has more emphasis on system-level chip architecture and integration issues, firmware-hardware trade-offs and co-design, selection and integration of processor cores and other IP cores, peripherals, buses and memories, etc. On the other hand, there will usually be some ASIC subsystems as part of an SOC chip and very often, the ASIC part is a critical one for the product. Using a DVB-x receiver SOC as an example, the physical-layer subsystem (from radio to bit stream) is commonly a custom designed ASIC subsystem.

While IP cores play a bigger role in SOC design, ASIC design is based mainly on standard cells. ASIC designers build circuits using general-purpose standard cells provided by the fabrication foundry (commonly referred to as "fab" for short) or third-party vendors. ASIC designers may also generate memory blocks as needed using a memory generation tool provided by the foundry or other third party vendors.

A standard-cell library contains logic gates with various sizes and speeds characterized at certain supply voltages, temperatures, and process conditions. After the system requirements are defined, algorithm, architecture, and implementation microarchitecture should be developed accordingly. Digital designers work on hardware solutions and create RTL (Register Transfer Level) code in a hardware description language (typically Verilog® or VHDL®). RTL synthesis maps RTL code to a standard-cell based netlist [1]. A place-and-route tool then places the cells on the floor-plan area reserved for the digital subsystem and makes routing connections according to netlist information and timing constraints.

During this process timing analysis is performed, clock trees are generated, and extra buffers are inserted to fix the timing relationship between data and clocks and also between clocks. Timing is to be verified in different ways including dynamic simulations and static timing analysis (STA). Dynamic timing simulation checks system functionality at the gate level using timing information obtained from the parasitic extraction after layout, which is also known as back-annotated simulation. STA is a thorough and efficient means of checking the timing on all paths without simulation. STA can be a separate step in the design flow or a built-in function of the physical synthesis, depending on EDA tools used.

Formal Verification (FV) can also be carried out to check logic equivalency between the RTL, the gate-level netlist after synthesis (prelayout), and the gate-level netlist generated after layout (which will include scan chains and extra buffers to fix timing). In addition, in order to achieve a quality design a large amount of engineering resources are spent on behavioral and functional verification and debugging, in system level, RTL level, and gate level. FPGA prototyping and emulation assists verification during and after this process.

The design engineering stage from concept to RTL is often referred to as front-end design, while back-end design refers to the flow from synthesis to "tape-out" (sending the design database, normally in GDS file format, to the foundry). DFT (Design for Test) features for production test, chip diagnosis features for chip evaluation, and plenty of spare cells for potential future metal fixes are also added at appropriate stages of the flow.

ECO (Engineering Change Order) fixes are sometimes preferred to perform small changes outside of the normal design flow either before a full-layer tape-out or after silicon evaluation and debugging, which will generate a tape-out with changes in metal-layer (routing) masks only. Physical verification is normally carried out on both the digital layout itself and the entire chip, and includes LVS (Layout Versus Schematic), DRC (Design Rule Check), antenna effect check, and other checks depending on the fabrication foundry requirements. Iterations of the design steps mentioned above are often necessary to meet a curtain design requirement or for optimization in die area, power consumption, functionality, and performance. Figure 13.1 illustrates a typical ASIC development flow.

In the earliest stage of designing an ASIC subsystem, either as a stand-alone chip or as a part of an SOC, one of the first tasks is to define what will be the clock rates, how the clocks will be generated, their relationship, and how they will be used across the chip. Microarchitecture, RTL implementation, synthesis, and timing analysis will all depend on the clock plan.

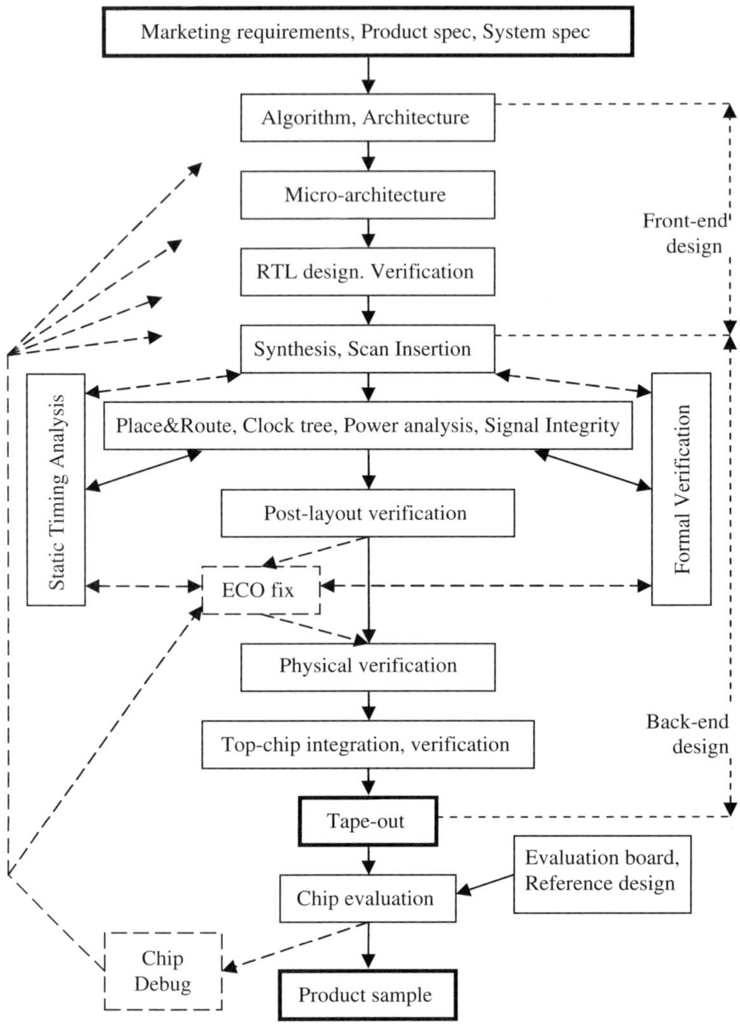

Fig. 13.1 Typical ASIC development flow

Clock rate definition is driven by data rate, process speed, supply voltage, availability and convenience of certain clocks, and the size-power trade-off. Often the targeted data rate brings the heaviest influence on major architecture choices, clock rates, and even process choice. For example, a high bps (bits per second) multimedia communications system implemented in a present mainstream ASIC process may clock the bit processing domain at a fraction of the bps rate and implement the operation in a parallel-processing circuit architecture. The same system may clock the video sample processing block at or near the video sampling rate and clock the audio sample processing block at a convenient rate that is much higher than the audio sample rate to save area using a serialized processing architecture.

13.3 Low Power ASIC Design

Longer battery life for portable devices requires lower chip power consumption. The emergence and development of digital multimedia broadcasting itself enlightens the trend of increasing system complexity in consumer electronic devices, which is growing at a much faster pace than battery technology advancement. The increasing gap between system algorithm complexity and battery capability is imposing an on-going challenge for low-power implementation of complex SOCs and ASICs.

Long battery life requires that the chip consume low operating current when the device is in use and low stand-by current when the device is not active but in a state ready for operation.

Power (heat) dissipation is a major driving-force for low-power design techniques in processor (CPU, DSP) design and may become an issue for very large-scale high-speed ASIC chips as well. The combination of decreasing semiconductor process geometries, increasing algorithm complexity, and increasing clocking speed is resulting in an increase of power density on chip, even at lower supply voltages.

For CMOS digital circuits total power consumption mainly consists of the following: dynamic power that charges and discharges capacitance when logic level changes, leakage power from current leaking through transistors and diodes, and short circuit power due to current flowing between supply and ground during gate switching. In custom (non-standard-cell-based) circuit design, designers have plenty of design options and various techniques to deal with all these types of power consumption [1–3]. Standard-cell-based ASIC designs can still use creative methods to reduce dynamic power and hence produce differentiating products while still having limited flexibility to control leakage power.

At the chip-architecture level, it is important to first understand the power budget, the primary product functionality, and the value and cost of potential add-on features; only then can one use sound judgment in defining the chip. Given limited engineering resources and schedule, more emphasis on product simplicity tends to be rewarded with lower power, smaller area, better design quality, and better chance of success. Care should be taken on formulating a chip-level power management scheme and on specifying power saving operation modes. Power consumption is also an important factor in making hardware–software partition and fixed-function ASIC versus programmable processor decisions.

Dynamic power is determined by

$$P_{\text{dynamic}} = \alpha C V^2 f$$

where α represents switching activity (probability); C is switching capacitance including gate capacitance, routing wire capacitance, and any other parasitic capacitance charged and discharged during logic transitions; V is the level of digital supply voltage; and f is the clock frequency. It follows that dynamic power can be controlled by balancing α, C, V, and f through architecture and logic design trade-offs.

Algorithm and architecture play key roles in determining dynamic power. A more efficient algorithm requires less gates (smaller$C$) and/or less switching activity (smaller α and/or f). For a given algorithm, implementation can be architected with proper area, power, and schedule/resources trade-offs. For example, a high-throughput data processing subsystem may use parallel processing and pipelined processing to enable a reduction in V and $\alpha \cdot f$, even with larger$C$, thereby reducing power consumption to a small fraction. In the next section an FIR filter will serve as an example to demonstrate hardware architecture options in more detail.

ASIC implementation typically employs synchronous design, meaning that all sequential circuit elements are only triggered at a clock edge. Therefore, dynamic current can be separated into two parts: the "clock current" consumed by clock trees to provide synchronous clocks to registers and the "data current" consumed by combinational logic performing data processing and registers performing data buffering. Clock-gating – turning off clock when it is not needed – is an effective way to save clock power.

Clock-gating can be implemented anywhere from system level, block level, to individual registers, and it is a good practice to have multiple levels of clock-gating. Microprocessor IP cores usually have clock-gating implemented inside the core where clocks are internally shut down as much as possible when no instruction is executed. In a packet-data communications system clock-gating can be implemented in the system level controlled by the power management firmware. For instance, the receiver subsystem clock is turned on only while receiving valid data packets. In implementing clock-gating circuitry care should be taken to ensure that no voltage glitches are induced on the clock lines during switching.

At a lower level the synthesis tool can perform automatic clock-gating using register enable information in the RTL code. Figure 13.2 demonstrates an example of automatic clock-gating, which represents a byte register bank with synchronous reset (to 0) and data enable. The amount of power saving due to clock-gating will depend on the duty cycle of the enable signal. As a byproduct, the area will become smaller with clock-gating due to the removal of the data multiplexers for holding data.

Subthreshold leakage current has become significant with technology scaling to deep submicron and now reaches a level comparable with dynamic current. Lower and lower threshold voltage (V_t), which is scaled down as supply voltage is scaled, results in high leakage current. Most standard cell library providers provide two or more library versions with different V_t levels so that ASIC designers have the option to use higher V_t cells to save leakage power or lower V_t cells for higher circuit speed. EDA tool venders also provide multi-V_t synthesis tools that can mix high- and low-V_t cells to optimally balance speed and power for given circuit timing constraints. The result from synthesis is usually that a small percentage of the cells are low-V_t to handle critical timing paths and a large percentage of the cells are high-V_t to save leakage power.

To many people's surprise, RTL coding styles also significantly impact area and power consumption, especially when it is not done right or with care. RTL design should be planned to take advantage of the logic optimization capability of the synthesis tool. Besides synthesis register based clock-gating mentioned above, logic synthesis strives to shrink chip area through expression merging and resources sharing.

```
parameter N = 8;
wire [N-1:0] d;

reg [N-1:0] q;
always @(posedge clk)begin
    if (~reset_n) q <= {N{1'b0}};
    else if (enable) q <= d;
end
```

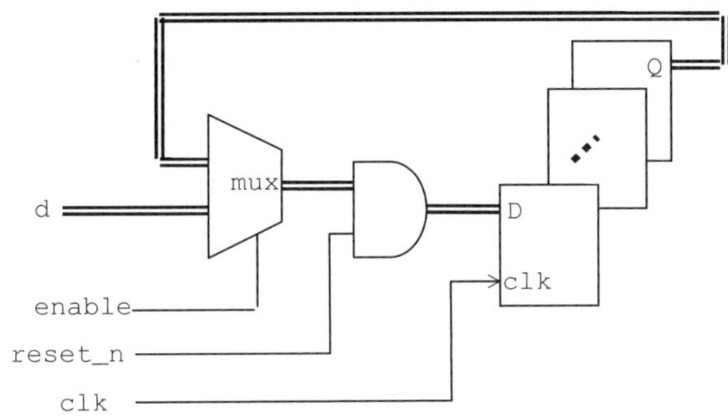

(**a**) Example of design without clock gating

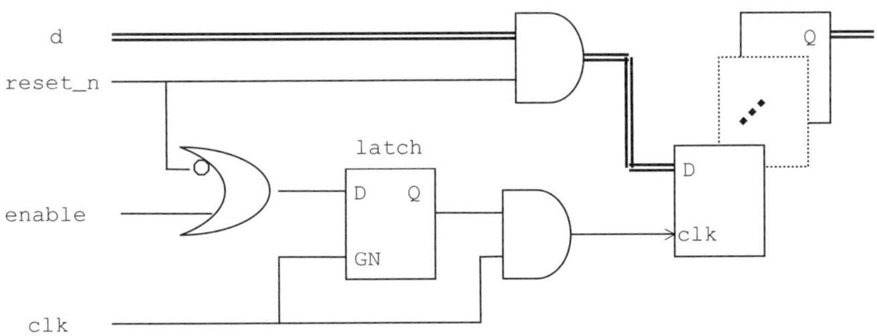

(**b**) Expected synthesis logic with clock gating enabled

Fig. 13.2 Example of register-level clock-gating (double line for byte bus): (**a**) example of design without clock-gating; (**b**) expected synthesis logic with clock-gating enabled

Memory types used in ASIC design include RAM's, ROM's, register files, and flip-flops. Flip-flops can be configured as shift registers or as individually accessible registers. The choice has an impact on area and power and is sometimes a trade-off between the two. To save power, care should be taken to enable memory only when needed – such as when accessing valid data – which may involve carefully generating control signals such as chip select, read enable, or write enable in RTL.

Some power reduction techniques are effective but heavily increase design complexity and risk. These may include multiple supply voltages, supply voltage/current shut down, variable supply voltage, and variable clock rate.

In summary, there are many design engineering aspects that contribute to ASIC power consumption, die area, or design quality in general, including architecture and algorithm, hardware microarchitecture, back-end design options, and RTL realization. A high quality design only comes from doing everything right.

13.4 Filter Implementation Choices under Power and Speed Context

As mentioned earlier, architecture plays a key role in the efficiency of hardware implementation. As an example, it was briefly mentioned that parallel processing and pipelining techniques could be exploited to reduce power consumption for high-throughput data processing. In general, how a certain algorithm should be implemented is determined by a combination of product requirements and engineering resources. During the stage of block-level architecture development, many factors are already predecided, such as process technology, supply voltage, standard cell timing characteristics, and block performance requirements. With the given constraints architecture-level trade-offs are critical for silicon area (manufacturing cost), power consumption (usage cost), design engineering cost, as well as development schedule and even chance of success.

Digital filtering is one of the most widely used functional blocks in multimedia systems, especially in communication modems and in media signal processing. This section will outline digital filter architecture options for ASIC implementation and then will use a digital filter as an example to demonstrate architecture variations with an emphasis on power and speed. An FIR filter implemented in the direct form [4] (the multiplier coefficients are the original ones in the transfer function) will serve as an example to outline implementation options from an ASIC design perspective. The concepts and methods outlined for the FIR filter can also be applied to other types of data-path functions and other varieties of digital filters such as multirate filters, IIR filters, and other forms (nondirect) of FIR filters.

Consider an L-tap FIR filter with impulse response $\{h_0, h_1, \ldots, h_{L-1}\}$ to be implemented in fixed-point arithmetic with N-bit 2's complimentary code representing the data samples, input $x(n)$, and output $y(n)$ (where n is time index), and M-bit 2's complimentary code representing filter coefficients $\{h_0, h_1, \ldots, h_{L-1}\}$.

13.4.1 *Tapped Delay Line*

The first hardware implementation structure to consider is the straightforward tapped delay line structure shown in Fig. 13.3a, which is also referred to as a transversal filter [4]. Here the sampling rate is the same as or lower than the clock rate, which is common when filtering modulated signals in a communication modem such as in a channel filter.

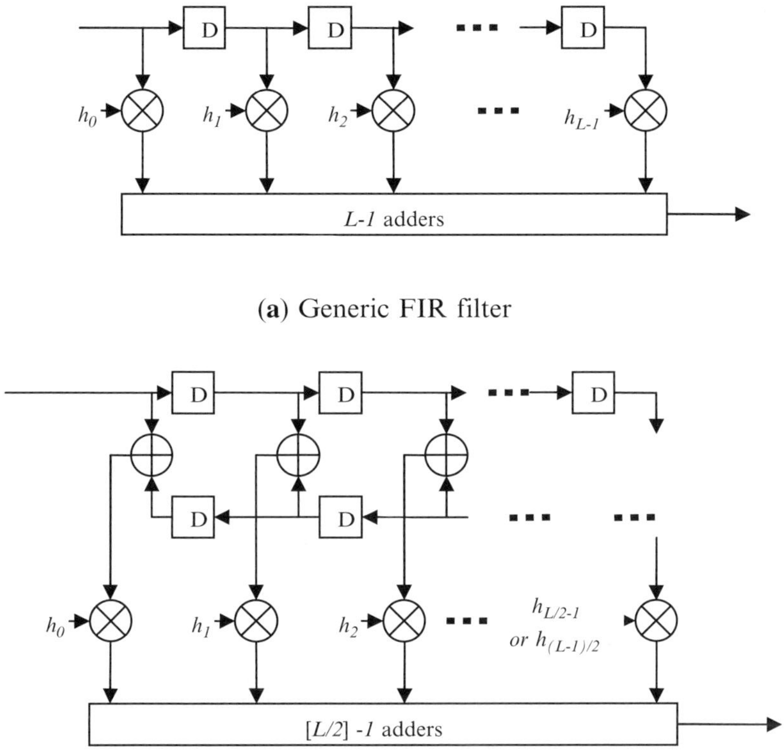

(**a**) Generic FIR filter

(**b**) Symmetric FIR filter

Fig. 13.3 Tapped delay line structure: (**a**) generic FIR filter; (**b**) symmetric FIR filter

This structure might be just right if the sampling rate (data throughput rate), clock rate, the process speed, and power consumption all fit. If the filter coefficients are fixed values, which is often the case, the fixed-point coefficients should have been carefully selected such that only a small number of adders are needed to perform the coefficient multiplications, as each "1" bit (in binary form) in the coefficients requires an addition operation.

In this structure the longest timing path (the combinational logic between registers, assuming the registers at the block boundaries are default) includes a coefficient multiplication and the entire *L*-to-1 addition tree. The total delay of a path with so many stages of arithmetic operations could easily sum up to more than a clock cycle duration for a reasonable filter length, data word length, and clock rate. If so, this could violate synthesis timing constraints and so it is wise to consider other implementation options which may also save power.

It is often desirable to have a symmetric coefficient sequence in an FIR filter in order to have a linear-phase response (the frequency components have the same latency and hence there is no waveform distortion). The filtering computation can take advantage of the coefficient symmetry by pairing the delayed taps with the same coefficients and adding them first to share a multiplier, as shown in Fig. 13.3b.

13.4.2 Transposing

A mathematically equivalent structure comes from the transposed version of the original signal flow. Figure 13.4 shows the transposed realization of the generic and symmetric FIR structures shown in Fig. 13.3. In the transposed realization the critical timing path is reduced to a small fraction: only one 2-to-1 adder plus a coefficient multiplication. As for area (gate count), the size of the data buffer is increased from N-bit wide to $(N+M-1)$-bit or so, possibly $N+M$-bit if both $-2^{(N-1)}$ and $-2^{(M-1)}$ are used as valid codes for data and coefficient, respectively, or less than $(N+M$-1)-bit if some word-length reduction (rounding or truncation) is performed after the coefficient multiplication operation.

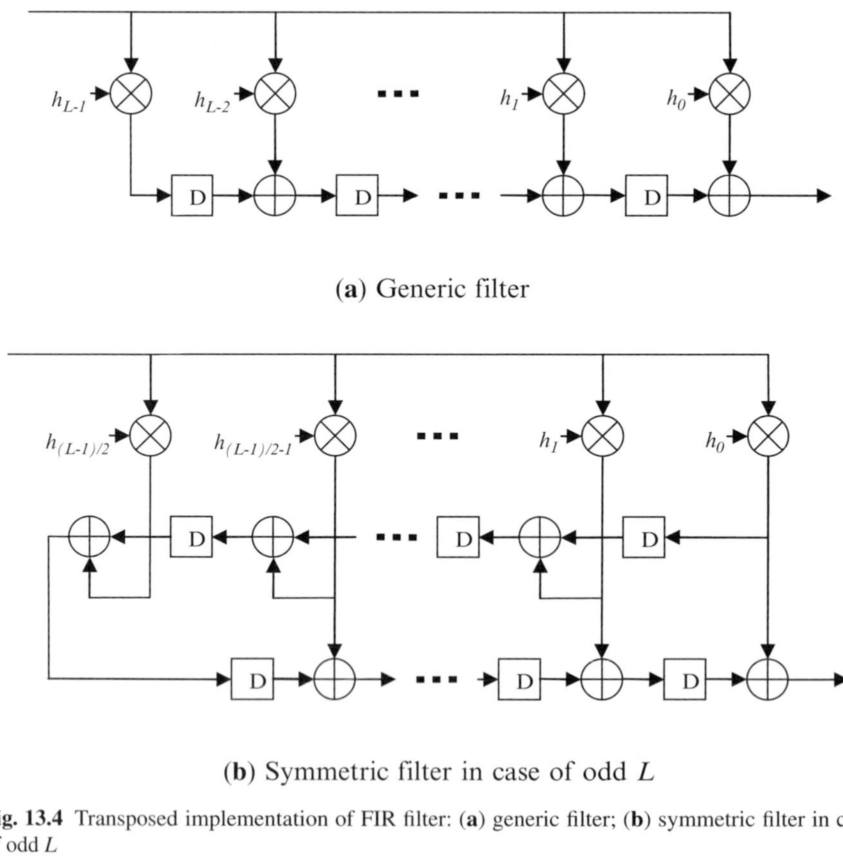

Fig. 13.4 Transposed implementation of FIR filter: (**a**) generic filter; (**b**) symmetric filter in case of odd L

13.4.3 Pipelining

A generic approach to reduce the delay of the critical timing path is to cut the data path into shorter sections by inserting registers as illustrated in Fig. 13.5. In the first example (Fig. 13.5a), the long timing path (1 multiplier plus $L-1$ adders) is broken into two separate timing paths. The second example (Fig. 13.5b) is a case of timing-critical high-speed processing, where it is of interest to further reduce path-delay from the original 1 multiplier plus 1 adder. Assuming that the delay from multiplier h_1 is much larger than the delay from the adder one can factor multiplier h_1 into two parts: $h_1 = h_{1A} \cdot h_{1B}$ and assign them into separate timing paths to better balance the delays on all paths.

This practice is referred to as pipelining. It is quite straightforward to apply pipelining to FIR filters and other data-path functions without feedback or recursive computation. Functions with recursive operations, such as an IIR filter, require more skill and thought [5]. For the sake of better design separation, modularity, and convenience, it is a common practice to insert registers at the boundaries of major design blocks and modules. Signal latency is increased when extra stages of registers are inserted along the data path.

Chip area may grow due to transposing and pipelining. When using flip-flops for data buffering, clock tree size and clock tree power will grow due to their increased number. On the other hand, there may be fewer voltage glitches in the combinational logic due to the reduced path delays, which in turn reduces the power consumed by the combinational logic. Due to the increased timing margin smaller cells tend to be used during synthesis, which results in smaller area and lower power than using the larger and faster cells. It should be emphasized that reducing critical timing path delays results in even more significant potential power savings when considering the possibilities of reducing supply voltage, reducing clock rate, or reducing processing duty cycle.

13.4.4 Retiming

The delay in the timing critical path can also be reduced by local adjustment of the location of delay elements without affecting data-path latency and input–output system behavior. For example, operation $(ax+b)z^{-1}$ (multiply, add, then delay) can be implemented as $(ax)z^{-1} + bz^{-1}$ (add after separate delays), as shown in Fig. 13.6. Assuming that the adder takes 1 time unit, the multiplier h_0 takes 4 time units, and the multiplier h_1 takes 5 time units, then the illustrated retiming modification reduces the largest path delay from 6 to 5 time units. Finer retiming can also be done by breaking the larger multiplier in a similar way to what was done for pipelining in Fig. 13.5b.

13.4.5 Parallel Processing

In the discussion above it was assumed that the signal sampling rate (data throughput rate) is the same as or lower than the clock rate. In some scenarios it is necessary to clock the data processing circuit at a slower rate than the signal sampling rate.

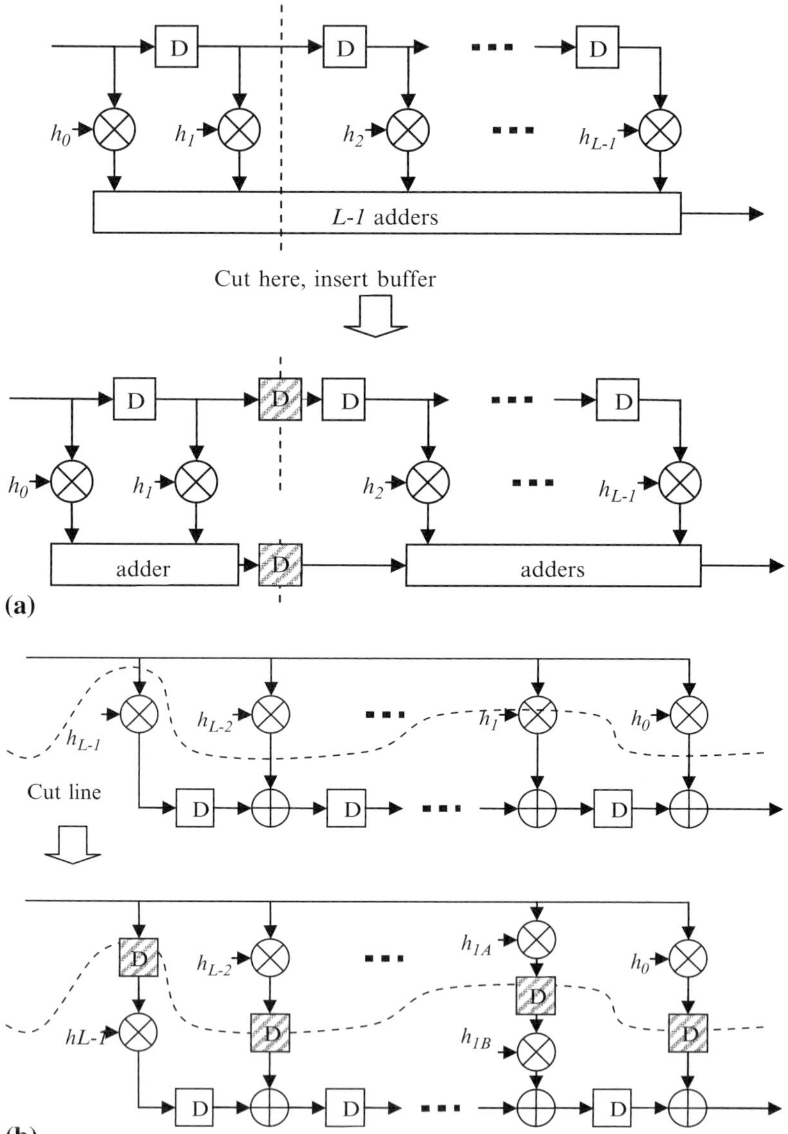

Fig. 13.5 Examples of pipelined processing: **(a)** tapped delay line FIR filter; **(b)** transposed FIR filter

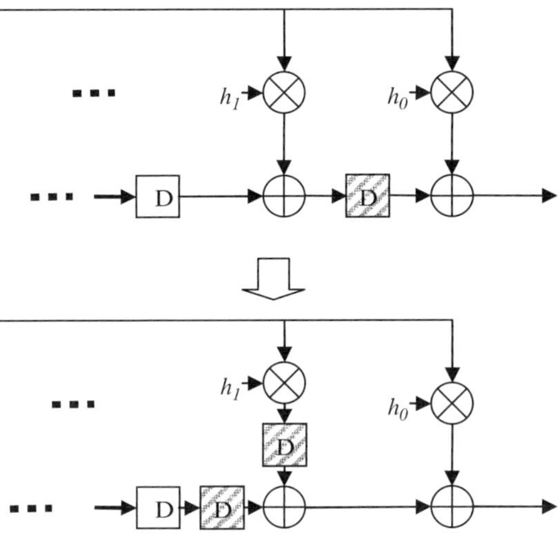

Fig. 13.6 Example of retiming

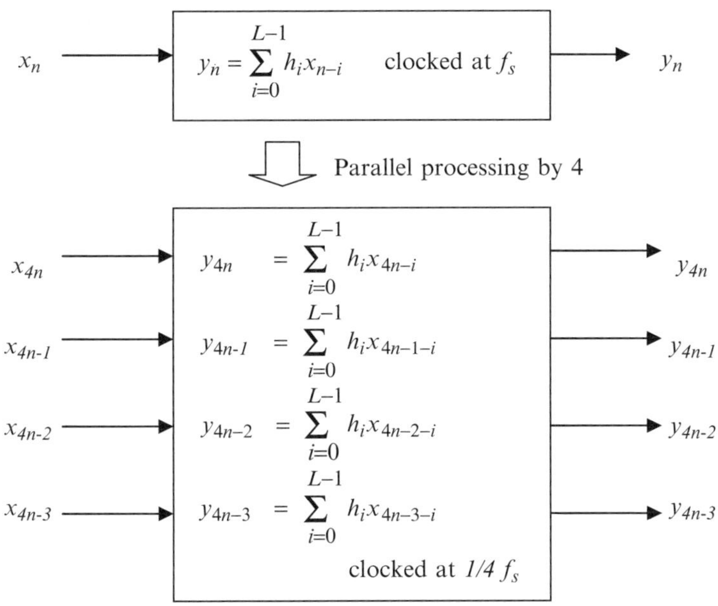

Fig. 13.7 Parallel processing implementation for high data throughput

For example, in a wireless communications system delivering uncompressed high-definition (HD) video, the communications signal bandwidth is close to 1 GHz or higher, and therefore it may be appropriate to make the clock rate, for example, one fourth $\left(\frac{1}{4}\right)$ of the sampling rate. The decision to do this is driven by the relatively slow process and standard cell speed compared to the data sample rate. The resulting filter has four parallel paths as shown in Fig. 13.7.

This multiple-input–multiple-output (MIMO) parallel filter replaces the original single-input–single-output (SISO) filter. The previously discussed structure options, such as pipelining and retiming, can also be applied to the MIMO filter. The MIMO system runs at a lower clock rate, processes four samples per clock cycle, and maintains the same total data throughput rate. This approach trades off space for time and power. Each filter coefficient is used four times in the MIMO filter and hence the combinational logic is four times larger. However, the size of the sequential logic does not necessarily change.

In high-rate communications systems, in order to keep the clock rate reasonable and clock plan simple, it may be desirable to apply the parallel processing technique to any high-rate bit operations such as scrambling, descrambling, interleaving, encoding or decoding, especially when considering giga-bit-per-second systems.

13.4.6 Low-Throughput Serialized Processing

So far this discussion has only covered the cases of data sampling rate close to or higher than the clock rate, which is likely the case for communication signals and video signals. For digital audio signal processing the clock rate is usually a few orders of magnitude higher than the audio sampling rate (32, 44.1, 48, 96, or 192 kHz) and often requires a very high filter order and wide word length to achieve a signal-to-noise ratio (SNR) of 100 dB or greater.

To save area a single generic multiplier and an adder can perform the computation in a serialized fashion. During each data sample period all the filtering calculations for one sample are carried out sequentially, which may involve coefficient multiplication, reading from and writing back to the memory, while cycling through all the coefficients. The procedure involves memory address incrementing (or decrementing) and hence one may find it convenient to address the memory in a circular way. This is similar to the way that software, which runs on DSP's or generic microprocessors, accesses memory.

13.4.7 Serial Processing of Multiple Channels

There are cases where the same operation is applied to multiple parallel data channels. Examples are: the same real-valued filter is applied to both in-phase (I) and quadrature (Q) components of a modulated signal, the same rate-converter is applied to both left (L) and right (R) components of a stereo audio signal, and the same processing is applied to all three components of a digital video signal. If speed allows, the signal components can be serialized to share the same combinational circuitry as illustrated in Fig. 13.8. The total number of delay units stays unchanged. The area saving comes from combinational computing hardware reuse because the identical operation is used on multiple data streams.

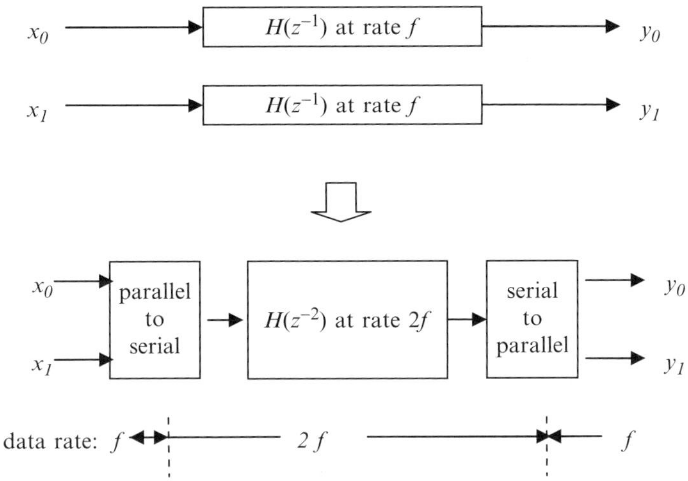

Fig. 13.8 Serial processing of multiple channel signals

13.4.8 Cascading

The discussion up to this point has been limited to direct-form filter realizations where the filtering function is realized without changing the multiplication coefficients from the original filter coefficients in the transfer function. Nondirect form filter structures, such as lattice structure realizations, are beyond the scope of this book, but there is one example that is simple and particularly convenient for high-speed implementation.

The transfer function of a larger filter can always be factored into several smaller filters. The subfilters can then be cascaded with data registers inserted as desired between filter stages. This technique is similar to pipelining where the subfilter stages are made timing independent. It should be mentioned that IIR filters are often implemented in second-order sections, also known as bi-quad filters, with optimally paired zeros and poles for numerical performance reasons.

13.5 Finite Number of Gates for Finite Field Operations

High data-rate encoding and decoding requires ASIC hardware implementation while lower data rate encoding and decoding can be implemented in software on a generic or special processor. For linear block codes based on GF algebra [6,7], GF algebraic operations need to be implemented in binary logic gates. For codes defined in GF(2^m), such as the widely used Reed-Solomon codes, both the encoder and decoder require computations in GF(2^m) operations. For codes defined in GF(2), such as the BCH codes, the decoder may also perform GF(2^m) operations.

While current ASIC synthesis tools are capable of performing a good job in synthesizing regular signed or unsigned integer adders and multipliers from simple RTL code lines in the form of $a+b$ and $a^*\ b$, it is often necessary to manually design lower-level RTL modules for the GF correspondence. For GF(2) the operations are straightforward: exclusive-or (XOR) for addition or subtraction and Boolean AND for multiplication. For $\mathrm{GF}(2^m)$, however, it is not always that straightforward. $\mathrm{GF}(2^m)$ addition and subtraction is simply a bit-wise XOR, similar to in GF(2). Division can be implemented as an inversion (reciprocal) followed by a multiplication.

There are a few ways to implement the $\mathrm{GF}(2^m)$ reciprocal. For high speed, one suitable approach is a straightforward m-bit to m-bit look-up table. The same technique also applies to logarithm and exponential operations in $\mathrm{GF}(2^m)$. While the sizes of these look-up tables for unary operations are acceptable, a straightforward look-up table for binary (two input variables) operations, such as multiplication, would be too large. The following section discusses how to realize high-speed multiplication in combinational logic with a small number of gates.

13.5.1 Multiplication

For prime p and nonzero m, field $\mathrm{GF}p^{(m)}$ consists of p^m elements, and is an extension field of $\mathrm{GF}(p)$. A root of a primitive irreducible polynomial $P(x) = x^m + p_{m-1}x^{m-1} + \cdots + p_1 x + p_0$ in $\mathrm{GF}p^{(m)}$, denoted as α, generates all nonzero elements of $\mathrm{GF}p^{(m)}$. From $P(\alpha) = 0$, $\alpha^m = p_{m-1}\alpha^{m-1} + \cdots + p_1\alpha + p_0$ follows, which suggests that any element in the field can be expressed as a polynomial of α with degree up to $m-1$, or as a vector of length m on basis $\{1, \alpha, \alpha^2, \ldots, \alpha^{m-1}\}$.

Multiplication of two elements can be carried out by multiplying them using polynomial representation of the field elements in two steps. First, the two polynomials of degree $m-1$, $A(x)$ and $B(x)$, are multiplied to yield a product polynomial $C(x) = A(x) \cdot B(x)$ of degree $2m-2$. Second, since the product element can also be expressed as a polynomial of α with degree up to $m-1$, the product polynomial $C(x)$ of degree $2m-2$ is reduced to polynomial $D(x)$ of degree $m-1$ with arithmetic modulo the field generator polynomial $P(x)$:

$$
\begin{aligned}
D(x) &= A(x) \cdot B(x) \bmod P(x) \\
&= \left(\sum_{i=0}^{m-1} a_i \alpha^i\right) \cdot \left(\sum_{i=0}^{m-1} b_i \alpha^i\right) \bmod P(x) \\
&= \left(\sum_{i=0}^{m-1} c_i \alpha^i\right) + \left(\sum_{i=m}^{2m-2} c_i \alpha^i\right) \bmod P(x) \\
&= \left(\sum_{i=0}^{m-1} c_i \alpha^i\right) + \left(\sum_{i=m}^{2m-2} c_i \left(\alpha^i \bmod P(x)\right)\right)
\end{aligned}
$$

where $c_0 = a_0 b_0$, $c_1 = a_0 b_1 + a_1 b_1$, and so on.

For efficient combinational circuit implementation, the modulo $P(x)$ operation is precomputed on individual terms as α^i mod $P(x)$, $i = m, m+1, \ldots, 2m-2$. Figure 13.9 illustrates an example of multiplication logic in GF(2^8), described in Verilog® syntax, where character "&" represents Boolean AND and "^" represents exclusive-OR (XOR).

In this approach, m^2 AND gates and roughly the same number of XOR gates are needed to implement the GF(2^m) multiplication in combinational logic. For a field with a larger dimension m, the subfield representation can be used to reduce the gate count to lower m^2.

It should be noted that it is not efficient to implement GF(2^m) multiplication by the procedure of logarithm table look-up, addition, and then exponential table look-up.

In case of multiplying by a constant, for a given constant B, the products between the basics and constant B should be precomputed. Then each vector element of product AB can be implemented by a linear combination of the vector elements of A, resulting in a much simpler logic.

13.5.2 Square and Square Root

GF(2^m) squaring can be implemented more efficiently than by using a generic multiplier. First, as a special case of multiplication, replace b with a in Fig. 13.9. As a result, the terms $c1$, $c3$, and so on become 0, term $c0$ is reduced to $a0$, term $c2$ is reduced to $a1$, and so on. It follows that the logic described as a generic multiplier in Fig. 13.9 will be reduced to the much smaller squaring logic shown in Fig. 13.10a.

The same squaring logic can also be derived from the basis representation. For element A in the field GF(2^m)

$$
\begin{aligned}
D(x) &= A(x) \cdot A(x) \bmod P(x) \\
&= \left(\sum_{i=0}^{m-1} a_i \alpha^i\right) \cdot \left(\sum_{i=0}^{m-1} a_i \alpha^i\right) \bmod P(x) \\
&= \left(\sum_{i=0}^{1m-1} a_i \alpha^{2i}\right) \bmod P(x) \\
&= \left(\sum_{i=m}^{2m-2} a_i \left(\alpha^{2i} \bmod P(x)\right)\right)
\end{aligned}
$$

The formula rearrangement suggests that the squaring operation on a field element can be obtained by applying the squaring operation to all the basis vectors $\{1, \alpha, \alpha^2, \ldots, \alpha^{m-1}\}$, which yields $\{1, \alpha^2, \alpha^4, \ldots, \alpha^{2m-2}\}$. An example of logic coded directly from this representation is shown in Fig. 13.10b, which is in fact same as the logic shown in (a).

```
wire [7:0] a, b;

wire c0 = (a[0]&b[0]);
wire c1 = (a[0]&b[1]) ^ (a[1]&b[0]);
wire c2 = (a[0]&b[2]) ^ (a[1]&b[1]) ^ (a[2]&b[0]);
wire c3 = (a[0]&b[3]) ^ (a[1]&b[2]) ^ (a[2]&b[1]) ^ (a[3]&b[0]);
wire c4 = (a[0]&b[4]) ^ (a[1]&b[3]) ^ (a[2]&b[2]) ^ (a[3]&b[1])
        ^ (a[4]&b[0]);
wire c5 = (a[0]&b[5]) ^ (a[1]&b[4]) ^ (a[2]&b[3]) ^ (a[3]&b[2])
        ^ (a[4]&b[1]) ^ (a[5]&b[0]);
wire c6 = (a[0]&b[6]) ^ (a[1]&b[5]) ^ (a[2]&b[4]) ^ (a[3]&b[3])
        ^ (a[4]&b[2]) ^ (a[5]&b[1]) ^ (a[6]&b[0]);
wire c7 = (a[0]&b[7]) ^ (a[1]&b[6]) ^ (a[2]&b[5]) ^ (a[3]&b[4])
        ^ (a[4]&b[3]) ^ (a[5]&b[2]) ^ (a[6]&b[1]) ^ (a[7]&b[0]);
wire c8 = (a[1]&b[7]) ^ (a[2]&b[6]) ^ (a[3]&b[5]) ^ (a[4]&b[4])
        ^ (a[5]&b[3]) ^ (a[6]&b[2]) ^ (a[7]&b[1]);
wire c9 = (a[2]&b[7]) ^ (a[3]&b[6]) ^ (a[4]&b[5]) ^ (a[5]&b[4])
        ^ (a[6]&b[3]) ^ (a[7]&b[2]);
wire ca = (a[3]&b[7]) ^ (a[4]&b[6]) ^ (a[5]&b[5]) ^ (a[6]&b[4])
        ^ (a[7]&b[3]);
wire cb = (a[4]&b[7]) ^ (a[5]&b[6]) ^ (a[6]&b[5]) ^ (a[7]&b[4]);
wire cc = (a[5]&b[7]) ^ (a[6]&b[6]) ^ (a[7]&b[5]);
wire cd = (a[6]&b[7]) ^ (a[7]&b[6]);
wire ce = (a[7]&b[7]);

wire [7:0] x7 = {c7, c6, c5, c4, c3, c2, c1, c0};
wire [7:0] x8 = {8{c8}} & 8'b00011101;
wire [7:0] x9 = {8{c9}} & 8'b00111010;
wire [7:0] xa = {8{ca}} & 8'b01110100;
wire [7:0] xb = {8{cb}} & 8'b11101000;
wire [7:0] xc = {8{cc}} & 8'b11001101;
wire [7:0] xd = {8{cd}} & 8'b10000111;
wire [7:0] xe = {8{ce}} & 8'b00010011;

wire [7:0] d = x7 ^ x8 ^ x9 ^ xa ^ xb ^ xc ^ xd ^ xe;
```

Fig. 13.9 Example of combinational logic for GF(256) multiplication

The GF(2^m) squaring operation described above is in fact a linear transform. The square root operation is simply the inverse of the squaring transform, and is also linear. Inverting the logic in Fig. 13.10 yields the square root logic shown in Fig. 13.11.

13.6 Efficient Implementation of Complex Number Operations

In modern wireless communications systems, modulated signals are complex-valued in nature. In the transmitters, the complex signal is generated from modulation – assigning data bits to the phase and magnitude values or the real and imaginary part values. In the receiver, a mixer produces in-phase (I) and quadrature (Q) components of the received modulated signal. Down stream of the receiver

```
wire [7:0] a;

wire [7:0] x7 = {1'b0, a[3], 1'b0, a[2],
                 1'b0, a[1], 1'b0, a[0]};
wire [7:0] x8 = {8{a[4]}} & 8'b00011101;
wire [7:0] xa = {8{a[5]}} & 8'b01110100;
wire [7:0] xc = {8{a[6]}} & 8'b11001101;
wire [7:0] xe = {8{a[7]}} & 8'b00010011;

wire [7:0] d = x7 ^ x8 ^ xa ^ xc ^ xe;
```

(**a**) Reduced from generic multiplier logic

```
wire [7:0] a;

wire [7:0] d = {a[6],
                a[3] ^ a[5] ^ a[6],
                a[5],
                a[2] ^ a[4] ^ a[5] ^ a[7],
                a[4] ^ a[6],
                a[1] ^ a[4] ^ a[5] ^ a[6],
                a[7],
                a[0] ^ a[4] ^ a[6] ^ a[7]};
```

(**b**) Same logic, obtained from squaring basic vectors

Fig. 13.10 Example of combinational logic for GF(256) square: (**a**) reduced from generic multiplier logic; (**b**) same logic, obtained from squaring basic vectors

```
wire [7:0] d;

wire [7:0] sqrt = {d[1],
                   d[7],
                   d[5],
                   d[3] ^ d[7],
                   d[5] ^ d[6] ^ d[7],
                   d[1] ^ d[3] ^ d[4] ^ d[5] ^ d[7],
                   d[2] ^ d[3] ^ d[5],
                   d[0] ^ d[1] ^ d[3]};
```

Fig. 13.11 Example of combinational logic for GF(256) square root

chain the I/Q signals are digitized, processed, and analyzed one way or another, which may include rotating the signal in the I–Q (real–imaginary) plane, calculating signal magnitude $A = |I + jQ| = \sqrt{|I|^2 + |Q|^2}$ or phase $\theta = \arctan(I/Q)$. This section discusses some techniques for efficient implementation of these operations.

13.6.1 Magnitude Estimation

Magnitude estimation is often involved in physical layer functions such as RSSI (received signal strength indication) measurement, AGC (automatic gain control) logic, and channel SNR (signal to noise ratio) evaluation. Without performing squaring and square root, the magnitude can be estimated as the follows:

$$A = |I + \mathrm{j}Q| = \sqrt{|I|^2 + |Q|^2} \approx \max(|I|, |Q|) + k \min(|I|, |Q|), \quad k \approx 0.4$$

where the absolute value of a negative number in 2's complimentary form can be implemented either in the mathematically correct way (inverse plus 1) or approximated by simply bit-wise inverse in order to save some time on the path. The value of k can be conveniently chosen to be 11/32, 3/8, or 1/2 for implementation simplicity.

Figure 13.12 shows the traces of magnitude estimation for a unit circle and lists the maximum and average magnitude errors in dB for several different $k$values. As shown, among the k values listed, 11/32 yields the lowest maximum magnitude error (0.49dB) while 13/32 leads to lowest average error (0.21dB).

Less complicated magnitude approximations that do not require comparators can also be used when accuracy is less of a concern, including simply $|I| + |Q|$ which has

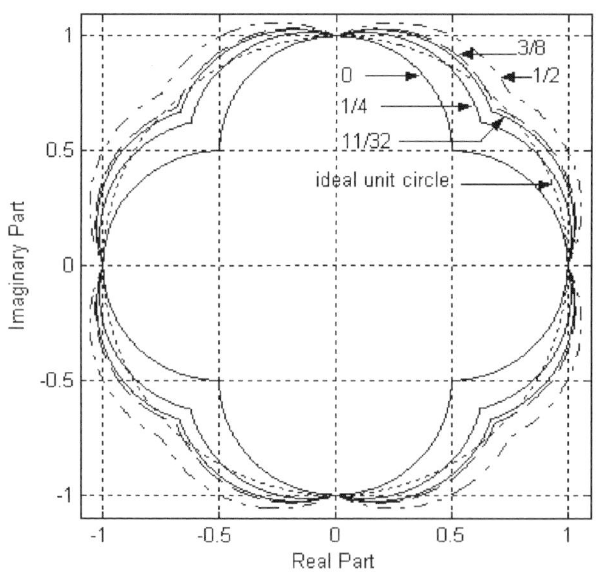

k	0	1/4	11/32	3/8	0.4	13/32	1/2
Max error (dB)	3.0	1.07	0.49	0.57	0.64	0.66	0.97
RMS error (dB)	3.0	0.37	0.24	0.22	0.21	0.21	0.25

Fig. 13.12 Magnitude estimation approximation

a maximum error of 3 dB, which is the same as the special case of the approximation with $k = 0$. A more accurate magnitude estimation algorithm featuring iterative approximation will be introduced later in this section.

13.6.2 Complex Multiplication

Multiplication between two complex numbers takes four real multipliers if done in a straightforward way, or three real multipliers and three adders as the following:

$$\begin{aligned} y &= (a+\mathrm{j}b)\cdot(c+\mathrm{j}d) \\ &= (ac-bd)+\mathrm{j}(bc+ad) \\ &= [(a+b)c-b(c+d)]+\mathrm{j}\,[(a+b)c+a(-c+d)] \end{aligned}$$

where the real and imaginary parts of the final line of the product in the last line above share the same first term $(a+b)c$. The size of the real multipliers in the 3-multiplier approach may grow by 1-bit on one input due to the addition/subtraction. The net area saving can be expected around 15% depending on word length.

13.6.3 CORDIC Algorithm and Implementation

CORDIC (COordinate Rotation DIgital Computer) is an iterative algorithm with only simple shift and addition/subtraction operations required and can be configured to implement a wide variety of elementary functions including trigonometric, linear, and other functions [8, 9]. In communications systems CORDIC can be applied to efficiently implement carrier frequency offset removal, phase or phase error measurement, I/Q conversion to magnitude/phase, magnitude/phase conversion to I/Q, multiplication by $\exp(\pm\mathrm{j}(2\pi/N)k)$ in FFTs and Direct Digital Synthesizers (DDS). In multimedia broadcasting systems it is also used to implement a Discrete Cosine Transform (DCT) as an element of video or audio image compression algorithms [10].

Rotation of a complex number $C = x+\mathrm{j}y$, or a vector (x, y) in the Cartesian plane by angle θ is represented by $C' = C\exp(\mathrm{j}\theta)$ or

$$\begin{aligned} x' &= x\cos\theta - y\sin\theta \\ y' &= y\cos\theta + x\sin\theta \end{aligned}$$

which can be rearranged as

$$\begin{aligned} x' &= \cos\theta(x - y\tan\theta) \\ y' &= \cos\theta(y + x\tan\theta) \end{aligned}$$

Table 13.1 CORDIC angle series (in degree)

step i	0	1	2	3	4	5	6	7	8	...
$\|\theta_i\|(^o)$	45	26.57	14.04	7.13	3.58	1.79	0.90	0.45	0.22	...

assuming that $C = x + jy$ is located on the right half plane, i.e. $-\pi/2 < \theta \leq \pi/2$. In cases where this is not true one can first apply a rotation by π or $\pi/2$, which is simply data negation and/or swapping. Next, any given angle θ can be approximated by the sum of a series of fixed angles θ_i defined by $\tan\theta_i = \pm 2^{-i}$, $i = 1, 2, 3, \ldots$, as listed in Table 13.1.

For a given rotation step i, multiplication by $(\tan\theta_i = \pm 2^{-i})$ can be implemented as a shift operation with possible negation; the common factor term $(\cos\theta_i)$ is a fixed positive scaling factor that maintains constant magnitude after rotation. Allowing the signal magnitude to be increased by $(1/\cos\theta_i)$, which is not a problem in most applications, cancels the term $(\cos\theta_i)$ and hence each rotation step can be expressed as

$$
\begin{aligned}
x' &= x - y \cdot 2^{-i} \cdot d_i \\
y' &= y + x \cdot 2^{-i} \cdot d_i
\end{aligned}
$$

where $d_i = \pm 1$ represents the direction of rotation. After the series of rotation steps the output vector will settle to the desired angle with its magnitude increased by

$$
A = \prod_i (1/\cos\theta_i) = \prod_i \sqrt{1 + 2^{-2i}} \approx 1.647
$$

Now look at configurations of CORDIC for different application scenarios. The first case is rotation by a predetermined angle or a small set of predetermined angles, such as in the case of an FFT. For a predetermined total phase of rotation, the direction (sign) of each step $d_i = \pm 1$ can be precalculated and then stored or hardwired in the design.

The next case is to rotate by a phase determined on the fly, such as an I/Q signal de-rotation to remove the carrier frequency offset. In this case it is necessary to have an angle accumulator or pipelined subtraction circuit initialized with the desired angle θ. Subtracting θ_i for each step, the direction of the next rotation step can then be determined by checking the sign of the residual angle, i.e.

$$
\begin{aligned}
a_{i+1} &= a_i - d_i \arctan(2^{-i}) \\
d_i &= \text{sign}(a_i)
\end{aligned}
$$

Another application is to measure phase and magnitude, such as for Cartesian to polar coordinate conversion. This is done by rotating the given vector towards the real axis (angle 0) and noting that the direction of next step is determined by the sign of the current residual imaginary part. The magnitude estimation is the final real part, subject to the constant gain of 1.647. The phase estimate can be obtained either by accumulating rotation angles of all steps on the fly, or by using a table look-up from a recorded direction sequence $\{d_0, d_1, \ldots, d_N\}$.

Polar to Cartesian coordinate conversion can also be realized similarly by rotation starting from the real axis and ending at the given angle. The resultant vector is the Cartesian coordinates. To remove the rotation gain of 1.647 the initial real part can be set as magnitude multiplied by (1/1.647) or 0.607. If starting from (0.607, 0) the rotation will generate the functions $(\cos\theta)$ and $(\sin\theta)$.

In some applications, it may be necessary to have one CORDIC rotation following another CORDIC with the later one determining the angle for the former one. In such cases, the rotation angles do not need to be calculated explicitly and only the 1-bit direction indicator for each step needs to be passed over. Besides saving the hardware for phase calculation on both sides, the rotation operation on the two sides can be pipelined and performed concurrently, leading to a saving in processing latency.

Most applications require 6–12 rotation steps to reach a reasonable phase accuracy. Depending on speed requirements and clocking scheme the rotation steps can be realized either serially (iterative operation on a set of hardware implementing a single rotation step) or in pipelined hardware stages. The implementation choices discussed in Sect. 13.4 can also be used for optimal implementation of a CORDIC operation.

13.7 Conclusions

This chapter discussed hardware implementation of multimedia communications systems from both ASIC design and architecture perspectives. Highly optimized design work is often the result of collaboration from both sides. The entire chapter is presented with emphasis on low-power and high-speed techniques, first by covering ASIC design flow and low-power ASIC design techniques and then by focusing on architectures for efficient ASIC implementation of common elements in multimedia communications systems including architecture choices for filters and general data paths alike, finite field operations, and complex number operations. To probe further, the books and articles listed next are good places to start.

References

1. N. E. Weste, K. Eshraghian (1993) *Principles of CMOS VLSI Design*, Addison-Wesley
2. J. M. Rabaey, A. Chandrakasan, and B. Nikolic (2002) *Digital Integrated Circuits – A Design Perspective*, Prentice-Hall
3. L. Benini, G. D. Micheli, and E. Macii (2001) "Designing Low-Power Circuits: Practical Recipes," *IEEE Circuits and Systems Magazine*, Vol. 1, No. 1, First Quarter 2001
4. S. K. Mitra (1998) *Digital Signal Processing – A Computer-Based Approach*, McGraw-Hill
5. K. Parhi (1999) *VLSI Digital Signal Processing Systems – Design and Implementation*, John Wiley & Sons, Inc.

6. I. S. Reed, X. Chen (1999) *Error-Control Coding for Data Networks*, Kluwer
7. S. B. Wicker (1995) *Error Control Systems for Digital Communication and Storage*, Prentice Hall
8. J. E. Volder (1959) "The CORDIC Trigonometric Computing Technique," *IRE Transactions on Electronic Computers*, Vol. EC-8, No. 3, pp. 330–334
9. Y. H. Hu (1992) "CORDIC-Based VLSI Architectures for Digital Signal Processing," *IEEE Signal Processing Magazine*, Vol. 9, No. 4, pp. 16–35
10. V. Bhaskaran and K. Konstantinides (1997) *Image and Video Compression Standards – Algorithms and Architectures*, Kluwer

Part III
Video Coding in Mobile Multimedia Broadcasting

Chapter 14
Objective and Subjective Assessment Methods of Video Quality in Multimedia Broadcasting

Harilaos G. Koumaras

14.1 Introduction

Current digital broadcasting network throughput rates are insufficient to handle raw video data in real time, even if low spatial and temporal resolution (i.e. frame size and frame rate) has been selected. Towards alleviating the network bandwidth requirements for efficient transmission of audiovisual content, coding/compression techniques have been applied on raw video data, performing compression on both temporal and spatial redundancy of the content.

More specifically, coding applications that are specialized and adapted in broadcasting digitally encoded audiovisual content have known an explosive growth in terms of development, deployment, and provision. Video coding is defined as the process of compressing and decompressing a digital video sequence, which results in lower data volumes, besides enabling the transmission of video signals over bandwidth-limited means, where uncompressed video signals would not be possible to be transmitted.

In this multievolutionary environment, the new era of digital video broadcasting has arrived and the beyond typical analog-based transmission for broadcasting services is a fact, setting new research challenges for the assessment of Perceived Quality of Service (PQoS) under the latest video encoding and broadcasting standards.

The majority of the compression standards have been proposed by the International Telecommunication Union (ITU) and the International Organization for Standardization (ISO) bodies, by introducing the following standards H.261, H.263, H.263+, H.263++, H.264, MPEG-1, MPEG-2, MPEG-4 and MPEG-4 Advanced Video Coding (AVC). Some of the aforementioned standards were developed in

H.G. Koumaras
NCSR Demokritos, Institute of Informatics and Telecommunications, Athens, Greece
e-mail: koumaras@iit.demokritos.gr

F.-L. Luo (ed.), *Mobile Multimedia Broadcasting Standards: Technology and Practice*,
DOI: 10.1007/978-0-387-78263-8_14,

partnership of ITU with Moving Pictures Expert Group (MPEG), under the venture name Joint Video Team (JVT), exploiting similar coding techniques developed by each body separately.

Each standard was designed and targeted a specific service and application, featuring therefore specific parameters and characteristics. For example, MPEG-1 was proposed by MPEG in order to be used by the Video Compact Disc (VCD) medium, which stores digital video on a Compact Disc (CD) with a quality almost similar to that of an analog VHS video. In 1994 MPEG-2 was proposed for encoding audiovisual content for broadcast signals, exploiting interlace format. MPEG-2 is also the coding format used by the widely successful commercial Digital Versatile Disc (DVD) medium, while the latest H.264, or MPEG-4 Part 10 AVC, aims at providing high broadcasting video quality at very low bit rates on a wide variety of applications, networks and systems or high definition resolution through Blu-Ray Discs and Multimedia content.

Video compression standards exploit in their algorithms the high similarity of the depicted data in the spatial, temporal, and frequency domain within and between subsequent frames of a video sequence [1]. Correlating this redundancy in these three domains, it has achieved data compression against a certain amount of visual data loss, which from the one hand cannot be retrieved but on the other hand it is not perceived by the viewers, since it is not conceived by the mechanisms of the Human Visual System (HVS).

Therefore, MPEG-based coding standards are characterized as lossy techniques, since they provide efficient video compression at the cost of a partial loss of the data and subsequently video quality degradation of the initial signal. Due to the fact that the parameters with strong influence on the video quality are normally those which play the most important role in the bit rate during the encoding/compression process, the issues of the user satisfaction and video quality assessment in correlation with the encoding parameters have been raised.

One of the future visions is the provision of audiovisual content at various quality and price levels. There are many approaches to this issue, one being the PQoS concept. The evaluation of the PQoS for audiovisual content will provide a user with a range of potential choices, covering the possibilities of low, medium, or high quality levels. Moreover, the PQoS evaluation gives the service provider and network operator the capability to minimize the storage and network resources by allocating only the resources that are sufficient to maintain a specific level of user satisfaction.

The evaluation of the PQoS is a matter of procedures, each time taking place after the encoding process (postencoding evaluation). The methods and techniques that have been proposed in the bibliography mainly aim at:

- Determining the encoding settings (i.e. resolution, frame rate, bit rate) that are required in order to carry out successfully a communication task of a multimedia application (i.e. video conference)
- Evaluating the quality level of a media clip based on the detection of artifacts on the signal caused by the encoding process

A content/service provider, depending on the content dynamics, must decide for the configuration of the appropriate encoding parameters that satisfy a specific level of user satisfaction.

Currently, the determination of the encoding parameters that satisfy a specific level of video quality is performed by recurring subjective or objective video quality assessments, each time taking place after the encoding process (repetitive postencoding evaluations). However, subjective quality evaluation processes of video streams require large amount of human resources, establishing it as an impractical procedure for a service provider. Similarly, the repetitive use of objective metrics on already encoded sequences may require numerous test encodings for identifying the specified encoding parameters, which is also time consuming and financially unaffordable from a business perspective.

Once the broadcaster has encoded appropriately the offered content at the preferred quality level, then the provision of the service follows. Digitally video encoded services, due to their interdependent nature, are highly sensitive to transmission errors (e.g. packet loss, network delay) and require high transmission reliability in order to maintain between sender and receiver devices their stream synchronization and initial quality level. Especially, in video broadcasting, which is performed over wireless environments, each transmitted from one end video packet can be received at the other end, either correctly or with errors or get totally lost. In the last two cases, the perceptual outcome is similar, since the decoder at the end-user usually discards the packet with errors, causing visual artifact on the decoded frame and therefore quality degradation.

In this context, this chapter discusses the various PQoS-related aspects that are involved in the end-to-end video quality assessment of MPEG-based broadcasting services, focusing on:

1. The assessment methods of the encoded broadcasting service, which aim at specifying a specific video quality level in terms of encoding bit rate and content dynamics
2. The impact of transmission impairments, such as the packet loss ratio during the transmission, on the respective cross QoS-related layers and delivered video quality level, respectively

The rest of the chapter is organized as follows: The next section presents and discusses both on the subjective and objective video quality assessment methods that have been proposed and are applied on digitally encoded and compressed video signals. Afterwards, the impact of various transmission conditions and impairments, such as packet loss occurred by bad transmission conditions, on the deduced perceptual quality is discussed. In this context, how this impact can be modeled in a deterministic way across the various QoS layers of the broadcasting service in terms of video quality degradation over error-prone transmission channel is described. Finally, the last section concludes the chapter.

14.2 Video Quality Assessment Methods at the Encoding/Generation Phase

The advent of quality evaluation was the application of pure error-based sensitive framework between the encoding and the original/uncompressed video sequence. Although these primitive methods provided a quantitative approach of the quality degradation between the encoded and original signals, they did not provide reliable measurements of the perceived quality, because they miss out the characteristics and sensitivities of the HVS. In this context, the most widely used primitive methods and quality metrics that used an error sensitivity framework are the peak signal to noise ratio (PSNR) and the mean square error (MSE).

Over the last years, emphasis has been put on developing methods and techniques for evaluating the perceived quality of digital video content mainly during the encoding process. These methods are categorized into two classes [2]: the subjective and objective ones.

- The subjective test methods involve an audience of people, who watch a video sequence and score its quality as perceived by them, under specific and controlled watching conditions.

The subjective assessment methods are further classified into classes depending on the test procedure, which may include simultaneous viewing of both degraded and original video signal (double stimulus methods) or of only one signal at a time (single stimulus methods).

- The objective evaluation methods, which successfully emulate the results that are derived from subjective quality assessments, based on criteria and metrics that can be measured objectively

The objective assessment methods are further classified into classes, according to the procedure of the quality evaluation, depending on the requirement of the initial uncompressed and nondegraded content into the evaluation process. Based on this categorization, the assessment methods are named as full reference, reduced reference, or no-reference methods, representative of the requirement or not of the initial uncompressed signal. The next sections discuss and present the most popular subjective and objective assessment method categories.

14.2.1 Subjective Assessment Methods

The subjective test methods have been mainly proposed by International Telecommunications Union (ITU) and Video Quality Experts Group (VQEG), involving an audience of people, who watch and score the quality of a video sequence as perceived by them, under specific and controlled watching conditions. Afterwards, the statistical analysis of the collected data is used for the evaluation of the perceived quality, usually exploiting the Mean Opinion Score (MOS) as the most reliable and typical metric of quality measurement.

Subjective test methods are described in ITU-R Rec. BT.500-11 [3] and ITU-T Rec. P.910 [4], suggesting specific viewing conditions, criteria for observers and test material selection, assessment procedure description and statistical analysis methods. The ITU-R Rec. BT.500-11 described subjective methods that are specialized for television applications, whereas ITU-T Rec. P.910 is intended for multimedia applications.

The subjective methods depending on the number of the simultaneous sequences under test are mainly classified as single or double stimulus when one or two signals are used, respectively [5].

In the single stimulus (SS) methods multiple separate scenes are shown simultaneously and the viewers are asked to evaluate each one separately. Depending on the playback order of the test signals, there SS methods are classified into two approaches: SS with no repetition of test scenes and SS where the test scenes are repeated multiple times. Three different scoring methods are used: adjectival, numerical, and noncategorical (i.e. a continuous scale with no numbers). Representative single stimulus methods are the following:

- Single Stimulus Method (SSM)
- Absolute Category Rating (ACR)
- Single Stimulus Continuous Quality Evaluation (SSCQE)

In the double stimulus methods, the observers watch multiple references and degraded scene pairs. The order of the reference scene relative to the degraded one may differ depending on the implemented method. Also the viewers may not be aware of which signal is the reference and/or the degraded one. Scoring is usually on an overall impression scale of impairment either using adjectival or noncategorical scale. Representative double stimulus methods are the following:

- Double Stimulus Continuous Quality Evaluation (DSCQE)
- Double Stimulus Impairment Scale (DSIS)
- Degradation Category Rating (DCR)
- Pair Comparison Method (PC)

14.2.2 Objective Assessment Methods

The preparation and execution of subjective tests is costly and time consuming and its implementation today is mainly limited to scientific purposes, especially at VQEG experiments.

For this reason, a lot of effort has recently been focused on developing cheaper, faster, and easier applicable objective evaluation methods. These techniques successfully emulate the subjective quality assessment results, based on criteria and metrics that can be measured objectively. The objective methods are classified according to the availability of the original video signal, which is considered to be of high quality.

The majority of the proposed objective methods in the literature require the undistorted source video sequence as a reference entity in the quality evaluation process, and due to this they are characterized as full reference methods [6]. The methods perform multiple channel decomposition of the video signal, where the proposed objective method is applied on each channel, which features a different weigh factor according to the characteristics of the HVS. The basic block diagram of the full reference methods with multiple channels is depicted in Fig. 14.1. These methods emulate characteristics of the HVS using Contrast Sensitivity Functions (CSF),

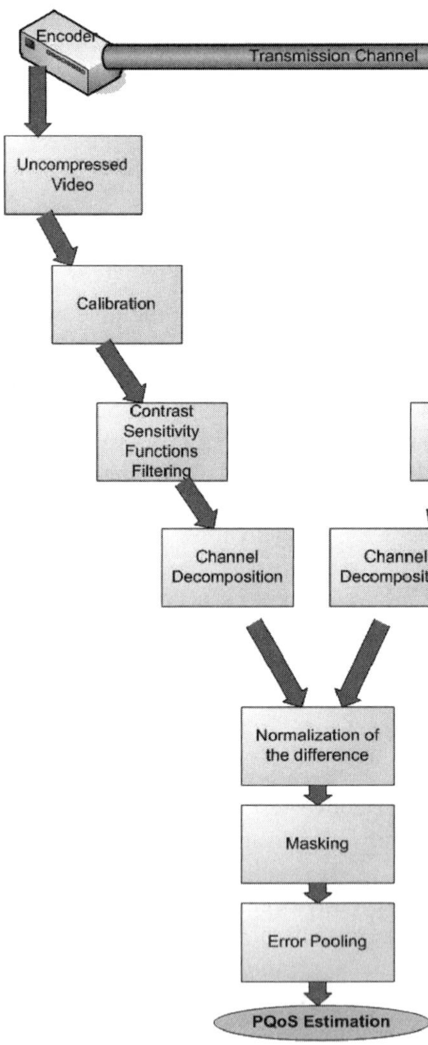

Fig. 14.1 Full reference methods with multiple channels

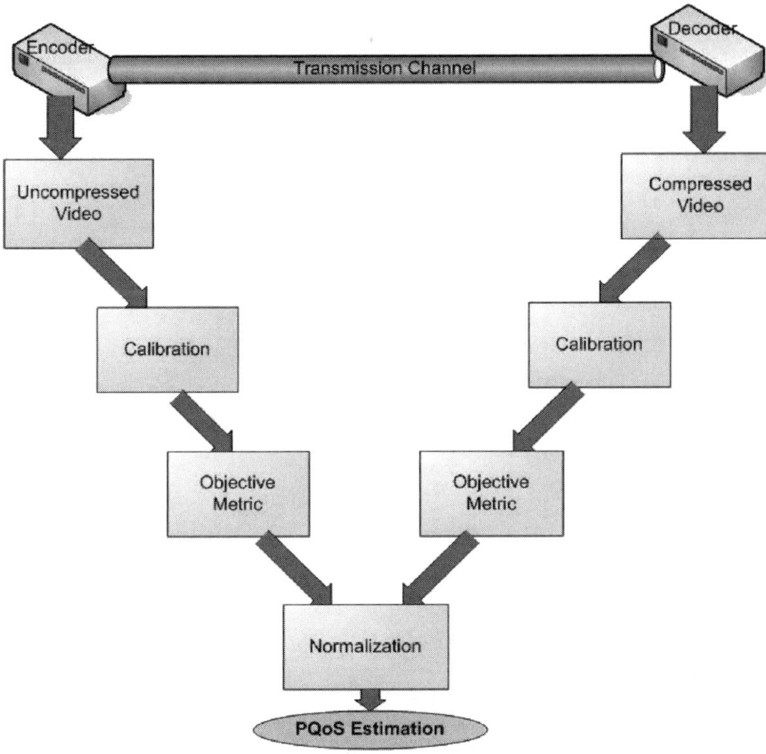

Fig. 14.2 Full reference methods with single channel

Channel Decomposition, Error Normalization, Weighting and finally Minkowski error pooling for combining the error measurements into single perceived quality estimation [7].

Similarly, in the bibliography, some full reference methods of single channel have been proposed, where the proposed objective metric is applied homogeneously on the video signal, without considering varying weight functions. The block diagram of these methods is depicted in Fig. 14.2. However, it has been reported [8,9] that in specific cases these complicated methods do not provide more accurate results than the simple mathematical measures. Due to this some new full reference metrics that are based on the video structural distortion, and not on error measurement, have been proposed [7, 10–13].

On the other hand, the fact that these methods require the original video signal as reference deprives their use in commercial video service applications, where the initial undistorted clips are not always accessible. Moreover, even if the reference clip is available, then synchronization predicaments between the undistorted and the distorted signal (which may have experienced frame loss) make the implementation of the full reference methods difficult and impractical.

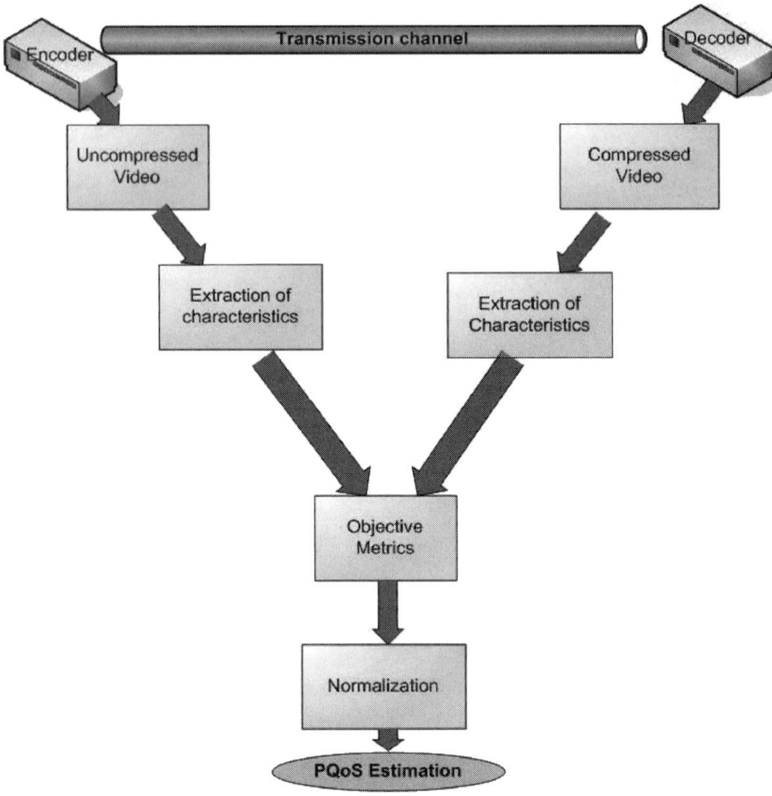

Fig. 14.3 Reduced reference methods

Due to these reasons, the recent research has been focused on developing methods that can evaluate the PQoS level based on metrics, which use only some extracted structural features from the original signal (reduced reference methods) [14, 15]. The block diagram of the reduced reference methods is depicted in Fig. 14.3.

Finally, some methods and techniques have been proposed in the bibliography that do not require any reference video signal (no-reference methods) [16, 17].

Nevertheless, due to the fact that the future vision is the provision of audiovisual content at various quality and price levels [18], there is a great need for developing methods and tools that will help service providers to predict quickly and easily the PQoS level of a media clip. These methods will enable the determination of the specific encoding parameters that will satisfy a certain quality level. All the previously mentioned postencoding methods may require repeating tests in order to determine the encoding parameters that satisfy a specific level of user satisfaction. This procedure is time consuming, complex, and impractical for implementation on the broadcasting services.

Towards this, recently research has been performed in the field of preencoding estimation and prediction of the PQoS level of a multimedia service as a function of the selected resolution and the encoding bit rate [19–24]. These methods provide fast and quantified estimation of the PQoS, taking into account the instant PQoS variation due to the spatial and temporal activity within a given encoded sequence. Quantifying this variation by the mean PQoS (MPQoS) as a function of the video encoding rate and the picture resolution, the MPQoS is finally used as a metric for preencoding PQoS assessment based on the fast estimation of the spatiotemporal activity level of a video signal.

14.3 Video Quality Issues During Service Transmission: Translation Between PQoS, AppQoS, and NQoS

This section discusses how the transmission errors and impairments of the transmission channel are mapped to the various QoS-related layers of the broadcasting service. More specifically the following subsections refer to the Perceived QoS (PQoS), Application QoS (AppQoS), and Network QoS (NQoS) layers, discussing the various aspects of their cross mapping.

Once the broadcaster has encoded appropriately the offered content at the preferred quality level, then the provision of the service follows. Digitally video encoded services, due to their interdependent nature, are highly sensitive to transmission errors (e.g. packet loss) and require high transmission reliability in order to maintain between sender and receiver devices their stream synchronization and initial quality level. Especially, in video broadcasting, which is performed over wireless environments, each transmitted from one end video packet can be received at the other end, either correctly or with errors or get totally lost. In the last two cases, the perceptual outcome is similar, since the decoder at the end-user usually discards the packet with errors, causing visual artifact on the decoded frame and therefore quality degradation.

The issue of mapping the perceptual impact of transmission errors (like packet loss) during the broadcasting on the delivered perceptual video quality at the end-user is a fresh topic in the field of video quality assessment since the relative literature appears to be limited with a small number of relative published works.

In this framework, Kanumuri et al. [25] proposed a very analytical statistical model of the packet-loss visual impact on the decoding video quality of MPEG-2 video sequences, specifying the various factors that affect the perceived video quality and visibility (e.g. maximum number of frames affected by the packet loss, on what frame type the packet loss occurs, etc.). However, this study focuses mainly on the pure study of the MPEG-2 decoding capabilities, without considering the parameters of the digital broadcasting or the latest encoding standards.

Similarly, in [26] is presented a transmission distortion model for real-time video streaming over error-prone wireless networks. In this work, an end-to-end video distortion study is performed, based on the modeling of the impulse propagation error (i.e. the visual fading behavior of the decoding artifact).

Table 14.1 Metrics of each QoS level

Service QoS level	Application Qos level	Network QoS level
User satisfaction	Decodable frame rate	Packet loss ratio
PQoS level	Decoding threshold	Packet loss scheme
Terminal specifications	Encoding parameters	Packet size

The deduced model, although it is very accurate and robust, enabling the media service provider to predict the transmission distortion at the receiver side is not a generic one. On the contrary, it is highly dependent on the video content dynamics and the selected encoder settings. More specifically, an initial quantification of the spatial and temporal dynamics of the content, which will allow the appropriate calibration of the model, is required. This prerequisite procedure (i.e. adapting the impulse transmission distortion curve based on the least mean square error criteria) is practically inapplicable by an actual content creator/provider. Moreover, the strong dependence of the proposed model on the spatiotemporal dynamics of the content deprives its implementation on sequences with long duration and mixed video dynamics, since not a unique impulse transmission distortion will be accurate for the whole video duration.

Regarding the mapping between the various discrete QoS layer (i.e. PQoS/AppQoS/NQoS), Table 14.1 defines the representative metrics of each level, which must be used and considered into any relative mapping process or model.

At the service QoS level, the critical metric is the user satisfaction (i.e. PQoS level). The evaluation of the PQoS for audiovisual content will provide a user with a range of potential choices, covering the possibilities of low, medium, or high quality levels (i.e. gold, silver, and bronze services). Moreover, the PQoS evaluation gives the service provider and network operator the capability to minimize the storage and network resources by allocating only the resources that are sufficient to maintain a specific level of user satisfaction.

As it has been already mentioned and explained in the previous sections, the evaluation of the PQoS is a matter of objective and subjective evaluation procedures, each time taking place after the encoding process (postencoding evaluation). Subjective quality evaluation processing of video streams (PQoS evaluation) requires large amount of human resources, establishing it as a time-consuming process (e.g. large audiences evaluating video/audio sequences). Objective evaluation methods, on the other hand, can provide PQoS evaluation results faster, but require large amount of machine resources and sophisticated apparatus configurations. Towards this, objective evaluation methods are based and make use of multiple metrics, which are related to the content's artifacts (i.e. tilling, blurriness, error blocks, etc.) resulting from the quality degradation due to the encoding process (see Figure).

At the AppQoS level, given that during the encoding process quality degradation of the initial video content (see Fig. 14.4) is incurred, the values of the encoding parameters (i.e. bit rate, resolution) play a major role in the resulting PQoS. Thus, the various encoding parameters must be used as metrics in quantifying the deduced

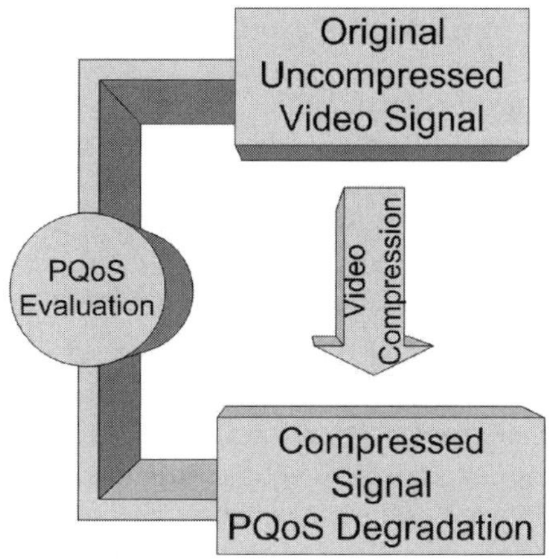

Fig. 14.4 Concept of the PQoS evaluation

PQoS level. If we also consider additional degradation due to transmission problems (i.e. limited bandwidth, network congestion), which finally result in packet loss at the video packet receiver during the service transmission, then the decodable frame rate can be considered as a metric for quantifying this phenomenon. The decodable frame rate Q is an application-level metric, with values ranging from 0 to 1.0. The larger the value of Q, the higher the successful decoding rate at the end-user. Q is defined as the fraction of decodable frame rate, which is the number of decodable frames (i.e. frames that are theoretically able to be decoded without considering the postfiltering or error concealment abilities of each decoder) over the total number of frames sent by a video source.

Since different codec and transmission techniques have different tolerance to packet loss, the theoretically expected decoding threshold will be also used as a metric in order to define the impact of the packet loss ratio on the frame loss ratio. A theoretical decoding threshold equal to 1.0 means that only one packet loss results in unsuccessful decoding of the corresponding frame, to which the missing packet is a part of.

Finally, at the NQoS level the metrics packet loss ratio, packet loss scheme and packet size may be considered as key parameters. Although, it is obvious that other network statistics and phenomena may be present over a broadcasting network (e.g. jitter, delay), however all these parameters are quantified into the packet loss effect, since this is the final outcome of all these network QoS-sensitive parameters at the video packet receiver. Otherwise, if no packet loss occurs due to these phenomena, then sophisticated buffer techniques may eliminate their impact. Thus, with the appropriate approach the packet loss ratio can be considered as adequate parameter and used as a network metric to the PQoS–NQoS and NQoS–PQoS mapping. Regarding the various packet loss schemes (e.g. unified, bursty, etc.), due to the stochastic nature of the PQoS degradation over an error-prone broadcasting channel, for reference purposes focus must be given on identifying the packet loss scheme, which provides the worst-case scenario in terms of affecting the decodable frame rate (i.e. the delivered PQoS level) for specific packet loss ratio.

More detailed explanation of each metric and description of its scope and role is presented in the following subsections following two discrete directions from PQoS down to NQoS and the opposite one from the NQoS up to the PQoS.

14.3.1 PQoS to AppQoS and NQoS Mapping

The mapping of the PQoS to the AppQoS covers the relationship between the service and the application level. Based on a predefined perceptual quality at PQoS, then the appropriate parameters at the application level (frame rate, bit rate, codec) are determined. The mapping is based on empirical data that are derived from subjective or objective quality assessments for different genres of content.

Concerning the initial preparation of the content at the requested/targeted PQoS level, a method for mapping the content dynamics/genre of the video to the encoding parameters that satisfy the requested/specific level of user satisfaction is necessary. Taking into account the instant PQoS variation due to the spatiotemporal activity within a given MPEG encoded content, the respective mean PQoS (MPQoS) as a function of the video encoding rate can be exploited as a metric for objective video quality assessment. Based on the proposed metric, this task will derive a reference table containing the encoding bit rate that satisfies specific quality (i.e. PQoS) levels depending on the spatiotemporal activity of the requested content.

Towards the specification of these reference MPQoS vs. bit rate rules, an objective quality meter tool of the PQoS level may be used, providing objective PQoS assessment for each frame within a video clip. The graphical representation of these results vs. time demonstrates the instant PQoS of each frame within the video clip, besides indicating the mean PQoS (MPQoS) of the entire video (for the whole clip duration). Similar experiments will be conducted for the MPQoS calculation of the same video content, each time applying different encoding parameters. The results of these experiments will be used to draw up experimental curves of the MPQoS of the given video content, as a function of the encoding parameters. The same procedure will be repeated for a set of video sequences, each one with different spatiotemporal activity level.

More specifically, considering three discrete spatiotemporal categories (i.e. high, medium, low) and their respective MPQoS vs. bit rate equations, the service provider should be able to specify the bit rate that satisfies a specific perceptual quality level. Figure 14.5 depicts the concept of this approach and the expected form of the PQoS vs. bit rate dependence as it has been reported in the relative literature [21, 27] due to the logarithmic sensitivity of the HVS.

As depicted in Fig. 14.5, curve (A) represents a video clip with low temporal and spatial dynamics, i.e. whose content has "poor" movements and low picture complexity. Such a curve can be derived, for example from a talk show. Curve (C) represents a short video clip with high dynamics, such as a football match. Curve (B) represents an intermediate case (a music video clip). Each curve and therefore each video clip can be characterized by: (a) the low bit rate (BR_L), which corresponds to

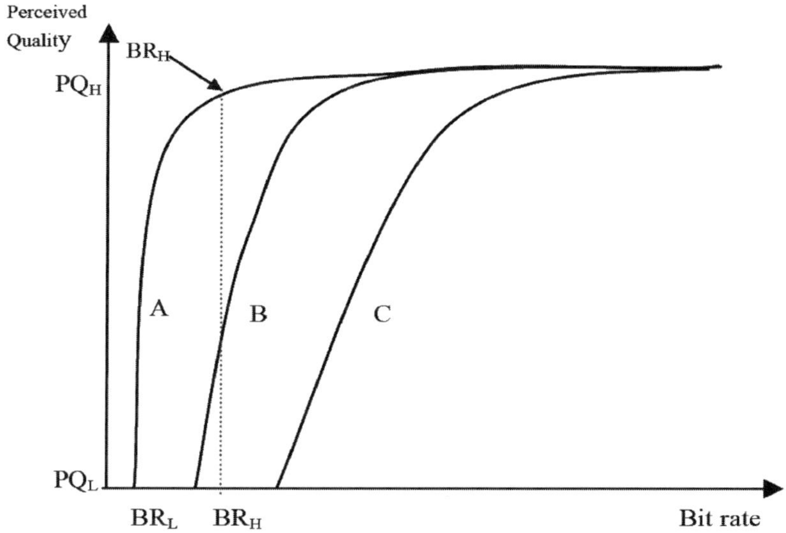

Fig. 14.5 PQoS vs. bit rate curves for various spatiotemporal contents

the lower value of the accepted PQoS (PQ_L) by the end-user (i.e. bronze service), (b) the high bit rate (BR_H), which corresponds to the minimum value of the bit rate for which the PQoS reaches its maximum value (PQ_H) (i.e. gold service, and (c) the shape of the curve, which is defined by the content dynamics. These parameters can be experimentally derived for reference purposes and further used for defining a generic equation for describing the MPQoS vs. bit rate curves. Based on relative published research [21], the respective MPQoS vs. bit rate curves are successfully modeled as follows:

$$\mathrm{MPQoS} = [\mathrm{PQ_H} - \mathrm{PQ_L}](1 - \mathrm{e}^{-\alpha[\mathrm{BR}-\mathrm{BR_L}]}) + \mathrm{PQ_L}, \ \alpha > 0 \text{ and } \mathrm{BR} > \mathrm{BR_L}$$

where the parameter α is the time constant of the exponential function, which determines the shape of the curve and BR the encoding bit rate of the service.

The preencoding PQoS assessment nature of the described procedure alleviates both the machine resource requirements and the time consumption of the already existing postencoding video quality assessment methods, making PQoS evaluation quick, easy, and economically affordable for commercial implementations.

The mapping of AppQoS to NQoS deals with the translation of application level parameters to parameters of the underlying network level. Multimedia services, especially broadcasting applications, tend to impose great demands on the communication networks concerning bandwidth, maximum tolerable delay, jitter, and packet loss. A pessimistic estimation of the required network resources might lead to over-provisioning of resources for a single multimedia service, resulting in bad link utilization and a waste of network resources. On the other hand, a too optimistic mapping bears the risk of congestion within the network resulting in packet loss that decreases the end-to-end QoS. Therefore, a trade-off between these extremes has to be envisaged.

In this context, the most relevant parameter at the upper application layer with direct impact to the NQoS is the video bit rate. As it has been already mentioned earlier, the selection of the appropriate video bit rate is influenced by a variety of factors, such as the content dynamics, the video codec, and the fidelity of the encoding.

Based on the selected video bit rate at the application level, the required network bandwidth can be derived, based on the overhead that is introduced by the protocol stack. In a typical digital broadcasting scenario, a considerable amount of headers is added to the actual media payload. This results in an overhead for each MPEG transport stream packet that is transmitted over the network.

14.3.2 NQoS to AppQoS and PQoS Mapping

Concerning the mapping of the Network QoS sensitive parameters (like delay, packet loss, etc.) to perceived video quality (i.e. PQoS) some approaches have already been proposed in the literature, which perform a very analytical statistical model of the packet-loss visual impact on the decoding video quality for MPEG-2 video sequences, specifying the various factors that affect perceived video quality and visibility (e.g. maximum number of frames affected by the packet loss, on what frame type the packet loss occurs, etc.). Similarly, a transmission/distortion modeling for real-time video streaming over error-prone wireless networks has also been presented, where a modeling of the impulse transmission distortion (i.e. the visual fading behavior of the transmission errors) is performed.

However, all the already proposed models are very codec and content specific, while they do not also provide any end-to-end video quality estimation, namely the degradation during the encoding process and the transmission/streaming procedure. In this framework, once the content has been prepared for delivery at the requested PQoS level, according to the reported monitored network conditions (e.g. packet loss rate) and the reference look-up tables/rules of the NQoS to PQoS mapping, the worst-case degradation to the application QoS (i.e. undecoded or lost frames) and to the service QoS (percentage of the total duration for which the end-user will experience degraded PQoS, i.e. delivered PQoS $<$ requested PQoS) could be able to predicted. According to this representation, a preprovision assessment of the end-to-end PQoS degradation will be performed. Thus, it is necessary to develop mapping rules between the application level (e.g. decodable frame rate) and the network level (i.e.. packet loss, packet size, packet loss scheme) parameters. Of course, the decodable frame rate metric of the application level may be extended as a metric to the service level, representing the duration percentage of the requested or no-degraded PQoS level.

It must be noted that during this whole mapping procedure, the sophisticated delay and delay variation phenomena may be not taken under consideration, since it can be supposed that they are successfully managed by efficient play-out buffer structures or they will eventually result in packet loss. So, the ultimate goal of the decision-taking is to find parameters for both the application and network levels, which do not violate the constraints that were imposed at the service level in terms of PQoS.

More specifically, regarding the application QoS and network QoS mapping, the translations between network packet loss ratio and decodable frame rate (Q), as well as packet size and decodable frame rate, can be exploited. Q is an application-level metric, with values ranging from 0 to 1.0. The larger the value of Q, the higher the successful decoding rate at the end-user. Q is defined as the fraction of decodable frame rate, which is the number of decodable frames over the total number of frames sent by a video source.

$$Q = \frac{N_{\text{dec}}}{(N_{\text{total}-\text{I}} + N_{\text{total}-\text{P}} + N_{\text{total}-\text{B}})}$$

where N_{dec} is the sum of number of theoretically expected to be successfully decoded I, P, B frames, i.e., $N_{\text{dec}-\text{I}}$, $N_{\text{dec}-\text{P}}$, and $N_{\text{dec}-\text{B}}$, without taking under consideration the postfiltering and error concealment techniques of the codec.

Due to the fact that the frames in an MPEG video sequence are interdependent, considering a packet loss, the visual distortion due to this packet loss will not be limited only to the frame, on which the specific lost packet belongs to. On the contrary, spatial error propagation will take place, which will infect all the frames that are interdependent to the specific frame, on which a packet loss occurred. Thus, in order to calculate the theoretically expected error propagation due to a packet loss, one must take under consideration the impulse transmission of the distortion.

From the hierarchical structure of MPEG encoding stream, a video frame may be considered theoretically undecodable directly or indirectly:

- Directly undecodable when the packet loss occurred in a group of packets that carry the data of the specific frame
- Indirectly undecodable, when the packet loss occurred in a group of packets that carry the data of another frame, from which the current frame is directly depended and its successful decoding depends on the successful decoding of the corrupted frame

For specifying the theoretically worst-case scenario in order to avoid the stochastic nature of the packet loss effect on the PQoS degradation, we will not consider any concealment method. So the decodable threshold (DT) is 1.0 (i.e. even one direct or indirect packet loss causes an undecodable frame). Therefore, our analysis will provide the worst case of video transmission quality degradation.

Due to the very specific structure of an MPEG stream (i.e. GOP type), which is specified by successive I, P, and B frames, given the deterministic fact that one packet loss results in the corresponding frame loss (i.e. DT $= 1.0$), then the whole mapping between AppQoS and NQoS can be mathematically modeled. More specifically, given a GOP structure, which is described by two parameters GOP(N, M), where N defines the GOP length (i.e. the number of frames of each GOP) and the $M - 1$ is the number of B frames between I–P or P–P frames, and taking under consideration the decoding inter-dependencies among the three frame types, then the impact of the packet loss ratio will be mathematically and deterministically formulated. In [28], it is presented the described theoretical mathematical model, which for given packet loss ratio and packet loss scheme provides the worst-case theoretically expected decodable frame rate, without considering the decoding and error concealment capabilities of each decoder. The proposed model of the theoretically expected is summarized in the following equation:

$$Q=\frac{N_{\text{dec}}}{(N_{\text{total}-\text{I}}+N_{\text{total}-\text{P}}+N_{\text{total}-\text{B}})}=\frac{N_{\text{dec}-\text{I}}+N_{\text{dec}-\text{p}}+N_{\text{dec}-\text{B}}}{(N_{\text{total}-\text{I}}+N_{\text{total}-\text{P}}+N_{\text{total}-\text{B}})}$$

$$=\frac{(1-p)^{C_\text{I}}*N_{\text{GOP}}+(1-p)^{C_\text{I}}*\sum_{j=1}^{N_\text{P}}(1-p)^{jC_\text{P}}*N_{\text{GOP}}+\left[(1-p)^{C_\text{I}+N_\text{P}C_\text{P}}+\sum_{j=1}^{N_\text{P}}(1-p)^{jC_\text{P}}\right]*(M-1)*(1-p)^{C_\text{I}+C_\text{B}}*N_{\text{GOP}}}{(N_{\text{total}-\text{I}}+N_{\text{total}-\text{P}}+N_{\text{total}-\text{B}})}$$

where $C_\text{I}C_\text{P}C_\text{B}$ are the mean number of packets that transport the data of each frame type, p is the packet loss rate, N_{GOP} is the total number of GOPs in the video flow, N_{dec} is the total number of decodable frames in the video flow, $N_{\text{dec}-\text{I}}N_{\text{dec}-\text{P}}N_{\text{dec}-\text{B}}$ are the number of decodable frames in each type and $N_{\text{total}-\text{I}}N_{\text{total}-\text{P}}N_{\text{total}-\text{B}}$ are the total number of each type of frames.

The validity of this theoretical and mathematical framework has been examined in [28] by performing experiments using the ns-2 simulation platform for uniform packet loss distribution. As it has been deduced and shown on the cited paper, the dependence of the theoretically expected decodable frame rate and packet loss rate can be successfully described by an equation of the following form:

$$Q=C_1\,\text{Ln}(p)-C_2$$

Especially for the case of packet size equal to 1000 bytes and based on the performed simulation, the above equation is specialized to

$$Q=-0.3211\,\text{Ln}(p)-0.4094 \text{ with } R^2=0.9971$$

For the derivation of the above equation the random uniform model has been used, which provides the distributed losses with the mean loss rate (p) and corresponds to the worst-case packet loss scenario, given that we have considered DT equal to 1.0. Regarding the AppQoS to the PQoS mapping, the application-layer Q metric can be extended to the service level, by expressing it in terms of duration percentage for error-free video transmission.

14.4 Conclusions

Digital video coding techniques have already prevailed in the upcoming broadcasting services and applications, enabling the provision of digital video content over various bandwidth-limited means and computationally low terminals. Video compression algorithms exploit the redundancy that a video signal contains in the spatial, temporal, and frequency domain. Thus, by removing this redundancy in these three different domain types, it has achieved high compression of the data with cost the perceptual degradation of the content.

This chapter outlines the various PQoS evaluation methods and comments their efficiency. These methods can be mainly categorized into two major classes: The subjective and objective ones. The subjective test methods involve an audience of people, who watch a video sequence and evaluate its quality as perceived by them, under specific and controlled watching conditions. The objective methods successfully emulate the subjective quality assessment results, based on criteria and metrics that can be measured objectively. These objective methods are classified, according to the availability of the original video signal to full reference, reduced reference, and no reference.

Finally, the chapter discusses how the transmission errors and impairments of the transmission channel are mapped to the various QoS-related layers of the broadcasting service. More specifically, it is examined how the PQoS, AppQoS, and NQoS layers are cross-related when transmission predicaments are present. In this context, the effect of the packet loss ratio on the theoretically expected ratio of decodable frames is discussed, describing how the interdependencies of the encoded frames create error propagation.

Acknowledgments Part of the work in this chapter has been supported and was carried out within the framework of the Information Society Technologies (IST) Integrated Project ENTHRONE phase II/ FP6–38463.

References

1. Richardson I E G (2004) *H.264 and MPEG-4 Video Compression*, ed., Wiley, ISBN: 978-0-470-84837-1, October 2003, USA.
2. Engelke U, Zepernick H-J (2006) Perceptual quality measures for image and video services. *Euro-NGI Workshop on Socio-Economic Aspects of Next Generation Internet*, Lyngby, Denmark, October 9–10, 2006.
3. ITU (2000) *Methodology for the Subjective Assessment of the Quality of Television Pictures.* Recommendation ITU-R BT.500–10.
4. ITU (1999) *Subjective Video Quality Assessment Methods for Multimedia Applications.* Recommendation ITU-T P.910.
5. Arriba C (2007) *Subjective Video Quality Evaluation and Estimation for H.264 Codec and QVGA Resolution Sequences.* Institut fur Nachrichtentechnik und Hochfrequenztechnik Fakultat fur Elektrotechnik und Informationstechnik, Technischen Universitat Wien.
6. Silva E A, Panetta K, Agaian S (2007) Quantifying image similarity using measure of enhancement by entropy. In *Mobile Multimedia/Image Processing for Military and Security Applications 2007*, Sos S Agaian, Sabah A Jassim, Editors, 65790U, Proceedings of SPIE 6579.
7. Wang Z, Sheikh H R and Bovik A C (2003) Objective video quality assessment. In *The Handbook of Video Databases: Design and Applications*, B Furht and O Marqure, Editors, CRC Press. p. 1041–1078, USA.
8. VQEG (2000) *Final Report from the Video Quality Experts Group on the Validation of Objective Models of Video Quality Assessment.*
9. Wang Z, Bovik A C, Lu L (2002) Why is image quality assessment so difficult? In*IEEE International Conference on Acoustics, Speech, and Signal Processing*, May 2002.

10. Wang Z, Lu L, Bovik A C (2004) Video quality assessment based on structural distortion measurement. *Signal Processing: Image Communication*, special issue on Objective Video Quality Metrics 19(2): p. 121–132.
11. Wang Z, Bovik A C, Sheikh H R, Simoncelli E P. Image quality assessment: from error visibility to structural similarity. *IEEE Transactions on Image Processing*, 13(4): p. 1–14, Apr. 2004.
12. Watson A B (1998) Toward a perceptual video quality metric. *Proceedings of SPIE Human Vision and Electronic Imaging*, 3299: p. 19–147.
13. Lu L, Wang Z, Bovik A C, Kouloheris J (2002) Full-reference video quality assessment considering structural distortion and no-reference quality evaluation of MPEG video. In *IEEE International Conference on Multimedia and Expo*, Aug. 2002.
14. Gunawan I P, Ghanbari M (2003) Reduced-reference picture quality estimation by using local harmonic amplitude information. In *London Communications Symposium 2003*, 8th-9th September 2003, London.
15. Ries M, Crespi C, Nemethova O, Rupp M (2007) Content based video quality estimation for H.264/AVC video streaming. *Proceedings of IEEE Wireless and Communications & Networking Conference*, Hong Kong.
16. Lu L, Wang Z, Bovik A C, Kouloheris J (2002) Full-reference video quality assessment considering structural distortion and no-reference quality evaluation of MPEG video. In *IEEE International Conference on Multimedia*, Volume: 1, p. 61–64, Lausanne, Switzerland.
17. Ries M, Nemethova O, Rupp M (2007) Motion based reference-free quality estimation for H.264/AVC video streaming. *International Symposium on Wireless Pervasive Computing*, San Juan, Poerto Rico, 5–7 February.
18. Seeling P, Reisslein M, Kulapala B (2004) Network performance evaluation using frame size and quality traces of single layer and two layer video: a tutorial. *IEEE Communications Surveys & Tutorials*, 6(3): p. 58–78.
19. Koumaras H, Kourtis A, Martakos D (2005) Evaluation of video quality based on objectively estimated metric. *Journal of Communications and Networks*, **7**(3): p. 235–242.
20. Koumaras H, Pallis E, Xilouris G, Kourtis A, Martakos D, Lauterjung J (2004) Pre-encoding PQoS assessment method for optimized resource utilization. In *HET-NETs04* Ilkley, West Yorkshire, U.K.
21. Koumaras H, Kourtis A, Martakos D, Lauterjung J (2007) Quantified PQoS assessment based on fast estimation of the spatial and temporal activity level. *Multimedia Tools and Applications*, 34(3): p. 355–374.
22. Koumaras H, Kourtis A (2007) Video quality prediction based on the spatial and temporal classification of the uncompressed content. *The 18th Annual IEEE International Symposium on Personal, Indoor and Mobile Radio Comm. (PIMRC)*, Athens, Greece, 3–7 September.
23. Koumaras H, Pliakas T, Kourtis A (2007) A novel method for pre-encoding video quality prediction. *IST Mobile Summit*, Budapest, Hungary.
24. Koumaras H, Gardikis G, Kourtis A (2003) Objective evaluation of the perceived quality of video content. *IST Mobile Summit 2003*, Aveiro, Portugal.
25. Kanumuri S, Cosman P C, Reibman A R, Vaishampayan V A (2006) Modeling packet-loss visibility in MPEG-2 video. *IEEE Transactions on Multimedia* , 8(2): p. 341–355.
26. He Z, Xong H (2006) Transmission distortion analysis for real-time video encoding and streaming over wireless networks. *IEEE Transactions on Circuits and Systems for Video Technology*, 16(9): p. 1051–1062.
27. Lee W, Srivastava J (2001) An algebraic QoS-based resource allocation model for competitive multimedia applications. *International Journal of Multimedia Tools and Applications*, 13(197–212).
28. Koumaras H, Kourtis A, Lin C-H, Shieh C-K (2007) A theoretical framework for end-to-end video quality prediction of MPEG-based sequences. *The Third International Conference on Networking and Services – ICNS07*, Athens, Greece.

Chapter 15
Video Coding Using the H.264/AVC Standard

Kun-Bin Lee

15.1 Introduction

H.264/AVC is the latest international video coding standard [1, 2] jointly developed by the Video Coding Experts Group (VCEG) of ITU-T and Moving Picture Experts Group (MPEG) of ISO/IEC. This standard is entitled "Advanced Video Coding" (AVC) and is published jointed as Part 10 of the MPEG-4 and ITU-Recommendation H.264. H.264/AVC uses state-of-the-art coding tools and provides enhanced coding efficiency for a wide range of applications including mobile multimedia broadcasting, video conferencing, internet protocol television (IPTV), digital cinema, IP multimedia subsystem (IMS), surveillance, etc. H.264/AVC can provide approximately 50% bit rate savings for equivalent perceptual quality relative to the performance of prior standards. Furthermore, H.264/AVC supports flexibilities in coding as well as organization of coded data that can improve error resilience. As might be expected, the increase in coding efficiency and coding flexibility comes at the expense of an increase in complexity with respect to earlier standards.

Since there are already plenty of overviews of this topic (e.g., [3–11]) and some works of H.264/AVC standard are not completed yet, the purpose of this chapter is to provide a quick start to understanding this standard. First, we will provide an overview of H.264/AVC applications and current development status. A brief introduction of H.264/AVC video coding layer structure will also be given. Then we will introduce some kernel coding tools of H.264/AVC, including intra-frame and inter-frame prediction, in-loop deblocking filter, transform and quantization, and entropy coding. Design optimization techniques for these coding tools are reviewed in another chapter of this book by the same author.

K.-B. Lee
MediaTek Inc., No.1, Dusing RD.1, Science-based Industrial Park, Hsinchu, Taiwan 300, R.O.C.
e-mail: kb.lee@mediatek.com

F.-L. Luo (ed.), *Mobile Multimedia Broadcasting Standards: Technology and Practice*,
DOI: 10.1007/978-0-387-78263-8_15,

15.2 Overview of H.264/AVC Coding Standard

The scope of H.264/AVC covers two layers: network abstraction layer (NAL) and video coding layer (VCL). Figure 15.1 shows the usage of H.264/AVC standard in transport environments. The VCL consists of the main compression engine for enhancing the coding efficiency, and comprises syntactical levels commonly known as the block-, macroblock-, and slice level. The VCL is designed to be as network independent as possible. The NAL adapts the bitstreams generated by the VCL and covers all syntactical levels above the slice level. The NAL defines the interface between the VCL and the transport environments, and is appropriate for the adaptation of H.264/AVCover RTP/UDP/IP, H.324/M, MPEG-2 transport, and H.320 [12,13]. Both the VCL and the NAL are designed in such a way that in heterogeneous transport environments, no VCL transcoding is necessary. In this chapter, we will focus on the discussion of VCL.

15.2.1 Applications

The video coding design problem can be posed as follows: given a maximum allowed delay and a maximum allowed complexity, achieve an optimal trade-off between bit rate and distortion for the range of environments envisioned in the scope of the applications [7]. Therefore, even in the same standard, different *profiles* are provided to support different coding tools for solving the aforementioned trade-off problem. In many applications, it might be neither practical nor economic to implement a decoder capable of dealing with all hypothetical uses of the syntax within a particular profile. Hence, *levels* are specified within each profile to specify sets of constraints imposed on values of the syntax elements in the bitstream. For example, Table 15.1 shows the adoption of video codecs in prerecorded Blu-ray

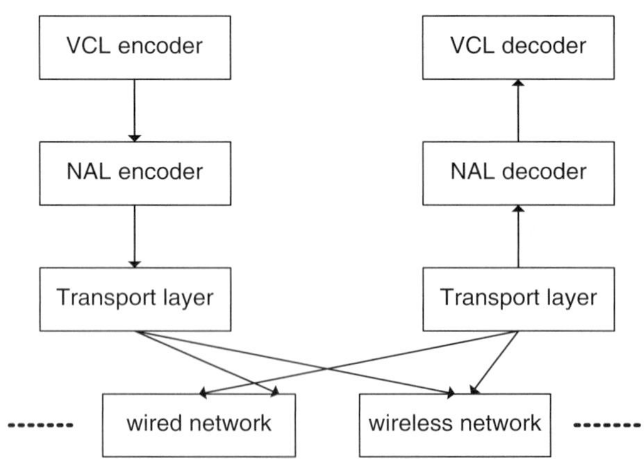

Fig. 15.1 The usage of H.264/AVC standard in transport environments

Table 15.1 Video codecs for BD-ROM

Video codec	H.264/AVC Main Profile @ Level 4.1 H.264/AVC High Profile @ Level 4.1 MPEG-2 Main Profile @ High Level SMPTE VC-1 Advanced Profile@ Level 3
Video format	1920×1080 HD (50i, 60i and 24p) 1440×1080 HD (50i, 60i and 24p) 1280×720 HD (50p, 60p and 24p) $720 \times 576/480$ SD (50i, 60i)

Table 15.2 Applications that adopt H.264/AVC as video codec

Application	Broadcast	(3G) mobile multimedia	Storage format
Standard/service	DVB ATSC DirecTV Euro1080	3GPP 3GPP2	BD HD DVD AVCHD

disc (BD-ROM). Only a subset of H.264/AVC standard is supported in this next-generation optical video disc format. A quick reference to coding tools for each profile specified in H.264/AVC can be found in [11].

H.264/AVC has been very strongly embraced by industry and new demands for different applications were still requested within a few years of the completion of the first version of this standard. For example, both of the next-generation optical video disc formats, the HD DVD and the BD, have included H.264/AVC as a mandatory player feature. The Digital Video Broadcasting (DVB) standards also include the use of H.264/AVC for broadcast television in Europe. Table 15.2 lists some examples of services or standards that already adopt H.264/AVC as their video codec. In summary, H.264/AVC may be used in the following application areas:

- Broadcast, subscription, and pay-per-view services over cable, DSL, satellite, and terrestrial transmission channels
- Real-time conversational services over wire-line and wireless communications
- Video-on-demand or multimedia streaming services over ISDN, LAN, wireless networks, etc.
- Storage formats on magnetic devices, optical disc
- High quality applications such as studio camcorders, editing systems, large-screen digital imagery, high fidelity display systems, etc.
- Surveillance applications, in which video sources not only need to be viewed on multiple devices ranging from high-definition monitors to videophones, but also need to be stored and archived
- Visual media such as 3DTV and free viewpoint TV (FTV) that provides rich 3-D information of a scene

15.2.2 H.264/AVC Structure

The VCL design follows the conventional block-based motion-compensated hybrid video coding concepts. Each picture is represented in block-shaped units of associated luma (Y) and chroma (Cb/Cr) samples. The basic video coding algorithm is a hybrid of inter-picture prediction to utilize temporal redundancy and transform coding of the prediction residual to exploit spatial statistical dependencies.

Figure 15.2 shows the subdivision of a picture. Macroblocks (MBs) are the basic building blocks for which the decoding process is specified. A picture may also be split into one or several slices, which represent regions of a given picture that their syntax elements can be parsed independent of each other. There are three basic slice types, i.e., *I*, *P*, or *B* type. In *I* slices, all MBs are coded using Intra Prediction (IP) which uses some prediction from the neighbors within the same picture but has no dependence on other pictures. In addition to the coding tools of the *I* slice, MBs of a *P* slice can also be coded using interprediction with at most one motion-compensated prediction signal per prediction block. MBs of a *B* slice can be coded using the coding tools of *P* slices or using interprediction with two motion-compensated prediction signals per prediction block. In addition to the aforementioned slice types, H.264/AVC also supports *SP* and *SI* slice types in extended profile for efficient and exact switching between different video streams [14].

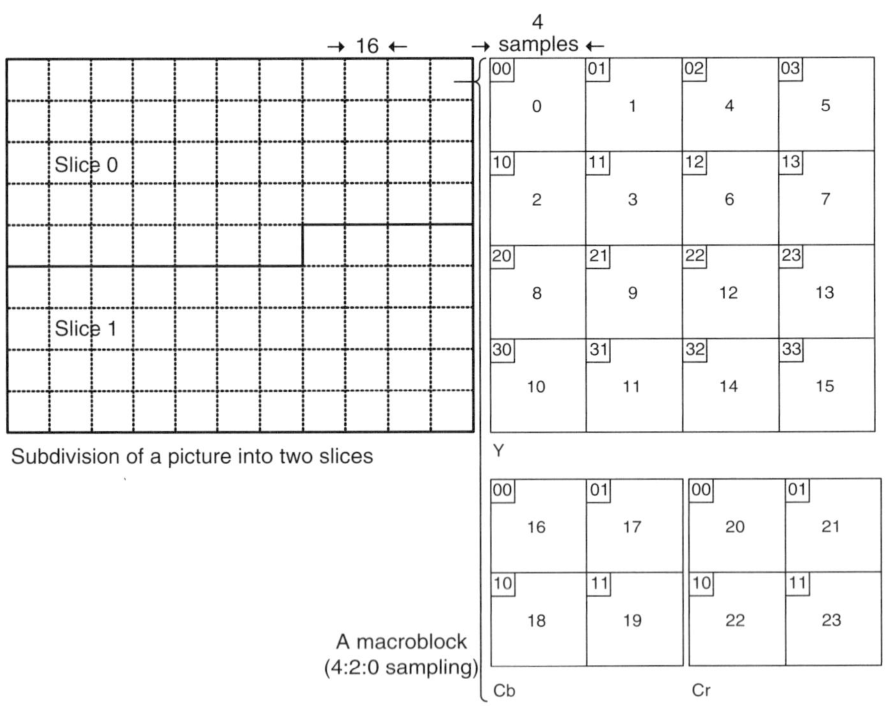

Fig. 15.2 Subdivision of a picture

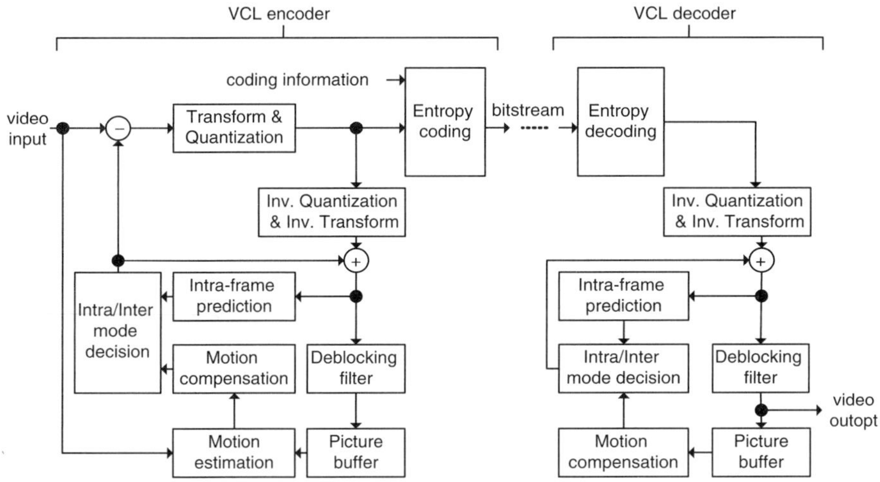

Fig. 15.3 Basic VCL codec structure of H.264/AVC

Data within a macroblock are transmitted in a predetermined order. Figure 15.2 shows the data order in an MB with 4:2:0 sampling format when 4×4 coding mode is selected. The order in this example also indicates the processing order of IP and entropy coding of residual blocks, which will be explained later.

Figure 15.3 shows basic VCL codec structure of H.264/AVC. A basic H.264/AVC video encoding process proceeds as follows. Each picture is split into blocks, and each block is encoded based on rate-distortion optimization. The first or a random-accessible picture of a video sequence is typically encoded in intramode. For all remaining pictures of the sequence, typically inter-picture coding modes are selected for most of blocks. The encoding process for interprediction consists of using motion estimation to decide motion data comprising the selected reference picture indexes and motion vectors to be applied for all samples of each inter-coded block. Encoding an MB by using IP or interprediction depends on the slice type and the mode decision algorithm. The motion and mode decision data are transmitted as prediction coding information.

The residual, which is the difference between the original block and its prediction, is transformed by a frequency transform. The transform coefficients are then scaled (quantized), entropy coded, and packed together with the prediction coding information in the bitstream. If a macroblock is coded by 16×16 intramode, then a block containing the transformed DC coefficient of each 4×4 luma block (i.e., samples 00–33 in Fig. 15.2) is transmitted first before transmitting luma AC blocks. Otherwise, only the luma residual blocks 0–15 are transmitted in the order shown in Fig. 15.2. Then transformed DC coefficients 00–11 of Cb and Cr blocks are transmitted if necessary. Finally, chroma residual blocks 16–23 are sent.

The encoding process also contains a model of the picture reconstruction process so that it can compute the same prediction values computed in the decoder for the prediction of subsequent blocks in current picture or subsequent coded pictures. When the deblocking filter is enabled within the coding loop to improve the visual quality and coding efficiency, the filtered frames are used as reference frames for motion compensation in subsequent coded pictures.

To reconstruct a coded picture, an inverse procedure is applied. The coded bitstream is entropy decoded to obtain the coding information and the quantized transform coefficients of each macroblock. Then prediction process, either intra or IP, is performed to obtain the prediction of a macroblock. The quantized transform coefficients are processed by inverse scaling and inverse transform to generate the residual of this macroblock. The residual is added to the prediction, and the addition result may then be fed into a deblocking filter to smooth out block-edge discontinuities induced by the block-wise processing. The reconstructed picture is then stored for the prediction of subsequent pictures (may also be displayed).

15.2.3 Profiles and Levels

Table 15.3 shows the publication stages of ITU Recommendation H.264 [2]. In Stage 1, three profiles were finalized. The baseline profile is intended for real-time conversational applications, such as videoconferencing and wireless communications. Potential applications of Main profile include TV broadcasting and stored digital video. However, the importance of this profile faded when the High profiles were defined for these applications. This is because the High profiles add more coding efficiency to what was previously defined in the Main profile without adding a significant amount of implementation complexity. Extended profile is in particular useful for streaming video applications. In addition to all coding tools of baseline profile, it includes *SP* and *SI* slices to facilitate switching between different coded streams, and Data Partitioning feature to improve performance in error-prone transmission environments. Extended profile also has all coding tools of Main profile except context-adaptive binary arithmetic coding (CABAC) tool.

Table 15.3 Evolution of H.264/AVC standard

Publication stage	Evolution
Stage 1 (May 2003)	Baseline, Extended, and Main profiles
Stage 2 (Mar. 2005)	Fidelity Range Extensions: a suite of four High profiles (High, High 10, High 4:2:2, and High 4:4:4 profiles)
Stage 3 (Jun. 2006)	H.264 (2005) Amendment 1: removal of the High 4:4:4 profile
Stage 4 (Apr. 2007)	H.264 (2005) Amendment 2: High 4:4:4 Predictive and four Intra-only profiles (High 10 Intra, High 4:2:2 Intra, High 4:4:4 Intra, and CAVLC 4:4:4 Intra)
Stage 5 (Nov. 2007)	H.264 (2005) Amendment 3: Scalable video coding

In Stage 2, Fidelity Range Extensions (FRExt) were added to provide further improvements in coding efficiency for higher-fidelity video material. Average bit-rate savings for the High profile of 10% relative to Main profile (both using CABAC tool) have been observed for a set of HD sequences [8].

In Amendment 1 published in Stage 3, High 4:4:4 profile is removed and replaced by professional profiles finalized in Stage 4. In this amendment, support of additional color spaces is also specified.

In Stage 4, to respond to strong demands for high quality applications such as studio camcorders, editing systems, digital cinema/large-screen digital imagery, high fidelity display systems, etc., the joint video team (JVT) of ISO/IEC MPEG and ITU-T VCEG has developed five new profiles, two new supplemental enhancement information (SEI) messages, and two new extended gamut color space indicators [10]. These new profiles are High 10 Intra, High 4:2:2 Intra, High 4:4:4 Intra, CAVLC 4:4:4 Intra and High 4:4:4 Predictive profiles. The enhancements include improved-efficiency 4:4:4 video format coding, improved-efficiency lossless macroblock coding with transform bypass technique, coding 4:4:4 video pictures using three separately coded color planes, and bit depths up to 14 bits per sample. Intra-only coding profiles have been defined to enable simple and quick editing of video without imposing the burden of implementing inter-picture decoding functionality. This feature is especially useful for content creation and postproduction applications.

In Stage 5, scalable video coding (SVC) was finalized. SVC addresses the needs of using video coding in highly time-varying environments, such as Internet networks and mobile communication systems, by enabling the transmission and decoding of partial bitstreams to provide video services with lower temporal or spatial resolutions or reduced fidelity [17]. SVC is also highly suitable for surveillance applications, in which video sources not only need to be viewed on multiple types of devices ranging from high-definition monitors to videophones, but also need to be stored and archived. By using SVC, high-resolution or high-quality parts of a bitstream can be deleted after a certain expiration time, so that only low-quality copies of the video are kept for long-term archive.

Currently, multiview video coding (MVC) is under development in the JVT [18, 19]. Since the efficient compression of MultiView Video (MVV) is a key enabling factor for 3DTV and FTV applications, MVC explores the interview statistical dependencies in addition to the temporal ones. MVC is scheduled to be released in 2008 and it will be an extension of H.264/AVC (Amendment 4).

15.3 Video Coding Layer

H.264/AVC achieves a considerable improvement and flexibility over previous video coding standards through a rich set of coding tools. In this section, we focus on some kernel coding tools shown in Fig. 15.3.

15.3.1 Intra-Frame Prediction

In all slice-coding types, two primary types of intracoding are supported: luma IP and chroma IP. A third type of intracoding, called I_PCM, is also provided to allow the encoder to simply bypass the prediction and transform processes and directly send the values of the encoded samples. Therefore, I_PCM enables lossless coding of selected regions. In addition, I_PCM mode also places a hard limit on the number of bits a decoder must handle for a macroblock.

In H.264/AVC, IP is used to reduce the high amount of bits coded by original video signal itself. For encoding a block in this mode, a prediction block is formed based on previously reconstructed (but unfiltered by deblocking filter) blocks. The residual signal between current block and the prediction is finally coded. For luma samples, the prediction block may be formed for each 4×4 subblock, each 8×8 block, or a 16×16 macroblock. Table 15.4 shows the complete IP modes in H.264/AVC. One case is selected from a total of nine prediction modes for each 4×4 or 8×8 luma block (I4MB/I8MB); four modes for a 16×16 luma block (I16MB); and four modes for each chroma block. Figure 15.4 illustrates the eight directional prediction modes mentioned in Table 15.4.

Basically, the picture reconstruction flow is to generate prediction value from neighboring pixels and then sum up the prediction value with the corresponding

Table 15.4 IP modes

I4MB	I16MB	I8MB	Chroma
Vertical	DC	Vertical	DC
Horizontal	Vertical	Horizontal	Vertical
DC	Horizontal	DC	Horizontal
Diagonal down-left	Plane	Diagonal down-left	Plane
Diagonal down-right		Diagonal down-right	
Vertical-right		Vertical-right	
Vertical-down		Vertical-down	
Horizontal-down		Horizontal-down	
Horizontal-UP		Horizontal-UP	

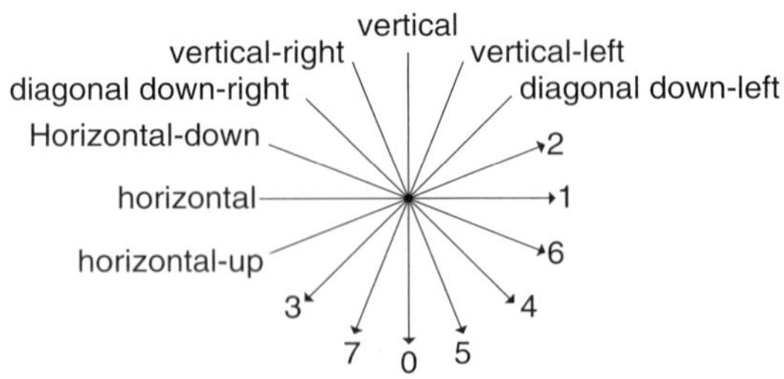

Fig. 15.4 Available IP directions

residual in the co-located position. The summed up value is the reconstructed pixel. Prediction across slice boundaries is not used, in order to keep all slices independent of each other.

For vertical and horizontal IP mode, the prediction value is exactly the neighboring pixel and therefore the proper neighboring pixel is selected to sum up with its corresponding residual. For DC modes, one value generated from its neighboring samples is used to predict all samples in the entire block. For other IP modes except the plane mode, proper neighboring samples and filters are selected to derive the prediction samples. The following gives two examples shown in Fig. 15.5:

- For I4MB diagonal down-right mode, the prediction samples ***a*** and ***d*** can be calculated as

$$\begin{aligned} \mathrm{a} &= (\mathrm{I}+2\times\mathrm{Z}+\mathrm{A}+2) \gg 2, \\ \mathrm{d} &= (\mathrm{B}+2\times\mathrm{C}+\mathrm{D}+2) \gg 2. \end{aligned}$$

- For I4MB horizontal-up mode, the prediction samples ***a*** and ***d*** can be calculated as

$$\begin{aligned} \mathrm{a} &= (\mathrm{I}+\mathrm{J}+1) \gg 1, \\ \mathrm{d} &= (\mathrm{J}+2\times\mathrm{K}+\mathrm{L}+2) \gg 2. \end{aligned}$$

For 8×8 luma IP modes, the prediction values are generated in a similar way to 4×4 modes except that a low-pass filter is performed on the neighboring samples first.

The prediction of chroma samples and luma in 16×16 prediction mode is similar. For I16MB and chroma plane prediction modes, parameters *pA*, *pB*, *pC*, *pH*, *pV* are derived first, and then prediction samples *pred[x, y]* are generated based on these parameters. Figure 15.6 shows the generation of prediction values for chroma plane IP mode, where chroma_format_idc indicates chroma format sampling structure and can be 0, 1, 2, or 3 to represent monochrome, 4:2:0, 4:2:2, or 4:4:4, respectively.

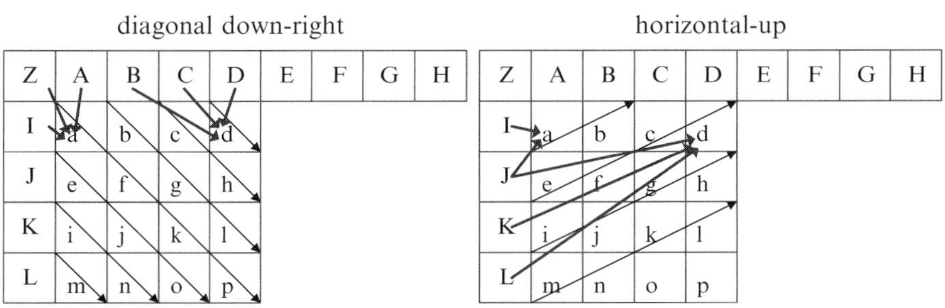

Fig. 15.5 Examples of generation of prediction values for I4MB

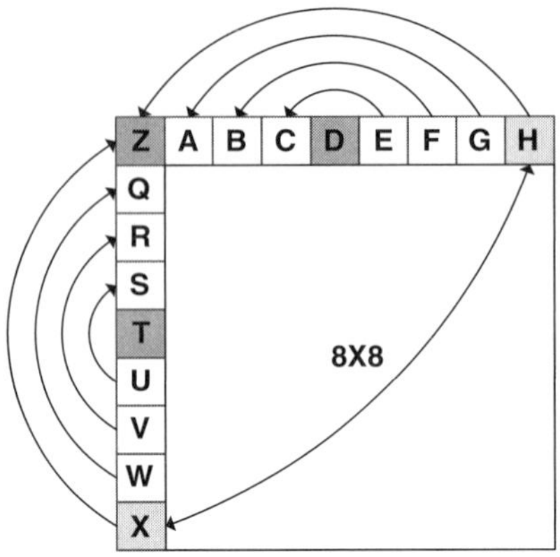


```
pred[x, y]= Clip1(pA +(x-7)*pB+ (y-7)*pC + 16)>>5)

pA =16(X + pH)
pB = (5 * pH+32) >> 6, if (chroma_format_idc == 3)
     (17 * pH+16) >> 5  otherwise
pC = (5 * pV+32) >> 6, if (chroma_format_idc != 1)
     (17 * pV+16) >> 5  otherwise

pH =     (E-C)     pV =     (U-S)
     + 2(F-B)           + 2(V-R)
     + 3(G-A)           + 3(W-Q)
     + 4(H-Z)           + 4(X-Z)
```

Fig. 15.6 Generation of prediction values for chroma plane IP mode

15.3.2 Inter-Frame Prediction

H.264/AVC uses several enhanced motion-prediction techniques to improve inter-frame prediction. In this section, weighted prediction (WP), variable block-size motion compensation, multiple reference picture motion compensation, and quarter-sample-accurate motion compensation will be discussed in more detail.

15.3.2.1 Weighted Prediction

In previous standards, biprediction has typically been performed with a simple averaging of the two prediction signals, and the prediction in the *P* macroblock types has not used weighting. In H.264/AVC, a WP coding tool is included to improve coding efficiency for scenes containing fades. In H.264/AVC WP coding tool, a multiplicative weighting factor and an additive offset are applied to the motion compensated prediction. In explicit mode, a weighting factor and offset may be coded in the slice header for each allowable reference picture index. In implicit mode, the weighting

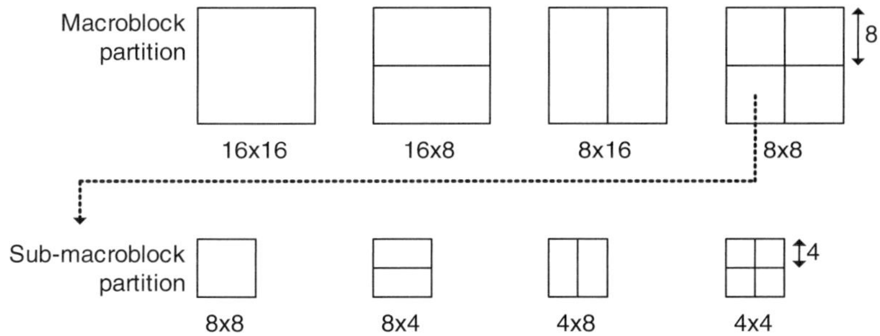

Fig. 15.7 Macroblock and submacroblock partitions

factors are not coded but are derived based on the relative picture order count (POC) distances of the two reference pictures. When coding fade-to-black sequences, bit-rate reductions of up to 67% were achieved [20].

15.3.2.2 Variable Block-Size Motion Compensation

H.264/AVC supports more flexibility in the selection of motion compensation block sizes and shapes than previous standards, with a minimum luma motion compensation block size as small as 4×4. As shown in Fig. 15.7, each macroblock can be partitioned into smaller regions with luma block sizes of 16×16, 16×8, 8×6, and 8×8 samples. When 8×8 macroblock partitioning is chosen, an additional syntax element is transmitted for each 8×8 partition to specify whether the 8×8 partition is further partitioned into smaller regions of 8×4, 4×8, or 4×4 luma samples.

15.3.2.3 Multiple Reference Picture Motion Compensation

In MPEG-2 standard and its predecessors, inter-coded blocks in *P* and *B* slices use only one and two previous reference pictures, respectively, to predict the samples in an incoming picture. H.264/AVC extends the enhanced reference picture selection technique proposed in H.263++ [21] to enable efficient coding by allowing an encoder to select among a larger number of pictures that have been reconstructed and stored in a multipicture buffer. This buffer is also maintained in the decoder according to memory management control operations specified in the bitstream. Unless the size of the multipicture buffer is set to one picture, the index at which the reference picture is located inside the multipicture buffer is signaled. The reference index parameter is transmitted for each motion-compensated 16×16, 16×8, 8×16, or 8×8 luma block. Submacroblock partitions within the 8×8 region use the same reference index for prediction. Figure 15.8 illustrates the concept of multiframe motion compensation and this concept can be extended to *B* slices.

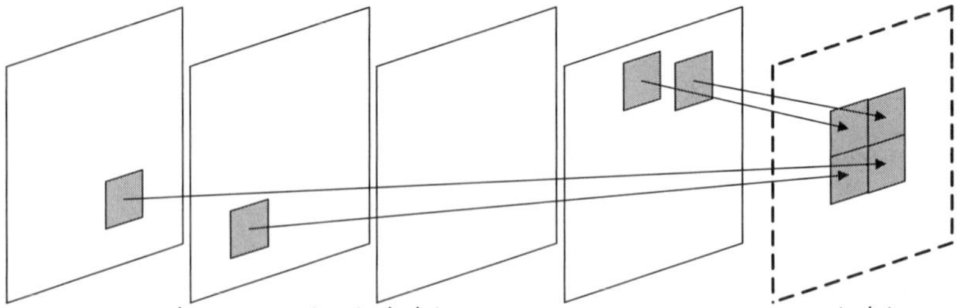

Fig. 15.8 Multiframe motion compensation

15.3.2.4 Quarter-Sample-Accurate Motion Compensation

The purpose of fractional pixel resolution is to get more accurate definition of picture content displacement. Most prior standards enable half-sample motion vector accuracy at most. The use of displacement vectors with higher resolution did not improve the coding efficiency and therefore has not been considered in those standards. By mathematical analysis, it is shown that aliasing components are deteriorating the prediction efficiency [22, 23]. Prediction errors containing camera noise and other signal components can be observed. Werner [22] supposed that these additional components originate from aliasing and proposed a Wiener interpolation filter for reducing the impact of aliasing. For H.264/AVC, the original filter was modified to a 6-tap filter in order to allow an easier hardware implementation. As a result, quarter-sample motion vector accuracy was incorporated into H.264/AVC.

Figure 15.9 shows integer sample positions (shaded blocks with upper-case letters) and fractional sample positions (un-shaded blocks with lower-case letters) for luma interpolation. Given the luma samples "A" to "U" at integer-sample locations, the luma samples "a" to "s" listed in Table 15.5 are derived by the following rules [1]. The luma prediction values at half-sample positions are derived by applying a 6-tap filter with tap values (1, −5, 20, 20, −5, 1). The luma prediction values at quarter-sample positions are derived by averaging samples at full and half-sample positions. The process for each fractional position is described below.

The samples at half-sample positions labeled b and h are derived by first calculating intermediate values $b1$ and $h1$ by applying the 6-tap filter to the nearest integer position samples in the horizontal direction and the vertical direction, respectively:

$$\begin{aligned} \mathrm{b1} &= (\mathrm{E} - 5 \times \mathrm{F} + 20 \times \mathrm{G} + 20 \times \mathrm{H} - 5 \times \mathrm{I} + \mathrm{J}), \\ \mathrm{h1} &= (\mathrm{A} - 5 \times \mathrm{C} + 20 \times \mathrm{G} + 20 \times \mathrm{M} - 5 \times \mathrm{R} + \mathrm{T}). \end{aligned}$$

The final prediction values b and h are derived using:

$$\begin{aligned} \mathrm{b} &= \mathrm{Clip1Y}((\mathrm{b1} + 16) \gg 5) \\ \mathrm{h} &= \mathrm{Clip1Y}((\mathrm{h1} + 16) \gg 5), \end{aligned}$$

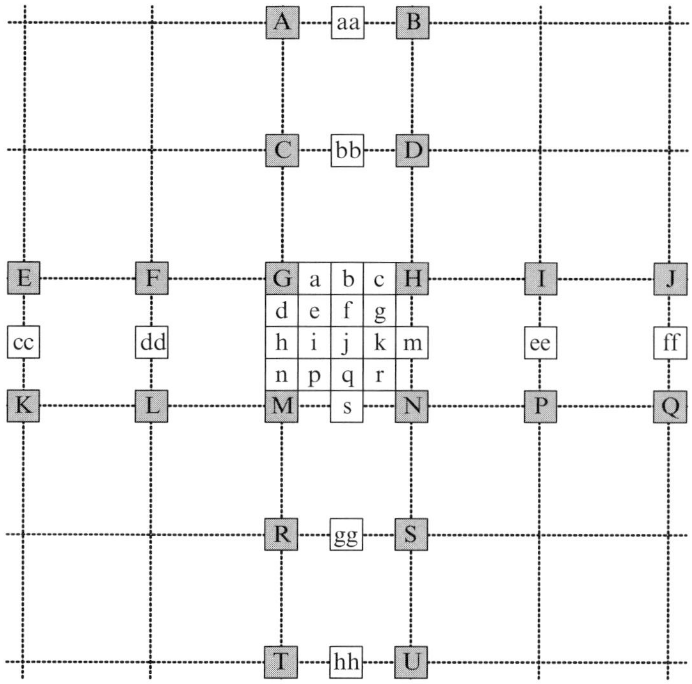

Fig. 15.9 Integer and fractional sample positions for luma interpolation

Table 15.5 Sample positions in luma interpolation

Sample position type	Sample position
Half-sample position	b, h, s, m
Half-sample position	j
Quarter-sample positions	a, c, d, n, f, i, k, q
Quarter-sample positions	e, g, p, r

where Clip1Y is a clipping function clip that clips a value to the range of 0–255 for 8-bit samples. The final prediction values s and m are derived from $s1$ and $m1$ in the same manner as the derivation of b and h.

The samples at half-sample position labeled as j can be derived by first calculating intermediate value denoted as $j1$ by applying the 6-tap filter to the intermediate values of the closest half-sample positions in either the horizontal or vertical direction because both yield an equal result, i.e.,

$$\mathrm{j1} = \mathrm{cc} - 5 \times \mathrm{dd} + 20 \times \mathrm{h1} + 20 \times \mathrm{m1} - 5 \times \mathrm{ee} + \mathrm{ff}, \text{ or}$$
$$\mathrm{j1} = \mathrm{aa} - 5 \times \mathrm{bb} + 20 \times \mathrm{b1} + 20 \times \mathrm{s1} - 5 \times \mathrm{gg} + \mathrm{hh},$$

where intermediate values aa, bb, gg, hh, and $s1$ are derived by applying the 6-tap filter horizontally in the same manner as the derivation of $b1$ and intermediate values *cc*, *dd*, *ee*, *ff*, and $m1$ are derived by applying the 6-tap filter vertically in the same manner as the derivation of $h1$. The final prediction value j is derived using:

$$\mathrm{j} = \mathrm{Clip1Y}((\mathrm{j1} + 512) \gg 10).$$

The samples at quarter-sample positions labeled as a, c, d, n, f, i, k, and q are derived by averaging with upward rounding of the two nearest samples at integer and half-sample positions. For example:

$$\mathrm{a} = (\mathrm{G} + \mathrm{b} + 1) \gg 1,$$
$$\mathrm{q} = (\mathrm{j} + \mathrm{s} + 1) \gg 1.$$

The samples at quarter-sample positions labeled as e, g, p, and r are derived by averaging with upward rounding of the two nearest samples at half-sample positions in the diagonal direction. For example:

$$\mathrm{e} = (\mathrm{b} + \mathrm{h} + 1) \gg 1,$$
$$\mathrm{r} = (\mathrm{m} + \mathrm{s} + 1) \gg 1.$$

The predicted chroma sample values are obtained by bilinear interpolation. Since the sampling grid of chroma (for 4:2:0 sampling format) has lower resolution than the sampling grid of the luma, the displacements used for chroma have one-eighth sample position accuracy. Given the chroma neighboring samples A, B, C, and D at integer sample locations, the predicted chroma sample value a at each fractional sample position shown in Fig. 15.10 is derived as follows:

$$\begin{aligned}\mathrm{a} = ((8 - \mathrm{dx}) \times (8 - \mathrm{dy}) \times \mathrm{A} + \mathrm{dx} \times (8 - \mathrm{dy}) \times \mathrm{B} \\ + (8 - \mathrm{dx}) \times \mathrm{dy} \times \mathrm{C} + \mathrm{dx} \times \mathrm{dy} \times \mathrm{D} + 32) \gg 6.\end{aligned}$$

15.3.3 In-Loop Deblocking Filter

One annoying characteristic of block-based coding is the production of visible block artifacts, which is due to the block-based transforms and motion compensated prediction. Overlapped block motion compensation (OBMC) [24] is a good solution to this problem and is adopted into H.263 standard. In H.264/AVC, although the small

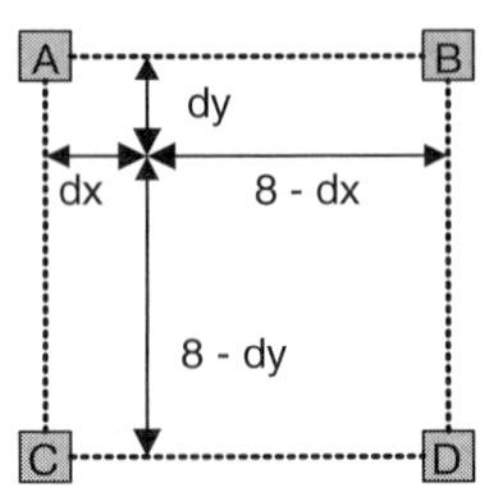

Fig. 15.10 Fractional sample position dependent variables in chroma interpolation

4×4 sample transform size somewhat reduces the problem, the adaptive deblocking filter is defined to maximize coding performance. The filter reduces the bit rate typically by 5–10% with significantly improved visual quality while producing the same objective quality as the nonfiltered video [25].

Prior to H.264/AVC, the use of a deblocking filter as an optional postprocessing technique in a decoder side to improve the visual quality of decoded pictures was already a well-known technology. Significant blocking artifacts appearing in low bit rate applications, such as video conferencing, could be substantially reduced by such deblocking filters. In contrast with previous standards, the H.264/AVC deblocking filter is not an optional feature in the decoder. When the deblocking filter is enabled within the coding loop, the filtered frames are used as reference frames for motion compensation in subsequent coded frames. When the deblocking filter is enabled, the filter strength is determined by coding modes of adjacent blocks, quantization step size, and the steepness of the luminance gradient between blocks, etc.

In H.264/AVC, the deblocking filter is adaptive on three levels: slice level, block-edge level, and sample level. On the slice level, the global filtering strength can be adjusted to the individual characteristics of the video sequence and the deblocking parameters are transmitted in the slice header. On the block-edge level, the filtering strength is decided based on inter/intra predictions, motion vector difference, and residuals in the two participating blocks. On the sample level, sample values and quantizer-dependent thresholds can turn off filtering for a particular sample to avoid distortions of real edges.

The filtering process is performed on a macroblock basis after the completion of the picture reconstruction process prior to deblocking filter process. For each MB and each color component, vertical edges are filtered first, starting with the edge on the left-hand side of the MB proceeding through the edges towards the right-hand side of the MB in their geometrical order, and then horizontal edges are filtered, starting with the edge on the top of the MB proceeding through the edges towards the bottom of the MB in their geometrical order. Figure 15.11 shows edges of a luma macroblock. A conditional filtering process is applied to all $N \times N$ (where $N = 4$ or $N = 8$ for luma, and $N = 4$ for chroma) block edges of a picture, except edges at the boundary of the picture and any edges for which the deblocking filter process is disabled. When 8×8 transforms are applied, $8\times$ block edges are filtered.

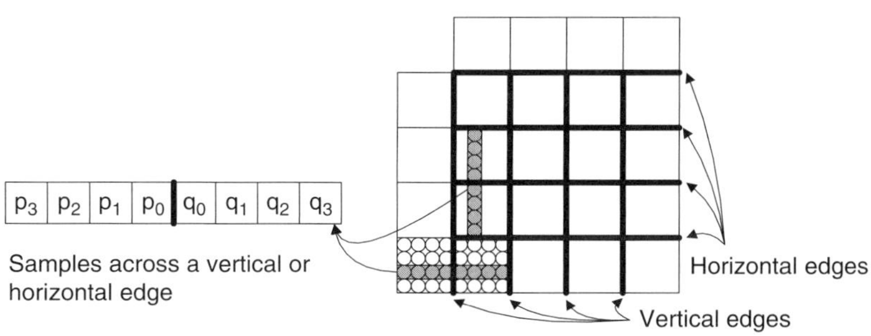

Fig. 15.11 Filtering operations on vertical and horizontal edges

The filtering process is applied to a set of eight samples across a horizontal or vertical edge denoted as p_i and q_i with $i = 0{-}3$ for the edge lying between p_0 and q_0. The input sample values p_i and q_i are conditionally replaced by the corresponding filtered result sample values p'_j and q'_j based on the definition of the standard. The good performance is based on reliable detection of real and artificially created edges and efficient filtering of the latter ones.

15.3.4 Transform and Quantization

Transforms used in H.264/AVC standard are characterized by integer transform, adaptive transform size, hierarchical transform, and its low complexity which can be implemented in 16-bit arithmetic computation without multiplication. Similar to previous video coding standards, H.264/AVC utilizes block-based transform coding for spatial redundancy removal. Unlike the popular 8×8 discrete cosine transform (DCT) used in previous standards, separable integer transforms in H.264/AVC can be computed exactly in integer arithmetic, thus avoiding inverse transform mismatch problems [26, 27].

Inverse transform mismatch problems could cause significant issues in H.264/AVC because H.264/AVC is more sensitive to prediction drift. In prior standards, prediction drift accumulation occurs once per inter-coded frame. In H.264/AVC, however, prediction drift can occur much more frequently. For example, 4×4 blocks can be predicted from their neighboring blocks with various IP modes. At each stage prediction drift can be accumulated. Thus, it is clear that as a result of the extensive use of various predictions in H.264/AVC, the residual coding must be drift-free. The drift arises from the fact that the inverse transform is not fully specified in integer arithmetic. The new transforms used in H.264/AVC are a scaled integer approximation to the DCT, which allows computation of the direct or inverse transform with just additions and a minimal number of shifts, but no multiplications.

H.264/AVC uses an adaptive transform block size, 4×4 or 8×8 [28]. The 4×4 transform size is preferred because that with similar objective compression capability, the smaller 4×4 transform has visual benefits resulting in less ringing artifacts. The similar objective compression capability is due to the improved inter and IP process. Consequently the residual signal has less spatial correlation. In H.264/AVC, the 8×8 transform is used in High profiles. Traditionally, 8×8 transforms have been used for image and video coding. The 8×8 size has the advantage of being dyadic, and large enough to capture and preserve trends and periodic structures. Compression performance gains have been reported at Standard Definition (SD) and High Definition (HD) resolutions when larger than 4×4 transforms are used [29]. The average bit-rate reduction is more significant for the coding of progressive material [28]. The use of adaptive block transform achieves significant rate-distortion benefits. More importantly, from the point of subtle texture preservation (such as keeping film details), adaptive transform also provides significant subjective quality benefits [30].

To further improve compression efficiency, H.264/AVC also employs a hierarchical transform structure [26, 31]. For blocks with mostly flat sample values, there is significant correlation among transform DC coefficients of neighboring blocks. Therefore, DC coefficients of a macroblock (e.g., luma samples 00–33 in Fig. 15.2) may undergo a second $M \times N$ transform to improve coding efficiency. When the 16×16 IP mode is used with the 4×4 integer transform, the DC coefficients of the sixteen 4×4 luma blocks in the macroblock are further transformed by a secondary 4×4 Hadamard transform. The DC coefficients of the 4×4 blocks of chroma samples are transformed using a secondary Hadamard transform as well. For 4:2:0 video format, this requires a 2×2 Hadamard transform; and for 4:4:4 video format, the chroma DC coefficients use the same 4×4 Hadamard transform used for luma.

After the residue is transformed, the transformed coefficient is quantized. In H.264/AVC, a quantization parameter (QP) ranged from 0 to 51 is used for determining the quantization of transform coefficients. The quantization step size is controlled logarithmically by QP rather than linearly as in previous standards, in a manner designed to reduce decoding complexity and enhance bit rate control capability. Each increase of 6 in the value of QP causes a doubling of the quantization step size, so each increase of 1 in the value of QP increases the step size by approximately 12.5%. Often a change of step size by approximately 12.5% also means roughly a reduction of bit rate by approximately 12.5%.

As illustrated in [3,27], the 4×4 forward transform equation 15.1 can be approximated by using Eq. 15.2.

$$Y = AXA^{\mathrm{T}} = \begin{bmatrix} a & a & a & a \\ b & c & -c & -b \\ a & -a & -a & a \\ c & -b & b & -c \end{bmatrix} \begin{bmatrix} x_{00} & x_{01} & x_{02} & x_{03} \\ x_{10} & x_{11} & x_{12} & x_{13} \\ x_{20} & x_{21} & x_{22} & x_{23} \\ x_{30} & x_{31} & x_{32} & x_{33} \end{bmatrix} \begin{bmatrix} a & b & a & c \\ a & c & -a & -b \\ a & -c & -a & b \\ a & -b & a & -c \end{bmatrix} \quad (15.1)$$

where A is the transform matrix whose entries are

$$a = \frac{1}{2}, \quad b = \sqrt{\frac{1}{2}}\cos\left(\frac{\pi}{8}\right), \quad c = \sqrt{\frac{1}{2}}\cos\left(\frac{3\pi}{8}\right).$$

$$Y = \left(CXC^{\mathrm{T}}\right) \otimes E_{forw}$$

$$= \left(\begin{bmatrix} 1 & 1 & 1 & 1 \\ 2 & 1 & -1 & -2 \\ 1 & -1 & -1 & 1 \\ 1 & -2 & 2 & -1 \end{bmatrix} \begin{bmatrix} x_{00} & x_{01} & x_{02} & x_{03} \\ x_{10} & x_{11} & x_{12} & x_{13} \\ x_{20} & x_{21} & x_{22} & x_{23} \\ x_{30} & x_{31} & x_{32} & x_{33} \end{bmatrix} \begin{bmatrix} 1 & 2 & 1 & 1 \\ 1 & 1 & -1 & -2 \\ 1 & -1 & -1 & 2 \\ 1 & -2 & 1 & -1 \end{bmatrix} \right) \otimes \begin{bmatrix} a^2 & \frac{ab}{2} & a^2 & \frac{ab}{2} \\ \frac{ab}{2} & \frac{b^2}{4} & \frac{ab}{2} & \frac{b^2}{4} \\ a^2 & \frac{ab}{2} & a^2 & \frac{ab}{2} \\ \frac{ab}{2} & \frac{b^2}{4} & \frac{ab}{2} & \frac{b^2}{4} \end{bmatrix}, \quad (15.2)$$

$$a = \frac{1}{2}, \quad b = \sqrt{\frac{2}{5}}, \quad c = \sqrt{\frac{1}{10}},$$

where $\otimes$ denotes element-by-element multiplication instead of normal matrix multiplication. Since multiplying with elements of matrix E_{forw} consists of simply scaling the output with constants, this scaling can be combined with the quantization tables in the encoder, and with de-quantization tables in the decoder. The transform matrix elements of C are all equal to ± 1 or ± 2, and the final scaling is absorbed by the quantization step. Since a multiplication by two in the aforementioned transform can be implemented as simply a left shift, the transform used in H.264/AVC is *multiplier-free*.

For a given step Q_s, usually an integer, the encoder can perform quantization by

$$Y_Q(i,j) = \operatorname{sign}(Y_{ij})\,\frac{|Y_{ij}| + f(Q_s)}{Q_s},$$

where i and j are the row and column indices and $f(Q_s)$controls the quantization width near the origin (the "dead zone"). To let all operations be computed in 16-bit arithmetic for input data with 9-bit dynamic range, H.264/AVC uses the following optimized quantization formula:

$$Y_Q(i,j) = \operatorname{sign}(Y_{ij})\left\{|Y_{ij}|\,Q(Q_M,i,j) + f2^{17+Q_E} \gg (17+Q_E)\right\}, \quad i,j = 0,\ldots,3,$$

where $Q_M = QP\%6$ and $Q_E = QP/6$. The parameter f is chosen by the encoder, and is typically in the range 0–0.5. The quantization factor $Q(Q_M,i,j)$ depends on the transform coefficient position $(i,\ j)$ inside the block. The quantization matrix for the quantization factors can be represented as

$$Q(k) = \begin{bmatrix} x_k & y_k & x_k & y_k \\ y_k & z_k & y_k & z_k \\ x_k & y_k & x_k & y_k \\ y_k & z_k & y_k & z_k \end{bmatrix},$$

where $k = Q_M$. Take $k = 0$ as the example, the default 4×4 quantization matrix for luma defined in H.264/AVC is

$$Q(0) = \begin{bmatrix} 13107 & 8066 & 13107 & 8066 \\ 8066 & 5243 & 8066 & 5243 \\ 13107 & 8066 & 13107 & 8066 \\ 8066 & 5243 & 8066 & 5243 \end{bmatrix}.$$

In this case, the relationship between elements of E_{forw} and elements of Q is

$$a^2 : ab/2 : b^2/4 \cong 0.25 : 0.1581 : 0.1 \cong 13107 : 8066 : 5243 = x_0 : y_0 : z_0.$$

Table 15.6 shows the matrix elements of $Q(k)$. As aforementioned, each increase of one in QP increases the step size by approximately 12.5%. This can be explained by

$$\begin{aligned} & x_5 && : x_4 && : x_3 && : x_2 && : x_1 && : x_0 \\ &= 13107 && : 11916 && : 10082 && : 9362 && : 8192 && : 7282 \\ &\cong 1.125^5 && : 1.125^4 && : 1.125^3 && : 1.125^2 && : 1.125 && : 1 \end{aligned}$$

Table 15.6 Quantization matrix elements

k	x_k	y_k	z_k
0	13,107	8,066	5,243
1	11,916	7,490	4,660
2	10,082	6,554	4,194
3	9,362	5,825	3,647
4	8,192	5,243	3,355
5	7,282	4,559	2,893

15.3.5 Entropy Coding

Two innovative entropy coding modes are supported in H.264/AVC, referred to as Context-Based Adaptive Variable-Length Coding (CAVLC) and as Context-Based Adaptive Binary Arithmetic Coding (CABAC). Both of them utilize context-based adaptability to improve coding efficiency. In earlier coding methods, it was implicitly assumed that the statistics of syntax elements are stationary, which however in practice is seldom the case. Especially residual in a motion-compensated video coder shows a highly nonstationary statistical behavior, depending on the video content, the coding conditions, and the accuracy of the prediction model. By incorporating context modeling in entropy coding framework, both entropy coding methods of H.264/AVC offer a high degree of adaptation to the underlying source. CABAC typically provides a average reduction in bit rate between 9% and 14% compared to CAVLC for typical test sequences in broadcast applications in a range of video quality of about 30–38 dB [32].

In addition to these two entropy coding modes, many syntax elements are coded using Exponential-Golomb coding, which is a simple infinite-extent codeword set. Thus, instead of defining a different variable-length coding (VLC) table for each syntax element, only the mapping methods (Table 15.9) to this single codeword table are customized to the syntax element statistics.

15.3.5.1 Exponential-Golomb Coding

Exponential-Golomb (Exp-Golomb) codes are variable-length codes with a regular construction [3, 33, 34]. The codeword is constructed as

$$[\text{M zeros}][1][\text{INFO}].$$

M zeros are a prefix code having a number of zeros before the first "1," whereas INFO is the following suffix code carrying information, as shown in Tables 15.7 and 15.8. The length of each Exp-Golomb codeword for its index codeNum is (2M + 1) bits and each codeword can be constructed as

$$\text{M} = \text{floor}(\log_2[\text{codeNum} + 1])$$
$$\text{INFO} = \text{codeNum} + 1 - 2^{\text{M}}$$

Table 15.7 Exponential-Golomb codeword and its index range

Codeword	Range of codeNum
1	0
0 1 x_0	1–2
0 0 1 x_1 x_0	3–6
0 0 0 1 x_2 x_1 x_0	7–14
0 0 0 0 1 x_3 x_2 x_1 x_0	15–30
...	...

Table 15.8 Exponential-Golomb codeword and its index in explicit form

Codeword	codeNum
1	0
010	1
011	2
00100	3
00101	4
00110	5
00111	6
00010000	7
...	...

Table 15.9 Utilization of Exponential-Golomb coding

Descriptor	Description
me(v)	Mapped Exp-Golomb-coded syntax element
se(v)	Signed integer Exp-Golomb-coded syntax element
	Syntax element value = $(-1)^{codeNum+1}$ Ceil (codeNum /2)
	Ceil(x): the smallest integer greater than or equal to x
te(v)	Truncated Exp-Golomb-coded syntax element
ue(v)	Unsigned integer Exp-Golomb-coded syntax element
	Syntax element value = codeNum

For index codeNum = 0, M and INFO are zero. Table 15.8 shows the first few codewords. Then, a syntax element v to be encoded is mapped to codeNum in one of the ways described in Table 15.9.

15.3.5.2 Context-Based Adaptive Variable-Length Coding

In a baseline profile of H.264/AVC, its basic entropy coding method consists of Exp-Golomb codes, which by means of individually customized mappings is applied to all syntax elements except those related to quantized transform coefficients. The quantized transform coefficients are compressed by five syntax elements to convey information including the number of nonzero quantized transform coefficients, the value of these coefficients, and the positions of these coefficients in a coded block [3, 35]. CAVLC encoding of a block of quantized transformed coefficients proceeds as follows.

Encoding the Number of Nonzero Coefficients (TotalCoeff) and Trailing 1s (TrailingOnes)

"Trailing 1s" indicate the number of coefficients with absolute value equal to 1 at the end of the scanning order of a coded block. For coding efficiency, TrailingOnes can be 0–3 in each coded block. If there are more than three Trailing 1s, the rest of Trailing 1s are coded as normal nonzero coefficients. The range of TotalCoeff depends on the block size. For example, TotalCoeff can have a value from 0 to 16 for a 4×4 block. Note that when a coding information (syntax coded_block_pattern) indicates an 8×8 block contains nonzero coefficients, there may be 4×4 blocks in this 8×8 block that do not contain any nonzero coefficients. In this case, TotalCoeff is 0 for these 4×4 blocks. As the example shown in Fig. 15.12, TotalCoeff = 5, TrailingOnes = 3, and the rest of nonzero coefficients are 6 and 1.

TotalCoeff and TrailingOnes are coded as a combined event coeff_token, which is coded by one of six tables [1]. The selection of tables depends on the variable nC. For chroma DC blocks, nC depends on the chroma format sampling structure, which is specified through chroma_format_idc, and may have a value of 0, -1, and -2. For other block types, nC is calculated according to its neighboring blocks, i.e., it is context adaptive. Then nC is classified as the following four groups: $0 <= nC < 2, 2 <= nC < 4, 4 <= nC < 8$, and $8 <= nC$. Finally, a choice between three VLC tables and one fixed-length code (FLC) table is made for this block according to nC.

Encoding the Value of Nonzero Coefficients

Based on the observation that the statistics of coefficient values has less spread for high-frequency coefficients than for low-frequency ones, the coefficients values are coded in a reverse scan order, i.e., from high-frequency coefficients to DC

0	6	1	0
0	-1	1	0
1	0	0	0
0	0	0	0

(a)

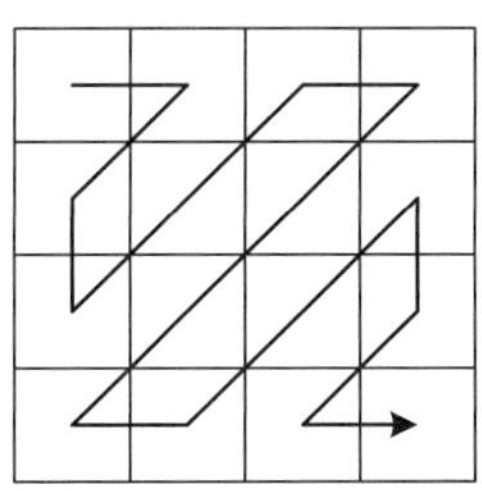

(b)

Fig. 15.12 (**a**) A 4×4 block and (**b**) zig-zag scan order for a 4×4 block coefficient. Firstly, for each TrailingOne starting with the highest frequency, the sign is encoded with a single bit. Then each remaining nonzero coefficient, called level, in the block is encoded by using several structured VLC tables. For improved coding efficiency, the selections of these tables are changed along with the coding process based on the QP, the number of coefficients, and the size of the previously coded level value. After each level is encoded, the VLC table number is updated. In this way, adaptation is obtained in the use of VLC tables.

Encoding Positions of Nonzero Coefficients

Position information coding is separated into two syntax elements total_zeros and run_before for separately indicating the total number of zeros (i.e. the number of zeros located before the last nonzero coefficient) and run (of zeros) before the nonzero coefficients. The reason for sending a separate total_zeros information is that usually a number of nonzero low-frequency coefficients exist in a coded block. By using this approach, zero-runs for these coefficients need not be encoded. This works especially well at high bit rates because the lower-frequency coefficients typically have no run of zeros.

The number of zeros preceding a nonzero coefficient, run_before, is also encoded in a reverse scan order. If there are no more zeros left (zeros_left = 0) to encode, it is not necessary to encode any more run_before values. Therefore, it is not necessary to encode run_before for the last nonzero coefficient. The selection of a VLC table for each run_before depends on the number of zeros that have not yet been encoded (i.e., zeros_left) and current run_before.

15.3.5.3 Context-Adaptive Binary Arithmetic Coding

For coding residual data, CAVLC exploits inter-symbol redundancies by switching VLC tables for various syntax elements according to the already coded symbols. However, CAVLC cannot provide an adaptation to the actually given conditional symbol statistics. Furthermore, coding events with symbol probabilities greater than 0.5 cannot be efficiently coded due to the fundamental lower limit of 1 bit/symbol imposed on variable-length codes. It is well known that arithmetic coding is especially useful when dealing with source data with symbols having highly skewed probabilities. By applying sophisticated processing steps, CABAC has better coding efficiency than CAVLC.

Figure 15.13 shows the simplified CABAC encoder. The encoding process consists of three elementary steps: binarization, context modeling, and binary arithmetic coding [32]. In the first step, a given nonbinary valued syntax element is uniquely mapped to a binary sequence, a so-called *bin string* or*bins*, which can also be interpreted in terms of a binary code tree. Binarization is the preprocessing step to reduce the alphabet size of the syntax elements. When a binary valued syntax element is given, this initial step is bypassed.

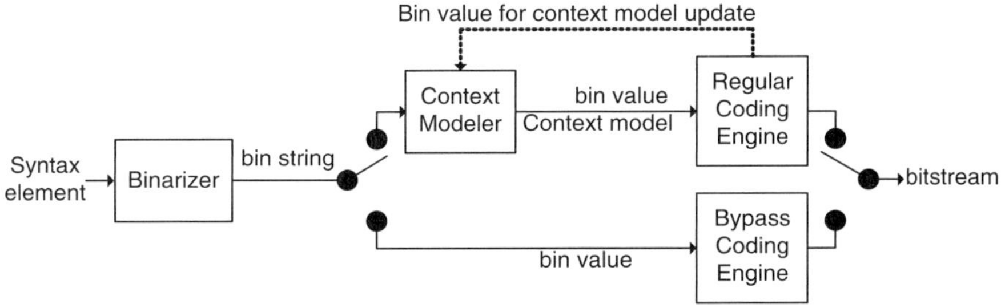

Fig. 15.13 Simplified CABAC encoder

Next, further processing of each bin depends on the associated coding mode, which can be either the *regular* or the *bypass* mode. The bypass coding mode is chosen for selected bins in order to allow a speedup of the whole encoding (and decoding) process. Bins related to the sign information or bins which are assumed to be uniformly distributed are processed in this mode. In the regular coding mode, prior to the actual binary arithmetic coding process the selected binary bin enters the context modeling stage, where the associated probability model is either selected by a fixed choice without any context model or adaptively chosen depending on the related context model. The latter case is generally applied only to the most frequently observed bins. In this way, CABAC enables adaptive context modeling on a subsymbol level and efficient exploitation of inter-symbol redundancies. In the final step of CABAC, each bin enters the binary arithmetic encoder, either in regular or bypass mode.

Among low-complexity binary arithmetic coding methods, the Q coder [36, 37] and its derivatives QM and MQ coder [38] have attracted great attention, especially in the context of the still image coding standardization groups JPEG and JBIG. Although the MQ coder represents the state of the art in fast binary arithmetic coding, it considerably degrades coding efficiency in the application scenario of H.264/AVC video coding [39]. An alternative multiplication-free binary arithmetic coding scheme, the so-called *modulo coder* (M coder), is used in the regular coding engine [32]. For coding in bypass mode, the computationally expensive probability estimation is completely omitted since the selected bins are assumed to be nearly uniformly distributed.

Coding of residual data in CABAC involves specifically designed syntax elements that are different from those used in the traditional run-length coding approach. Figure 15.14 shows the encoding flow for blocks with at least one nonzero (nz) quantized transformed coefficient. The first part of this flow is the coding of position information by encoding a binary-valued *significant* (i.e., nonzero) *map* to indicate the occurrence and the location of nonzero quantized transformed coefficients within a scan path of a given block. The second part of this flow is to encode the magnitude and sign of each nonzero level in a reverse scanning order. Context models for coding of binarized level magnitudes are based on the number of previously coded transmitted level magnitudes in a reverse scan order. Encoding

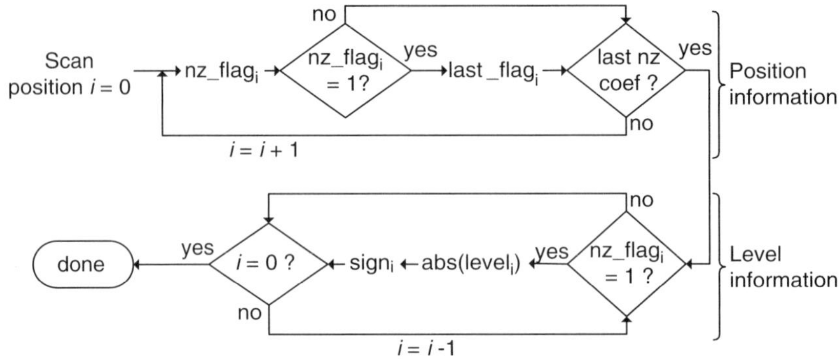

Fig. 15.14 Simplified CABAC encoding flow of transformed coefficients

nonzero levels in the reverse scanning order is motivated by the statistical observation that levels with the magnitude equal to one are dominate at the end of the scanning path.

15.4 Summary

H.264/AVC has a rich set of coding tools to provide coding efficiency, error/loss robustness, and flexibility for effective use over a wide variety of application domains. More new features are still under development. Currently, the MPEG/VCEG JVT is developing MVC that explores the interview statistical dependencies in addition to the temporal ones for 3DTV and FTV applications. In addition to technical features and theoretical performance, there are many considerations for practical applications. These compelling advantages of H.264/AVC come at the price of an increase of complexity in both encoding and decoding. Realization of these applications relies on VLSI for cost effective implementation. The design challenges include real-time processing and low power consumption under cost constraints. Review on implementation issues and design techniques of H.264/AVC VCL decoding systems is presented in another chapter of this book by the same author.

References

1. ITU-T Rec. H.264 and ISO/IEC 14496-10 (MPEG4-AVC), *Advanced Video Coding for Generic Audiovisual Services*, March 2005.
2. ITU Recommendation H.264 page, available on: http://www.itu.int/rec/T-REC-H.264/e
3. I. Richardson, *H.264 and MPEG-4 Video Compression: Video Coding for Next Generation Multimedia*. Wiley, March 2003.

4. T. Wiegand, G. J. Sullivan, G. Bjontegaard, and A. Luthra, "Overview of the H.264/AVC video coding standard," *IEEE Trans. Circuits Syst. Video Technol.*, vol. 13, pp. 560–576, July 2003.
5. J. Ostermann et al., "Video coding with H.264/AVC: tools, performance, and complexity," *IEEE Mag. Circuits Syst.*, vol. 4, pp. 7–28, First quarter 2004.
6. G. J. Sullivan, P. Topiwala, and A. Luthra "The H.264/AVC advanced video coding standard: Overview and introduction to the fidelity range extensions," *SPIE*, vol. 5558, pp. 53–74, Aug. 2004.
7. G. Sullivan and T. Wiegand, "Video compression – from concepts to the H.264/AVC standard," *Proc. IEEE*, vol. 93, no. 1, pp. 18–31.
8. D. Marpe, T. Wiegand, and S. Gordon, "H.264/MPEG-4 AVC fidelity range extensions: tools, profiles, performance and application areas," *Proc. ICIP 2005*, pp. 593–596, Sept. 2005.
9. G. Sullivan, "The AVC video coding standard," *New Multimedia Technology Workshop*, Oct. 2007.
10. G. Sullivan, et al., "New standardized extensions of MPEG4-AVC/H.264 for professional-quality video applications," *Proc. IEEE Intl. Conf. on Image Proc. (ICIP 2007)*, San Antonio, TX, USA, Sept. 2007.
11. Wikipedia H.264/MPEG-4 AVC page, available on: http://en.wikipedia.org/wiki/H.264
12. S. Wenger, "H.264/AVC over IP," *IEEE Trans. Circuits Syst. Video Technol.*, vol. 13, pp. 645–656, July 2003.
13. T. Stockhammer, M. Hannuksela, and T. Wiegand, "H.264/AVC in wireless environments," *IEEE Trans. Circuits Syst Video Technol.*, vol. 13, pp. 657–673, July 2003.
14. M. Karczewicz and R. Kurçeren, "The SP and SI frames design for H.264/AVC," *IEEE Trans. Circuits Syst. Video Technol.*, vol. 13, pp. 637–644, July 2003.
15. S. Wittmann and T. Wedi, "Transmission of post-filter hints for video coding schemes," *Proc. IEEE Intl. Conf. Image Proc. (ICIP 2007)*, San Antonio, TX, USA, Sep. 2007.
16. A. Segall, L. Kerofsky, and S. Lei, "Tone mapping SEI message," JVT document JVT-T060, Joint Video Team (JVT) of ISO/IEC MPEG & ITU-T VCEG, Klagenfurt, Austria, July 2006.
17. H. Schwarz, D. Marpe, and T. Wiegand, "Overview of the scalable video coding extension of the H.264/AVC standard," *IEEE Trans. Circuits Syst. Video Technol.*, vol. 17, pp. 1103–1120, Sept. 2007.
18. A. Vetro, Y. Su, H. Kimata, and A. Smolic, "Joint draft 1.0 on multiview video coding," in Joint Video Team, Hangzhou, China, Oct. 2006, Doc. JVT-U209.
19. A. Smolic et al. "Coding algorithms for 3DTV – A survey," *IEEE Trans. Circuits Syst. Video Technol.*, vol. 7, no. 11, pp. 1606–1621.
20. J. Boyce, "Weighted prediction in the H.264/MPEG4 AVC video coding standard," in *Proc ISCAS*, pp. 789–792, May 2004.
21. T. Wiegand, X. Zhang, and B. Girod, "Long-term memory motion-compensated prediction," *IEEE Trans. Circuits Syst. Video Technol.*, vol. 9, no. 2, pp. 70–84, 1999.
22. O. Werner, "Drift analysis and, drift reduction for multiresolution hybrid video coding," *Signal Processing: Image Commun.*, vol. 8, no. 5, 1996.
23. T. Wedi and H. G. Musmann, "Motion- and aliasing-compensated prediction for hybrid video coding," *IEEE Trans. Circuits Syst. Video Technol.*, vol. 13, no. 7, pp. 577–586,2003.
24. M. T. Orchard and G. J. Sullivan, "Overlapped block motion compensation: An estimation theoretic approach," *IEEE Trans. Image Proc.*, vol. 3, no. 5, pp. 693–699, 1994.
25. P. List et al., "Adaptive deblocking filter," *IEEE Trans. Circuits Syst. Video Technol.*, vol. 13, no. 7, pp. 614–619, 2003.
26. H. Malvar, A. Hallapuro, M. Karczewicz, and L. Kerofsky, "Low-complexity transform and quantization in H.264/AVC," *IEEE Trans. Circuits Syst. Video Technol.*, vol. 13, no. 7, pp. 598–603, 2003.
27. A. Hallapuro, M. Karczewicz, and H. Malvar, "Low complexity transform and quantization – Part I: basic implementation," Joint Video Team (JVT) of ISO/IEC MPEG & ITU-T VCEG 2nd Meeting, JVT-B038, Feb. 2002.
28. S. Gordon, D. Marpe, and T. Wiegand, "Simplified use of 8x8 transform – Updated proposal & results," Joint Video Team (JVT) of ISO/IEC MPEG and ITU-T VCEG, doc. JVT-K028, Munich, Germany, March 2004.

29. S. Gordon, "ABT for film grain reproduction in high definition sequences," Joint Video Team (JVT) of ISO/IEC MPEG and ITU-T VCEG, doc. JVT-H029, Geneva, Switzerland, May, 2003.
30. M. Wien, "Variable block size transforms for H.264/AVC," *IEEE Trans. Circuits Syst. Video Technol.*, vol. 13, no. 7, pp. 604–613, 2003.
31. H. S. Malvar, *Signal Processing with Lapped Transforms*. Boston, MA: Artech House, 1992.
32. D. Marpe, H. Schwarz, and T. Wiegand, "Context-based adaptive binary arithmetic coding in the H.264/AVC video compression standard," *IEEE Trans. Circuits Syst. Video Technol.*, vol. 13, no. 7, pp. 620–636,2003.
33. J. Teuhola, "A compression method for clustered bit-vectors," *Information Process. Lett.*, vol. 7, pp. 308–311, Oct. 1978.
34. D. Marpe et al., "Improved CABAC," VCEG-O18, ITU-T Q6/16, WP 3/16, Thailand, Dec. 4–6, 2001.
35. G. Bjontegaard and K. Lillevold, "Context-adaptive VLC coding of coefficients," JVT-C028, May 6, 2002.
36. W. B. Pennebaker, J. L. Mitchell, G. G. Langdon, and R. B. Arps, "An overview of the basic principles of the Q-coder adaptive binary arithmetic coder," *IBM J. Res. Dev.*, vol. 32, pp. 717–726, Nov. 1988.
37. J. L. Mitchell and W. B. Pennebaker, "Optimal hardware and software arithmetic coding procedures for the Q-Coder," *IBM J. Res. Dev.*, vol. 32, no. 6, pp. 727–736, 1988.
38. D. Taubman and M. W. Marcellin, *JPEG2000 Image Compression: Fundamentals, Standards and Practice*. Boston, MA: Kluwer, 2002.
39. D. Marpe and T. Wiegand, "A highly efficient multiplication-free binary arithmetic coder and its application in video coding," in *Proc. ICIP*, Barcelona, Spain, Sept. 2003.

Chapter 16
H.264/AVC Error Concealment for DVB-H Video Transmission

Susanna Spinsante, Ennio Gambi, and Damiano Falcone

Digital Video Broadcasting for handheld terminals (DVB-H) is a broadcast system designed for high-speed data transmission in highly dispersive mobile channel conditions. This so-called "TV on Mobile" technology uses the Internet Protocol (IP) to carry highly compressed audio/video information; the ITU-T H.264 Advanced Video Coding (AVC) standard has been selected by the European Telecommunications Standards Institute (ETSI) as the video compression enabling technology. This choice, motivated by the good trade-off between coding efficiency and video quality, provides the further benefit of robustness against loss of parts of the received stream. As a matter of fact, transmission errors on the mobile channel can lead to IP datagram corruptions or losses; if not recovered by the prescribed Forward Error Correction (FEC) techniques, they can determine audio and video quality degradation. To face video degradation, Error Concealment strategies at the decoder can be efficiently tailored to the available hardware and computational resources of the receiving device, and do not require any changes in the encoding technique, nor any redundancy added to the transmitted data. Error Concealment refers to post-processing techniques employed by the video decoder, which basically exploits the remaining spatial and temporal redundancy in the received video data to compensate the lost information.

This chapter discusses the main features that make the H.264/AVC standard particularly suited to wireless video communications, and provides an overview of possible Error Concealment strategies that could be efficiently implemented in DVB-H devices, to compensate for errors and data losses due to the channel impairments.

S. Spinsante (✉), E. Gambi, and D. Falcone
Dipartimento di Elettronica, Intelligenza artificiale e Telecomunicazioni,
Università Politecnica delle Marche, Via Brecce Bianche 12, I-60131 Ancona, Italy
e-mail: s.spinsante@univpm.it

F.-L. Luo (ed.), *Mobile Multimedia Broadcasting Standards: Technology and Practice*,
DOI: 10.1007/978-0-387-78263-8_16,

16.1 The H.264/AVC Standard in Wireless Video Communication

At present, video transmission in wireless environments is a well-consolidated and quite familiar communication technology. Despite its appeal and easiness to use, however, wireless video transmission represents a very challenging task, which requires high-compression efficiency plus a network friendly design.

Among the video compression techniques developed during the last decades, the ITU-T H.264/AVC standard [13] emerged as the one able to provide the best facilities to address these two fundamental issues, i.e. high-compression efficiency and network oriented design. As a matter of fact, H.264/AVC is recognized as an optimal solution both for conversational and nonconversational video communication services, thanks to its wide range of specific levels and profiles, that can be selected and tuned properly, to meet each different application context requirements [19,29].

The H.264/AVC distinguishes between two different conceptual layers, the Video Coding Layer (VCL) and the Network Abstraction Layer (NAL). The VCL specifies an efficient representation for the coded video signal; the NAL defines the interface between the video codec itself and the communication system protocol architecture. Outside of the standard scope is the definition of transport and encapsulation techniques for the NAL units: they depend on the underlying protocol architecture implemented by the system. As an example, for IP-based packet switched communications, the 3GPP consortium [11] has chosen to use Session Initiation Protocol (SIP) and Session Description Protocol (SDP) for call control, and Real time Transport Protocol (RTP) for media transport. In this specific application context, a NAL unit is typically encapsulated in RTP/UDP/IP data structures. The H.264 VCL, that deals with the actual video coding processing and data, can provide the best trade-off between compression efficiency and quality degradation; it permits a significant bit rate reduction, if compared to alternative standards, such as ITU-T H.263 [14] and ISO/IEC MPEG-4 [12]. At the same time, the H.264 NAL component allows a seamless and easy integration of the coded video into multiple protocol and network architectures, a network oriented optimization of the coding parameters, and provides a set of tools designed to meet the requirements of packet oriented video transportation over wired and wireless networks [26].

When dealing with wireless communications, in general, the available bandwidth is the most critical issue. The bit rate on a radio link is usually limited, so that low bit rate communications are likely to be typical; this makes compression efficiency become the main requirement for a video coding technology to be preferred in a mobile environment. In this sense, a number of improvements in the coding strategy make H.264/AVC largely more efficient than all the other hybrid video coding solutions. The main features responsible for an increased coding efficiency are multiframe motion compensated prediction, adaptive block size for motion compensation, generalized B-pictures concept, quarter-pel motion accuracy, intracoding using prediction in the spatial domain, in-loop deblocking filter, and efficient entropy coding methods. The main components of an H.264/AVC encoder are shown in Fig. 16.1.

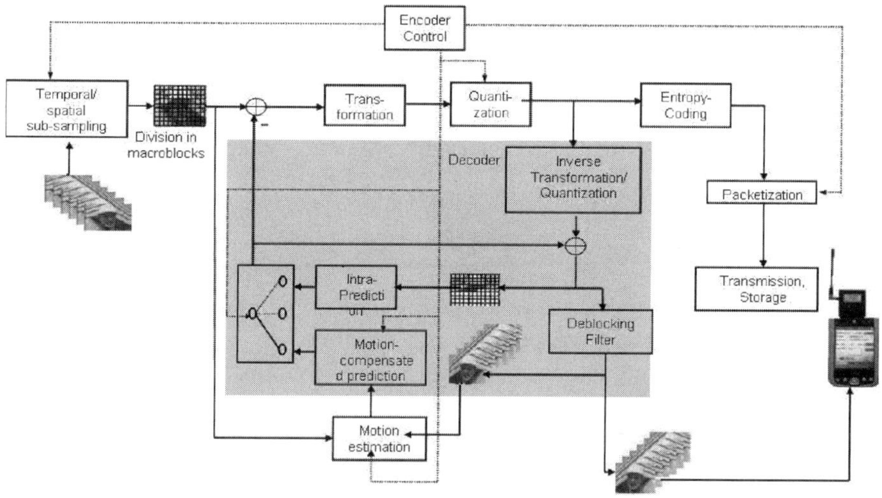

Fig. 16.1 H.264/AVC encoder main components

In addition to bandwidth limitations, the mobile radio channel represents a harsh communication environment, due to attenuation, shadowing, fading, and multiuser interference, which result in global channel conditions that are not only time varying, but also location dependent. The frequency of channel variations cannot be described deterministically, so, even if sophisticated transmission techniques and signal processing algorithms have been conceived and applied to their compensation, residual errors have to be tolerated, and faced somehow. Again, H.264/AVC proves to be an efficient tool, when integrated by basic strategies conceived to face residual errors on the channel, namely Error Resilience and Error Concealment. Additional features, besides those enabling an increased compression efficiency, have been included in the H.264/AVC; among them, the insertion of regular Intraframes with Instantaneous Decoder Refresh (IDR) and Rate Control strategies (see, for example, [16,31]). They may help either in implementing Error Resilience and Concealment techniques, by providing a number of anchor points to recover the video quality, and improving adaptation to the network conditions, by setting a suited target bit rate for the encoded video signal.

All the solutions applicable at the encoder side, to increase the video bitstream robustness against errors and losses, fall inside the definition of Error Resilience, which is usually related to the introduction of some kind of redundancy in the encoded data. As an example, Error Resilience methods for compressed videos consist of adding extra parameters, or more synchronization points. Error Resilience can be a powerful and effective tool; however, on one hand it requires to change the basic encoder structure, and sometimes this can avoid its integration in existing systems, and, on the other hand, packet losses and resulting reference frame mismatches between encoder and decoder are still unavoidable. As a consequence, the effects of spatio-temporal error propagation during decoding remain in general quite severe: a limited amount of corrupted data can cause a remarkable decrease of quality over a huge portion of the decoded video content.

An alternative, or complementary, solution to face these issues is represented by Error Concealment algorithms. They can be applied at the decoder side only, in a way that is not dependent on the encoder implementation; this way, little or no extra information at all is needed, and thus there is little cost on the bandwidth, and at the transmitter. The H.264/AVC standard defines how a compliant decoder reacts to an error free encoded bitstream; appropriate concealment algorithms on the detected error or lost macroblocks (MBs) and frames can be applied within the decoder, to mitigate the degradation effects, and to improve the final perceived video quality, ideally making the damages subjectively imperceptible.

Two well-known and classical concealment algorithms, i.e. weighted pixel value averaging for Intrapictures [21] and boundary-matching-based motion vector recovery for Interpictures [17], were initially tailored for H.264, and provided as a non-normative feature in the H.264 test model [27]. Later on, many other advanced concealment strategies have been proposed and implemented, facing specific requirements for a given application. A general overview of concealment solutions is presented in a next section.

16.2 DVB-H Channel Impairments: Errors and Losses

The Digital Video Broadcast (DVB) Project started research work related to mobile reception of DVB–Terrestrial (DVB-T) signals in 1998, together with the introduction of commercial terrestrial digital TV services in Europe. In 2000, an EU-sponsored project concluded that mobile reception of DVB-T is possible, but it implies dedicated broadcast networks, as mobile services are more demanding in robustness (i.e., constellation and coding rate) than broadcast networks planned for fixed DVB-T reception.

In early 2002, the DVB community was asked to provide technical specifications to allow delivery of rich multimedia contents to handheld terminals, thus making it possible to receive TV-type services in a small, handheld device like a mobile phone. Being services delivered in an environment suffering severe mobile multipath and high levels of man-made noise, among the specific features targeted to this kind of devices, the transmission system shall offer suited means to mitigate the channel effects on the receiver capabilities. The answer to these requirements came from the DVB-H system specification, which was published as ETSI Standard (EN 302 304), in November 2004 [4].

Broadcast transmission systems shall offer a simple way to cope with the multiple signal replicas reaching the receivers. For receivers in motion, complexity comes not only from the multiplicity of received echoes delayed in the time domain, but also from the frequency shift affecting such echoes. Signals received in motion are affected by a frequency Doppler shift, which is in relation with the receiver speed, and the relative angle between the motion direction and the signal incoming

Table 16.1 TU6 channel model

Tap number	Delay (μs)	Power (dB)	Doppler spectrum
1	0.0	−3	Rayleigh
2	0.2	0	Rayleigh
3	0.5	−2	Rayleigh
4	1.6	−6	Rayleigh
5	2.3	−8	Rayleigh
6	5.0	−10	Rayleigh

direction. The incoming angle provides a sign and weighting factor to the Doppler frequency shift; both the carrier frequency and the speed of the receiver will proportionally increase the Doppler shift value. Echoes affected by Doppler frequency shifts are perceived as a noise contributing to Inter Carrier Interference (ICI) [15]. ICI can be mitigated, in receivers using dedicated signal processing techniques, until a level at which the orthogonality of the subcarriers is broken, making demodulation impossible.

Research activities on channel propagation models suited to simulate the DVB-H mobile communication showed that the Typical Urban 6-paths model (TU6), described in Table 16.1 [5], can be considered representative for the typical mobile reception, with Doppler frequency above 10 Hz. The TU6 model has been heavily used both for simulation and laboratory test, and results from numerous field trials highly correlate with those obtained from simulations. Nevertheless, concerns remain in regard to the TU6 suitability for reception with Doppler frequency below 10 Hz (i.e., the pedestrian and indoor reception), suggesting further modelling work.

In the case of DVB-T services delivered to mobile receivers, the Erroneous Seconds Ratio (ESR) parameter, i.e. the amount of seconds with errors over the observation period, has proven to be a suited mean for Quality of Restitution (QoR) evaluation. For a service continuously delivered, a 5% ESR (denoted as ESR5) has shown to be highly correlated with the subjectively perceived reception quality (ESR = 5% corresponds to 1 s with errors over a 20 s observation period).

In the case of DVB-H, where services are delivered through FEC for Multiprotocol Encapsulation (MPE-FEC), which is the standard DVB way of carrying IP datagrams in MPEG-2 Transport Stream protected time slice bursts, other evaluation figures have been defined: Frame Error Ratio (FER) and MPE FER (MFER). In this context, FER is defined as the ratio of Application Data Tables (ADTs) containing errors, without MPE-FEC error correction being applied, during a given observation period. Consequently, FER5 corresponds to 5% ADTs containing errors, counted over the observation period. MFER is the ratio of uncorrected MPE-FEC frames during an observation period; MFER5 therefore corresponds to 5% uncorrected MPE-FEC frames over the observation period. FER and MFER showed to be extremely good indicators of the QoR for each service. Moreover, FER constitutes a mimic of a DVB-T like transmission, while MFER highlights the improvement brought by the DVB-H transmission. FER5 and MFER5 have been used during laboratory test sessions where their drawbacks (i.e., the need to wait for the reception of a large number of MPE-FEC frames) are tolerable; sufficient QoR for streaming video applications is achieved with MFER smaller than 5%.

If all erroneous frames were discarded at the decoder, the user could experience long times of blackout during a service. This is undesired, especially for streaming video services, where absolutely error free reception is not required, and can be compensated by means of proper concealment techniques.

The approach adopted to overcome the effects of receiver mobility is a Reed-Solomon link layer FEC code, denoted with RS(255, 191) [18]; moreover, to facilitate error detection in the receiver, a Cyclic Redundancy Check (CRC) is also performed, for sections of encoded RS data [8]. Several decoding methods have been tested for application in the DVB-H link layer FEC; among them, Section Erasure (SE), conventional Non Erasure (NE), Transport Stream Erasure (TSE), Hierarchical Section Erasure (HSE), and Hierarchical Transport Stream Erasure (HTS). The advantage of hierarchical decoding and the drawback of erasure decoding are clearly demonstrated when measuring ESR. Figure 16.2 [8] presents the ratios of visible erroneous seconds, for a video stream bit rate of 350 kbps and a frame rate of 25 fps, in the case of a 10 Hz Doppler frequency. A visible error is assumed to occur when at least one of the 25 images is erroneous; the error is assumed to be visible when at least half of the bytes of the image are incorrect. The QoR criterion ESR5 is marked as a solid line in the figure. For simulation purposes, a model that determines whether a TS packet contains errors or not is sufficient as a tool for FER evaluation. In practice, when a TS packet contains eight erroneous bytes or less, the packet can be labelled error free. This is due to the eight-byte correction capability of the RS code used in the DVB-T modulators [20].

RS coding is further reinforced, in DVB-H, by virtual time interleaving. For mobile conditions above 10 Hz Doppler, the MPE-FEC protection scheme lowers

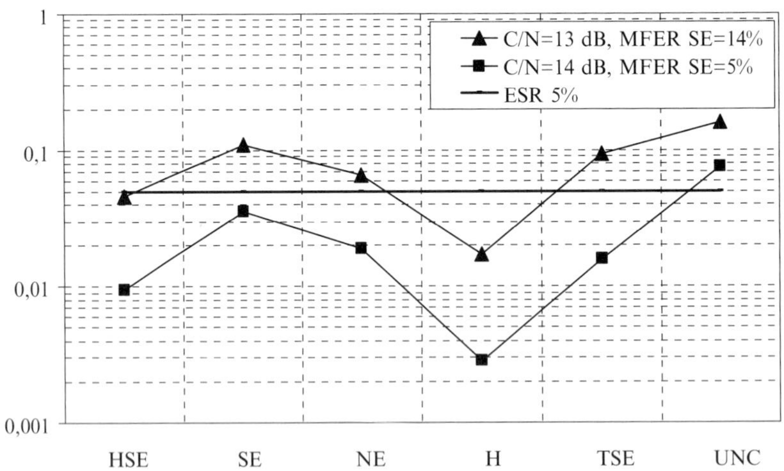

Fig. 16.2 Ratios of visible erroneous seconds, for a Doppler frequency of 10 Hz, when different decoding techniques are applied (UNC stands for the Uncoded option)

the required C/N, while the maximum tolerable receiver speed increases. For pedestrian situations, below 10 Hz Doppler, the effects of the virtual time interleaver are less efficient, and DVB-H transmissions benefit mainly from the ruggedness brought by the RS protection, and DVB-T in general. But, in this area, the absolute duration of the service bursts is expected to bring further advantages. In brief, the use of MPE-FEC in DVB-H transmissions nicely makes the service availability independent of the receiving speed, while canceling a large part of the C/N penalty suffered by the receiver in motion. Further details about the DVB-H system and architecture can be found in [4, 6].

16.3 Overview of Error Concealment Strategies in H.264/AVC

In the framework of Motion Estimation based video codecs, like H.263 and H.264, which adopt variable length coding techniques, even a single error bit can cause the loss of synchronism at the decoder, thus making impossible the right processing of the subsequently received bits, until the arrival of the next synchronization codeword. In order to avoid this "avalanche" effect, each frame is divided into slices; a slice consists of a number of MBs and, at the beginning of each slice, a synchronization code is inserted. This way, errors in the decoding of a slice will not have effects on the correct decoding of the following slices.

When one or more MBs within a frame are lost, the video decoder can restore the damaged frame, by using past and/or future frames, or the other MBs of the same frame, correctly received. According to these options, temporal or spatial Error Concealment, respectively, are defined; a third set of Error Concealment solutions are called hybrid, as they mix available temporal and spatial information to recover the lost data. In the framework of H.264 video coding, it is possible to adopt spatial Error Concealment for Intra (I) pictures and temporal Error Concealment for Predicted (P) and Bi-predicted (B) pictures. Algorithms restoring Intraframes, in general, interpolate the missing pixels on the basis of the available neighboring pixels in the same frame. Algorithms used for Interframes can be a little bit more involved, as they can adopt different approaches to restore the lost MBs: as an example, if the frame is characterized by low motion, MBs can be copied from one or more previously decoded frames; if the frame belongs to a high motion sequence, a missing MB can be restored by searching and copying the most likely one, among those correctly received.

Temporal Error Concealment methods estimate the motion of missing MBs and use this estimate to perform motion compensated temporal replacement of the lost MBs. First, a list of Motion Vector (MV) candidates is formed, for estimating the motion of the missing MB. Then, each of the MV candidate is tested by a matching error measure. The MV that minimizes the matching error is selected as a replacement for the missing MV of the lost MB. The performance of temporal Error Concealment methods is strongly affected by the matching error measure used. Two different criteria for selecting a suited matching measure are the smoothness criterion and the uniform motion criterion. Smoothness is measured as the Boundary Matching Error (BME), defined as the sum of absolute differences along the one-pixel boundary between the recovered MB and the surrounding ones. Motion uniformity is measured as the External Boundary Matching Error (EBME), which is defined as the sum of absolute differences between the multiple pixel boundary of MBs adjacent to the missing one, in the current frame, and the same boundary of MBs adjacent to the replacement MB in the reference frame. Tests [1] show that using EBME can lead to substantial improvements in concealment performance, with EBME being favored over BME thanks to its lower complexity.

Concealment in the spatial domain is typically based on mean interpolation or fast Discrete Cosine Transform (DCT) methods, and it is usually applied to MBs of I frames, which are used as a reference for P and/or B frames. Considerable compression ratios can be achieved through DCT, by ignoring most of the high frequency DCT coefficients, which does not lead to perceivable quality degradation. The 2-D DCT transform is defined by

$$y(k,l) = \sum_{i=0}^{M-1}\sum_{j=0}^{N-1} \cos\left(\frac{\pi(2i+1)k}{2M}\right)\cos\left(\frac{\pi(2j+1)l}{2N}\right)x(i,j) \tag{16.1}$$

where $x(i,j)$ and $y(k,l)$ denote an input and output matrix of the same size, $M \times N$, respectively.

The transformed block, whose bidimensional DCT coefficients are computed according to (16.1), contains both the lost pixels and the reliable pixels; if the number of high frequency coefficients set to zero equals the number of missing pixels, the lost area can be fully determined by solving a determined system of equations [28]. This procedure can be viewed as a kind of interpolation using border pixels to restore the missing ones, with the further advantage of being able to preserve edge information, if present in border pixels. Thanks to this feature, DCT based interpolation can outperform mean interpolation, which yields visual artifacts when applied to sharp blocks, or blocks containing edge information. Several other approaches to spatial concealment can be found in the technical literature; among them, it is worth citing the Projection Onto Convex Sets (POCS) method [30] and Maximum A Posteriori (MAP) estimator based methods (see, for example, [22,23]).

Results and quality performance obtainable through hybrid Concealment techniques are discussed, among the others, in [3]. The frame motion content, which determines the concealment strategy to adopt, can be computed as the average number of MVs in the frame. This value is compared against a threshold, above which the missing MBs are restored through the MVs of the nearby MBs, thus minimizing the difference among the edge pixels. This way, a uniform image can be obtained, as justified by the expectation that the movement of a close area in the image is strongly correlated. Intraframes are recovered by interpolating the lost pixels from the adjacent ones, inside the same frame. The lost MBs of a P frame are recovered from those of the previous frame, if they belong to a low movement video sequence, or from the near ones correctly received inside the same frame, if they belong to a high motion video sequence.

(a) (b)

Fig. 16.3 (**a**) Corrupted Stefan CIF Intraframe and (**b**) recovered Stefan CIF Intraframe

An introductory example is shown in Fig. 16.3, by means of a corrupted Stefan CIF sequence, generated by simulating the loss of two RTP packets, each one transporting one slice of 22 MBs (a single line of a CIF picture) and belonging to the first I frame. In Fig. 16.3a the degradation of the video content is clearly visible, when the decoder does not perform Error Concealment. Comparison with Fig. 16.3b, corresponding to the activation of the Reference Decoder JM 7.3 [9] Error Concealment module, shows that, since the corrupted frame is of I type, the lost MBs have been interpolated from the adjacent ones, and the recovered slices look softened.

If the lost RTP packet contains MBs belonging to an I or a P frame, this causes error propagation also in the subsequent P frames. As a consequence, the other frames show a residual error that, however, gets lower and lower as the sequence evolves in time. In the example of Fig. 16.3, where the Stefan sequence is encoded according to an IPPP... P pattern, the effects of the errors in frame #1, which is the only I frame, become weaker and weaker, and just after 50 frames the recovered sequence is very similar to the original one.

To get an objective evaluation of the quality degradation experienced by concealed video sequences, it is useful to refer to the Peak Signal to Noise Ratio (PSNR) values of the image components, Luma (Y), Chroma blue (U), and Chroma red (V). PSNR is defined as follows:

$$\mathrm{PSNR} = 10\log\left(\frac{255^2}{\frac{1}{NM}\sum_{i=0}^{N-1}\sum_{j=0}^{M-1}\left(x(i,j)-x_{\mathrm{R}}(i,j)\right)^2}\right) \tag{16.2}$$

where $x(i,j)$ refers to the original color component in pixel position (i,j) and $x_{\mathrm{R}}(i,j)$ refers to the color component after decoding; N and M specify the image size.

An example of evaluation of the Y, U, and V PSNR values as a function of a parameter expressing the amount of RTP packets lost during transmission (namely

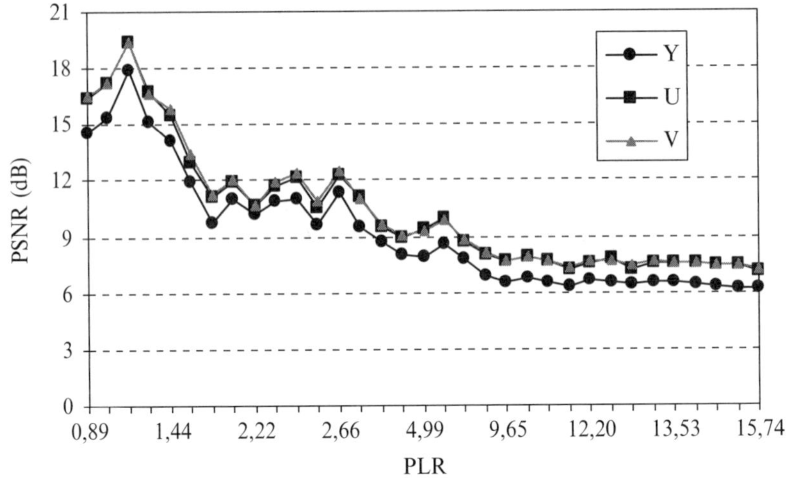

(a)

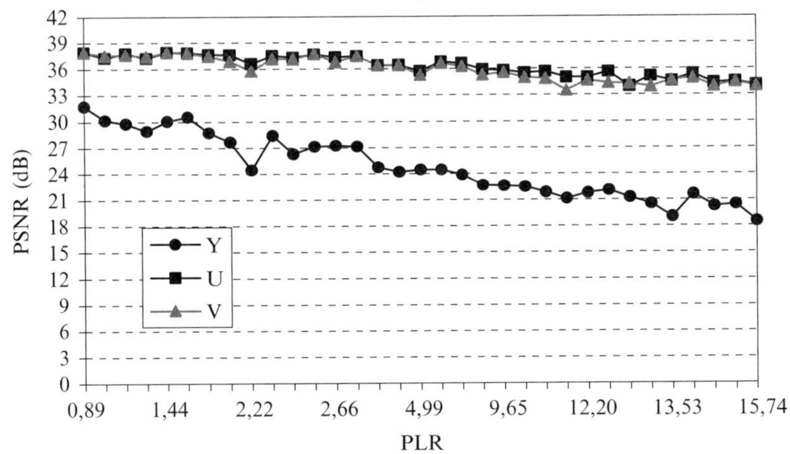

(b)

Fig. 16.4 Examples of PSNR evaluation (Stefan CIF sequence) as a function of the packet loss (**a**) without and (**b**) with activation of the Reference Decoder JM 7.3 Error Concealment algorithm

packet loss rate, PLR) is shown in Fig. 16.4, for the Stefan CIF sequence. By comparing the curves obtained without (Fig. 16.4a) and with (Fig. 16.4b) Error Concealment, it is possible to see that, in the former case, PSNR decreases almost exponentially, whilst, after activation of the concealment algorithm, degradation for increasing PLR occurs according to a much more favorable, roughly linear, law. It is wise to remind that the objective evaluation provided by PSNR is usually a preliminary step in judging the quality of the concealed image. A further, fundamental, step is the subjective evaluation that can be realized via formal subjective tests, or expert viewing experiments.

Further tests can be performed by setting different values of the Intraperiod (IP) parameter, which influences the encoder Error Resilience capability: as a matter of fact, I frames in the encoded video sequence represent "refresh points" of the sequence itself, which help an increase of PSNR. Actually, when IP is not zero, the Error Concealment algorithm is more effective, but only when errors or slice losses do not affect the I frames. On the contrary, if the reference I frames are damaged, PSNR values may also decrease.

The example in Fig. 16.5 refers to IP = 0; in Fig. 16.6, instead, IP = 15 is assumed. As shown in Fig. 16.6a, any time a packet is lost, the PSNR degrades, but the quality level is restored by the arrival of the I frame (with IP = 15). The only exception (for the considered example) is at frame #15 where the I frame is simultaneous with a loss. The situation is very different when the concealment algorithm is active (see Fig. 16.6b): the PSNR values for the three components are high and almost constant, thus testifying the effectiveness of the procedure for the considered sequence.

As discussed before, it is possible to adopt spatial Error Concealment for I pictures and temporal Error Concealment for P and B pictures. Accordingly, suited procedures were implemented in the JM7.3 Reference Software. Based on experimental tests, there is some evidence that the efficiency of the latter algorithm is superior to that of the former one, most of all if quality perception is measured though a subjective evaluation. As a matter of fact, for both techniques, the application of the restoration algorithm permits to increase the average PSNR (objective evaluation), but when using spatial Error Concealment a "soft effect" generally appears, with severe implications on the image quality. Such an effect is not present when using the temporal replacement algorithm, for P frames.

Following experimental evidences, the temporal Error Concealment technique could be applied to any frame in error, either Intra or Inter, with the obvious exception of the first frame, where spatial Error Concealment must be necessarily applied (temporal replacement does not work). This change has been easily introduced in the reference software, acting on the condition which permits to choose the algorithm for recovering a damaged frame. The results obtainable through such an easy modification are very interesting.

Obviously, in order to test the efficiency of the modification, the IP parameter must be set at a value different from 0. For IP = 0, the modification is ineffective. An example is shown in Fig. 16.7, where IP = 15 and frame #60 of the Akiyo sequence is considered. The Akiyo sequence is characterized by low motion, basically localized in the central area of the frame. In this condition, use of the temporal Error Concealment is particularly effective, since subsequent frames exhibit strong correlation. It is therefore possible to maintain high quality for the image, simply copying the missing MBs from the previous frame. Conversely, the spatial Error Concealment would restore the missing MBs on the basis of the nearby pixels correctly received, often altering completely the scene, most of all when it contains marked edges or details. The proposed widespread adoption of the temporal replacement technique might show its limits in the case of sequences with high motion, like

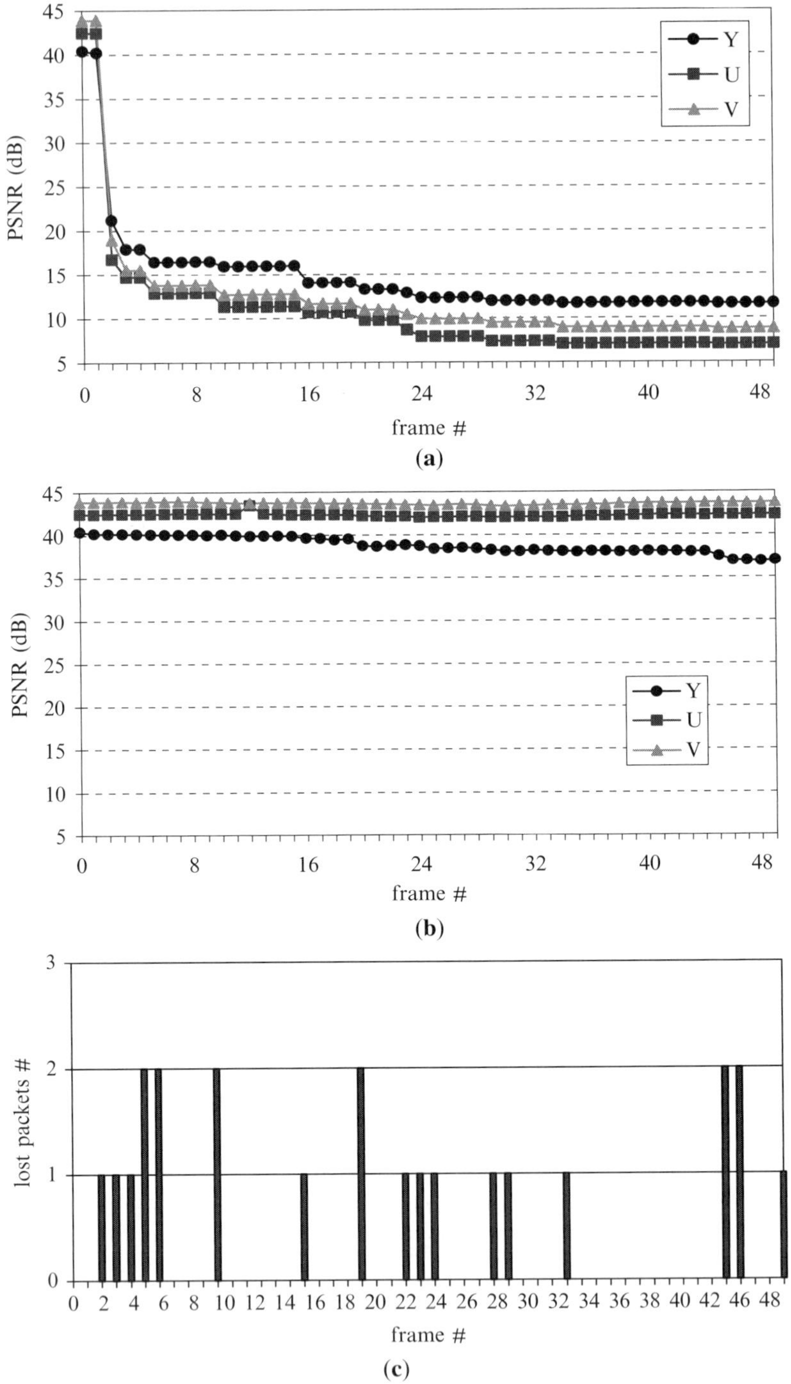

Fig. 16.5 Examples of PSNR evaluation (Akiyo CIF sequence, IP = 0): (**a**) without and (**b**) with activation of the Error Concealment algorithm and (**c**) simulated packet loss distribution (PLR = 2.77%)

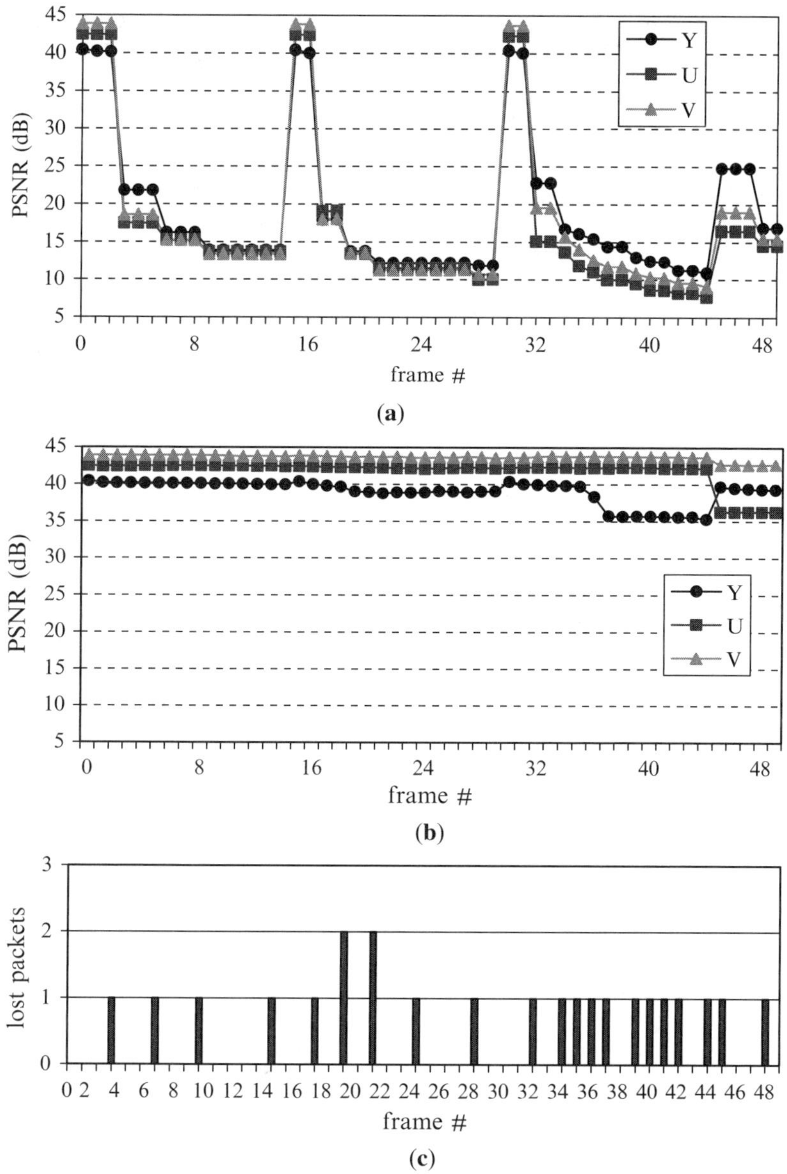

Fig. 16.6 Examples of PSNR evaluation (Akiyo CIF sequence, IP = 15, PLR = 2.55%): (**a**) without and (**b**) with activation of the Error Concealment algorithm and (**c**) simulated packet loss distribution

Stefan. Actually, the performance achievable in this case, by the modified algorithm, is generally worse than that in the case of low motion sequences.

Anyway, it must be said that the annoyance resulting from an incomplete temporal replacement is often more tolerable than the softening effect induced by the

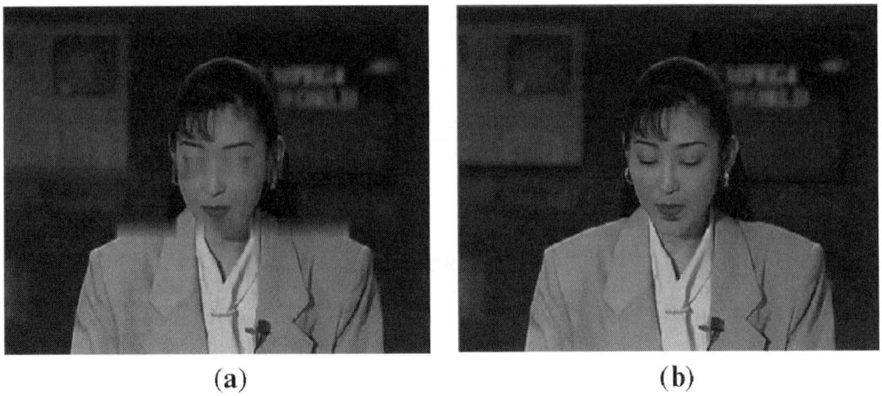

Fig. 16.7 Comparison between the result of (**a**) the Error Concealment algorithm in JM7.3 and (**b**) its modified version, for frame #60 of the Akiyo sequence

spatial restoration. Local (i.e., in correspondence of specific frames) evaluation of the PSNR is often ambiguous, not providing a good measure of the subjective quality of the recovered image. In this sense, values averaged on the ensemble of frames are more significant.

A more flexible solution for an effective concealment of video sequences featuring different properties can exploit scene change detection to adjust the recovering strategy, in such a way as to better fit each frame characteristic. This is the basic idea on which the concealment solution presented in the next section is built, which makes it possible to obtain an effective strategy under several conditions, without a remarkable computational overhead on the decoding device.

Before moving to the discussion of a scene change detection based concealment solution suited to DVB-H devices, it is worth citing a further family of Error Concealment algorithms that are able to face whole frame losses, a condition that can be quite common in wireless environments. As a matter of fact, when dealing with low bit rate packet based video communications, video frames may have a very small size; in many cases, a single frame fills the payload of a single network packet. If a packet gets lost, a whole frame loss occurs. In order to face whole frame losses, a novel technique has been proposed in [2]: optical flow (OF) estimation is applied to Error Concealment, together with multiframe estimation that exploits the multiple reference frame buffer featured by H.264/AVC codecs. The loss of a whole video frame makes the assumptions that underlie most Error Concealment algorithms not verified; this way, classical spatial or temporal solutions cannot be applied. The capability of estimating entire missing frames of a video sequence is built upon the OF constraint, i.e. finding an approximate solution of the OF equation, whose unknowns are both the MVs and the intensity information of the missing frame. The proposed algorithm divides the estimation problem into two subtasks: the estimation of the missing frame OF, based on the previous received frames, and the projection of the previous frame onto the missing one, based on this estimated information. The complexity of the solution is due to the fact that the relationship between the

Table 16.2 Average PSNR (dB) for several video sequences (QCIF, YUV 4:2:0, frame rate = 10 fps, bit rate = 64 kbps) at various PLRs, obtained by the OF based EC method, and the repetition of the last correct frame EC method

Sequence	PLR	OF based EC	Repetition based EC
Foreman	5%	23.95	21.36
Foreman	10%	20.82	18.60
Foreman	20%	20.39	19.14
Table Tennis	5%	28.80	26.39
Table Tennis	10%	29.07	26.35
Table Tennis	20%	23.10	20.94
Coastguard	5%	28.09	26.09
Coastguard	10%	25.18	23.00
Coastguard	20%	24.46	22.39

OF and the intensity information is not known. A generally used model for such a relation is given by

$$\frac{dx(s_H, s_V, t)}{dt} = 0 \tag{16.3}$$

being s_H and s_V the horizontal and vertical spatial coordinates, t the temporal coordinate, and $x(s_H,\ s_V,\ t)$ the intensity distribution of the video sequence. The proposed algorithm has been extensively tested on several sequences, with both independent and correlated packet loss models, using loss parameters typical of the wireless context. It resulted to outperform the only alternative technique for whole frame losses, i.e. the repetition of the last correctly received frame, as shown in Table 16.2, for some of the tested configurations. The proposed solution exhibits very high PSNR gains, provides good visual quality, and results to be an attractive choice for real-time applications, thanks to its limited complexity.

To conclude this section, a huge amount of technical literature exists on the topic of Error Concealment, and new solutions are continuously proposed to the scientific community; details about the schemes herein discussed, or novel strategies, can be found in the cited references.

16.4 An Error Concealment Algorithm for H.264/AVC Video Transmission to DVB-H Devices

As previously discussed, the available bandwidth on a wireless channel, and therefore the bit rate over the radio link, is limited and likely to be low, so that the achievable compression is one of the main features motivating the adoption of H.264/AVC in a wireless environment.

In addition to high-compression efficiency, the video coding standard for real-time services in wireless environments has to be error resilient. However, this feature, which mainly affects the encoder, relies on the activation of coding options that are usually beyond the H.264/AVC baseline profile, and require an increased complexity in the encoder operations. On the other hand, the baseline profile is the one able to better handle real-time video communications. Loss of data is a consequence of the errors and impairments due to the harsh radio environment: it is necessary to implement suited strategies able to manage data cancelations. Moreover, it is worth noting that new directions in the design of wireless systems do not necessarily aim at error rate minimization, but rather at throughput maximization. If we assume that bit errors are detected by the lower layer entities, any remaining transmission error results in a packet loss. Therefore, it is necessary to ensure that the decoder is able to react to data losses, by applying appropriate concealment strategies.

The concealment algorithm originally implemented within the Reference Software version JM9.7 [10] is a hybrid scheme: if the block to conceal belongs to an I frame, the missing pixels are obtained as a weighted interpolation of the pixels located in the available adjacent MBs, according to a suboptimum approach based on a MAP estimate. In the case of a lost MB in a P frame, the algorithm adopts two different strategies, depending on the motion degree of the video content: low motion MBs are simply replaced by copying the corresponding MBs from the last reference frame, whereas high motion MBs require further computation. Both the strategies are based on the so-called status map of the MBs, in which every MB is marked as correctly received, lost, or concealed. In order to recover a current MB marked as lost, not only the available correctly received MBs, but also the previously concealed ones, can be used in the concealment process.

As any hybrid concealment solution, the algorithm provided within Reference Software JM9.7 shows some drawbacks, basically related to the following issues:

- The restoring strategy applied to I frames is not effective on wide damaged areas, or if the image is rich in details; under these conditions, spatial interpolation determines a shading effect with loss of details.
- The restoring strategy conceived for P frames is effective in the case of null or very low motion video sequences, but it produces wrong effects when scene change events occur.

With the aim of improving the behavior of the JM9.7 concealment scheme, and, more in general, of a hybrid solution, the integration of a low complexity scene change detection algorithm is proposed [24, 25], so that concealment of the damaged frames can be performed on the basis of detected scene change events, instead of referring to the coding features of the frames (i.e., I or P frame). The results presented in the following refer to the JM9.7 Reference Software; even if other and more recent versions of the Reference Software have been released, their Error Concealment features do not differ significantly from those present in JM9.7 that can be assumed as a representative implementation.

16.5 Error Concealment Based on Scene Change Detection

When processing P frames, the H.264/AVC encoder exploits the temporal correlation among them to find a good predictor for a block, using motion compensation techniques. When a scene change occurs, the correlation among blocks gets lost and the majority of the MBs in the frame is coded in Intramode. According to this consideration, if the number of MBs encoded as Intra within an Interframe gets higher than a threshold, it is reasonable to assume that a scene change has occurred. In this case, even in presence of an Interframe, it is better to perform a spatial concealment at the decoder, by interpolation from adjacent MBs. For experimental purposes, the threshold has been set to 30%; actually, this value is not always appropriate, especially in some specific situations related to panning effects of the camera. Besides a threshold for detecting scene change events, it is also important to set a threshold for the minimum number of correctly received MBs. In the presence of wide damaged areas, copying the missing MBs from the previous frame is better than restoring them by interpolation, at least from the subjective perception point of view.

Temporal correlation is not available in the case of I frames, so that a simple scene change detection can be performed by properly sampling four distinct MBs, one within each quadrant of the frame. Once the sample MBs are selected, the corresponding blocks in the most recent available reference image are located, for SAD (Sum of Absolute Difference) computation. The resulting SAD value is compared to a threshold, whose value is obtained experimentally; if higher, a scene change is revealed.

A block diagram of the concealment scheme resulting from the application of the simple scene change detector is plotted in Fig. 16.8. As shown in the diagram,

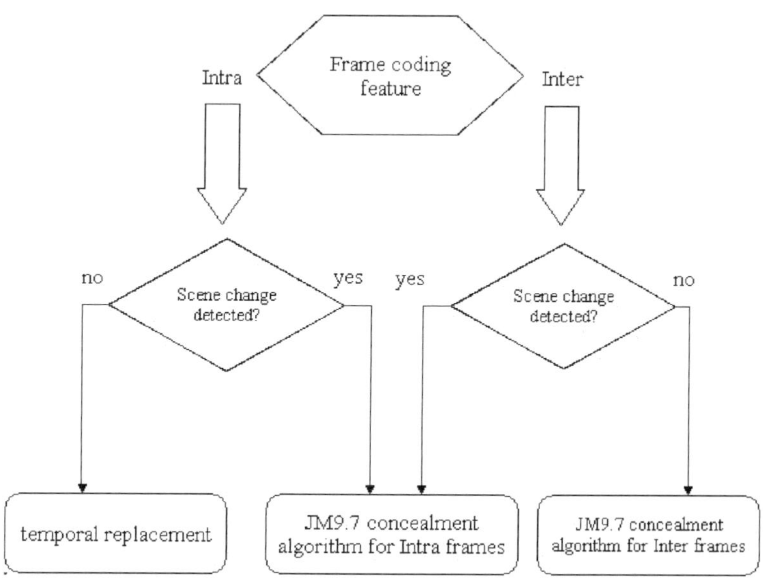

Fig. 16.8 Block diagram of the Error Concealment scheme based on a simple scene change detector

either in the case of an Intra or Interframe, a scene change detection is performed. If no scene change is revealed within an Intraframe, then a plain temporal replacement strategy [3] is adopted, to restore the damaged blocks, otherwise the concealment algorithm for Intraframes, already provided by the JM9.7 Reference Software, is invoked. The same JM9.7 concealment scheme for Intraframes is activated also in the case of a scene change detected within an Interframe, whereas the JM9.7 concealment solution, specifically conceived for Interframes, is activated, when no changes are detected.

16.6 Experimental Results and Performance Evaluation

In order to test the effectiveness of the proposed low complexity algorithm based on scene change detection, a modified version of the JM9.7 Reference Software decoder has been implemented, and tested off-line, on a 300 frames QCIF video sequence obtained as an ensemble of different, shorter, sub-sequences. This way, true scene changes occur at any time a given sub-sequence ends and a new one starts; moreover, all sub-sequence features, like panning effects in Stefan or Foreman, are maintained. The sub-sequences have been chosen according to their motion properties, ranging from standard static sequences (i.e., Akiyo, Mother& Daughter), to dynamic ones (like Stefan, Gladiator). A quantization parameter (QP) equal to 28 has been selected for both Inter and Intraframes; five frames are available for reference during the encoding process, and an IP equal to 15 has been configured. This way, every 15 frames, a frame is always encoded as Intra, in order to ensure a refresh point in the sequence, which can help recovering from eventual errors in the decoding process.

In order to simulate an error prone transmission environment, the encoder has been suitably modified, so that the resulting encoded sequence shows a PLR of 14.35%: in a set of 2702 RTP packets used to encapsulate the compressed video data, 388 are randomly deleted to simulate their loss. Each packet carries a single slice of a QCIF frame, i.e. a row of 11 MBs. In Fig. 16.9 the percent number of MBs that are correctly received within each frame of the sequence is shown, according to the MB classification provided by the status map at the decoder.

The ensemble video sequence is given as input to the modified version of the decoder, to which the following configuration applies:

- Scene change detection for Interframes: percent number of Intra-MBs >30%, percent increment of the number of Intra-MBs between subsequent frames >30%
- Concealment of Interframes with detected scene change: spatial concealment, as the original JM9.7
- Concealment of Interframes with no scene change detected: temporal concealment, as the original JM9.7
- Scene change detection for Intraframes: four sample MBs, SAD threshold equal to 5000

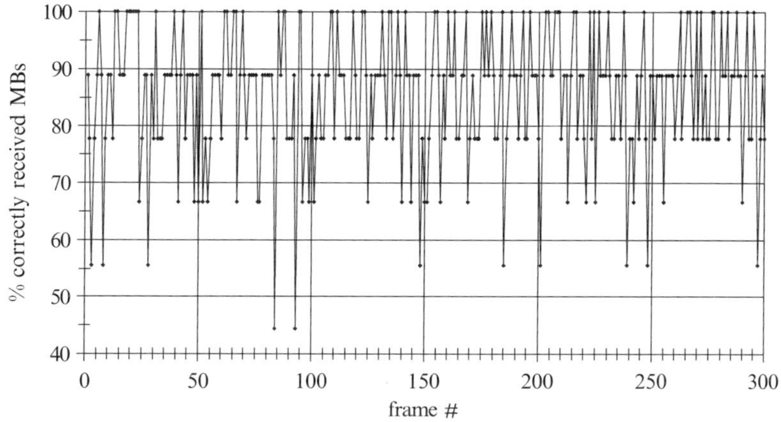

Fig. 16.9 Percent number of MBs correctly received, for each frame

Table 16.3 Average PSNR values resulting from the JM 9.7 decoder and the modified decoder integrating a simple scene change detector

	PSNR Y (dB)	PSNR U (dB)	PSNR V (dB)
Original JM 9.7 decoder	27.33	42.39	42.67
Modified decoder	30.79	46.48	47.02

– Concealment of Intraframes with detected scene change: spatial concealment, as the original JM9.7
– Concealment of Intraframes with no scene change detected: temporal replacement

The performance obtained by means of the modified decoder is compared to those provided by the original Reference Software decoder, in terms of objective quality of the decoded sequence, expressed by the average PSNR values of its color components. The results are summarized in Table 16.3.

The comparison among the results in Table 16.3 shows that the modified decoder has, on average, a better behavior, which is actually confirmed also by the results related to each single frame of the sequence, as shown in Fig. 16.10.

Besides an objective evaluation of the two decoding schemes, their performance has been subjectively tested also, by considering a visual comparison between corresponding frames decoded by the two different tools. In the following, results related to the case of a scene change between Interframes and Intraframes are shown, which again confirm a better behavior of the modified scheme, with respect to the original one.

In Fig. 16.11b it is evident that frame no. 201 is restored by spatial interpolation, even if it is an Interframe. The result obtained is better than the one shown in Fig. 16.11a, where slices not belonging to the correct video sequence are used anyway to recover the same frame (201). Better performance is also confirmed in the case of Intraframes, as shown in Fig. 16.12.

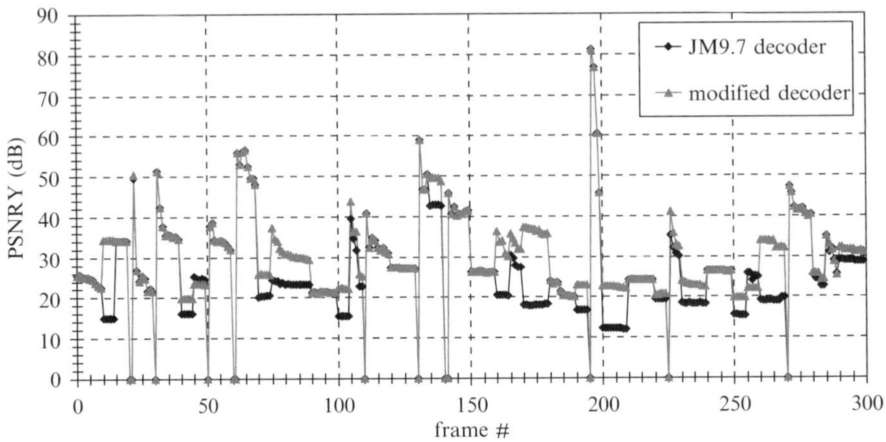

Fig. 16.10 Comparison of the frame PSNR Y values obtained by means of the two decoders

Fig. 16.11 Visual comparison in the case of a scene change between Interframes: (**a**) JM 9.7 decoder and (**b**) modified decoder

Fig. 16.12 Visual comparison in the case of a scene change between Intraframes: (**a**) JM 9.7 decoder and (**b**) modified decoder

As previously described, in order to detect a scene change in the case of Intraframes, four sample MBs are selected within the QCIF frame, and their SAD values compared to the ones of the corresponding MBs co-located in the following frame. Actually, this strategy fits well to bigger image formats, like CIF, whereas it can be considered redundant in the case of a QCIF frame. Moreover, the SAD computation performed over four different MBs could be resource consuming at the receiver. According to these considerations, the scene change detector can be modified, in the case of Intraframes, by considering a single scanning procedure over the whole QCIF frame, i.e. a single reference MB for the SAD computation.

Comparing the new SAD values resulting from a single MB evaluation to the ones obtained by testing up to 4 MBs, it is possible to see that even if the values are different, the scene change detector shows the same behavior, because the relationship of the SAD values towards the threshold set to detect a scene change event has not been modified. This is reported in Fig. 16.13. As a final remark, the computational time required by the off-line application of the original JM9.7

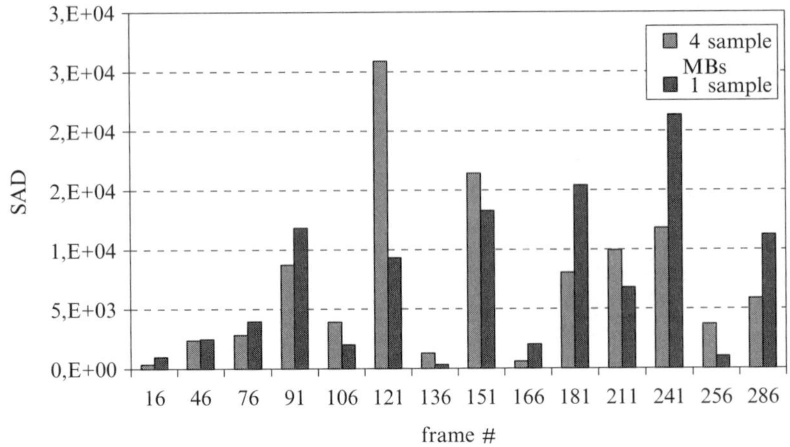

Fig. 16.13 SAD values obtained over each frame by considering 1 and 4 sample MBs

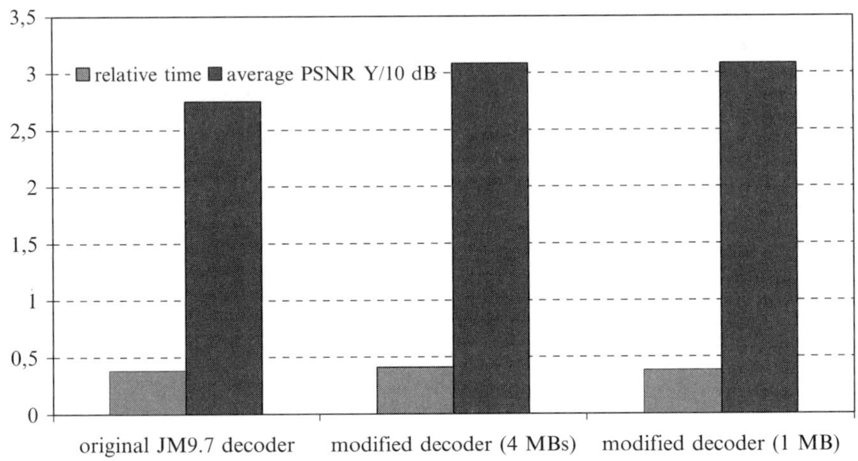

Fig. 16.14 Computational time (in relative units) and quality comparison among the original JM 9.7 decoder, the modified decoder using 4 MBs, and the modified decoder using 1 MB for SAD computation

and the two alternative decoders, together with the corresponding output quality, in terms of the average PSNR Y value of the video sequence, are evaluated. The last version of the modified decoder, that uses only 4 MB for SAD computation, gives the best quality result, and a computational time equal to the one required by the original JM9.7 decoder, as reported in Fig. 16.14. At a parity of required computational time and objective quality provided, the modified decoder performs an additional SAD computation with respect to the original JM 9.7 decoder, which means the modified concealment approach has a reduced complexity, with respect to the original one. Finally, the two modified decoders have been tested against the original JM9.7 one, for a number of different PLRs (ranging from 5.75% to 18%), according to the test conditions suggested by the JVT Packet Loss Simulator [7], always providing better performance.

16.7 Conclusions

This chapter discussed the main features of the H.264/AVC standard that make it particularly suited to wireless video communications, and provided an overview of the main issues related to the complexity and variability of the wireless DVB-H channel. In order to face channel errors and impairments, suited Error Concealment algorithms can be conceived and implemented at the decoder, to recover the lost video information and improving the final quality perceived by the mobile user.

After a detailed overview of the most relevant concealment solutions available in the technical literature, a specific strategy based on a simple scene change detector has been presented, which could be effectively adopted within DVB-H devices, given the design constraints they have to face, in terms of available computing power and memory resources. The suggested solution can improve the final quality delivered to the user, without increasing the required computing time.

References

1. Agrafiotis D, Bull DR, Canagarajah CN (2006) Enhanced error concealment with mode selection. IEEE Trans Circ Syst Video Technol 16: 960–973
2. Belfiore S, Grangetto M, Magli E, Olmo G (2005) Concealment of whole-frame losses for wireless low bit rate video based on multiframe optical flow estimation. IEEE Trans Mult 7: 316–328
3. Chiaraluce F, Ciccarelli L, Gambi E, Spinsante S (2004) Performance evaluation of error concealment techniques in H.264 video coding. In: Proc PCS04, San Francisco
4. European Telecommunications Standards Institute (2004) Digital video broadcasting (DVB); transmission system for handheld terminals (DVB-H). ETSI EN 302 304 V1.1.1
5. Failly M (1989) Digital land mobile radio communications (final report). Commission of the European Communities, Directorate General Telecommunications, Information Industries and Innovation, pp 135–147
6. Faria G, Henriksson JA, Stare E, Talmola P (2006) DVB-H: digital broadcast services to handheld devices. In: Proc IEEE, Volume 94, Issue 1, pp 194–209
7. Guo Y, Li H (2005) SVC/AVC loss simulator. Input Document to JVT (ISO/IEC JTC1/SC29/WG11 and ITU-T SG16 Q.6), 17th Meeting, France
8. Himmanen H, Hazmi A, Paavola J (2006) Comparison of DVB-H link layer FEC decoding strategies in a mobile fading channel. In: Proc IEEE PIMRC
9. http://iphome.hhi.de/suehring/tml/download/jm73.zip
10. http://iphome.hhi.de/suehring/tml/download/jm97.zip
11. http://www.3gpp.org
12. International Organisation for Standardization ISO/IEC JTC1/SC29/WG11 (2002) Coding of moving pictures and audio: MPEG-4 overview
13. International Standard ISO/IEC 14496-10 (2005) H.264 Advanced Video Coding for generic audiovisual services

14. International Telecommunication Union (2005) H.263 video coding for low bit rate communications
15. Jingyu H, Xiaomin X, Limin M (2007) An adaptive doppler shift estimator in mobile communication systems. Ant Wir Prop Lett 6: 117–121
16. Kwon DK, Shen MY, Kuo CCJ (2007) Rate Control for H.264 video with enhanced rate and distortion models. IEEE Trans Circ Syst Video Technol 17: 517–529
17. Lam WM, Reibman AR, Liu B (1993) Recovery of lost or erroneously received motion vectors. In: Proc ICASSP, pp 417–420
18. Lin S, Costello DJ (2004) Error Control Coding, 2nd Edition. Prentice-Hall, Pearson Prentice-Hall, Upper Saddle River, NJ 07458
19. Ostermann J, Bormans J, List P, Marpe D, Narroschke M, Pereira F, Stockhammer T, Wedi T (2004) Video Coding with H.264/AVC: tools, performance, and complexity. IEEE Circ Syst Mag: 7–28, Volume 4, Issue 1
20. Poikonen J, Paavola J (2006) Error models for the transport stream packet channel in the DVB-H link layer. In: Proc IEEE Int Conf Comm, pp 1861–1866
21. Salama P, Shroff NB, Delp EJ (1998) Error concealment in encoded video. In: Image Recovery Techniques for Image Compression Applications. Norwell, Kluwer, 1998
22. Shirani S, Erol B, Kossentini F (2000) A concealment method for shape information in MPEG-4 coded video sequence. IEEE Trans Mult 2: 185–190
23. Shirani S, Kossentini F, Ward R (2000) A concealment method for video communications in an error-prone environment. IEEE J Sel Areas Comm 18: 1122–1128
24. Spinsante S, Gambi E, Chiaraluce F (2006) An improved error concealment strategy driver by scene motion properties for H.264/AVC decoders. In: Proc EUSIPCO06, Italy
25. Spinsante S, Gambi E, Falcone D, Morichetti S (2006) Low complexity concealment of H.264/AVC encoded video for DVB-H devices. In: Proc MobiMedia, Italy
26. Stockhammer T, Hannuksela MM, Wiegand T (2003) H.264/AVC in wireless environments. IEEE Trans Circ Syst Video Technol 13: 657–673
27. Varsa V, Hannuksela MM, Wang YK (2001) Non-normative error concealment algorithms. ITU-T VCEG-N62
28. Wang H, Lv J (2005) A novel error concealment scheme for intra frames of H.264 video. In: Proc IEEE Int Work VLSI Design & Video Tech, China: pp 300–303
29. Wiegand T, Sullivan GJ, Bjøntegaard G, Luthra A (2003) Overview of the H.264/AVC video coding standard. IEEE Trans Circ Syst Video Technol 13: 560–576
30. Yu GS, Liu MK, Marcellin MW (1998) POCS-based error concealment for packet video using multiframe overlap information. IEEE Trans Circ Syst Video Technol 8: 422–433
31. Zhou SM, Li J, Fei J, Zhang Y (2007) Improvement on rate-distortion performance of H.264 rate control in low bit rate. IEEE Trans Circ Syst Video Technol 17: 996–1006

Chapter 17
AVS-M: Mobile Video Standard

Liang Fan

17.1 Introduction

AVS-M standard is the seventh part of the standards developed by the Audio Video Coding Standard (AVS) Workgroup of China with focusing on video coding for mobility systems and devices with limited computation capability and power consumption. AVS-M standard can cover a broad range of applications including mobile multimedia broadcasting, IP multimedia subsystem (IMS), multimedia mailing, multimedia services over packet networks, video conferencing, video phone, and video surveillance.

AVS Workgroup was authorized and established by Science and Technology Department under Ministry of Information Industry of China in June, 2002. To meet the requirement of information industry and realize the union of the native enterprises and research institutions, the mandate of the group is to establish (or edit) such general technical standards as the compression, decoding, processing, and the representation of digital audio–video and meanwhile provide the digital audio–video equipments and systems with high-efficient and economic coding/decoding technologies. This standard is applied in the field of significant information industry such as high-resolution digital broadcast, high-density laser-digital storage media, wireless broad-band multimedia communication, and internet broad-band stream media.

The AVS standard is the series of "Information Technology Advanced Coding of Audio and Video" standard abbreviation. There are 10 parts of the AVS standard, as shown in Table 17.1.

Liang F.
Department of Electronics and Communication Engineering, Sun Yat-sen University, Guangzhou 510275, P. R. China
and
Media Processing and Communication Lab, Sun Yat-sen University, Guangzhou 510275, P. R. China
e-mail: isslf@mail.sysu.edu.cn

F.-L. Luo (ed.), *Mobile Multimedia Broadcasting Standards: Technology and Practice*, 485
DOI: 10.1007/978-0-387-78263-8_17,

Table 17.1 The AVS standard

Part no.	Name
1	System
2	Video
3	Audio
4	Conformance test
5	Reference software
6	Digital media right management
7	Mobile video
8	Transmit AVS via IP network
9	AVS file format
10	Mobile speech and audio coding

AVS-M standard is the seventh part of the AVS standard, which was finished by the group in 2006 [1].

The rest of this chapter will give an overview on AVS-M standard with being organized into the following five sections. Section 17.2 presents the overall architecture of AVS-M codec and its features. In Sect. 17.3, the structure and syntax of AVS-M bitstream will be defined. Section 17.4 is devoted to the key techniques employed in AVS-M standard. Definitions of the profile and level used in AVS-M standard will be presented in Sect. 17.5. The last section will give some conclusions and further discussions on AVS-M standard with its applications and implementation.

17.2 Architecture of AVS-M Codec and Its Features

Figures 17.1 and 17.2 depict the block diagram of AVS-M encoder and decoder, respectively. Because of the same modules, the architecture of AVS-M looks like that of H.264 [2].

Each input macroblock needs to be predicted (intra predicted or inter predicted) in AVS-M. In an AVS-M encoder, the S_0 is used to select the proper prediction method for current macroblock. In an AVS-M decoder, the S_0 is controlled by the macroblock type of current macroblock.

The intra predictions are derived from the neighboring pixels in left and top blocks. The unit size of intra prediction is 4×4 because of the 4×4 integer cosine transform (ICT) used by AVS-M. There are nine intra prediction modes for luma blocks, and three modes for luma blocks. A detail description of AVS-M intra prediction can be found in Sect. 17.4.3.

The inter predictions are derived from the decoded frames. Seven types of block sizes, i.e., 16×16, 16×8, 8×16, 8×8, 8×4, 4×8, and 4×4 are supported in AVS-M. The precision of motion vector in inter prediction is up to 1/4 pixel. Section 17.4.4 gives more detail about AVS-M inter prediction.

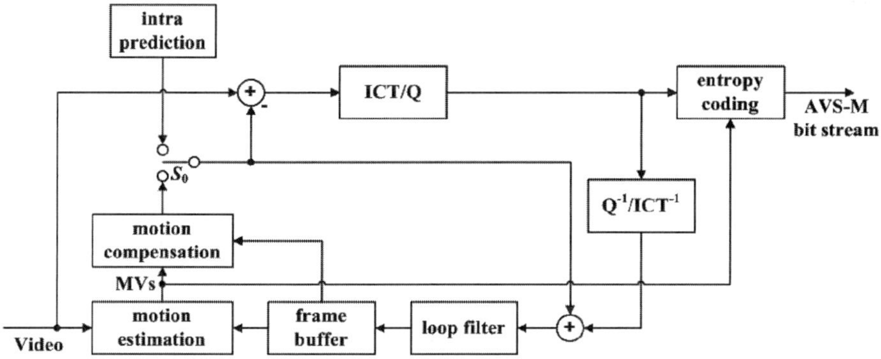

Fig. 17.1 The block diagram of AVS-M encoder

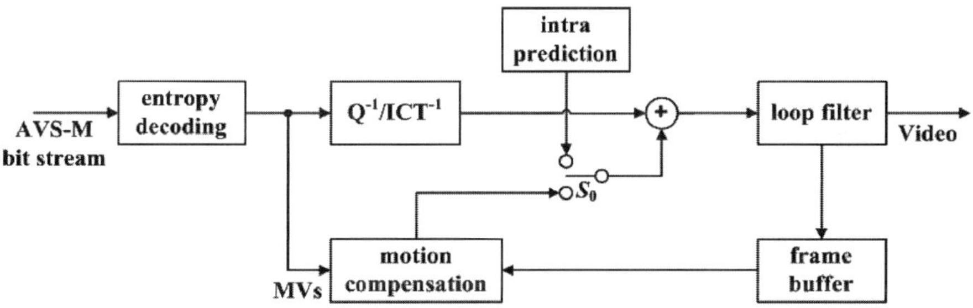

Fig. 17.2 The block diagram of AVS-M decoder

The prediction residues are transformed with 4×4 ICT. The ICT coefficients are quantized using a scale quantizer. More detail of the transform and quantization can be found in Sect. 17.4.2. The scanning order of quantized coefficients used in AVS-M is still zig-zag similar to that used in MPEG-1 [3] and MPEG-2 [4].

AVS-M employs an adaptive VLC coding technique. There are two different types of Exp-Golomb codebook corresponding to different distributions, as described in Sect. 17.4.1. Some mapping tables are defined to map coded symbol into a special codebook and its elements.

The sum of prediction and current reconstructed error image forms the reconstructed reference. AVS-M uses the deblocking filter in motion compensation loop. The deblocking process directly acts on the reconstructed reference first across vertical edges and then across horizontal edges. Obviously, different image regions and different bit rates need different smoothes. Therefore, the deblocking filter is automatically adjusted in AVS-M depending on activities of blocks and QP parameters. Section 17.4.5 has the detail of AVS-M loop filter.

17.3 Structure and Syntax of AVS-M Bitstream

17.3.1 AVS-M Bitstream Format

AVS-M specifies the network abstraction layer (NAL) unit stream and the byte stream. The NAL unit stream consists of a sequence of NAL units. The byte stream is constructed from the NAL unit streams by prefix each NAL unit with a start code "0000 0000 0000 0000 0000 0001". The nal unit type specified in AVS-M is in Table 17.2.

The syntax of AVS-M NAL units is shown in Fig. 17.3. Figure 17.4 gives the syntax of sequence parameter set NAL unit and picture parameter set NAL unit. Figure 17.5 shows the syntax of supplemental enhancement information NAL unit. Figures 17.6 and 17.7 give the syntax of random access point indicator NAL unit and RBSP (raw byte sequence payload) trailing bits, respectively.

17.3.2 Video Sequence

Video sequence is the highest syntactic structure of AVS-M bitstream. An AVS-M bitstream consists of one or more video sequences. An AVS-M video sequence consists of one or more access units. The first access unit of each video sequence should be an IDR access unit. All subsequent access units in the video sequence are non-IDR access units.

Table 17.2 NAL unit type

nal unit type	Content of NAL unit and RBSP syntax
0	Unspecified
1	Picture header of a non-IDR picture
2	Picture header of an IDR picture
3	Slice
4	Sequence parameter set (SPS)
5	Picture parameter set (PPS)
6	Supplemental enhancement information (SEI)
7	Random access point indicator
8 .. 23	Reserved
24 .. 31	Unspecified

```
nal_unit( NumBytesInNALunit ) {
    forbidden_zero_bit
    nal_ref_idc
    nal_unit_type
    for ( i = 0; i < NumBytesInNALunit-1; i++ ) {
        rbsp_byte[i]
    }
}
```

Fig. 17.3 Syntax of NAL unit

```
seq_parameter_set_rbsp( ) {
    profile_idc
    level_idc
    seq_parameter_set_id
    delta_time_picture_distance_1
    num_ref_frames
    horizontal_size_minus1
    vertical_size_minus1
    aspect_ratio
    frame_cropping_flag
    if ( frame_cropping_flag == '1' ) {
        frame_cropping_left_offset
        frame_crop_right_offset
        frame_crop_top_offset
        frame_crop_bottom_offset
    }
    hrd_parameters_present_flag
    if ( hrd_parameters_present_flag == '1' )
        hrd_parameters( )
    rbsp_trailing_bits( )
}
```

(**a**) Sequence parameter set.

```
pic_parameter_set_rbsp( ) {
    pic_parameter_set_id
    seq_parameter_set_id
    fixed_picture_qp_flag
    picture_reference_flag
    if ( ! picture_reference_flag )
        sliding_window_size_flag
    skip_mode_flag
    loop_filter_disable_flag
    if ( ! loop_filter_disable_flag )
        loop_filter_parameter_flag
    constrained_intra_pred_flag
    if ( more_rbsp_data( ) )
        half_pixel_mv_flag
    rbsp_trailing_bits( )
}
```

(**b**) Picture parameter set.

Fig. 17.4 Syntax of parameter set

If an access unit contains a random access point indicator NAL unit, it either is an IDR access unit or it contains gradual_random_access SEI message. Every access unit can have at most one random access point indicator NAL unit. When random access point indicator NAL unit is present, it shall be the first NAL unit of the access unit in decoding order. The decoding of the access units following it shall not refer to any sequence parameter set NAL unit, picture parameter set NAL unit or SEI NAL unit preceding it.

In gradual random access SEI message, recovery_frame_cnt specifies the frame number of the recovery picture containing this SEI message, with range of 0–31. If decoding process starts from the coded picture containing gradual random access SEI message, all pictures are indicated to be correct in content starting at the position of the picture having the frame number equal to current frame number frame_num incremented by recovery_frame_cnt in modulo 32 arithmetic.

17.3.3 Pictures

AVS-M only supports progressive video sequence. So in AVS-M, one picture is one frame. AVS-M supports the 4:2:0 format, as shown in Fig. 17.8.

AVS-M specifies two types of pictures, which is I picture and P picture. In AVS-M, P picture can have a maximum of two reference frames for forward prediction. Motion vectors can exceed the boundaries of the reference picture. In this case,

```
sei_rbsp( ) {
    do {
        sei_message( )
    } while( more_rbsp_data( ) )
}
```

(a) SEI RBSP.

```
sei_message( ) {
    payloadType = 0
    while ( next_bits(8) == 0xFF ) {
        ff_byte /* equal to 0xFF */
        payloadType += 255
    }
    last_payload_type_byte
    payloadType += last_payload_type_byte
    payloadSize = 0
    while ( next_bits(8) == 0xFF ) {
        ff_byte /* equal to 0xFF */
        payloadSize += 255
    }
    last_payload_size_byte
    payloadSize += last_payload_size_byte
    sei_payload( payloadType, payloadSize )
}
```

(b) SEI message.

```
sei_payload( payloadType, payloadSize ) {
    if ( payloadType == 0 )
        user_data( payloadSize )
    else if ( payloadType == 1 )
        full_frame_freeze( payloadSize )
    else if ( payloadType == 2 )
        full_frame_freeze_release( payloadSize )
    else if ( payloadType == 3 )
        scalable_layer_profile_level( payloadSize )
    else if ( payloadType == 4 )
        hrd_buffering_parameters( payloadSize )
    else if ( payloadType == 5 )
        gradual_random_access( payloadSize )
    else
        reserved_sei_message( payloadSize )
}
```

(c) SEI payload.

```
gradual_random_access( payloadSize ) {
    recovery_frame_cnt
    rbsp_trailing_bits( )
}
```

(d) Gradual random access SEI message.

Fig. 17.5 Syntax of SEI

Fig. 17.6 Syntax of random access point indicator

```
random_access_point_indicator_rbsp( ) {
}
```

Fig. 17.7 Syntax of RBSP trailing bits

```
rbsp_trailing_bits( ) {
    rbsp_stop_one_bit /* equal to 1 */
    while( ! byte_aligned( ) )
        rbsp_alignment_zero_bit /* equal to 0 */
}
```

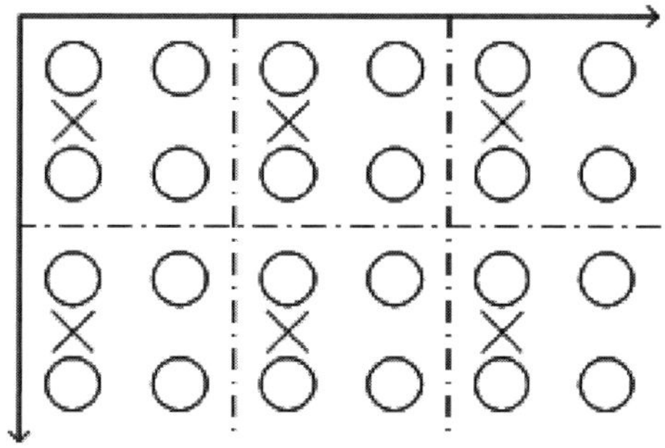

Fig. 17.8 Nominal vertical and horizontal locations of 4:2:0 luma and chroma samples in a frame

Fig. 17.9 Syntax of picture header

```
picture_header_rbsp( ) {
    picture_coding_type
    picture_distance
    pic_parameter_set_id
    if ( nal_unit_type == 2 )
        picture_distance_gap_minus1
    frame_num
    picture_qp
    if ( loop_filter_parameter_flag != 0 ) {
        alpha_ci_offset
        cp_offset
        loopfilter_qp_offset
    }
    if ( sliding_window_size_flag )
        sliding_window_size_minus1
    rbsp_trailing_bits( )
}
```

the nearest pixel within the frame shall be used to extend the boundary. A reference pixel block shall not exceed the picture boundary beyond 16 pixels for luma samples and 8 pixels for chroma samples, in both horizontal director and vertical direction.

The syntax of picture header is shown in Fig. 17.9. Pictures are ordered consecutively in the bitstream, whose order is the same as decoding order. An access unit consists of one primary coded picture and other NAL units except the picture header NAL units and slice NAL units.

17.3.4 Slices

A slice is a sequence of macroblocks ordered consecutively in the raster scan. Macroblocks within a slice shall not overlap and slices shall also not overlap. Except in loop filtering process, the decoding process for macroblocks within a slice shall not use the data from other slices of the picture. Figure 17.10 gives an example of slice structure.

The syntax of slice is shown in Fig. 17.11. Slice NAL units shall be ordered in ascending order of the value of first_mb_in_slice.

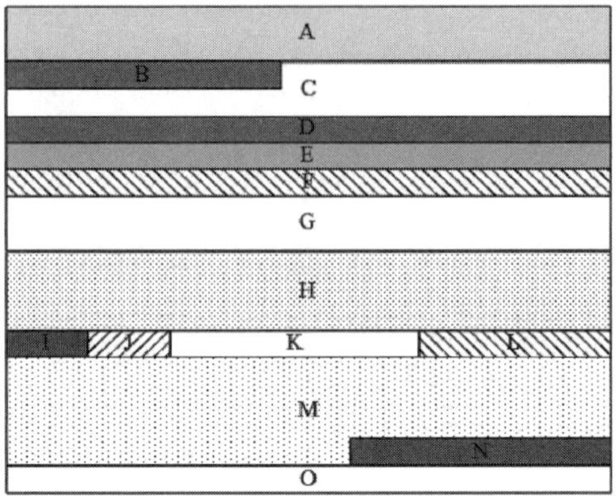

Fig. 17.10 Slice structure

```
slice_layer_ rbsp( ) {
    slice_header( )
    slice_data( )
    rbsp_trailing_bits( )
}
```

(a) Slice layer RBSP.

```
slice_header( ) {
    first_mb_in_slice
    if ( fixed_picture_qp_flag == '0' ) {
        fixed_slice_qp_flag
        slice_qp_delta
    }
    if ( ! loop_filter_disable_flag )
        disable_loop_filter_slice_flag
}
```

(b) Slice header.

```
slice_data() {
    do {
        if ( PictureType != 0 ) {
            if ( skip_mode_flag == '1' )
                mb_skip_run
        }
        if ( MbIndex < MbWidth × MbHeight && more_rbsp_data( ) )
            macroblock( )
    } while ( more_rbsp_data( ) )
}
```

(c) Slice data.

Fig. 17.11 Syntax of slices

17.3.5 Macroblocks and Blocks

Picture is divided into macroblocks. The upper-left sample of each macroblock shall not exceed picture boundary.

Macroblock partitioning and submacroblock partitioning are shown in Figs. 17.12a and b, respectively. The partitioning is used for motion compensation. The number in each rectangle specifies the order of appearance of motion vectors and reference indices in a bitstream.

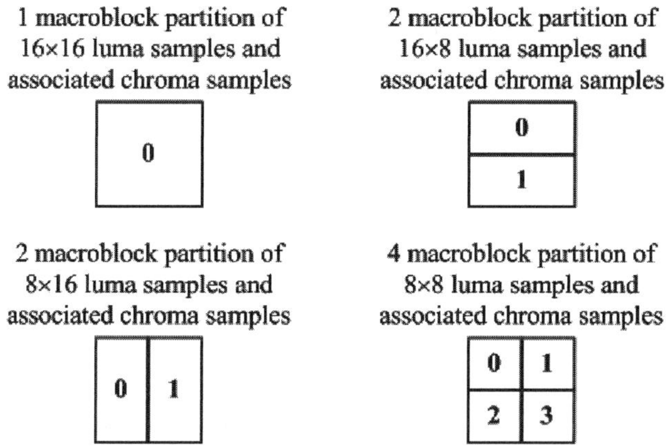

(a) Macroblock partitioning.

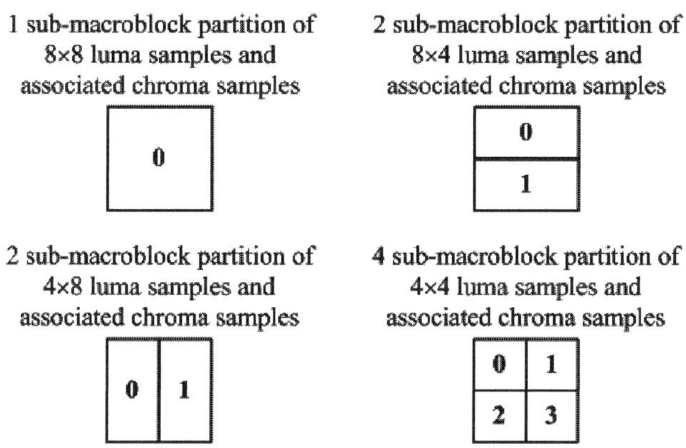

(b) Sub-macroblock partitioning.

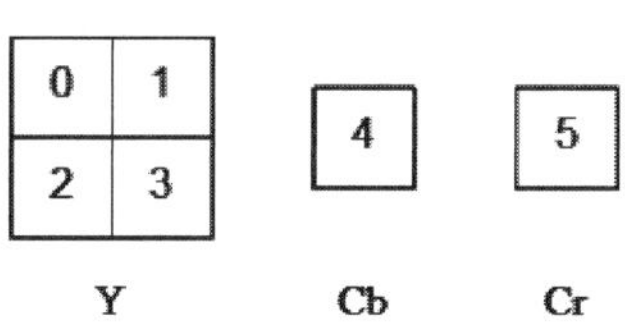

(c) 8 × 8 partitions.

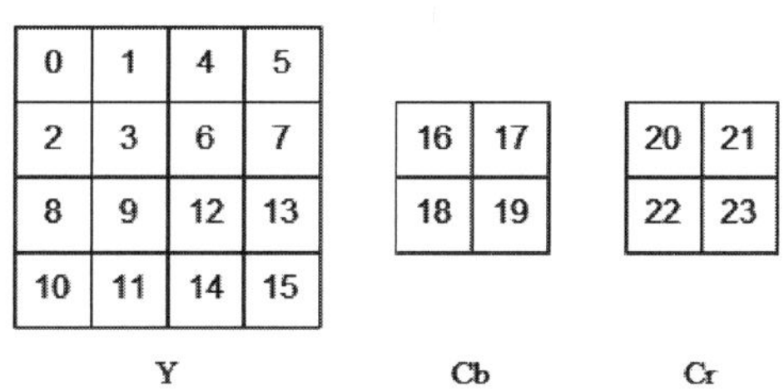

(d) 4 × 4 partitions.

Fig. 17.12 Macroblocks

A macroblock can be partitioned into six 8 × 8 blocks (four luma blocks and two chroma blocks) or twenty-four 4 blocks (16 luma blocks and 8 chroma blocks). The coding order of these blocks is shown in Figure 17.12c and d, respectively.

There are two macroblock types of I-picture specified by AVS-M. If mb_type is 1, the type of current MB is I_4x4; otherwise, the type is I_Direct.

Table 17.3 Macroblock types of P picture

MbTypeIndex	Macroblock type	Number of MVs
0	P_Skip	0
1	P_16x16	1
2	P_16x8	2
3	P_8x16	2
4	P_8x8	4 .. 16
5	I_4x4	0

Table 17.4 Submacroblock types of P picture

sub_mb_type	Submacroblock type	Number of MVs
0	Sub_8x8	1
1	Sub_8x4	2
2	Sub_4x8	2
3	Sub_4x4	4

The macroblock types of P-picture are shown in Tables 17.3 and 17.4. If skip_mode_flag is 1, then MbTypeIndex is equal to mb_type plus 1; otherwise, MbTypeIndex is equal to mb_type. If MbTypeIndex is greater than or equal to 5, MbTypeIndex is set to 5.

The syntax of macroblock and block is shown in Fig. 17.13.

17.4 The Key Techniques in AVS-M Standard

17.4.1 Entropy Coding

AVS-M uses Exp-Golomb code, as shown in Table 17.5, to encode syntax elements such as quantized coefficients, CBP, macroblock coding type, and motion vectors. Eighteen coding tables are used in quantized coefficients encoding. The encoder uses the run and the absolute value of the current coefficient to select the table.

17.4.2 Transform and Quantization

Like H.264, AVS-M also uses a 4×4 integer cosine transform (ICT). Equation (17.1) gives the 4×4 inverse transform matrix of AVS-M. The operations of the transform/quantization can be completed within 16 bits. AVS-M uses a prescaled integer transform (PIT) technology, all of the scale-related operations has been done in the encoder. The decoder does not need any scale operations.

```
macroblock() {
    mb_type
    if ( MbType != 'P_Skip') {
        if ( MbType == 'I_4x4' ) {
            for ( i=0; i<16; i++ ) {
                pred_mode_flag
                if ( pred_mode_flag )
                    intra_luma_pred_mode
            }
            intra_chroma_pred_mode
        }
        else if (MbType == 'I_Direct' ) {
            intra_chroma_pred_mode
        }
        else {
            if ( IsRef0anIdrPic == 0 && picture_reference_flag == 0 ) {
                for ( i = 0; i <MvNum; i++ )
                    mb_reference_index
            }
            if ( MbType != 'P_8x8' ) {
                for ( I = 0; i <MvNum; i++ ) {
                    mv_diff_x
                    mv_diff_y
                }
            }
            else {
                for ( i = 0; i < 4; i++ )
                    sub_mb_type
                for ( i = 0; i < 4; i++ ) {
                    for ( j = 0; j < SubMvNum[sub_mb_type[ i ]]; j++ ) {
                        mv_diff_x
                        mv_diff_y
                    }
                }
            }
        }
        if ( ! (MbTypeIndex >= 5) )
            cbp
        for ( i = 0; i < 6; i++ ) {
            if ( MbCBP & ( 1 << i ))
                cbp_4x4
        }
        if ( MbCBP > 0 && ! fixed_slice_qp_flag )
            mb_qp_delta
        for ( i = 0; i < 6; i++ )
            block( i )
    }
}
```

(a) Macroblock.

Fig. 17.13 Syntax of macroblock and block

```
block( i ) {
    if ( i < 6 && (MbCBP & ( 1 << i ))) {
        for ( j=0; j<4; j++ ) {
            if ( cbp_4x4 & (1 << j ) ) {
                do {
                    trans_coefficient_4
                    if (trans_coefficient_4 >= 71)
                        escape_run
                } while ( trans_coefficient_4 != "EOB" && !LastCoefficient )
            }
        }
    }
}
```

(**b**) Block.

Fig. 17.13 *continued*

Table 17.5 *k*-th order Exp-Golomb code

Exponential	Code structure	Range of code number
	1	0
	0 1 x_0	1 .. 2
$k = 0$	0 0 1 x_1 x_0	3 .. 6
	0 0 0 1 x_2 x_1 x_0	7 .. 14
	......	......
	1 x_0	0 .. 1
	0 1 x_1 x_0	2 .. 5
$k = 1$	0 0 1 x_2 x_1 x_0	6 .. 13
	0 0 0 1 x_3 x_2 x_1 x_0	14 .. 29
	......	......

$$\mathbf{T}_4 = \begin{bmatrix} 2 & 3 & 2 & 1 \\ 2 & 1 & -2 & -3 \\ 2 & -1 & -2 & 3 \\ 2 & -3 & 2 & -1 \end{bmatrix} \tag{17.1}$$

For a 4×4 block, the decoded levels w'_{ij} are dequantized with (17.2).

$$x_{ij} = \left(x'_{ij} \times \mathrm{d}(QP) + 2^{s(QP)-1} \right) \gg s(QP), \tag{17.2}$$

where x_{ij} is the dequantized coefficient, QP is the quantization parameter, $\mathrm{d}(QP)$ is the inverse quantization table, and $s(QP)$ is the varied shift value for inverse quantization. The range of x'_{ij} is $[-2^{11}, 2^{11} - 1]$. The range of QP is $[0, 63]$.

The horizontal inverse transform is performed as follows

$$\mathbf{H}' = \mathbf{X} \times \mathbf{T}_4^{\mathrm{T}}, \tag{17.3}$$

where $\mathbf{X}$ is the 4×4 dequantized coefficient matrix and $\mathbf{H}'$ is the intermediate result after the horizontal inverse transform.

Then the vertical inverse transform is performed

$$\mathbf{H} = \mathbf{T}_4 \times \mathbf{H}', \qquad (17.4)$$

$\mathbf{H}$ is the 4×4 matrix after transform. The range of the elements h_{ij} of $\mathbf{H}$ shall be $[-2^{15}, 2^{15} - 17]$.

The elements r_{ij} of the residual matrix is calculated as

$$r_{ij} = \left(h_{ij} + 2^4\right) \gg 5 \qquad (17.5)$$

17.4.3 Intraframe Prediction

AVS-M uses intraframe prediction to improve the performance of intracoded macroblocks. The intraframe prediction in AVS-M is conducted for each 4×4 luma/chroma block in the spatial domain. AVS-M defines 9 modes for 4×4 luma block and 3 modes for 4×4 chroma block, as shown in Tables 17.6, and 17.7 and Fig. 17.14.

Each 4×4 luma block has a predicted intra prediction mode, which is derived from the left and upper neighbouring blocks. AVS-M defines two intra-coded macroblock coding types, I_4x4 and I_Direct. Intra-coded macroblocks in P picture can be I_4x4 only.

Table 17.6 Luma intra prediction mode

Luma mode	Name
0	Intra_Luma_Down_Left
1	Intra_Luma_Vertical_Left
2	Intra_Luma_Vertical
3	Intra_Luma_Vertical_Right
4	Intra_Luma_Down_Right
5	Intra_Luma_Horizontal_Down
6	Intra_Luma_Horizontal
7	Intra_Luma_Horizontal_Up
8	Intra_Luma_DC

Table 17.7 Chroma intra prediction mode

Chroma mode	Name
0	Intra_Chroma_DC
1	Intra_Chroma_Horizontal
2	Intra_Chroma_Vertical

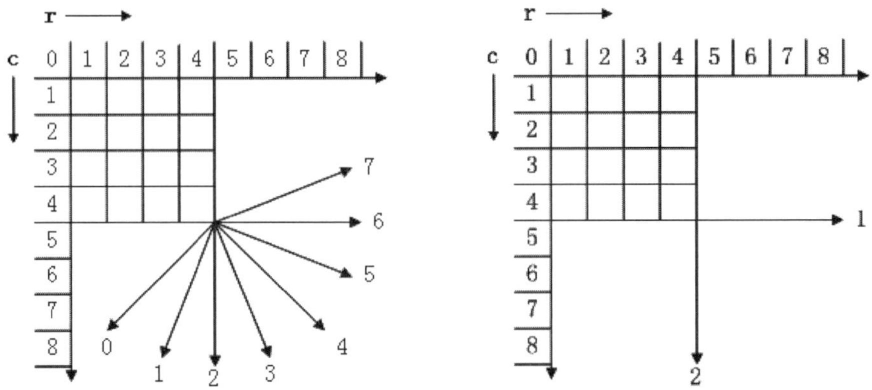

(a) Luma prediction. (b) Chroma prediction.

Fig. 17.14 Intraframe prediction

If a macroblock is I_4x4 macroblock, the intraprediction mode of 4×4 luma blocks in the macroblock is determined as follows:

1. If pred_mode_flag is '0', intra prediction mode is equal to predicted intra prediction mode.
2. If pred_mode_flag is "1", and intra_luma_pred_mode is less than predicted intraprediction mode, then intraprediction mode is equal to intra_luma_pred_mode; otherwise, intraprediction mode is equal to intra_luma_pred_mode plus 1.

If a macroblock is I_Direct macroblock, the intraprediction mode of the sixteen 4×4 luma blocks in the macroblock is equal to predicted intra prediction mode.

If current block is a chroma block, the intraprediction mode is equal to intra_chroma_pred_mode. Every 4×4 chroma block of the macroblock uses the same intra prediction mode.

According to Figs. 17.14a and b, the number of reference samples used in intra prediction is 17 (samples marked with $r[i]$ or $c[i]$, $i = 0, \ldots, 8$). Only 9 of 17 reference samples are get from current picture. Reference samples $r[5]$ to $r[8]$ is equal to $r[4]$, and $c[5]$ to $c[8]$ is equal to $c[4]$.

17.4.4 Interframe Prediction

AVS-M defines I picture and P picture. P pictures use forward motion compensated prediction. The maximum number of reference pictures used by a P picture is two. To improve the error resilience capability, one of the two reference pictures can be a I/P pictures far away from current picture.

AVS-M also specifies nonreference P pictures. If the nal_ref_idc of a P picture is equal to 0, the P picture shall not be used as a reference picture. The nonreference P pictures can be used for temporal scalability.

The reference pictures are identified by the reference picture number, which is 0 for IDR picture. The reference picture number of a non-IDR reference picture is calculated as follows:

$$\text{refnum} = \begin{cases} \text{refnum}_{\text{prev}} + \text{num} - \text{num}_{\text{prev}}, & \text{num}_{\text{prev}} \leq \text{num} \\ \text{refnum}_{\text{prev}} + \text{num} - \text{num}_{\text{prev}} + 32, & \text{otherwise} \end{cases} \tag{17.6}$$

where num is the frame_num value of current picture, num_{perv} is the frame_num value of the previous reference picture, and $\text{refnum}_{\text{perv}}$ is the reference picture number of the previous reference picture.

After decoding current picture, if nal_ref_idc of current picture is not equal to 0, then current picture is marked as "used for reference". If current picture is an IDR picture, all reference pictures except current picture in decoded picture buffer (DPB) shall be marked as "unused for reference". Otherwise, if nal_unit_type of current picutre is not equal to 0 and the total number of reference pictures excluding current picutre is equal to num_ref_frames, the following applies:

1. If num_ref_frames is 1, reference pictures excluding current picutre in DBP shall be marked as "unused for reference."
2. If num_ref_frames is 2 and sliding window size is 2, the reference picture excluding current picutre in DPB with smaller reference picture number shall be marked as "unused for reference."
3. Otherwise, if num_ref_frames is 2 and sliding window size is 1, the reference picture excluding current picture in DBP with larger reference picture number shall be marked as "unused for reference."

The size of motion compensation block can be 16×16, 16×8, 8×16, 8×8, 8×4, 4×8 or 4×4. If the half_pixel_mv_flag is equal to '1', the precision of motion vector is up to 1/2 pixel, otherwise the precision of motion vector is up to 1/4 pixel. When half_pixel_mv_flag is not present in the bitstream, it shall be inferred to be "11."

Figure 17.15 gives the positions of integer, half and quarter pixel samples. Capital letters indicate integer sample positions, while small letters indicate half and quarter sample positions. The interpolated values at half sample positions can be obtained using 8-tap filter $F_1 = (-1,4,-12,41,41,-12,4,-1)$ and 4-tap filter $F_2 = (-1,5,5,-1)$. The interpolated values at quarter sample position can be obtained by linear interpolation.

According to Fig. 17.15, half sample b is calculated as follows:

$$b' = -C + 4D - 12E + 41F + 41G - 12H + 4I - J \tag{17.7}$$

$$b = (b' + 32) \gg 6 \tag{17.8}$$

and half sample h is calculated as follows:

$$h' = -A + 5F + 5N - S \tag{17.9}$$

$$h = (h' + 4) \gg 3. \tag{17.10}$$

Both b and h should be cliped to the range $[0, 255]$.

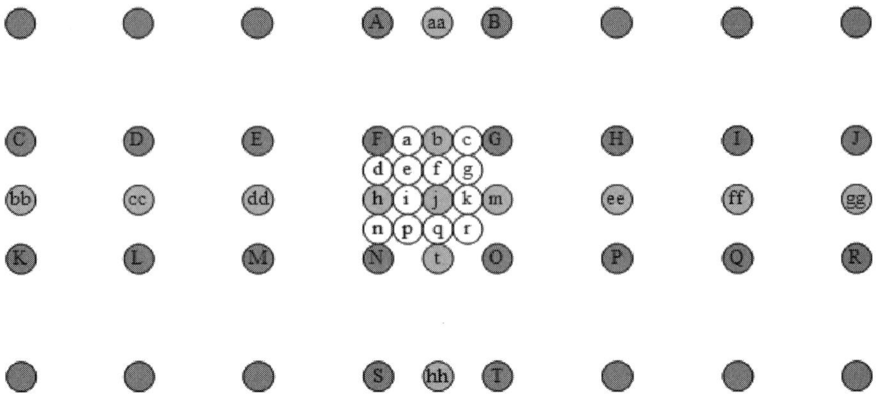

Fig. 17.15 The positions of integer, half and quarter pixel samples

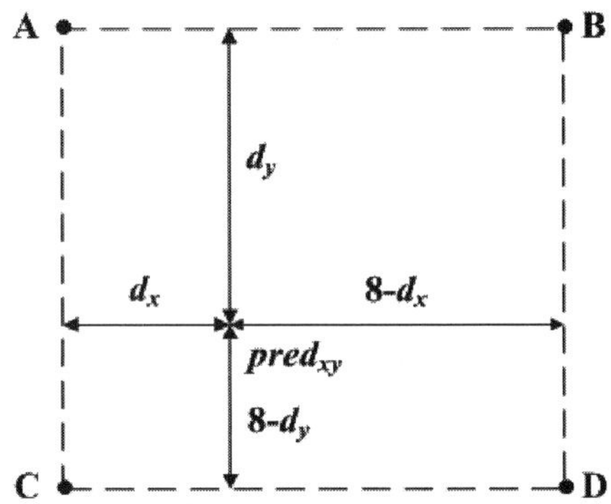

Fig. 17.16 Relationship between variable positions and reference sample

Quarter sample a is calculated as

$$a = (F + b + 1) \gg 1. \tag{17.11}$$

Interpolation of chroma sample values is shown in Fig. 17.16. A, B, C, D are the integer sample values around the interpolated sample. d_x and d_y are the horizontal and vertical distances from predicted sample to A, respectively.

The predicted sample $\mathrm{pred}_{\mathrm{xy}}$ is calculated as

$$\mathrm{pred}_{\mathrm{xy}} = \left((8 - d_\mathrm{x})(8 - d_\mathrm{y})A + d_\mathrm{x}(8 - d_\mathrm{y})B + (8 - d_\mathrm{x})d_\mathrm{y}C + d_\mathrm{x}d_\mathrm{y}D + 32\right) \gg 6, \tag{17.12}$$

$\mathrm{pred}_{\mathrm{xy}}$ should be cliped to the range $[0, 255]$.

17.4.5 Loop Filter

Block-based video coding often produces blocking artifacts especially at low bit rates. AVS-M defines an adaptive in-loop de-blocking filter to improve the decoded

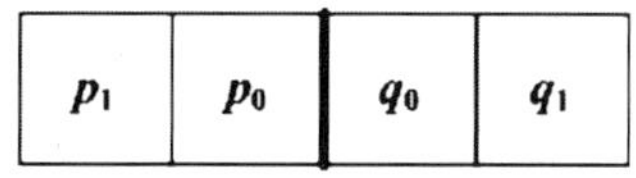

Fig. 17.17 Horizontal or vertical edge of 4×4 block

visual quality. The filtering is applied to the boundaries of luma and chroma blocks except for the boundaries of picture or slice. The filtering process uses four samples positions on two vertical or horizontal sides. The filtering mode of AVS-M loop filter is selected as follows:

1. If current macroblock is intra coded, use intra macroblock filtering mode.
2. Otherwise, if current macroblock is inter coded but not coded in skip mode, or QP of current macroblock is greater than or equal to a threshold, use inter macroblock filtering mode.
3. Otherwise, current macroblock is not filtered.

Figure 17.17 shows four sample positions on two vertical or horizontal sides of p and q (edge is expressed in bold line).

If current macroblock is to be filtered and the (17.13) is true, then samples around the edge are filtered.

$$|p_0 - q_0| < \alpha, \tag{17.13}$$

where α is the threshold value for block edge, which can be obtained from IndexA. IndexA can be calculated using the average quantization parameter of the two corresponding blocks and AlphaCIOffset:

$$QP_{\text{av}} = (QP_p + QP_q + 1) \gg 1 \tag{17.14}$$

$$\text{IndexA} = QP_{\text{av}} + \text{AlphaCIOffset} \tag{17.15}$$

IndexA should be cliped to the range $[0, 63]$. The value of AlphaCIOffset is equal to alpha_ci_offset in the bitstream. The range of alpha_ci_offset is -8 to 8. If alpha_ci_offset is not present in the bitstream, it shall be inferred to be 0.

17.5 The Profile and Level in AVS-M Standard

The purpose of defining profiles and levels is to facilitate interoperability among streams from various applications. A profile is a subset of syntax, semantics, and algorithms defined by AVS-M, where a level puts constraints on the parameters of the stream. In AVS-M, a Jiben Profile has been defined. The profile_idc of Jiben Profile is 16. There are 9 levels (1.0, 1.1, 1.2, 1.3, 2.0, 2.1, 2.2, 3.0, and 3.1) defined in AVS-M, as shown in Table 17.8. The limits of each level are shown in Tables 17.9, 17.10, and 17.11.

Level 1.0 and Level 1.1 are defined for SQCIF (128×96) video and QCIF (176×144) video with max bitrate 64 kbit/s and 128 kbit/s. Both of Level 1.0 and Level 1.1 supports max frame rate of 30 fps for SQCIF video, and 15 fps for QCIF video.

Table 17.8 Levels of AVS-M standard

level_idc	Level
0	Forbidden
10	1.0
12	1.1
14	1.2
16	1.3
30	2.0
32	2.1
34	2.2
50	3.0
52	3.1
Others	Reserved

Table 17.9 Limits of Level 1.x

Parameter	Level 1.0	Level 1.1	Level 1.2	Level 1.3
Max number of MBs per second	1,485	1,485	6,000	11,880
Max number of MBs per frame	99	99	396	396
Max bits per MB	4,096	4,096	4,096	4,096
Max bitrate (bit/s)	64,000	128,000	384,000	768,000
Max decoded picture buffer size (byte)	114,048	114,048	456,192	456,192
Max coded picture buffer size (bit)	175,000	350,000	1,000,000	2,000,000
Max range of horizontal MVs	$[-2048, +2047.75]$	$[-2048, +2047.75]$	$[-2048, +2047.75]$	$[-2048, +2047.75]$
Max range of vertical MVs	$[-32, +31.75]$	$[-32, +31.75]$	$[-32, +31.75]$	$[-32, +31.75]$

Levels 1.2, 1.3, and 2.0 are defined for CIF (352×288) video, and the max bitrate is 384 kbit/s, 768 kbit/s, and 2 Mbit/s, respectively. The max frame rate of Level 1.2 for CIF video is 15 fps, and the max frame rate of Levels 1.3 and 2.0 for CIF video is 30 fps.

Levels 2.1 and 2.2 can support video resolution up to 352×480 or 352×576 with the frame rate 30 fps of 25 fps. Level 3.0 supports VGA video (640×480, 30 fps) video, and Level 3.1 supports D1 video (720×576, 25 fps or 720×480, 30 fps).

The limits of the range of horizontal and vertical motion vector are useful for implementation of AVS-M codec on devices with limited computation capability and power consumption.

Table 17.10 Limits of Level 2.x

Parameter	Level 2.0	Level 2.1	Level 2.2
Max number of MBs per second	11,880	19,800	20,250
Max number of MBs per frame	396	792	1,620
Max bits per MB	4,096	4,096	4,096
Max bitrate (bit/s)	2,000,000	4,000,000	4,000,000
Max decoded picture buffer size (byte)	456,192	912,384	1,555,200
Max coded picture buffer size (bit)	2,000,000	4,000,000	4,000,000
Max range of horizontal MVs	$[-2048, +2047.75]$	$[-2048, +2047.75]$	$[-2048, +2047.75]$
Max range of vertical MVs	$[-32, +31.75]$	$[-256, +255.75]$	$[-256, +255.75]$

Table 17.11 Limits of Level 3.x

Parameter	Level 3.0	Level 3.1
Max number of MBs per second	36,000	40,500
Max number of MBs per frame	1,620	1,620
Max bits per MB	4,096	4,096
Max bitrate (bit/s)	6,000,000	8,000,000
Max decoded picture buffer size (bytes)	1,555,200	1,555,200
Max coded picture buffer size (bits)	6,000,000	8,000,000
Max range of horizontal MVs	$[-2048, +2047.75]$	$[-2048, +2047.75]$
Max range of vertical MVs	$[-256, +255.75]$	$[-256, +255.75]$

17.6 Conclusions

AVS-M standard is the seventh part of the AVS standard (Information Technology Advanced Coding of Audio and Video) developed by the AVS Workgroup of China. AVS-M standard uses of state-of-the-art video coding technologies to achieve higher coding efficiency. AVS-M standard mainly focus on video coding for mobility systems and devices with limited computation capability and power consumption. So the implementation complexity is an very important issue that the AVS-M standard should consider about.

Table 17.12 gives the comparison between AVS-M Jiben Profile and H.264 Baseline Profile [5] [6].

Compare with H.264 Baseline Profile, AVS-M Jiben profile has lower implementation complexity in both encoder and decoder.

Figures 17.18–17.24 gives the performance comparison between AVS-M Jiben profile and H.264 Baseline profile [7]. In these figures, "AVS-M" represents AVS-M Jiben Profile, "H.264" represents H.264 Baseline profile, "ref1" represents one

Table 17.12 Comparision between AVS-M Jiben profile and H.264 Baseline profile

Module	AVS-M Jiben profile	H.264 Baseline profile
Intraluma prediction	9 4 × 4 modes direct mode	9 4 × 4 modes 4 16 × 16 modes
Intrachroma prediction	3 4 × 4 modes	4 4 × 4 modes
Intraprediction Reference samples	9 samples	17 samples
Interprediction	16 × 16 to 4 × 4	16 × 16 to 4 × 4
Subpixel interpolation	8-tap (1/2 horizontal) 4-tap (1/2 vertical) linear (1/4)	6-tap (1/2) linear (1/4)
Transform and quantization	4 × 4 ICT (3,2) without scale in decoder	4 × 4 ICT (2,1) with scale in decoder
entropy coding	2D-VLC Exp-Golomb code	CAVLC Huffman/Exp-Golomb code
Loop filter	each pixel filtering once fewer pixels need filtering	each pixel filtering once or 2 times

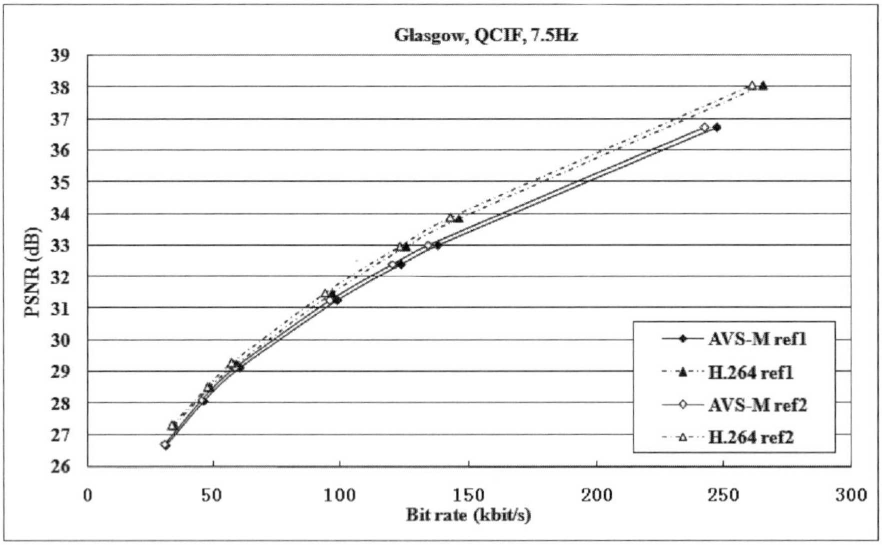

Fig. 17.18 Experimental results of Glasgow (QCIF, 7.5 Hz) where the *four curves* from *bottom* to *upper* are obtained from AVS-M ref1, AVS-M ref2, H.264 ref1, and H.264 ref2, respectively

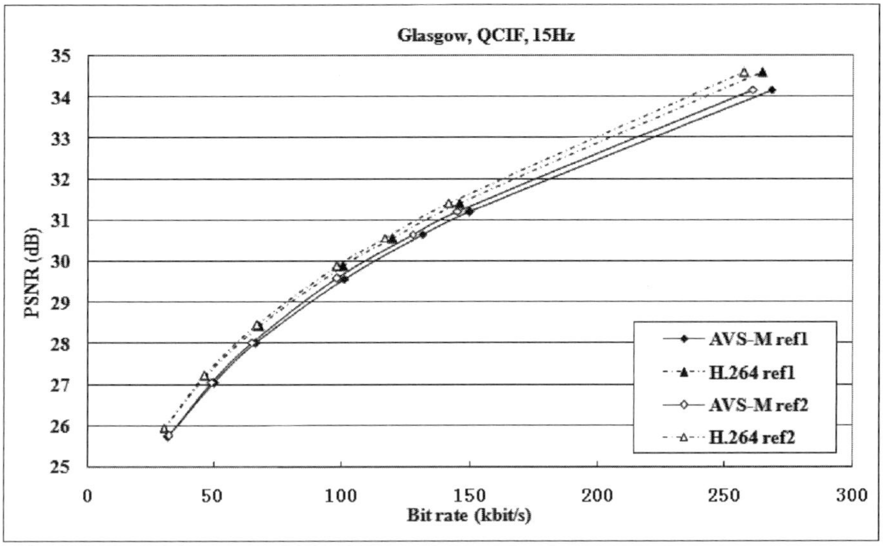

Fig. 17.19 Experimental results of Glasgow (QCIF, 15 Hz) where the *four curves* from *bottom* to *upper* are obtained from AVS-M ref1, AVS-M ref2, H.264 ref1, and H.264 ref2, respectively

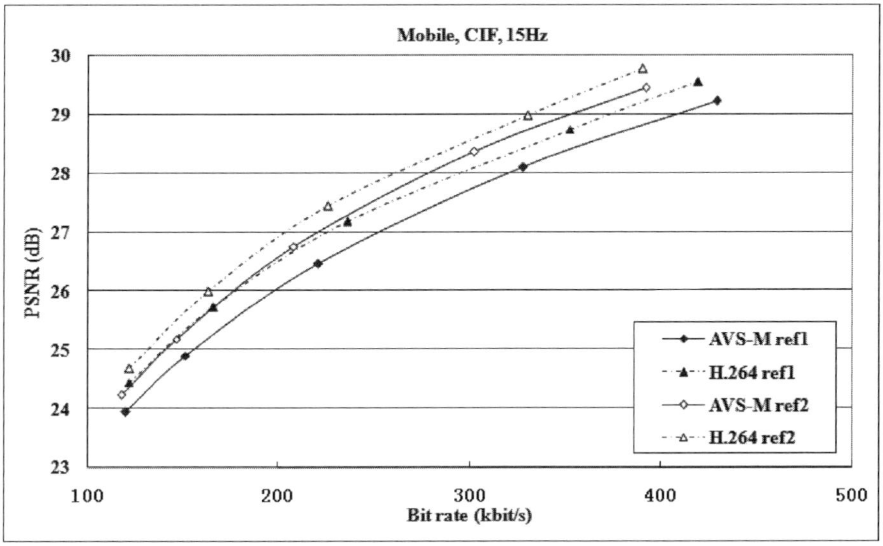

Fig. 17.20 Experimental results of Mobile (CIF, 15 Hz) where the *four curves* from *bottom* to *upper* are obtained from AVS-M ref1, H.264 ref1, AVS-M ref2, and H.264 ref2, respectively

reference frame, and "ref2" represents two reference frames. AVS-M uses reference software WM3.3a, and H.264 uses reference software JM12.0. The test conditions are shown in Table 17.13 [8].

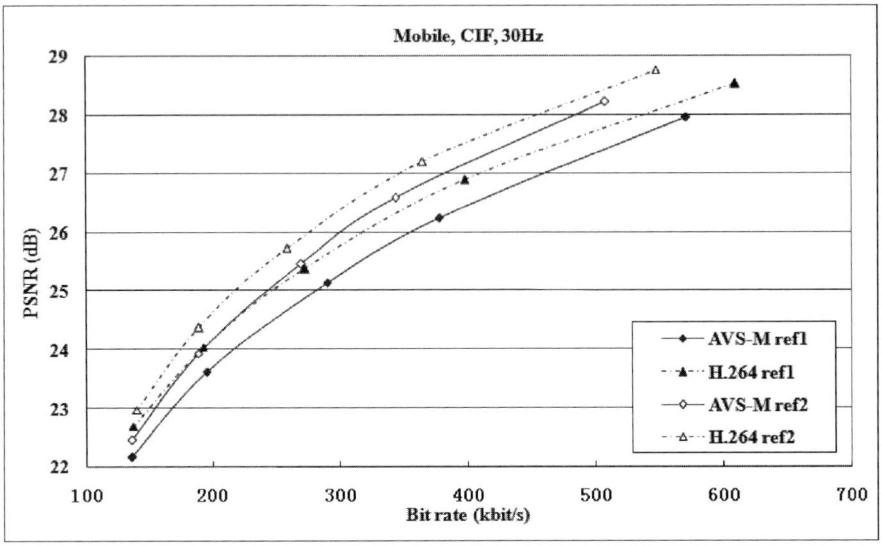

Fig. 17.21 Experimental results of Mobile (CIF, 30 Hz) where the *four curves* from *bottom* to *upper* are obtained from AVS-M ref1, H.264 ref1, AVS-M ref2, and H.264 ref2, respectively

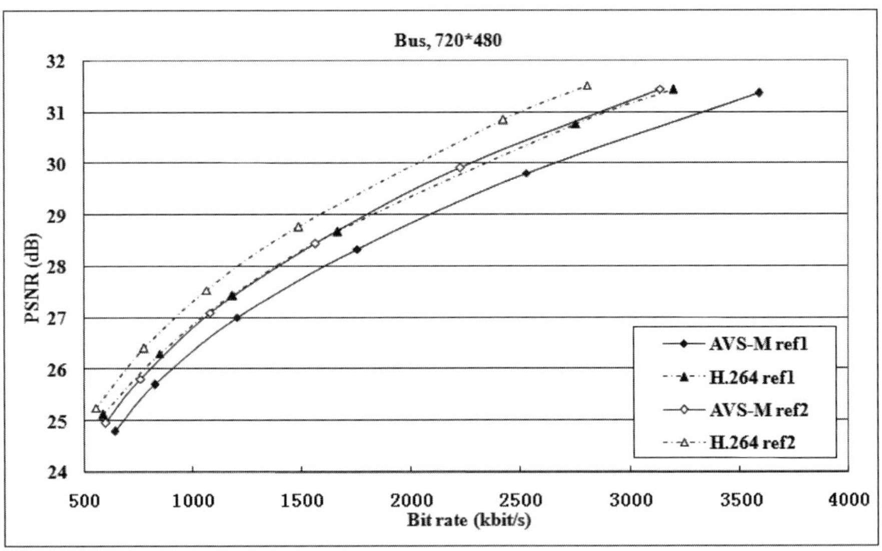

Fig. 17.22 Experimental results of Bus (720 × 480) where the *four curves* from *bottom* to *upper* are obtained from AVS-M ref1, H.264 ref1, AVS-M ref2, and H.264 ref2, respectively

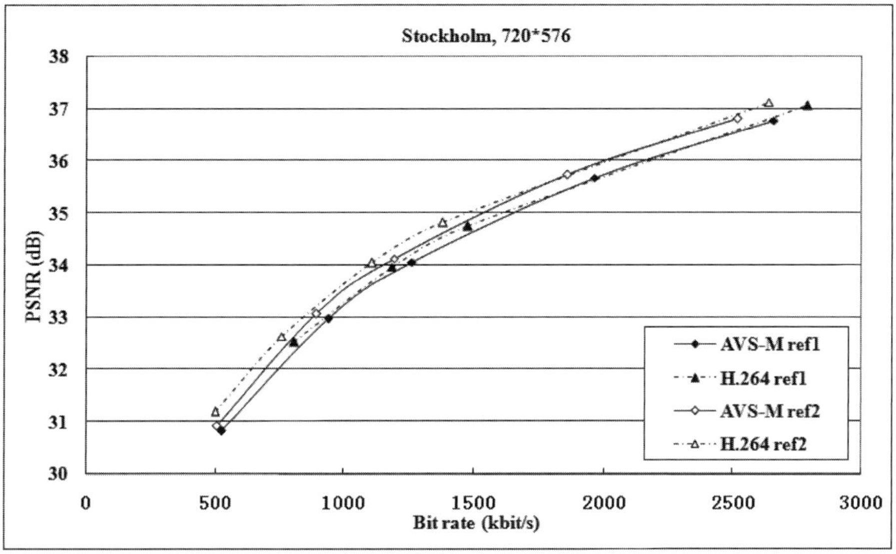

Fig. 17.23 Experimental results of Stockholm (720 × 576) where the *four curves* from *bottom* to *upper* are obtained from AVS-M ref1, H.264 ref1, AVS-M ref2, and H.264 ref2, respectively

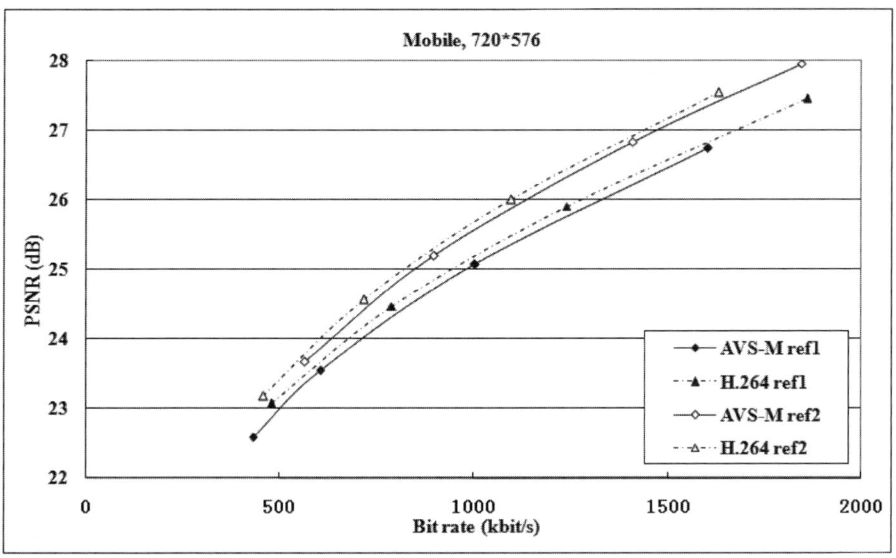

Fig. 17.24 Experimental results of Mobile (720 × 576) where the *four curves* from *bottom* to *upper* are obtained from AVS-M ref1, H.264 ref1, AVS-M ref2, and H.264 ref2, respectively

Table 17.13 Test conditions of AVS-M and H.264 performance test

Resolution	CIF	QCIF
Sequence	Bus, Football, Mobile, Paris	Foreman, Glasgow, News, Tempete
Frame rate (fps)	15, 30	15, 7.5
Bit rate (kbit/s)	128, 144, 192, 256, 384 512, 768	32, 48, 64, 96, 128, 144, 256
Resolution	720×576	720×480
Sequence	Mobile, Parkrun, Stockholm	Football, Tempete, Bus
Frame rate (fps)	25	30
Bit rate (kbit/s)	512, 768, 1000, 1500, 2000 3000, 4000	512, 768, 1000, 1500, 2000 3000, 4000

References

1. (2006) Final Draft GB/T 20090.7 Information Technology - Advanced Coding of Audio and Video - Part 7: Mobile Video, AVS N1294
2. (2003) Draft ITU-T recommendation and final draft international standard of joint video sepcfication (ITU-T Rec. H.264/ISO/IEC 14496-10 AVC), JVT-G050
3. (1993) ISO/IEC 11172-2, Coding of Moving Pictures and Associated Audio for Digital Storage Media at up to about 1.5 Mbit/s - Part 2: Video
4. (2000) ISO/IEC 13818-2, Generic Coding of Moving Pictures and Associated Audio Information - Part 2: Video
5. Wiegand T, Schwarz H, Joch A, Kossentini F, Sullivan G J (2003) IEEE Trans Circuits Syst Video Technol 13:688–703
6. Puri A, Chen X, Luthra A (2004) Signal Processing Image Communicaiton 19:793–849
7. (2007) AVS-P2 and AVS-P7 Performance Test Report, AVS M1979
8. (2006) AVS-P2 and AVS-P7 Performance Test, AVS N1353

Chapter 18
Design and Implementation of H.264/AVC Decoder

Kun-Bin Lee

18.1 Introduction

Recently multimedia and wired/wireless communication technologies have fundamentally changed the way we create, communicate, and consume audiovisual information. H.264/AVC, the latest international video coding standard [1], uses state-of-the-art coding tools and provides enhanced coding efficiency for a wide range of applications including mobile multimedia broadcasting, video conferencing, internet protocol television (IPTV), digital cinema, IP multimedia subsystem (IMS), surveillance, etc. As might be expected, the increase in coding efficiency and coding flexibility comes at the expense of an increase in complexity with respect to earlier standards. Realization of these applications relies on VLSI for cost-effective implementation.

In this chapter, we will discuss and review the issues and design techniques of H.264/AVC decoding systems. First, we will discuss the development issues of H.264/AVC decoding systems. The performance of software-based and hardware-based solution will also be reviewed. At the second part, we will review the latest designs of key coding tools of an H.264/AVC decoder, including intraprediction (IP), variable block-size motion compensation (MC), in-loop deblocking filter, inverse transform and inverse quantization, context-adaptive variable length decoding (VLD), and context-adaptive binary arithmetic decoding (CABAD). As the increasing integration density of various intellectual properties (IPs) into the system-on-a-chip (SoC) systems, the memory system becomes a dominant role to determine the final performance, area, and power consumption of SoC chips [2]. At the third part, memory usage and some related optimization techniques for H.264/AVC will also be reviewed.

K.-B. Lee
MediaTek Inc., No.1, Dusing RD.1, Science-based Industrial Park, Hsinchu, Taiwan 300, R.O.C.
e-mail: kb.lee@mediatek.com

F.-L. Luo (ed.), *Mobile Multimedia Broadcasting Standards: Technology and Practice*, DOI: 10.1007/978-0-387-78263-8_18,

18.2 Hardware/Software Partition

The design of embedded VLSI systems poses many challenges in areas such as design complexity, low power versus high speed, and hardware/software codesign and coverification. An embedded system architecture comprises software programmable modules, such as general-purpose processors or DSPs, and dedicated hardware modules. These modules can be common-off-the-shelf components or customized design components. They can also be integrated into a single chip, as called SoC. Besides maintaining most of the efficiency of dedicated hardware to optimize the system performance and power consumption, the programmability is introduced in the system to offer the desired flexibility in the development process. In addition to the trade-off between flexibility and efficiency, the system designers are currently faced with the enormously difficult work of integrating these heterogeneous components. To effectively conquer these challenges, we have to understand the prosperities of multimedia processing and SoC development issues.

In addition to the characters of real-time processing and high volume data, the characteristics of multimedia processing also differ from those of the traditional applications. To understand the characteristics of multimedia processing, a multimedia benchmark MediaBench [3] and MediaBench II [4] are proposed and the workload evaluations on this benchmark are performed [3, 5, 6]. Table 18.1 shows the benchmark programs among the 15 application packages. As opposed to conventional microprocessor applications, multimedia applications typically use small integer data types of 16 bits or less. This is particularly true of most video and image. It can quantitatively be stated that multimedia applications use 16-bit or smaller data types nearly 70% of the time, offering ample opportunity for subword parallelism or potentially even narrower data paths [5]. In comparison with general-purpose applications, the ratio of data types and sizes is significant. As reported by Hennessy

Table 18.1 Application packages and benchmark programs in MediaBench I & II

Media type	Application package	Programs
Video	MPEG-2	mpeg2dec, mpeg2enc
	H.263	h263dec, h263enc
	MPEG-4	mpeg4dec, mpeg4enc
	H.264/AVC	ldecod, lencod
Image	JPEG2000	Jasperdec, jasperenc
	JPEG	cjpeg, djpeg
	EPIC	epic, unepic
	Ghostscript	gs
Graphics	Mesa	mipmap, osdemo, texgen
Audio	ADPCM Coder	Rawcaudio, rawdaudio
Speech	GSM	gsmdecode, gsmencode
	G.721	g721dec, g721enc
	Rasta	rasta
Security	PGP	pgpdecode, pgpencode
	Pegwit	pegwitdec, pegwitenc

and Patterson [7], the percentage for byte, halfword, word, and double word data sizes is 7%, 19%, 74%, and 0% for SPECint92, and 0%, 0%, 31%, and 69% for SPECfp92, respectively. Similarly, a lot of studies have shown that the dramatically different nature and demands of media processing force profound changes in processor designs [8–11], including significant fine-grained data parallelism lends itself to machines with SIMD architectures, extensive data reorganization enables SIMD architectures to adapt to a variety of input data stream layouts, high instruction-reference locality to small loops yields a much higher degree of correlation between overall application speedup and loop/kernel speed speedup, etc.

In addition to the aforementioned media processing characteristics that make changes in designs of programmable processors, the general characteristics of the multimedia signal processing algorithms also change the designs of memory architectures [12–14], VLSI design [15–18], system-level exploration [19–22], and functional verification [23, 24], etc. The aforementioned real-time multimedia applications involve a combination of complex data- and control-flow, where also complex data types are manipulated and transferred. Therefore, multimedia SoC platforms are usually heterogeneous in nature. These systems combine high-performance data processing as well as slow-rate control processing, synchronous and asynchronous part, analog versus digital, guaranteed data bandwidth and minimized access latency, deterministic and random processing events, and so on.

Figure 18.1 shows a typical heterogeneous platform design for multimedia systems. The platform usually contains one or more programmable components (general-purpose processors, application-specific instruction-set processors, or DSPs), augmented with some specialized IPs for data paths or co-processors (accelerators). The programmable components run software components, being slow to medium speed algorithms, while time-critical parts are executed on dedicated hardware accelerators. Realization of cost-effective, power-efficient platforms relies on sophisticated design of these key components.

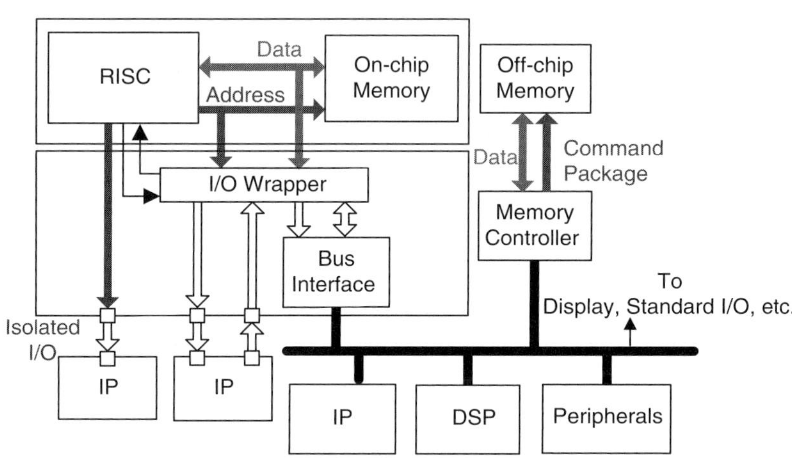

Fig. 18.1 A typical heterogeneous platform for multimedia systems

Next to these programmable components and dedicated hardware accelerators, the platform contains large amounts of on- and off-chip memories and peripherals. An important design issue is that not only these components, but also the data transfer schemes among these components shall be carefully designed. Increasing levels of integration are leading to a growing volume and diversity of data and control traffic exchanged among these components. As a result, a poorly designed on-chip communication architecture could become a severe impediment to system performance, power consumption, and cost. In order to support the large volume of heterogeneous communication traffic, techniques to efficiently achieve system's communication requirements need to be included as an integral part of any system design. These techniques include communication architectures, interface protocols, bus topologies, arbitration mechanisms, buffer sizing, coding schemes, etc.

In addition to the key components to perform multimedia processing and the data transfer schemes to meet communication requirements of multimedia systems, the access to off-chip memory is more and more a limiting factor for the performance of embedded systems. Previous researches have shown that data transfer and storage of multidimensional array signals dominate the performance and power consumption in SoCs [19, 25]. In particular, the data transfer to off-chip memory is especially important due to the scarce resource of off-chip bandwidth. As many studies have shown, the off-chip memory system is one of the primary performance bottlenecks in current systems. For example, Hennessy and Patterson [26] showed that while microprocessor performance improved 35% per year until 1986, and 55% per year since 1987, the access time to DRAM has been improving about 5% per year. Rattner illustrated that whereas a Pentium Pro required 70 instruction cycles for a DRAM access, a Pentium 4 running at 2 GHz takes 500–600 cycles [27]. Even as the performance of DRAM is ever-improved, the system overheads like turnaround time and request queuing account for a significant portion of inefficiencies in memory access. Schumann [28] reported that, in Alpha workstations, 30–60% of primary memory latency is attributable to system overhead rather than to latency of DRAM components. For multicore SoC designs, the performance of memory subsystem is even worse, due to the share of memory bus with different access requirements of these cores. Therefore, memory access management is essential for heterogeneous and data-intensive applications. It becomes especially critical when the same background memory in a unified memory architecture (UMA) must be used. In multimedia systems for instance, signal processing applications (H.264/AVC decoding, display processing, etc.) must obtain a more or less guaranteed bandwidth, while the CPU and possibly some peripherals require low latency for the best performance. The interface and arbitration to background memory is therefore of growing importance.

Before we discuss the design of H.264/AVC decoders, we first take a look at a real-world multimedia SoC platform, Philips' Nexperia Digital Video Platform (DVP) [29, 30]. The EDA Consortium selected Nexperia silicon design team as the recipient of its annual 2002 Design Achievement Award for the Nexperia-based Home Entertainment Engine pnx8500 (a.k.a. Viper), which along with its successor pnx8525 are the first instantiations of the Nexperia DVP [29]. Viper, a design with a size about 35 million transistors, receives, optionally decrypts, decodes, converts, and displays multiple media streams having different data formats. Besides MPEG-2 transport streams, the chip typically handles live video, audio, and various other stream types, in compressed or uncompressed formats. The video engine inside Nexperia PNX4103 is a 350 MHz very long instruction word (VLIW) TriMedia DSP core. It can encode baseline H.264 up to VGA or decode main profile H.264 up to D1 resolution, both at 30 fps. This VLIW DSP core contains instruction-set extensions for H.264 context-adaptive binary arithmetic decoding operations.

One of Viper's most challenging tasks is controlling all media streams to decode and produce the right data at the right time. Therefore, a high-performance point-to-point, 64-bit memory management interface (MMI) bus is used for bandwidth- and latency-critical access to the external SDRAM. In Nexperia DVP, the most critical resource, regarded as the most "scarce" or "expensive," is off-chip memory bandwidth [31]. The second and third critical resources are CPU cycle and off-chip memory size, respectively. To regulate the distribution of available bandwidth over the requesting blocks, Nexperia-DVP utilizes a sophisticated arbitration scheme to achieve two goals:

- Fair distribution of bandwidth over requesting blocks
- Shortest possible transaction latency for CPUs

Table 18.2 shows the three most computation-intensive subfunctions of H.264 main and baseline decoders. The performance of the baseline decoder [32] is evaluated on a Pentium 3 platform and the results are from the averages of all sequences in a test set. The performance of the main profile H.264/AVC decoder [33] is evaluated on an IBM PowerPC platform and the results are from the averages of four sequences, each has 720×576, 1280×720, and 1920×1088 resolution. Note that some kernels of this decoder are optimized by using Altivec media instructions. In addition, CABAD is used in this performance evaluation. When running at 1.6 GHz, the platform can perform H.264/AVC main profile decoding on D1 for an average frame rate of 26.88 fps. To perform H.264 main profile high definition (HD) decoding, it is expected that a general-propose processor PowerPC needs to run at 3.0 GHz [33]. Therefore, a pure software solution for H.264/AVC decoder might not be feasible for some applications. To lower the working frequency of the decoder, e.g., for the purpose of low-cost, low-power designs, or feasible designs of HD decoding, hardware accelerators are essential.

Table 18.3 shows H.264/AVC decoding performance on different platforms. Be aware of the experimentation conditions may significantly affect the decoding

Table 18.2 Most computation-intensive subfunctions of H.264/AVC decoders

Baseline profile	Main profile
Deblocking (33%)	Deblocking (49.01%)
Interpolation (25%)	Chroma interpolation (19.98%)
Bitstream parsing and entropy decoding (13%)	Entropy decoding (12.08%)

Table 18.3 Performance of different H.264/AVC decoders

Platform	Resolution	Frame rate (fps)	Clock rate required
IBM PowerPC[33]	D1	26.88	1.6 GHz
TI OMAP2420[34]	QVGA	30	220 MHz (C55x)
Tensilica Diamond [35]	D1	30	172 MHz
Dedicated hardware[36]	2048 × 1024	30	200 MHz
Dedicated hardware[37]	1920 × 1088	30	120 MHz

performance, especially the memory subsystem and the test sequences. In OMAP2420 [34], a 330 MHz ARM1136 processor, a 220 MHz TMS320C55xDSP, and hardware accelerators are used to perform H.264 decoding. It is shown in [38] that OMAP2420 can perform H.264/AVC decoding on QVGA, 30 fps. The hardware accelerators used in this platform at least include two modules to perform parts of H.264/AVC decoding:

- VLCD: entropy decoding and inverse quantization
- iMX: MC, IDCT, reconstruction, deblocking

Tensilica's Diamond processor family is based on highly efficient configurable and extensible Xtensa processor architecture. When running at 172 MHz, this processor can perform main profile H.264/AVC decoding on D1, 30 fps [35]. To perform main profile HD decoding, dedicated hardware designs [36,37] are more feasible at much lower clock frequencies. In general, to realize a feasible H.264/AVC decoder (e.g., a low-power or a high-performance decoder), the programmable processor performs high-level syntax decoding, hardware accelerator control, and system managements. The selections of performing subfunctions of H.264/AVC by the programmable processors or hardware accelerators highly depend on the applications and available resources of the processors. In spite of software-based, hardware-based, or hybrid decoding systems, an elegant memory subsystem is a key factor to have a successful design.

18.3 Implementation of Coding Tools

18.3.1 Intra Prediction (IP)

In H.264/AVC, IP is used to reduce the high amount of bits coded by original input signal itself. One case is selected from a total of nine prediction modes for each 4×4 or 8×8 luma block (I4MB/I8MB); four modes for a 16×16 luma block (I16MB); and four modes for each chroma block. To reduce the computation complexity, the regularities of these modes are investigated [39–41]. For example, Fig. 18.2 shows I4MB diagonal down-right mode and Table 18.4 shows one possible processing flow that common terms are computed at the same time for those prediction

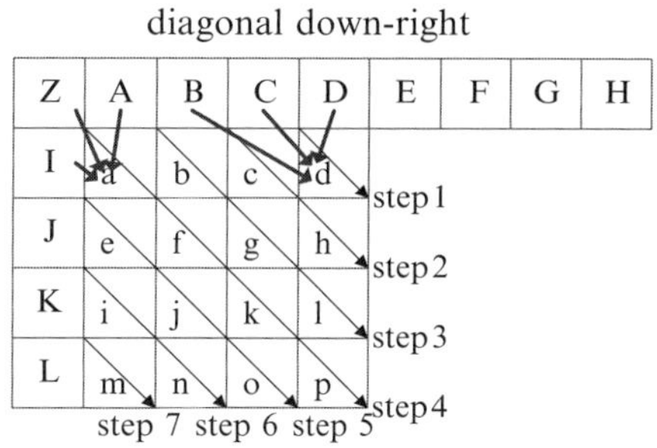

Fig. 18.2 Generation of prediction values for I4MB diagonal down-right mode

Table 18.4 Processing steps of luma 4×4 diagonal down-right mode

Step	Prediction sample	Final value	Immediate value
0			$P0 = C + D$
1	d	$(P0 + P1 + 2) >> 2$	$P1 = B + C$
2	c, h	$(P1 + P2 + 2) >> 2$	$P2 = A + B$
3	b, g, l	$(P2 + P3 + 2) >> 2$	$P3 = Z + A$
4	a, f, k, p	$(P3 + P4 + 2) >> 2$	$P4 = I + Z$
5	e, j, o	$(P4 + P5 + 2) >> 2$	$P5 = J + I$
6	i, n	$(P5 + P6 + 2) >> 2$	$P6 = K + J$
7	m	$(P6 + P7 + 2) >> 2$	$P7 = L + K$

samples requiring these common terms. In this case, different steps (and therefore the processing time) are required for each prediction mode. Other processing flows can be exploited to have a constant processing time while the computation complexity can still be reduced by the regularities of each mode. Furthermore, the regularities among all modes can be exploited to reduce the complexity of multiplexers for dedicated hardware designs or reduce the instruction usages for programmable processors.

18.3.2 Variable Block-Size MC

The main design issues of H.264/AVC variable block-size MC [42] are the computation complexity and the memory bandwidth. As shown in [32] and Table 18.2, interpolation might occupy 25% computation in a baseline decoder. To improve the performance, media instructions are used or designed for software-based decoder. On the other hand, multiplication-free interpolations are explored to reduce hardware cost. The worst-case memory access of MC might occupy 75% of total memory bandwidth [43]. To reduce memory bandwidth, various aspects of MC are investigated, including the processing flow and architecture of MC to have better data reuse, precise data fetch to reduce redundant memory access, different frame memory organizations to improve memory efficiency [43–46].

As aforementioned, multiplication-free interpolations are explored to reduce MC hardware cost. We take chroma interpolation as the example. Given the chroma neighboring samples A, B, C, and D at full-sample locations, the predicted chroma sample value ***a*** at each subsample position is derived as follows:

$$a = ((8-\mathrm{d}x)\times(8-\mathrm{d}y)\times \mathrm{A} + \mathrm{d}x\times(8-\mathrm{d}y)\times \mathrm{B} \\ +(8-\mathrm{d}x)\times \mathrm{d}y\times \mathrm{C} + \mathrm{d}x\times \mathrm{d}y\times \mathrm{D} + 32) \gg 6.$$

This prediction value can be derived without multiplication operations by rearranging the original equation, for example, as

$$a = ((8-\mathrm{d}x)\times[(8-\mathrm{d}y)\times \mathrm{A} + \mathrm{d}y\times \mathrm{C}] \\ +\mathrm{d}x\times[(8-\mathrm{d}y)\times \mathrm{B} + \mathrm{d}y\times \mathrm{D}] + 32) \gg 6.$$

Let $\mathrm{F}(\alpha, \beta, \mu) = (8-\mu)\times\alpha + \mu\times\beta = 8\times\alpha + \mu\times(\alpha-\beta)$

$$\text{Then } a = ((8-\mathrm{d}x)\times \mathrm{F}(\mathrm{A}, \mathrm{C}, \mathrm{d}y) + \mathrm{d}x\times \mathrm{F}(\mathrm{B}, \mathrm{D}, \mathrm{d}y) + 32) \gg 6 \\ = (\mathrm{F}(\mathrm{F}(\mathrm{A}, \mathrm{C}, \mathrm{d}y), \mathrm{F}(\mathrm{B}, \mathrm{D}, \mathrm{d}y), \mathrm{d}x) + 32) \gg 6.$$

The value of $8\times\alpha$ can be obtained by a left shift of α, whereas the value of $\mu\times(\alpha-\beta)$ can be derived from combinational operations of shift/add/sub on $(\alpha-\beta)$. For example, $5\times(\alpha-\beta) = (\alpha-\beta) + [(\alpha-\beta) \ll 2]$. By using this approach, no multiplication is required and thus the hardware cost can be reduced. The multiplication-free computation of predicted luma samples can be obtained in the similar approach used to generate the predicted chroma samples.

To improve memory efficiency when performing MC, one popular method is to store Cb and Cr samples in an interleaved method. This method is widely used in video decoding systems, such as MPEG-2, MPEG-4 decoders. In a straightforward design, frame buffer for Y, Cb, and Cr is stored individually. For 4:2:0 sampling format, the block size of chroma component is a quarter of corresponding luma component partition. Therefore, if Cb and Cr can be accessed in an interleaving approach, such as {Cb, Cr, Cb, Cr} in a 32-bit DRAM data word, memory bandwidth utilization can be improved due to the elimination of memory page miss penalty. Luma component, however, is rarely stored with chroma component in such interleaving approach, especially in 4:2:0 sampling format. One reason of not adopting this approach is the different frame size of luma and chroma component. Another reason is that different neighboring pixels are required for luma and chroma interpolation. Therefore, no significant performance gain is expected when storing luma/chroma in the interleaving approach. As reported in [46], only 2.47% memory bandwidth in average is improved when such interleaving approach is applied. On the other hand, more than 6.31% improvement on memory bandwidth is obtained when Cb and Cr can be accessed in an interleaving approach [43].

Another useful method to improve memory efficiency is to fetch data precisely. This method is based on the observation that the number of reference samples for one $M\times N$ partition varies with the interpolation position. For example, $M\times(N+5)$

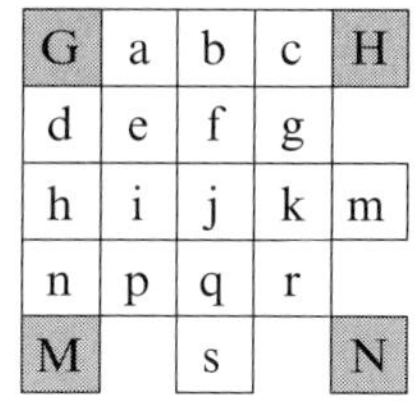

Fig. 18.3 Sample positions of luma interpolation. Shaded blocks are integer sample

and $(M+5) \times (N+5)$ samples are required for half sample position b and j shown in Fig. 18.3, respectively, in a single prediction direction. By precisely controlling the exact size of reference samples, a memory bandwidth reduction 7.36%–40.12% can be obtained [43].

To support the seven different block sizes of variable block-size MC in H.264/AVC standard, from a size of 16×16 to 4×4, simple 4×4-based MC processing flow is widely adopted. The problem of this implementation is the redundant access to the reference data in the overlapped region, which in turn leads to overhead in the memory bandwidth. For example, the maximum number of the valid reference samples for a 8×8 luma block in a single direction prediction is $(8+5) \times (8+5) = 169$. When the 4×4-based MC processing flow is applied, the number of valid reference samples is $4 \times (4+5) \times (4+5) = 324$. That is, 155 samples are accessed redundantly. By identifying the reference samples in overlapped regions and excluding redundant access to these samples, a memory bandwidth reduction from 17.3% to 41.4% was reported in [43].

18.3.3 In-loop Deblocking Filter

The deblocking filter in H.264/AVC is computation and memory intensive due to its highly adaptive mode decision and small 4×4 block size. It is reported that even with highly optimized filtering algorithm, the deblocking operation still occupies one third of the computational complexity of a decoder [47]. In addition, the memory usage of deblocking processing is also considerable to have a low-cost, high-performance and low-power design. A direct deblocking processing flow is to process all vertical edges first, then transpose the intermediate results, and finally process all horizontal edges. The major drawback of this direct approach is that intermediate data storage is as large as a whole 16×16 macroblock (MB) size. Besides, this data flow accesses the original data and intermediate results of a 4×4 block four times. Adopting this coding flow requires a larger buffer size and higher memory bandwidth [48]. Most researches explore the processing order to optimize the processing speed under the constraint of available local memory [48–51]. This section briefly introduces an interleaved horizontal–vertical filtering flow [49] that reuses intermediate data as soon as the data is available. With such processing flow, both memory bandwidth and intermediate storage can be reduced.

Figure 18.4 shows the 4×4 block index *blki*. In the following discussion, neighboring blocks *blk1–blk4* and *blk5, blk10, blk15, blk20* are retrieved from external

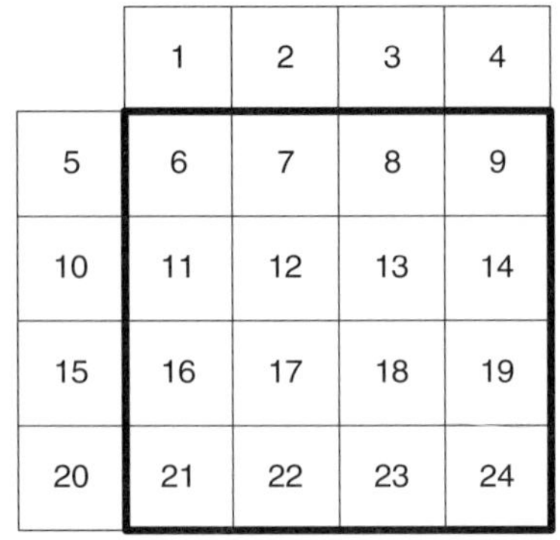

Fig. 18.4 Block index of 4 × 4 blocks of Luma component

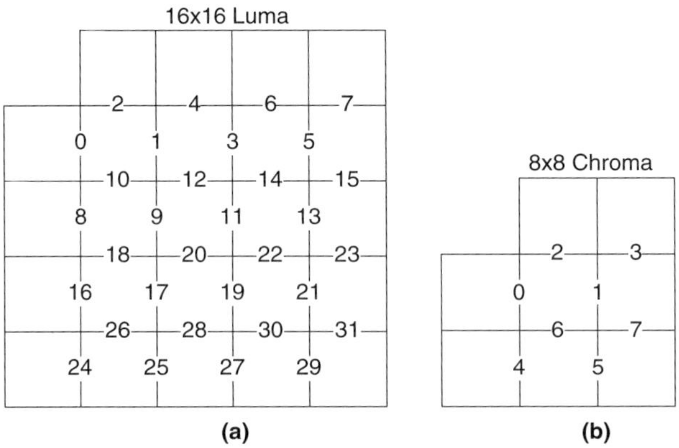

Fig. 18.5 The numbered edges in the processing order of the interleaved horizontal–vertical filtering flow for (**a**) Luma component and (**b**) Chroma component

memory. Figure 18.5 shows the numbered edges in the processing order of the interleaved horizontal–vertical filtering flow. First, the horizontal filtering is applied for vertical edge 0 and then edge 1. The input sample values of *blk5–blk7* are therefore conditionally replaced by the corresponding filtered result sample values. Since all required samples (*blk1* and the updated *blk6*) are available for the horizontal edge 2, the vertical filtering can be applied to edge 2. This horizontal–vertical interleaved filtering approach is repeated for each 4 × 4 block in a raster scan order.

With this interleaved horizontal–vertical filtering flow, the filtered results can be used immediately. Therefore, the number of memory access and the size of local buffer required to process the left, top, and right edge of a 4 × 4 block can be reduced. The bottom horizontal edge of this 4 × 4 block will be processed when its below neighboring block is available. Therefore, only the four 4 × 4 blocks above the current filtering block row are required to be stored (e.g., storing *blk6–blk9* when processing *blk11–blk14*). With this data flow management, the size and the access number of the internal memory are both greatly reduced.

By using the interleaved horizontal–vertical filtering flow, only a buffer with a size of four 4 × 4 blocks (4 × 4 × 4 × 8 bits) is required for storing blocks above

Table 18.5 Comparison of different architectures of H.264/AVC deblocking filter

	[49]	[48]	[52]
Technology	0.25 μm	0.25 μm	0.25 μm
Cycle/MB	300	614	446
Gate count	13.41K	20.66K	24K
Memory size (bits)	16×32	160×32	$64 \times 32 \times 1$ $96 \times 32 \times 2$
Memory type	Two port	Two port	Dual port $\times$ 1 Two port $\times$ 2

the current filtering block row. Based on this filtering flow, two hardwired designs were proposed in [49, 50]. The resulting designs can easily meet the real-time high-resolution requirement (2048×1024@30 fps) with low-cost hardware. Table 18.5 shows a simple comparison of different H.264/AVC deblocking filter designs. Note that the evaluated performance may be different under different input and output data assumptions. For example, reading of neighboring pixels (e.g., *blk1–blk4*) from another extra local buffer or from external memory may affect the overall decoding performance.

To further enhance the performance, simultaneous processing of the horizontal filtering on a vertical edge and the vertical filtering on a horizontal edge can be considered. A window architecture hardwired design based on such processing flow was proposed [51]. In such processing flow, two sets of deblocking filter are required and the internal memory is also larger. Furthermore, joint optimization of deblocking and MC can further reduce the system memory bandwidth and improve the overall H.264/AVC decoding performance.

18.3.4 Inverse Transform and Inverse Quantization

Transforms used in H.264/AVC standard are characterized by integer transform, adaptive transform size, and hierarchical transform [53]. These novel transform algorithms also make the implementation complexity much less than those transforms used in the previous standards. The new transforms include $8 \times 8/4 \times 4$ integer transforms and $4 \times 4/2 \times 2$ Hadamard transforms. All transforms have a form similar to $Y = CXC^T$. Therefore, previous architecture designs for traditional discrete cosine transform (DCT) can be easily applied to transforms used in H.264/AVC. In addition, due to the use of integer arithmetic, the computation of the forward or inverse transform can be completed with simply additions and a minimal number of shifts, but no multiplications. In terms of either hardware or software implementation, transforms in H.264/AVC are significantly lower complexity than those transforms defined in prior standards.

There are plenty of architecture designs for traditional DCT. The basic idea is to exploit the even-symmetric and odd-symmetric characteristics of the transform

Table 18.6 Forward/inverse 4 × 4 transforms used in H.264/AVC

Forward transform	Inverse transform	Forward/inverse Hadamard transform
$\begin{bmatrix} 1 & 1 & 1 & 1 \\ 2 & 1 & -1 & -2 \\ 1 & -1 & -1 & 1 \\ 1 & -2 & 2 & -1 \end{bmatrix}$	$\begin{bmatrix} 1 & 1 & 1 & 1/2 \\ 1 & 1/2 & -1 & -1 \\ 1 & -1/2 & -1 & 1 \\ 1 & -1 & 1 & -1/2 \end{bmatrix}$	$\begin{bmatrix} 1 & 1 & 1 & 1 \\ 1 & 1 & -1 & -1 \\ 1 & -1 & -1 & 1 \\ 1 & -1 & 1 & -1 \end{bmatrix}$

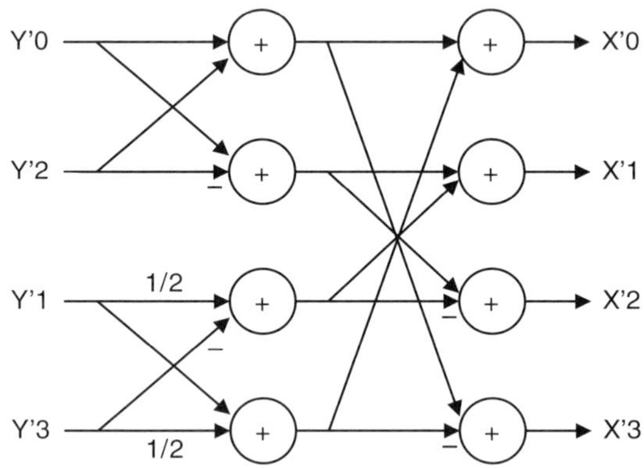

Fig. 18.6 One-dimensional inverse transform

matrix to reduce the computation complexity and enable the parallel processing [54]. Due to the similar operations in both forward and inverse transform, the architecture can be easily shared with both transforms. To further reduce the hardware cost and enhance the computing speed, techniques include direct two-dimensional (2-D) DCT algorithm, the bit-level adder-based distributed arithmetic, and common subexpression sharing are investigated in depth [55]. These techniques can also be applied to design a shared architecture for all forward/inverse transforms with different transform size.

Table 18.6 shows all 4 × 4 forward and inverse transforms used in H.264/AVC. The inverse and forward Hadamard transforms are the same due to the orthogonal feature of this transform. A direct mapping of 1-D inverse transform in Table 18.6 to hardware implementation is shown in Fig. 18.6. Obviously, various 1-D and 2-D hardware architectures can be easily shared to compute these transforms by appropriately adjusting the corresponding coefficients. Furthermore, the hardware can also be easily shared by 8 × 8 forward and inverse transforms. Most designs focus on pursuing high throughput because of the simplicity of H.264/AVC transforms. Some high-throughput designs are shown in Table 18.7. It is easy to have a high-throughput transform processing unit with a low-cost design. However, the cost to have a high-throughput H.264/AVC quantization/dequantization is considerable because of multiplication operations, even though implementation of quantization/dequantization is quite straightforward.

Table 18.7 Throughput of different H.264/AVC transform designs

Solution	[56]	[57]	[58]
samples/cycle	4	8	16

Table 18.8 CAVLD for bit pattern (0001 0011 1100)

Codeword	Syntax	Value	Output array
000100	coeff_token	TotalCoeff = 2, TrailingOnes = 1	Empty
1	Sign of T1	−1	−1
1	Level	2	2, −1
110	total_zeros	1	2, −1
0	run_before	1	2, 0, −1

18.3.5 Context-Adaptive Variable Length Decoding

In VLD process, the codeword length is detected from a coded data stream and this codeword is used to determine the actual symbol with the aid of predefined codeword values, i.e., codeword table. The input stream is then aligned for the next decoding iteration. In general, there is no explicit boundary information for detecting the end or beginning of the codeword in the coded data stream. Therefore, the length of the current codeword should be known before the next codeword can be decoded. This data dependency complicates the decoder design substantially and limits the decoding performance. For context-adaptive VLD, the data dependency can be even more complex due to the selection of a codeword table for the next codeword depends on the decoding result of current codeword.

There are several works for CAVLD by using programmable processors. In [59], a pattern-search method before CAVLD is proposed, which is based on the observation that 60% of 4×4 blocks are coded within only 15 bits. For the example shown in Table 18.8, the bit pattern (0001 0011 1100) represents the coded block of the zig-zag ordered array (2, 0, −1, 0, 0, 0, 0, 0, 0, 0, 0, 0, 0, 0, 0). If this bit pattern is matched in the proposed pattern look-up table, the standard CAVLD procedure can be skipped and the reconstruction of this 4×4 block is completed. Otherwise, the standard CAVLD procedure is performed for this block. Based on the ARM920T embedded system experiment environment, this pattern-search method has a hit rate range of 53–81.28% and a performance improvement range of 6.12–14.86% for QCIF test sequences. This approach requires extra loop-up tables for bit pattern matching.

For overcoming the heavy memory access in table look-up-based CAVLD, several works focus on using arithmetic operations (AOs) to replace these memory accesses [60–62]. Take the decoding of run_before as the example, Table 18.9 shows the original table for zeros_left > 6 and its corresponding arithmetic operations (in the AO > 6 column). In this example, when zeros_left > 6, a 3-bit bit string $I[2{:}0]$ is checked first. If $I[2{:}0] > 0$, then run_before $= 7 - I[2:0]$, otherwise run_before $= 4 + m$, where m is the number of zeros followed by the first 1.

Table 18.9 Tables for run_before

run_before	zeros_left							
	1	2	3	4	5	6	>6	AO >6
0	1	1	11	11	11	11	111	$7-I[2:0]$
1	0	01	10	10	10	000	110	
2	–	00	01	01	011	001	101	
3	–	–	00	001	010	011	100	
4	–	–	–	000	001	010	011	
5	–	–	–	–	000	101	010	
6	–	–	–	–	–	100	001	
7	–	–	–	–	–	–	0001	
8		–	–	–	–	–	00001	
9	–	–	–	–	–	–	000001	
10	–	–	–	–	–	–	0000001	$4+m$
11	–	–	–	–	–	–	00000001	
12	–	–	–	–	–	–	000000001	
13	–	–	–	–	–	–	0000000001	
14	–	–	–	–	–	–	00000000001	

In [60], Moon et al. define arithmetic equations for partial entries of coeff_token VLC tables and all entries of run_before VLC tables. The proposed algorithm achieves an approximate 65–88% savings in memory access as compared to the conventional CAVLC decoding. In [61], Moon proposes a new coeff_token decoding method that the higher probable codewords are decoded with the integer arithmetic operations whereas the less probable ones are decoded with a new codeword structure which is designed to remove the redundant memory accesses. This new algorithm achieves an approximately 95% memory access saving compared with the conventional approach. Inspired by [60], Kim et al. first analyze correlation of VLC codes, then classify codes into multiple groups and perform arithmetic operations on some of these groups instead of the conventional table look-up method [62]. Arithmetic equations for all entries of *coeff_token* and *run_before* VLC table and partial entries of *total_zeros* table are defined. Compared with the conventional approach, this method achieves about 94% and 50% reduction in memory access rate and average CAVLD processing time, respectively. The conventional approach mentioned in [60–62] is JM software with different versions.

For hardwired designs, more optimizations on table space reduction and parallel processing are explored. The easiest implementation is to partition codeword tables to save memory space. Tables for run_before can also be replaced by arithmetic operations due to their simple structures. An architecture for run_before decoder based on arithmetic operations is proposed in [63].

To improve decoding performance, multisymbol decoding schemes dedicated to H.264/AVC CAVLD [64–67] have been widely exploited. The major design issue of multiple-symbol decoding is to break the data dependencies between codewords under the management of the increasing hardware and control complexity, especially when large codeword tables and long codewords are used. In H.264/AVC

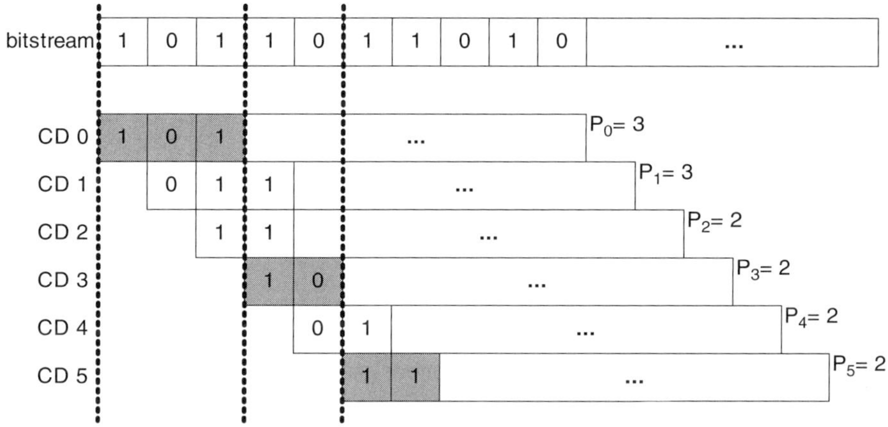

Fig. 18.7 Parallel multiple-symbol decoding scheme

CAVLD, since there is only one symbol of coeff_token and total_zeros in each block, optimization on costs of tables for these two syntax elements is more important. On the other hand, the easiest syntax to realize multisymbol decoding is the sign of trailing ones (T1s.) Once coeff_token is decoded, the number (and the length) of T1s is known. Thus the signs of all T1s can be decoded in one clock cycle instead of one sign in one clock cycle.

Multisymbol decoding for run_before has also been exploited. Based on the parallel multiple-symbol decoding scheme proposed in [64], a bit-position decoding approach is proposed for CAVLD run_before in [65]. In this approach, all run_before of a block can be decoded using less than three clock cycles. The basic idea is to detect all possible codewords in the local bitstream buffer and the sum of the valid codeword lengths is provided to a shifter aligning the encoded input bitstream for a new decoding cycle. The implementation is then a trade-off between the maximum number of output codewords and the critical path. An example of detecting multiple codewords for run_before is shown in Fig. 18.7. At the start of the decoding cycle, zeros_left $= 7$ in this example. At position 0, 3, and 5 in the local bitstream buffer, codewords for run_before $= 2$, 1, and 0 are detected, respectively, by codeword detector (CD) unit CD 0, CD 3, and CD 5. When a match is found, the length of the codeword at position i is also returned by the variable p_i to select the next valid CD. The updated zeros_left is also required for the next CD to select correct codeword table.

There are some other works of decoding multiple symbols for run_before. In [66], two symbols of run_before can be decoded within a clock cycle when zeros_left ≤ 6. In this case, two levels will be re-positioned to the correct locations in the reconstructed block. In [67], multiple symbols can be decoded within one clock cycle when a group of run_before equal to zero. For high quality video streams with low QPs, most coefficients in a $4 \times 4/2 \times 2$ block would be nonzero. In this case, it might consume lots of bits for run_before equal to zero, which in turn lead to more decoding cycles for single symbol decoding methods.

Since multiple symbols for level (including T1s) and run_before can be decoded in a clock cycle, the buffer(s) for storing these data should be carefully designed. Furthermore, nonzero coefficients must be used in the correct order for the next stage, i.e. inverse quantization (IQ). Joint consideration of CAVLD and IQ can further improve overall performance and reduce total buffer usage.

18.3.6 Context-Adaptive Binary Arithmetic Decoding

CABAD is the inverse process of CABAC. It consists of similar three elementary steps: binary arithmetic decoding, context modeling, and de-binarization, as shown in Fig. 18.8. While context modeling in the decoding process is almost identical to that in the encoding process, binary arithmetic decoding and de-binarization are the inverse processes of binary arithmetic coding and binarization in the encoding process. In decoding process, binary arithmetic decoding and context modeling are the two most time-consuming parts. Because the algorithm has complicated decoding procedure and strong data dependency, it is hard to exploit its pipeline and parallel facilities in these two parts. Therefore, CABAD is typically not suitable for software implementation by general-purpose processors, particularly for high-resolution video. To further improve decoding performance, methods of multisymbol decoding within one clock cycle might be required.

Since *modulo coder* (M coder) [68] used as the binary arithmetic codec in the H.264/AVC standard is inspired by the Q coder [69,70] and MQ coder [71], previous design concepts for these coders might be applied to M coder. To improve the performance of the arithmetic coding, some designs focus on the improvement of clock rate of the multiplication-free arithmetic coding [72–74], while other designs focus on parallel processing multiple symbols within one clock cycle [75–77]. Most parallel designs focus on the improvement of the multiplication-free arithmetic coder, Q coder [70]. In the original Q coder proposed in [70], either a single more probable symbol (MPS) or less probable symbol (LPS) is processed in one iteration. In [75], Feygin et al. modify Q coder architecture by employing loop unrolling and speculative execution of the inner loop of the arithmetic coding algorithm to achieve either a single LPS or multiple MPS's coding in one iteration. In [76], Jiang incorporates multiplication into the Q coder coding operation to achieve parallel arithmetic coding. Jiang's architecture has a tree-structure PE array and each PE has its

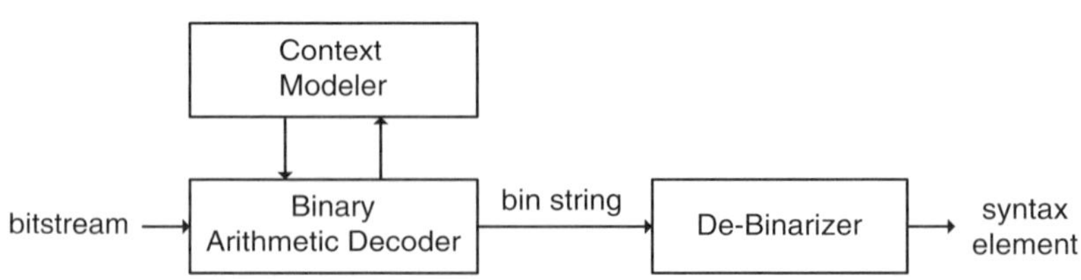

Fig. 18.8 CABAD flow

own renormalization operation. Furthermore, Jiang proposes a similar tree-structure parallel design for bilevel image to encode a group of input symbols within each cycle [77]. For input symbols with even number $2N$, their design requires $4N$ PE's with a latency of $2+\log_2(2N)$ cycles. In [78], Lee et al. propose a multisymbol context-based arithmetic coding architecture for MPEG-4 shape coding by cascading multiple range update units to handle multiple symbols with particular contexts. In MPEG-4 shape coding, it is necessary to initialize the arithmetic encoder prior to encoding the first pixel of each binary alpha blocks and to flush/terminate it after encoding the last pixel of the binary alpha block. Because these frequent initializations and terminations of the arithmetic encoder are a source of inefficiency, a multiplicative arithmetic codec was chosen in preference to the less efficient shift-subtract type [79].

To improve the performance of binary arithmetic decoding used in H.264/AVC, the characteristics of data dependency must be exploited. For either pipeline designs [80,81] or multisymbol decoding designs [82–84], data dependency of context fetch, range update, and renormalization is investigated. For pipeline designs, the design issues are to reduce/balance delay path in each pipeline stage and to solve pipeline data and structure hazards [26, 80]. For multisymbol decoding designs, the challenge is to solve data dependency under the management of the increasing hardware and control complexity. In general, the data dependency of context is exploited first to enhance the decoding performance. For example, inter-stage registers in pipeline design [81] or forwarding logic in multisymbol design [82] are usually introduced to eliminate memory access conflict, which occurs when two successive bins/bits have the same context. For another example, decoding of *coded_block_flag* syntax element usually requires more clock cycles because the generation of its context needs more clock cycles to get neighboring data. By parallel processing of decoding of quantized coefficients and fetching of neighboring data, an 8% performance improvement was reported in [82].

Figure 18.9 shows the multibin decoding method using a cascade of binary arithmetic decoders (BADs). In this method, the number of bins that can be decoded within one clock cycle is constrained by the working clock frequency. Therefore, the less time required for decoding a symbol, the more symbols can be decoded in one clock cycle. Figure 18.10 shows the range update (or called internal subdivision) of binary arithmetic decoding. For the regular binary arithmetic decoding, the computation time for decoding an MPS without renormalization is much less than all other cases. Therefore, the processing of next bin might start early. Another case which can be utilized is the bypass decoding, in which no probability estimation

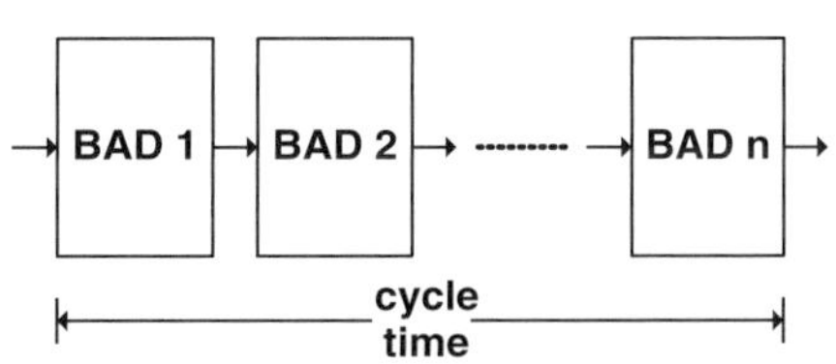

Fig. 18.9 H.264/AVC multibin decoding in a cascade of BADs

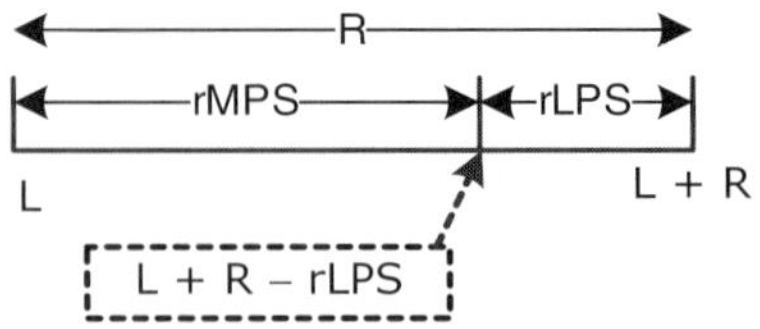

L: The lower bound of the interval
R: The range of the interval
LPS: Value of the least probable symbol
rLPS: Range corresponding to the LPS

```
if (symbol == LPS) {
        R' = rLPS
        L' = L + R –rLPS
} else {
        R' = R –rLPS
        L' = L
}
```

Fig. 18.10 Range update of binary arithmetic decoding

and context fetching are required, thus its decoding time is also quite short. In these two cases, either the pipeline latency can be shorter [81] or multiple bins can be decoded within one clock cycle [82, 83]. Considering that the renormalization does not occur in some cases, the design in [84] performs two-bin decoding only when the first bin is an MPS without renormalization. On the other hand, renormalization units are also combined in the two cascaded BADs in the designs [82, 83] to decode two regular coded bins in one clock cycle.

18.4 Memory Usage and Data Transfer

In H.264/AVC, the significant improved coding efficiency mainly relies on improving the prediction for the current MB based on previously transmitted and decoded data. Consequently, previously decoded results need to be stored. Therefore, the method of storing these data significantly affects the overall performance, cost and power consumption of the decoding system.

Researches on architecture design have shown that 50–80% of the area cost in (application-specific) architectures for real-time multidimensional signal processing is due to *memory components*, e.g., embedded SRAMs and register files [85]. Also the power consumption is heavily dominated by memory access both in custom hardware [86] and in processors [87]. Table 18.10 shows the relative energy of different operations [19, 86]. Data transfer and memory access operations consume much more power than a data-path operation. For example, fetching an operand from an off-chip memory for an addition operation consumes 33 times more power than the addition itself in case of a processor. Furthermore, memory access performance, including latency and bandwidth, could significantly affect system performance. This is especially true in high-performance, memory-intensive applications, such as those for multimedia processing. Obviously, a promising avenue for further optimization of H.264/AVC decoders under various design constraints must take data transfer and memory subsystems into consideration.

Table 18.10 Relative energy per operation at a 1.5 V supply in 0.8 μm CMOS technology

Operation	Relative energy/op
16b carry–select adder	1
16b multiplier	3.6
$8 \times 128 \times 16$ SRAM (read)	4.4
$8 \times 128 \times 16$ SRAM (write)	8.9
External I/O access	10
16b memory transfer	33

Table 18.11 Upper neighboring pixels needed for deblocking filter and IP

Usage	Size (bits)	1080HD@ 4:2:0 (Kbits)
Upper neighboring pixels for deblocking	$4 \times w \times cf \times 8$	122
Upper neighboring pixels for intra prediction	$w \times cf \times 8$	30

18.4.1 Memory Usage

The storage required by an H.264/AVC decoder can be divided into four classes.

- Interframe information: This type of information is required for decoding for multiple frames, includes information for whole current frame. Information belonging to this type includes current reconstructed frame, reference frames, MB information required for B-direct MB decoding, slice group map.
- Inter-MB information: This type of information is required for decoding the following MBs in current frame. For example, pixels or MB mode information from upper neighboring MBs are required for processing current MB.
- Intra-MB information: This type of storage is used to store intermediate results when processing an MB. The buffer for this purpose includes storing entropy decoded residual information, inverse transformed coefficients, etc.
- Constant data: This includes variable-length decoding tables, default quantization tables, etc.

In current semiconductor technology, interframe information (except slice group map) is generally stored in off-chip memory, especially in off-chip DRAMs. Intra-MB information and constant data are stored in local storages. Storing of inter-MB information in on-chip or off-chip memory needs more consideration of performance, cost, and power consumption. Although the size of this information is not as large as interframe information, it is still considerable. In addition, the performance of access this information is also crucial to the overall decoding performance. Table 18.11 shows the upper neighboring pixels above current MB row needed for deblocking filter and IP. In the table, w is the width of the picture and *cf* is chroma format, where in H.264/AVC, *cf* can be 1, 2, 2, and 3 for monochrome, 4:2:0, 4:2:2, and 4:4:4, respectively. Note that for decoders supporting MB adaptive frame/field (MBAFF) coding, the size is double.

18.4.2 Memory Design Optimization

There are already a lot of memory design optimization/trade-off schemes for previous video coding standards. One of them is the interleaved storing of Cb and Cr, which is already mentioned in design of MC. For storing of reference frames, another popular scheme is tile-based mapping of frame data to multiple banks of DRAMs [88–92]. This scheme improves DRAM access performance and reduces power consumption by exploiting the page-access mode and multibank architecture features of most types of DRAMs, including SDRAM and mobile DRAM.

Figure 18.11 shows a simplified architecture of a two-bank SDRAM. All memory banks share the data and address bus, whereas each bank has its own row decoder, column decoder, and row buffer. The mode register stores several SDRAM operation modes such as burst length, burst type, CAS (column address strobe) latency, etc. An m-bank SDRAM has a similar architecture. A complete SDRAM access may consist of several commands including row-activation, column-access (read/write) and precharge, as shown in Fig. 18.12. A row-activation command, together with the row and bank address, is used to open (or called activate) a specific row (or called page) in a particular bank, and copy all data in the selected row into the selected bank's row buffer for the subsequent column accesses. After accepting this command, SDRAM needs a latency called t_{RCD} (ACTIVE to column-access delay) to accomplish the command. No other commands can be issued to this bank during this latency. However, commands to other banks are permissible due to the independent parallel processing capability of each bank. Once a row of a particular

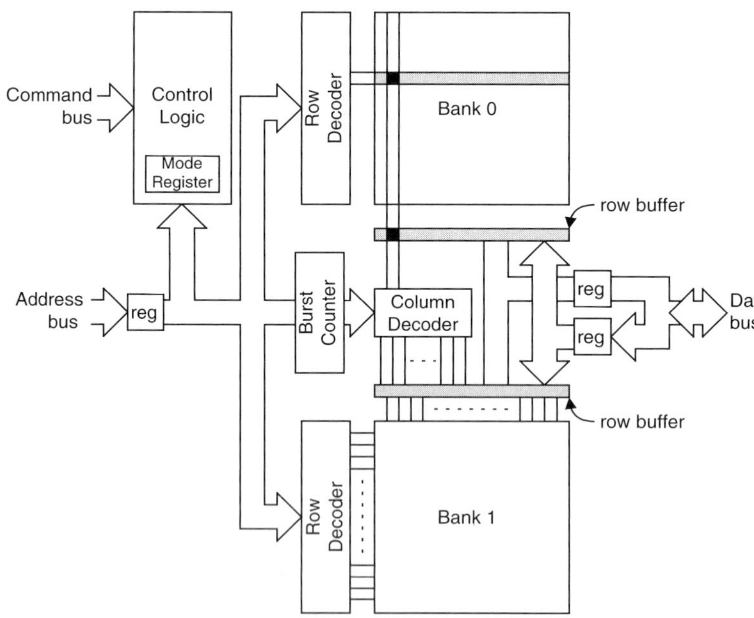

Fig. 18.11 A simplified architecture of a two-bank SDRAM

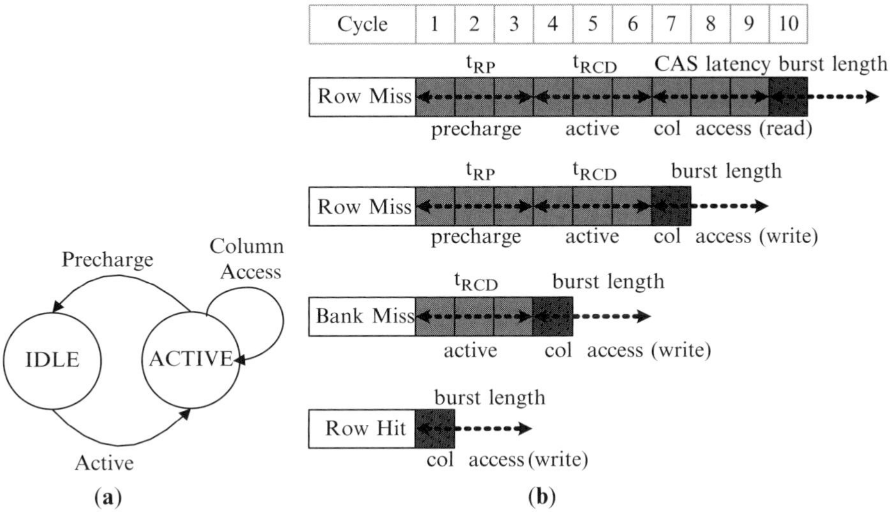

Fig. 18.12 (**a**) Simplified bank state diagram and (**b**) access latencies of different access statuses

bank has been opened, a column-access command can be issued to read/write data from/to the addressed word(s) within the row buffer. To issue either a read or write column-access command, DRAM address bus is also required to indicate the column address of the open row in that bank. For a write access, DRAM data bus is needed to transfer write data from the time the command is issued until the whole burst transfer is completed. As for a read access, DRAM data bus is used to transfer data after a latency called CAS, which is the latency from the read column-access command, is registered to the time the first read datum is available. The precharge command, together with the information on address bus, can be used to deactivate a single open row in a particular bank or all rows in all banks. While processing the precharge command, the addressed bank or banks are not allowed to accept any other commands during a time called t_{RP} (PRECHARGE command period).

SDRAM bandwidth utilization and latency is lower and longer, respectively, when more commands are required for an SDRAM access. The number of commands needed for a complete SDRAM access deeply depends on the state of the bank addressed by the SDRAM access. Figure 18.12a and b shows a simplified bank state diagram and the access latencies due to different access statuses: bank miss, row miss, and row hit. In a bank miss status, an incoming access is addressed to a bank in the IDLE state, therefore it must first activate the target row and then issue the column-access command. For a row miss status, the addressed bank is in ACTIVE state and the row address of its activated row is not identical to that of an incoming access. In this case, the incoming access has to first precharge the bank, then activate the target row, and finally issue column-access commands. As for a row hit status, the addressed bank is in ACTIVE state and the row address of its activated row is the same as that of the incoming access. Hence, column-access commands can be directly issued. The above discussion is only based on a simplified condition.

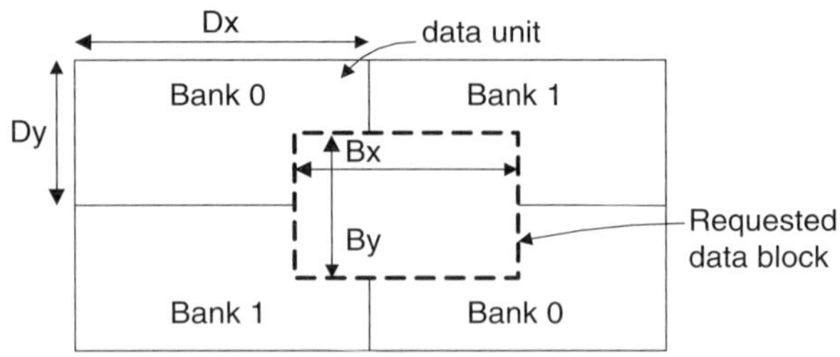

Fig. 18.13 An example of tile-based mapping to two banks of DRAM

Based on the aforementioned access characteristics, various tile-based mapping of frame data to multiple banks of DRAMs were proposed to minimize miss penalty. Figure 18.13 shows one possible mapping. In H.264/AVC, the size of requested data block $B_x \times B_y$ depends on the $M \times N$ partition size, which ranges from a size of 16×16 to 4×4. The design parameters D_x and D_y denote the horizontal and vertical size of the data unit, which will be mapped into a single DARM page. If a requested data block is allocated within a data unit, then no row miss occurs, resulting in the best memory access performance. Otherwise, either row miss or bank miss might occur. If the misses do happen, DRAM controllers might have the ability to hide the miss penalty [90, 93]. However, the extra power consumption for miss penalty cannot be avoided.

Compared with accessing to on-chip memory, accessing to off-chip DRAMs requires more power consumption and latency. In addition, the performance of accessing off-chip DRAMs is more unpredictable than that of accessing on-chip SRAMs, due to probable row/bank miss penalty and DRAM access arbitration. As aforementioned, storing of inter-MB information in on-chip or off-chip memory needs more consideration of performance, cost, and power consumption. For performance consideration, a Row Store Buffer with a size of 21KB is used to store pixel data from previous MBs for main profile HD decoding in [36]. A line-pixel-lookahead (LPL) scheme [94] is proposed to use a size of 19.2kbits internal memory for storing upper neighboring pixels for IP and deblocking, leading to 11% power consumption improvement in their three-level memory hierarchy environment. Compared with [36], a trade-off for this inter-MB information buffer that uses 0.5KB of internal memory with about 10% increase in external memory bandwidth is proposed in [37].

18.5 Summary

H.264/AVC standard is designed for a wide range of applications and has been very strongly embraced by industry. For different applications and market segments, the key design issues of an H.264/AVC decoder can be low-cost, low-power consumption, or the best rendering quality. For example, lower-power consumption might be more crucial to the H.264/AVC decoder designs in handheld devices. The overall design space is large and complex because each coding tool might be implemented in a programmable processor, a co-processor, or a dedicated hardware accelerator. In this chapter, we have reviewed design techniques of some kernel coding tools for video coding layer of H.264/AVC. The most important design factor is the data management and flow control. In particular, the data transfer to off-chip memory is especially important due to the scarce resource of off-chip bandwidth. In fact, although the tremendous progress in VLSI technology provides an ever-increasing number of transistors and routing resource on a single chip, and hence allows integrating heterogeneous control and computing functions to realize SoCs, the improvement on off-chip communication is limited due to the number of available I/O pins and the physical design issues of these pins. On one hand the application is characterized by a variety of access patterns. On the other hand, new memory devices and organizations provide a set of features. To find the best match between the application characteristics and the memory organization features, the designer needs to explore different memory configurations in combination with different design architectures.

References

1. ITU-T Rec. H.264 and ISO/IEC 14496-10 (MPEG4-AVC), Advanced Video Coding for Generic Audiovisual Services, Mar.2005.
2. K.-B. Lee and T.-S. Chang, "Chapter 4: SoC memory system design," *Essential Issues in Soc Design: Designing Complex Systems-on-chip*, pp. 73–118, Nov. 2006, Springer, Netherlands.
3. C. Lee, W. Mangione-Smith, and M. Potkonjak, "MediaBench: A tool for evaluating multimedia and communication systems," in *Proc. Int. Symp. Microarchitectures*, pp. 330–335, Dec. 1997.
4. J. E. Fritts, F. W. Steiling, and J. A. Tucek, "MediaBench II Video: expediting the next generation of video systems research," in *Proc. SPIE*, pp. 79–93, Jan. 2005.
5. J. Fritts, *Architecture and compiler design issues in programmable media processors*, Ph.D. Thesis, Dept. of Electrical Engineering, Princeton University, 2000
6. B. Bishop, T.P. Kelliher, and M.J. Irwin, "A detailed analysis of MediaBench," in *Proc. Signal Processing Systems (SiPS 99)*, pp. 448–455, Oct. 1999.
7. J.L. Hennessy and D.A. Patterson, *Computer Architecture: A Quantitative Approach*, 2nd ed., Morgan Kaufmann Publishers, San Francisco, 1996.
8. K. Diefendorff and P.K. Dubey, "How multimedia workloads will change processor design," *IEEE Computer*, vol. 30, no. 9, pp. 43–45, Sep. 1997.
9. T. M. Conte et al., "Challenges to combining general-purpose and multimedia processors," *IEEE Computer*, pp. 33–37, Dec. 1997.
10. I. Kuroda and T. Nishitani, "Multimedia processors," *Proc. IEEE*, vol. 86, pp. 1203–1221, Jun. 1998.
11. S. Rixner et al., "A bandwidth-efficient architecture for media processing," in *Proc. ACM/IEEE Int. Symp. Microarchitecture*: IEEE CS'98, Nov.–Dec. 1998, pp. 3–13.
12. Y. Oshima, B.J. Sheu, and S.H. Jen, "High-speed memory architectures for multimedia applications," *IEEE Circ. Devices Mag.*, vol. 13, pp. 8–13, Jan. 1997.
13. B. Prince, *High Performance Memories: New Architecture DRAMs and SRAMs*, John Wiley & Sons, 2nd ed., Jul. 1999, England.
14. Y.-H. Park, S.-H. Han, J.-H. Lee, and H.-J. Yoo, "A 7.1-GB/s low-power rendering engine in 2-D array-embedded memory logic CMOS for portable multimedia system," *IEEE J. Solid-State Circuits*, vol. 36, pp. 944–955, Jun. 2001.

15. Ackland, "The role of VLSI in multimedia," *IEEE J. Solid-State Circuits*, vol. 29, pp. 381–388, Apr. 1994.
16. P. Pirsch, N. Demassieux, and W. Gehrke, "VLSI architectures for video compression—A survey," *Proc. IEEE*, vol. 83, pp. 220–246, Feb. 1995.
17. V. Baskaran and K. Konstantinides, *Image and Video Compression Standards: Algorithms and Architecture*, Kluwer Academic, Norwell, MA, 1995.
18. P. Pirsch and H.-J. Stolberg, "VLSI implementations of image and video multimedia processing systems," *IEEE Trans. Circuits Syst. Video Technol.*, vol. 8, pp. 878–891, Nov. 1998.
19. F. Catthoor et al., *Custom Memory Management Methodology: Exploration of Memory Organisation for Embedded Multimedia System Design*, Kluwer Academic Publishers, Boston, Sep. 1998.
20. Francky Catthoor, *Unified Low-power Design Flow for Data-dominated Multi-media and Telecom Applications*, Kluwer Academic Publishers, Dordrecht, Jul. 2000.
21. D. Laneer, M. Cornero, G. Goosens, and H. De Man, "Data routing: A paradigm for efficient data-path synthesis and code generation," in *Proc. 7th Int. Symp. High-Level Synthesis*, 1994, pp. 17–22.
22. S. Tarafdar and M. Leeser, "A data-centric approach to high-level synthesis," *IEEE Trans. Computer-Aided Design*, vol. 19, no. 11, Nov. 2000, pp. 1251–1267.
23. H. Samsom, F. Franssen, F. Catthoor, and H. De Man, "Verification of loop transformations for real time signal processing applications," in *VLSI Signal Process. VII*, 1994, pp. 208–217.
24. M. Cupak, F. Catthoor, and H.J. De Man, "Efficient system-level functional verification methodology for multimedia applications," *IEEE Des. Test. Comput.*, vol. 20, no. 2, pp. 56–64, Mar.–Apr. 2003.
25. P.R. Panda, N. Dutt, A. Nicolau, *Memory Issues in Embedded Systems-on-chip: Optimizations and Exploration*, Kluwer Academic, Boston, 1999.
26. J.L. Hennessy and D.A. Patterson, *Computer Architecture: A Quantitative Approach*, 3rd ed., Morgan Kaufmann Publishers, San Francisco, 2002.
27. Anthony Cataldo, MPU designers target memory to battle bottlenecks, EE Times, (10/19/01, available on http://www.siliconstrategies.com/story/OEG20011019S0125)
28. R.C. Schumann, "Design of the 21174 memory controller for DIGITAL personal workstations," *Digital Technical J.*, vol. 9, no. 2, pp. 57–70, 1997.
29. R. Goering, "Philips design team wins EDAC award," EEdesign, May 30, 2002.
30. S. Dutta, R. Jensen, and A. Rieckmann, "Viper: A multiprocessor SoC for advanced set-top box and digital TV systems," *IEEE Des. Test. Comput.*, vol. 18, no. 5, pp. 21–31, Sep.–Oct. 2001.
31. G. Martin and H. Chang, *Winning the SoC Revolution: Experiences in Real Design*, Kluwer Academic Publishers, Boston, Jun. 2003.
32. M. Horowitz, A. Joch, and F. Kossentini, "H.264/AVC baseline profile decoder complexity analysis," *IEEE Trans. Circuits Syst. Video Technol.*, vol. 13, no. 7, pp. 704–716, Jul. 2003.
33. M. Alvarez, E. Salamí, A. Ramirez, and M. Valero, "A performance characterization of high definition digital video decoding using H.264/AVC," *IEEE International Symposium on Workload Characterization*, pp. 24–33, Oct. 2005.
34. K. Ramkishor, "Media processor architectures from TI," TI Developer Conference, Nov. 2004.
35. Diamond Standard VDO Video Engines Product Brief, ver. 2–2007, available on: http://www.tensilica.com/pdf/video.pdf
36. Y. Hu, A. Simpson, K. McAdoo, and J. Cush, "A high definition H.264/AVC hardware video decoder core for multimedia SoCs," in *Proc. IEEE Int. Symp. Consumer Electronics*, pp. 385–389, Sept. 2004.
37. C.C. Lin, et al, "A 160K gates/4.5 kb SRAM H.264 video decoder for HDTV applications," *IEEE J. Solid-State Circuits*, vol. 42, no. 1, Jan 2007.
38. OMAP2420, http://www.ti.com/omap2420
39. Y.-W. Huang, B.-Y. Hsieh, T.-C. Chen, and L.-G. Chen, "Analysis, fast algorithm, and VLSI architecture design for H.264/AVC intra frame coder," *IEEE Trans. Circuits Syst. Video Technol.*, vol. 15, no. 3, pp. 378–401, Mar. 2005.

40. J.-W. Chen, C.-C. Lin, J.-I. Guo, and J.-S. Wang, "Low complexity architecture design of H.264 predictive pixel compensator for HDTV application," in *Proc. ICASSP*, pp. 932–935, May 2006.
41. E. Sahin and I. Hamzaoglu, "An efficient intra prediction hardware architecture for H.264 video decoding," in *Proc. DSD*, pp. 448–454, 2007.
42. M. Wien, "Variable block-size transforms for H.264/AVC," *IEEE Trans. Circuits Syst. Video Technol.*, vol. 13, no.7, pp. 604–613, 2003.
43. R. Wang, M. Li, J. Li, and Y. Zhang, "High throughput and low memory access sub-pixel interpolation architecture for H.264/avc HDTV decoder," *IEEE Trans. Consumer Electron.*, vol. 51, no. 3, pp. 1006–1013, 2005.
44. S.Z. Wang, T.A. Lin, T.M. Liu and C.Y. Lee, "A new motion compensation design for H.264/AVC decoder," in *Proc. ISCAS*, pp. 4558–4561, 2005.
45. J.-W. Chen, C.-C. Lin, J.-I. Guo, and J.-S. Wang, "Low complexity architecture design of H.264 predictive pixel compensator for HDTV applications," in *Proc. ICASSP*, vol. 3, pp. 932–935, May 2006.
46. A. Azevedo, B. Zatt, L. Agostini, and S. Bampi, "MoCHA: a bi-predictive motion compensation hardware for H.264/AVC decoder targeting HDTV," in *Proc. ISCAS*, pp. 1617–1620, May 2007.
47. P. List et al., "Adaptive deblocking filter," *IEEE Trans. Circuits Syst. Video Technol.*, vol. 13, no. 7, pp. 614–619, Jul. 2003.
48. Y.-W. Huang et al, "Architecture design for deblocking filter in H.264/JVT/AVC," in *Proc. Multimedia Expo.*, Jul. 2003, vol. 1, pp. 693–696.
49. C.-C. Cheng, T.-S. Chang, and K.-B. Lee, "An in-place architecture for the deblocking filter in H.264/AVC," *IEEE Trans. Circuits Syst. II: Express Briefs*, vol. 53, no.7, pp. 530–534, Jul. 2006.
50. K.Y. Min and J. W. Chong, "A memory and performance optimized architecture of deblocking filter in H.264/AVC," *Int. Conf. MultimediaUbiquitous Engineering*, pp. 220–225, Seoul Korea, Apr. 2007.
51. C.M. Chen and C.H. Chen, "Window architecture for deblocking filter in H.264/AVC," *IEEE Int. Symp. Signal Processing Information Technol.*, pp. 338–342, Vancouver Canada, Aug. 2006.
52. B. Sheng, W. Gao, and D. Wu, "An implemented architecture of deblocking filter for H.264/AVC," in *Proc. Int. Conf. Image Processing*, vol. 1, pp. 665–668, Oct. 2004,
53. H. Malvar, A. Hallapuro, M. Karczewicz, and L. Kerofsky, "Low-complexity transform and quantization in H.264/AVC," *IEEE Trans. Circuits Syst. Video Technol.*, vol. 13, no. 7, pp. 598–603, Jul. 2003.
54. A. Madisetti and A.N.Willson, Jr., "A 100 MHz 2-D 8x8 DCT/IDCT processor for HDTV applications," *IEEE Trans. Circuits Syst. Video Technol.*, vol. 5, pp. 158–165, Apr. 1995.
55. T.S. Chang, C.S. Kung, and C.W. Jen, "A simple processor core design for DCT/IDCT," *IEEE Trans. Circuits Syst. Video Technol.*, vol. 10, pp. 439–447, Apr. 2000.
56. T.C. Wang, Y.W. Huang, H.C. Fang, and L.G. Chen, "Parallel 4 x 4 2-D transform and inverse transform architecture for MPEG-4 AVC/H. 264," in *Proc. IEEE ISCAS*, pp. 800–803, May 2003.
57. K.H. Chen, J.I. Guo, and J.S. Wang, "A high-performance direct 2-D transform coding IP design for MPEG-4AVC/H.264," *IEEE Trans. Circuits Syst. Video Technol.*, vol. 16, no. 4, pp. 472–483, Aug. 2006.
58. L. Agostini, R. Porto, J. Guntzel, and I.S. Silva, "High throughput multitransform and multiparallelism IP for H.264/AVC video compression standard," in *Proc. IEEE ISCAS*, pp. 5419–5422, May 2006.
59. S.Y. Tseng and T.W. Hsieh, "A pattern-search method for H.264/AVC CAVLC decoding," in *Proc. ICME*, July 2006, pp. 1073–1076.
60. Y.H. Moon, G.Y. Kim, and J.H. Kim, "An efficient decoding of CAVLC in H.264/AVC video coding standard," *IEEE Trans. Consumer Electron.*, vol. 51, no. 3, pp. 933–938, Aug. 2005.

61. Y.H. Moon, "A new coeff-token decoding method with efficient memory access in H.264/AVC video coding standard," *IEEE Trans. Circuits Syst. Video Technol.*, vol. 17, no. 6, pp. 729–736, Jun. 2007.
62. Y.H. Kim, Y.J. Yoo, J. Shin, B. Choi, and J. Paik, "Memory-efficient H.264/AVC CAVLC for fast decoding," *IEEE Trans. Consum. Electron.*, vol. 52, pp. 943–952, Aug. 2006.
63. H.-C. Chang, C.-C. Lin, and J.-I. Guo, "A novel low-cost high-performance VLSI architecture for MPEG-4 AVC/H.264 CAVLC decoding,"in *Proc. IEEE Int. Symp. Circuits Syst. (ISCAS)*, 2005, pp. 6110–6113.
64. J. Nikara, S. Vassiliadis, J. Takala, and P. Liuha, "Multiple-symbol parallel decoding for variable length codes," *IEEE Trans. Very Large Scale Integration Systems*, vol. 12, no. 7, pp. 676–685, Jul. 2004.
65. Y.N. Wen, G.L. Wu, S.J. Chen, and Y.H. Hu, "Multiple-symbol parallel CAVLC decoder for H.264/AVC," in *Proc. IEEE APCCAS*, Dec. 2006, pp. 1240–1243.
66. Guo-Shiuan Yu, and Tian-Sheuan Chang, "A zero-skipping multi-symbol CAVLC decoder for MPEG-4 AVC/H.264," in *Proc. ISCAS*, pp. 21–24, May 2006.
67. T.-H. Tsa, D.-L. Fang, and Y.-N. Pan, "A hybrid cavld architecture design with low complexity and low power considerations," in *Proc. ICME*, pp. 1910–1913, Jul. 2007.
68. D. Marpe and T. Wiegand, "A highly efficient multiplication-free binary arithmetic coder and its application in video coding," in *Proc. ICIP*, Barcelona, Spain, pp. 263–266, Sept. 2003.
69. W.B. Pennebaker, J.L. Mitchell, G.G. Langdon, and R.B. Arps, "An overview of the basic principles of the Q coder adaptive binary arithmetic coder," *IBM J. Res. Dev.*, vol. 32, pp. 717–726, Nov. 1988.
70. J.L. Mitchell and W.B. Pennebaker, "Optimal hardware and software arithmetic coding procedures for the Q coder," *IBM J. Res. Dev.*, vol. 32, no. 6, pp. 727–736, Nov. 1988.
71. D. Taubman and M.W. Marcellin, *JPEG2000 Image Compression: Fundamentals, Standards and Practice*, Kluwer, Boston, MA, 2002.
72. M. Tarui, M. Oshita, T. Onoye, and I. Shirakawa, "High-speed implementation of JBIG arithmetic coder," in *Proc. IEEE TENCON*, vol. 2, 1999, pp. 1291–1294.
73. Y.-T. Hsiao, H.-D. Lin, K.-B. Lee, and C.-W. Jen, "High-speed memory-saving architecture for the embedded block coding in JPEG2000," in *Proc. Int. Symp. Circuits and Systems (ISCAS'02)*, Phoenix, USA, vol. 5, pp. 133–136, May 2002.
74. K.-K. Ong, W.-H. Chang, Y.-C. Tseng, Y.-S. Lee, and C.-Y. Lee, "A high throughput low cost context-based adaptive arithmetic codec for multiple standards," in *Proc. Int. Conf. Image Processing*, 2002, vol. 1, pp. 872–875.
75. G. Feygin, P.G. Gulak, and P. Chow, "Architectural advances in the VLSI implementation of arithmetic coding for binary image compression," in *Proc. Data Compression Conference (DCC '94)*, 1994, pp. 254–263.
76. J. Jiang and S. Jones, "Parallel design of arithmetic coding," *IEE Proceedings-E: Computer and Digital Techniques*, vol. 141, pp. 327–333, Nov. 1994.
77. J. Jiang, "Parallel design of Q coders for bilevel image compression," in *Proc. Int. Conf. Parallel and Distributed Systems*, 1994, pp. 230–235.
78. K.-B. Lee, J.-Y. Lin, and C.-W. Jen, "A multisymbol context-based arithmetic coding architecture for MPEG-4 shape coding," *IEEE Trans. Circuits Syst. Video Technol.*, vol. 15, no. 2, pp. 283–295, Feb. 2005.
79. N. Brady, F. Bossen, and N. Murphy, "Context-based arithmetic encoding of 2D shape sequences," in *Proc. Int. Conf. Image Processing*, Santa Barbara, CA, vol. I, pp. 29–32, Oct. 1997.
80. Y. Yi and I.-C. Park, "High-speed H.264/AVC CABAC decoding," *IEEE Trans. Circuits Syst. Video Technol.*, vol. 17, no. 4, pp. 490–494, Apr. 2007.
81. L. Li, Y. Song, T. Ikenaga, and S. Goto, "Hardware architecture design of CABAC codec for H.264/AVC," in *Proc. IEEE VLSI-DAT*, Hsinchu, Taiwan, pp. 1–4, Apr. 2007.
82. J.-W. Chen and Y.-L. Lin, "A high-performance hardwired CABAC decoder," in *Proc. ICASSP*, Honolulu, Hawaii, USA, pp. 37–40, Apr. 15–20, 2007.
83. W. Yu and Y. He, "A high performance CABAC decoding architecture," *IEEE Trans. Consumer Electron.*, vol. 51, no. 4, pp. 1352–1359, Nov. 2005.

84. Y.-C. Yang et al., "A high throughput VLSI architecture design for H.264 context-based adaptive binary arithmetic decoding with lookahead parsing," *IEEE International Conference on Multimedia & Expo (ICME)*, pp. 357–360, Jul. 2006.
85. G. Goossens et al, "Synthesis of flexible IC architectures for medium throughput real-time signal processing," *J. VLSI Signal Processing*, vol. 5, no. 4, 1993.
86. T.H. Meng, B. Gordon, E. Tsern, and A. Hung, "Portable video-on-demand in wireless communication," in *Proc. IEEE*, vol. 83, no. 4, pp. 659–680, Apr. 1995.
87. V. Tiwari, S. Malik, and A. Wolfe, "Power analysis of embedded software: a first step towards software power minimization," in *Proc. ICCAD*, pp. 384–390, Nov. 1994.
88. M. Winzker, P. Pirsch, and J. Reimers, "Architecture and memory requirements for stand-alone and hierarchical MPEG2 HDTV-decoders with synchronous DRAMs," in *Proc. Int. Symp. Circuits Systems*, pp. 609–612, Apr. 1995.
89. H. Kim and I.-C. Park, "Array address translation for SDRAM-based video processing applications," *Electron. Lett.*, vol. 35, pp. 1929–1931, Oct. 1999.
90. H. Kim and I.-C. Park, "High-performance and low-power memory-interface architecture for video processing applications," *IEEE Trans. Circuits Syst. Video Technol.*, vol. 11, no. 11, pp. 1160–1170, Nov. 2001.
91. E.G.T. Jaspers and P.H.N. de With, "Bandwidth reduction for video processing in consumer systems," *IEEE Trans. Consum. Electron.*, vol. 47, no. 4, pp. 885–894, Nov. 2001.
92. C.-H. Li, W.-H. Peng, and T. Chiang, "Design of memory sub-system with constant-rate bumping process for H.264/AVC decoder," *IEEE Trans. Consum. Electron.*, vol. 53, no. 1, pp. 209–217, 2007.
93. K.-B. Lee, T.-C. Lin, and C.-W. Jen, "An efficient quality-aware memory controller for multimedia platform SoC," *IEEE Trans. Circuits Syst. Video Technol.*, vol. 15, no. 5, pp. 620–633, May 2005.
94. T.-M. Liu et al., "A 125 mW, fully scalable MPEG-2 and H.264/AVC video decoder for mobile applications," *IEEE J. Solid-State Circuits*, vol. 42, no. 1, pp. 161–169, Jan. 2007.

Part IV
Audio Processing in Mobile Multimedia Broadcasting

Chapter 19
Audio Coding and Classification: Principles and Algorithms

Karthikeyan Umapathy and Sridhar Krishnan

Copyright Information[1, 2]

19.1 Introduction

A normal human can hear sound vibrations in the range from 20 Hz to 20 kHz. Signals that create such audible vibrations qualify as an audio signal. Creating, modulating, and interpreting audio clues were among the foremost abilities that differentiated humans from the rest of the animal species. Over the years, methodical creation and processing of audio signals resulted in the development of different forms of communication, entertainment, and even biomedical diagnostic tools. With the advancements in the technology, audio processing was automated and various enhancements were introduced. The current digital era furthered the audio processing with the power of computers. Complex audio processing tasks were easily implemented and performed in blistering speeds. The digitally converted and formatted audio signals brought in high levels of noise immunity with guaranteed quality of reproduction over time. However, on the one hand, the benefits of digital audio

K. Umapathy and S. Krishnan (✉)
Ryerson University, Canada
e-mail: karthi(krishnan)@ee.ryerson.ca

[1] Portions of this chapter reprinted, with permission, from ([1] K. Umapathy, S. Krishnan and S. Jimaa. Multi-group Classification of Audio Signals using Time-frequency Parameters. IEEE Transactions on Multimedia, vol. 7, Issue 2, Pages 308 - 315, Apr 2005. [2] K. Umapathy, S. Krishnan and Raveendra K. Rao. Audio Signal Features Extraction and Classification using Local Discriminant Bases. IEEE Transactions on Audio, Speech and Language Processing, vol. 15, Issue 4, Pages 1236 - 1246, May 2007.) ©[2005, 2007] IEEE.

[2] Portions of this chapter are reproduced from the open source article: (K. Umapathy and S. Krishnan. Perceptual coding of Audio Signals using Adaptive Time-frequency Transformation. EURASIP Journal on Audio, Speech and Music Processing. Volume 2007 (2007), Article ID 51563.)

F.-L. Luo (ed.), *Mobile Multimedia Broadcasting Standards: Technology and Practice*, 539
DOI: 10.1007/978-0-387-78263-8_19,

format came with the penalty of huge data rates and difficulties in protecting copyrighted audio content over Internet. On the other hand, the ability to use computers brought in great power and flexibility in analyzing and extracting information from audio signals. This contrasting pros and cons of digital audio inspired the development of variety of audio processing techniques.

With relevance to the main theme of this book, audio processing, in particular audio compression, plays an important role in mobile multimedia broadcasting. Multimedia content delivery over cellular networks are receiving a lot of attention recently with the current 3GPP standard defining it as a part of its standard. The cost involved (bandwidth related) with the usage of these services, low processing capability, and limited memory capacity of the wireless mobile receivers demand very low-bit rate high quality audio compression approaches. Specifically it would be ideal to use a less complex audio coder that can work satisfactorily for all kinds of audio at low bit-rates. In the near future, the multimedia content requests to service providers will be driven by actual content of the audio and video patterns then the text based searches. Audio processing will also play an central role in sensing the audio environment and adaptively switching the performance of the multimedia-enabled wireless devices in different environments. As it would not be possible to cover the complete array of audio processing techniques in a chapter, we limit our presentation to the basics of two important audio processing tasks for multimedia broadcasting viz. audio compression (Coding) and audio feature extraction (for Classification) using a time–frequency (TF) approach.

The rest of the chapter is organized as follows: Sect. 2 is devoted to the theories and the algorithms related to TF analysis. Section 3 will deal with the use of TF analysis in audio coding and also will present the comparisons among some of the audio coding technologies including adaptive time–frequency transform (ATFT) coding, MPEG-Layer 3(MP3) coding, and MPEG Advanced Audio Coding (AAC). In Sect. 4, TF analysis-based audio classification will be covered. Summary of this chapter will be given in Sect. 5.

19.2 Time–Frequency Analysis

Signals can be classified into different classes on the basis of their characteristics. One such classification is deterministic and random signals. Deterministic signals are those, which can be represented mathematically or in other words all information about the signals are known a priori. Random signals take random values and cannot be expressed in a simple mathematical form like deterministic signals, instead they are represented using their probabilistic statistics. When the statistics of such signals vary over time, they qualify to form another subdivision called nonstationary signals. Nonstationary signals are associated with time-varying spectral content and most of the real world (including audio) signals fall into this category. Because of the time-varying behavior, it is challenging to analyze nonstationary signals.

Early signal processing techniques were mainly using time domain operations such as correlation, convolution, inner product, and signal averaging. Although the time-domain operations provided some information about the signal, they were limited in their ability to extract the frequency content of a signal. Introduction of Fourier theory addressed this issue by enabling the analysis of signals in the frequency domain. However, Fourier technique provided only the global frequency content of a signal and not the time occurrences of those frequencies. Hence neither time-domain nor frequency domain analysis were sufficient enough to analyze signals with time-varying frequency content. To over come this difficulty and to analyze the nonstationary signals effectively, techniques that could give joint time and frequency information were needed. This gave birth to the TF transformations.

In general, TF transformations can be classified into two main categories on the basis of: (1) signal decomposition approaches and (2) Bilinear TF distributions (also known as Cohen's class). In decomposition-based approach, the signal is approximated into small TF functions derived from translating, modulating, and scaling a basis function having a definite time and frequency localization. Distributions are two-dimensional energy representations primarily used for visualization purposes and cannot be efficiently used for parameterization of the signal. For the applications in hand (audio compression, audio classification), the automatic choice would be a parametric decomposition approach. There are variety of TF decomposition techniques with different TF resolution properties. Some examples in the increasing order of TF resolution superiority are STFT (Short-Time Fourier Transform), Wavelets, Wavelet packets, Pursuit-based algorithms [1]. The next subsection will give the theory behind some of these techniques.

19.2.1 Time–Frequency Theory

TF representation is all about how well the frequency content of a signal can be defined at infinitely small time or instantaneous time. The success of any TF transformation lies in how well it can transform the signal on to a TF plane with optimal TF resolution. The ideal case would be to have both time and frequency resolution as high as possible or in other words to have infinitely small TF tile in the TF plane (TF tiling is a two-dimensional display with time and frequency as its axes to show the TF resolution achieved by different techniques). However, the ideal TF resolution cannot be achieved because of the Heisenberg's uncertainty principle, which states that infinitely small widths in both time and frequency domains cannot exist simultaneously. The TF tiling has to satisfy the condition $\sigma_t \sigma_\omega >= \frac{1}{2}$, where σ_t and σ_ω are the respective time width and frequency width of the TF structure.

Figure 19.1 shows the possible TF tiling of two TF atoms or structures with time (σ_t) and frequency (σ_ω) widths satisfying Heisenberg's principle. It can be clearly observed from the condition that one cannot go below the lower bound. It is proved that only Gabor functions or atoms (Gaussian) satisfy the lower bound condition

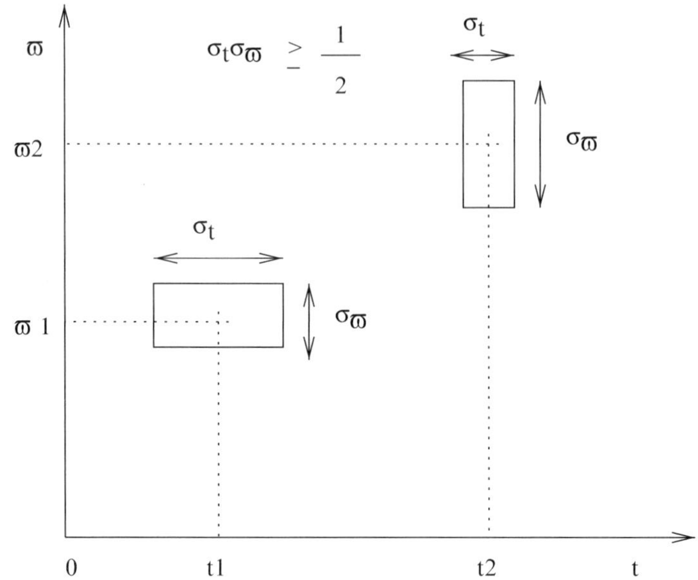

Fig. 19.1 TF tiling demonstrating Heisenberg's principle. ω - angular frequency and t - time

meaning they have the best TF localization properties or in other words have the best possible TF resolution [1]. Now with this uncertainty principle governing the optimal TF resolution achievable, we will discuss few joint TF techniques evolved over years in addressing the TF resolution requirement of nonstationary signal analysis.

The first technique developed for joint TF analysis was by Gabor in 1940s. He used a translating window function to segment the signals into overlapping small time segments and then applied Fourier on them. It is given by the following equation,

$$X(\tau,\omega) = \int x(t)g(t-\tau)\exp(-j\omega t)\mathrm{d}t, \tag{19.1}$$

where $x(t)$ is the signal, and $g(t-\tau)$ is the sliding window function. This technique is called the STFT (short-time Fourier transform). Because of the segmentation, frequency content for each time segment was estimated individually thereby giving the tie between time and frequency. The TF resolution provided by this technique was uniform through out the TF plane and is decided by the width of the window function. The assumption is that during those small time segments, signals remain stationary (quasi-stationary). This approach suffered from the frequency smearing or oscillatory effect associated with the type of windowing function used.

Figures 19.2 and 19.3 demonstrate two situations of TF tilings, for short duration and a long duration window function. The fixed TF resolution provided by this technique could not answer all the TF resolution required for processing complex signals with varying TF resolutions over the complete TF plane. However, this technique is

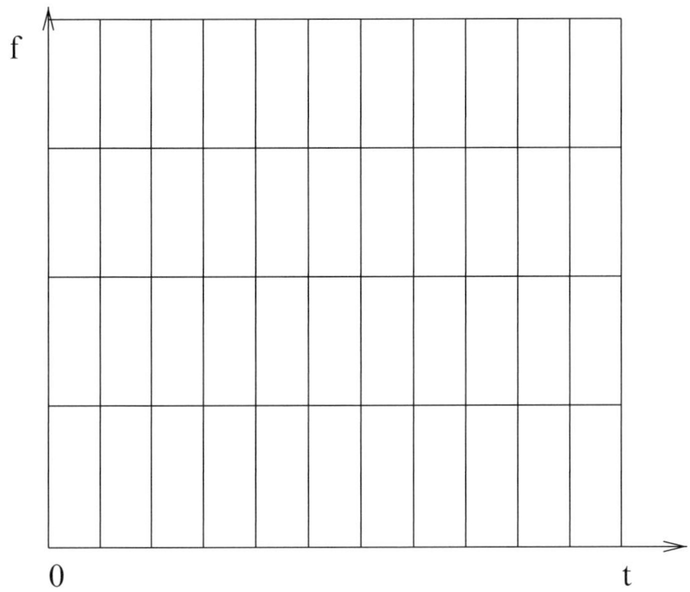

Fig. 19.2 TF tiling of STFT technique with short time window. f - frequency and t - time

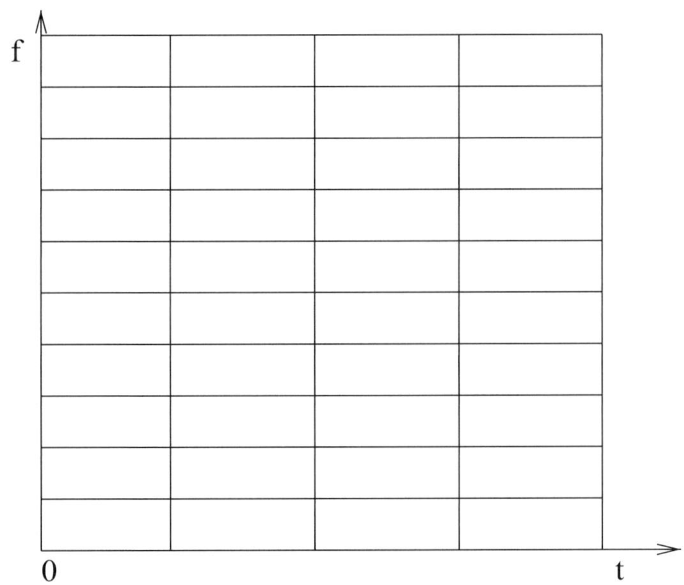

Fig. 19.3 TF tiling of STFT technique with long time window. f - frequency and t - time

widely used to visualize the TF plane to get a coarser idea. Spectrogram, which is nothing but the squared modulus of the STFT, is generally used to display the TF energy distribution over the TF plane.

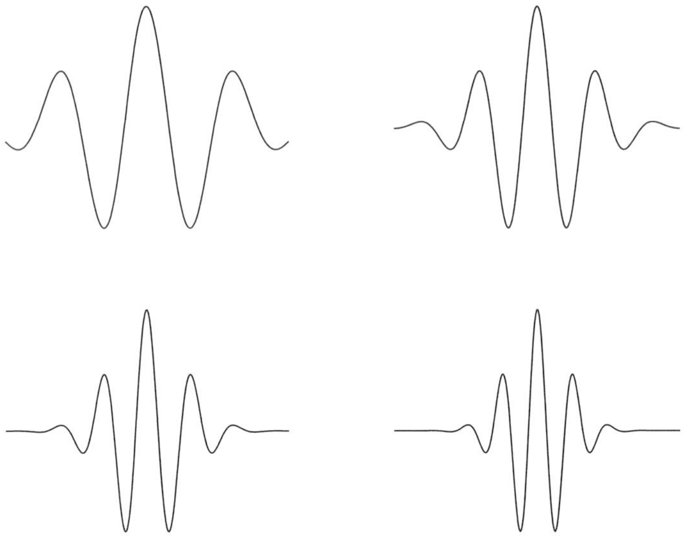

Fig. 19.4 Morlet wavelet with different scaling

To solve the varying TF resolution requirements, an adaptive technique was needed. Wavelets were introduced to address this adaptive TF resolution requirements. In wavelets, the basis function used are small waves called mother wavelets, which satisfy few mathematical conditions. These waves took different sizes by stretching and compressing themselves to different scaled versions of the mother wavelet. Figure 19.4 demonstrates the different scaled version of morlet wavelet.

These different scaled versions of the mother wavelet are slided across the signals and the similarity measure is checked to model the localized signal structures with the wavelet of a particular scaling. Because, wavelets signals are analyzed using scaled versions of mother wavelet translating across the signal, it is often called as time-scale analysis. It is given by the following equation,

$$CWT_x(\tau,s) = \frac{1}{\sqrt{s}} \int x(t) g(\frac{t-\tau}{s}) \mathrm{d}t, \tag{19.2}$$

where $g(\frac{t-\tau}{s})$ is the mother wavelet, τ being the translational parameter, and s being the scaling parameter. Unlike STFT where the time width of the window function is fixed, wavelets have an adaptive varying time width defined by the scaling parameter. This scaling parameter that stretches and compresses the wavelets contribute to the change in the center frequency of the wavelets. Small scale factors corresponds to higher frequencies and larger scale factor corresponds to the lower frequencies. In other words, wavelets uses short time scales to capture the high frequency structures and a long time scale to capture the low frequency structures in a signal. In practice for most of the applications, discrete wavelet transformations follow a dyadic scaling as shown in Fig. 19.5 to simplify numerical calculations. However, M band scaling also exist for specific applications but the design and implementation of

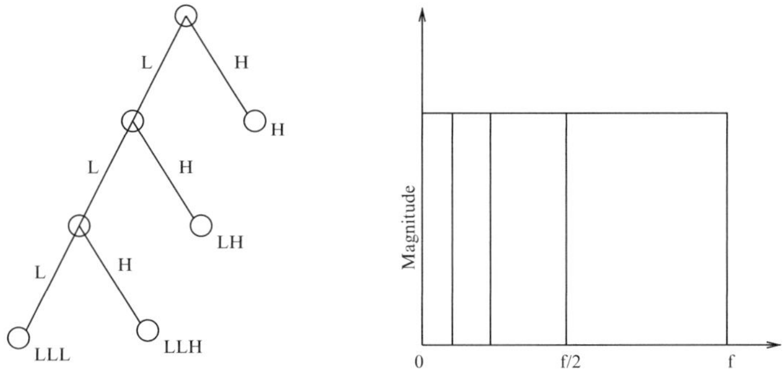

Fig. 19.5 Dyadic scaling of wavelet and its frequency division. f - frequency

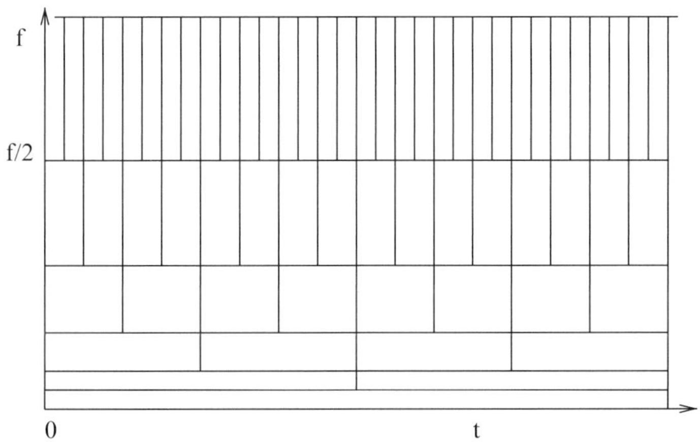

Fig. 19.6 TF tiling for 2 band wavelet technique. f - frequency and t - time

these systems are much complex and expensive in terms of computational effort. It can be proved that if the frequency axis is completely covered by dilated dyadic wavelets then it defines a complete and stable representation. This dyadic scaling divides the frequency axis into two halves equally as high and low frequencies. Each time the low frequency part is further divided into two halves and this process repeats. Now applying the uncertainty principle it can be noted that the TF resolution varies for each frequency division over the frequency axis. For one particular frequency interval of fixed size, the time resolution remains the same. But over the complete frequency axis both time and frequency resolutions vary based on Heisenberg's condition. Figure 19.6 demonstrates the TF tiling achieved for a two band wavelet basis. It can be noted that at high frequencies wavelets provide high time resolution but poor frequency resolution and vice versa at low frequencies.

Even though we were successful in achieving varying TF resolutions over TF plane, these adaptive TF tiling are fixed for every interval of frequency. We cannot expect a nonstationary signal to have same time resolution over a particular

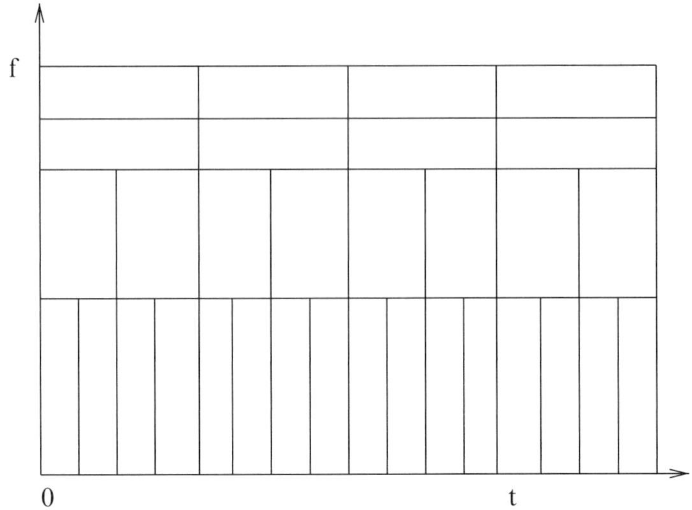

Fig. 19.7 TF tiling for wavelet packet. f - frequency and t - time

frequency interval. Signals containing components with same frequency support but different time supports cannot be possibly modeled efficiently using wavelets.

Wavelet packet bases and local cosine bases were introduced to address this problem to some extent [1]. Unlike discrete wavelet bases where high frequencies have poor frequency resolution, in wavelet packet the frequency axis can be divided into any required intervals. By this way any frequency resolution can be achieved at any frequencies; however, still for each frequency interval the time resolution remains the same. Figures 19.7 and 19.8 demonstrate the two sample representations of wavelet packet bases TF tiling. It can be noted wavelet packet bases provide adaptive varying frequency resolutions unlike wavelets that have fixed varying frequency resolutions. But in both cases the time resolution remains the same for a particular frequency interval.

In local cosine bases [1] the opposite of wavelet packet bases is achieved, That is we can have any time resolution but for a particular time resolution we have a fixed frequency resolution. Wavelet packet bases and local cosine bases are duals. The combination of varying time resolutions as in the case of local cosine bases and varying frequency resolutions as in the case of wavelet packet bases would be an ideal choice to model nonstationary signals. To achieve this nonorthogonal basis functions were needed i.e., the basis functions should have an over-complete and redundant combinations of bases for all possible translations, modulations, and scalings. The collection of over complete combinations of a TF basis function can be called as a redundant TF dictionary. Applying a pursuit algorithm (such as matching pursuit algorithm [2]) with efficient search techniques on this TF dictionary, optimal TF resolutions can be achieved at the cost of computational complexity. The basis functions chosen to form the redundant dictionary determine the nature of the modeling/decomposition. The matching pursuit algorithm when used with

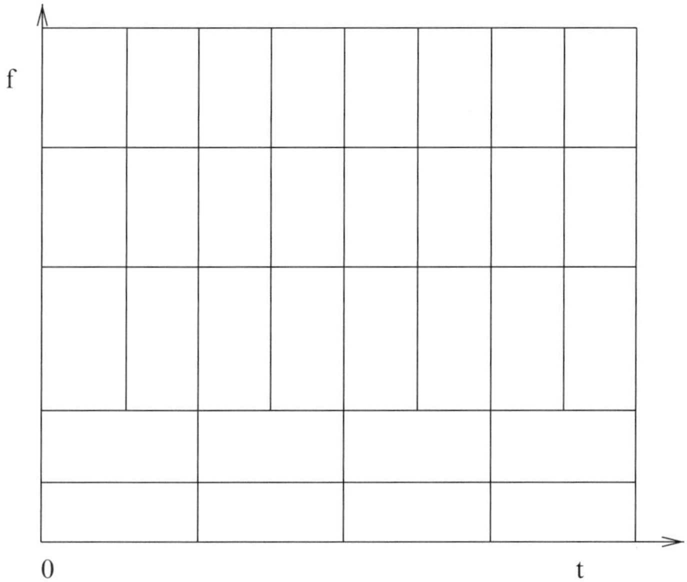

Fig. 19.8 TF tiling for wavelet packet. f - frequency and t - time

a TF dictionary yields adaptive time-frequency transform (ATFT) [2]. Because of the adaptive nature and the over-complete dictionary any TF resolution satisfying the Heisenberg's condition can be achieved at any part of the TF plane. One possible ATFT TF tiling is shown in the Fig. 19.9. Many variants of matching pursuits based algorithms can be chosen to suit the specific needs of coding or classification applications; however, since we chose to demonstrate both coding and classification using a single TF decomposition approach, we used ATFT for the analysis of audio signals. A detailed explanation of the ATFT algorithm follows in the next subsection.

19.2.2 ATFT Algorithm

As explained in the previous subsection, the ATFT technique is based on the matching pursuit algorithm with TF dictionaries. ATFT has excellent TF resolution properties (better than Wavelets and Wavelet Packets) and due to its adaptive nature (handling nonstationarity), there is no need for signal segmentations. Flexible signal representations can be achieved as accurately as possible depending upon the characteristics of the TF dictionary.

In the ATFT algorithm, any signal $x(t)$ is decomposed into a linear combination of TF functions $g_{\gamma_n}(t)$ selected from a redundant dictionary of TF functions [2]. In this context, redundant dictionary means that the dictionary is over-complete and contains much more than the minimum required basis functions i.e., a collection of

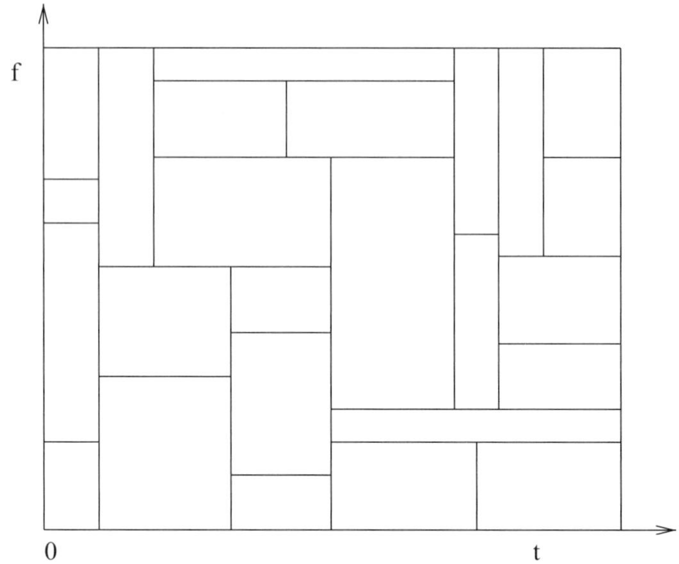

Fig. 19.9 One representation of TF tiling provided by the ATFT algorithm. f - frequency and t - time

nonorthogonal basis functions that is much larger than the minimum required basis functions to span the given signal space. Using ATFT, we can model any given signal $x(t)$ as

$$x(t) = \sum_{n=0}^{\infty} a_n g_{\gamma_n}(t), \tag{19.3}$$

where

$$g_{\gamma_n}(t) = \frac{1}{\sqrt{s_n}} g\left(\frac{t-p_n}{s_n}\right) \exp\{j(2\pi f_n t + \phi_n)\} \tag{19.4}$$

and a_n are the expansion coefficients.

The scale factor s_n, also called as octave parameter, is used to control the width of the window function, and the parameter p_n controls the temporal placement. The parameters f_n and ϕ_n are the frequency and phase of the exponential function, respectively. The index γ_n represents a particular combination of the TF decomposition parameters (s_n, p_n, f_n, and ϕ_n). The signal $x(t)$ is projected over a redundant dictionary of TF functions with all possible combinations of scaling, translations, and modulations. The dictionary of TF functions can either suitably be modified or selected, based on the application in hand. When $x(t)$ is real and discrete, like the audio signals in the presented technique, we use a dictionary of real and discrete TF functions. Because of the redundant or overcomplete nature of the dictionary it gives extreme flexibility to choose the best fit for the local signal structures (local optimization) [2]. This extreme flexibility enables to model a signal as accurately as possible with the minimum number of TF functions providing a compact approximation of the signal.

In our technique, we used the Gabor dictionary (Gaussian functions), which has the best TF localization properties [3]. At each iteration, the best correlated TF function was selected from the Gabor dictionary. The remaining signal called the residue was further decomposed in the same way at each iteration subdividing them into TF functions. After M iterations, signal $x(t)$ could be expressed as

$$x(t) = \sum_{n=0}^{M-1} \langle R^n x, g_{\gamma_n} \rangle g_{\gamma_n}(t) + R^M x(t), \tag{19.5}$$

where the first part of (19.5) is the decomposed TF functions until M iterations, and the second part is the residue, which will be decomposed in the subsequent iterations. This process is repeated till all the energy of the signal is decomposed. At each iteration, some portion of the signal energy was modeled with an optimal TF resolution in the TF plane. Over iterations it can be observed the captured energy increases and the residue energy falls. On the basis of the signal content, the value of M could be very high for a complete decomposition (i.e. residue energy = 0). Examples of Gaussian TF functions with different scale and modulation parameters are shown in Fig. 19.10. The order of computational complexity for one iteration

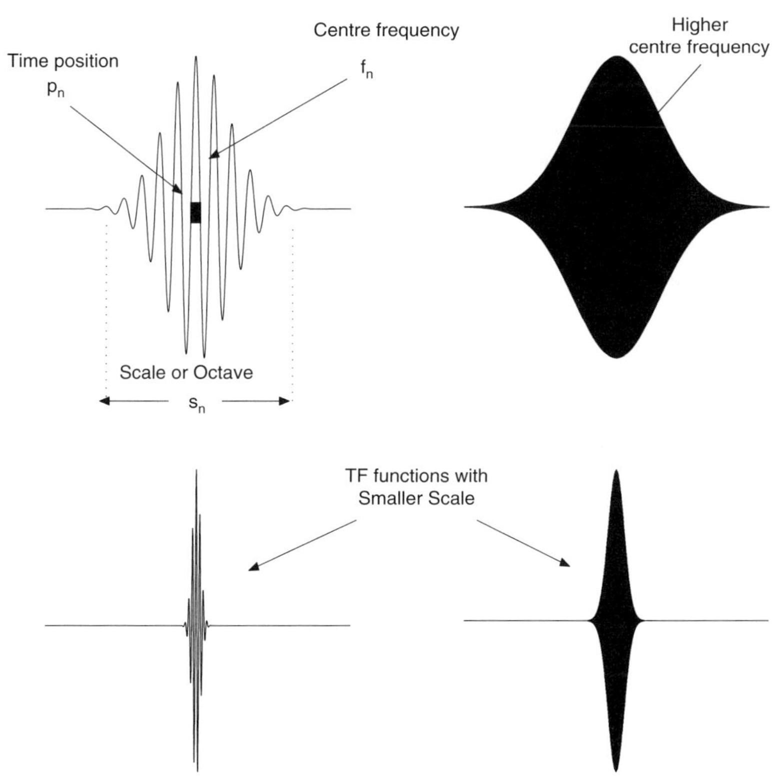

Fig. 19.10 Gaussian TF function with different scale and modulation parameters

of the ATFT algorithm is given by $O(N \log N)$, where N is the length of the signal samples. The time complexity of the ATFT algorithm increases with the increase in the number of iterations required to model a signal, which in turn depends on the nature of the signal. Compared with this the computational complexity of modified discrete cosine transform (MDCT) used in few of the state-of the-art audio coders is only $O(N \log N)$ (same as FFT).

Once the signal is modeled accurately or decomposed into TF functions with definite time and frequency localization, the TF parameters governing the TF functions could be analyzed for extracting application specific information. In our case, we process the TF decomposition parameters of the audio signals to perform both audio compression and classification as will be explained in the following sections.

19.3 Audio Coding

Until the invention of Gramophone in 1880s, it was a dream to record and reproduce sounds. With the advancements in science, slowly the audio recording and storage techniques have improved: migrating from the records, to magnetic tape media in 1950s, to digital format on a CD in 1980s, to today's much sophisticated application oriented Internet driven audio formats. Considering only the time-frame after the digital era began, audio processing has undergone tremendous amount of development and refinement. The invention of CD made a big change in digital audio representation and storage. Audio in digital format on a CD brought in many advantages such as robustness, dynamic range, and unprecedented high fidelity. Conventional CD audio are typically sampled at 44.1 or 48 kHz with 16 bits/sample (pulse code modulation) resolution to maintain high quality reproduction of sound. Because of the high sampling rates, the amount of data associated with the audio sequences is high in terms of 700 kbps, and 1.41 Mbps for stereo. The high rate is acceptable as long as the audio is listened by a single listener on his home CD player or a computer. When the same data have to be transmitted from one place to another over the network or carried in portable miniature audio devices, these high bit-rates increased the demand for expensive bandwidth/storage requirements. Although the leap in the nonlinear growth of hardware technologies have addressed the storage issues to some extent by producing miniaturized high capacity storage devices, still the need for audio compression has not diminished.

To address this high demand for audio compression, many techniques were introduced to reduce the bit-rates without sacrificing much of the audio quality. Several algorithms/coders were developed to exploit the redundancy information contained in the audio signal in time or frequency or joint TF domains. In general, they can be subdivided into temporal waveform coders, subband coders, and transform coders. Earliest of all were the temporal waveform coders using techniques such as PCM (pulse code modulation) and ADPCM (adaptive differential pulse code modulation), where the time domain redundancy is exploited to achieve reduction in bit-rates [4]. In subband coders, the signal is divided into multiple subbands of frequency and the psychoacoustics (science of hearing) model is used for performing a variable quantization to achieve reduction of bit-rates [4]. In transform coders, the audio data are first transformed into another domain such as frequency or joint TF. The transform or decomposition parameters obtained in the process of transformation is analyzed, and variably quantized based on their degree of contribution in reconstructing the audio signal. Psychoacoustics models are used to further filter the perceptually irrelevant parameters so as to achieve compact representation of the signal in either frequency or joint TF domain. Since it is out of scope of this chapter to cover all of the existing audio compression methodologies, the authors recommend the work of Ted Painter [4] for a comprehensive review of most of the existing audio compression techniques.

Audio signals are highly nonstationary in nature and the best way to analyze them is to use a joint TF approach. The presented coding methodology is based on ATFT and falls under the transform-like coder category. The usual methodology of a transform based coding technique involves the following steps : (1) Transforming the audio signal into frequency or TF domain coefficients, (2) processing the coefficients using psychoacoustic models and computing the audio masking thresholds, (3) controlling the quantizer resolution using the masking thresholds, (4) applying intelligent bit allocation schemes, and (5) enhancing the compression ratio with further lossless compression schemes. The ATFT-based coder nearly follows the above general transform coder methodology; however, unlike the existing techniques, the major part of the compression was achieved by exploiting the joint TF properties of the audio signals. The block diagram of the ATFT coder is shown in Fig. 19.11. The ATFT approach provides higher TF resolution than the existing TF techniques such as wavelets and wavelet packets [2]. This high resolution sparse decomposition enables us to achieve a compact representation of the audio signal in the transform domain itself. Also, due to the adaptive nature of the ATFT, there was no need for signal segmentation.

Psychoacoustics was applied in a novel way on the TF decomposition parameters to achieve further compression. In most of the existing audio coding techniques, the fundamental decomposition components or building blocks are in the frequency domain with corresponding energy associated with them. This makes it much easier for them to adapt the conventional, well-modeled psychoacoustics techniques into their encoding schemes. In contrast, in ATFT, the signal was modeled using TF functions which have a definite time and frequency resolution (i.e., each individual TF function is time limited and band limited), hence the existing psychoacoustics models need to be adapted to apply on the TF functions.

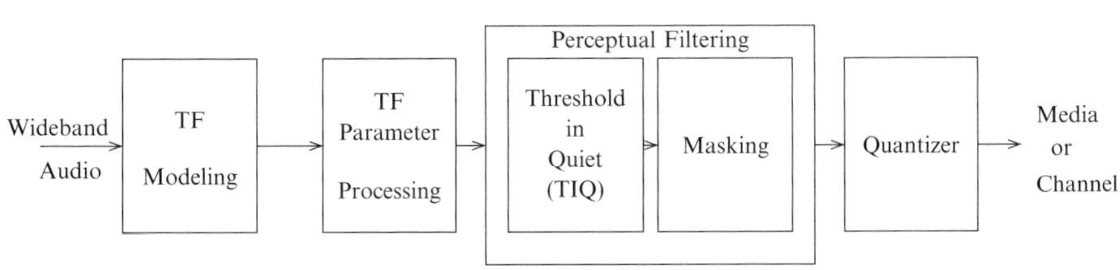

Fig. 19.11 Block diagram of ATFT audio coder

19.3.1 ATFT of Audio Signals

Any signal could be expressed as a combination of coherent and noncoherent signal structures. Here the term coherent signal structures means those signal structures that have a definite TF localization (or) exhibit high correlation with the TF dictionary elements. On the one hand, the ATFT algorithm models the coherent signal structures well within the first few 100 iterations, which in most cases contribute to >90% of the signal energy. On the other hand, the noncoherent noise like structures cannot be easily modeled since they do not have a definite TF localization or correlation with dictionary elements. Hence these noncoherent structures are broken down by the ATFT into smaller components to search for coherent structures. This process repeats until the whole residue information is diluted across the whole TF dictionary [2]. From a compression point of view, it would be desirable to keep the number of iterations ($M <<< N$), as low as possible and at the same time sufficient enough to model the audio signal without introducing perceptual distortions. Considering this requirement, an adaptive limit has to be set for controlling the number of iterations. The energy capture rate (signal energy capture rate per iteration) could be used to achieve this. By monitoring the cumulative energy capture over iterations, we could set a limit to stop the decomposition when a particular amount of signal energy was captured. The minimum number of iterations required to model an audio signal without introducing perceptual distortions depends on the signal composition and the length of the signal. In theory, due to the adaptive nature of the ATFT decomposition, it is not necessary to segment the signals. However, due to the computational resource limitations (Pentium III, 933 MHZ with 1 GB RAM), we decomposed the audio signals in 5 s durations. The larger the duration decomposed, the more efficient is the ATFT modeling. This is because if the signal is not sufficiently long, we cannot efficiently utilise longer TF functions (highest possible scale) to approximate the signal. As the longer TF functions cover larger signal segments and also capture more signal energy in the initial iterations, they help to reduce the total number of TF functions required to model an audio signal. Each TF function has a definite time and frequency localization, which means all the information about the occurrences of each of the TF functions in time and frequency of the signal is available. This flexibility helps us later in our processing to group the TF functions corresponding to any short time segments of the audio signal for computing the psychoacoustic thresholds. In other words, the complete length of the audio signal can be first decomposed into TF functions and later the TF functions corresponding to any short time segment of the signal can be grouped together. In comparison, most of the DCT and MDCT-based existing techniques have to segment the signals into time frames and process them sequentially. This is needed to account for the nonstationarity associated with the audio signals and also to maintain a low signal delay in encoding and decoding.

In the presented technique for a signal duration of 5 s, the limit was set to be the number of iterations needed to capture 99.5% of the signal energy or to a maximum of 10,000 iterations. For a signal with less noncoherent structures, 99.5% of signal energy could be modeled with a lower number of TF functions than a signal

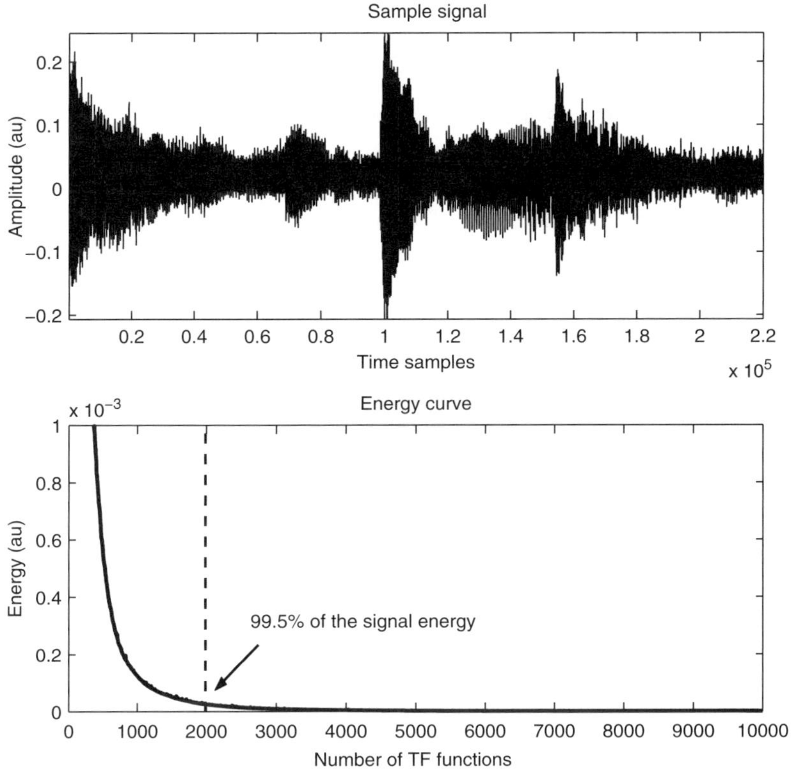

Fig. 19.12 Energy cutoff of the sample signal in panel 1. au - arbitrary units

with more noncoherent structures. In most cases, a 99.5% of energy capture nearly characterises the audio signal completely. The upper limit of the iterations is fixed to 10,000 iterations to reduce the computational load. Figure 19.12 demonstrate the number of TF functions needed for a sample audio signal. In the figure, the lower panel shows the energy capture curve for the sample audio signal in the top panel with number of TF functions in the *X* axis and the normalised energy in the *Y* axis. On average, it was observed that 6,000 TF functions are needed to represent a signal of 5 s duration sampled at 44.1 kHz.

19.3.2 Psychoacoustics

Psychoacoustics is the science of hearing. This plays an important role in audio compression techniques. A transform coder provides only mathematically nonredundant coefficients and has no perceptual criteria attached to it. Applying perceptual filtering to the transform coefficients helps us in achieving perceptually lossless coders.

The two main psychoacoustics effect that is of importance to digital audio compression are threshold-in-quiet (TiQ) and audio masking. Much work has been done on the area of psychoacoustics right from 1930s. Many experiments were performed in arriving at a human auditory model over years.

19.3.2.1 Threshold-in-Quiet

TiQ can be defined as the minimum detectable level of sound in the absence of any external sounds. Our ear has a non-linear frequency response. We need different sound pressure level (SPL) to hear different frequencies. In general humans are sensitive to the frequencies between 1 to 5 kHz even with very low SPLs. Frequencies below and above this range need high SPL to be perceived by our ears. The SPL has a dynamic range of 60 dB over the audible range (0-20 kHz). Figure 19.13 depicts the TiQ curve measured as a function of frequency.

The TIQ can be well approximated by the non-linear function [4]

$$T_q(f) = 3.64\left(\frac{f}{1000}\right)^{-0.8} - 6.5e^{-0.6}\left(\frac{f}{1000} - 3.3\right)^2 + 10^{-3}\left(\frac{f}{1000}\right)^4 \text{ dBSPL}, \tag{19.6}$$

where f is the frequency variable. In audio coding, any signal component with an SPL below this curve need not be coded or in other words the signal component will not be perceived by ears. In existing coders, the quantization noise introduced by the coders are kept below this TiQ curve so that it is not perceived by the ears. The TiQ concept also provides a convenient and perceptually least disturbing mechanism for regulating the output bit-rates.

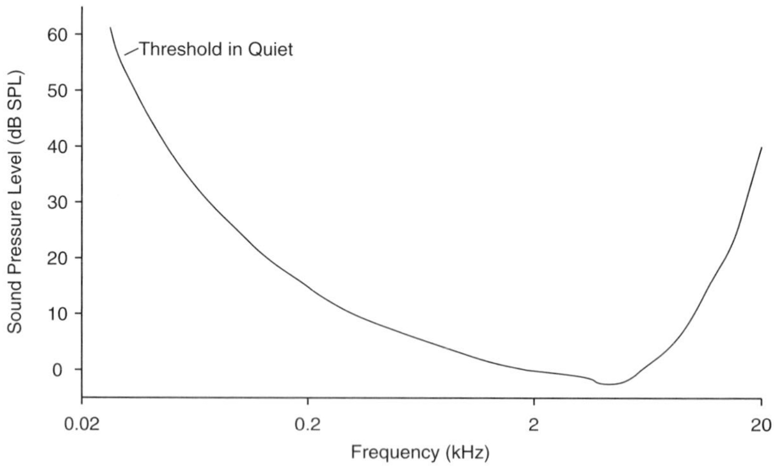

Fig. 19.13 Threshold-in-quiet curve. Reproduced from [5]

19.3.2.2 Audio Masking

Audio masking is a process in which a strong signal renders a weak signal inaudible. This phenomenon is observed by us in everyday life. We do not hear someone talking when a fast moving train passes by or a helicopter that flies low, making lot of noise. The strong signal is referred as masker and the weak signal as maskee. The amount of masking depends on the energy of the masker, the frequency separation, the time occurrence, and duration of both masker and maskee.

Our ear can be modeled as an overlapping band pass filters in the audible range. A particular area in the bascillar membrane is excited for a particular band of frequencies. On the basis of this phenomena, audible range is broken into 25 critical bands with varying bandwidths corresponding to different areas of the bascillar membrane [6, 7]. A space to frequency translation is achieved by dividing the audible frequencies into critical bands. The critical bandwidths can be approximated as [4]

$$BW_c(f) = 25 + 75\left(1 + 1.4\left(\frac{f}{1000}\right)^2\right)^{(0.69)} \text{ Hz}, \tag{19.7}$$

where f is the frequency variable. Figure 19.14 depicts the critical band bandwidths and center frequencies. A strongly occurring signal with frequency in a particular critical band increases the audibility threshold of the corresponding area in the bascillar membrane. This increase in the audible threshold makes the bascillar

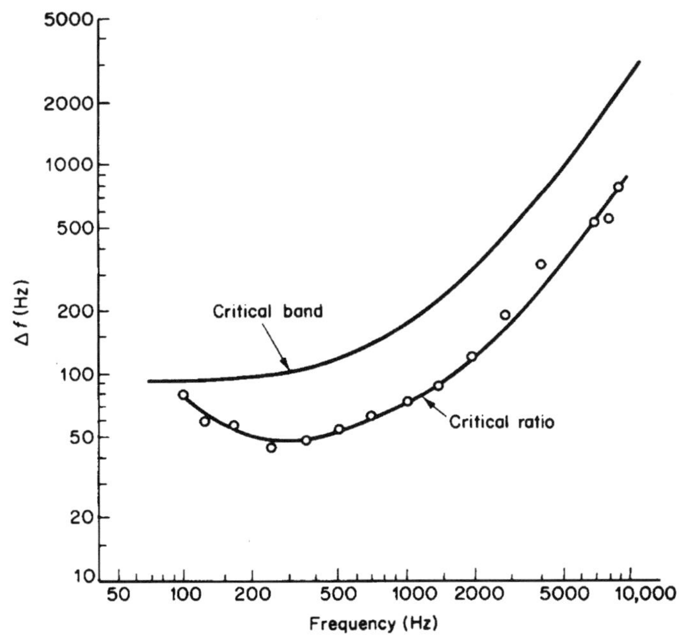

Fig. 19.14 Critical band bandwidths and center frequencies. Reproduced from [5]

membrane insensitive to weak signals with frequencies in the same critical band and also affects the neighboring bands by raising their audible threshold. The level of masking produced by a masker located at the center of the critical band is approximately same across the critical band.

There are two types of masking that happens with audio signals.

- Simultaneous masking
- Temporal masking

Simultaneous masking: This is a frequency domain phenomena. A strong signal (masker) simultaneously occurring with a weak signal (maskee) will mask the weak signal within a critical band. This type of masking within the critical band is called intraband masking. The effect of the masker also extends to the neighboring bands and vice-versa. However, this effect of masking is much lower. This type of masking is called interband masking. Simultaneous masking is effective when the masker frequency is lower than the maskee's frequency. Figure 19.15 shows the simultaneous masking phenomena in the frequency domain.

The shaded area represents the cumulative auditory shadow that results from the simultaneous presence of tonal components S1, S2, and S3. The signal S2 is the highest power tonal component and therefore dominates the total masking threshold. S3 is completely masked, whereas S1 is only partially masked. The component S3 can be removed completely because it will not be audible due to the strong presence of S2. In existing coders, S1 and S2 are allocated with corresponding bits so that the introduced quantization error falls well below the masking threshold. Basically masking locally increases the threshold of audibility allowing the possibility to have imperceptible quantization noise with reduced bits. Figure 19.16 illustrates the combined TiQ with masking threshold. However, the presented technique uses the masking property in a novel way when compared with the existing coders.

The nature of the masker also influences the level of masking. Tonal masking is the masking generated by sinusoidal-like signals. Noise masking is the masking

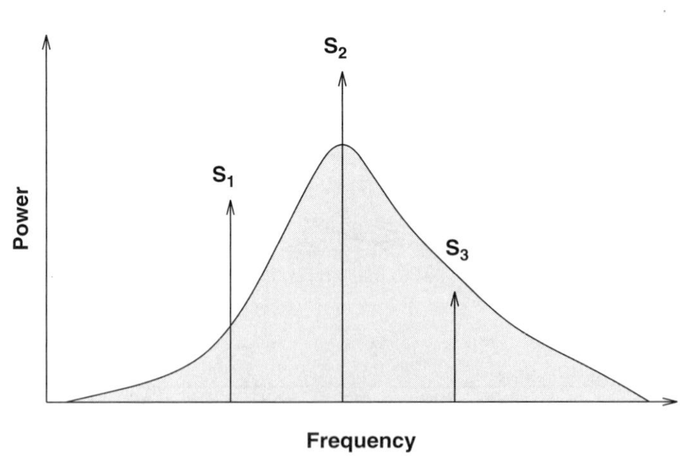

Fig. 19.15 Simultaneous masking. Reproduced from [5]

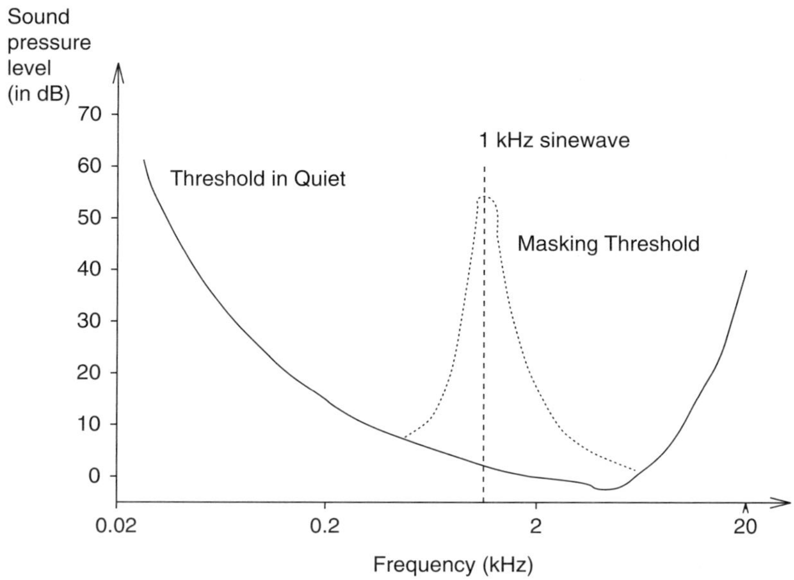

Fig. 19.16 Combined TiQ and masking threshold. Reproduced from [5]

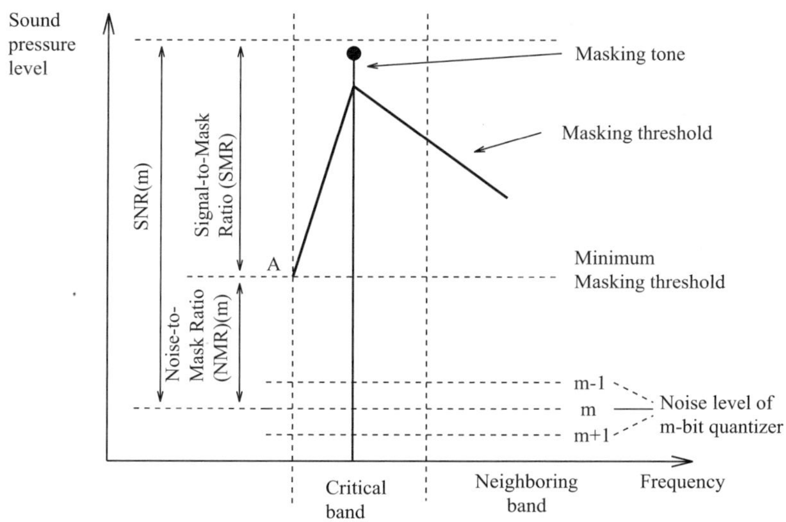

Fig. 19.17 Masking threshold and signal to mask ratio. Reproduced from [5]

generated by band-limited noise. The calculation of noise masking requires the signal power in each critical band. In tonal masking, the presence of tonal components are identified within the critical band. For each tonal component, the masking threshold is calculated. Figure 19.17 illustrates the masking pattern for a tonal simultaneous masker. The masking effect within the critical band (intraband) and the neighboring band (interband) are shown. Signal to mask ratio and noise to mask ratio denote the log distances from minimum masking threshold to the masker and the noise level introduced due to quantization. By combining both the noise masking and the tonal masking thresholds a global masking threshold is calculated, which will be applied on the audio segment to either remove the perceptually irrelevant components completely or quantize it in such a way the noise level remains below the masking thresholds.

Temporal masking: This is a time domain phenomena. Here a strongly occurring signal (masker) in time domain masks the weak signals that are preceding it or following it. Because of the retention capabilities of our ear, the increased audibility threshold due to the occurrence of a strong signal remains for a short period during which weak signals following or preceding are made inaudible. Figure 19.18 shows the typical masking pattern prior to the occurrence (premasking), during the occurrence (simultaneous) and after the occurrence (postmasking) of a masker. Premasking lasts only about 5 ms whereas postmasking extends from 50 to 300 ms. Premasking phenomenon is efficiently used with adaptive block size transform to compensate for pre-echo [4, 8] distortions. The masking that happens to the preceding signals is called backward masking and the masking that happens to the following signal is called forward masking. Figure 19.18 illustrates the premasking region, simultaneous masking and the postmasking region.

In general masking depends on the following factors:

- Strength of the masker (dB level)
- Frequency separation of the masker and the maskee (critical band)
- Time occurrence of the masker (simultaneous or temporal)
- Time duration of the masker (simultaneous)

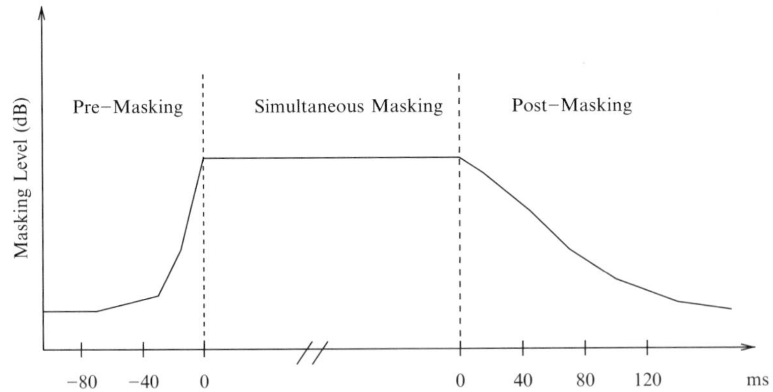

Fig. 19.18 Typical masking pattern. Reproduced from [5]

19.3.3 Implementation of Psychoacoustics

In the conventional coding methods, the signal is segmented into short time segments and transformed into frequency domain coefficients. These individual frequency components are used to compute the psychoacoustic masking thresholds and accordingly their quantization resolutions are controlled. In contrast, in our approach we computed the psychoacoustic masking properties of individual TF functions and used them to decide whether a TF function with certain energy was perceptually relevant or not based on its time occurrence with other TF functions. TF functions are the basic components of the presented technique and each TF function has a certain time and frequency support in the TF plane. So their psychoacoustical properties have to be studied by taking them as a whole to arrive at a suitable psychoacoustical model.

19.3.3.1 Threshold-in-Quiet (TIQ)

TF functions form fundamental building blocks of the presented coder and they can take all possible combinations of time duration and frequency. However, in the ATFT algorithm implementation, they could take any time-width between 2^2 samples (90 μs) and 2^{14} samples (0.4 s) in steps with any frequency between 0 and 22,050 Hz (max frequency). The time support of a frequency component also plays an important role in the hearing process. From our experiments, we observed that longer duration TF functions were heard much better even with lower energy levels than the shorter duration TF functions. Hence out of all the possible durations of the TF functions, the highest possible time duration of 16,384 samples corresponding to the octave 14 (the term octave is from the implementation nomenclature i.e., the scale factor doubles in each step) was the most sensitive TF function for different combinations of frequencies. This forms the worst case TF function in our modeling for which our ears are more sensitive. So it is obvious that this TF function has to be used to obtain the worst case threshold in quiet (TIQ) curve for our model. The curve obtained in this way will hold good for all other TF functions with all possible combinations of time-widths and center frequencies. Figure 19.19 demonstrates the different modulated versions of the TF function with maximum time-width (octave 14).

The TF functions (duration 0.4 s) with different center frequencies were played to each of the listeners. It should be noted that the "frequency" here means the center frequency of the TF function and not the absolute frequency as used in regular psychoacoustics experiments. In general, each of the TF functions will have a center frequency and a frequency spread based on the time-width they can take. For this experiment as we are using only the TF function with the longest width (duration 0.4s), the frequency spread is fixed. For each frequency, setting the amplitude of the TF function was reduced in steps until the listener could no longer hear the TF function anymore. Once this point is reached, the amplitude of the TF function is increased and played back to the listener to confirm the correct point of

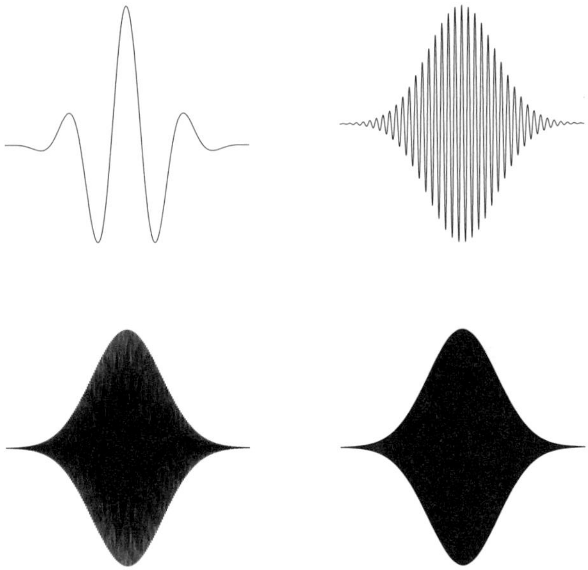

Fig. 19.19 TF function with time-width of 16,384 samples modulated at different center frequencies

minimum audibility. This is repeated for the following values of center frequencies: 10 Hz, 100 Hz, 500 Hz, 1 kHz, 2 kHz, 4 kHz, 6 kHz, 8 kHz, 10 kHz, 12 kHz, 16 kHz, and 20 kHz. The minimum audible amplitude level for each frequency setting was recorded. The values obtained from five listeners were averaged to obtain the absolute threshold of audibility for TF functions.

To reduce the computational complexity, the frequency range was divided into three bands of low frequency (500 Hz and below), sensitive frequencies (from 500 Hz to 15 kHz) and high frequencies (15 kHz and above). The experimental values were averaged to get uniform thresholding for the low and high frequency bands. In the middle or sensitive band, the lowest averaged experimental value was selected as threshold of audibility throughout the band. Figure 19.20 illustrates the averaged TIQ curve superimposed on the actual TIQ curve. The TF functions were grouped into the above mentioned three frequency groups. Amplitude of the TF functions were calculated from their energy and octave values. These amplitude values were checked with the TIQ average values. The TF functions whose amplitude values fell below the averaged TIQ values were discarded.

19.3.3.2 Audio Masking Applied to TF Functions

Similar to TIQ, the existing masking techniques cannot be used directly on the presented coder for the same reasons explained earlier. So masking experiments were conducted to arrive at masking thresholds for TF functions with different time-widths. The possible time duration of TF functions varies between 2^2 and 2^{14} in

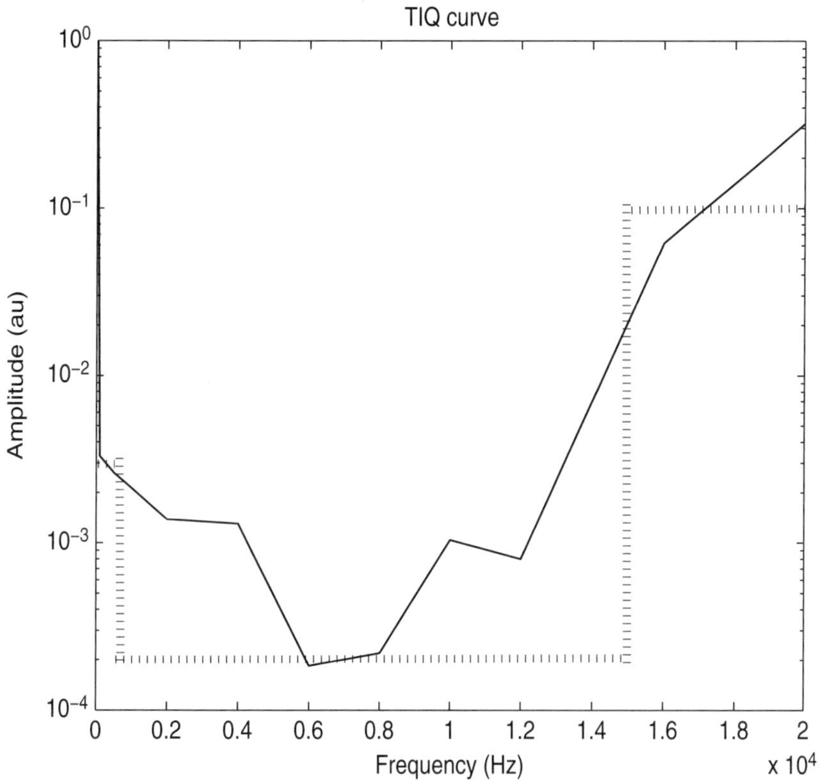

Fig. 19.20 Average thresholding applied to TIQ curve. *Solid line* denotes the actual TIQ curve and *dashed line* denotes the applied threshold. au - arbitrary units

steps of powers of 2, each of the time-width TF function was examined for its masking properties. Each of this different duration TF functions can occur at any point in time with frequencies between 20 Hz and 20 kHz. Out of the possible durations of the TF functions the shorter durations $(2^2–2^7)$ are transient-like structures, which have larger bandwidths but little time support. Removing these TF functions in the process of masking will introduce more tonal artifacts in the reconstructed signal. This happens because the complex frequency pattern of the signal is disturbed to some extent. Hence, these functions were preserved and not used for masking purposes.

The remaining TF functions with time-widths $(2^8–2^{14})$ were used for the masking experiments. TF functions with each of these time-widths (durations from 256 to 16,384 samples) were tested for their masking capabilities with other time-width TF functions at various energies and frequencies. The TF functions were first grouped into equivalents of 400 time samples (10 ms). This is possible as each of the TF functions have the precise information about their time occurrence. Once they were grouped into time slots equivalent to 10 ms, the TF functions falling in each time slot were divided into 25 critical bands, based on their center frequencies. In each critical band, the TF function with highest energy was located. Relative energy difference of this TF function with the remaining TF functions in the same critical band was computed. Using a look up table, each of the remaining TF functions was verified if it would be masked by the relative energy difference with the TF function having the highest energy. The experimental procedure for computing the look up table of masking thresholds will be explained in subsequent paragraphs. The TF functions, which fell below the masking threshold defined by the look up tables, were discarded.

As shown in Fig. 19.21a within the 10 ms duration, the location of masker and maskee TF functions can occur anywhere. The worst case situation would be when the masker TF function occurs at the beginning of the time slot, and the maskee TF function occurs at the end of the time slot or vice versa. So all of our testing was done for this worst case scenario by placing the masker TF function and the maskee TF function at the maximum distance of 10 ms.

On the basis of the duration of masker and maskee TF functions, one of the following could occur as depicted in Fig. 19.21b.

- Masker and maskee are apart in time within the 10 ms, in which case they do not occur simultaneously. In this situation, masking is achieved due to temporal masking effects where a strong occurring masker masks the preceding, and following weak signals in time domain.
- Masker duration is large enough that the maskee duration falls within the masker (two scenarios shown in Fig. 19.21b) even after a 400 samples shift. In this case simultaneous masking occurs.
- Masker duration is shorter than the maskee duration. In this case, both simultaneous and temporal masking is achieved. The simultaneous masking occurs during the duration of the masker when the maskee is also present. Temporal masking occurs before and after the duration of the masker.

Four sets of experiments were conducted with masker TF function (normalized in amplitude) taking center frequency of 150 Hz, 1 kHz, 4.8 kHz, and 10.5 kHz (critical band center frequencies) and the maskee TF function taking center frequency of 200 Hz, 1.1 kHz, 5.3 kHz, and 12 kHz (corresponding critical band upper limits), respectively. As the masking thresholds depend also on the frequency separation of masker and maskee, maximum separation from the critical band center frequency was taken for our experiments for maskee TF functions. TF functions of each time-width were used as maskers to measure their masking capabilities on the remaining of each time-width TF functions for all the above four different frequency sets. Both (masker and maskee TF functions) were placed apart with 10 ms duration and played to the listeners. Each time the amplitude of the maskee TF function was reduced till the listener perceived only the masker TF function, or in other words, until there was no difference observed between the masker TF function played individually or played together with the maskee TF function. At this point, the masker TF function's energy was sufficient to mask the maskee TF function. The difference in their energies is calculated in dB and used as the masking threshold for the particular time-width maskee TF function when occurring simultaneously with that

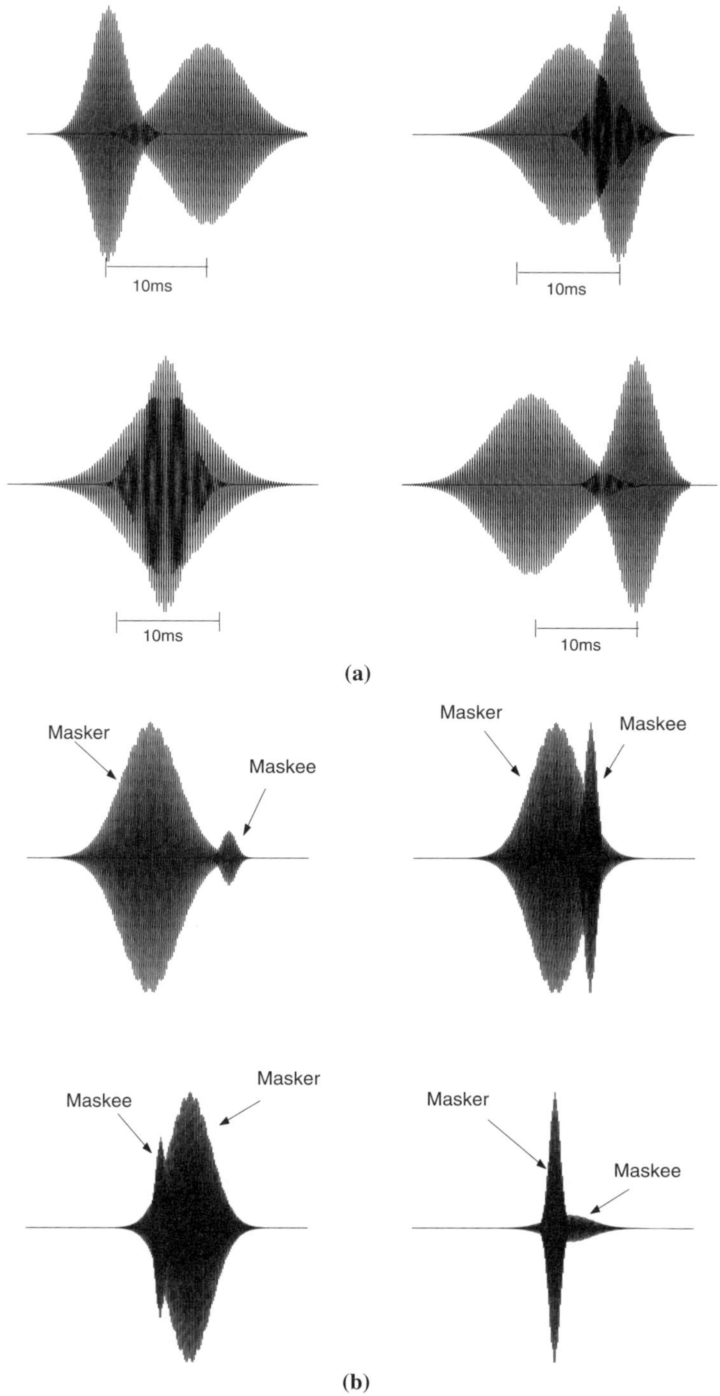

Fig. 19.21 (**a**) Illustration of few possible time occurrences of two TF functions as masker and maskee, (**b**) Possible masking conditions that can occur within the 10 ms time slot

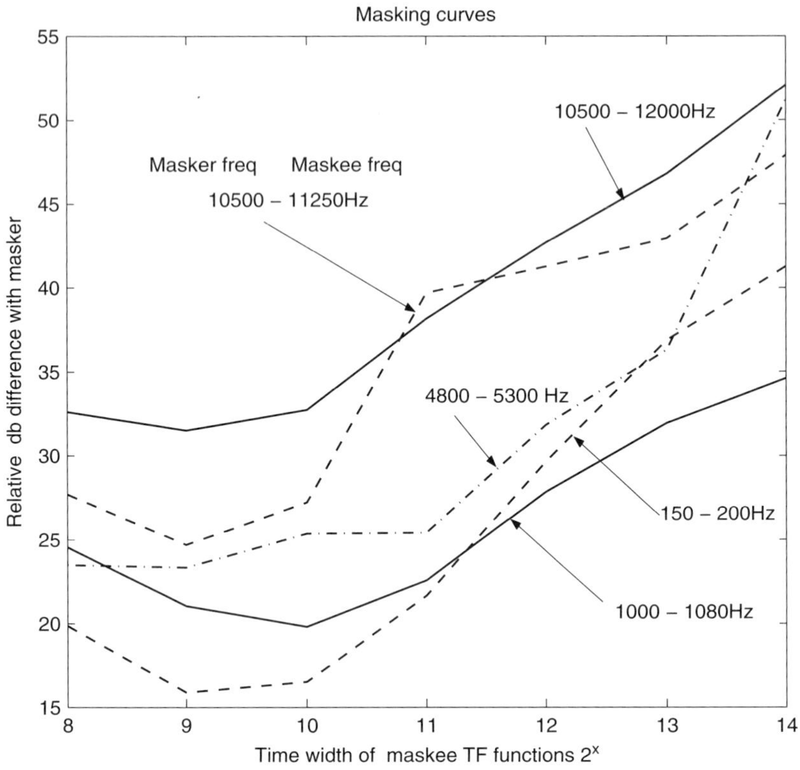

Fig. 19.22 Masking curves for different time-width of TF functions

particular time-width masker TF function. Once all the measurements were finished, each time-width TF function was analyzed as a maskee against all the remaining time-width TF functions as masker. An average energy difference was computed for each time-width TF function below which they will be masked by any other time-width TF functions. Five different listeners participated in the test and their average masking curves for each time-width of TF functions were computed. Figure 19.22 shows the different masking curves obtained for different durations of TF functions. The *X* axis represents the different time-width TF functions and the *Y* axis represents the relative energy difference with the masker in dB.

The masking curve obtained for critical band center frequency 10.5 kHz deviates from the remaining curves considerably. This is because the frequency separation between the masker and the maskee becomes very high at this band. This is because we used for all our experiments the upper limit of the critical band as the maskee frequency to simulate the worst case scenario. To demonstrate this frequency separation dependence on masking performance, a second masking curve was obtained for the critical band with a center frequency of 10.5 kHz for masker but this time the frequency separation between masker and maskee was reduced by half. The curve dropped down explaining the increase in masking performance i.e., when the frequency separation between the masker and maskee was reduced, the average relative dB difference required for masking also reduces.

From these curves, it could be observed the masking curves of critical bands with center frequencies 150 Hz, 1 kHz, and 4.8 kHz remain almost the same. Hence, the masking curve obtained for 1 kHz was used as the lookup table for the first 20 critical bands. The remaining five critical bands use the masking curve obtained for the critical band with a center frequency of 10.5 kHz (with 12 kHz upper limit) as the lookup table. These lookup tables were used to verify if a TF function will be masked by the relative dB difference of it with the TF function having highest energy within the same critical band.

The flow chart in Fig. 19.23 gives an overview of the masking implementation used in the presented coder.

19.3.4 Quantization

Most of the existing transform-based coders rely on controlling the quantizer resolution, based on psychoacoustic thresholds to achieve compression. Unlike this, the presented technique achieves a major part of the compression in the transformation itself followed by perceptual filtering. That is, when the number of iterations M needed to model a signal is very low compared with the length of the signal, we just need $M \times L$ bits, where L is the number of bits needed to quantize the 5 TF parameters that represent a TF function. Hence, we limited our research work to scalar quantizers as the focus of the research mainly lies on the TF transformation block and the psychoacoustics block rather than the usual subblocks of the data compression application.

As explained earlier each of the five parameters energy (a_n), center frequency (f_n), time position (p_n), octave (s_n) and phase (ϕ_n) are needed to represent a TF function and thereby the signal itself. These five parameters were to be quantized in such a way that the quantization error introduced was imperceptible while obtaining good compression. Each of the five parameters has different characteristics and dynamic range. After careful analysis of them the following bit allocations were made. In arriving at the final bit allocations informal mean opinions score (MOS) tests were conducted to compare the quality of the audio samples before and after quantization stage.

In total, 54 bits are needed to represent each TF function without introducing significant perceptual quantization noise in the reconstructed signal. The final form of data for M TF functions will contain

- Energy parameter (Log companded) = $M \times 12$ bits
- Time position parameter = $M \times 15$ bits
- Center frequency parameter = $M \times 13$ bits
- Phase parameter = $M \times 10$ bits
- Octave parameter = $M \times 4$ bits

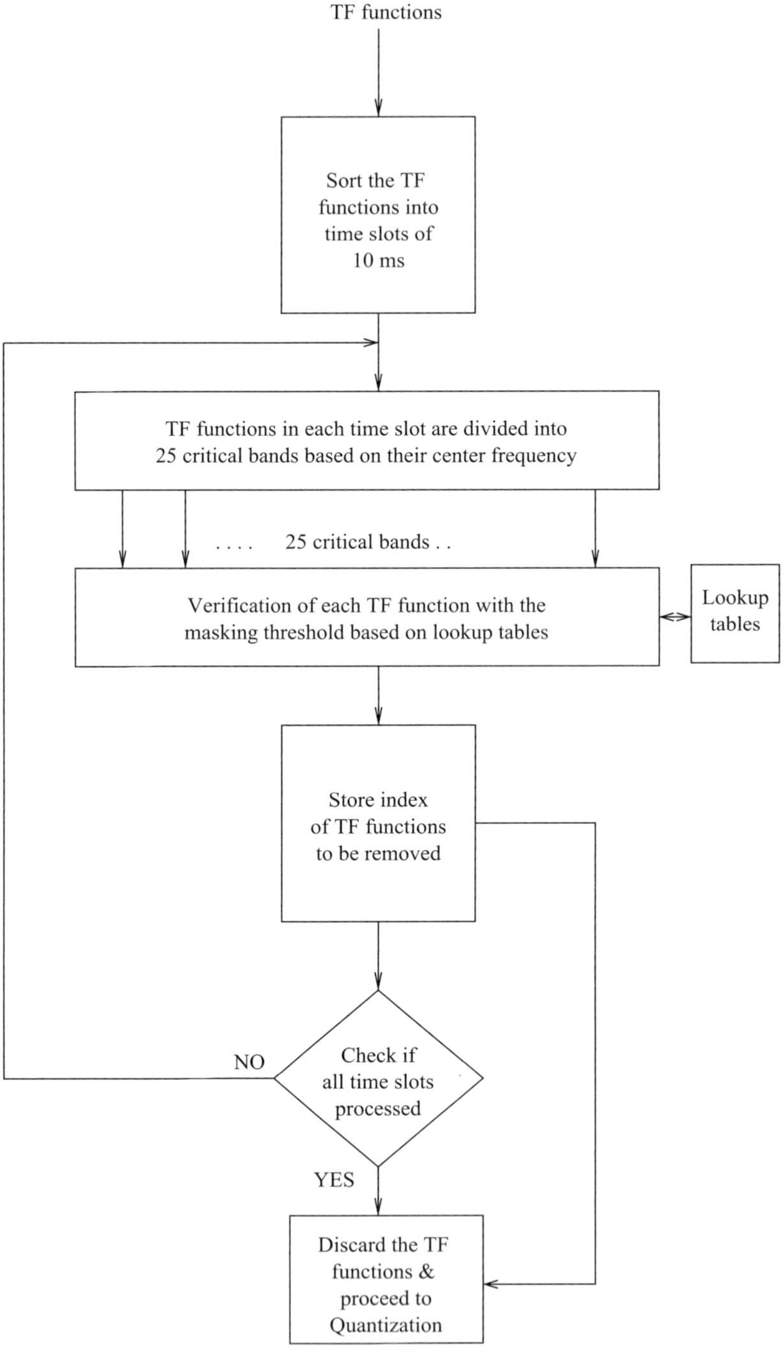

Fig. 19.23 Flow chart of the masking procedure

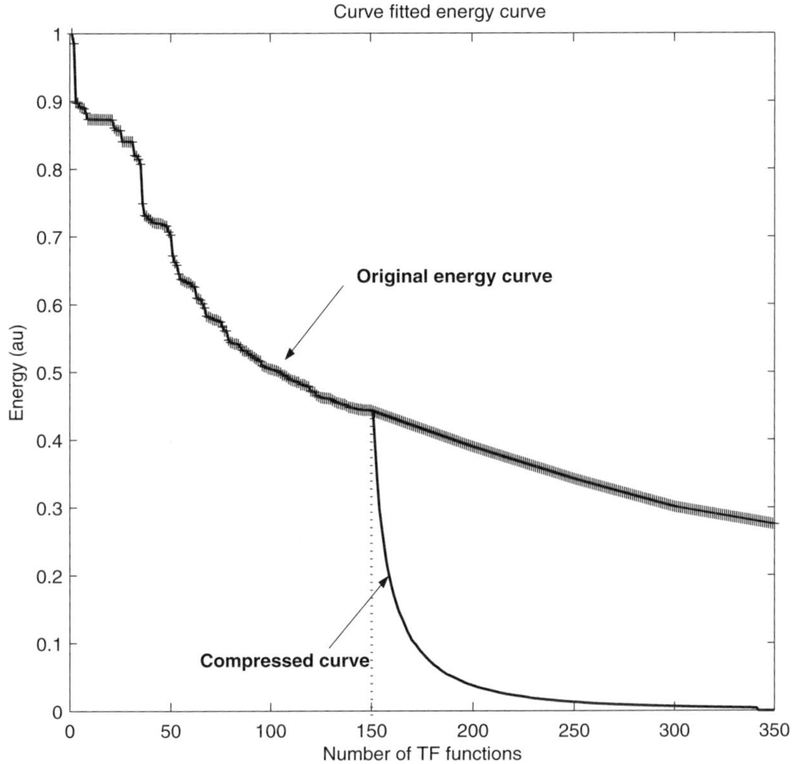

Fig. 19.24 Log companded original and curve fitted energy curve for a sample signal. au - arbitrary units

The sum of all the above ($= 54 \times M$ bits) will be the total number of bits transmitted or stored representing an audio segment of duration 5 s. The energy parameter after log companding was observed to be a very smooth curve as shown in Fig. 19.24. Fitting a curve to the energy parameter further reduces the bit rate. Nearly 90% of the energy is present in the first few 100 TF functions and hence they are not used for curve fitting. The remaining number of TF functions were divided into equal lengths of 50 points on the curve. Only the values corresponding to these 50 points need to be sent with the first few original 100 values. The distance between these 50 points can be treated as linear comparing the spread of total number of TF functions. In the reconstruction stage, these 50 points can be interpolated linearly to the original number of points. The error introduced in this procedure was very small due to the smooth slope of the curve. Moreover, this error was introduced only in the 10% energy of the signal that was not perceived. To better explain the benefit of the presented curve fitting approach in reducing the bit rate, let us take an example of transmitting 5,000 TF functions. To transmit the energy parameter for 5,000 TF functions (without applying curve fitting) will require $5{,}000 \times 12$ bits = 60,000 bits. With curve fitting, say we preserve the energy parameter for the first

150 TF functions and thereafter select the energy parameter from every 50th TF function in the remaining 4,850 TF functions. This will result in [150 + (4850/50 = 97)] = 247 values of the energy parameter requiring only $247 \times 12 = 2,964$ bits for transmission. We see a massive reduction in bits due to curve fitting. Figure 19.24 demonstrates the original curve super imposed with the fitted curve. Every kth point in the compressed curve corresponds to actually the $(3+k) \times 50$th point in the original curve. A correlation value of 1 was achieved between the original curve and the interpolated reconstructed curve.

With just a simple scalar quantizer and curve fitting of the energy parameter, the presented coder achieves high compression ratios. Although a scalar quantizer was used to reduce the computational complexity of the presented coder, sophisticated vector quantization techniques can be easily incorporated to further increase the coding efficiency. The five parameters of the TF function can be treated as one vector and accordingly quantized using predefined codebooks. Once the vector is quantized, only the index of the codebook needs to be transmitted for each set of TF parameters resulting in a large reduction of the total number of bits. However, designing the codebooks would be challenging as the dynamic ranges of the 5 TF parameters are drastically different. Apart from reducing the number of total bits, the quantization stage can also be utilized to control the bit rates suitable for CBR (constant bit rate) applications.

19.3.5 Compression Ratios

Compression ratios achieved by the presented coder were computed for eight sample wideband audio signals (of 5 s duration) as described below. These eight sample signals (viz. ACDC, DEFLE, ENYA, HARP, HARPSICHORD, PIANO, TUBULARBELL, and VISIT) were representative of wide range of music types.

- As explained earlier, the total number of bits needed to represent each TF function is 54.
- The energy parameter is curve fitted and only the first 150 points in addition to the curve fitted point need to be coded.
- So the total number of bits needed for M iterations for a 5 s duration of the signal is $TB_1 = (M \times 42) + ((150 + C) \times 12)$, where C is the number of curve fitted points, and M is the number of perceptually important functions.
- The total number of bits needed for a CD quality 16 bit PCM technique for a 5 s duration of the signal sampled at 44,100 Hz is $TB_2 = 44,100 \times 5 \times 16 = 3,528,000$.
- The compression ratio can be expressed as the ratio of number of bits needed by the presented coder to the number of bits needed by the CD quality 16 bit PCM technique for the same length of the signal, i.e,

$$\text{Compression ratio} = \frac{\text{TB}_2}{\text{TB}_1}$$

- The overall compression ratio for a signal was then calculated by averaging all the 5 s duration segments of the signal for both the channels.

The presented coder is based on an adaptive signal transformation technique, i.e., the content of the signal and the dictionary of basis functions used to model the signal plays an important role in determining how compact a signal can be represented (compressed). Hence, VBR (variable bit rate) is the best way to present the performance benefit of using an adaptive decomposition approach. The inherent variability introduced in the number of TF functions required to model a signal and thereby the compression is one of the highlights of using ATFT. Although VBR would be more appropriate to present the performance benefit of the presented coder, CBR mode has its own advantages when using with applications that demand network transmissions over constant bitrate channels with limited delays. The presented coder can also be used in CBR mode by fixing the number of TF functions used for representing signal segments; however, due to the signal adaptive nature of the presented coder, this would compromise the quality at instances where signal segments demand a higher number of TF functions for perceptually lossless reproduction. Hence, we choose to present the results of the presented coder using only the VBR mode.

We compared the presented coder with two existing popular and state-of-the-art audio coders viz MP3 (MPEG 1 layer 3) and MPEG-4 AAC/ HE-AAC. Advanced audio coding (AAC) is the current industrial standard that was initially developed for multichannel surround signals (MPEG-2 AAC [9]). The transformation technique used is the modified discrete cosine transform (MDCT). Compared with mp3, which uses a polyphase filter bank and an MDCT, new coding tools were introduced to enhance the performance. The core of MPEG-4 AAC is basically the MPEG-2 AAC but with added tools to incorporate additional coding enhancements and MPEG-4 features so that a broad range of applications are covered. There are many application-specific profiles that can be chosen to adaptively configure the MPEG-4 audio for the user needs. It is claimed that at 128 kbps the MPEG-4 AAC is indistinguishable from the original audio signal [10]. As there are ample studies in the literature [9, 11–15] available for both MP3 and MPEG-2/4 AAC, more details about these techniques are not provided in this chapter. The average bit-rates were used to calculate the compression ratio achieved by MP3 and MPEG-4 AAC as described below.

- Bitrate for a CD quality 16 bit PCM technique for 1 s stereo signal is given by $TB_3 = 2 \times 44100 \times 16$.
- The average bit rate/s achieved by (MP3 or MPEG-4 AAC) in VBR mode = TB_4.
- Compression ratio achieved by (MP3 or MPEG-4 AAC) = $\frac{TB_3}{TB_4}$.

The 2nd, 4th, and 6th columns of Table 19.2 show the compression ratio (CR) achieved by the MP3, MPEG-4 AAC, and the presented ATFT coders for the set of 8 sample audio files. It is evident from the table that the presented coder has better compression ratios than MP3. When comparing with MPEG-4 AAC, 5 out of 8 signals are either comparable or have better compression ratios than the MPEG-4

AAC. It is noteworthy to mention that for slow music (classical type) the ATFT coder provides 3–4 times better comparison than MPEG-4 AAC or MP3.

The compression ratio alone cannot be used to evaluate a audio coder. The compressed audio signals has to undergo a subjective evaluation to compare the quality achieved with respect to the original signal. The combination of the subjective rating and the compression ratio will provide a true evaluation of the coder performance.

Before performing the subjective evaluation, the signal has to be reconstructed. The reconstruction process is a straight forward process of linearly adding all the TF functions with their corresponding five TF parameters. To do that, first the TF parameters modified for reducing the bit rates have to be expanded back to their original forms. The log compressed energy curve was log expanded after recovering back all the curve points using interpolation on the equally placed 50 length points. The energy curve was multiplied with the normalization factor to bring the energy parameter as it was during the decomposition of the signal. The restored parameters (energy, time-position, center frequency, phase, and octave) were fed to the ATFT algorithm to reconstruct the signal. The reconstructed signal was then smoothed using a 3rd order Savitzky-Golay [16] filter and saved in a playable format.

Figure 19.25 demonstrates a sample signal (/'HARP'/) and its reconstructed version and the corresponding spectrograms. It can be clearly observed from the reconstructed signal spectrogram compared with the original signal spectrogram, how accurately the ATFT technique has filtered out the irrelevant components from the signal (evident from Table 19.2 - (/'HARP'/) - high compression ratio vs. acceptable quality). The accuracy in adaptive filtering of the irrelevant components is made possible by the TF resolution provided by the ATFT algorithm.

19.3.6 Subjective Evaluation of ATFT Coder

Subjective evaluation of audio quality is needed to assess the audio coder performance. Even though there are objective measures such as SNR, total harmonic distortion (THD), and noise-to-mask ratio [17], they would not give a true evaluation of the audio codec particularly if they use lossy schemes as in the presented technique. This is because say for example in a perceptual coder, SNR is low; however, audio quality is claimed to be perceptually lossless. In this case, SNR measure may not give the correct performance evaluation of the coder.

We used the subjective evaluation method recommended by ITU-R standards (BS. 1116). It is called a "double blind triple stimulus with hidden reference" [4, 17]. In this method, listeners are provided with three stimuli A, B, and C for each sample under test. A is the reference/original signal, B and C are assigned to either of the reference/original signal or the compressed signal under test. Basically, the reference signal is hidden in either B or C and the other choice is assigned to the compressed (or impaired) signal. The choice of reference or compressed signal for B and C is completely randomized. Figure 19.26 explains the choices A, B, and C. For each sample audio signal, listeners listen to all three (A, B, C) stimuli, and

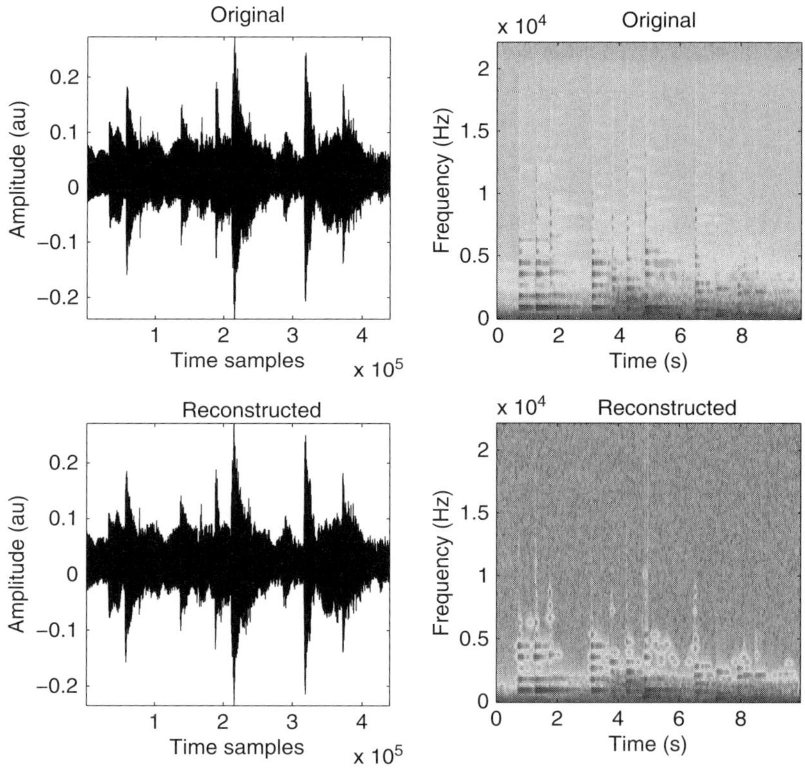

Fig. 19.25 Example of a sample original (/'HARP'/) and the reconstructed signal with their respective spectrograms. *X*-axes for the original and reconstructed signal are in time samples, and *X*-axes for the spectrogram of the original and the reconstructed signal are in equivalent time in seconds. Note that the sampling frequency = 44.1 kHz. au - arbitrary units

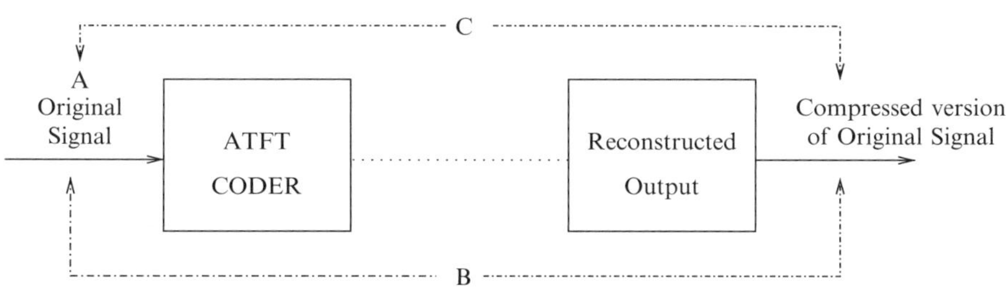

Fig. 19.26 Block diagram explaining MOS choices A, B, and C

compare A with B and A with C. After each comparison of A with B, and A with C, they grade the quality of the B and C signals with respect to A in 5 levels from 1 to 5 as shown in Table 19.1. The levels 1–5 corresponds to (1) unsatisfactory or very annoying, (2) poor or annoying, (3) fair or slightly annoying, (4) good or perceptible but not annoying, and (5) excellent or imperceptible [4, 17].

Table 19.1 Description of the ratings used in the mean opinion score

MOS	Audio quality	Level of distortion
5	Excellent	Imperceptible
4	Good	Just perceptible but not annoying
3	Fair	Perceptible and slightly annoying
2	Poor	Annoying but not objectionable
1	Unsatisfactory	Very annoying and objectionable

Table 19.2 Compression ratio (CR) and subjective difference grades (SDG)

Samples	MP3		AAC		ATFT	
	CR	SDG	CR	SDG	CR	SDG
ACDC	7.5	0.067	9.3	−0.067	8.4	−0.93
DEFLE	7.7	−0.2	9.5	−0.067	8.3	−1.73
ENYA	9	0	9.6	−0.133	20.6	−0.8
HARP	11	−0.067	9.4	−0.067	36.3	−1
HARPSICHORD	8.5	−0.067	10.2	0.33	9.3	−0.73
PIANO	13.6	0.067	9.6	−0.2	40	−0.8
TUBULARBELL	8.3	0	10.1	0.067	10.5	−0.53
VISIT	8.4	−0.067	11.5	0	11.6	−2.27
AVERAGE	9.3	−0.03	9.9	−0.02	18.3	−1.1

MP3 moving picture experts group I layer 3, *AAC* MPEG-4 AAC, *Moving Picture* Experts Group 4 Advanced audio coding VBR Main LTP profile, ATFT Adaptive time-frequency transform

A subjective difference grade (SDG) [4] was computed by subtracting the absolute score assigned to the hidden reference from the absolute score assigned to the compressed signal. It is given by 19.8

$$\text{SDG} = \text{Grade}_{\{\text{compressed}\}} - \text{Grade}_{\{\text{reference}\}} \tag{19.8}$$

Accordingly the scale of SDG will range from (−4 to 0) with the following interpretation, (−4) unsatisfactory or very annoying, (−3) poor or annoying, (−2) fair or slightly annoying, (−1) good or perceptible but not annoying, and (0) excellent or imperceptible. Fifteen listeners (randomly selected) participated in the MOS studies and evaluated all the three audio coders (MP3, AAC, and ATFT in VBR mode). The average SDG was computed for each of the audio sample. The 3rd, 5th, and 7th columns of the Table 19.2 show the SDGs obtained for MP3, AAC, and ATFT coders, respectively. MP3 and AAC SDGs fall very close to the Imperceptible (0) region, whereas the proposed ATFT SDGs are spread out between −0.53 and −2.27.

19.3.7 Results and Discussion

The compression ratios (CR) and the SDG for all three coders (MP3, AAC, and ATFT) are shown in Table 19.2. All the coders were tested in the VBR mode. For the presented technique, VBR was the best way to present the performance benefit of using an adaptive decomposition approach. In ATFT, the type of the signal and the characteristics of the TF functions (type of dictionary) control the number of transformation parameters required to approximate the signal and thereby the compression ratio. The inherent variability introduced in the number of TF functions required to model a signal is one of the highlights of using ATFT. Hence, we choose to present comparison of the coders in the VBR mode.

The results show that the MP3 and AAC coders perform well with excellent SDG scores (imperceptible) at a compression ratio around 10. The presented coder does not perform well with all of the eight samples. Out of the 8 samples, 6 samples have an SDG between -0.53 and -1 (imperceptible - perceptible but not annoying) and 2 samples have SDG below -1. Out of the 6 samples with SDGs between (-0.53 and -1), 3 samples (ENYA, HARP, and PIANO) have compression ratios 2–4 times higher than MP3 and AAC, and 3 samples (ACDC, HARPSICHORD, and TUBULARBELL) have comparable compression ratios with moderate SDGs.

Figure 19.27 shows the comparison of all three coders by plotting the samples with their SDGs in *X* axis and compression ratios in the *Y* axis. If we can virtually divide this plot in segments of SDGs (horizontally) and the compression ratios (vertically), then the ideal desirable coder performance should be in the right top corner of the plot (high compression ratios and excellent SDG scores). This is followed next by the right bottom corner (low compression ratios and excellent SDG scores) and so on as we move from right to left in the plot. Here the terms "Low" and "High" compression ratios are used in a relative sense on the basis of the compression ratios achieved by all the three coders in this study. From the plot, it can be seen that the MP3 and AAC coders occupy the right bottom corner, whereas the samples from ATFT coder are spread over. As mentioned earlier, 3 out the 8 samples of the ATFT coder occupy the right top corner, however, only with moderate SDGs that are much less than the MP3 and the AAC. Three out of the remaining five samples of the ATFT coder occupy the right bottom corner, however, again with only moderate SDGs that are less than MP3 and AAC. The remaining two samples perform the worst occupying the left bottom corner.

We analyzed the poorly performing ATFT-coded signals DEFLE and VISIT. DEFLE is a rapidly varying rock like signal with minimal voice components, and VISIT is a signal with dominant voice components. We observed that the symmetrical and smooth Gaussian dictionary used in this study does not model the transients well, which are the main features of all rapidly varying signals like DEFLE. This inefficient modeling of transients by the symmetrical Gaussian TF functions resulted in the poor SDG for the DEFLE. A more appropriate dictionary would be a damped sinusoids dictionary [8], which can better model the transient as decaying structures in audio signals. However, a single dictionary alone may not be sufficient to model all types of signal structures. The second signal VISIT has significant amount(s) of

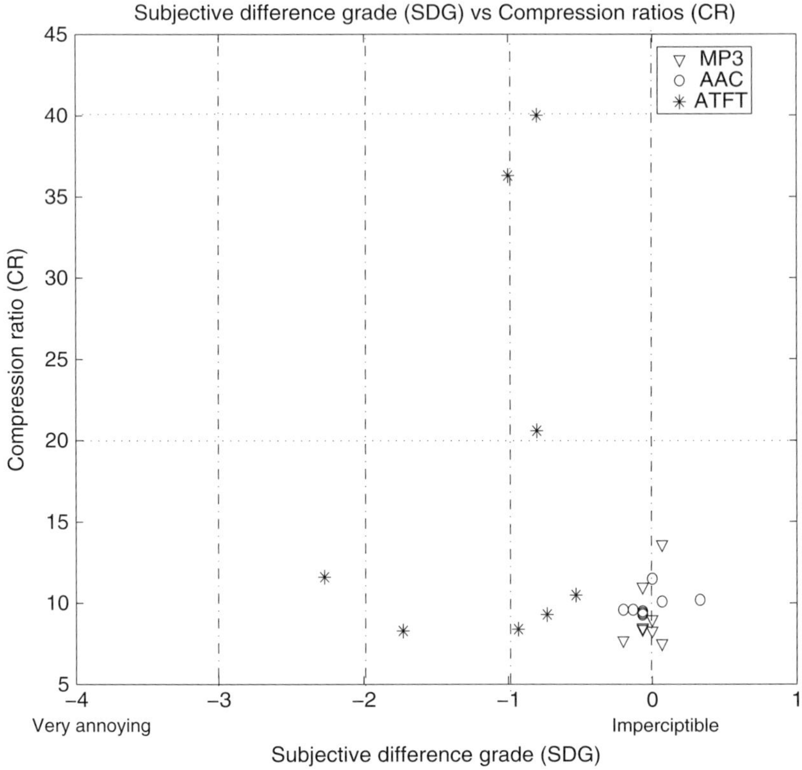

Fig. 19.27 Subjective difference grade (SDG) vs. compression ratios (CR)

voice components. Even though the main voice components are modeled well by the ATFT, the noise such as hissing and shrilling sounds (noncoherent structures) could not be modeled within the decomposition limit of 10,000 iterations. These hissing and shrilling sounds actually add to the pleasantness of the music. Any distortion in them is easily perceived, which could have reduced the SDG of the signal to the lowest of the group -2.27. The poor performances with the two audio sample cases could be addressed by using a hybrid dictionary of TF functions and residue coding the noncoherent structures separately. However, this would increase the computational complexity of the coder and reduce the compression ratios.

We have covered most details involved in a stage by stage implementation and evaluation of a transform-based audio coder. The approach demonstrated the application of ATFT for audio coding and the development of a novel psychoacoustics model adapted to TF functions. The compression strategy was changed from the conventional way of controlling quantizer resolution to achieving majority of the compression in the transformation itself. Listening tests were conducted and the performance comparison of the presented coder with MP3 and AAC coders were presented. From the preliminary results, although the proposed coder achieves high compression ratios, its SDG scores are well below the MP3 and AAC family of coders. The proposed coder, however, performs moderately well for slowly varying classical like signals with acceptable SDGs. The proposed coder is not as refined as the state-of-the-art commercial coders, which to some extent explains its poor performance.

From the results presented for the ATFT coder, the signal adaptive performance of the coder for a specific TF dictionary is evident i.e. with a Gaussian TF dictionary the coder performed moderately well for slow-varying classical like signals than fast varying rock like signals. In other words, the ATFT algorithm demonstrated notable differences in the decomposition patterns of classical like and rock like signals. This is a valid clue and a motivating factor that these differences in the decomposition patterns if quantified using TF decomposition parameters could be used as discriminating features for classifying audio signals. We apply this hypothesis in extracting TF features for classifying audio signals for a content-based audio retrieval application as will be explained in the following section.

19.4 Audio Classification

Audio feature extraction plays an important role in analyzing and characterizing audio content. Auditory scene analysis, content-based retrieval, indexing, and fingerprinting of audio are few of the applications that require efficient feature extraction. The general methodology of audio classification involves extracting discriminatory features from the audio data and feeding them to a pattern classifier. Different approaches and various kinds of audio features were proposed with varying success rates. Audio feature extraction serves as the basis for a wide range of applications in the areas of speech processing [18], multimedia data management and distribution [19–22], security [23], biometrics and bioacoustics [24]. The features can be extracted either directly from the time domain signal or from a transformation domain depending upon the choice of the signal analysis approach. Some of the audio features that have been successfully used for audio classification include mel-frequency cepstral coefficients (MFCC) [21, 22], spectral similarity [25], timbral texture [22], band periodicity [19], LPCC (linear prediction coefficient derived cepstral coefficients) [26], zero crossing rate [19, 26], MPEG-7 descriptors [27], entropy [28], and octaves [20]. Few techniques generate a pattern from the features and use it for classification by the degree of correlation. Few other techniques use the numerical values of the features coupled to statistical classification methods.

In this section we present a content-based audio retrieval application employing audio classification and explain the generic steps involved in performing successful audio classification. The simplest of all retrieval techniques is the text-based searching where the information about the multimedia data is stored with the data file. However, the success of these type of text-based searches depend on how well they are text indexed by the author and they do not provide with any information on the real content of the data. To make the retrieval system automated, efficient, and intelligent, content-based retrieval techniques were introduced. The presented work

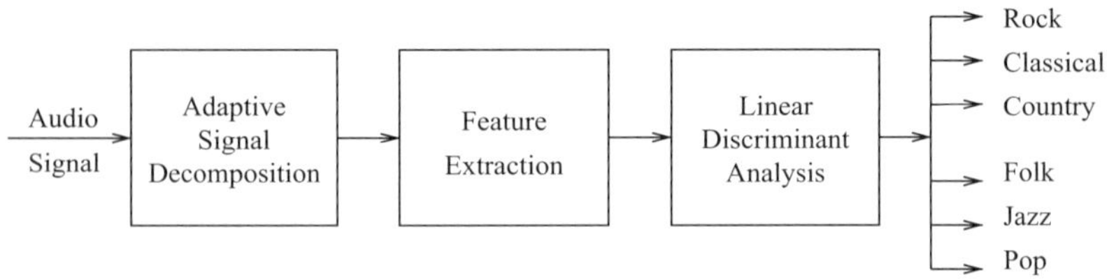

Fig. 19.28 Block diagram of the proposed audio classification scheme

focuses on one such ways for automatic classification of audio signals for retrieval purposes. The block diagram of the proposed technique is shown in Fig. 19.28.

In content-based retrieval systems, audio data are analyzed, and discriminatory features are extracted. The selection of features depends on the domain of analysis and the perceptual characteristics of the audio signals under consideration. These features are used to generate subspaces dividing the audio signal types to fit in one of the subspaces. The division of subspaces and the level of classification vary from technique to technique. When a query is placed, the similarity of the query is checked with all subspaces, and the audio signals from the highly correlated subspace are returned as the result. The classification accuracy and the discriminatory power of the features extracted determine the success of such retrieval systems.

Most of the existing techniques do not take into consideration the true nonstationary behavior of the audio signals while deriving their features. The presented approach uses the same ATFT transform that was discussed in the previous audio coding section. ATFT approach is one of the best ways to handle nonstationary behavior of the audio signals and also due to its adaptive nature, does not require any signal segmentation techniques as used by most of the existing techniques. Unlike many existing techniques where multiple features are used for classification, in the proposed technique only one TF decomposition parameter is used to generate a feature set from different frequency bands for classification. Because of its strong discriminatory power, just one TF decomposition parameter is sufficient enough for accurate classification of music into six groups.

19.4.1 Audio Database

A database consisting of 170 audio signals were used in the proposed technique. Each audio signal is a segment of 5 s duration extracted from individual original CD music tracks (wide band audio at 44,100 samples/second) and no more than one audio signal (5 s duration) was extracted from the same music track. The 170 audio signals consist of 24 rock, 35 classical, 31 country, 21 jazz, 34 folk, and 25 pop signals. As all signals of the database were extracted from commercial CD music tracks, they exhibited all the required characteristics of their respective music genre, such as guitars, drumbeats, vocal, and piano. The signal duration of 5 s was arrived at using the following rationale that the longer the audio signal analyzed, the better the extracted feature, which exhibits more accurate music characteristics. As the ATFT algorithm is adaptive and does not need any segmentation, theoretically there is no limit for the signal length. However, considering the hardware (Pentium III @ 933 MHz and 1.5 GB RAM) limitations of the processing facility, we used 5 s duration samples. In the proposed technique, first all the signals were chosen between 15 and 20 s of the original music tracks. Later by inspection those segments, which were inappropriately selected were replaced by segments (5 s duration) at random locations of the original music track in such way their music genre are exhibited.

19.4.2 Feature Extraction

All the signals were decomposed using the ATFT algorithm. The decomposition parameters provided by the ATFT algorithm were analyzed, and the octave s_n parameter was observed to contain significant information on different types of music signals. In the decomposition process, the octave or scaling parameter is decided by the adaptive window duration of the Gaussian function that is used in the best possible approximation of the local signal structures. Higher octaves correspond to longer window durations and the lower octaves correspond to shorter window duration. In other words, combinations of these octaves represent the envelope of the signal. The envelope (temporal structures) [29] of an audio signal provides valid clues such as rhythmic structure [22], indirect pitch content [22], phonetic composition [6], tonal and transient contributions. Figure 19.29 demonstrates a sample piece of a music signal and its reconstructed version using 10 TF functions. The relation between the octave parameter and the envelope of the signal is clearly seen. On the basis of the composition of different structures in a signal, the octave mapping or distribution varies significantly. For example, more lower order octaves are needed for signals containing lot of transient like structures and in contrast more higher order octaves are needed for signal containing rhythmic tonal components. As an illustration, from Fig. 19.30 it can be observed that signals with similar spectral characteristics exhibit a similar pattern in their octave distribution. Signals 1 and 2 are rock-like music, whereas Signals 3 and 4 are instrumental classical. Comparing the spectrograms with the octave distributions, one can observe that the octave distribution reflecting the spectral similarities for the same category of signals.

To further improve the discriminatory power of this parameter, the distribution of this parameter is grouped into three frequency bands 0–5, 5–10, and 10–20 kHz. This is done since analyzing the audio signals in subbands will provide more precise information about their audio characteristics [30]. The bounds for frequency bands were arrived considering the fact that most of the audio content lies well within 10 kHz range so this band needs to be looked more in detail hence broken further into 0–5 kHz and 5 kHz to 10 kHz and the remaining as one band between 10 and 20 kHz. By this frequency division, we get an indirect measure of signal envelope

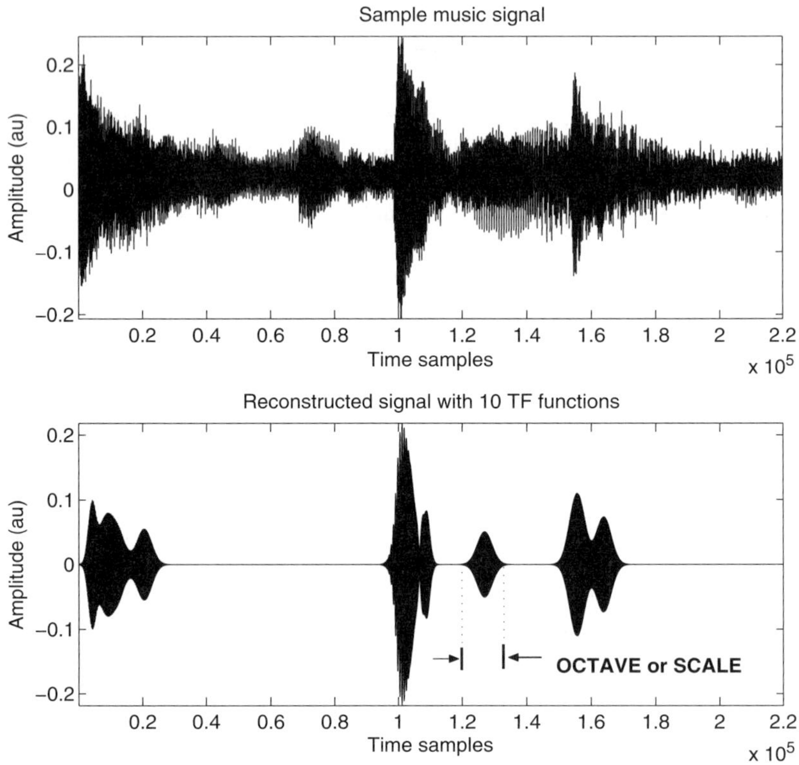

Fig. 19.29 A sample music signal, and its reconstructed version with 10 TF functions

contribution from each frequency band. From Fig. 19.30 even though we see difference in the distribution of octaves between rock-like and classical-like music, it becomes more evident when the distribution is divided into three frequency bands as shown for a sample rock and a classical signal in Figs. 19.31 and 19.32. Dividing the octave distribution into frequency bands basically reveal the pattern in which the temporal structures occur over the range of frequencies. As music is the combination of different temporal structures with different frequencies occurring at same or different time instants, each type of music exhibit an unique average pattern. On the basis of the subtle differences between patterns to be detected, the division of octave distribution over fine frequency intervals and the dimension of the feature set can be controlled.

After decomposing all the audio signals using ATFT, the TF functions were grouped into three frequency bands on the basis of their center frequencies f_n. Then the distribution of each of the 14 octave parameter s_n values were calculated over the three frequency bands to get a total of $14 \times 3 = 42$ different distribution values. All these 42 values of each audio segment was used as a feature set for classification. As an illustration, in Figs. 19.31 and 19.32 the X-axis represents the 14 octave

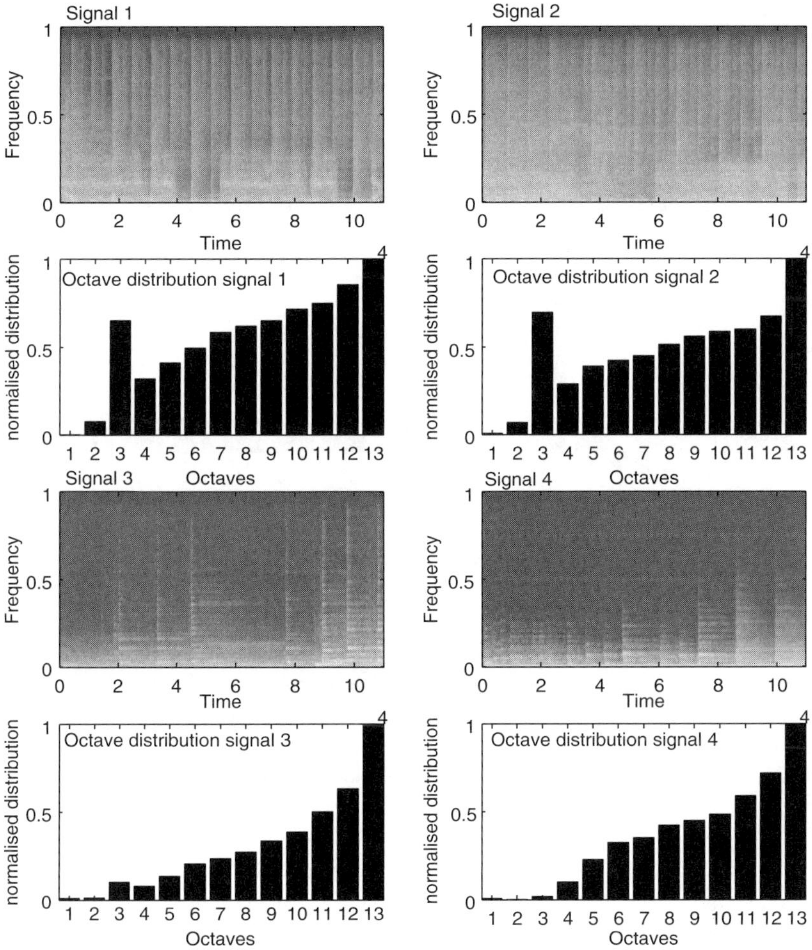

Fig. 19.30 Comparison of octave distributions. Signals 1 & 2 : Rock-like signals, and Signals 3 & 4 : Classical-like signals

parameters and the *Y*-axis represents the distribution of the octave parameters over three frequency bands for 10,000 iterations. Each of the distribution value forms one of 42 elements in the feature set.

19.4.3 Pattern Classification

The motivation for the pattern classification is to automatically group audio signals of same characteristics, using the discriminatory features derived as explained in previous subsection.

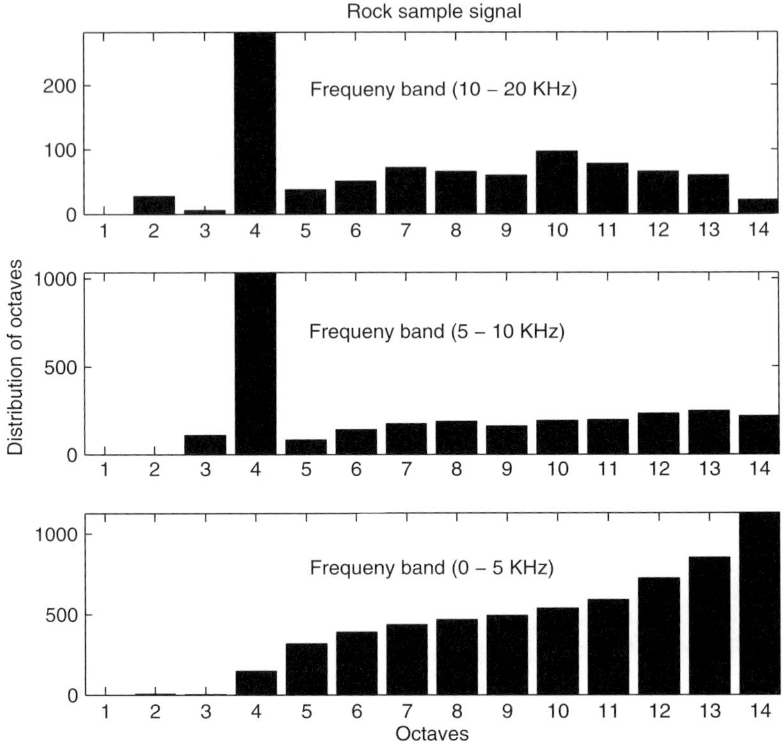

Fig. 19.31 Octave distribution over three frequency bands for a rock signal

Pattern classification was carried out by linear discriminant analysis (LDA)-based classifier-using the SPSS software [31]. In discriminant analysis, the feature vector derived as explained above were transformed into canonical discriminant functions such as

$$f = u_1 b_1 + u_2 b_2 + \ldots\ldots. + u_q b_q + a, \tag{19.9}$$

where $\{\mathbf{u}\}$ is the set of features, $\{\mathbf{b}\}$ and a are the coefficients and constant, respectively. The feature dimension q represents the number of features used in the analysis. Using the discriminant scores and the prior probability values of each group, the posterior probabilities of each sample occurring in each of the groups were computed. The sample was then assigned to the group with the highest posterior probability [31].

The classification accuracy was estimated using the leave-one out method, which is known to provide a least bias estimate [32]. In the leave-one-out method, one sample is excluded from the dataset, and the classifier is trained with the remaining samples. Then the excluded signal is used as the test data, and the classification accuracy is determined. This is repeated for all samples of the dataset. Since each signal is excluded from the training set in turn, the independence between the test and the training set are maintained.

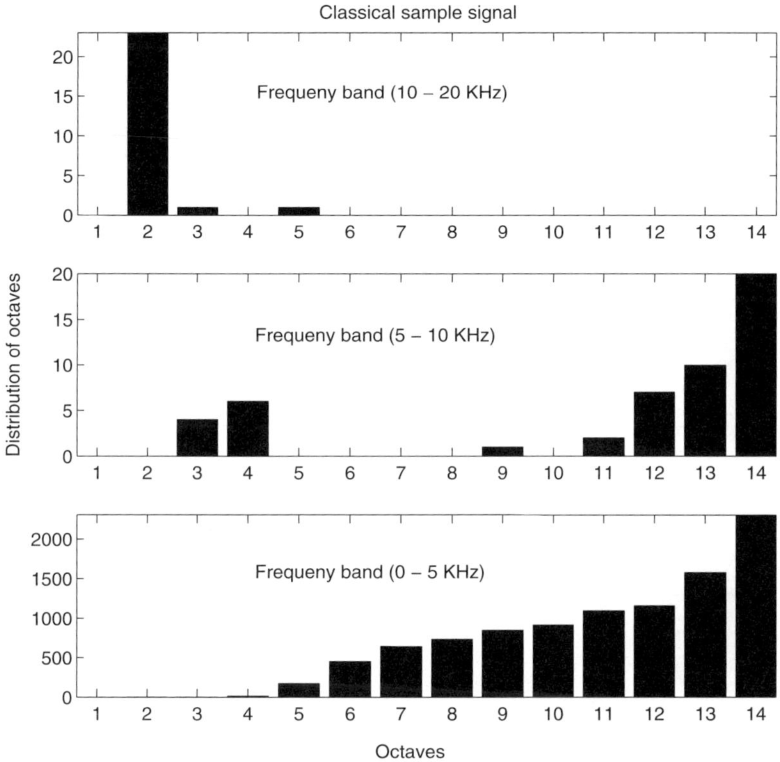

Fig. 19.32 Octave distribution over three frequency bands for a classical signal

19.4.4 Results and Discussion

A database of 170 audio signals consisting of 24 rock, 35 classical, 31 country, 21 jazz, 34 folk, and 25 pop each of 5 s duration was used. All the 170 audio signals were decomposed and the feature set of 42 octave distribution values were extracted. The extracted feature sets for the entire 170 signals were fed to the classifier based on LDA. Six-group classification was performed (rock, classical, country, jazz, folk, and pop). Table 19.3 shows the confusion matrices for different classification procedures. An overall classification accuracy of 97.6% is achieved by the regular LDA method and 91.2% with the leave-one-out based LDA method. In the regular LDA method, all the 24 rock, 35 classical, 31 country, and 25 pop were correctly classified with 100% classification accuracy. Two out of 21 jazz and 2 out of 34 folk signals were misclassified with an correct classification accuracy of 90.5% and 94.1%, respectively. The classification accuracy of 91.2% with the leave-one-out method proves the robustness of the proposed technique and independence of the achieved results irrespective of the dataset size. Figure 19.33 shows the all-groups scatter plot with the first two canonical discriminant functions. One can clearly observe the

Table 19.3 Classification results

Method	Gr	Ro	Cl	Co	Ja	Fo	Po	*CA%*
Regular	Ro	24	0	0	0	0	0	100
	Cl	0	35	0	0	0	0	100
	Co	0	0	31	0	0	0	100
	Ja	0	2	0	19	0	0	90.5
	Fo	1	0	0	1	32	0	94.1
	Po	0	0	0	0	0	25	100
			Overall					97.6
Cross-Validated	Ro	23	0	1	0	0	0	95.8
	Cl	0	34	0	1	0	0	97.1
	Co	1	0	29	0	1	0	93.5
	Ja	0	3	0	18	0	0	85.7
	Fo	1	1	0	2	30	0	88.2
	Po	2	0	2	0	0	21	84
			Overall					91.2

Method: Regular - Linear discriminant analysis; Cross - validated - Linear discriminant analysis with leave-one-out method, *CA%* Classification accuracy rate; *Gr* Groups; *Ro* Rock; *Cl* Classical; *Co* Country; *Ja* Jazz; *Fo* Folk; *Po* Pop

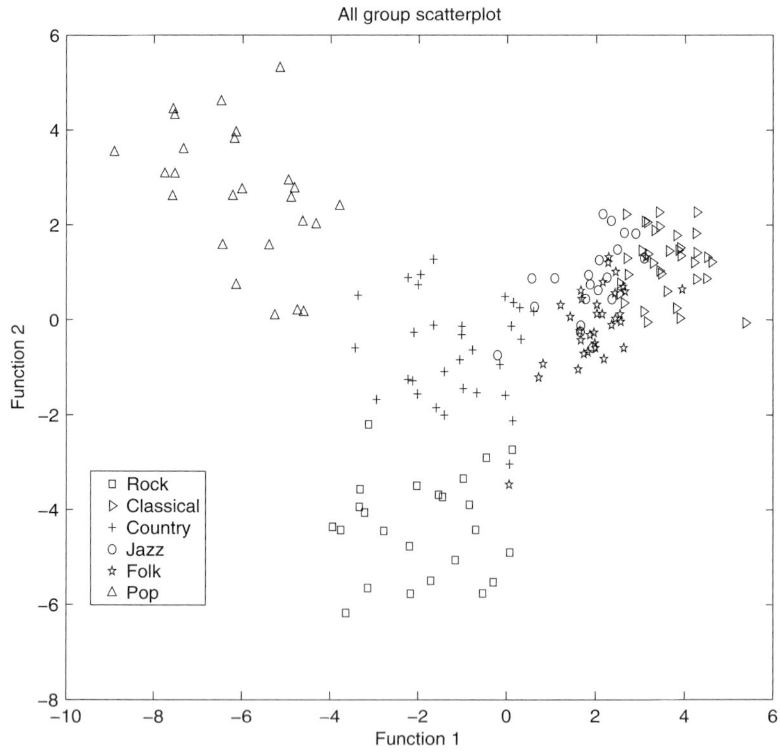

Fig. 19.33 All-groups scatter plot with the first two canonical discriminant functions

significant separation between the group spaces explaining the high discriminatory power of the feature set on the basis of the octave distribution.

The misclassified signals were analyzed but could not identify a clear auditory clue to why they were misclassified. However, their differences are observed in the feature set. Considering the known fact that no music genre has clear hard line boundaries and the perceptual boundaries are often subjective (For example, rock and pop often have overlaps and likewise jazz and classical too have overlaps), we may attribute the classification error of these signals on the natural overlap of the music genre and the amount of knowledge imparted to the classifier with the given database.

In this section, we have covered details involved in a simple audio classification task using a time-frequency approach. The high classification accuracies achieved by the proposed technique clearly demonstrates the potential of a true nonstationary tool in the form of a joint TF approach for audio classification. More interestingly, a single TF decomposition parameter is used for feature extraction proving the high discriminatory power provided by TF approach compared with the existing techniques.

19.5 Summary

In this chapter, we presented a stage-by-stage implementation of two important audio processing tasks viz. (1) audio compression and (2) audio classification, using a TF approach. The methodology used is based on a joint time-frequency (TF) approach that is best suited for analyzing highly nonstationary audio signals. Unlike many existing works, we have shown that a single adaptive TF approach can perform well for both audio compression and classification. Although the audio compression results were not on par with the state-of-the-art coders, we introduced a novel way of performing audio compression. Moreover, the proposed coder is not as refined as the state-of-the-art commercial coders, which to some extent explains its poor performance. A content-based audio retrieval application was presented to explain the basic blocks of audio classification. TF features were extracted from the audio signals and were segregated into six groups using a pattern classifier. High classification accuracies of $>90\%$ (cross validated) were reported, which proves the robustness of the proposed technique and the power of TF approach in processing audio signals.

References

1. Mallat S (1998) A wavelet tour of signal processing. Academic press, San Diego, CA.
2. Mallat S G and Zhang Z (1993) "Matching pursuit with time-frequency dictionaries", *IEEE Trans. Signal Processing*, 41(12): 3397–3415.

3. Cohen L (1989) "Time-frequency Distributions – a review", *Proceedings of the IEEE*, 77(7): 941–981.
4. Painter T and Spanias A (2000) "Perceptual Coding of Digital Audio", *Proceedings of the IEEE*, 88(4): 451–513.
5. Black S M (1995) "Wavelet packet coding of wideband audio signals", *M.E.Sc Thesis, University of Western Ontario*.
6. Moore B C J (1992) An Introduction to the Psychology of Hearing. Academic Press., Toronto, ON.
7. Scharf B (1970) Critical bands – Foundations of Modern Auditory Theory, vol. 1. Academic press, New York, NY.
8. Goodwin M M (1998) Adaptive Signal Models: Theory, Algorithms and Audio Applications. Kluwer Academic Publishers, Norwell, MA.
9. Brandenburg K and Bosi M (1997) "MPEG-2 Advanced Audio Coding: Overview and Applications", 103rd *Audio Engineering Society Convention, New York*, Preprint 4641.
10. http://www.apple.com/mpeg4/aac/
11. Eberlein E et al. (1993) "Layer-3, a flexible coding standard", 94th *Audio Engineering Society Convention, Berlin*, Preprint 3493.
12. Herre J et al. (1995) "Second generation ISO/MPEG audio layer-3 coding", 98th *Audio Engineering Society Convention, Paris*.
13. JTC1/SC29/WG11 I (2002) "Overview of the MPEG-4 Standard", *International Organization for Standardization*.
14. http://www.iis.fraunhofer.de/amm/techinf/index.html .
15. Meltzer S and Moser G (2006) MPEG-4 HE-AAC v2 audio coding for todays digital media world, *EBU Technical Review*.
16. Orfanidis S J (1996) Introduction to Signal Processing. Prentice-hall, New Jersey, NJ.
17. Ryden T (1996) "Using Listening Tests to Assess Audio Codecs", *Collected Papers on Digital Audio Bit-Rate Reduction, AES*, 115–125.
18. Campbell-Jr. J P (1997) "Speaker recognition: a tutorial", *Proceedings of the IEEE*, 85(9): 1437–1462.
19. Lu L and Zhang H-J (2002) "Content Analysis for Audio Classification and Segmentation", *IEEE Transactions on Speech and Audio Processing*, 10(7): 504–516.
20. Umapathy K, Krishnan S and Jimaa S (2005) "Multi-group classification of audio signals using time-frequency parameters", *IEEE Transactions on Multimedia*, 7(2): 308-315.
21. Guo G and Li S Z (2003) "Content-based audio classification and retrieval by support vector machines", *IEEE Transactions on neural networks*, 14(1): 209–215.
22. Tzanetakis G and Cook P (2002) "Music Genre Classification of Audio Signals", *IEEE Transactions on Speech and Audio Processing*, 10(5): 293–302.
23. Burges C J, Platt J C and Jana S (2003) "Distortion discriminant analysis for audio fingerprinting", *IEEE Transactions on Speech and Audio Processing*, 11(3): 165–174.
24. Dugelay J L et al. (2002) "Recent advances in biometric person authentication", *IEEE International Conference on Acoustics, Speech, and Signal Processing, 2002. Proceedings. (ICASSP '02).*, 4: 4060–4063.
25. Cooper M and Foote J (2003) "Summarizing popular music via structural similarity analysis", *IEEE Workshop on Applications of Signal Processing to Audio and Acoustics, 2003*, 127–130.
26. Xu C, Maddage N C and Shao X (2005) "Automatic music classification and summarization", *IEEE Transactions on Speech and Audio Processing*, 13(3): 441–450.
27. Kim H G, Moreau N and Sikora T (2004) "Audio classification based on MPEG-7 spectral basis representations", *IEEE Transactions on circuits and systems for video technology*, 14(5): 716–725.
28. Esmaili S, Krishnan S and Raahemifar K (2004) "Content based audio classification and retrieval using joint time-frequency analysis", *IEEE International Conference on Acoustics, Speech and Signal Processing (ICASSP)*, V: 665–668.
29. Soltau H, Schultz T, Westphal M and Waibel A (1998) "Recognition of music type", *Proc. IEEE International Conference on Acoustics, Speech and Signal Processing*, 1137–1140.

30. Allamanche E et al. (2001) "Content-based Identification of Audio Material Using MPEG-7 Low Level Description", *Proc. 2nd Annual International Symposium on Music Information Retrieval 2001*, 197–204.
31. Inc. S (1990) "SPSS Advanced Statistics User's Guide", *User manual, SPSS Inc., Chicago, IL.*
32. Fukunaga K (1990) Introduction to Statistical Pattern Recognition. Academic Press, Inc., San Diego, CA.

Chapter 20
DRA Audio Coding Standard: An Overview

Yu-Li You and Wenhua Ma

20.1 Introduction

A lossy audio coding or compression algorithm explores statistical redundancy and perceptual irrelevance of an input audio signal, which may include multiple channels, to obtain a compact representation suitable for efficient transmission or storage. Figure 20.1 is a generic architecture designed to achieve this and is the basis for most audio coding algorithms or standards. The following is a brief description of its major components:

Time-Frequency Analysis: Frequently referred to as an analysis filter bank, it transforms each channel of the input audio signal into a set of time-frequency parameters suitable for quantization and encoding so that their statistical redundancy and perceptual irrelevance can be readily exploited. It may come in the form of Fourier transform, discrete cosine transform(DCT), linear prediction, or subband filter banks. Modified discrete cosine transform (MDCT) is a filter bank widely used in audio coding standards.

Joint Channel Coding: It exploits the statistical redundancy and/or perceptual irrelevance between two or more audio channels. The most widely used techniques include sum/difference coding and joint intensity coding.

Perceptual Model: It computes the masking threshold below which audio signals or quantization noise is not audible, hence quantifies the maximum amount of distortion that the quantization unit of an audio coder can introduce without producing audible artifacts.

Global Bit Allocation: It allocates bits to groups of time-frequency parameters, based on the masking thresholds provided by the perceptual model, usually using

Y.-L. You (✉) and W. Ma
Guangdong Provincial Key Lab for Digital Audio Technologies, Digital Rise Technology Co. Ltd., 6th Floor, Bldg. 2, Science and Tech Park, South China University of Technology, Guangzhou, Guangdong, China
e-mail: yuliyou@hotmail.com

F.-L. Luo (ed.), *Mobile Multimedia Broadcasting Standards: Technology and Practice*,
DOI: 10.1007/978-0-387-78263-8_20,

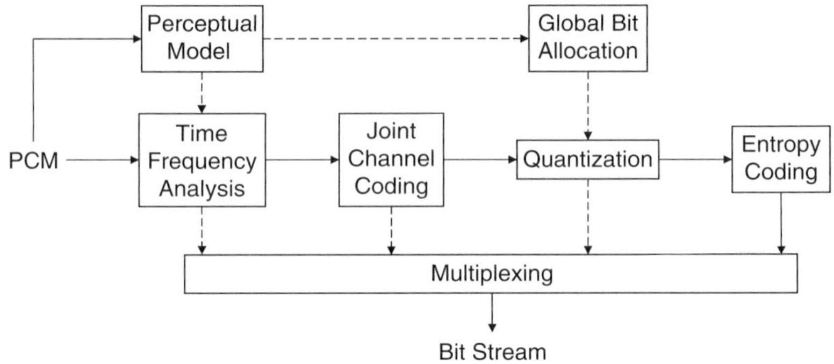

Fig. 20.1 A generic audio coder. The *solid lines* represent data flow and the *dashed lines* represent control signal or side information

a procedure in which the bit resource is iteratively allocated to the group of time-frequency parameters whose quantization noise is most audible.

Quantization: It exploits the perceptual irrelevancy by quantizing the time-frequency parameters using a step size that is usually adjusted by the global bit allocation unit. Statistical redundancy can also be exploited using techniques such as linear prediction (DPCM). The quantization can be uniform, nonlinear, or PDF (probability density function) optimized. It can be performed on either scalar or vector data.

Entropy Coding: It exploits the remaining statistic redundancy in the quantized time-frequency parameters called quantization indexes. A variety of entropy coding techniques, such as Huffman and arithmetic coding, can deployed.

Multiplexing: The entropy codes for all quantization indexes as well as side informations are packed together to form an integral bit stream.

DRA (Dynamic Resolution Adaptation) multichannel digital audio coding algorithm [1], adopted as China's national standard for its electronics industry, is essentially a bare-bone implementation of the above generic audio coder to produce a codec with low decoder complexity, which is becoming more prominent in an era of mobile devices. As shown in Fig. 20.2a, it uses transient-localized MDCT to provide improved pre-echo suppression with small bit and computation overheads. It uses statistic allocation of Huffman codebooks to enhance the coding efficiency of Huffman codes. Its quantization unit and Huffman codebooks are designed in such a way that 24-bit signal path is allowed throughout the codec so that highest audio quality can be delivered if bit rate allows. As shown in Fig. 20.2b, reconstruction of the number of quantization units and short/brief window sequencing are the only decoding blocks that would be extra when compared with a decoder derived from the generic audio coder. Since both of them can be implemented using short procedures, the complexity of DRA decoder is very low, barely more than what is necessary. Although simple, DRA standard delivers state-of-art coding efficiency as is shown by the five ITU-R BS.1116 compliant subjective listening tests.

This chapter is organized into the following sections. Section 20.2 is devoted to transient-localized MDCT and its use in DRA audio coding standard. Section 20.3

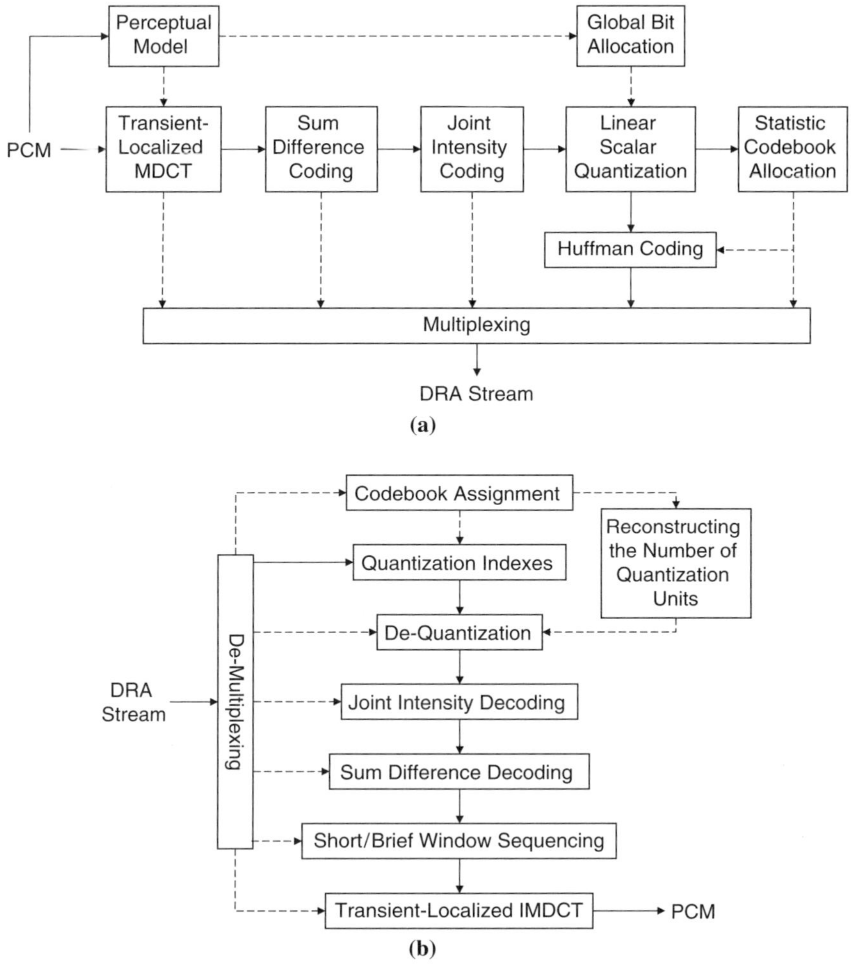

Fig. 20.2 DRA encoder (**a**) and decoder (**b**). The *solid lines* represent data flow and the *dashed lines* represent control signal or side information

addresses all other processing blocks deployed in DRA encoder and Sect. 20.4 provides an overview to DRA decoder. Results of subjective listening tests are reported in Sect. 20.5 and this chapter is concluded in Sect. 20.6.

20.2 Transient-Localized MDCT and its Use in DRA Encoder

An audio signal often consists of quasistationary episodes that are interrupted by dramatic transients. Therefore, it is desirable for an audio coding algorithm to employ a filter bank that can adapt its temporal-frequency resolution to this piecewise quasistationary nature of audio signals, i.e., having high frequency resolution during the quasistationary episodes and high temporal resolution around the transients.

One widely used filter bank is MDCT (modified cosine transform) [2, 3], which can be described by the following basis function:

$$h(k,n) = w(n)\sqrt{\frac{2}{M}}\cos\left[\frac{\pi}{M}\left(n+\frac{M+1}{2}\right)\left(k+\frac{1}{2}\right)\right], \tag{20.1}$$

where $k = 0, 1, ..., M-1$; $n = 0, 1, ..., 2M-1$; and $w(n)$ is a window function of size $2M$. A widely used window function is the following sine window:

$$w(n) = \sin\left[\left(n+\frac{1}{2}\right)\frac{\pi}{2M}\right]. \tag{20.2}$$

The temporal-frequency resolution of an MDCT is largely determined by M, which may be referred to as block size. A large M means low temporal resolution but high frequency resolution, while a small M means high temporal resolution but low frequency resolution.

To adapt its temporal-frequency resolution to the piecewise quasistationary nature of audio signals, many widely used audio coding algorithms deploy an MDCT that switches between two block sizes [3]. Figure 20.5a is a typical window sequence that frequently occurs under such a scheme [3]. This window-switching strategy was carried further in Xiph.Org Foundation's Vorbis, which can switch between two block sizes that are power-of-two from 64 to 8,192 samples [10] and in Microsoft WMA whose block sizes may be 64, 128, 256, 512, 1,024, or 2,048 samples [11]. Table 20.1 is a partial list of the audio coding algorithms utilizing this window-switching strategy.

Let us focus on the MDCT that switches between two block sizes. The window functions corresponding to these two block sizes are illustrated in Fig. 20.3a and b. In order for the MDCT to be able to properly switch between these two block sizes, the perfect reconstruction conditions [2, 3] call for the use of the three

Table 20.1 Block sizes of switched-window MDCT used by various audio coding algorithms

Algorithm	MDCT block sizes in samples
Dolby AC-2A [4]	128/512
Dolby AC-3 [5]	128/256
Sony ATRAC [6]	32/128 and 32/256
Lucent PAC [7]	128/1024
MPEG 1/2 Layer 3 [8]	6/18
MPEG 2/4 AAC [9]	128/1024
Xiph.Org Vorbis [10]	Any two from 64,128,256,512,1,024,2,048,4,096,8,192
Microsoft WMA [11]	64,128,256,512,1,024,or 2,048
Digirise DRA [1]	128/1,024

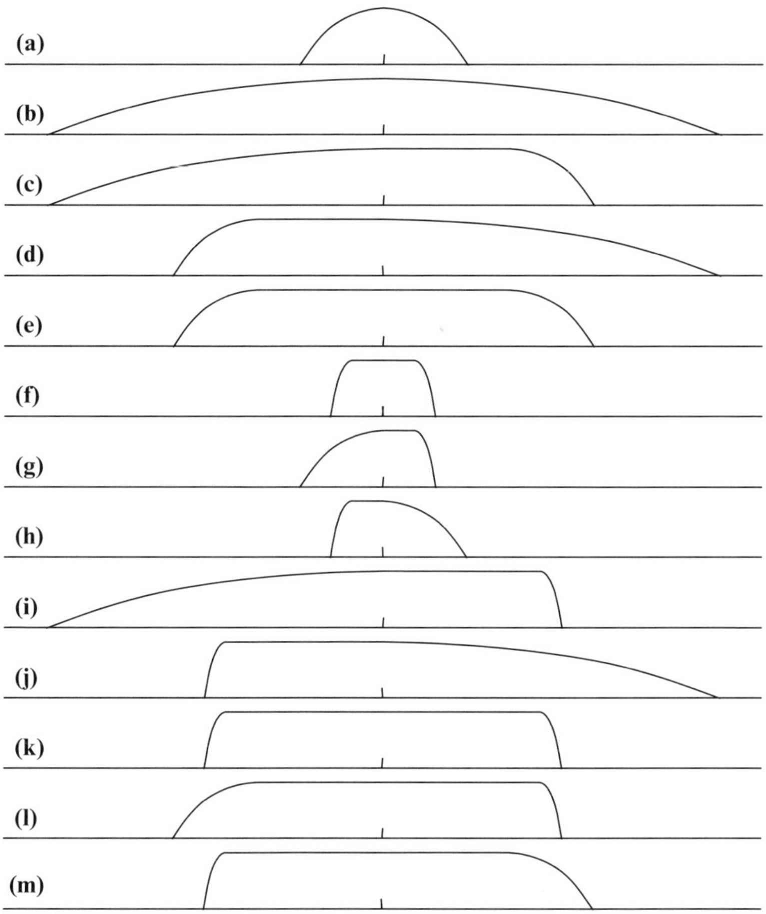

Fig. 20.3 Window functions. (**a**) WS_S2S, (**b**) WL_L2L, (**c**) WL_L2S, (**d**) WL_S2L, (**e**) WL_S2S, (**f**) WS_B2B, (**g**) WS_S2B, (**h**) WS_B2S, (**i**) WL_L2B, (**j**) WL_B2L, (**k**) WL_B2B, (**l**) WL_S2B, and (**m**) WL_B2S

transitional window functions illustrated in Fig. 20.3c–e. Window function (e) may not be necessary if certain restriction on the application of the short window is imposed.

Because transients in an audio signal typically consist of no more than a few samples, the short block size for a frame of detected transient should be a few samples as well, thereby matching the filter's temporal resolution to the transient. Unfortunately, such a choice results in extremely poor frequency resolution within the frame and, therefore, is inappropriate for the rest of the samples in the same frame because such other samples, provided they are sufficiently far away from the transient, are quasistationary and therefore are better processed using high frequency resolution. This conflict conventionally has resulted in a compromise short block size that is neither optimal for the transient samples nor for the quasistationary samples in the same frame.

Since quantization noise for all MDCT coefficients in a MDCT block spreads uniformly across the whole MDCT window in the time domain, the part of the quantization noise that appears before a transient attack can be easily heard because it is not masked by the transient. This is referred to pre-echo artifacts.

Temporal noise shaping (TNS) [12], used by AAC [9], is one of the pre-echo control methods [3]. It deploys linear prediction over the MDCT coefficients for the block with transient, to boost the coding gain for the block and also to shape some of the pre-echo quantization noise to behind the transient. TNS is computationally intensive due to the linear prediction and the overhead for transferring the description of the predictor, including the prediction filter coefficients, is remarkable.

DRA audio coding standard [1] uses a very simple method for pre-echo suppression. As presented here, this method reduces the effective size of the short window applied to the transient samples, thereby providing transient localization while leaving frequency resolution unchanged. Pre-echo is better suppressed because transient localization reduces the spread of both quantization noise and high bit rates associated with transients.

20.2.1 Brief Window Functions

DRA audio coding algorithm deploys a MDCT that switches its window sizes between 256 and 2,048 points. All DRA window functions are based on the sine window function (20.2) and constructed via the perfect reconstruction conditions [2, 3]. What makes DRA audio coding algorithm different is the introduction of a new "brief window function", illustrated in Fig. 20.3f. Labeled as WS_B2B, it is still a short window of size 256, the same size as other windows within the frame. Unlike those other windows, however, brief window WS_B2B uses only a central portion of its overall length for signal shaping, employing a number of leading and trailing zeros to improve its temporal resolution.

In particular, the brief window WS_B2B is designed such that it is nonzero within the central 160 samples, with the first 16 and last 16 of such samples overlapping the respective transition windows that are adjacent to it, and with zeros for the first 48 and the last 48 samples of the window. Obviously, its effective window size is reduced from 256 to 160.

To switch to/from this brief window from/to the long (WL_L2L) and short (WS_S2S) windows, the perfect reconstruction conditions [2, 3] call for the addition of transitional windows which are illustrated as (g), (h), (i), (j), (k), (l), and (m) in Fig. 20.3.

Note that this brief window is applied only to the block of samples containing a transient, while the short and/or the appropriate transition windows are applied to the quasistationary samples in the remainder of the frame.

20.2.2 Enhanced Preecho Control

The significantly reduced effective size of the brief window offers better preecho suppression for the following reasons:

1. High bit rates associated with transients are constrained to fewer samples and finer temporal resolution is deployed to transient samples.
2. The worst spread of quantization noise is reduced from 256 samples for the regular short window to only 160 samples for the brief window. For a typical sample rate of 48 kHz, for example, they amount to 5.33 and 3.3 ms, respectively. Given that significant premasking tends to last about 1–2 ms before a transient attack [3], the worst spread of quantization noise that may cause preecho is reduced from 4.3–3.3 (Fig. 20.4a) ms to 2.3–1.3 ms (Fig. 20.4b). This is obviously a significant reduction.

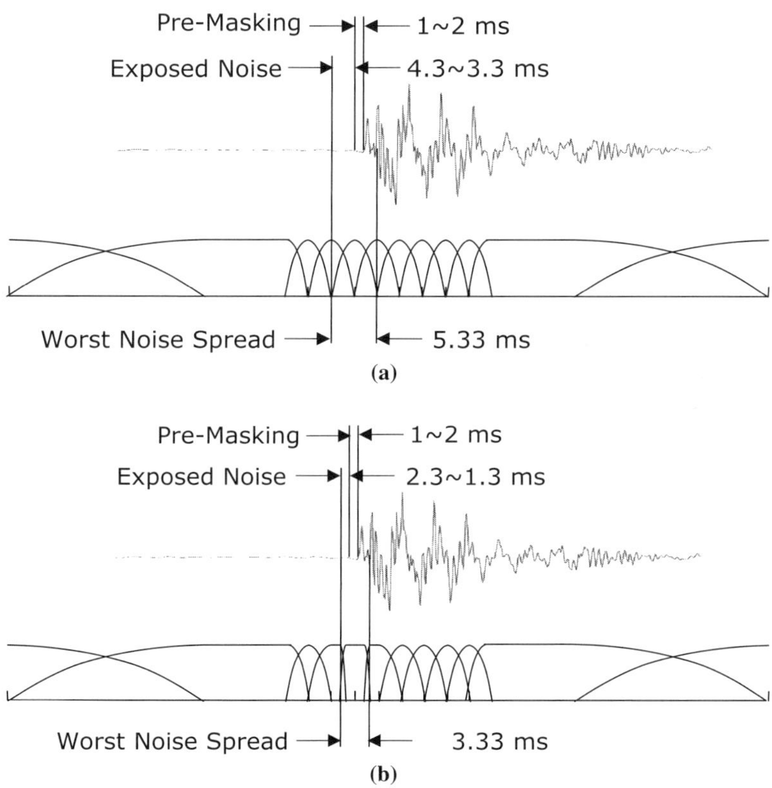

Fig. 20.4 Worst exposed quantization noise for the conventional switched-window MDCT (**a**) and transient-localized switched-window MDCT (**b**)

20.2.3 *Window Sequencing*

Because of the increased number of windows when compared with the conventional approach, the determination of appropriate window sequence is more involved for DRA audio coding algorithm, but a short procedure suffices that applies the perfect reconstruction conditions [1]. This is outlined as follows.

Before we start, let us first define transient segments. For a frame with detected transient(s), there are eight short MDCT blocks, only one or two of them may be allowed to contain a transient. The first transient segment is defined as the MDCT blocks starting from the first MDCT block of the frame to before the first transient MDCT block. The second transient segment is defined as the MDCT blocks starting from the first transient MDCT block to the last MDCT block of the frame if there is only one transient in the frame. If there are two transients in the frame, the second transient segment is defined as the MDCT blocks starting from the first transient MDCT block to before the second transient MDCT block and the third transient segment is defined as the MDCT blocks starting from the second transient MDCT block to the end of the frame. For a frame without transient, however, there is only one long MDCT block. For completeness, this frame may be considered as consisting of only one transient segment.

20.2.3.1 Long Windows

If there is no transient detected within the current frame, select a long window, the specific shape of which depending on the existence and location of any transient in the previous and the subsequent frame, as shown in Table 20.2.

20.2.3.2 Short Windows

If a transient has been detected in the current frame, the selection of windows is a little bit more involved. First of all, identify the location(s) of the transient(s). Around a transient, select a series of short windows according to the following principles:

Table 20.2 Determination of long windows for a frame without detected transient

Previous Frame	Current frame	Subsequent frame
	WL_L2L	No transient
No transient	WL_L2S	Transient, but not in the first block
	WL_L2B	Transient in the first block
Transient, but	WL_S2L	No transient
not in the last	WL_S2S	Transient, but not in the first block
block	WL_S2B	Transient in the first block
Transient in	WL_B2L	No transient
the last block	WL_B2S	Transient, but not in the first block
	WL_B2B	Transient in the first block

Table 20.3 Allowed placement of windows around a block with a detected transient

Pretransient	Transient	Posttransient
WL_L2B		WL_B2L
WL_S2B		WL_B2S
WL_B2B	WS_B2B	WL_B2B
WS_S2B		WS_B2S
WS_B2B		WS_B2B

- WS_B2B is applied to the block where the transient occurs, to improve the temporal resolution of the MDCT.
- The window for the block that is immediately before this transient block has a designation of the form "...2B".
- The window for the block that is immediately after this transient block has a designation of the form "..._B2...".

Consequently, the allowed placement of windows is summarized in Table 20.3:

For the rest of the frame (away from the transient), short window WS_S2S should be deployed, except for the first and last blocks of the frame, whose windows are assigned as follows:

1. For the first block of the frame, if there is no transient in the last block of the previous frame, short window WS_S2S should be used; otherwise, short window WS_B2S should be used.
2. For the last block of the frame, if there is no transient in the first block of the subsequent frame, short window WS_S2S should be used; otherwise, short window WS_S2B should be used.

Some example window sequences are shown in Fig. 20.5. (a) is an example for the conventional approach. (b) shows that a transient occurs in the first block of the frame, so brief window WS_B2B is deployed for this block, WL_L2B is deployed for the previous frame and WS_B2S is deployed for the second block of the frame. (c) shows that a transient occurs in the third block of the frame. (d) shows that two transients occur in the third and sixth blocks, so two brief windows are placed, respectively. (e) shows that a transient occurs in the last block of the frame.

20.2.4 Indication of Window Sequence to Decoder

The encoder needs to indicate to the decoder the window(s) that it used to encode the current frame so that the decoder can use the same window(s) to decode the frame. This can be accomplished using a window index with the conventional approach. This is also true for DRA audio coding algorithm.

For a frame without a detected transient, one label from those in Table 20.2 needs to be transferred to the decoder.

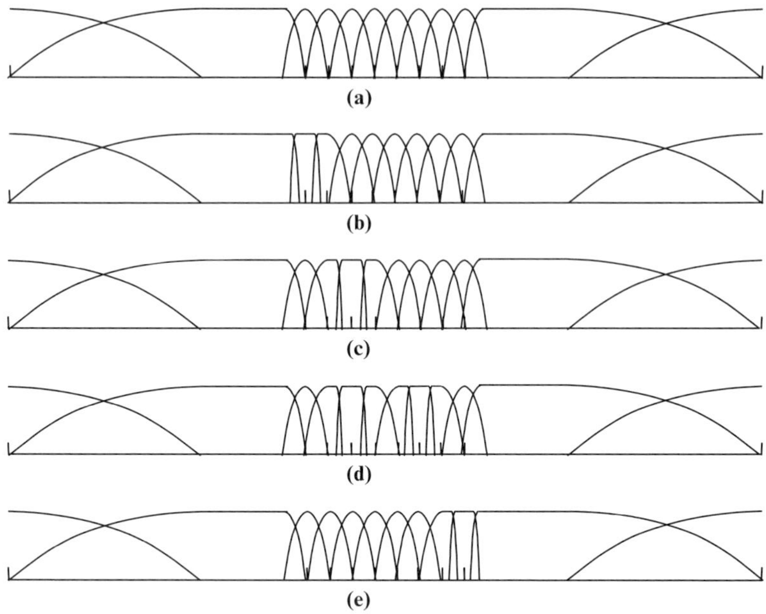

Fig. 20.5 Window sequence examples. (**a**) Window sequence for the conventional method. (**b**) Window sequence for the proposed method when a transient occurs in the first block. (**c**) When a transient occurs in the third block. (**d**) When two transients occur in the third and sixth blocks, respectively. (**e**) When a transient occurs in the last block

For a transient frame, the window sequencing procedure outlined in Sect. 20.2.3 needs to know:

1. Transient location(s), which the conventional method also needs and is(are) conveyed by the length(s) of the transient segment(s)
2. Whether there is a transient in the first block of the current frame and of the subsequent frame, respectively

This later information may be conveyed by the nomenclature WS_CurrSubs, where

1. Curr (S=no, B=yes) identifies if there is transient in the first block of current frame
2. Subs (S=no, B=yes) identifies if there is transient in the first block of the subsequent frame

This is shown in Table 20.4. Obviously, the first column of Table 20.4 is the same set of labels used for the short windows, i.e., (a), (f), (g), and (h) in Fig. 20.3. Combining the labels in Tables 20.2 and in 20.4, we get the complete set of window labels shown in Table 20.5 and Fig. 20.3.

The total number of windows is now 13, while that number is 5 or 4 for the conventional approach (depending on whether or not window WL_S2S is forbidden), so one or two more bits (4 vs. 3 or 2) are needed to transfer the window sequencing information. And that is the only overhead.

Table 20.4 Encoding of the existence or absence of transient in the first block of the current and subsequent frames

Label	Transient in the first block of	
	Current frame	Subsequent frame
WS_B2B	Yes	Yes
WS_B2S	Yes	No
WS_S2B	No	Yes
WS_S2S	No	No

Table 20.5 Window indexes and their corresponding lables

Window index	Window label
0	WS_S2S
1	WL_L2L
2	WL_L2S
3	WL_S2L
4	WL_S2S
5	WS_B2B
6	WS_S2B
7	WS_B2S
8	WL_L2B
9	WL_B2L
10	WL_B2B
11	WL_S2B
12	WL_B2S

20.3 DRA Encoder

Transient-localized MDCT described earlier above is an important component of DRA encoder and the other components of it are described as follows.

20.3.1 Linear Scalar Quantization

Linear scalar quantization is used to quantize MDCT coefficients. One quantization step size is shared by all MDCT coefficients boxed by the critical bands in the frequency domain and by the transient segments in the time domain. This box of MDCT coefficients is referred to as a quantization unit.

The quantization step size itself is logarithmically quantized and spans a range from 1 to 2^{23}.

When the quantization step size is one, the maximum allowed quantization index is $\pm 2^{23}$ and the Huffman codebooks are designed to accommodate this. Consequently, a 24-bit signal path is allowed throughout the codec so that audio quality exceeding the perceptual capability of the human ear can be delivered if the bit rate suffices.

20.3.2 Sum/Difference Coding

Sum/difference coding may be optionally (usually at low bit rates) deployed by the encoder. If this is the case, the left and right channels can be reconstructed in the decoder from the sum/difference encoded channels as follows

$$\text{LeftChannel} = \text{SumChannel} + \text{DifferenceChannel}, \tag{20.3}$$

$$\text{RightChannel} = \text{SumChannel} - \text{DifferenceChannel}. \tag{20.4}$$

Note that sum/difference coding is applied to the front and any surround left/right pairs when activated.

20.3.3 Joint Intensity Coding

At very low bit rates, joint intensity coding may be optionally deployed by the encoder. Instead of joining stereo pairs, all normal channels (other than the LFE channels) are joined into the left channel, thereby providing significant bit rate reduction when surround sounds are involved.

At the decoder side, all normal channels (other than the left) are reconstructed from the left channel as follows:

$$\text{JointChannel} = \text{ScaleFactor} \times \text{LeftChannel}, \tag{20.5}$$

where ScaleFactor is the scale factor calculated and embedded in the bit stream by the encoder.

20.3.4 Perceptual Model

The perceptual model computes the masking threshold below which audio signals or quantization noise is not audible. This threshold is used by the global bit allocation unit to adjust the quantization step size so that the resultant quantization noise is below this threshold. Although it is a necessary component of DRA encoder, it is not part of the decoder and there is essentially no restriction on its implementation. Therefore, most perceptual models can be adapted for a DRA encoder.

20.3.5 Global Bit Allocation

It allocates bits to all quantization units based on the masking thresholds provided by the perceptual model. Although it is a necessary component of DRA encoder, it is not part of the decoder and there is essentially no restriction on its implementation. Therefore, most bit allocation technologies can be adapted for a DRA encoder.

A widely used bit allocation algorithm involves the following iterative procedure:

1. Find the quantization unit whose quantization noise is most audible, i.e., the ratio of quantization noise to masking threshold is the highest
2. Decrease the quantization step size for this unit
3. Repeat steps 1 and 2 until either
   - The bit pool is exhausted
   - The quantization noise for all quantization units are below their respective masking thresholds

20.3.6 Statistic Allocation of Codebooks

With conventional approach to Huffman codebook allocation, all quantization indexes in a quantization unit usually share one Huffman codebook, as is shown in Fig. 20.6a. The assigned codebook is usually the smallest one that can accommodate the maximum quantization index within the quantization unit. Therefore, once the quantization step size is fixed, the Huffman codebook is also fixed, there is no room for optimization.

As the statistic properties of the quantization indexes are not necessarily segmented by the boundaries of quantization units, the traditional approach does not provide a good match, if any, between the statistic properties of the Huffman codebooks and those of the quantization indexes. This motivates a statistic-adaptive approach to codebook assignment.

As shown in Fig. 20.6b, quantization units are ignored when it comes to codebook selection. Instead, the following steps are taken to adaptively match codebooks to the local statistic properties of quantization indexes:

1. Assign to a quantization index the smallest codebook that can accommodate it, thereby converting quantization indexes into codebook indexes
2. Segments these codebook indexes into large segments based on their local statistic properties
3. Select the largest codebook index for each segment as the codebook index for the segment

The advantage of this statistic-adaptive approach to codebook assignment vs. the traditional approach is shown in Fig. 20.6. Since the largest quantization index falls into quantization unit 4 in Fig. 20.6a, a large codebook (Codebook 6) is assigned to quantization unit 4 using the traditional method, which is obviously not a good match because most of the indexes in the unit are much smaller. Using the DRA approach, however, the largest quantization index is segmented into segment 2 as shown in Fig. 20.6b, so share a codebook with other large quantization indexes. Also, all quantization indexes in segment 3 are small, so a small codebook (Codebook 2) is selected. This obviously results in fewer bits for coding the quantization indexes.

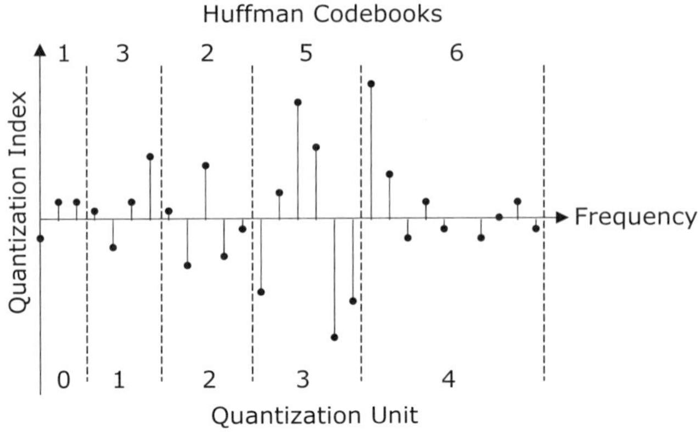

(a)

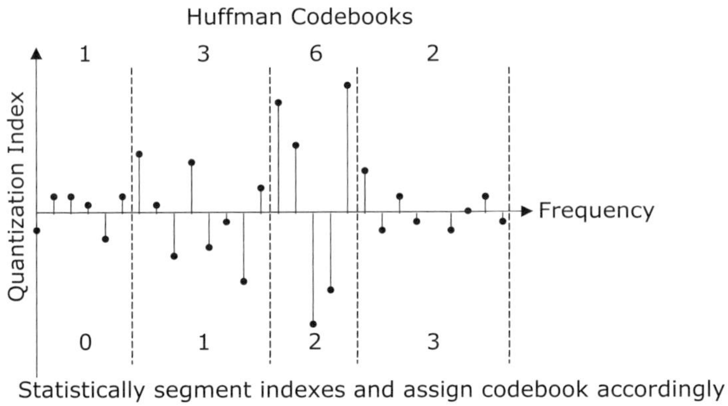

(b)

Fig. 20.6 Traditional (**a**) and statistic-adaptive (**b**) approaches to Huffman codebook assignment

With the conventional approach, only the codebook indexes need to be transfered to the decoder as side information, because the application ranges of codebooks are the same as the quantization units, which are predetermined. The DRA approach, however, need to transfer the application ranges of codebooks as side information, in addition to the codebook indexes, since they are independent of the quantization units. This overhead must be contained in order for the new approach to offer any advantage. This can be accomplished by imposing a limit to the number of segments.

20.3.7 Huffman Coding

Two sets of Huffman codebooks, shown in Table 20.6, are provided, one for transient frames and the other for quasistationary frames. Note that the last two codebooks, namely HuffDec18_256x1 and HuffDec27_256x1, are unsigned, so their

Table 20.6 Huffman codebooks for quantization indexes

Codebook index	Dimension	Code index range	Signed	Quasistationary codebooks	Transient codebooks
0	0	0	N/A	N/A	N/A
1	4	[-1,1]	Yes	HuffDec10_81x4	HuffDec19_81x4
2	2	[-2,2]	Yes	HuffDec11_25x2	HuffDec20_25x2
3	2	[-4,4]	Yes	HuffDec12_81x2	HuffDec21_81x2
4	2	[-8,8]	Yes	HuffDec13_289x2	HuffDec22_289x2
5	1	[-15,15]	Yes	HuffDec14_31x1	HuffDec23_31x1
6	1	[-31,31]	Yes	HuffDec15_63x1	HuffDec24_63x1
7	1	[-63,63]	Yes	HuffDec16_127x1	HuffDec25_127x1
8	1	[-127,127]	Yes	HuffDec17_255x1	HuffDec26_255x1
9	1	[-255,255]	No	HuffDec18_256x1	HuffDec27_256x1

Table 20.7 Frame structure

Sync word	0x7FFF
Frame header	Bits describing the audio signal, such as its sample rate, the numbers of normal and LFE channels, etc.
Normal channel(s)	Bits representing 1 to 64 normal channels.
LFE channel(s)	Bits representing 0 to 3 LFE channels.
Error detection	Bits for error detections.
Auxiliary data	Bits for auxiliary data such as time code.

actual range of code index is [0,255]. This is extended to $[-255,255]$ with the transmission of a sign bit for each quantization index encoded with either of the two codebooks.

The maximum code index of HuffDec18_256 $\times$ 1 and HuffDec27_256 $\times$ 1, namely 255, is an escape indicating that the quantization index is out of the range that the two codebooks can represent. In case of this, recursive indexing is deployed, which represents a quantization index q as follows

$$q = m \times 255 + r, \tag{20.6}$$

where m is the quotient and r is the reminder. The reminder r is still encoded with either HuffDec18_256x1 or HuffDec27_256x1, but the quotient m is packed directly into the bit stream. Obviously, this can easily accommodate a signal path of 24 bits.

20.3.8 Multiplexer

The Huffman codes for the quantization indexes of all MDCT coefficients and side information are packed by the multiplexer to form a frame of DRA audio data and a sequence of such frames form a DRA bit stream. The major components of a DRA frame are arranged as shown in Table 20.7. The major components of normal and LFE channels are further arranged as shown in Table 20.8 and 20.9, respectively.

Table 20.8 Data structure for a normal channel

Window sequence	Index for MDCT window functions Number of transient segments Length of each transient segment
Huffman codebook assignment	Number of codebooks for each transient segment Application range of each codebook Codebook index for each application range
Quantization	Huffman codes for quantization indexes of all MDCT coefficients
Step size	Huffman codes for quantization step sizes
Sum/diff	Decisions for sum/difference coding
Joint intensity	Decisions and scale factors for joint intensity coding

Table 20.9 Data structure for an LFE channel

Huffman codebook assignment	Number of codebooks for each transient segment Application range of each codebook Codebook index for each application range
Quantization	Huffman codes for quantization indexes of all MDCT coefficients
Step size	Huffman codes for quantization step sizes

20.4 Decoder

The DRA decoder is shown in Fig. 20.2b and its components are described as follows.

20.4.1 Codebook Assignment

As shown in Tables 20.8 and 20.9, all the information for codebook assignment is embedded in the bit stream, ready for the decoder to reconstruct the complete codebook assignment used by the encoder.

20.4.2 Quantization Indexes

Using the codebook assignment structure reconstructed in the last step, all the quantization indexes are decoded from the bit stream.

20.4.3 Reconstructing the Number of Quantization Units

The number of quantization units needs to be reconstructed for each transient segment because it is not embedded in the bit stream. For each transient segment, this can be accomplished as follows:

1. Get the range of nonzero quantization indexes by adding the application ranges of all codebooks in that segment. The upper limit of this range indicates the valid bandwidth of the transient segment.
2. The quantization unit that accommodates this upper limit is the last one with nonzero quantization indexes, so the index corresponding to this quantization unit represents the number of quantization units in the transient segment.

20.4.4 DeQuantization

De-quantization is performed for all quantization indexes by multiplying each quantization index with its respective quantization step size unpacked from the bit stream.

20.4.5 Joint Intensity Decoding

If the bit stream indicates that joint intensity coding was deployed in the encoder, the decoder performs decoding as shown in (20.5).

20.4.6 Sum/Difference Decoding

If the bit stream indicates that sum/difference coding was deployed in the encoder, the decoder performs decoding as shown in (20.3) and (20.4).

20.4.7 Short/Brief Window Sequencing

The window sequencing information is conveyed in the bit stream using an index shown in Table 20.5. Once this index is unpacked from the bit stream, the window label is directly looked up from Table 20.5.

If the label indicates a long window, the window function corresponding to the label is directly supplied to Transient Localized IMDCT and Short/Brief Window Sequencing is not performed.

On the otherhand, if the label indicates a short window, the label is used to look up Table 20.4 to determine whether there is a transient in the first block of the current frame as well as in the subsequent frame. This information, along with the boundaries of the transient segments is used to determine the short/brief window sequencing in the current frame using the procedure outlined in Sect. 20.2.3.

20.4.8 Transient Localized IMDCT

The window function(s) determined in the last step is(are) supplied to a standard IMDCT procedure to reconstruct the PCM audio samples from the reconstructed MDCT coefficients.

20.5 Subjective Listening Tests

During its standardization process, DRA audio coding algorithm went through five rounds of ITU-R BS.1116 [13] compliant subjective listening test. The test grades are shown in Table 20.10.

The bit rate was configured in such a way that it is the upper limit absolutely not to be exceeded in any frame. For example, if the sample rate is 48 kHz, a bit rate of 128 kbps translates into 2,730 bits per frame because a DRA frame consists of 1,024 samples. No frame can use more than 2,730 bits and no bit reservoir is allowed.

The first test was conducted in August 2004 by National Testing and Inspection Center for Radio and TV Products (NTICRT) under the Ministry of Information Industry of China. Ten stereo sound tracks selected mostly from SQAM CD [14] and five 5.1 surround sound tracks were used in the test. The test subjects were all expert listeners consisting of conductors, musicians, recording engineers, and audio engineers.

The other four tests were all performed by the State Lab for DTV System Testing (SLDST) under the State Administration for Radio, Film, and TV of China. Its state-of-the-art listening room deploys all Genelec speakers, including three 1,037 and two 1032A, and a Tascam DM24 mixer.

Some of the most famous Chinese conductors, musicians, and recording engineers were among the test subjects, but senior and graduate students from the School of Recording Arts, Communication University of China, and the Department of Sound Recording, Beijing Film Academy, were found to possess better acuity due to their age and training, so became the majority in the later tests.

Other than a few Chinese sound tracks, most of the test materials were selected from the SQAM CD and a group of surround sound tracks used by EBU and MPEG, including Pitch_pipe, Harpsichord, and Elloit1.

Table 20.10 Grades for ITU-R BS.1116 compliant subjective listening tests

Lab	Date	Stereo @ 128 kbps	5.1 @ 320 kbps	5.1 @ 384 kbps
NTICRT	08/2004	4.6	–	4.7
SLDST	10/2004	4.2	–	4.0
SLDST	01/2005	4.1	–	4.2
SLDST	07/2005	4.7	–	4.5
SLDST	08/2006	–	4.6	4.9

The last test, while conducted by SLDST, was actually ordered and supervised by China Central TV (CCTV) as part of its DTV standard evaluation program. CCTV was only interested in surround sounds, so DRA was tested at 384 kbps as well as 320 kbps. This test was conducted in comparison with two major international codecs, DRA came out as the clear winner.

The test results from SLDST are more consistent because the same group of engineers performed the tests in the same lab, and the listener pool was largely unchanged. The later the test, the more difficult the test became, because the test administrators and many listeners had learned the characteristics of DRA-processed audio and knew where to look for distortions. The gradually increasing test grades reflect the continuous improvement to the encoder, especially to its perceptual model and transient detection unit.

20.6 Conclusions

This chapter provides an overview to China's DRA audio coding standard, including its encoder and decoder. It shows that DRA standard is essentially a bare-bone adaptive transform coder that deploys transient-localized MDCT for improved pre-echo suppression and statistic allocation of codebooks for enhanced Huffman coding efficiency. Its quantizer and Huffman codebooks are designed in such a way that a 24-bit signal path is allowed throughout the codec so that highest audio quality can be delivered if the bit rate suffices. Although its decoder complexity is very low, it delivers state-of-art coding performance that is documented by the five ITU-R BS.1116 compliant subjective listening tests.

References

1. Yu-Li You, Weixiong Zhang, Mao Xu, and Subin Zhang, *Electronics Industry Standard: Multichannel Digital Audio Coding Technology, SJ/T11368-2006*, Ministry of Information Industry, People's Republic of China, 2007.
2. J. P. Princen and A. B. Bradley, "Analysis/synthesis filter bank design based on time domain aliasing cancellation," *IEEE Transactions on ASSP*, vol. 34, no. 5, pp. 1153–1161, 1986.
3. T. Painter and A. Spanias, "Perceptual coding of digital audio," *Proceedings of the IEEE*, vol. 88, no. 4, pp. 451–513, 2000.
4. G. Davidson and M. Bosi, "AC-2: high quality audio coding for broadcasting and storage," *46th Annual Broadcasting Engineering Conference*, pp. 98–105, April 1992.
5. Dolby Laboratories, *Digital Audio Compression Standard A/52B*, Advanced Television Systems Committee (ATSC), 2005.
6. K. Tsutsui, "ATRAC (adaptive transform acoustic coding) and ATRAC 2," *The Digital Signal Processing Handbook, V. Madisetti and D. Williams, Editors, CRC Press*, pp. 43.16–43.20, 1998.
7. J. Johnston, D. Sinha, S. Dorward, and S. Quackenbush, "AT&T perceptual audio coding (PAC)," *Collected Papers on Digital Audio Bit-Rate Reduction*, pp. 73–81, 1996.
8. MPEG, *Coding of moving pictures and associated audio for digital storage media at up to about 1.5 Mbit/s – Part 3: Audio*, vol. 11172-3, ISO/IEC, 1992.

9. MPEG, *Information technology: Generic coding of moving pictures and associated audio information Part 7: Advanced Audio Coding (AAC)*, vol. 13818-7, ISO/IEC, 1997.
10. *Vorbis I specification*, Xiph.org Foundation, 2004.
11. Wikipedia, *Windows Media Audio*, http://en.wikipedia.org/wiki/Windows_Media_Audio, October 2007.
12. J. Herre and J.D. Johnston, "Enhancing the performance of perceptual audio coders by using temporal noise shaping (TNS)," *101st AES Convention*, 1996 Reprint #4384.
13. International Telecommunication Union, *Recommendation ITU-R BS.1116 - Methods for the subjective assessment of small impairments in audio systems, including multichannel sound systems*, 1994.
14. EBU, *Sound Quality Assessment Material Recordings for Subjective Tests, Tech. 3253*, April 1988.

Chapter 21
Audio Coding Standard Overview: MPEG4-AAC, HE-AAC, and HE-AAC V2

Yujie Gao

21.1 Introduction

Nowadays, Advanced Audio Coding (AAC) becomes one of the most popularly adopted audio formats in mobile society. In this chapter, brief history of MPEG4 AAC decoder family will be introduced, followed by more details for MPEG4-AAC, HE-AAC, and HE-AAC V2 systems.

In April 1997, MPEG-2 Advanced Audio Coding (MPEG-2 AAC) [1] compressing algorithm, which takes advantages of some new spectrum processing and compression tools like temporal noise shaping (TNS) [1, 2], became an international standard. Compared to previously existing audio compressing algorithms, the new standard provides outstanding audio quality and exceptional compression ratio and thus achieves lower bit rate in the encoded bitstreams, and gradually becomes one of the new choices of audio codec standards for broadcasting, internet services, and mobile applications.

MPEG-4 AAC standard [2] was adopted by the MPEG community in 1999. It is based on MPEG-2 AAC standard and keeps maximum compatibility with existing MPEG-2 AAC standard from bitstream syntax point of view. In other words, generally speaking, MPEG-4 AAC decoder should be able to decode MPEG-2 AAC bitstreams. On the other hand, the new standard adopts further improvements on scalability, error resilience and some additional spectral processing features including Perceptual Noise Substitution (PNS), Long-Term Predictor (LTP), etc. [2]. Therefore, MPEG-2 AAC decoder might face issues while decoding MPEG-4 AAC stream if MPEG-4 specific tools or features are used.

In 2003, the MPEG community further standardized High Efficiency AAC (HE-AAC) [2], an extension of AAC algorithm targeting on low bit-rate applications with higher coding efficiency. HE-AAC adopts a new tool called Spectral Band

---

Y. Gao
Qualcomm Inc., San Diego, California, USA
e-mail: jgao@qualcomm.com

F.-L. Luo (ed.), *Mobile Multimedia Broadcasting Standards: Technology and Practice*, 607
DOI: 10.1007/978-0-387-78263-8_21,

Replication (SBR) [2, 3], which can reconstruct high-frequency band output based on low-frequency band data and some side information.

In 2004, HE-AAC Version 2 (HE-AAC V2) [4] was standardized by the MPEG community. It uses Parametric Stereo tool (PS) [4] on the basis of SBR (HE-AAC), which can reconstruct stereo audio signals based on monaural downmixed signals and limited number of additional stereo parameters.

In summary, MPEG-4 AAC, HE-AAC, and HE-AAC V2 make up the AAC decoder family. Based on different compression efficiency requirements, the most appropriate one from them can be chosen to achieve the best compression ratio with required audio quality.

The concept of audio object type and profile has been playing a very important role in AAC coding standards. As specified in the MPEG-2 and MPEG-4 ISO specifications, AAC comes in different "flavors" which gives maximum flexibility to different applications and usage models. In MPEG-2 standard, they are called Profiles [1]. While in MPEG-4, they also include Audio Object Types or AOT in short [5]. The AAC standards support tens of these flavors, by adopting different optional tools in encoding or decoding process. AAC compressors family consists of all these different flavors and therefore is suitable for a broad range of different applications. However, those different audio object types are not necessarily compatible to each other.

In MPEG-2 AAC spec, three profiles can be supported: Main Profile, Low Complexity Profile (LC), and Scalable Sampling Rate Profile (SSR) [1]. While in MPEG-4 AAC, multiple audio object types are supported. Some of them are almost the counterpart of corresponding MPEG-2 AAC profiles, for example, AAC Main object, AAC-LC object, and AAC-SSR object [5]. The bitstream syntax of the above MPEG-4 audio object types is very similar to their corresponding counterpart of the MPEG-2 profiles, except that MPEG-4 bitstream might have PNS-related data. Thus, an MPEG-4 AAC decoder which supports above audio object types can parse and decode corresponding MPEG-2 profile bitstream, while a specific MPEG-2 profile decoder can also parse its MPEG-4 counterpart object only if the stream does not contain any PNS information.

Some popularly used MPEG-4 AAC audio object types in mobile society include MPEG-4 AAC LC object, MPEG-4 AAC LTP object, MPEG-4 ER AAC LC (Error Resilient AAC Low Complexity) object, and SBR-related objects [5]. All of them are targeting on one or multiple goals key important to mobile applications, such as: low complexity and low power with good audio quality, low bit rate, suitable channel settings, and error robustness, etc.

The MPEG-4 AAC LC object is the basic audio object type and minimum requirement for AAC decoding, just like the MPEG-2 AAC LC profile.

The MPEG-4 LTP object type adds long-term prediction (LTP) on top of basic MPEG-4 LC object. LTP takes advantages of the redundancy between successive frames in a clear pitched audio signal, to achieve lower bit rate in encoding such audio source. As it is based on LC object, the MPEG-4 AAC LTP object compatible decoder can smoothly decode both MPEG-2 LC profile and MPEG-4 LC object type.

The ER AAC LC object type is the combination of error resilience functionality and the AAC LC object. In some mobile services like streaming and broadcasting, bitstreams often contains bit errors after going through transmission channels. The error resilience tools provide more protection on important side information or spectrum information encoded in thc bitstreams and thus can improve audio quality for bitstream error cases. As additional information needs to be packed into bitstreams to achieve error resilience, the bitstream syntax of ER object type is very different from the others and thus not compatible with non-ER object either. Therefore, a non-ER object decoder is not able to decode ER bitstreams.

The SBR-related objects can cover multiple audio object types with different AOT values. Basically, they are adopting SBR tool on top of other MPEG-4 AAC audio object types, like Main, LC, and LTP, etc., respectively [5].

The rest of this chapter is organized into as follows. The next three sections will present more details of MPEG-4 AAC, HE-AAC, and HE-AAC V2, respectively. At the end of this chapter, some conclusions and further discussions will be made.

21.2 MPEG-4 AAC

The schematic diagrams of the encoder and decoder of MPEG-AAC are shown in Fig. 21.1a and b [1, 2]. Based on different audio profiles or audio object types the system supports, it can have different optional tools in spectral processing part for

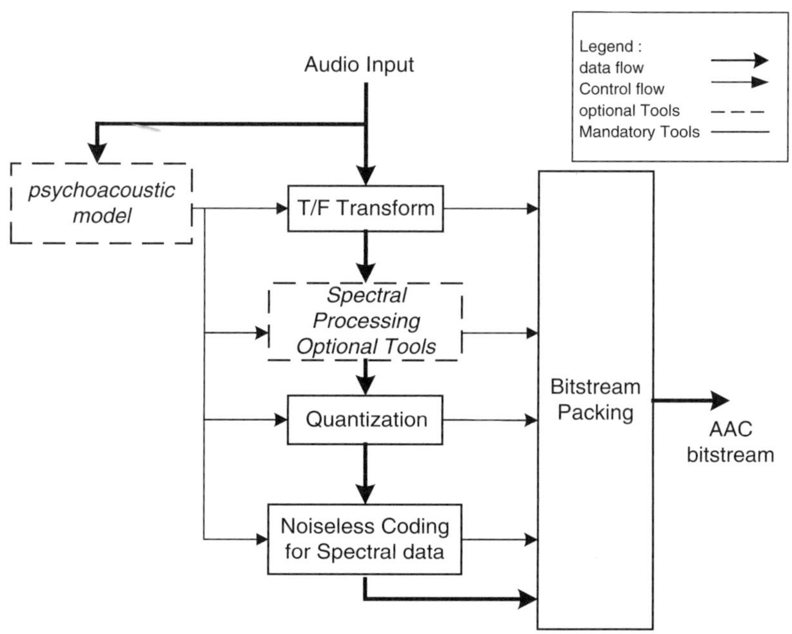

Fig. 21.1 (a) Schematic diagram for MPEG-AAC encoder

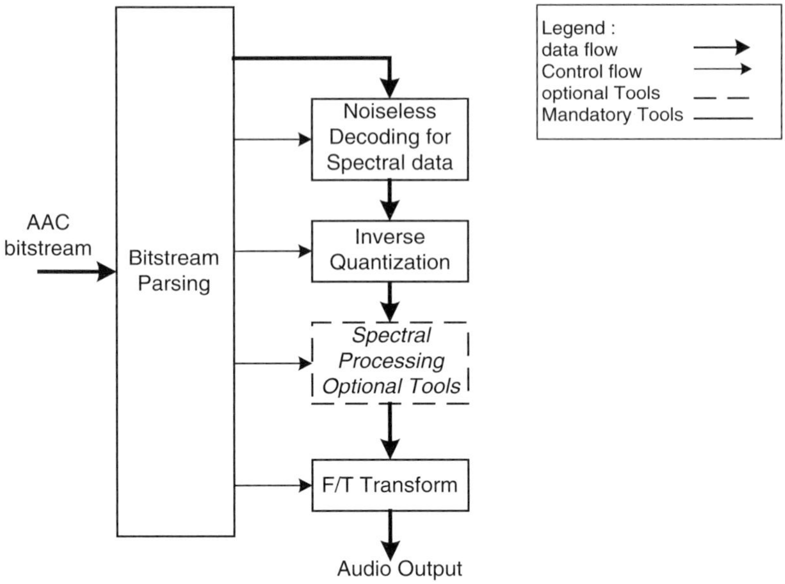

Fig. 21.1 (b) Schematic diagram for MPEG-AAC decoder

both encoder and decoder. Thus, here only mandatory tools are clearly specified in these two diagrams. In latter part of this section, some popularly used optional spectral processing tools will be described in more details with a particular AAC decoding system.

Just like many other transform-based audio encoders, the central part of AAC encoder is time to frequency domain transform. The encoder runs on the scaled quantized spectral data and chooses one of the noiseless coding algorithms to pack the audio information into final bitstreams. Most of these processing could be controlled by information generated from corresponding psychoacoustic model. The psychoacoustic model is based on some key important characteristics of human perceptual system, such as mask effect. For example, distortions in spectral below the so-called "masking threshold" might not be perceived by human auditory system. Thus, psychoacoustic model can process audio signal in such a way which skips the inaudible information and only keeps the audible signals. More basics of perceptual audio coding can be found in [6, 7].

In AAC encoder, the psychoacoustic model will calculate the masking threshold and scalefactor band mapping and make decisions on window information like window length [1, 2]. All these information will further control the whole encoding process. Some of them might also be packed into the bitstream and finally guide the decoding process for corresponding bitstreams in decoder side.

AAC decoder, on the other hand, is a counterpart of AAC encoder, like illustrated in Fig. 21.1 (b). In mobile society, for end user listening experience, audio content playback is the critical part. Thus, the rest of this section will focus more on decoding system and more details on each module in a particular AAC decoder will follow. Encoding process might be described in some decoder tools just to help better understanding of the decoder system.

21.2.1 Sampling Rates Supported in MPEG-4 AAC Decoder

Theoretically, AAC encoder should be able to support absolute sampling rates. However, to improve compression efficiency, only fixed number (12, in AAC specification) of sampling rate related tables are referred to, they are 8,000, 11,025, 12,000, 16,000, 22,050, 24,000, 32,000, 44,100, 48,000, 64,000, 88,200, and 96,000 Hz. Thus, sampling frequency mapping might be necessary if the original signal sampling rate is not matching one of above sampling rates being picked. MPEG-AAC standard specifies guidelines for frequency mapping [2]. For example, any frequency between 37,566 and 46,009 Hz will use tables for sampling rate at 44100 Hz.

21.2.2 Bit Rate Allowed in MPEG-4 AAC Decoder

Generally speaking, from the encoder side, AAC algorithm allows a maximum of 6 bits per sample [1, 2]. As such, the maximum bit-rate AAC can support is sampling rate and total number of channels dependent. And for each channel, the maximum bit rate allowed can be up to (6* sampling frequency). For example, for bitstream with sampling rate 24,000 Hz, the maximum bit rate allowed is 6*24,000, i.e., 144 kbps. As AAC can support up to 96,000 Hz, the maximum allowed bit rate will be 576 kbps per channel. However, as AAC is adopted targeting on high compressing efficiency, most of AAC bitstreams have much lower bit rate than the theoretical upper bound, for example, 128 kbps AAC bitstream is considered state of art [8,9]. And some popularly used AAC bit rates (depending on corresponding sampling rates) include 128, 96, 64 kbps or even lower. MPEG-AAC can also support variable bit rate (VBR) [1, 2].

21.2.3 MPEG-4 AAC Decoder Program Flow and Major Modules

Like most of other transform-based audio decoders, AAC decoder is also to reconstruct the audio PCM samples in time domain based on the reconstruction guiding information and quantized spectrum domain data encoded in the bitstream payload. It has a bunch of tools or functional modules. Some of them are always required such as noiseless decoding, inverse quantization and rescaling, and filter bank [2]. Others are optional, including M/S stereo processing, PNS, prediction, intensity stereo processing, LTP, dependently switched coupling, TNS, gain control, etc. [2]. Some optional tools might be only applicable to some particular AAC audio object types.

For better illustration, here one particular AAC decoder system is picked up for detailed discussion. It can support some of the most popularly adopted audio object types in mobile society, such as MPEG-4 AAC LC object type, MPEG-4 AAC LTP object type, and MPEG-4 ER AAC LC object type. Figure 21.2 is its program flow. As optional tools used in AAC system are related to audio object types it supports, only those optional tools adopted in this particular system are plotted in Fig. 21.2. Furthermore, even all allowed optional tools for this particular system are plotted, some of them can be mutually exclusive to each other. In other words, one tool might

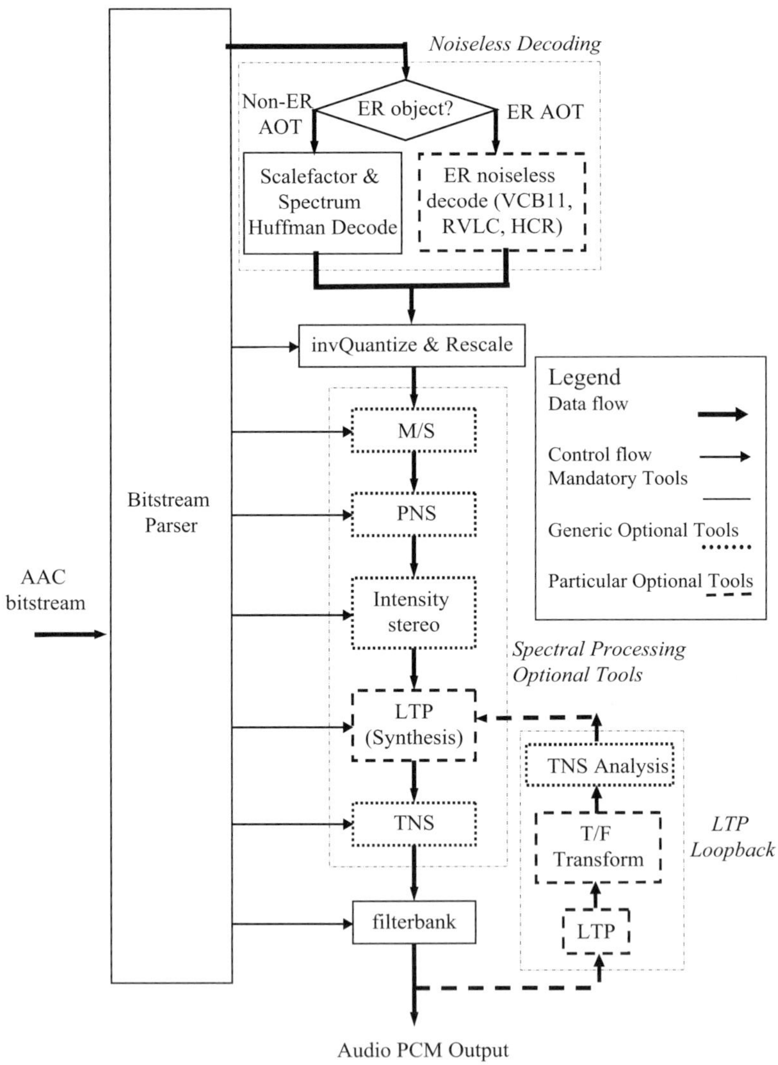

Fig. 21.2 Program flow for one particular AAC decoder system supporting AAC LC, ER AAC LC, and AAC LTP audio object types

take priority while it presents with other modules. And some tools related information specific to one tool which is embedded in the bitstream might have different or "modified" meanings other than what is specified in its original definition while its mutually exclusive counterpart tool presents. A high-level review for each functional module in this particular system and related mutually exclusive information are described below. The complete processing details and algorithm information for each module in the diagram as well as those tools not necessary for this particular system (but might be required for a complete MPEG-4 AAC decoding system) can be found in MPEG-4 AAC Specification [2].

21.2.3.1 Bitstream Parsing

AAC bitstream can come with different formats, for example, ADIF Format (Audio Data Interchange Format) [1, 5], ADTS Format (Audio Data Transport Stream) [1, 5], and MPEG-4 systems format [5], etc.

ADIF format contains only one header at the start of the bitstream which contains all decoding necessary information, followed by the raw data. As there is no audio frame boundary information embedded in the bitstream, bitstream with ADIF format can only be decoded starting from the very beginning of it. Thus, there is no way to perform fast forward, rewind, or re-sync if any bit error happens.

On the other hand, unlike the ADIF file which only contains one ADIF header at the very beginning of the file, ADTS file contains multiple ADTS headers, one before each frame of the raw data. An ADTS header starts with the fixed header, which consists of the byte-aligned sync-word followed by all other header information not changing from frame to frame. The sync word gives decoder the capability to re-sync to the start of a frame, and the fixed header pattern can be used for validating the sync word. Thus, ADTS format is more error robust to bitstream error compared with ADIF format. The variable ADTS header comes after the fixed header, containing header data which changes from frame to frame.

As both sync words and frame length are embedded in ADTS header, the cross-check of both along with decoding process can enhance the error robustness of AAC decoding on ADTS bitstreams. The known information of frame boundary in ADTS header are also very helpful while re-sync the bitstream in case of fast forward, rewind, or other kind of repositioning.

MPEG-4 format, commonly referred by file extension of ".mp4," is a true multimedia format, which can hold different media types such as audio, video, animation, etc. [10, 11]. The AAC bitstream, being one audio standard supported in MPEG-4, can also be packed in such MPEG-4 file format. In this case, additional MPEG-4 compliant parser might be required to strip out AAC audio track which is then passed through AAC decoder to get all AAC specific information controlling audio reconstruction. If multiple media types present in one MPEG-4 bitstream, different kinds of synchronizations such as Audio and Video Synchronizations might also have to be maintained in a real-time system if such concurrency requirement is needed.

The major functionality of bitstream parsing module is to unpack all AAC specific parameters from audio raw data, such as global gain, window and group information, section data, scalefactor data, pulse data, TNS data, etc. [2]. Those parameters are then passed to different decoder modules/tools as control information, to help reconstruction of the final decoded samples.

21.2.3.2 Noiseless Decoding

The first step of reconstructing audio samples starts from noiseless decoding of all scalefactors and quantized spectral data. To better understand how data is packed, viewing it from encoder side could be helpful. For AAC encoding, spectrum coefficients are divided into different bands, called scalefactor band, within each a common scalefactor will be applied as gain to all spectral data to shape the quantization noise. The scalefactor band partition is related to critical band in human perception system. It is also sampling rate dependent. To achieve maximum compression, both scalefactors and four-tuples or two-tuples quantized spectral coefficients are entropy-encoded via different Huffman codebooks, starting from low frequency and moving towards high frequency. Huffman codebook selection is packed in the section information within the bitstream along with compressed spectrum data. Within each section, same Huffman codebook is used. The total number of section allowed could be up to the number of scalefactor bands. There are 11 Huffman codebooks for spectrum coefficients, and one additional Huffman codebook for scalefactor. Each spectrum Huffman codebook has its own sign feature (i.e., can represent either signed or unsigned coefficients), dimension feature (i.e., can represent coefficients either in four-tuples or two-tuples) and its own allowed largest absolute value (LAV). Beyond spectrum and scalefactor codebooks, there are special codebooks like "zero" codebook, "intensity" codebook, or "noise substitution" codebook which can be used when no actual quantized spectral data is transmitted [2]. The latter two kinds of codebooks will be further discussed in particular tools/modules using them.

In decoder side, after Huffman decoding of scalefactors and spectrum, indexes for four-tuples or two-tuples spectral data are obtained. Further index unpacking process is necessary to generate quantized spectrum data. The unpacking process is based on the sign feature of corresponding Huffman codebook and its dimension as well as its LAV. The unpacked index might be grouped or interleaved based on the window and group information packed in the bitstream. Thus, de-interleave might be necessary.

21.2.3.3 Error Resilience Tools in MPEG-4 AAC

Currently, AAC is becoming the choice of audio compression formats for a broad range of applications including internet, broadcasting, and streaming, etc. During transmission, bit errors with different patterns might be injected to the bitstream based on actual transmission channel characteristics. Thus, error resilience tools become more and more important in AAC. Some popularly used error resilience tools in MPEG-4 AAC include VCB11 (Virtual Codebook 11), RVLC (Reversible Variable Length Coding), and HCR (Huffman Codeword Reordering) [2]:

Virtual Codebook 11

Huffman codebook 11 is actually an ESC codebook for spectral data, which can encode much larger value than its LAV via escape sequence [2]. Thus, it is possible to encode huge spectral data into the bitstream while necessary. However, if only one ESC codebook is used, it is very hard to tell whether a decoded huge spectral data is what got actually encoded originally or due to bitstream errors. To resolve this, MPEG-4 AAC introduces 17 different codebook indices and these indices will replace the codebook information of codebook number "11" in the section data. All these 17 indices are still referring to the same ESC codebook. However, different indices can allow different maximum spectral values. Thus, if a decoded spectral data is larger than the LAV corresponding virtual index allows, it can be concluded that it is due to bitstream errors [2].

Reversible Variable Length Coding

Scalefactors are important information for reconstructing audio data, thus any scheme adding more protection to them can help with error resilience. RVLC is one of such options which can replace normal Huffman coding while encoding scalefactors. RVLC enables backward decoding if forward decoding faces errors. On the other hand, to accommodate backward decoding features, additional information or overhead has to be transmitted in the bitstream, like the reversible global gain, the length of RVLC payload, and additional information on PNS or intensity stereo if they present [2]. RVLC improves error robustness at the cost of choosing more redundant Huffman codebook and packing additional information in the bitstream which slightly reduce the coding efficiency.

Huffman Codeword Reordering

The basic idea of HCR is to give different levels of protection to codewords according to their importance and minimize error propagation. The scheme places codewords with high importance or priority (called priority codewords, or PCWs) [2] at the known position in the bitstream, i.e., at the beginning of known-length segments. Therefore, as long as there are no bitstream errors in PCWs, they can be correctly decoded even bitstream corruption happens in other codewords (called non-PCWs). As PCWs are scattered at specific locations in the bitstream, there are gaps between PCWs. And a special reordering algorithm is adopted to sort non-PCWs according to their particular importance and then reposition them into those gaps. It is very clear that HCR itself does not increase the encoded bitstream size as it is only reordering the codewords. However, it does pay the price of a more complicated encoding and decoding process in bitstream packing and parsing portion [2].

From the above descriptions on these error resilience tools, it can be easily found out that error resilience audio object type has different bitstream syntax, thus ER object is not compatible to non-ER AAC decoder.

21.2.3.4 Inverse Quantization and Rescaling

After Huffman decoding, the quantized spectrum data goes through nonuniform inverse quantization block to get the un-scaled de-quantized spectra. Then corresponding scalefactor is translated to gain value to be applied on top of inverse quantization results.

For long window case, inverse quantization and rescaling is more straightforward. While for short window case, if multiple groups present, the same scalefactor will be applied to all spectral coefficients in the scalefactor window bands within the same group [2]. Therefore, inverse quantization and rescaling module might also require data de-interleaving and is also controlled by side information embedded in the bitstreams.

21.2.3.5 M/S Stereo Processing

M/S tool is popularly used in channel pairs to achieve further coding efficiency when common window is used for AAC Main, LC, SSR, and LTP object types, etc. To support this feature, Mid/Side decision information has to be packed in the bitstream, so the de-quantized channel pair spectra can be converted to left and right information. Whether M/S coding is used or not is indicated by a set of one-bit flag according to window group and scalefactor band in the encoded data [2]. Sometimes M/S related information embedded in the bitstream might be used for some other optional tools which are "mutually exclusive" to M/S. Such information will be reiterated in the description for those particular tools.

21.2.3.6 Perceptual Noise Substitution

Sometimes it can achieve more coding efficiency to derive spectrum coefficients in some scalefactor bands from random numbers rather than quantized spectral data. In such cases, no spectral data needs to be transmitted. Instead, noise energy data will be packed in the bitstream. And "noise substitution" codebook can be used by setting PNS flag on scalefactor band and group basis. In such scenario, i.e., while PNS is used, M/S tool should not be applied. However, to achieve further coding efficiency, in case of a channel pair, some M/S related bits in the bitstream syntax can be reused for noise correlation control if both channels have PNS data in the same scalefactor band and group [2].

For error resilience AAC object, if reverse variable length coding tool is used, additional information need to be packed in the bitstream in order to accommodate decoding PNS-related data from backwards. And the PNS information reconstructing process is slightly different too [2].

21.2.3.7 Intensity Stereo Processing

This is another popular tool used for channel pairs. It can be activated by selecting either one of the two intensity Huffman codebooks on selective scalefactor band in the right channel of a channel pair. Unlike M/S stereo, for intensity stereo, outputs of both left and right channels are derived from one set of spectral coefficients with different panning, etc. As such, it has the restriction that common window information should be shared in both channels. Two special codebooks can be used for intensity stereo, one for in-phase, the other one for out-of-phase. However, the actual phase applied finally can be switched from in-phase to out-of-phase or vice versa, based on M/S activation flag information packed in the bitstream. This is based on the mutually exclusive characteristic between M/S tool and intensity stereo tool. In other words, in any particular scalefactor band and group, when intensity stereo tool is used, M/S stereo tool should not be applied. Thus, M/S related bits can be reused in case of intensity stereo presents to achieve maximum coding efficiency [2].

21.2.3.8 LTP Processing

For signals with many tonal components, LTP, using 1 tap of IIR filter in time domain for forward adaptive prediction, is very effective to reduce the redundancy between successive frames. LTP can only be used for long window case, and can be turned on or off on frame basis based on the LTP flag setting. LTP coefficients are transmitted as part of the side information in bitstream [2]. For channel pair, left channel and right channel could have different LTP settings.

Attention needs to be paid while both LTP and PNS data present in one particular band. As PNS data indicates the corresponding spectral data is derived from random vector instead of any decoded or prediction-reconstructed spectrum, accordingly, LTP should not take effect in such scenario.

It should be kept in mind that LTP is different from frequency domain prediction, another optional tool which can only be used for AAC Main audio object type (or AAC Main profile in MPEG-2 AAC) [1, 2], and thus beyond the scope of this particular AAC system.

21.2.3.9 TNS Processing

TNS is an effective tool to shape quantization noise in time domain by applying all-pole filter in frequency domain on selected spectral data. All TNS-related information are packed in the bitstream, including number of filters used, resolution of coefficients, frequency region it should be applied, filter order, filtering direction, and coefficients information. The range of values in above fields could have dependency on window information. Some of them might be also related to audio object type. For example, maximum TNS filter order is only 7 for short window, while for long window it can be up to 20 for Main object and up to 12 for others [2]. Some other parameters, like maximum number of scalefactor bands TNS filter can be applied, can be also sampling rate dependent [2]. All these varieties should be taken into account at the system design stage while considering memory usage efficiency, to guarantee a robust decoding system implementation.

For audio object types with LTP tool adopted, for example, MPEG-4 AAC LTP object, TNS can also be used. In other words, LTP and TNS are not mutually exclusive features. However, in such case that both present, as LTP is using time domain prediction with corresponding synthesis part in frequency domain to reconstruct predicted spectrum, there is a feedback path for LTP (LTP loopback in Fig. 21.2), and TNS analysis filter is necessary in the loopback. At the same time, LTP synthesis has to come before TNS synthesis filter in the main decoder path. Both restrictions are to guarantee both tools working on proper signals they are designed to apply on, as illustrated in Fig. 21.2.

21.2.3.10 Filter Bank

Filter bank is the key module to transform the time/frequency domain signal back to PCM data. Its core is the inversed modified discrete cosine transform (IMDCT), plus windowing and overlap-adding. Window length is 2048 in case of long window vs. 256 for short window. The actual window shape could be either Kaiser-Bessel derived or sine window, depending on the window sequence and shape information packed in the bitstream. IMDCT is performed on half of the window-length time–frequency domain values to generate time domain data with window-length. Then windowing is applied. Finally, the first half of the resulted sequence is added to the second half of windowed sequence in previous block (overlap-adding) to generate the final output [2].

Beyond what we described here in this particular AAC decoder system, there are some other optional tools available in main MPEG4-AAC decoder program flow if complete audio object types are required to be supported, such as dependently switched coupling, independently switched coupling, and gain control, etc. [2]

21.2.4 Computational Complexity of MPEG-4 AAC

MPEG-AAC has several computational complex modules which in total take most of implementation MIPs (millions of instructions per second), such as Huffman decoding and inverse quantization, and IMDCT. For some cases, TNS could be part of the group too. The final implementation MIPs for functional modules and the entire decoder system can depend on lots of factors, including bit rate, sampling rate and actual audio signals' characteristics. Generally speaking, to optimize the actual decoding system, modules like IMDCT can be part of the focuses. A lot of fast algorithms on IMDCT have been proposed which can achieve high efficiency while being implemented on DSP platforms [12–15].

21.3 HE-AAC

21.3.1 Overview

Generally speaking, for human listening experience, low frequency component is critical and more important than high frequency component. Thus, algorithms and models emphasizing low frequency importance are commonly used in audio encoders. However, while talking about audio quality, high frequency components could play critical role. That is why lots of traditional audio encoders have to use a lot of bits encoding high frequency components to achieve required audio quality which reduces their coding efficiency. A new tool, called SBR [2], comes in focusing on improving coding efficiency of high frequency components. The principles of SBR processing have been thoroughly discussed in lots of papers [3, 16–18], and the algorithm details for SBR processing in MPEG-4 HE-AAC can be found in [2]. In this chapter, just a high-level review is given. The basic idea behind SBR is that audio signal between low frequency and high frequency could have strong correlation and so it is possible to reconstruct high band signal based on low band signal and some additional critical information about the similarity and/or relationship between spectrum envelopes from high band and low band. High Efficiency Advanced Audio Coding (HE-AAC) is actually composed of MPEG-4 core AAC algorithm plus SBR processing. And SBR tool can improve coding efficiency for high frequency portion of an audio signal significantly compared with traditional audio compression tools.

21.3.2 HE-AAC, The Superset of MPEG-4 AAC

In MPEG-4 standard, SBR is an additional tool which can be combined with AAC. Furthermore, SBR part can act as a pure "Plus" or a "postprocessor" of AAC, which reconstructs high band from low band based on SBR-related side information. As such, SBR can be applied or combined with different MPEG-4 AAC audio object types, such as LC object, LTP object, and ER AAC LC objects. Therefore, HE-AAC is a superset of AAC [5, 16]. It can improve AAC performance by either increasing decoded audio bandwidth and thus improving audio quality with a given bit rate, or reducing the encoded bit rate for a given quality requirement as only the low band

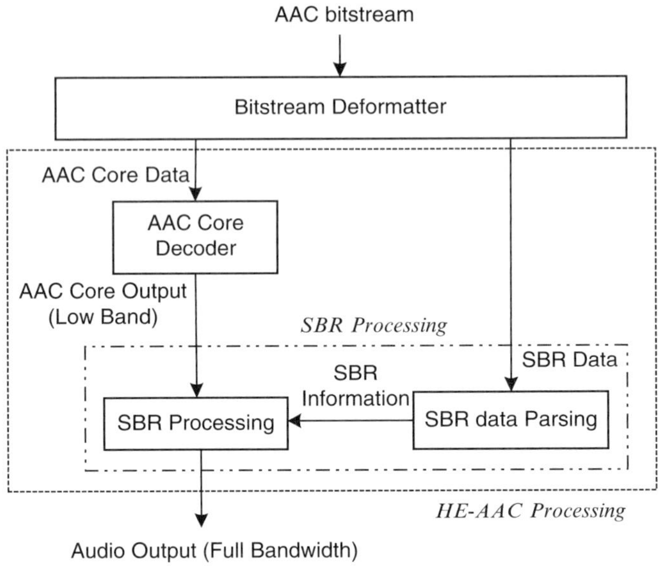

Fig. 21.3 HE-AAC is the superset of AAC with SBR as "post-processor"

spectral data is actually AAC encoded and for high band, only limited information to reconstruct high band from low band are transmitted as side information.

The relationship between SBR and AAC in an HE-AAC decoding system can be illustrated as in Fig. 21.3.

21.3.3 HE-AAC, A Multirate System

Based on Fig. 21.3 and above description of HE-AAC decoder, it can be easily told that HE-AAC system is actually running on different rates for core decoder part and SBR part. In other words, HE-AAC is a dual rate system [3, 18]. The sampling rate of the final output in the HE-AAC system is actually doubled compared with the core AAC decoder sampling rate within the system. Thus, attention needs to be paid while trying to figure out the final output sampling rate of an HE-AAC compliant decoding system. Details will be further discussed in the following section along with SBR signaling.

21.3.4 SBR Compatibility and Signaling

As HE-AAC is a superset of AAC, HE-AAC decoder should be able to decode AAC bitstreams without restrictions. In other words, AAC only bitstream is compatible with HE-AAC decoding system. On the other hand, for HE-AAC bitstreams, to achieve maximum backward compatibility with existing MPEG-4 AAC system, it can pack the additional SBR data into AAC bitstream in an AAC-backward-compatible way, i.e., packing in AAC extension payload. As such, HE-AAC bitstream syntax can be compliant with AAC only decoder. For example, an MPEG-4 AAC only decoder can decode HE-AAC bitstream by decoding core AAC-related bitstream portion only and skipping SBR-related payload and actual SBR processing. Clearly, by doing this, the final audio output quality will be reduced as the audio bandwidth is limited to the core AAC portion only and no high band of audio signals can be reconstructed. In other words, even HE-AAC and AAC bitstreams have maximum backward and forward compatibility with each other, HE-AAC-compliant decoders are still required while decoding HE-AAC bitstreams in order to achieve full output bandwidth and full audio quality.

There are different ways of signaling SBR content in the bitstream:

21.3.4.1 Implicit Signaling [5]

SBR content could be embedded in AAC extension payload with proper "ID" specific to SBR. In such cases, due to dual-rate feature of the HE-AAC system, the final output sampling frequency of the system might be different from (for example, might be doubled of) AAC core decoder sampling frequency (for applicable sampling rates). Thus, for system settings of the final output sampling rate, it might be required to perform precheck for SBR data in the bitstream. The precheck requires overhead. And sometimes the overhead could be significant as SBR data is not necessarily embedded in the bitstream from the first audio frame. In other words, it might require partial or most of the bitstream to be predecoded to tell whether SBR presents. In a real-time system, this might be difficult to achieve or sometimes even impossible. An alternate way to it is to always assume SBR presents, i.e., to always assume the system final output sampling rate is doubled of the original bitstream for applicable sampling rates. In the latter case, if SBR data does not present, SBR tool can be used as an upsampler with the upsampling factor of 2 to guarantee the final resulting sampling rate matching the setting of the system. The disadvantage of this method is that SBR processing is generally computationally complex, and it is actually "wasting" mips by using SBR tool just to do upsampling. Thus, there is a trade-off between system efficiency and compatibility.

The actual solution for a particular system supporting implicit signaling should be based on its own application requirement and system functionality. However, it has to be taken into account at the design stage as it is mandatory for HE-AAC compliance decoder to support implicit signaling.

21.3.4.2 Explicit Signaling [5]

Explicit signaling is done by explicitly defining SBR audio object type in the bitstream in either backward compatible way or in hierarchical way. While explicit signaling is used, implicit signaling is not applied. And there is no ambiguity of the system setting in this case.

21.3.5 SBR Decoding and Processing

Understanding the processing in encoder side can always help with a better understanding of decoder side. For an SBR encoder, audio samples are carefully analyzed and crossover frequency between low band and high band is chosen based on actual sampling frequency and target bit rate. AAC core encoder can run on low band and get good coding efficiency as it is originally designed for. The similarity and dissimilarity between high band spectrum envelope and low band spectrum envelope are transformed into SBR data to be further encoded and packed into bitstream multiplexing with AAC encoded data [3]. Different Huffman codebooks are used for various SBR data encoding.

On the decoder side, the bitstream deformatter will spit out AAC core decoder data and SBR data. AAC core decoder data will then be fed into core AAC decoder to generate an audio frame with one audio frame length of samples, which will be further fed into analysis quadrature mirror filter (QMF) bank. On the other path, SBR data goes through SBR bitstream parsing, Huffman decoder and inverse quantization. The resulting SBR information will guide high frequency band generator and envelope adjustor. High frequency band generator does transposition of low band spectrum envelope to high band, based on QMF analysis filter output running on low band audio frame generated by core decoder. The reconstructed high frequency band spectrum then goes through envelope adjuster, which is also guided by decoded SBR data. Finally synthesis QMF filter bank convert both the output of analysis QMF bank (for low band) and output of envelope adjuster to get the final audio PCM data [2].

21.3.6 Computational Complexity of SBR Processing

There are two versions of SBR processing available in MPEG-4 standard with different complexities: High Quality (HQ) SBR tool [2] and Low Power (LP) SBR tool [2]. There is no difference between the two versions of SBR processing and bitstream packing part from SBR encoder side, as such both share the same bitstream syntax. However, from decoder side, different versions adopt different processing algorithms with different complexities and thus yield different audio outputs.

The major difference between HQ SBR and LP SBR is the feature of filter bank, including both analysis QMF bank and synthesis QMF bank. They are just the counterpart of each other. HQ analysis QMF filter bank splits time domain signal into 32 subbands by generating 32*32 complex subband samples from 1,024 core decoded time domain samples. Its synthesis part consists of a 64 subband QMF bank which transforms QMF domain signals into real-valued sample in time domain. On the other hand, LP version of SBR processing only runs on real-valued samples for both analysis QMF bank and corresponding synthesis bank. Thus, all imaginary valued processing are skipped and computational complexity can be significantly reduced. However, alias effect might present during pure real-valued processing, and hence additional functional modules like alias detector and alias reduction needed [2]. Generally speaking, LP SBR is less complex in computation and consumes less MIPs than HQ SBR. However, the HQ SBR offers higher audio quality. In a particular system, which version of SBR processing is used should be again based on the actual application requirement such as demanded audio quality, computation ability available and complexity or power consumption requirement on the particular platform.

In SBR implementation, the transposer and QMF filter bank for both analysis and synthesis sides are good examples of computational complex modules [3].

21.4 HE-AAC V2

21.4.1 Overview

High Efficiency Advanced Audio Coding V2 (HE-AAC V2) is standardized via applying another new tool called PS tool, which exploits more features from stereo audio signals, on top of HE-AAC [4]. In other words, HE-AAC V2 is combining AAC core algorithm, SBR tool, and PS tool together, with the latter two act as "postprocessing" modules of AAC, to achieve further low bit rate for stereo signal encoding on the basis of already-highly-coding-efficient AAC and HE-AAC system. PS bitstream might maintain good audio quality at very low bit rate, for example, at bit rates even below 10 kbps [19].

21.4.2 HE-AAC V2, A Superset of HE-AAC and MPEG-4 AAC

Just like HE-AAC adopts SBR on top of AAC core to make itself a superset of MPEG-4 AAC, HE-AAC V2 applies PS on top of HE-AAC and becomes a superset of HE-AAC, and hierarchically again a true superset of MPEG-4 AAC [5]. Therefore, an HE-AAC V2 compliant decoder can decode both HE-AAC and AAC bitstreams without restrictions. The relationship among AAC, SBR, and PS in an HE-AAC V2 decoding system is illustrated in Fig. 21.4. It is very clear that HE-AAC V2 system, as it completely contains HE-AAC system inside, is a multiple-rate system. And the final channel settings of the audio output could be different from what it presents in the original bitstream for core AAC portion too, due to PS reconstruction of stereo information.

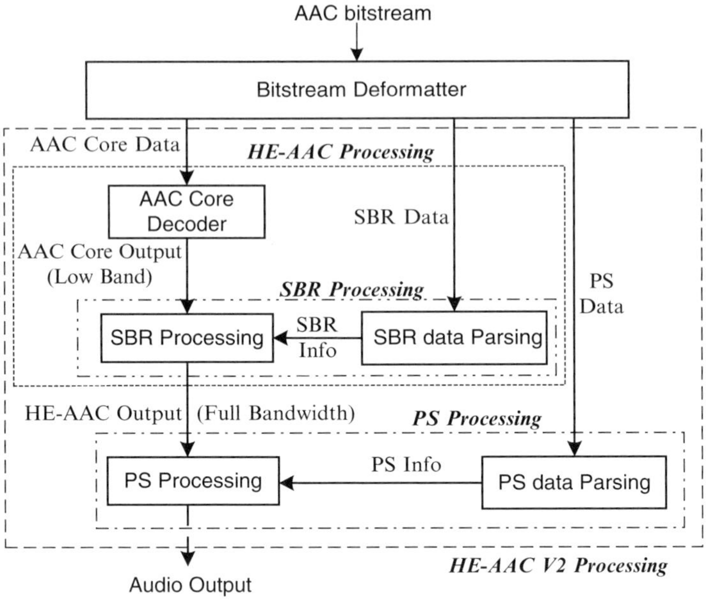

Fig. 21.4 Relationship among AAC, SBR, and PS in an HE-AAC V2 system

21.4.3 Compatibility of HE-AAC V2

As HE-AAC V2 is proposed on top of HE-AAC and thus AAC core too, to achieve maximum compatibility with existing HE-AAC system and MPEG-AAC only system, from bitstream syntax point of view, HE-AAC V2 bitstream can pack the additional PS data in a backward compatible way, i.e., packing PS data into the SBR extension payload of a mono HE-AAC stream. In the other words, a non-PS compliant HE-AAC decoder system can decode HE-AAC V2 bitstream by focusing on HE-AAC decoding and skipping PS-related information and processing, and finally generates an audio output at full bandwidth, but with only mono signal. And a core AAC only decoder system, which does not support either SBR or PS capability, can decode HE-AAC V2 bitstream by skipping both SBR and PS-related information and processing and generate final audio output with only half of the bandwidth and mono signal. On the other hand, HE-AAC V2 compliant decoder can decode HE-AAC bitstream or MPEG-AAC only bitstream and generate output at full bandwidth and appropriate channel settings [4, 5].

21.4.4 Signaling for PS

Similarly to SBR, PS can be either implicitly signaled or explicitly signaled [4, 5].

For implicit signaling, PS data presents in SBR extension with proper "ID" specified to PS. In this case, there is no way to tell whether PS data is embedded or not without actual decoding or checking the bitstream in advance. Similarly as SBR implicit signaling, for final channel settings, system might be required to either do prechecking of PS data, or always assume nonmono output mode. In latter case, if it turns out PS data not present in a bitstream, additional effort might be needed to match the actual bitstream scenario with the assumed system presetting.

For explicit signaling of PS, audio object type for PS is explicitly specified in the bitstream. There will be no ambiguity for channel settings in such case.

21.4.5 PS Processing and Complexity

Related studies conducted in PS algorithm and processing can be found in papers like [19–21], and PS processing specific to MPEG-4 AAC standard is described in [4]. This section only gives a simple overview as below: In encoder side, PS tool is used to extract the stereo feature of audio signals into a small number of parameters, such as inter-channel intensity difference (IID), inter-channel phase difference (IPD), overall phase differences (OPD), and inter-channel coherence (ICC). These PS data can cover the relative level, phase behavior, and the similarity between left and right channels. They are the basis for reconstructing stereo signals from the downmixed input. PS data is then encoded and packed with coded monaural signal. In PS decoder side, monaural samples go through QMF analysis and low frequency filtering to achieve high frequency resolution, and are further passed into stereo processing together with stereo parameters after de-correlation. Thus, stereo information in QMF domain is generated. These QMF domain signals further pass through QMF synthesis filter bank and finally generate output samples for both left and right channels in time domain.

Attention needs to be paid for implementing PS compliant audio decoder on top of HE-AAC decoding system, as PS processing requires HQ SBR [4]. This affects PS decoding complexity. However, as HE-AAC V2 system performs most decoding and quite some computation-complex modules (for example, the entire HE-AAC processing embedded in the system) on single channel, the overall computational complexity is acceptable.

21.5 Conclusions

MPEG-4 AAC decoder family consists of three members: AAC, HE-AAC, and HE-AAC V2, with each of the latter two being a true superset of its predecessor. Each of the latter two members can achieve an overall higher coding efficiency than its predecessor too. Furthermore, the bitstream syntax for AAC decoder family is designed in a way to guarantee maximum compatibility. Thus, AAC decoder family codecs are commonly adopted in a broad range of current mobile TV, portable media players, and internet applications.

Moreover, related compression and coding techniques with the backward compatibility are still being employed in the development and standardization for audio transmission, storage, and broadcasting. A good representative is the MPEG Spatial Coding standard which allows the compression of multichannel audio down to bit rates which so far are used for the compression of mono-channel or stereo-channel through the above MPEG AAC family codec. More importantly, this backward compatibility can be used to upgrade existing distribution infrastructures for stereo or mono audio content (radio channels, Internet streaming, music downloads, etc.) towards the delivery of multichannel audio while retaining full compatibility with existing receivers [22–28].

References

1. ISO/IEC JTC 1/SC 29/WG 11 MPEG International Standard, "Advanced Audio Coding (AAC)", ISO/IEC IS 13818-7:2003 Subpart 7.
2. ISO/IEC JTC 1/SC 29/WG 11 MPEG International Standard, "General Audio Coding (GA) – AAC, TwinQV, BSAC", ISO/IEC IS 14496-3:2005(E) Subpart 4.
3. A. Ehret, M. Dietz, K. Kjörling, "State-of-the-Art Audio Coding for Broadcasting and Mobile Applications", in 114th AES Convention, Amsterdam, March 22–25, 2003.
4. ISO/IEC JTC 1/SC 29/WG 11 MPEG International Standard, "Technical Description of Parametric Coding for High Quality Audio", ISO/IEC IS 14496-3:2005(E) Subpart 8.
5. ISO/IEC JTC 1/SC 29/WG 11 MPEG International Standard, "Main", ISO/IEC IS 14496-3:2005(E) Subpart 1.
6. K. Brandenburg, "Perceptual Coding of High Quality Digital Audio", in Applications of Digital Signal Processing to Audio and Acoustics, M. Kahrs, K. Brandenburg, Eds., Chapter 2, pp. 39–83. Kluwer, Boston, 1998.
7. T. Painter, A. Spanias, "Perceptual Coding of Digital Audio", Proceedings of IEEE, 88 (4), pp. 451–513, 2000.
8. ISO/IEC JTC 1/SC 29/WG 11 MPEG-4, "Audio Verification Test Results: Audio on the Internet", ISO/IEC JTC 1/SC 29/WG 11 N2425, Oct 1998.
9. Gilbert A. Soulodre, Theodore Grusec, Michel Lavoie, Louis Thibault, "Subjective Evaluation of State-of-the-Art 2-Channel Audio Codecs", in 104th AES Convention, Amsterdam, May 16–19, 1998.
10. ISO/IEC "Coding of Audio-visual Objects – Part 1: Systems (MPEG-4 Systems, 2nd edition)", ISO/IEC IS 14496-1:2001, 1999.
11. C. Herpel, G. Franceschini, D. Singer, "Transporting and Storing MPEG-4 Content", in the MPEG-4 Book, F. Pereira, T. Ebrahimi, Eds., Chapter 7. Prentice Hall, Englewood Cliffs, NJ, USA, 2002.
12. Gi Lee, "A New Algorithm to Compute the Discrete Cosine Transform", IEEE Transactions on Acoustics, Speech, and Signal Processing, 32(6), pp. 1243–1245, 1984.
13. Liu, Wen-Chieh Lee, "A Unified Fast Algorithm for Cosine Modulated Filter Banks in Current Audio Coding Standards", Journal of the Audio Engineering Society, 47(12), pp. 1061–1075, 1999.
14. Y.H. Fan, V.K. Madisetti, R.M. Mersereau, "On Fast Algorithms for Computing the Inverse Modified Discrete Cosine Transform", IEEE Signal Processing Letters, 6(3), pp. 61–64, 1999.
15. V.B. Britanak, K.R. Rao, "An Efficient Implementation of the Forward and Inverse MDCT in MPEG Audio Coding", IEEE Signal Processing Letters, 8(2), pp. 48–51, 2001.
16. Martin Wolters, Kristofer Kjörling, Daniel Homm, Heiko Purnhagen, "A Closer Look into MPEG-4 High Efficiency AAC", in 115th AES Convention, New York, Oct 10–13, 2003.

17. Martin Dietz, Lars Liljeryd, Kristofer Kjörling, Oliver Kunz, "Spectral Band Replication, a Novel Approach in Audio Coding", in 112th AES Convention, Munich, May 10–13, 2002.
18. Andreas Ehret, Kristofer Kjörling, Jonas Rödén, Heiko Purnhagen, Holger Hörich, "aacPlus, Only a Low-bitrate Codec?", in 117th AES Convention, San Francisco, Oct 28–31, 2004.
19. Jeroen Breebaart, Steven van de Par, Armin Kohlrausch, Erik Schuijers, "High-Quality Parametric Spatial Audio Coding at Low Bitrates", in 116th AES Convention, Berlin, May 8–11, 2004.
20. Erik Schuijers, Jeroen Breebaart, Heiko Purnhagen, Jonas Engdegård, "Low Complexity Parametric Stereo Coding", in 116th AES Convention, Berlin, May 8–11, 2004.
21. Jonas Engdegård, Heiko Purnhagen, Jonas Rödén, Lars Liljeryd, "Synthetic Ambience in Parametric Stereo Coding", in 116th AES Convention, Berlin, May 8–11, 2004.
22. Jeroen Breebaart, Christof Faller, "MPEG Surround", in Spatial Audio Processing: MPEG Surround and Other Applications, Chapter 6. Wiley, England, Jan 2008.
23. C. Faller, F. Baumgarte, "Binaural Cue Coding Applied to Stereo and Multi-channel Audio Compression", in 112th AES Convention, Munich, May 10–13, 2002.
24. J. Herre, K. Kjoerling, J. Breebaart, C. Faller, S. Disch, H. Purnhagen, J. Koppens, J. Hilpert, J. Roeden, W. Oomen, K. Linzmeier, K.S. Chong, "Mpeg Surround – the iso/mpeg Standard for Efficient and Compatible Multi-channel Audio Coding," in Preprint 122nd AES Convention, May 2007.
25. C. Faller, "Parametric Multi-channel Audio Coding: Synthesis of Coherence Cues", IEEE Transactions on Speech and Audio Proceedings, 14(1), pp. 299–310, 2006.
26. J. Herre, C. Faller, C. Ertel, J. Hilpert, A. Hoelzer, C. Spenger, "MP3 Surround: Efficient and Compatible Coding of Multi-channel audio", in Preprint 116th AES Convention, May 2004.
27. C. Faller, "Coding of Spatial Audio Compatible with Different Playback Formats", in Preprint 117th AES Convention, Oct 2004.
28. J. Breebaart, G. Hotho, J. Koppens, E. Schuijers, W. Oomen, and S. van de Par, "Background, Concept and Architecture for the Recent mpeg Surround Standard on Multi-channel Audio Compression," Journal of Audio Engineering Society, 55(5), pp. 331–351, 2007.

Chapter 22
Spatial Audio Coding and MPEG Surround

Christof Faller

22.1 Introduction

Surround sound has been widely utilized in cinemas for decades, but wide adoption of home cinema surround was only enabled recently by the digital video disc (DVD). While a compact disc (CD) stores its stereo audio content uncompressed as 16-bit PCM at a sampling rate of 44.1 kHz, the DVD stores its six multichannel surround channels (five main audio channels plus a low-frequency effects channel) compressed with the perceptual audio coder AC-3 from Dolby Laboratories. The multichannel surround signal would require too much storage space on the DVD and thus it is compressed.

The most prominent perceptual audio coder is MP3 (MPEG-1 Layer 3). Being a relatively old audio coder it only supports up to two audio channels. Many perceptual audio coders can also code multichannel surround signals and achieve roughly a compression ratio of 10. *Spatial audio coding* is motivated by the need to code multichannel audio at lower bitrates. It enables coding of multichannel audio at bitrates comparable with mono or two-channel stereo bitrates. Thus, spatial audio coding can enable multichannel surround for applications where the bitrate is constrained to a mono or stereo bitrate. Further, spatial audio coding can be compatible with mono or stereo, facilitating upgrading of existing mono and stereo services to multichannel surround.

This chapter is organized as follows. Section 22.2 discusses multichannel surround and its benefits compared with the widely used stereo system. In Sect. 22.3, spatial audio coding is discussed and motivated in the context of conventional perceptual audio coding and matrix surround. In order to explain and motivate spatial audio coding, basic spatial hearing knowledge is presented in Sect. 22.4. The application of these principles to audio coding and spatial audio coding is explained in

C. Faller
Audiovisual Communications Laboratory, EPFL, Lausanne, Switzerland
e-mail: christof.faller@epfl.ch

F.-L. Luo (ed.), *Mobile Multimedia Broadcasting Standards: Technology and Practice*,
DOI: 10.1007/978-0-387-78263-8_22,

Sect. 22.5. Aspects of the commercially available MP3 Surround audio coder and the standards-based MPEG Surround coder are discussed in Sect. 22.6. Conclusions are presented in Sect. 22.7.

22.2 Multichannel Surround

A stereo audio playback system is illustrated in Fig. 22.1. Usually the loudspeakers are located in the front, 30° to the left and right from the optimal listening position. The listener and the two loudspeakers form a regular triangle. When a stereo signal is appropriately recorded or mixed, a front sound stage can be realistically reproduced. Different objects or instruments appear as virtual sources somewhere between the two loudspeakers. Not only can the stereo system reproduce virtual sources at any direction between the loudspeakers, but also ambience and a spatial impression can be reproduced. Section 22.4 discusses in detail how stereo signal properties relate to virtual source positions and other perceived aspects.

The stereo playback system has three weaknesses, which are overcome or partially overcome by a surround audio playback system:

- Positions of virtual sources are limited to directions between the two front loudspeakers.
- The reproduction of spatial impression, i.e., the impression of being in a room or concert hall is limited.
- When the listener position is not centered between the two loudspeakers center virtual sources move to the nearer loudspeaker.

The third weakness is illustrated in Fig. 22.2. Panel (a) illustrates a perceived front stage sound image for a listener in the optimal listening position ("sweet spot"). Panel (b) illustrates how the center virtual sources are located close to the nearer loudspeaker when the listener is not in the sweet spot.

For home music listening, often not much attention is paid to virtual source positions and many consumers do never or only seldomly listen to a stereo audio playback system from a sweet spot position. The afore mentioned weaknesses are more

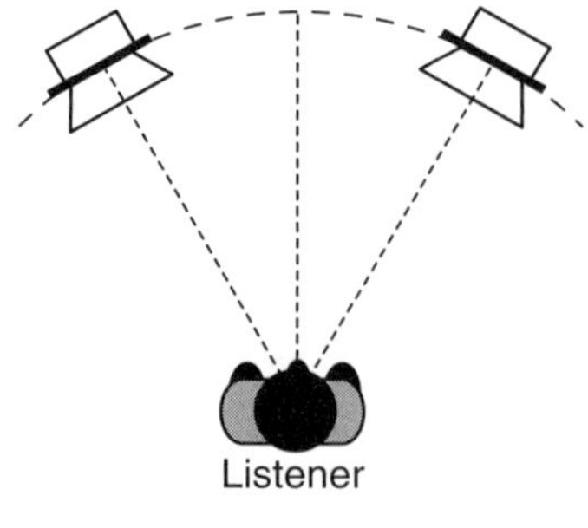

Fig. 22.1 Stereo audio playback system

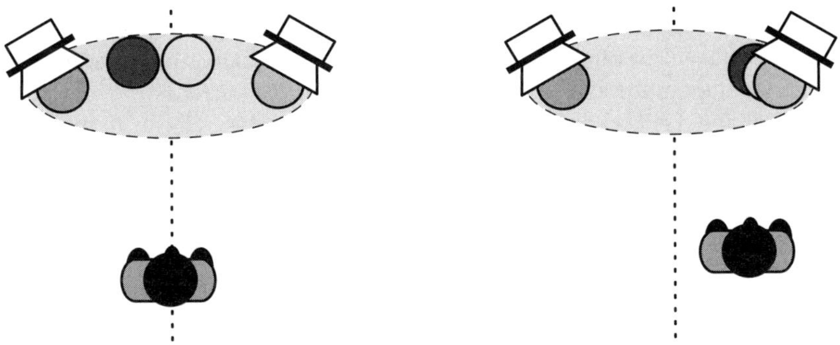

Fig. 22.2 The sweet spot problem: (**a**) listener in an optimal listening position, (**b**) listener off center

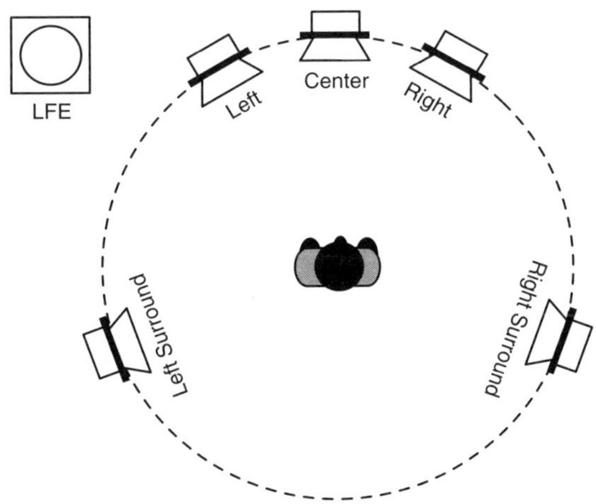

Fig. 22.3 5.1 surround audio playback system

severe for a cinema application. In films usually the dialogue is located in the center and thus should be perceived from the center of the movie screen. For all off-center listeners the dialogue appears from the side when a stereo playback system is used. This motivates why in cinemas an additional center loudspeaker has been used for long. When the dialogue is given to the center loudspeaker all listeners perceive it as being located in the center as is desired. Also it is desirable for a cinema audio system to be able to play sound effects from different directions than only from front directions. This motivates the use of surround (rear) loudspeakers. Surround loudspeakers have the additional benefit that they can be used to improve also spatial impression. By emitting simulated lateral reflected sound from the rear side concert hall acoustics can be simulated.

Figure 22.3 illustrates a multichannel surround audio playback system. The front left and right loudspeakers are at similar 30° left and right positions as for the stereo audio playback system, motivated by standard stereo backward compatibility. The front center channel has the purpose to improve virtual source localization for listeners who are not in the sweet spot. According to the 5.1 surround standard [1] the surround left and right loudspeakers are at directions $\pm 110°$ relative to the forward axis. This is a compromise between loudspeakers on the side (important for spatial impression) and loudspeakers in the rear (for rear sound effects). The LFE channel is for *low-frequency effects* for which the main loudspeakers usually can not build enough sound pressure. A designated low-frequency loudspeaker (subwoofer) is used to play back the LFE channel, which contains frequencies up to 120 Hz. The LFE signal is boosted by 10 dB prior to amplification and playback to support higher effective signal levels.

22.3 From Audio Coding and Matrix Surround to Spatial Audio Coding

The goal of this section is to provide background and prior art knowledge for surround audio signal representation and coding. First, conventional perceptual audio coding is described as is used for storing surround audio on DVD. Second, matrix surround, the original cinema and home surround audio system, is described. Matrix surround has some conceptual similarities to spatial audio coding. Third, spatial audio coding is briefly described in the context of perceptual audio coding and matrix surround.

22.3.1 Conventional Perceptual Audio Coding

A major breakthrough in audio coding was enabled by incorporating a receiver model into subband audio coders. The receiver model, usually denoted perceptual model, determines the masked threshold as a function of frequency and time, i.e., the level of noise that can be added at each frequency and time such that it is just not noticeable. Coding efficiency is improved by controlling the quantization error such that it is just below the masked threshold and thus not perceptible. Audio coders employing this principle are denoted *perceptual audio coders*. This coding principle was introduced for speech coding by Zelinski [2]. Brandenburg [3] applied this technique to wideband audio signals.

When several audio channels need to be encoded, such as when coding stereo or multichannel surround audio signals, redundancy between the audio channels may be explored for reducing the bitrate. *Mid/side* (M/S) coding [4] reduces the redundancy between a correlated channel pair by transforming it to a sum/difference channel pair prior to quantization and coding. The masked threshold depends on the interaural signal properties of the signal (masker) and quantization noise (maskee). When coding stereo or multichannel audio signals, the perceptual model needs to consider this interaural dependence of the masked threshold. This dependence is often described by means of the *binaural masking level difference* (BMLD) [5].

Today many proprietary perceptual audio coders [6–8] and international standards-based perceptual audio coders [9–14] are commercially available. When requiring that the coded audio signal can not be distinguished from the original audio signal, state-of-the-art perceptual audio coders are able to reduce the bitrate of CD audio signals by a factor of about 10. When higher compression ratios are required, the audio bandwidth needs to be reduced or coding distortions will exceed the masked threshold.

For many applications a compression factor of 10 is not sufficient. If lower bitrate is to be achieved with perceptual audio coders, the coding distortions exceed the masked threshold resulting in annoying artifacts. To avoid temporal and noise-like artifacts at low bitrates, perceptual audio coders have been extended with parametric coding principles. One such parametric technique applicable for stereo or multichannel audio coding is *intensity stereo coding* (ISC) [15]. ISC is a parametric technique for coding of audio channel pairs. It can be viewed as a technique applying simple spatial audio coding principles. Since it considers only interchannel level differences and operates in the critically sampled audio coder spectral domain its performance is however very limited compared with spatial audio coding. Another parametric technique that has been applied within the framework of perceptual audio coders is denoted as *spectral bandwidth replication* (SBR) [16]. SBR applies perceptual audio coding at frequencies below a certain frequency (e.g., $<4-12$ kHz) and parametric coding above that frequency. Very little information is used to parameterize the upper spectrum, and the decoder uses information from the lower spectrum to generate a perceptually meaningful upper spectrum. Ideas related to SBR were already introduced by Makhoul [17] in the context of speech coding.

22.3.2 Matrix Surround

Matrix surround was the technology that enabled wide adoption of multichannel surround sound in cinemas in the early 1970s. Matrix surround is backward compatible to stereo. A multichannel surround audio signal is downmixed to a stereo signal using a specific linear matrix. This stereo signal can be used as a conventional optical stereo track on movies. A matrix decoder, fed with this stereo signal regenerates a multichannel surround signal. Later, matrix surround was also widely used on VHS video tapes enabling early home cinema surround by means of using a matrix surround decoder fed with the VHS stereo audio track. A similarity between spatial audio coding and matrix surround is that both are based on a backward compatible downmix of the surround signal.

22.3.2.1 Matrix Encoding

The audio channels are encoded as

$$\begin{bmatrix} l_t(t) \\ r_t(t) \end{bmatrix} = \begin{bmatrix} 1 & 0 & \frac{1}{\sqrt{2}} & j\frac{\sqrt{3}}{2} & j\frac{1}{2} \\ 0 & 1 & \frac{1}{\sqrt{2}} & -j\frac{1}{2} & -j\frac{\sqrt{3}}{2} \end{bmatrix} \begin{bmatrix} l(t) \\ r(t) \\ c(t) \\ l_s(t) \\ r_s(t) \end{bmatrix}, \tag{22.1}$$

where $l(t)$, $r(t)$, $c(t)$, $l_s(t)$, $r_s(t)$ denote the front left, front right, front center, surround left, and surround right audio channels, respectively. The j denotes a 90° phase shift.

Alternatively, (22.1) can be written as

$$\begin{aligned} \begin{bmatrix} l_t(t) \\ r_t(t) \end{bmatrix} &= \begin{bmatrix} 1 \\ 0 \end{bmatrix} l(t) + \begin{bmatrix} 0 \\ 1 \end{bmatrix} r(t) + \begin{bmatrix} \frac{1}{\sqrt{2}} \\ \frac{1}{\sqrt{2}} \end{bmatrix} c(t) + j \begin{bmatrix} \frac{\sqrt{3}}{2} \\ -\frac{1}{2} \end{bmatrix} l_s(t) + j \begin{bmatrix} -\frac{1}{2} \\ \frac{\sqrt{3}}{2} \end{bmatrix} r_s(t) \\ &= \mathbf{v}_l l(t) + \mathbf{v}_r r(t) + \mathbf{v}_c c(t) + \mathbf{v}_{l_s} l_s(t) + \mathbf{v}_{r_s} r_s(t), \end{aligned} \tag{22.2}$$

where the five vectors are unit vectors mapping the five input channels to the two output channels. The vectors are shown relative to $l_t(t)$ and $r_t(t)$ in Fig. 22.4.

22.3.2.2 Passive and Active Matrix Decoding

The simplest way of decoding a matrix encoded signal is to apply a static decoder matrix. Formulated in terms of the previously defined vectors, the matrix decoded channels are as follows:

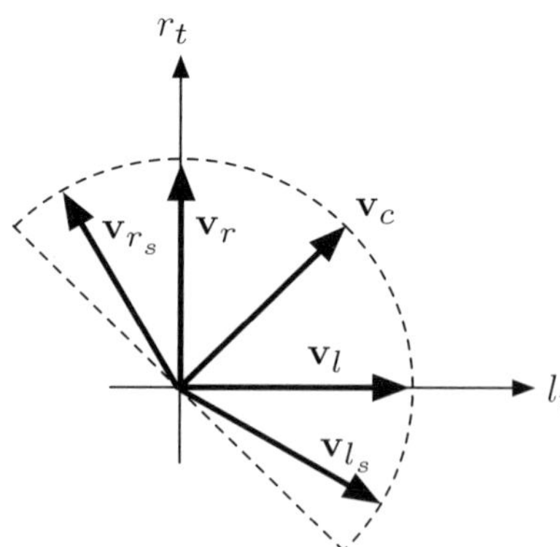

Fig. 22.4 The vectors for matrix encoding the five main channels of a 5.1 signal

$$
\begin{aligned}
\hat{l}(t) &= [l_t(t)\ r_t(t)]\,\mathbf{v}_\mathrm{l} \\
\hat{r}(t) &= [l_t(t)\ r_t(t)]\,\mathbf{v}_\mathrm{r} \\
\hat{c}(t) &= [l_t(t)\ r_t(t)]\,\mathbf{v}_\mathrm{c} \\
\hat{l}_\mathrm{s}(t) &= [l_t(t)\ r_t(t)]\,\mathbf{v}_{l_\mathrm{s}} \\
\hat{r}_\mathrm{s}(t) &= [l_t(t)\ r_t(t)]\,\mathbf{v}_{r_\mathrm{s}}\,.
\end{aligned} \tag{22.3}
$$

Each output channel is equal to the projection of the matrix encoded signal onto the corresponding matrix encoding vector.

As can easily be verified, the described encoding/decoding strategy perfectly reconstructs a channel in a scenario where only a single channel is active. However, the other channels are not zero in this case, i.e., there is a lot of crosstalk. The situation when only one channel is active can easily be detected. In such a situation one could set all the other audio channels to zero and thus improve the channel separation. This is a principle idea behind "active matrix decoders" [18].

Often the rear channels of matrix systems are low-pass filtered and delayed. The low-pass filtering mimics late reverberation, which has high frequencies attenuated due to more absorption of sound in air at high frequencies. The rear channels are delayed to prevent that signal components are "jumping" between front and back when the matrix decoder can not well separate the channels. Because of the precedence effect [5, 19] front/back correlated signal components will always be perceived from the front if the back channels are delayed. It is especially important for dialogue in movie sound tracks to firmly always stay in front.

22.3.2.3 Parametric Matrix Decoding

Parametric matrix decoding [20] separates the matrix stereo signal into directional and ambient/diffuse sound. For directional sound its direction in the original surround signal is estimated and multichannel amplitude panning is applied to render it at this direction. Ambient sound is mixed to the multichannel output signal with the goal of generating a high-quality multichannel spatial impression.

In the following, it is shown that the matrix stereo signal as generated with (22.1) indeed contains information to uniquely determine the direction of sound in the original surround signal. A surround signal with directional sound from a specific angle α is considered. It is assumed that the surround signal is generated by selecting the two audio channels whose corresponding loudspeakers enclose α and applying amplitude panning to this signal pair such that the virtual source is expected to appear at angle α. The "tangent amplitude panning law" [21] is used to determine the angle relative to the loudspeaker pair. The angle α is defined such that front center corresponds to $\alpha = 0°$.

To such signals matrix encoding (22.1) is applied and the real-valued right/left amplitude ratio r is determined. Figure 22.5 shows amplitude ratios as a function of angle α of the virtual source. The loudspeaker directions are indicated as vertical dotted lines. The horizontal dotted lines indicate values of -1, 0, and 1. Note that

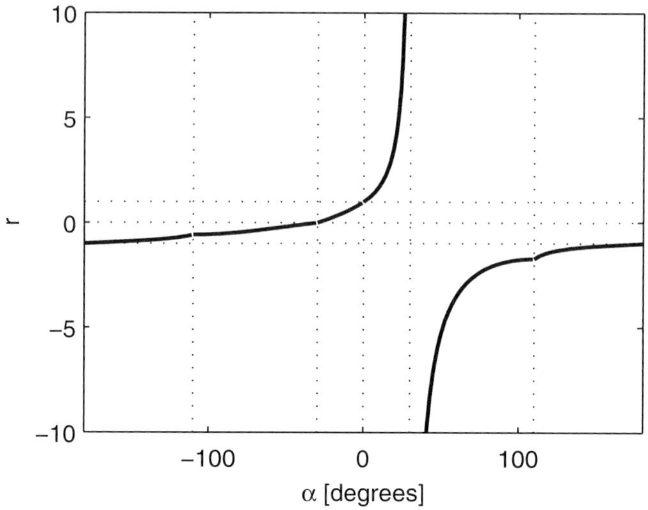

Fig. 22.5 The real-valued amplitude ratio as a function of direction of direct sound α in the original surround audio signal

the amplitude ratio uniquely determines the direction of the virtual source and thus the matrix stereo signal contains all information that is needed to render sources correctly to their original direction.

22.3.3 Spatial Audio Coding

A very efficient way to represent surround audio signals is to use a matrix encoder and then to apply a stereo perceptual audio coder. This not only allows to code surround audio signals at a bitrate of stereo audio coding, but also has the benefit of stereo backward compatibility (if no matrix surround decoder is used after the stereo decoder then the matrix stereo signal is played back). The described approach has the following drawbacks:

- The audio quality is limited by the modest performance of matrix encoding/decoding processes.
- A matrix surround downmix is worse than a downmix optimized for stereo quality (e.g., ITU downmix [1]) due to out-of-phase signal components in the downmix.

Spatial audio coding overcomes both of these limitations. It uses a downmix optimized for stereo quality. Some additional side information is used, which enables higher surround audio quality than matrix surround. Further, spatial audio coding can also be used with mono downmixes for even lower bitrates. On the other hand, also downmixes with more than two channels may be used.

22.4 Spatial Hearing

In this section, spatial hearing is reviewed. The focus is on phenomena that are relevant for spatial audio coding. Similarly to the way humans perceive a visual image, humans are also able to perceive an *auditory spatial image*. The different objects that are part of the auditory spatial image are denoted as *auditory objects*. When stereo or multichannel audio signals are played back over headphones or loudspeakers they evoke an auditory spatial image in the listener.

22.4.1 Spatial Hearing with One Sound Source

The simplest listening scenario is when there is one sound source in *free-field*. In this case, the ear input signals can be viewed as being filtered versions of the source signal. The filters modeling the path of sound from a source to the left and right ear entrances are commonly referred to as *head-related transfer functions* (HRTFs) [5]. For each source direction different HRTFs need to be used for modeling the ear entrance signals.

A more intuitive but only approximately valid view for the relation between the source angle ϕ and the ear entrance signals considers the difference in length of the paths from the source to the two ear entrances as a function of the source angle ϕ [5]. As a result of the different path lengths, there is a difference in arrival time between both ear entrances, denoted as *interaural time difference* (ITD). Additionally, the shadowing of the head results in an intensity difference of the left and right ear entrance signals, denoted as *interaural level difference* (ILD). For example, a source to the left of a listener results in a higher intensity of the signal at the left ear than at the right ear.

The following measures are used for ITD and ILD relative to the ear entrance signals $\tilde{x}_1(n)$ and $\tilde{x}_2(n)$:

- ITD [samples]:

$$\tau_{12}(n) = \arg\max_{d} \{\Phi_{12}(d,n)\}, \tag{22.4}$$

  with a short-time estimate of the normalized cross-correlation function

$$\Phi_{12}(d,n) = \frac{E\{\tilde{x}_1(n-d_1)\tilde{x}_2(n-d_2)\}}{\sqrt{E\{\tilde{x}_1^2(n-d_1)\}E\{\tilde{x}_2^2(n-d_2)\}}}, \tag{22.5}$$

  where

$$\begin{aligned} d_1 &= \max\{-d,0\} \\ d_2 &= \max\{d,0\}, \end{aligned} \tag{22.6}$$

  and $E\{.\}$ denotes time averaging.

- ILD [dB]:

$$\Delta L_{12}(n) = 10\log_{10}\left(\frac{E\{\tilde{x}_2^2(n-d_2)\}}{E\{\tilde{x}_1^2(n-d_1)\}}\right). \tag{22.7}$$

Diffraction, reflection, and resonance effects caused by the head, torso, and the external ears of the listener result in that ITD and ILD not only depend on the source angle ϕ but also on the source signal. Nevertheless, if ITD and ILD are considered as a function of frequency, it is a reasonable approximation to say that the source angle solely determines ITD and ILD as implied by data shown in [22]. When only considering frontal directions ($-90^\circ \leq \phi \leq 90^\circ$) the source angle ϕ approximately causally determines ITD and ILD. However, for each frontal direction there is a corresponding direction in the back of the listener resulting in a similar ITD-ILD pair. Thus, the auditory system needs to rely on other cues for resolving this front/back ambiguity. Examples of such cues are head movement cues, visual cues, and spectral cues (different frequencies are emphasized or attenuated when a source is in the front or back) [5]. The following discussion does not cover these other cues, since these are not considered explicitly in spatial audio coding. For audio playback systems with loudspeakers these other cues are automatically inherent in the ear entrance signals due to the physical location of the loudspeakers.

22.4.2 Ear Entrance Signal Properties and Lateralization

Figure 22.6(a) illustrates perceived auditory objects for different ITD and ILD [5] for two coherent left and right headphone signals. When left and right headphone signals are coherent, have the same level (ILD = 0), and no delay difference

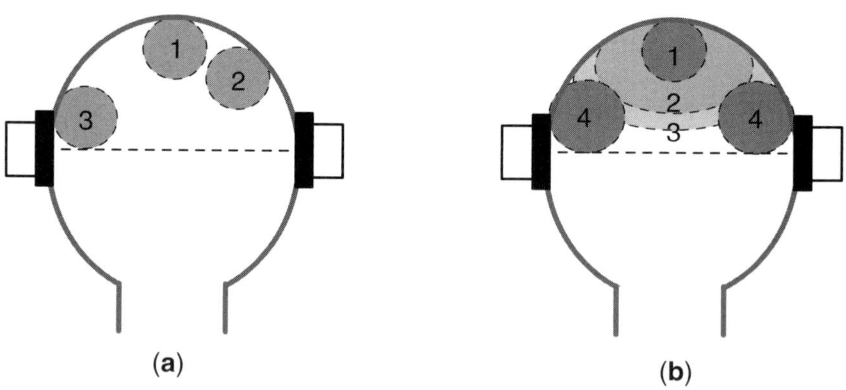

Fig. 22.6 (**a**) ILD and ITD between a pair of headphone signals determine the location of the auditory object, which appears in the frontal section of the upper head. (**b**): The width of the auditory object increases (1–3) as the interaural coherence (IC) between the left and right headphone signals decreases, until two distinct auditory objects appear at the sides (4).

(ITD = 0), an auditory object appears in the center between the left and right ears of a listener. More specifically, the auditory object appears in the center of the frontal section of the upper half of the head of a listener, as illustrated by region 1 in Fig. 22.6(a). By increasing the level on one side, e.g., right, the auditory object moves to that side as illustrated by region 2 in Fig. 22.6(a). In the extreme case, when only the signal on the left is active, the auditory object appears at the left side as illustrated by region 3 in Fig. 22.6(a). ITD can be used similarly to control the position of the auditory object.

Another ear entrance signal property that is considered in this discussion is a measure for the degree of "similarity" between the left and right ear entrance signals, denoted as *interaural coherence* (IC). IC here is defined as the maximum value of the normalized cross-correlation function,

$$c_{12}(n) = \max_{d} \Phi_{12}(d,n)\,, \tag{22.8}$$

where delays d corresponding to a range of ± 1 ms are considered. IC as defined has a range between 0 and 1. IC = 1 means that two signals are coherent (signals are equal with possibly a different scaling and delay) and IC = 0 means that the signals are independent.

When two identical signals (IC = 1) are emitted by the two transducers of the headphones, a relatively compact auditory object is perceived. For noise the width of the auditory object increases as the IC between the headphone signals decreases until two distinct auditory objects are perceived at the sides, as illustrated in Fig. 22.6(b) [23].

22.4.3 Two Sound Sources: Summing Localization

For two sources at a distance (e.g., loudspeaker pair), ITD, ILD, and IC are determined by the HRTFs of both sources and by the specific source signals. Nevertheless, it is interesting to assess the effect of cues similar to ITD, ILD, and IC, but relative to the source signals and not ear entrance signals. To distinguish between these same properties considered either between the two ear entrance signals or two source signals, respectively, the latter are denoted ICTD (interchannel time difference), ICLD (interchannel level difference), and ICC (interchannel coherence). For headphone playback, ITD, ILD, and IC are (ideally) the same as ICTD, ICLD, and ICC. In the following a few phenomena related to ICTD, ICLD, and ICC are reviewed for two sources located in the front of a listener.

Figure 22.7(a) illustrates the location of the perceived auditory objects for different ICLD for two coherent source signals [5]. When left and right source signals are coherent (ICC = 1), have the same level (ICLD = 0), and no delay difference (ICTD = 0), an auditory object appears in the center between the two sources as illustrated by region 1 in Fig. 22.7(a). By increasing the level on one side, e.g., right,

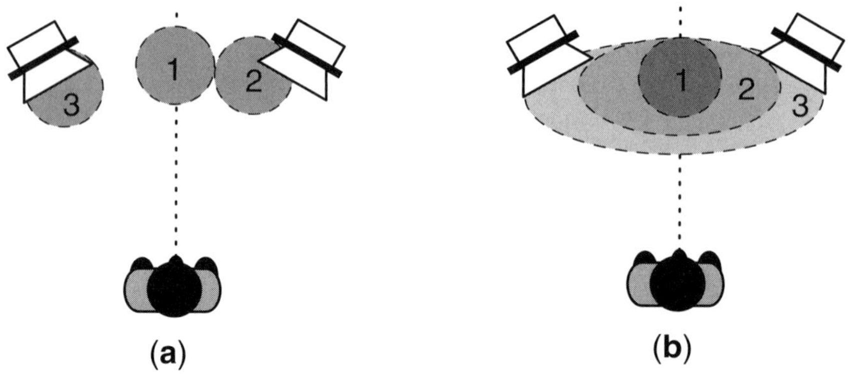

Fig. 22.7 (**a**) ICTD and ICLD between a pair of coherent source signals determine the location of the auditory object, which appears between the two sources. (**b**) The width of the auditory object increases (1–3) as the IC between left and right source signals decreases

the auditory object moves to that side as illustrated by region 2 in Fig. 22.7(a). In the extreme case, when only the signal on the left is active, the auditory object appears at the left source position as is illustrated by region 3 in Fig. 22.7(b). ICTD can be used similarly to control the position of the auditory object. This principle of controlling the location of an auditory object between a source pair is also applicable when the source pair is not in front of the listener. However, some restrictions apply for sources to the sides of a listener [24, 25]. There is an upper limit for the angle between such a source pair beyond which localization of auditory objects between the sources degrades.

When coherent wideband noise signals (ICC $= 1$) are simultaneously emitted by a pair of sources, a relatively compact auditory object is perceived. When the ICC is reduced between these signals, the width of the auditory object increases [5], as illustrated in Fig. 22.7(b).

The insight that when signals with specific properties are emitted by two sources the direction of the auditory object can be controlled is of high relevance for applications. It is this property that makes stereo audio playback possible. With two appropriately placed loudspeakers, the illusion of auditory objects at any direction between the two loudspeakers can be generated.

Another relevance of the described phenomena is that for loudspeaker playback and headphone playback similar cues can be used for controlling the location of an auditory object. This is the basis, which makes it possible to generate signal pairs that evoke related illusions in terms of relative auditory object location for both loudspeaker and headphone playback. If this were not the case, there would be a need for different signals depending on whether a listener uses loudspeakers or headphones.

22.4.4 Spatial Impression

So far the discussion mostly focused on the attribute of perceived direction or lateralization of auditory objects. One exception was the discussion of the role IC and ICC play for noise signals in determining the extent of the auditory object. In the following, other attributes related to auditory objects and the auditory spatial image are briefly discussed. These attributes mostly depend on the properties of reflections relative to the direct sound.

Spatial impression is defined as the impression a listener spontaneously gets about type, size, and other properties of an actual or simulated space [5]. Note that spatial impression is an attribute of auditory spatial images, which is not exclusively associated with auditory object properties. It also includes properties more associated with the auditory spatial image as a whole than with the single auditory objects. Spatial impression is largely determined by the relation between direct sounds and reflections, and number, strength, and directions of reflections. In the following, attributes related to spatial impression are briefly reviewed. More complete reviews are given in [5, 26].

Coloration: The first early reflections up to about 20 ms later than the direct sound can cause timbral colorization due to a "comb filter" effect, which attenuates and amplifies frequency components in a frequency-periodic pattern.

Distance of auditory object: In free-field, the following two ear entrance signal attributes change as a function of source distance: power of signal reaching the ears and high-frequency content (air absorption). For a source for which a listener knows its likely level of emitted sound, such as speech, the overall sound level at the ear entrances provides an absolute distance cue [27, 28]. However, in situations when a listener does not expect a source to have a certain emitting level, overall sound level at the ear entrances can not be used for judging absolute distance [29].

On the other hand, in a reverberant environment there is more information available to the auditory system. The reverberation time and the timing of the first reflections contain information about the size of a space and the distance to the surfaces, thus giving an indication about the expected range of source distances. For relatively distant sources the ratio of the power of direct to reflected sound is a reliable distance cue; see e.g. [27, 28, 30].

Width of auditory objects and envelopment: As implied by the results presented in Sects. 22.4.2 and 22.4.3 IC and ICC are related to the width of auditory objects. IC can be related to the width of auditory objects and *listener envelopment* [31, 32] by computing it for the early and late part of *binaural room impulse responses* (BRIRs) (e.g., up to 80 ms and later part). These two measures are often denoted early and late *interaural cross-correlation coefficient* (IACC) [33, 34]. A thorough review of IACC and related measures is given in [26].

Since IC and ICC are in many cases directly related, i.e., lower ICC between a loudspeaker pair results in lower IC between the ear entrance signals [35], also ICC can be related to the width of auditory objects and listener envelopment.

22.4.5 Limitations of the Auditory System and Discussion

The periphery of the auditory system decomposes the ear input signals into frequency bands [36]. It is often assumed that the frequency resolution at which binaural cues are processed by the auditory system is related to these frequency bands [37]. The temporal resolution at which the auditory system can track binaural cues is often referred to as "binaural sluggishness." The corresponding time constants are between 30 and 100 ms [37].

Localization blur [5] (amount of directional change that is just noticeable) for horizontal directions is smallest for sources in the front and back and largest for sources at the sides. It is about 1–2° for a source in the front. Just noticeable differences for ITD and ILD are related to localization blur and are smallest for front and rear sources and larger for side sources.

The sensitivity to changes in IC [38] is very high for a reference IC value of 1. In this case, changes as small as 0.002 can be perceived. The smaller the reference IC value the smaller is its sensitivity. For a reference value of 0, the change of coherence must be about 100 times larger than for a reference value of 1.

When a surround audio signal is played back, one listens to this signal without a reference. That is, small localization errors will not be noticed. Thus, for spatial audio coding it is not necessary to consider the precisions of binaural cues as so far discussed, but lower precision is often sufficient. Further, when there are multiple concurrently active sources, localization performance at best stays nearly the same as for a single source and in many cases decreases [5]. Usually localization performance slightly decreases in the presence of reflections [39].

An auditory model explaining aspects of source localization in complex listening scenarios (concurrent sound, reflections) has been presented in [40]. The model hypothizes how the auditory system obtains source localization information in complex listening scenarios. It assumes that the auditory system only processes ITD and ILD at time instants when these correspond to a source direction. Time instants when ITD and ILD are "corrupted" by concurrent sound or reflections are ignored. These time instants are identified using a time-adaptively computed IC. This model implies that the auditory system uses mostly only ITD, ILD, and IC for horizontal source localization in complex listening scenarios.

22.5 Spatial Audio Coding

In this section basic spatial audio coding principles are described. The case of using a single downmix channel is the fully perceptually motivated case of spatial audio coding and is thus described first. Then, the case when more than one downmix channel is used is discussed. Finally, complete spatial audio coding-based systems incorporating a conventional audio coder for coding of the downmix signal are described.

As implied by the discussion of the previous section, the interchannel cues of audio signals largely determine the aspects of the auditory spatial image that is perceived when listening to such signals. The specific interchannel cues are ICTD (interchannel time difference), ICLD (interchannel level difference), and ICC (interchannel coherence). The basic idea behind spatial audio coding is to represent a stereo or multichannel signals as a mono downmix signal plus interchannel cues. The interchannel cues are estimated from the audio signal to be coded. These estimated cues (denoted spatial cues) are quantized and coded and transmitted to the decoder along with the downmix signal. Given the downmix signal and the interchannel cues the decoder uses a *spatial cue synthesis* to obtain a stereo or multichannel audio signal, which is perceived as being similar to the original audio signal. Using of spatial cues for multichannel audio coding has been introduced in [41–44] as "binaural cue coding."

22.5.1 Spatial Audio Analysis

Figure 22.8 illustrates the analysis process used in spatial audio coding. The input channels are downmixed to a single audio channel. The downmix process may be implemented as a sum or weighted sum of the input channels. Alternatively, time and frequency adaptive equalization may be applied [45] to preserve total energy in the downmix. Practical experience showed that for virtually all cases there is no substantial benefit in using ICTD and thus usually only ICLD and ICC are estimated.

The estimation of ICLD and ICC between a channel pair is illustrated in Fig. 22.9. The input audio channels are converted to a time–frequency signal representation. Suitable time–frequency representations are the short time Fourier transform (STFT) or other complex filterbanks. The uniformly spaced spectral STFT coefficients are grouped to form subbands with a frequency resolution inspired by the frequency resolution of the peripheral human auditory system [36]. ICLD and ICC are estimated in each subband with a time resolution also motivated by the limits of the auditory system.

Time–frequency processing suitable for spatial audio coding is described in detail in [44,46,47]. Time and frequency resolution limitations of the auditory system have been discussed in Sect. 22.4.5. The time and frequency resolution for estimating ICLD and ICC is chosen as low as possible according to these limitations.

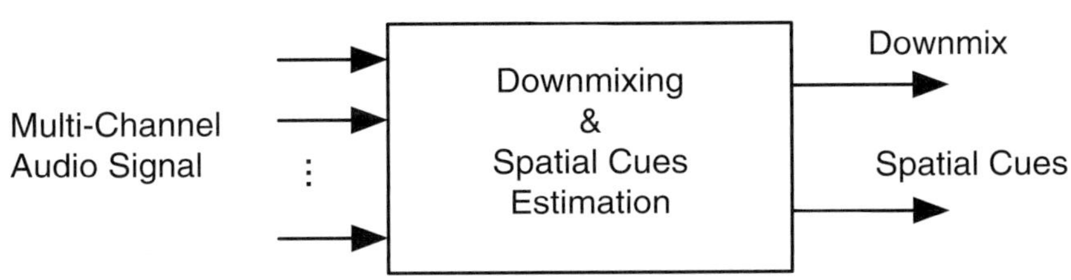

Fig. 22.8 Spatial audio analysis process

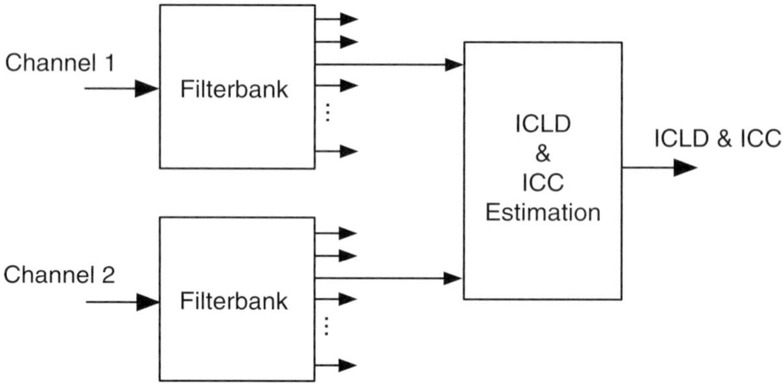

Fig. 22.9 Spatial audio analysis process. ICLD and ICC estimation is shown for a single subband pair. Such estimation is also applied to all other subbands

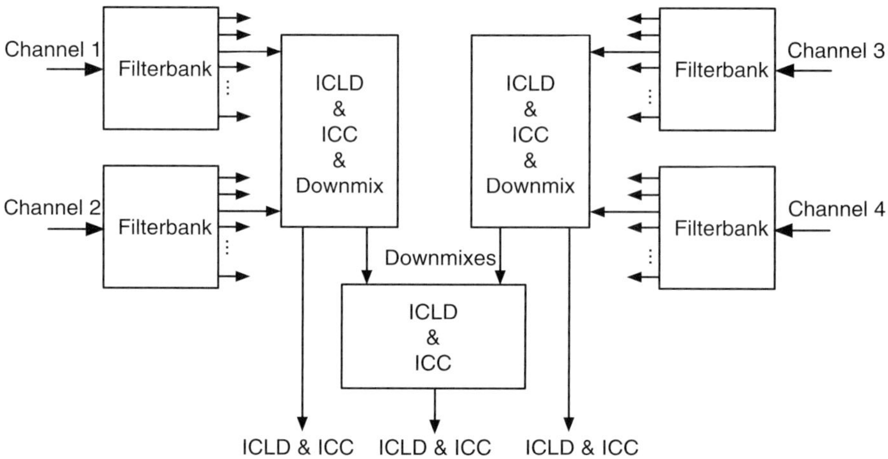

Fig. 22.10 Spatial audio analysis process

When there are more than two audio channels to be analyzed, there are numerous possibilities to estimate ICLD and ICC. For ICLD it is sufficient to estimate it between a reference channel and all the other channels. To completely describe the coherence properties between the audio channels one would have to estimate ICC between each possible audio channel pair. However, this is tedious and corresponding ICC synthesis would be very complicated. In [44] it was proposed to estimate at each time in each subband ICC only between the two strongest channels.

The MPEG Surround [48, 49] spatial audio coder simplifies the question on how to estimate ICLD and ICC for the multichannel case by cascading stereo analysis blocks. An example of such a cascade for four audio channels is shown in Fig. 22.10. ICLD and ICC are estimated between the first two and last two input channels. Further, ICLD and ICC are estimated between the downmixes of the two considered input channel pairs.

22.5.2 Spatial Audio Synthesis

The spatial audio synthesis process is illustrated in Fig. 22.11. The inputs of the spatial audio synthesis are the downmix signal and the spatial cues (as a function of time and frequency, estimated as described previously). The downmix signal is processed such that the interchannel cues of the resulting multichannel surround audio signal approximate the spatial cues estimated from the original audio signal.

An example for ICLD and ICC synthesis for two audio channels is shown in Fig. 22.12. The same filterbank is used as for cue analysis (Fig. 22.9) in order to match the time–frequency resolution of the estimated ICLD and ICC cues. The shown scheme conceptually corresponds to what is used for parametric stereo (PS) [46]. A slightly different cue synthesis was proposed in [50], using two filters as opposed to one filter. The filter in Fig. 22.12 generates an independent audio channel. The two output channel subbands are then generated as a weighted sum of the downmix and independent audio channel. The weights a_1, a_2, b_1, and b_2 are determined by the desired ICLD, ICC, and energy preservation (the sum of the two output channels need to have same power as the downmix channel).

When more than two channels need to be synthesized, a process symmetric to the ICLD and ICC estimation process is used. A general N-channel analysis and synthesis process was proposed in [50]. The MPEG Surround spatial audio coder, symmetric to its cue analysis cascade, uses a synthesis cascade with elements generating two channels from one channel [49].

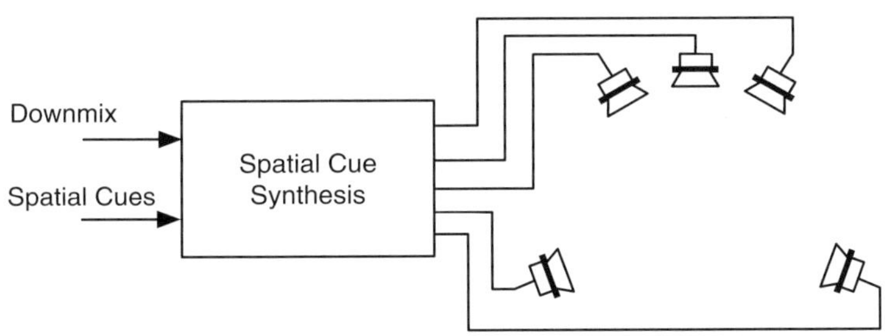

Fig. 22.11 Spatial audio synthesis process

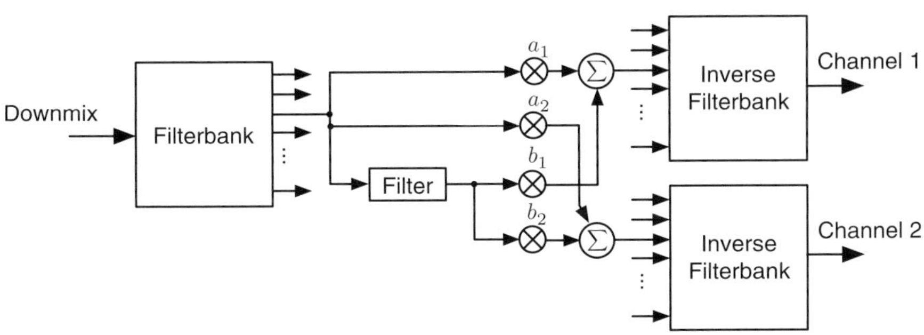

Fig. 22.12 Spatial audio synthesis process

22.5.3 Using More than One Downmix Channel

Similar to matrix surround, spatial audio coding principles can also be applied using a stereo downmix signal with advantages over matrix surround as discussed in Sect. 22.3.3. Using a stereo downmix has not only the advantage of stereo backwards compatibility, but also better audio quality is achieved since it is easier to generate a surround signal from two stereo channels than a single mono channel.

As opposed to the matrix surround encoding matrix (22.1), the ITU downmix matrix is usually used,

$$\begin{bmatrix} l_t(t) \\ r_t(t) \end{bmatrix} = \begin{bmatrix} 1 & 0 & \frac{1}{\sqrt{2}} & 1 & 0 \\ 0 & 1 & \frac{1}{\sqrt{2}} & 0 & 1 \end{bmatrix} \begin{bmatrix} l(t) \\ r(t) \\ c(t) \\ l_s(t) \\ r_s(t) \end{bmatrix}, \tag{22.9}$$

often yielding a higher quality stereo signal since there are no out-of-phase signal components in the stereo downmix. Time and frequency adaptive equalization may be used for energy preservation in the downmix [45].

MP3 Surround [51,52] and MPEG Surround [48,49] use (in most cases) a stereo downmix. More details on 5.1 surround coding with stereo downmix with MP3 Surround and MPEG Surround are given in the next section.

22.5.4 Combining Spatial Processing and a Conventional Audio Coder

Previously, principles were described that allow to reduce the number of channels of a surround audio signal to a downmix with usually only one or two channels plus spatial cues. It was also described how to recover a surround audio signal from the downmix by means of spatial cue synthesis.

Figure 22.13 illustrates a complete audio coder, employing spatial audio coding principles. The input surround audio signal is downmixed and the spatial cues are estimated. The estimated spatial cues are quantized and coded and the downmix signal is coded with a conventional perceptual audio coder. The coded spatial cues side information is added to the encoded downmix audio bitstream.

The corresponding decoder is shown in Fig. 22.14. The combined bitstream is parsed to obtain the bitstreams for the downmix audio decoder and spatial cues decoder. The decoded downmix signal is processed with spatial cue synthesis to recover the surround audio signal.

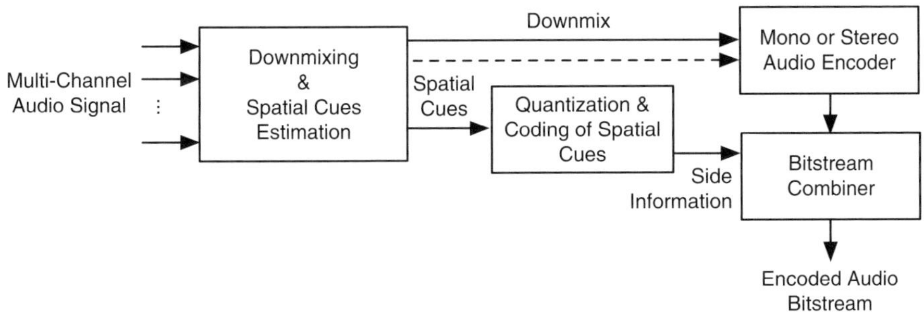

Fig. 22.13 A surround audio encoder employing a conventional perceptual audio encoder and spatial audio coding

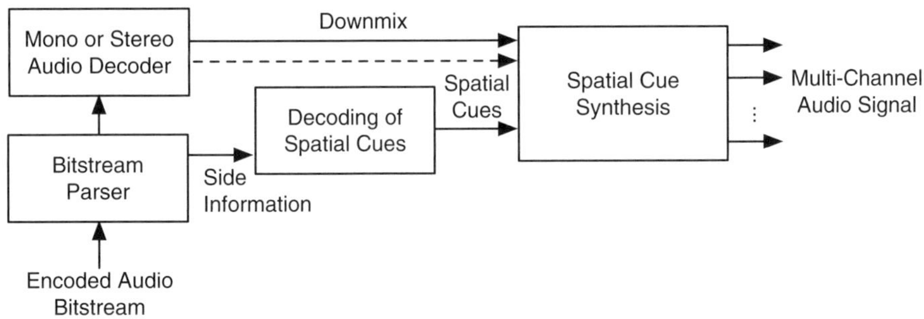

Fig. 22.14 A surround audio decoder employing a conventional perceptual audio decoder and spatial audio coding

22.6 MP3 Surround and MPEG Surround

MP3 Surround [51, 52] and MPEG Surround [48, 49] utilize a MP3 and any MPEG audio coder, respectively, for coding the stereo downmix signal (both coders also support mono downmix, which is however much less often used due to lack of stereo backward compatibility and lower audio quality). This section describes specific aspects of these coders and assumes that the reader is familiar with the general principles described in the previous section.

22.6.1 MP3 Surround

The most popular perceptual audio coder, MP3, supports only coding of mono and two-channel stereo signals. MP3 Surround is an extension to MP3 (pre and post-processor) which upgrades it to 5.1 surround. The spatial cues side information is added to the MP3 bitstream as ancillary data, invisible to legacy decoders. Only about 10–12 kb s^{-1} are needed in addition to the stereo downmix coding bitrate for 5.1 surround coding.

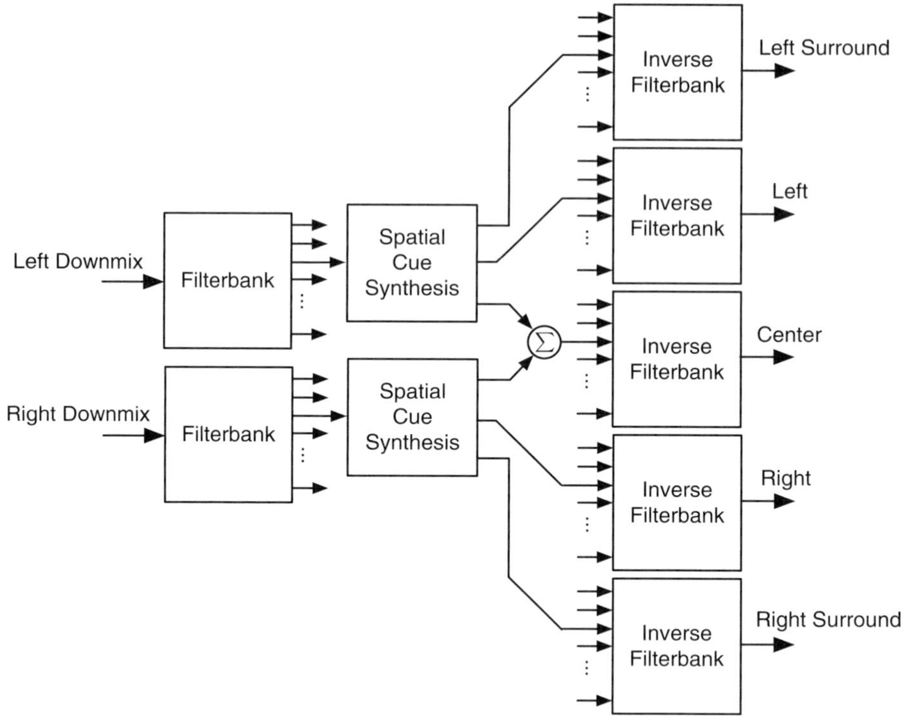

Fig. 22.15 A MP3 Surround decoder conceptually uses two three-channel spatial cue synthesis processes

An MP3 Surround decoder essentially applies two one-to-three channel spatial audio synthesis processes to generate the five main audio channels of the 5.1 surround audio signal from the stereo downmix. One spatial audio synthesis generates left, surround left, and half of the center channel and the other spatial synthesis generates right, surround right, and the other half of the center channel as illustrated in Fig. 22.15. Optionally, "center channel cancellation" is used [52] to reduce undesired correlations between the center and side channels. Not shown in the figure, the side information, extracted from the stereo MP3 bitstream, is decoded to obtain the spatial cues, which control the two-to-three channel synthesis processes.

Generation of 5.1 audio channels is much more demanding for a device than generating two stereo channels in terms of processing power and memory requirements. MP3 Surround was designed with low computational complexity in mind. Thus, only ICLD are synthesized in the spatial audio synthesis (no computationally intense ICC synthesis). To mitigate the effect that front and surround channels are potentially more correlated than desired, the rear channels are delayed.

The MP3 Surround encoder generates the downmix as described in Sect. 22.5.3. Further, the spatial cues are analyzed using conceptually two three-channel ICLD analysis processes in the subband domain.

22.6.2 MPEG Surround

Similarly to MP3 Surround, MPEG Surround adds the side information to the bitstream of the encoded downmix signal, invisibly to the downmix signal decoder for backward compatibility. MPEG Surround features a number of improvements and addition to the so far described spatial audio coding principles. These improvements enable higher audio quality and more versatility as is discussed in the following. The descriptions in the following are brief and many more details are given in [49] or in the following specific references on the various techniques.

22.6.2.1 Sophisticated Time–Frequency Processing

As opposed to a STFT, a quadrature mirror filter (QMF) bank with 32 bands is used. To increase frequency resolution, additional QMF banks are cascaded at the lower frequencies. The ICLD and ICC analysis and synthesis is carried out with a signal adaptive time–frequency resolution. For example, when transients occur the time resolution is increased.

22.6.2.2 Three-Channel Analysis and Synthesis

Synthesis of the center channel is particularly difficult. When the center channel is more correlated with the other channels than in the original surround signal, the spatial image appears as being too narrow. This is particularly noticeable in audio signals with virtually independent channels, e.g., signals with applause. As mentioned in Sect. 22.5.2, MPEG Surround uses a cascade of one-to-two channel synthesis blocks. Additionally, MPEG Surround incorporates a two-to-three channel synthesis block, based on prediction. Figure 22.16 illustrates the spatial synthesis process used in MPEG Surround for generating the five main audio channels of a 5.1 surround signal. The two-to-three synthesis block generates the center channel subbands and total left and total right audio signal subbands. The total left and total right subbands are further processed with one-to-two spatial synthesis processes to generate the front and surround channel subbands. On the analysis side, MPEG Surround uses a similar cascade to obtain the prediction gains, ICLD, and ICC. Details on the two-to-three channel synthesis and its corresponding analysis are given in [53].

22.6.2.3 Support for Various Surround Formats

MPEG Surround supports a wide range of configurations using different cascades of the described two-to-three channel and one-to-two channel analysis and synthesis processes, supporting various surround formats (e.g., 7.1, 10.1 surround) and different numbers of downmix channels.

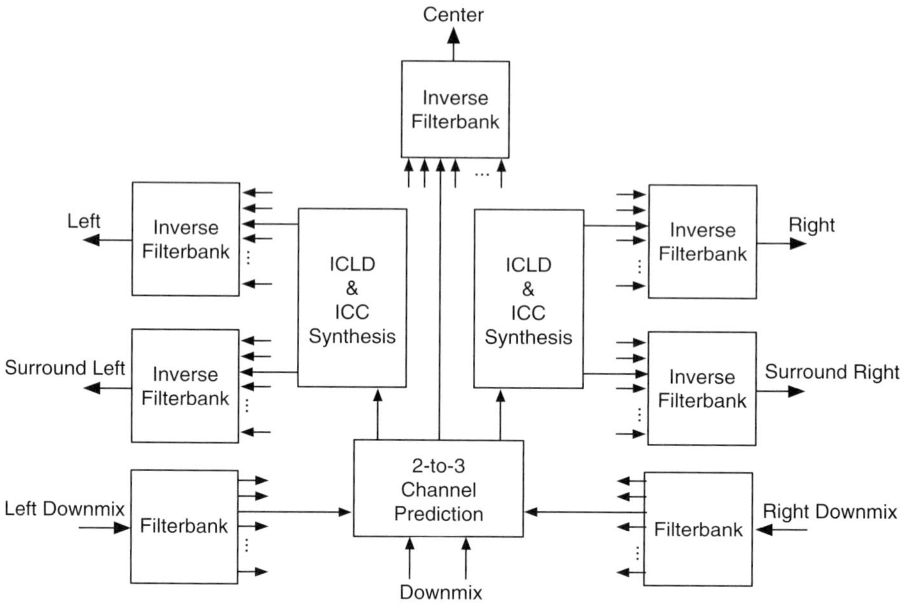

Fig. 22.16 A MPEG Surround decoder uses one two-to-three channel prediction process and two one-to-two channel spatial cue synthesis processes for 5.1 surround synthesis

22.6.2.4 Residual Coding

While spatial audio coding achieves good audio quality at low bitrates, it is usually not scalable to be transparent (not distinguishable from the original audio signal). To enable scalability, MPEG Surround features residual coding where residual signals are added to the spatial synthesis output to force the waveforms to be more similar to the original surround signal waveforms. Of course this is at the cost of a higher overall bitrate.

22.6.2.5 Matrix Decoding

The sophisticated spatial synthesis of MPEG Surround can also be applied for matrix surround (Sect. 22.3.2) decoding. The principle is the following. The ICLD and ICC between the matrix stereo input channels are estimated. A lookup table is used to generate all the cues needed for the MPEG Surround signal synthesis. These cues are applied (as opposed to using these cues from side information). The lookup table in MPEG Surround contains values of the multi-channel spatial cues that were determined by means of training.

22.6.2.6 Binaural Audio

While the low bitrate of MPEG Surround potentially enables multichannel surround on mobile devices, multiloudspeaker surround is in principle not mobile. Binaural audio (often also denoted 3D audio) [54] is a technology that enables playback of surround audio over conventional two-channel headphones. Each channel of the surround audio signal is filtered with HRTFs (measured anechoically or in a room) simulating the acoustic transfer of sound from that specific loudspeaker to the left and right ear entrances.

While such HRTF filtering can be applied to the MPEG Surround decoder output signals, the computational complexity of this process is high. MPEG Surround features a unique technology to convert the stereo downmix signal directly and with low computational complexity to a binaural audio surround signal for headphone playback. A two-by-two channel processor (complex two-by-two matrix) is applied in the MPEG Surround subband domain. This matrix is determined as a function of the surround ICLD and ICC spatial cues. This process is described in detail in [55].

22.7 Conclusions

Similarly as perceptual audio coding revolutionized mono and stereo audio coding by enabling high quality at low bitrates, spatial audio coding revolutionizes surround audio coding by enabling coding of multichannel audio at much lower bitrates than conventional techniques. This is made possible by consideration of human spatial hearing. Spatial audio coding is not based on accurately representing the waveform of each audio channel, but merely synthesizes interchannel properties of surround audio signals such that they are perceived similarly as the original signals.

This chapter described spatial audio coding, MP3 Surround, and MPEG Surround. The emphasis was on context and the goal was not to describe all mechanical details of these audio coders. The advantages of surround sound over stereo were discussed. Aspects of spatial hearing relevant for spatial audio coding were reviewed. The general concepts behind spatial audio coding were explained and specific aspects of MP3 Surround and MPEG Surround were discussed.

References

1. Rec. ITU-R BS.775, *Multi-Channel Stereophonic Sound System with or without Accompanying Picture*, ITU, 1993, http://www.itu.org.
2. R. Zelinski and P. Noll, "Adaptive transform coding of speech signals," *IEEE Trans. Acoust. Speech, and Signal Processing*, vol. 25, pp. 299–309, August 1977.
3. K. Brandenburg, G. G. Langenbucher, H. Schramm, and D. Seitzer, "A digital signal processor for real time adaptive transform coding of audio signal up to 20 khz bandwidth," in *Proc. ICCC*, 1982, pp. 474–477.

4. J. D. Johnston and A. J. Ferreira, "Sum-difference stereo transform coding," in *Proc. ICASSP-92*, 1992, pp. 569–572.
5. J. Blauert, *Spatial Hearing: The Psychophysics of Human Sound Localization*, The MIT Press, Cambridge, MA, revised edition, 1997.
6. L. D. Fielder, M. Bosi, G. Davidson, M. Davis, C. Todd, and S. Vernon, "AC-2 and AC-3: low-complexity transform-based audio coding," in *Collected Papers on Digital Audio Bit-Rate Reduction*, N. Gilchrist and C. Grewin, Eds., pp. 54–72. Audio Engineering Society, 1996.
7. K. Tsutsui, H. Suzuki, O. Shimoyoshi, M. Sonohara, K. Akagiri, and R. M. Heddle, "ATRAC: Adaptive transform acoustic coding for MiniDisc," in *Collected Papers on Digital Audio Bit-Rate Reduction*, N. Gilchrist and C. Grewin, Eds., pp. 95–101. Audio Engineering Society, 1996.
8. D. Sinha, J. D. Johnston, S. Dorward, and S. Quackenbush, "The perceptual audio coder (PAC)," in *The Digital Signal Processing Handbook*, V. Madisetti and D. B. Williams, Eds., Chapter 42. CRC Press, IEEE Press, Boca Raton, FL, 1997.
9. K. Brandenburg and G. Stoll, "ISO-MPEG-1 audio: a generic standard for coding of high-quality digital audio," *J. Audio Eng. Soc.*, pp. 780–792, Oct. 1994.
10. ISO/IEC, *Coding of moving pictures and associated audio for digital storage media at up to about 1.5 Mbit/s, Part 3: audio*, ISO/IEC 11172-3 International Standard, 1993, JTC1/SC29/WG11.
11. G. Stoll, "ISO-MPEG-2 audio: A generic standard for the coding of two-channel and multichannel sound," in *Collected Papers on Digital Audio Bit-Rate Reduction*, N. Gilchrist and C. Grewin, Eds., pp. 43–53. Audio Engineering Society, 1996.
12. M. Bosi, K. Brandenburg, S. Quackenbush, L. Fielder, K. Akagiri, H. Fuchs, M. Dietz, J. Herre, G. Davidson, and Y. Oikawa, "ISO/IEC MPEG-2 advanced audio coding," *J. Audio Eng. Soc.*, vol. 45, no. 10, pp. 789–814, 1997.
13. B. Grill, "The MPEG-4 general audio coder," in *Proc. AES 17th Int. Conf.: High-Quality Audio Coding*, Florence, Italy, September 1999, AES, pp. 147–156.
14. ISO/IEC, *Generic coding of moving pictures and associated audio information, Part 7: advanced audio coding*, ISO/IEC 13818-7 International Standard, 1997, JTC1/SC29/WG11.
15. J. Herre, K. Brandenburg, and D. Lederer, "Intensity stereo coding," *96th AES Conv., Feb. 1994, Amsterdam (preprint 3799)*, 1994.
16. M. Dietz, L. Liljeryd, K. Kjörling, and O. Kunz, "Spectral band replication – a novel approach in audio coding," in *Preprint 112th Conv. Aud. Eng. Soc.*, May 2002.
17. J. Makhoul and M. Berouti, "High-frequency regeneration in speech coding systems," in *Proc. ICASSP*, 1979, vol. 428–431.
18. K. Gundry, "A new active matrix decoder for surround sound," in *Proc. AES 19th Int. Conf.*, June 2001.
19. R. Y. Litovsky, H. S. Colburn, W. A. Yost, and S. J. Guzman, "The precedence effect," *J. Acoust. Soc. Am.*, vol. 106, no. 4, pp. 1633–1654, Oct. 1999.
20. C. Faller, "Matrix surround revisited," in *Proc. 30th Int. Conv. Aud. Eng. Soc.*, March 2007.
21. B. Bernfeld, "Attempts for better understanding of the directional stereophonic listening mechanism," in *Preprint 44th Conv. Aud. Eng. Soc.*, Feb. 1973.
22. W. Gaik, "Combined evaluation of interaural time and intensity differences: psychoacoustic results and computer modeling," *J. Acoust. Soc. Am.*, vol. 94, no. 1, pp. 98–110, July 1993.
23. R. I. Chernyak and N. A. Dubrovsky, "Pattern of the noise images and the binaural summation of loudness for the different interaural correlation of noise," in *Proc. 6th Int. Congr. on Acoustics Tokyo*, 1968, vol. 1, pp. A–3–12 (see Blauert 1997, Fig. 3.24).
24. G. Theile and G. Plenge, "Localization of lateral phantom sources," *J. Audio Eng. Soc.*, vol. 25, no. 4, pp. 196–200, 1977.
25. V. Pulkki, "Localization of amplitude-panned sources II: Two- and three-dimensional panning," *J. Audio Eng. Soc.*, vol. 49, no. 9, pp. 753–757, 2001.
26. R. Mason, *Elicitation and measurement of auditory spatial attributes in reproduced sound*, Ph.D. thesis, University of Surrey, 2002 (a review of existing measurements that relate to spatial impression).

27. D. H. Mershon and L. E. King, "Intensity and reverberation as factors in the auditory perception of egocentric distance," *Percept. Psychophys.*, vol. 18, no. 6, pp. 409–415, 1975.
28. D. H. Mershon and J. N. Bowers, "Absolute and relative cues for the auditory perception of egocentric distance," *Perception*, vol. 8, pp. 311–322, 1979.
29. P. D. Coleman, "Failure to localize the source distance of an unfamiliar sound," *J. Acoust. Soc. Am.*, vol. 34, pp. 345–346, 1962.
30. A. Bronkhorst and T. Houtgast, "Auditory distance perception in rooms," *Nature*, vol. 397, pp. 517–520, Feb. 1999.
31. M. Morimoto and Z. Maekawa, "Auditory spaciousness and envelopment," in *Proc. 13th Int. Congr. on Acoustics*, Belgrade, 1989, vol. 2, pp. 215–218.
32. J. S. Bradley and B. A. Soulodre, "Listener envelopment: an essential part of good concert hall acoustics," *J. Acoust. Soc. Am.*, vol. 99, pp. 22, Jan. 1996.
33. J. S. Bradley, "Comparison of concert hall measurements of spatial impression," *J. Acoust. Soc. Am.*, vol. 96, no. 6, pp. 3525–3535, 1994.
34. T. Okano, L. L. Beranek, and T. Hidaka, "Relations among interaural cross-correlation coefficient ($IACC_E$), lateral fraction (LF_E), and apparent source width (asw) in concernt halls," *J. Acoust. Soc. Am.*, vol. 104, no. 1, pp. 255–265, July 1998.
35. K. Kurozumi and K. Ohgushi, "The relationship between the cross-correlation coefficient of two-channel acoustic signals and sound image quality, and apparent source width (asw) in concernt halls," *J. Acoust. Soc. Am.*, vol. 74, no. 6, pp. 1726–1733, Dec. 1983.
36. B. C. J. Moore and B. R. Glasberg, "Suggested formula for caculating auditory-filter bandwidth and excitation patterns," *J. Acoust. Soc. Aneruca*, vol. 74, pp. 750–753, 1983.
37. I. Holube, M. Kinkel, and B. Kollmeier, "Binaural and monaural auditory filter bandwidths and time constants in probe tone detection experiments," *J. Acoust. Soc. Am.*, vol. 104, no. 4, pp. 2412–2425, Oct. 1998.
38. J. F. Culling, H. S. Colburn, and M. Spurchise, "Interaural correlation sensitivity," *J. Acoust. Soc. Am.*, vol. 110, no. 2, pp. 1020–1029, Aug. 2001.
39. William Morris Hartmann, "Listening in a room and the precedence effect," in *Binaural and Spatial Hearing in Real and Virtual Environments*, Robert H. Gilkey and Timothy R. Anderson, Eds., pp. 349–376. Lawrence Erlbaum Associates, Mahwah, NJ, 1997.
40. C. Faller and J. Merimaa, "Source localization in complex listening situations: selection of binaural cues based on interaural coherence," *J. Acoust. Soc. Am.*, vol. 116, no. 5, pp. 3075–3089, Nov. 2004.
41. C. Faller and F. Baumgarte, "Efficient representation of spatial audio using perceptual parametrization," in *Proc. IEEE Workshop on Appl. Sig. Proc. to Audio and Acoust.*, Oct. 2001, pp. 199–202.
42. C. Faller and F. Baumgarte, "Binaural cue coding applied to stereo and multi-channel audio compression," in *Preprint 112th Conv. Aud. Eng. Soc.*, May 2002.
43. F. Baumgarte and C. Faller, "Binaural cue coding, Part 1: psychoacoustic fundamentals and design principles," *IEEE Trans. Speech Audio Proc.*, vol. 11, no. 6, pp. 509–519, Nov. 2003.
44. C. Faller and F. Baumgarte, "Binaural cue coding, Part 2: schemes and applications," *IEEE Trans. Speech Audio Proc.*, vol. 11, no. 6, pp. 520–531, Nov. 2003.
45. A. Baumgarte, C. Faller, and P. Kroon, "Audio coder enhancement using scalable binaural cue coding with equalized mixing," in *Preprint 116th Conv. Aud. Eng. Soc.*, May 2004.
46. E. Schuijers, J. Breebaart, H. Purnhagen, and J. Engdegard, "Low complexity parametric stereo coding," in *Preprint 117th Conv. Aud. Eng. Soc.*, May 2004.
47. C. Faller, *Parametric Coding of Spatial Audio*, Ph.D. thesis, Ecole Polytechnique Fédérale de Lausanne (EPFL), Switzerland, July 2004, Thesis No. 3062, http://library.epfl.ch/theses/?nr=3062.
48. J. Herre, K. Kjörling, J. Breebaart, C. Faller, S. Disch, H. Purnhagen, J. Koppens, J. Hilpert, J. Rödén, W. Oomen, K. Linzmeier, and K. S. Chong, "Mpeg surround – the iso/mpeg standard for efficient and compatible multi-channel audio coding," in *Preprint 122th Conv. Aud. Eng. Soc.*, May 2007.
49. J. Breebaart and C. Faller, *Spatial Audio Processing: MPEG Surround and Other Applications*, Wiley, Jan. 2008.

50. C. Faller, "Parametric multi-channel audio coding: Synthesis of coherence cues," *IEEE Trans. on Speech and Audio Proc.*, vol. 14, no. 1, pp. 299–310, Jan. 2006.
51. J. Herre, C. Faller, C. Ertel, J. Hilpert, A. Hoelzer, and C. Spenger, "MP3 surround: efficient and compatible coding of multi-channel audio," in *Preprint 116th Conv. Aud. Eng. Soc.*, May 2004.
52. C. Faller, "Coding of spatial audio compatible with different playback formats," in *Preprint 117th Conv. Aud. Eng. Soc.*, October 2004.
53. J. Breebaart, G. Hotho, J. Koppens, E. Schuijers, W. Oomen, and S. van de Par, "Background, concept and architecture for the recent mpeg surround standard on multi-channel audio compression," *J. Audio Eng. Soc.*, vol. 55, no. 5, pp. 331–351, May 2007.
54. J. Huopaniemi, *Virtual Acoustics and 3D Sound in Multimedia Signal Processing*, Ph.D. thesis, Laboratory of Acoustics and Audio Signal Processing, Helsinki University of Technology, Finland, 1999, Rep. 53.
55. J. Breebaart, L. Villemoes, and K. Kjörling, "Binaural rendering in mpeg surround," *EURASIP J. Advances in Signal Processing*, 2008, doi: 10.1155/2008/732895.

Index

B

D

Printed in the United States of America